Genes, Cells

A·N·D

Organisms

GREAT BOOKS IN EXPERIMENTAL BIOLOGY

EDITED BY
John A. Moore

A Garland Series

17 TITLES THAT STAND AS
MONUMENTS OF BIOLOGICAL THOUGHT

GREAT BOOKS IN EXPERIMENTAL BIOLOGY

1. *THE CELL IN DEVELOPMENT AND HEREDITY*,
   by Edmund B. Wilson.
   1925.

2. *THE CELL THEORY: A RESTATEMENT, HISTORY, AND CRITIQUE*,
   by John R. Baker.
   1948–1955.

3. *SPECIES AND VARIETIES, THEIR ORIGIN BY MUTATION: LECTURES DELIVERED AT THE UNIVERSITY OF CALIFORNIA*,
   by Hugo De Vries.
   3rd edition, 1912.

4. *THE GENETICS OF DROSOPHILA*,
   by Thomas Hunt Morgan
   (with Calvin B. Bridges and A. H. Sturtevant).
   1925.

5. *THE THEORY OF THE GENE*,
   by Thomas Hunt Morgan.
   1926.

6. *AN INTRODUCTION TO GENETICS*,
   by A. H. Sturtevant and G. W. Beadle.
   1939.

7. *RECENT ADVANCES IN CYTOLOGY*,
   by C. D. Darlington.
   1937.

8. *PHYSIOLOGICAL GENETICS*,
   by Richard Goldschmidt.
   1938.

9. *INVERTEBRATE EMBRYOLOGY*,
   edited by Matazo Kume and Katsuma Dan.
   1968.

10. *EMBRYONIC DEVELOPMENT AND INDUCTION*,
    by Hans Spemann.
    1938.

11. *THE ORGANIZATION OF THE DEVELOPING EMBRYO*,
    by Ross G. Harrison.
    1969.

12. *PROBLEMS OF FERTILIZATION*,
    by Frank R. Lillie.
    1919.

13. *THE BIOLOGY OF THE CELL SURFACE*,
    by Ernest Just.
    1939.

14. *FEATURES IN THE ARCHITECTURE OF PHYSIOLOGICAL FUNCTION*,
    by Joseph Barcroft.
    1934.

15. *THE NEUROPHYSIOLOGICAL BASIS OF MIND; THE PRINCIPLES OF NEUROPHYSIOLOGY*,
    by John Carew Eccles.
    1953.

16. *RECOLLECTIONS OF MY LIFE*,
    by Santiago Ramon Y. Cajal.
    1937.

17. *THE DISCOVERY AND CHARACTERIZATION OF TRANSPOSABLE ELEMENTS: THE COLLECTED PAPERS OF BARBARA McCLINTOCK*.
    1987.

RECENT ADVANCES IN CYTOLOGY

C. D. Darlington

Garland Publishing, Inc.
NEW YORK & LONDON • 1988

This edition is reprinted by arrangement
with Churchill Livingstone.

LIBRARY OF CONGRESS CATALOGING-IN-PUBLICATION DATA

Darlington, C. D. (Cyril Dean), 1903–
Recent advances in cytology / C. D. Darlington.
p.   cm.—(Genes, cells, and organisms)
Reprint. Originally published: 2nd ed. Philadelphia :
The Blakiston Company, 1937.
Bibliography: p.
Includes index.
ISBN 0-8240-1376-X
1. Cytology.   I. Title.   II. Series.
[DNLM: 1. Chromosomes.   QH 605 D221r 1937a]
QH581.2.D36   1988
DNLM/DLC
for Library of Congress   88-21258

*All volumes in this series are printed on
acid free, 250-year-life paper.*

PRINTED IN THE UNITED STATES OF AMERICA

RECENT ADVANCES IN CYTOLOGY

By

C. D. DARLINGTON
D.Sc., Ph.D.

Head of the Cytological Department,
John Innes Horticultural Institution
London

With a Foreword by

J. B. S. HALDANE, M.A., F.R.S.

Professor of Genetics, University College, London;
Head of the Genetical Department, John Innes Horticultural Institution

SECOND EDITION

With 16 Plates, 160 Text-figures
and 81 Tables

THE BLAKISTON COMPANY
PHILADELPHIA
1937

To

F. A. JANSSENS
1863–1924

W. C. F. NEWTON
1894–1927

KARL BELAR
1895–1931

FOREWORD

I AM delighted, but not surprised, that a new edition of "Recent Advances in Cytology" has been called for, and am fortunate in being permitted to write a foreword for it. The book is concerned with those aspects of cytology which have a bearing on genetics, and therefore deals mainly with the nucleus. But a study of the mechanics of nuclear division inevitably involves that of extra-nuclear structures such as the centrosome; and a great deal of light is thrown on the nature of protoplasm by its peculiar behaviour when organised on the spindle.

Modern cytology has two very remarkable features. Its principles are the same for plant and animal cells. From a study of the chromosomes in the Liliaceæ we can clear up previously obscure phenomena in the nuclei of the Orthoptera, and conversely. And the remarkably uniform behaviour of nuclei makes deduction and prediction possible on a very much larger scale than in any other field of morphology and physiology. Thus the principles deduced largely from a study of the monocotyledons led to the prediction of phenomena which were verified in the genetics of man.

Further, the uniformity of the nuclear mechanisms can be extrapolated with great confidence into the past. We can be reasonably sure that an Acanthodian or a Pteridosperm nucleus was organised on modern lines. We can therefore deduce that the principles of genetics and the method of evolution were much the same in remote geological epochs as they are to-day. Just because the nuclear mechanism has apparently reached the limits of its own evolution it furnishes a basis for the evolution of other characters. An attempt to study the evolution of living organisms without reference to cytology would be as futile as an account of stellar evolution which ignored spectroscopy.

The first edition of this book was the object of numerous attacks. In fact, one of the sessions of the Sixth International Congress of

Genetics in 1932 was mainly occupied in disproving Dr. Darlington's conclusions. However, most of these objections have been quietly withdrawn in the four succeeding years. The most important correction to the views expressed in the first edition has been made by Dr. Darlington himself. It is the discovery that in the males of *Drosophila*, and doubtless of other Diptera, where there is no genetical crossing-over, the meiotic autosomal bivalents are held together not by chiasmata, but by attraction of a special character.

This book is indispensable not only because of the discoveries it describes, but almost equally on account of the coveries, to borrow a word from Samuel Butler. A fundamental covery is that the expressions " reductional division " and " equational division," those bogies of our schooldays, are meaningless. For a given section of a chromosome either meiotic division may be equational or reductional. A teacher of biology may, for the sake of simplicity, neglect some of the more recent discoveries in cytology. He cannot neglect such a covery as this.

It is perfectly possible that " Recent Advances in Cytology " marks a turning point in the history of biology. For some centuries the deductive method in the biological sciences has been very properly suspect. But first in genetics, and now in cytology, we have returned to it. General principles have been discovered of such wide validity that we can predict from them with considerable confidence, and on the rare occasions when the prediction is falsified, we are inclined to look for undetected causal agencies rather than to recast our first principles. This attitude has long been normal in chemistry and physics. Its introduction into biology, however unwelcome it may be to conservative biologists, is a sign of the growing unity of science.

J. B. S. HALDANE.

PREFACE TO
THE SECOND EDITION

CYTOLOGY began by describing what sort of things cells were. It continued by inferring what things happened inside them. Its last and longest task is to discover why these things happen. The purpose of this book is to explain what questions are raised by this last kind of enquiry, and how some of them may be answered.

The first stage of simple description survives in cytology, as it must needs survive until it has embraced the whole diversity of living organisms. The different stages of development therefore continue side by side to-day. None the less the distinction of method between the old and the new is profound. To many it is so profound as to be unintelligible. Finding out *why* things happen in the cell is an entirely different matter from finding out *how* they happen. The one problem is a matter of skill and common sense. The other takes us into a new element. We are plunged into inferences, often speculative inferences, which connect mechanics, physiology and genetics. We find that the cell is part of an interlocking system of growth and reproduction, heredity and variation. Everything that happens in the cell is related to everything else that happens in the organism, or indeed has happened in its ancestors.

It is impossible at one and the same time to deal with all these dialectical relationships. I can describe only those that seem to me most important at the moment. In the first edition I took the evolutionary point of view as being the most neglected and most necessary. I devoted a last chapter to reconsidering our knowledge of the cell from this point of view; I attempted to show cell-processes as the products of an evolution of the genetic system which in changing adapts itself to its own requirements. This theory can now be applied with greater rigour in the light of our increasing knowledge of genotypic control and in the light of the comparisons and experiments embodied in the chiasmatype theory

PREFACE TO SECOND EDITION

of crossing-over and the precocity theory of meiosis. In the present edition I have therefore recast the whole account in terms of evolution. I have shown mitosis as giving rise to meiosis and sexual reproduction, diploidy to polyploidy in one direction and to complex heterozygosis in another. I have represented structural hybridity as the basis of sexual differentiation and I have deduced the conditions of parthenogenesis from experimental observations of the breakdown of sexual reproduction.

Such a treatment leaves the other relationships of cell-behaviour in the background. Of these the most important at the present moment are the mechanical relationships. The special treatment that I previously gave to evolution I now therefore devote to arranging our knowledge of chromosome behaviour and nuclear and cell-division to reveal the laws of movement and enquire into their causes. Ten years ago, in approaching these problems, we could scarcely see the wood for the trees. Now order is beginning to appear. Causal relationships are being established. The result, I believe, is unmistakable. Cell structures, having a size intermediate between those of molecular and macroscopic systems, show properties of movement and development peculiar to their own level of integration. These properties depend on the physiologically active contents of the nucleus, and since they are co-extensive with heredity as we understand it, their study becomes a necessary part of the investigation not only of the mechanics, but also of the physiology of heredity. Thus in the cell, mechanics, physiology and physical chemistry are being brought together to show the unity of living processes.

I have to thank Professor Haldane, Dr. Catcheside, Dr. Mather and Miss Upcott for reading the proofs. I am indebted to Mr. La Cour and Mr. Osterstock for many microphotographs, and to the various authors acknowledged for the use of numerous text-figures.

C. D. DARLINGTON.

TABLE OF CONTENTS

PAGE

CHAPTER I. Cell Genetics
1. The Nucleus, the Cell and the Organism 1
2. Reproduction 4
3. Heredity 12

CHAPTER II. Mitosis : The Constancy of the Chromosomes
1. The Resting Nucleus : Nucleoli 19
2. The Movements of the Chromosomes 22
3. The Structure of the Chromosomes 31
4. Abnormal Forms of Mitosis 47

CHAPTER III. Mitosis : The Variation of the Chromosomes
1. The Mitotic Constants 52
2. Genotypic Control 53
3. Numerical Variation 60
4. Structural Variation 69
5. The Comparative Morphology of the Chromosome Complement 75

CHAPTER IV. Meiosis in Diploids and Polyploids
1. Introduction 85
2. Outline of Meiosis 86
3. Meiosis in Diploids 89
4. Meiosis in Polyploids 119
5. The Conditions of Meiosis 132

CHAPTER V. Structural Hybrids
1. Classification by Function 135
2. Meiosis in Structural Hybrids 139
3. The Conditions of Metaphase Pairing 163
4. Chromosome Pairing in Undefined Hybrids . . . 168
5. Permanent Prophase : the Salivary Glands . . . 175

CHAPTER VI. The Behaviour of Polyploids
1. The Origin of Polyploids in Experiment . . . 183
2. Pairing, Segregation and Fertility 194
3. Polyploid Species 197
4. Differential Affinity : Autosyndesis 198
5. Classification of the Polyploid Species 207
6. The Development of Polyploid Species . . . 220
7. The Inference of Polyploidy 229

CHAPTER VII. The Chromosomes in Heredity : Mechanical

1. The Chromosome Theory of Heredity 244
2. Segregation 247
3. The Theory of Crossing-over 249
4. Cytological Evidence of Crossing-over 254
5. Crossing-over in Structural Hybrids 260
6. The Parallel Investigation of Crossing-over . . . 281

CHAPTER VIII. The Chromosomes in Heredity : Physiological

1. The Two Methods of Enquiry 302
2. Differentiation of Behaviour in the Cell 303
3. Balance 315
4. Inert Chromosomes and Genes 329
5. The Gene 332

CHAPTER IX. Permanent Hybrids

1. Prior Conditions 335
2. Complex Heterozygotes : *Œnothera* 338
3. Sex Heterozygotes 356

CHAPTER X. The Breakdown of the Genetic System : Adventitious

1. The Breakdown of Mitosis 395
2. The Breakdown of Meiosis 399
3. The Effects of Structural Change 419

CHAPTER XI. The Breakdown of the Genetic System : Controlled

1. Definition and Classification 434
2. Apomixis in Relation to the Life Cycle 435
3. Non-Recurrent Apomixis 442
4. Meiosis in Relation to Parthenogenesis 451
5. The Origin of Apomixis in Experiment 464
6. Replacement of Sexual Reproduction 470

CHAPTER XII. Cell Mechanics

1. Introducing an Axiomatic 479
2. The Internal Mechanics of the Chromosomes . . . 483
3. The External Mechanics of the Chromosomes . . . 492
4. Ultra-Mechanics 547

APPENDICES

I. Interpretation 563
II. Technique 569
III. Glossary 572
IV. Bibliography 584

Index 648

LIST OF TABLES

PAGE

Chapter I
Table 1. Variation in the Haploid and Diploid Generations . . 6

Chapter II
Table 2. The Time Taken by Division of the Nucleus . . . 31
,, 3. Estimated Total Lengths of Chromosomes in Complement (in microns) 31
,, 4. Relatives Sizes of " Solid " and " Vesicular " Nuclei . . 46

Chapter III
Table 5. Lengths of Chromosomes in Species and their Hybrids . 56
,, 6. Types of Chromosome Constitution 61
,, 7. Instances of Syndiploidy in Male Germinal Cells . . 65
,, 8. Analysis of Variation in *Crepis* 70
,, 9. Organisms with an Extreme Range of Size in the Chromosomes 81
,, 10. Range of Size and Number of Chromosomes in Animals and Plants 83

Chapter IV
Table 11. Examples of a Diffuse Stage at Diplotene or Diakinesis . 94
,, 12. Classification of Organisms according to the Distribution of Chiasmata in their Longest Chromosomes at Metaphase 110–111
,, 13. Lagging of Bivalents with Interstitial Chiasmata . . 115
,, 14. Observation of Meiosis in Auto-Polyploids . . . 125
,, 15. Variation in Pairing of Chromosomes in Triploids and Tetraploids 126

Chapter V
Table 16. Examples of Supernumerary Fragment Chromosomes . 145
,, 17. Chiasma Frequencies of Fragments 146
,, 18. Pairing in Tetraploid Interchange Heterozygotes . . 159
,, 19. Classification of Interchange Heterozygotes . . . 159
,, 20. Classification of Unequal Bivalents 161
,, 21. Classified Observations of Pairing of Chromosomes in Undefined Hybrids 172
,, 22. The Fluctuation of Chromosome Pairing in Hybrids . . 174
,, 23. Chiasma Frequency of Hybrids Compared with that of their Parents 175
,, 24. Observations of Salivary Glands in Diptera . . . 178

Chapter VI

Table		Page
24A.	Chromosome Association in Polyploids	185
25.	Fertility showing the Origin of Polyploids after Hybridisation	187
26.	The Origin of Allopolyploids from Hybrids	190
27.	Polyploid Hybrids with Abnormalities of Chromosome Constitution	193
28.	Autotetraploids with Reduced Fertility	194
28A.	Chromosome Pairing in Polyhaploids	201
29.	Pairing of Chromosomes in Hybrids and Hybrid Derivatives of Polyploid Species	208–213
30.	Polyploid Species of an Intermediate Type	214
31.	Pairing in an Intermediate Octoploid Plant, *Bromus erectus* var. *eu-erectus*	215
32.	Autopolyploid Species and Mutants	216
33.	Polyploid Species with Variable Numbers of Chromosomes	220
34.	Observations of Cell Size in Haploids, Diploids and Polyploids	221
35.	Measurements of Cells in Haploids, Diploids and Polyploids	223
36.	Linear Measurements of Cells in Polyploid Derivatives	224
37.	Examples of Polyploidy within the Species	227
38.	Chromosome Numbers in the Phanerogama	229
39.	Allopolyploid Species whose Origin can be Determined	234
40.	Observations of Secondary Pairing at Meiosis	242

Chapter VII

Table		
41.	Interlocking of Chromosomes at Meiosis	258
41A.	Results of Dyscentric Crossing-over	269
42.	Dyscentric Structural Hybrids	273
43.	Secondary Structural Changes due to Crossing-over after Primary Changes	280
44.	Comparative Scope and Validity of Experimental and Direct Methods of Studying Crossing-over	298

Chapter VIII

Table		
45.	Observations of Nucleolar Behaviour	306
46.	Comparison of Cycles of Differential Condensation in Different Organisms	310
47.	Distribution of Chromosomes at Meiosis	318
48A.	Trisomic Organisms showing Differential Action of Chromosomes	319
48B.	Unbalanced Clonal Forms	319
49.	Transmission of Extra Chromosomes of Triploids to their Progeny	320
50.	Transmission of Extra Chromosomes of Trisomics to their Progeny	321
51.	Pollen Abortion in *Datura*	321
52.	Effects of Abnormal Chromosome Constitution on Viability	322
53.	Distribution of Extra Chromosomes of the Triploid *Nicotiana paniculata* ($n = 12$) × *N. rustica* ($n = 24$) in the Backcrosses to its Parents	327

LIST OF TABLES

Chapter IX

Table		Page
54.	Summary of Ring-forming Properties in Forms of *Œnothera*	344
55.	The Numbers of Species, Hybrids and Mutants in *Œnothera* having Different Types of Configurations Compared with their Frequency if Segments were Assorted at Random	349
56.	Representative Examples of Sex Chromosomes	366
57.	Examples of Haplo-Diploid Sex Differentiation	377
58.	Primitive and Advanced Types of Sex Determination	393

Chapter X

Table		Page
59.	Chromosome Pairing in Two Strains of *Crepis capillaris* and in their Hybrids by *C. tectorum*	400
60.	The Causal Relationships of Different Kinds of Non-pairing of Chromosomes at Meiosis	401
61.	Exceptional Failure of Pairing not Due to Hybridity	402
62.	Classification of Genotypically Conditioned Abnormalities of Meiosis	404
63.	Numbers of Cells in "Tetrads" of Haploid Plants	416
64.	Progeny of Haploids	417
65.	Cases of Non-Reduction	417

Chapter XI

Table		Page
66.	Cases of Non-Recurrent Parthenogenesis in the Angiosperms	444–446
67.	Methods of Suppressing Meiosis with Diploid Parthenogenesis	450
68.	Methods of Apomixis in *Artemisia nitida*	452
69.	Chromosome Numbers of Apomictic Species	468
70.	Cases of Polyploidy within the Species Associated with the Occurrence of Apomixis	469

Chapter XII

Table		Page
71.	Observations of Chromosome Coiling	491
72.	Examples of Arrest of Terminalisation by Change of Homology	511
73.	The Change in Chiasma Frequency from Diplotene to Metaphase with Partial and Complete Terminalisation	513
74.	Abnormal Spindles	535
75.	Names Applied to the Centromere and the Centric Constriction	536
76.	Relationship of the State of the Centromere to the Movements of the Chromosomes	540
77.	Mechanical Types of Chromosomes	554
78.	Analysis of Certain Simple Structural Changes in Terms of Breakage and Reunion	557

LIST OF PLATES

Microphotographs of living and fixed cells in sectioned and smear preparations.

		PAGE
I.	Root-tips of *Crepis dioscoridis* and *C. capillaris*	22
II.	Pollen-grains of *Paris quadrifolia*	46
III.	Mitosis in animals and plants. 1 and 2. Testes of *Chorthippus parallelus* 3. Testis of *Locusta migratoria* 4. Root-tip of *Eremurus spectabilis* 5 and 6. Pollen grains of *Tradescantia bracteata* and *Brodiaea uniflora* 7 and 8. Root-tips of *Puschkinia libanotica* and *Tricyrtis hirta*.	78
IV.	First metaphase of meiosis showing major spirals in *Hyacinthus orientalis*, *Tradescantia virginiana*, and *T. rubra*	118
V.	Pollen mother-cells of triploid plants. 1 and 2, zygotene in *Tulipa*. 3, First division in *Fritillaria*.	126
VI.	First metaphase in pollen moter-cells of interchange hybrids: *Campanula*, *Rhœo*, *Œnothera* and *Pisum*	152
VII.	Salivary gland chromosomes in Diptera, *Chironomus* and *Drosophila*	178
VIII.	Secondary pairing at first and second metaphase of meiosis in pollen mother-cells of polyploid plants. *Cydonia*, *Pentstemon*, *Dahlia*, *Prunus*, *Æsculus*	236
IX.	Results of Crossing-over in inversion heterozygotes, animal and plant: *Chorthippus*, *Tulipa*, *Pæonia*, *Podophyllum*	274
X.	1–5. Meiosis in *Phragmatobia* 6. Sex chromosomes in *Melandrium* 7. Meiosis in monosomic *Zea Mays* 8 and 9. Polymitosis in pollen grains of *Zea Mays*.	358
XI.	Sex chromosomes in *Chorthippus* (Orthoptera), *Hemerobius* (Neuroptera), *Putorius*, man and the rat (Mammalia)	388
XII.	Mitosis showing the formation of dicentric chromatids following X-ray treatment and crossing-over in inversions: *Tulipa*, *Crocus*, *Allium*, *Vicia*	430

LIST OF PLATES

		PAGE
XIII.	1 and 2. Mitosis in the pollen-tube of *Tulipa* 3–8. Natural and artificial uncoiling of spirals in *Paris* and *Tradescantia*	482
XIV.	1–3. Meiosis in the oöcytes of *Allolobophora* 4 and 5. First anaphase in *Kniphofia* 6. First metaphase in *Mecostethus* (♂) 7. Mitosis in the embryo of a fish, *Coregonus*	522
XV.	The centromere at meiosis in pollen mother-cells of *Agapanthus, Fritillaria, Alstrœmeria, Galtonia* and *Tradescantia*	540
XVI.	1–3. Living and fixed spermatocytes of *Stenobothrus* 4–7. Ultra-violet photographs of living spermatocytes of *Melanoplus*	564

RECENT ADVANCES IN CYTOLOGY

CHAPTER I

CELL GENETICS

The Cell and the Nucleus—Differentiation—Reproduction—Asexual and Sexual—Haploid and Diploid Phases—Sexual Differentiation—Genotype and Environment—Their Bearing on Chromosome Behaviour—And on Sexual Differentiation.

> quippe, ubi non essent genitalia corpora cuique,
> qui posset mater rebus consistere certa ? [1]
> LUCRETIUS, *De Rerum Natura*, I, 167—168.

1. THE NUCLEUS, THE CELL AND THE ORGANISM

(i) **The Problem.** The object of the present work is to describe the bodies responsible for heredity, to show how they live, move and have their being. We now know that the Atomists were right ; there are such bodies. But before considering them we must see how organisms are constructed, how they reproduce and how heredity expresses itself. Then we can examine the *genitalia corpora* and better understand their relations with heredity.

(ii) **Structure.** The living organism contains certain materials which for various reasons are held to be non-living. Such, for example, are the skeletal structures of plants and animals, consisting largely of cellulose, calcium carbonate or chitin, their storage products, such as starch, fats and glycogen, and their fluid contents, such as cell sap and blood plasma. The rest of the plant or animal body is described for convenience as the living substance or *protoplasm*. Various non-living substances, such as water, can often

---

[1] For, if each organism had not its own begetting bodies, how could we with certainty assign to each its mother ?

be separated from it without altering its essential character. Protoplasm is an organ of behaviour, not a chemical entity.

In certain organisms no differences of structure have yet been detected between different parts of the protoplasm. In most bacteria this is perhaps due to their small size. In the Cyanophyceæ and larger bacteria, differences have been found, but their meaning is still doubtful. Elsewhere it is always possible to distinguish between a small dense *nucleus* and the rest of the living substance, which is called *cytoplasm*. In the cytoplasm other bodies can be very generally made out ; they are known generically as chondriosomes and plastids. They are very widely recognised, but their relation with one another in different groups of organisms are not always clear. It is known, however, in the flowering plants that plastids differ in their potentialities for developing pigment and sometimes transmit these differences permanently and independently of other genetic influences. This can be shown by breeding experiments, but not by direct observation (Renner, 1934). The nucleus, on the other hand, can be seen in all stages of development. It is always carried by the germ cells of both parents in sexual reproduction and is the only structure of which this is known. It is recognised customarily by its characteristic method of propagation. All nuclei arise by division of a pre-existing nucleus into two. To do this, the mother-nucleus resolves itself into a number of double bodies, the chromosomes, whose halves separate to form two daughter nuclei which are exactly equivalent, and all the descendants of a nucleus derived in this way have the same complement of chromosomes. This process is known as *mitosis*.

All organisms may be said to arise from bodies of protoplasm with single nuclei. Each of these is described as a single *cell*. In the lowest organisms with nuclei, the Protista, growth of this cell is followed by its division, together with that of its nucleus, by mitosis, and the separation of the daughter cells. The organism thus remains *unicellular* and every mitosis is an act of reproduction, but instead of one individual begetting another, one individual so far as we can see simply becomes two. The external simplicity of its organisation is not always associated with simplicity of structure within the cell. Indeed, while the Protista embrace the simplest organisms they also include those with the most complex development within the cell. This range of cellular evolution is matched by a range of form of the nucleus and of its methods of division greater than that found in all the higher animals and plants (v. Ch. II), as is indeed appropriate in what must necessarily be the oldest group of organisms. In many of the lower organisms (Myxomycetes, Phycomycetes, some Chlorophyceæ) division of the nucleus is not accompanied by division of the cell, that is of the whole organism, which in consequence comes to contain many nuclei; it is *multinucleate* although we may still take it to be unicellular. Elsewhere, division of the cell into two compartments takes place at the same time as division of the nucleus. The daughter cells may be separated by non-living secretions of the cell with a connective or cementing function, or by differentiated parts of their own cytoplasm. The organism developing in this way is said to be *multicellular*. It reproduces by the separation of either single cells or groups of cells from the main body. Both unicellular and multicellular organisms may pass through a multinucleate stage of development, *e.g.*, at spore formation in many Protista, in the germ tubes of the Basidiomycetes and in the pollen tubes of flowering plants.

Apart from these special conditions the term "cell" is a convenient designation for a separate body of protoplasm containing a single nucleus.

(iii) **Differentiation.** Certain cells may reproduce themselves by mitosis without change indefinitely under constant conditions. This is true of many Protista, as well as of many young cells of the higher animals in tissue culture. Change of conditions leads to a change in appearance or behaviour affecting the whole life of the organism, such as encystation in the Protista. But even under constant conditions the products of cell division are usually dissimilar from the parent cell. In the unicellular Protista in which a series of different forms occur in regular or irregular sequence, making up the "life cycle," the daughter cells are usually like one another but unlike the parent cell. This may be spoken of as differentiation-in-time. In multicellular organisms one of the daughter cells is usually unlike either the parent or its sister; in this way the organism comes to consist of many cells of different forms and properties. This may be spoken of as differentiation-in-space-and-time. With higher organisation (both in the Protista and elsewhere) external conditions have less and less to do with the direction of differentiation of particular cells although they continue to modify its rate and consequently affect the result in detail. We may say that, in general, the greater the internal capacity for differentiation the less the special reaction to external changes.

2. REPRODUCTION

(i) **Asexual Reproduction.** Growth and reproduction equally depend on increase in size and in organisms having a differentiation of their substance into nucleus and cytoplasm, both are necessarily related to nuclear division.

In the unicellular Protista, in which each individual has a single nucleus, or a pair of complementary nuclei, division of the nucleus entails division of the whole organism: mitosis is an act of reproduction. In multicellular organisms, division is merely a concomitant of growth. In these under suitable conditions such growth, like reproduction in the Protista, may continue indefinitely. Such conditions occur in the higher plants where vegetative propagation can be carried on without limit, either naturally in the formation of bulbs, runners, viviparous shoots or apomictic seeds (to which special consideration will be given later) or artificially in propagation by cuttings and layers, buds and grafts. Many species of plants and animals depend for their preservation on these purely asexual or " vegetative " means of propagation. This is usually associated with great uniformity within the species, which in plants may be described as " clones."

Reproduction by mitosis alone can therefore continue indefinitely. Cases where it has failed to do so and where vitality can then be restored by sexual reproduction are due to special adaptation to conditions of sexual reproduction.

(ii) **Sexual Reproduction.** Sexual reproduction is known in all groups of living organisms with nuclei. In the Flagellata it has only been established in three genera (*cf.* M. Robertson, 1929) and in the Protista as a whole it is difficult as yet to estimate its importance. Elsewhere, however, whether normally or in some modification, it is an essential part of the life cycle of nearly all species.

Sexual reproduction consists superficially in the formation and separation from the rest of the organism of single cells, the germ-cells or *gametes* (constituted by the whole organism, in the Protista, and by a specialised part of it, in the higher plants and animals) and their subsequent fusion in pairs to give new cells, known as *zygotes*.

Its essential genetic characteristics are two. The first was seen by Oscar Hertwig (in 1875) to consist in the fusion of the nuclei of the two gametes. Since they carry the same number of chromosomes the product has a double or *diploid* number. This

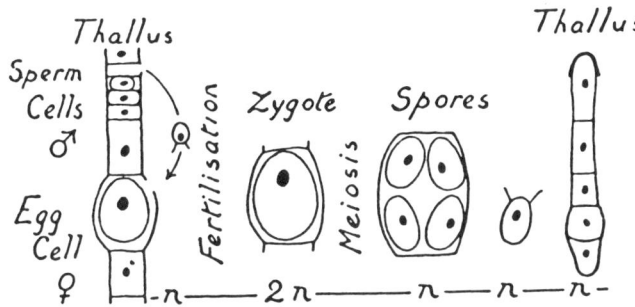

FIG. 1.—The life cycle of an alga such as *Œdogonium*. A haploid thallus produces haploid sperm and egg cells (n) which fuse to give a diploid zygote ($2n$). This undergoes immediate meiosis to give haploid spores from which a new thallus develops.

is *fertilisation*. The second was predicted by Weismann (in 1887). It consists in the compensating process of reduction or *meiosis*. In a *mother-cell* two nuclear divisions follow one another rapidly while the chromosomes only divide once, so that four nuclei are formed, to each of which a halved or *haploid* number of chromosomes is distributed.

The recurrence of fertilisation and meiosis gives two types of nuclei in each sexually reproducing species, the diploid or double nucleus of the zygote and the haploid or halved nucleus of the gamete. Either or both of these may, in different species, continue to multiply by mitotic divisions, giving a number of unicellular organisms or one multicellular one, so that we have three types of reproductive cycle, according to the importance of the diploid or haploid phase, as follows :

(a) Meiosis immediately follows fertilisation (" zygotic reduction "). The organism is then always haploid, apart from the zygote.

(b) Fertilisation immediately follows meiosis (" gametic

TABLE I
Variation in the Haploid and Diploid Generations

	Purely Haploid, " Zygotic Reduction."	Predominantly Haploid (phases dissimilar).	Equally Haploid and Diploid (phases similar).	Predominantly Diploid (phases dissimilar).	Purely Diploid, " Gametic Reduction."
PROTISTA.	Flagellata (some)[1] Sporozoa-Coccidia. (e.g., Aggregata Dobell, 1925). Sporozoa-Gregarina (some).	—	—	—	Rhizopoda (some, e.g., Actinophrys, Belar, 1922).[1] Ciliata (e.g., Paramœcium, Jennings, 1920). Sporozoa-Gregarina (some).[2]
PLANTS.	Charophyta. Chlorophyceæ (most). Phycomycetes. Ascomycetes.[4] Basidiomycetes.[4]	Bryophyta.[3]	Chlorophyceæ (e.g.,Cladophora, spp., Schussnig, 1930, 1931). Rhodophyceæ (most). Phæophyceæ (most).	Phæophyceæ (e.g., Laminaria). Myxomycetes (probably). Pteridophyta. Phanerogama.[2]	Chlorophyceæ (e.g., Cladophora, spp., Higgins, 1930). Phæophyceæ (e.g., Fucus).[6]
ANIMALS.			Icerya.		All Metazoa.[5]

[1] Most are non-sexual.
[2] Subordinate phase parasitic on predominant phase.
[3] Some have one mitosis in the haploid phase.
[4] v. Infra.
[5] Except males of those with haplo-diploid sex-differentiation (v. Ch. IX.).
[6] One (female) and four (male) mitoses in the haploid phase.

reduction "). The organism is then always diploid, apart from the gametes.

(c) A mitotic generation is intercalated after each process. There is then an alternation of haploid and diploid generations. The single cell from which the haploid generation is derived, following meiosis in the diploid, is called a *spore*. The spore mother-cell of the diploid plant, the *sporophyte*, produces haploid spores from which the haploid *gametophytes*, known as " prothallia " in the

Pteridophyta, are developed, and these bear the haploid gametes which fuse and give the new sporophyte. The gametophyte and sporophyte may be indistinguishable from one another morphologically (as in most Rhodophyceæ) save in the production of sexually differentiated germ-cells by the gametophyte and of non-motile spores by the sporophyte. Usually, however, the two generations are sharply distinct and the less important generation (as in the Bryophyta and in the flowering plants) may be parasitic on the more important one. Table I classifies the main groups of animals and plants in this respect (*cf*. Belar, 1926 *b* ; M. Robertson, 1929 ; Naville, 1931, for Protista ; Correns, 1928, for Phanerogama ; Hartmann, 1929, *a* and *b*, for lower plants ; Witschi, 1929, for Metazoa).

(iii) **Sexual Differentiation : The Gametes.** Sexual differentiation consists in the production by the species of two kinds of germ-cells which are complementary. It is determined by the same kinds of conditions as all other differentiation. But it has such special importance in reproduction and in the evolution of particular genetic mechanisms that it requires special consideration.

Sexual differentiation is not found in the Flagellata, or in many Phycomycetes and Chlorophyceæ, and it is negligible in the Rhizopoda, Gregarina (Sporozoa) and in the Basidiomycetes and other Fungi.

But in all other sexually reproducing organisms including many Algæ and Protozoa there is a differentiation between a small, active or motile *male* gamete, a spermatozoon or spermatozoid, and a large, inactive or non-motile *female* gamete, the egg cell, usually charged with nutritive material and having a larger nucleus.

When the male gamete, or the spore where there is a haploid generation, arises at meiosis, the cytoplasm of its mother cell is equally divided among the four daughter cells (except in certain Diptera, *cf*. Metz, 1926, and in the Cyperaceæ, Piech, 1928). The female gamete arises (except in certain Mollusca, Turbellaria and oögamous Algæ) by the suppression of all but one of the products of meiosis. This one becomes the egg nucleus, while the other products are extruded as the " polar bodies," which at once degenerate. Where a short haploid generation occurs (reduced and specialised for the production of the female germ-cells, as in the higher plants) an intermediate condition is found. The four potential spores are almost equal; the gametophyte may then develop from one (*cf.* *Œnothera*, Ch. IX), or from the collaboration of all four.

It is of particular importance in studying the mechanism and

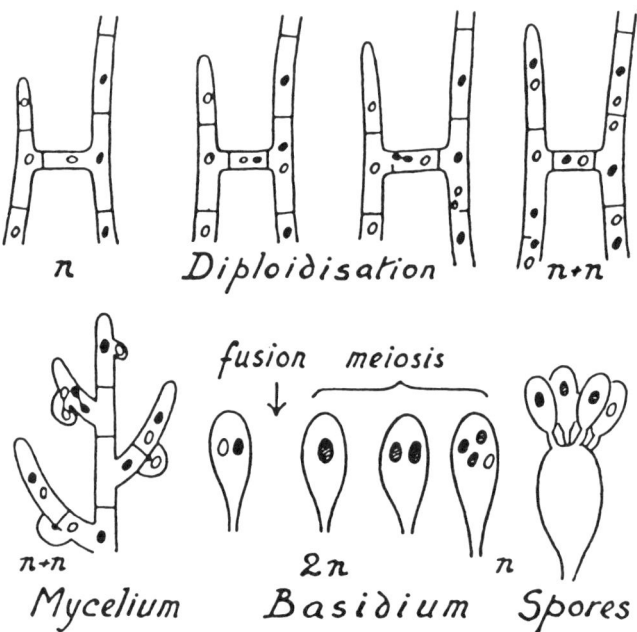

Fig. 2.—Life cycle of a basidiomycete, showing diploidisation, by which the haplophase becomes a diplophase, the division of conjugate nuclei and the cells containing them, and the occurrence of meiosis immediately after the fusion of these haploid nuclei (*cf.* Buller, 1931).

material basis of heredity to examine the male gamete—sperm, " antherozoid " or " generative nucleus " of the pollen grain. The female gamete is usually an over-developed cell, the male an under-developed one; that is to say, its constituents are reduced, and by studying their reduction we can see what is essential, and what unessential, in carrying the contribution of one parent in heredity.

Its development (v. Belar, 1928 b) shows that the simplest type of spermatozoon in the higher animals (e.g., *Stenobothrus*) consists of four constituents: (i) Nucleus, making the "head" of the sperm; it is dense and stains deeply like the mitotic chromosomes. (ii) Centrosome, a cytoplasmic body, lying at the base of the nucleus, forming the "middle-piece" and often divided; from it develops the axial filament of the "flagellum" or tail. (iii) Mitochondria, cytoplasmic bodies making the sheath of the axial filament. (iv) Undifferentiated cytoplasm making a thin coat over the head and flagellum—part of the cytoplasm of the parent cell having often been rejected. (*Cf.* Bowen, 1924; Schrader, 1929.)

Sharp (1920, *et al.*) has shown that the development of the male gamete, the "spermatozoid," in the Pteridophyta follows the same course. It is possible to trace the origin of the spermatozoid from its antecedents: the material of the nucleus and the "blepharoplast," which corresponds to the centrosome and the middle-piece of the animal spermatozoon; it, also, is concerned with the development of cilia on the coat of the spermatozoid. The cytoplasm from the mother-cell is left behind by the ripe gamete. In the flowering plants it is possible to show genetically and cytologically that in some species plastids are carried into the egg cell by the pollen tube, in others nothing but the nucleus passes from the male parent in fertilisation.

As a constituent of the male gamete the nucleus alone is constant in its contribution. All functional spermatozoa contain a nucleus which can be traced to one of the four bodies of chromosomes separated at meiosis. In certain organisms such as the flat-worm *Ancyracanthus* (Mulsow, 1912) and *Filaria* (Meves, 1915) the chromosomes can be counted as discrete bodies in the live and ripe sperms. The non-nuclear constituents, on the other hand, are variable in the proportion of their contribution. The same is true of the antherozoids of the lower plants. The centrosome, the only other permanent structure constantly carried by the sperm, is actually the one which is lost by the egg and transmitted on the male side only.

Both sperm and pollen grains are formed on occasion, particularly in hybrids with irregular meiosis, with a disorganised or incomplete nucleus or with a multiple nucleus. Spermatozoa with either too little (" oligopyrene " or " apyrene "), or too much, nucleus may develop the external apparatus of a sperm, or even function to give progeny. Kuhn (1929 b) found that sperms lacking certain chromosomes (the X and Y) functioned in fertilisation in *Drosophila*.

When pollen grains have too little nucleus, on account of their more complicated vegetative life, they do not produce functional gametes or even develop as far as their vegetative division (*cf.* Ch. VIII). When grains with too much nucleus function, they are of great importance in connection with the origin of polyploids and unbalanced forms (Chs. V, VIII, and X).

The fusion of the nuclei in fertilisation usually takes place directly, but the process shows great variation. The two bodies of chromosomes may first unite at the mitosis which begins immediately before the nuclei meet (*e.g.*, *Ascaris*).

Two special types of fertilisation require consideration. In some fungi, fusion of nuclei does not immediately follow the fusion of their cells. In the Basidiomycetes the haploid spore develops into a mycelium containing haploid nuclei. The mycelium is at first multinucleate, but later cell-walls are formed so that each cell contains one haploid nucleus ; this is the *haplophase*. Pairs of cells then fuse, derived from the same plant or mycelium in " homothallic " species or from different mycelia in " heterothallic " species. But their nuclei do not fuse ; they exist and divide at each cell division, cohabiting throughout the *diplophase*. Eventually these " conjugate nuclei " fuse in the reproductive organ, the basidium, and the next division is meiosis which gives the four haploid spores. Thus, while the *cells* are diploid in the diplophase, the *nuclei* are haploid. Similar conditions are found in the Ascomycetes (*cf.* Winge, 1935, on *Saccharomyces*). The significance of this remarkable behaviour will be considered later (Fig. 2).

In the flowering plants one kind of gametophyte, the pollen grain or microspore, produces the male gamete, its " generative nucleus," after undergoing two haploid mitoses (*cf.* Geitler, 1935), while the other, the embryo-sac cell or megaspore, produces the female gamete, usually after three mitoses, together with seven other nuclei. Fertilisation is a double process, as shown by Navashin and Guignard in 1899 (*cf.* Stenar, 1928; Jørgensen, 1928; Dahlgren,

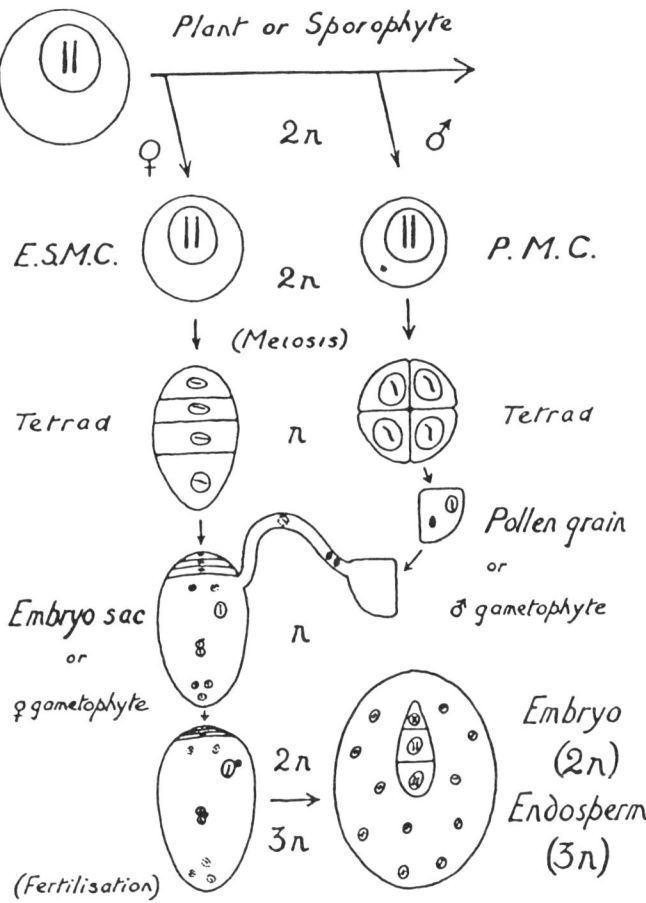

Fig. 3.—Life cycle of a flowering plant, an hermaphrodite dicotyledon, showing the history of one pair of chromosomes in the haploid, diploid and triploid phases.

1927, 1930; Gerassimowa, 1933 on *Crepis;* Poddubnaja-Arnoldi and Dianowa (1934) on *Taraxacum;* Hoare (1934) on *Scilla;* Mahony (1935) on *Fagopyrum; cf.* Shimamura (1935) on *Ginkgo).*

The male gamete migrates through the cytoplasm of pollen-tube and embryo-sac and fuses with the egg nucleus. A second, sister, generative nucleus fuses with the product of fusion of two " polar nuclei " in the embryo-sac to give the " triple fusion-nucleus " from which the nutritive endosperm develops, a new " triploid " organism with three nuclear complements, and independent of its sister embryo (Fig. 3). It has been shown that this second fusion may sometimes fail in *Zea*.

The female gametophyte, or embryo-sac, and the two new plants, embryo and endosperm derived from it, although morphologically mere organs of the parent sporophyte, are genetically three distinct individuals. Since the endosperm is preponderantly maternal in origin (as shown cytologically by its origin and genetically by the properties it inherits) and its development is essential to the development of the embryo, for which it provides nutritive material, it follows that differences in seed production in reciprocal crosses between species, such as are frequently observed in the angiosperms, may be referred to the different relationships between embryo and endosperm (*cf.* Thompson, 1930; Watkins, 1932; Kihara and Nishiyama, 1932) as well as to the relationship between the pollen and the maternal tissue which is concerned with incompatibility.

The new individual always arises from the fusion of the two germ-cells, and no others contribute normally. But it has been possible to show genetically that two fertilisations may take place side by side and the products join together to give a mosaic individual (*cf.* Ch. XI). In such cases probably another cell, the second polar body, functions as an auxiliary egg-cell. It is possible also that two pollen grains may function in fertilising the same egg-cell, both fusing with its nucleus, to give a triploid individual (Ishikawa, 1918, on *Œnothera*).

3. HEREDITY

(i) **Genotype and Environment.** Heredity is the property that organisms have of reproducing their like. It must be supposed, deterministically, that organisms carry something that they give to their offspring, and that this " something " determines the likeness. Under the simplest conditions, in the asexual reproduction of a clone or the sexual reproduction of an inbred race, it is always found that the likeness is imperfect ; further, the lack of likeness is not inherited, but is evidently due to the effect of different external conditions on the organism. This principle was first clearly recognised by Louis de Vilmorin. It was expressed by Weismann in the notion that the likeness was due to the constant properties of the *germ-plasm*, which was handed down unchanged from generation to generation. Finally, it was developed by Johannsen (1911, *et al.*) into the notion that the properties of an individual were the product of the reaction of its *genotype*, or type of germ-plasm, and the environment. On this notion genetical theory is built.

Environment requires a word of explanation. A fern may abnormally produce a prothallium vegetatively, and although the two must be supposed to have the same genotype their growth is entirely different. The reason is that although the two are separate individuals their development is conditioned by the circumstances of their origin, and the difference between them is simply a form of differentiation, the method of which is, like all other properties, a product of the genotype-environment reaction. So also with the various kinds of life cycle in the Protista : these are in different degrees subject to the variation in external conditions, but the different degrees of this subordination are to be considered a property of the genotype. Thus the visible properties of the organism (its phenotype) are the product of the reaction of its genotype with its environment. Further—and this is most important in the present study of cytology—differences in the properties of individual cells and nuclei of organisms may be referred to one of four determining conditions : (i) differences of genotype ; (ii) differences in relation to development or life-history (*i.e.*, in the differentiation in space or time) ; (iii) differences in environment. A fourth condition will later be seen to be important in the special circumstances of germ-cell formation in hybrids—the occurrence of differences between corresponding chromosomes. These conditions must constantly be borne in mind in attempting to analyse the behaviour of chromosomes.

When therefore a particular character in an organism is said to be *due* to heredity or *due* to the environment, or a particular problem is said to be " genetical " or " physiological," the statement is merely relative and depends on convenience of comparison with some ideal or " normal " heredity or environment. This implication is often forgotten because the " normal " condition seems so obvious. Yet it is not at all obvious in cases where a condition of revealing the character is, as is often the case in considering chromosome behaviour, differentiation within the organism. An abnormal property of part of the cells in one organism may be said to be determined by an environmental difference between these cells and the rest of the organism ; it may also with equal correctness be said to be determined by a genetic difference between this organism and other organisms. For example, the formation of testes and ovaries by different cells of the same hermaphrodite is primarily a genetic property, just as is the formation of testes and ovaries in different males and females of a species, but while the first is immediately determined by a developmental difference (differentiation) the second is immediately determined by a genetic difference.

Take another instance of great importance in cytology where, after a number of mitotic divisions, a diploid nucleus suddenly divides by meiosis. The difference in behaviour of the nucleus at different times shows it to be subject to differentiation, as are the larger structures. It is also subject to genetic control, as shown by comparison with other organisms having no meiosis. Where, on the other hand, meiosis occurs immediately after fertilisation, it is equally possible to regard it as determined directly by the diploid condition and not by genetically controlled differentiation since all the following mitoses might equally take the form of meiosis in a diploid nucleus.

Now, our task is to find the materials whose properties answer to the abstract description of the genotype. They must be embodied in " permanent " structures; that is to say, the structures themselves must be stable, and they must be capable of reproducing themselves identically.

These permanent structures are found in the nucleus at the resting stage and in the chromosomes during mitosis. The nucleus and its chromosomes alone amongst visible structures are handed down from cell to cell and from generation to generation in all organisms. In describing them it will be necessary to show their bearing on the genotype and *vice versâ*. For this purpose a point of view is necessary rather different from the usual one, for while the chromosomes contain the " something " which we identify with the genotype they themselves cannot be directly identified with it. Their form and behaviour have special properties of continuity and autonomy, but, as we have already seen in regard to meiosis, they are nevertheless controlled within certain limits by this "something." This also will have to be constantly borne in mind.

(ii) **Conditions of Sexual Differentiation.** Sexual differentiation consists in the production of two different kinds of germ cells by the sexually reproducing species, and this, as already shown, can be attributed to differences in one of three conditions: environment, position in development, genotype. The first two arise without genetic differences, the third by the segregation of such differences at meiosis. A few examples will provide an illustration of the characteristic way in which these three methods operate.

(a) *Environment*. When a fertilised egg of *Bonellia viridis* (Echiuroidea) falls free it develops into a female; when it falls on the proboscis of a female it develops into a male which is parasitic on the female (Baltzer). Differences of genetic constitution and of position in development therefore have no influence on the kind of gametes produced by the organism. The external environment alone is decisive. In the worm *Dinophilus*, a parental difference is decisive, for large eggs develop into females and small eggs into males (*cf.* Goldschmidt, 1920; Shen, 1936).

(b) *Development*. (a) Differentiation in space: in hermaphrodite plants and animals the kind of gamete to be produced by a given mother-cell is determined by its position in the body. In *Actinophrys sol* (Belar, 1926 b) the two products of one cell-division remain attached and undergo meiosis side by side, each expelling two " polar bodies." They then fuse, the one which completes its reduction first taking the initiative. This is the minimum of differentiation and is probably conditioned by difference of position as in a multicellular hermaphrodite. (b) Differentiation in time: in some plants or animals (*e.g.*, protandrous Mollusca, *Rana* sometimes and *Cannabis sativa*) the kind of gamete is determined by the period of its development, and an individual, which at one time produces exclusively male gametes, later produces exclusively female.

(c) *Genotype.* In *Sphærocarpus Donnellii* (Allen, 1919 ; Lorbeer, 1927) the diploid sporophyte produces two kinds of haploid spores. One develops into prothallia with only female gametes, the other into prothallia with only male gametes independently of any variation in environmental or developmental conditions.

Thus it will be seen that in the second type, differentiation of cells arises directly within the individual. It is a primary differentiation. In the first and third types it arises from a differentiation of the individuals which bear the cells. It is a secondary differentiation. The differentiations within and between individuals may equally be genetically determined, and this introduces a source of confusion. The genetic determination of sex within an individual depends on genetically determined differentiation within that individual, and does not differ in this from any other such differentiation. The genetic determination of sex between individuals depends on a special mechanism, the segregation of dissimilar genotypes in a heterozygous parent at meiosis. The investigation of this mechanism is an important part of chromosome studies. But it must be remembered that the chromosome mechanism of segregation is merely the switch which sets the organism on one of two alternative paths of development.

This classification is sufficient for a large number of cases, but for others a further analysis is convenient. Thus in the Metazoa, not the haploid but the diploid is genetically differentiated in regard to the sex of the gamete produced. And since the two sexes are equally produced by the fusion of the opposite kinds of gametes one of these must be composed genetically of two types. The kind of gamete formed is not therefore determined by its own genetic constitution, but by that of the diploid parent which bears it. The same principle applies to the flowering plants where the sexual properties of the whole haploid generation are similarly determined by the constitution of the parent, whether male, female, or hermaphrodite, on which it is, in effect, parasitic. Thus the immediate condition of this differentiation is an environmental difference, but a prior condition is a genetic difference.

The special conditions of haplo-diploid sex differentiation will be considered later.

Since sexual differentiation necessarily restricts the possibility of fusion to gametes of opposite character and therefore often of different origin, other types of differentiation having a *restrictive* effect on fusion are often classified with it. Thus in certain Protista, Fungi, and Algæ, fusion is restricted to cells having a different genetic constitution. This is described as " heterothallism " in the Phycomycetes and Basidiomycetes, and " physiological anisogamy " in the Chlorophyceæ and "relative sexuality" in *Ectocarpus* (Hartmann, 1929). Such behaviour is analogous in genetic cause and contraceptive effect with self-sterility in the flowering plants ; all these kinds of *incompatibility* demand fusion of gametes that are *genetically* different. The genetical types of gametes are usually of several different kinds, and the system is not alternative. Sexual differentiation, on the other hand, is strictly alternative, and its essential property is the provision of gametes so differentiated as to secure readier fusion and more successful development of the product. *Sexual differentiation* demands the fusion of gametes that are *morphologically* different.

Incompatibility and sexual differentiation are therefore distinct in their genetical basis and in their physiological function. When sexual differentiation comes secondarily to be determined by genetic differentiation, the two agree in securing, wholly or partly, the fusion of genetically distinct gametes, *i.e.*, hybridisation, in a broad sense. The importance of distinguishing them, however, remains, for incompatibility may be superimposed on sexual differentiation of gametes in hermaphrodites and serve the end of encouraging hybridisation where genetic determination of sex has failed to do so.

To conclude, botanists and zoologists who take too little account of one another's work have introduced a confusion into the use of the word sex. This confusion can be avoided by using the word with a consistent genetic meaning in all plants and animals. It will be applied to the two following relationships :—

1. *Sexual Reproduction, i.e., propagation co-ordinating meiosis and fertilisation in the life cycle.* It has originally the function of recombining genetic differences. Its antithesis is asexual reproduction, where the co-ordination does not occur either through the absence of the system or, as in Apomixis, through its breakdown, partial or complete.

2. *Sexual Differentiation, i.e., the production by a sexually reproducing group of organisms of larger (female, ♀) germ cells, which are fertilised by smaller (male, ♂) germ cells.* Two degrees of sexual differentiation occur ; first, differentiation between the cells themselves, or heterogamy ; and secondly, differentiation between the individual organisms that bear them, or diœcism. Heterogamy has originally the function of economy in allowing fertilisation by motile germ cells at a distance to be combined with the provision of food materials for the product by non-motile cells. It thus increases the scope of cross-fertilisation and enlarges the interbreeding group. Its antithesis is isogamy. Diœcism has originally the function of ensuring cross-fertilisation and of differentiating the work of reproduction. Its antithesis is monœcism or hermaphroditism. Incompatibility and heterothallism are restrictions of sexual reproduction, analogous in effect to sexual differentiation, but distinct from it inasmuch as they do not depend on a contrast in size of the germ-cells and may be found with it or without it.

CHAPTER II

MITOSIS: THE CONSTANCY OF THE CHROMOSOMES

The Resting Nucleus—The Behaviour of the Chromosomes—Abnormal Forms of Mitosis—The Structure of the Chromosomes—Constrictions—Spiral Structure—Permanence and Division of the Chromosomes.

> Make me to see't; or at the least so prove it
> That the probation bear no hinge or loop
> To hang a doubt on.
> SHAKESPEARE. *Othello*, Act III, Scene 3.

1. THE RESTING NUCLEUS : NUCLEOLI

THE nucleus is a body denser than the surrounding cytoplasm, for in centrifuged cells it passes to the periphery of the cell (Beams and King, 1935). It is also more viscous, and dehydration shrinks it less than the cytoplasm. It is usually globular, except in certain specialised tissues or when it degenerates. It is a smooth-surfaced body with a sharp boundary between it and the cytoplasm. There is no reason to suppose that it is enclosed by a membrane distinct from the materials within and without it, although such a membrane may be formed under the stimulus of micro-dissection. On the other hand the cyclical changes that will be described in the life of the nucleus are not related to similar changes in the cytoplasm. Their interface must, therefore, behave as a kind of semi-permeable membrane separating two independent systems (*cf*. Ch. XII).

The living nucleus is often optically homogeneous, but, as will be seen, this does not necessarily imply any physico-chemical homogeneity.

The size of the nucleus is subject to great variation, not only as between different organisms, but also within the same organism. In bulk it may best be compared with the chromosomes into which it resolves itself at mitosis, for these are of relatively constant size. The micronucleus of the infusorian is little if at all bigger than the mitotic chromosomes. The nucleus in the male gametes and in some body cells may be only two or three times their bulk. But in the egg cells of some of the higher animals and in the vegetative nuclei of pollen grains, for example, the nucleus may be 10^4 or 10^5 the size of the chromosomes (*e.g.*, *Malva*, Nemeč, 1910, Tischler, 1922).

Special genetic conditions may affect the relation of the bulk of the chromosomes to that of the nucleus, for in the male bee with half the number of chromosomes of the female the nuclei are of similar size in corresponding cells (Nachtsheim, 1913, *cf.* Schrader and Hughes-Schrader, 1931 ; Torvik, 1931 ; *v.* Ch. IX).

This variation in size bears some relation to the variation in size of the cell in the same organism (the Kern-plasma ratio). When different organisms are considered the volume relationship is apparently subject to genetic control (*e.g.*, *Tradescantia*, Ch. III, Fig. 12). Under the same genetic conditions doubling of the size of the nucleus (by doubling the number of its constituent chromosomes) leaves the volume relationship unchanged in most cases (Ch. VIII).

In almost all resting nuclei are embedded one or more amorphous, spherical or occasionally rod-shaped bodies of higher density and refractiveness than the surrounding ground substance (Fig. 4). These are called nucleoli. They occupy in all not more than one-twentieth to one-fiftieth of the total volume of the nucleus and characteristically lie away from the surface. With appropriate fixation the nucleoli stain deeply with basic dyes (*cf.* Zirkle, 1928, 1931). They do not, as a rule, give the thymo-nucleic acid reaction with Feulgen's stain which is characteristic of chromosomes (*cf.* Geitler, 1935, *b* on *Spirogyra*), but they may do so in some circumstances so that the distinction is not absolute. As a rule they break up and gradually disappear as the nucleus begins to divide, *i.e.*, at prophase, and are reconstituted at the beginning of the next resting stage. For these reasons they cannot be supposed to have any genetic significance or any physiological continuity with the chromosomes as a whole.

Owing to its greater density the nucleolus may be expelled from the nucleus by centrifuging (Andrews, 1915 ; Nemec, 1929 *a*). It may then persist in the cytoplasm as long as twenty-seven days,

NUCLEOLI

and only gradually disappear. Similarly, on account of its density, the nucleolus lies at the bottom of the large nucleus in an echinoderm egg. When the egg is rotated the nucleolus appears to fall to the bottom again (Gray, 1927). Presumably this is due to a movement of the nucleolus in the nucleus and not to a movement of the nucleus itself.

In the Protista, nucleoli vary greatly in form, position and behaviour (*cf.* Belar, 1926 *b*). In some species, nucleoli may lie on the periphery of the nucleus (*e.g.*, *Amœba terricola*). In other species they are numerous and rod-shaped, resembling chromosomes (*e.g.*, *Euglypha*). In others again an internal structure can be distinguished (*e.g.*, *Chilodon*). Nucleoli are not found in the small compact nuclei of the generative cells of animals, but they are sometimes found in those of plants (Upcott, 1936 *b*). As a rule they increase in size with the resting nucleus. In the alga *Acetabularia*, Hammerling (1932) found that the nucleus grew with the plant, without any division, and as it grew the nucleoli fused and developed into a single large branched body proportionate to the size of the nucleus.

The nucleoli are occasionally found to persist to metaphase of mitosis. Sometimes this is clearly abnormal, as in the gametophyte (pollen grain) of hybrid plants (D., 1929 *c*, in *Tradescantia* spp.). Sometimes it is characteristic of a species (*e.g.*, in *Tulipa*, Upcott, unpublished; *Fritillaria*, Frankel, 1936), and in the lower organisms the nucleoli may be extruded into the cytoplasm and lost, or even divide and pass into the daughter nuclei (Belar, 1926 *b*, *Chlamydophrys Schaudinni*). In some species of *Spirogyra* nucleolar behaviour is like that in higher plants. In others the chromosomes become embedded in the nucleolus during the prophase and remain so until telophase, the nucleolar mass dividing to the two poles. In others again the nucleolus behaves in an intermediate way; it is shed by the chromosomes at the end of prophase and the chromosomes then become larger (Geitler, 1935 *a* and *b*). Nucleolar and chromosome material thus seems to be spatially interchangeable. In the higher plants, *e.g.*, in *Zea Mays*, these bodies degenerate on reaching the poles and do not enter the new nuclei (Yamaha and Sinoto, 1925; Zirkle, 1928, 1931). Such bodies in *Amœba verrucosa* have been called "chromatin nucleoli," but we need not ascribe any special virtue to this persistence; a rapidity of prophase changes or a greater viscosity of the nucleus, as Belar suggests, may be held responsible.

2. THE MOVEMENTS OF THE CHROMOSOMES

(i) **Introduction.** Nuclei divide by the characteristic process of mitosis in the course of which the whole nucleus, apart from the nucleoli, resolves itself into longitudinally split threads, the chromosomes. Recent work, particularly on the Protista, has shown that certain features, precisely those of genetic significance, are universal. Mitosis can accordingly be defined by its essential properties as: *the separation of the identical halves of the split chromosomes into two identical groups from which two daughter nuclei are reconstituted.* Consequently it may also be said that mitosis is the process by which a nucleus gives rise to two daughter nuclei each morphologically equivalent to one another.

The longitudinal splitting of the chromosomes was first observed by Flemming in 1880. Its physiological meaning was pointed out by Roux in 1883: if equivalent nuclei can only be produced by such longitudinal splitting then the chromosomes must be longitudinally differentiated in their physiological properties. Experiment has now shown this conclusion to be valid (*v.* Ch. VIII).

There are two kinds of nuclear division: simple mitosis and double mitosis associated with the reduction in the number of the chromosomes; this is merely an abnormality of the simple mitosis, but it is convenient on account of its special character to refer to it as *meiosis* in contradistinction to mitosis, the normal type which we are now going to consider.

(ii) **Prophase.** The prophase of a mitotic division is first indicated by the appearance of coiled, contorted threads which as a rule, but not always, are evenly spaced throughout the whole body of the nucleus. These structures can be observed in living cells as well as after fixation and staining (Plate XVI, *cf.* Lucas and Stark, 1931) although in living material the nucleus first appears granular. The granulation is reasonably supposed to represent an

PLATE I

MICROPHOTOGRAPHS REPRODUCED BY KIND PERMISSION OF
DR. M. S. NAVASHIN

FIG. 1.—Section of root-tip of *Crepis dioscoridis*, $2n = 8$. Notice the four chromosome types in all metaphases, and also various stages of prophase and telophase of mitosis. Navashin-hæmatoxylin sectioned preparation. × 250.

FIG. 2.—Outer metaphase plate at 1 o'clock in Fig. 1. × 2300. The chromosome lying right and left at 6 o'clock in the plate is from another cell and is out of focus in this photograph.

FIG. 3.—Metaphase of mitosis in root-tip of triploid *Crepis capillaris* ($2n = 3x = 9$). Of the three D chromosomes one has a large trabant and two a small one. × 3000 (*v*. Text).

PLATE I

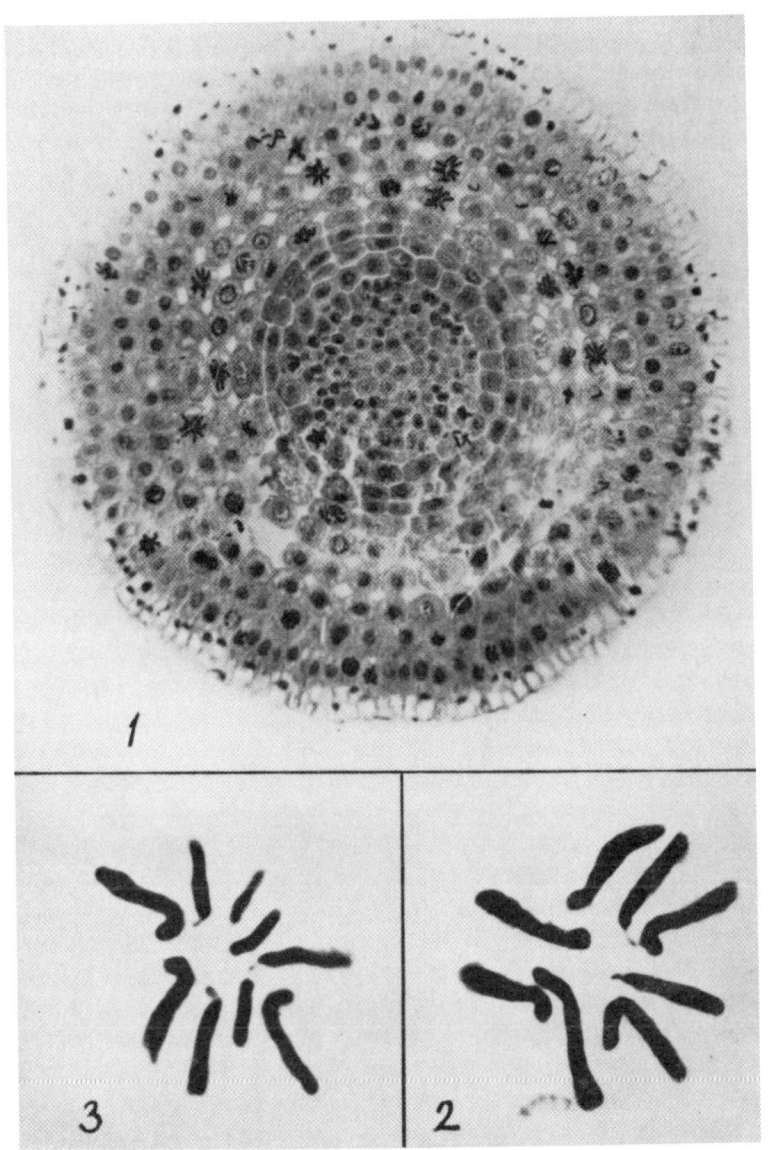

[To face p. 22.

optical section through the coiled threads which are later clearly distinguishable (Belar, 1929 b).

In the most favourable material it is possible to see at the earliest stage that the threads are double, and it is probable therefore that

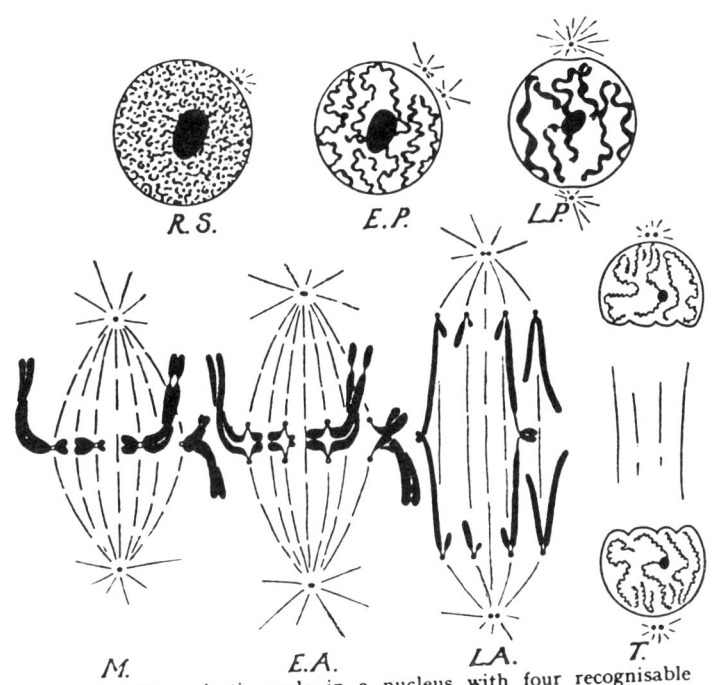

FIG. 4.—The mitotic cycle in a nucleus with four recognisable chromosomes, one having a subterminal centromere has a nucleolar constriction. *R.S.*, resting stage. *E.P.* and *L.P.*, early and late prophase (with dividing centrosome and disappearing nucleolus). *M.*, metaphase (with spindle). *E.A.* and *L.A.*, early and late anaphase. *T.*, telophase.

this is their universal characteristic throughout prophase (Ch. IX). The double threads can be traced in their development until metaphase, when they are recognised as chromosomes. The two single threads of which they are composed may be described as chromatids. This term was originally used by McClung to describe the single threads at meiosis, but since the single threads (being half-chromosomes) correspond in every way at the stages when they occur in the two types of nuclear division there can be no objection to calling a half-chromosome a chromatid both at meiosis and at mitosis.

Following their appearance the chromosomes contract lengthwise, and to a slight extent in gross bulk. They are evidently more rigid owing to loss of water, and they appear to suffer less from the action of fixatives. During the resting stage the nucleus has probably been uniformly filled with the material of the chromosomes and with water associated with them. Loss of fluid by the chromosomes during prophase therefore merely means that the nucleus becomes differentiated into chromosomes and " nuclear sap."

It was formerly held that at this stage the chromosomes were united end to end to form one " continuous spireme." It has now been shown in all clear cases (and especially at meiosis, on which great attention has been concentrated) that this is not so (*cf.* Fig. 4). Since ring chromosomes have been discovered (Ch. III) this structure is inconceivable. Evidently the free ends of chromosomes are particularly liable to collapse with defective treatment (*v.* Appendix I). The assumption of a continuous spireme is therefore unjustifiable on cytological grounds. On genetical grounds it is equally unjustifiable, for the constant production of a single continuous ring requires the secondary assumption of a temporary *specific* attraction between dissimilar pairs of ends of chromosomes. This assumption is contrary to the principles of chromosome mechanics (*v.* Ch. XII) and breaks down when applied to hybrids and polyploids and cases of interlocking at meiosis (*q.v.*). A genuine continuous spireme probably occurs under special conditions in the coccid *Icerya purchasi* (Hughes-Schrader, 1927). The chromosomes now usually have a uniform thread structure. But in certain nuclei (especially at the prophase of meiosis) a granular structure can be seen. Each chromosome then consists of a string of characteristic particles of unequal sizes at unequal distances apart. These particles or chromomeres as we shall see later have a permanent linear order.

This is most clearly observed at meiosis (*cf.* Wenrich, 1916 ; Gelei,

1921; and v. Ch. IV). The linear arrangement of the chromomeres in different nuclei is constant. It seems unlikely that these delicate structures are uninjured by fixation; no doubt their discontinuity is exaggerated. But their constancy shows them to be significant, as well as characteristic, artefacts (v. Appendix I). It is therefore difficult to regard them, as Kaufmann has suggested (1931), merely as twists in the chromatid. They demonstrate morphologically the linear differentiation of the chromosome. And it is important to notice that the two chromatids correspond exactly. Since, as we shall see, they must be supposed to arise from the longitudinal division of one, that division has been, so far as cytological evidence can show, into two identical halves. This identity is preserved from the time the doubleness is first seen at prophase throughout all stages of mitosis and all succeeding divisions of the chromatids at succeeding mitoses.

Further linear contraction of the chromosomes is accompanied by an increase in their thickness. There is no direct evidence as to what change takes place in the internal arrangement of the chromosomes, but indirect evidence suggests that it is due to the gradual assumption of a spiral form by the chromatid (v. infra). The increase in thickness obliterates the distinction between chromomeres and the chromatids gradually assume a rounded outline and eventually a highly uniform cylindrical structure, apart from constrictions, which will be considered later.

In most organisms contraction of the chromosomes reaches its maximum at the last stages of prophase. They are now probably one-tenth to one-twentieth of the length they were at the earliest stage of prophase. The two chromatids of each chromosome lie parallel or slightly coiled round one another and in close proximity.

(iii) **Metaphase : the Spindle.** At the end of prophase the chromosomes attain, as a rule, their greatest contraction. A new structure, the spindle, now develops. It varies in its method of origin. It may arise inside the nucleus; or when, as is usual in the higher organisms, the nuclear-cytoplasmic surface breaks down, it merely fills the space previously occupied by the nucleus. The spindle also varies in different organisms in its shape and in its appearance when treated with different reagents, but it can be

26 MITOSIS: THE CONSTANCY OF THE CHROMOSOMES

recognised everywhere by certain general characteristics. It lies between two more or less defined poles, and has more or less regularly the shape from which it takes its name. It is more rigid than the cytoplasm surrounding it and has less water to lose. When dehydrated in fixation it contracts crosswise more than lengthwise and frequently shows a lengthwise striation when contracted in this way. It is therefore evidently differentiated in its water content, which must be arranged in longitudinal channels to some of which the name of " spindle fibres " probably refers. Outside the nucleus, in most animals and in many of the lower plants, lies a body which is actively concerned in the division of the nucleus. This is the centrosome. It consists of a small granule which is sometimes capable of differential staining. It divides into two during telophase or prophase. There has often been difficulty in following its history, but it would appear from Boveri's studies that it is transmitted solely by the sperm of which it constitutes the " middle-piece," the centrosome of the egg-cell degenerating in sexual reproduction. Where centrosomes are present the spindle nearly always develops between them as an extension of a radial differentiation of the cytoplasm which appears around them during prophase.

During the prophase the chromosomes have been disposed evenly throughout the nucleus. As the spindle envelops the nucleus and its boundaries disappear the chromosomes come to lie midway between the two poles of the spindle in a flat " plate " on which they distribute themselves evenly. Comparison of different organisms shows that this distribution depends on one point in each chromosome lying on the equator of the spindle. This point of association with the spindle, or point of " attachment " to it, as it has been somewhat misleadingly described, is marked by a constriction or hiatus in the chromosome. This constriction marks the position of the *centromere*. This body is usually unstained at the metaphase of mitosis, and its behaviour will be described in detail later (Ch. XII). The constriction may be described as the centric constriction.

The parts of the chromosomes near the centromere are spoken of as " proximal," those further away as " distal." Where the chromosomes are short they may lie entirely in the equatorial plane as though in a flat plate. Where they are long, relative to the size of the spindle, the limbs of the chromosomes remote from the centromere lie at random—equatorially or axially. But owing to the spindle space being limited, the limbs of long chromosomes on the edge of the spindle lie in the outer cytoplasm, and the bends at their centromeres therefore point inwards. In many animals, all or most of the centromeres lie on the periphery of the spindle; the distal parts of the chromosomes then lie wholly outside the

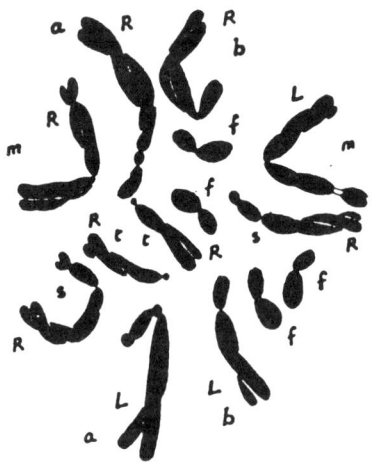

FIG. 5.—Metaphase plate in *Puschkinia libanotica* (2n = 10 + 4*ff*). *R.* and *L.*, relational coiling of chromatids, *f*, supernumerary fragments, *a*, *m* and *s* have secondary constrictions. (D. 1936 *b*.) × 3,000.

spindle, and in living cells (*cf.* Belar, 1929, *a* and *b*) can be seen to move in the streaming cytoplasm. In this way the chromosomes may fill up the greater part of the cell space.

Until metaphase, the chromatids are held together less closely elsewhere than at the centromere. The momentary condition of " full " metaphase is reached when the limbs of the chromosomes have come together as closely throughout their length as they were earlier at the centromere. It may be imagined that the stage with the distal parts of the chromosome separate follows this stage with the closest association, and is actually the first stage in the separation of the chromatids. This is not so. Separation always begins at the centromere, at a time when the distal ends are more closely united than earlier, and a wide separation of the distal ends shows an *early* metaphase stage, not a late one (*cf.* Fig. 5).

(iv) **Metaphase : the Plate.** While their centromeres lie on the equator of the spindle, the whole group of chromosomes lies in a " plate," the metaphase plate, which has been generally studied in polar view in order to make out the number, sizes and shapes of the chromosomes that are characteristic of the organism, the race or the species. In number and size the chromosomes are, as a rule, invariable at this stage throughout the organism. In shape they may vary slightly owing to different degrees of spiralisation. In very rapid gonial mitoses they may be less spiralised than usual (*e.g.* in the protozoan *Aggregata*, Belar, 1926), while in the mitoses immediately preceding meiosis in the Orthoptera and in that immediately following meiosis in the Angiosperms they are usually more spiralised (Janssens, 1924 ; White, 1935, on *Mecostethus* ; Darlington, 1929, on *Tradescantia* ; Janaki-Ammal, unpub., on *Saccharum*).

In distribution on the plate two main types are found in higher organisms. In the first, the so-called " central spindle " type of *Salamandra*, the centromeres lie at the edge of the spindle and the bodies of the chromosomes lie outside in the cytoplasm. In the second, which is usual in plants, the centromeres lie evenly in the equatorial plane of the spindle and the bodies of most of the chromosomes are therefore inside the spindle, flat on the plate when they are small, turned up on either side when they are longer. There is no absolute distinction between these types. Many Orthoptera are intermediate and abnormal pollen grains of *Fritillaria* and *Tulipa* have been found with the *Salamandra* type of plate. Such variations will probably be found to be widespread.

(v) **Anaphase.** The end of metaphase is marked by a change in the centromere or, where it is not visible, at the centric constriction. The two chromatids which have lain closest together at this point now separate (Figs. 4 and 19B). Under suitable conditions of fixation the centromere becomes visible in each chromatid at the point of separation (S. Nawaschin, 1912 ; Trankowsky, 1930).

This body is about 0·2 μ in diameter, in the largest chromosomes. It has received various names according to the stage or organism in which it has been seen. We shall find later that the centromere has important functions in the movements of chromosomes quite apart from the spindle, and since the same movements occur even when it is not visible with the treatment used, it seems necessary to assume that it is always present, but that, like the centrosomes, it is of variable staining capacity and in the smaller chromosomes is too small to see. The chromatids remain in contact at their ends until the anaphase separation finally draws them apart. It is common to refer to the separation of the chromatids as the "division" of the chromosome, but, as we have already seen, the division has taken place much earlier, and, throughout the prophase, chromatids have been merely held together either by a common pellicle surrounding them or, more probably, by a specific attraction existing between them. The only part of the chromosome that divides at this stage is the centromere, and when its division is inhibited artificially the division of the nucleus is itself arrested at metaphase (cf. Ch. X). At anaphase therefore it is an attraction which has to be overcome rather than an intricate connection which has to be severed. The relations of the chromatids at meiosis cannot be understood unless this principle is grasped.

When the separation is complete the corresponding groups of chromatids, now called "daughter chromosomes," pass towards opposite poles. This movement has been analysed by Belar with remarkable success, working with living cells (1927, 1929, a and b). In the first part of the movement the spindle undergoes no important change in shape or character, and it appears that the movement is independent of external variations that can be brought to bear on it experimentally. So far it may be said to be "autonomous." At late anaphase the situation is entirely different. There is little evidence that the chromosomes move any further through the spindle, but the spindle itself changes shape. The middle part, between the separated bodies of chromosomes, expands lengthwise and contracts correspondingly crosswise so that the chromosomes are pushed apart and the anaphase movement completed. Belar maintains that this is an inherent and "active" property of the spindle itself, more especially because with dehydration in hypertonic solutions the stretching of the spindle is exaggerated. Probably it is a property correlated with the special distribution of its water content. Where, in meiosis in *Stenobothrus*, one pair of chromosomes in separating offer, on account of their greater length, more resistance to separation than the rest, the whole spindle is tilted evenly. This argues not merely pressure from the middle part of the spindle but an independent coherence of the middle part and of the polar parts. The effect of the stretching of the spindle can be seen in pollen grain mitoses where one pole has been lying close to the wall. The spindle then continues to push the inner set inwards, while the movement of the outer set has been stopped by the wall in mid-anaphase (*cf.* Ch. XII). Where the chromosomes are very long, as in mitosis in *Tradescantia* and *Aggregata*, their separation is helped by their shortening still further at late anaphase.

(vi) **Telophase.** When the chromosomes reach the pole they contract longitudinally and form a compact mass which passes through stages in some respects comparable with the reverse processes of prophase. These appear to consist essentially in the imbibition of water. In some organisms the chromosomes form separate vesicles at first (Belar, 1928 *b*), and where the resting stage is short these may persist, separately identifiable, to the next division (*v. infra*).

(vii) **Duration of Mitosis.** The duration of successive stages of mitosis has been recorded from observations of living cells by Belar (1926 *b*, 1929 *b*, *Tradescantia*) and others. The following observations indicate the range of variation found, but do not include either the quickest or the slowest examples. The third case is an observation of meiosis, which will be specially considered later.

Apart from observations of the living cell the relative frequencies of different stages found in fixed material show great variation in their timing. Certain stages in the prophase of meiosis may be indefinitely prolonged, and in glandular tissue mitotic prophases have been found that are as permanent as the cells containing them (Ch. V). Telophase, on the other hand, is a uniformly short stage.

SPIRALISATION

TABLE 2

The Time Taken by Division of the Nucleus

	Prophase.	Beginning of Metaphase.	Metaphase.	Anaphase.	Beginning of Telophase.	Telophase.	Total.
Protozoan (mitosis):							
Rhogostoma schüssleri	6'	2'	2'	·5'	2'	20'	32·5'
Euglypha sp.	40'	7'	18'	6'	8'	100'	179'
Plant (mitosis):							
Tradescantia virginiana	181'	10'	4'	15'	130'		340'
Animal (meiosis, first division):							
Stenobothrus lineatus	—	—	6'	42'	39'	60'	—

3. THE STRUCTURE OF CHROMOSOMES

(i) Spiral Structure. At metaphase all chromosomes consist of two cylindrical rods, the chromatids, which are normally of even diameter. This diameter is a racial characteristic constant in most

TABLE 3

Estimated Total Lengths of Chromosomes in Complement (in Microns)

	Stages of First Division of Meiosis.					Mitotic Metaphase*
	Leptotene	Pachytene	Diplotene	Diakinesis	Metaphase	
Lilium species	—	1,469	257	152	152	—
$n = 12$, Belling, 1928, c†						
Aloë purpurascens	—	545	—	55	—	—
$n = 7$.						
Bellevalia romana	—	386	—	—	41	63
$n = 4$. Dark, 1934						
Dendrocœlum lacteum	421	—	—	—	—	86
$n = 7$. Gelei, 1921.						

* This length is halved to correspond with the haploid complement measured at meiosis.
† Belling states (1931 *a*) that leptotene threads in *Lilium* are six and a half times as long as at metaphase. This seems to disagree with the results recorded here.

tissues for all the chromosomes of a complement except for such as may be less in bulk than a sphere of the specific diameter (v. Fig. 7).

Variation in the apparent diameter of the chromosome is due to, first, the pairs of chromatids being seen sometimes endways-on, *i.e.*, in the same line of vision, as is always the case near the centromere in polar views, and sometimes sideways-on, when the chromosome will be just twice as broad ; and, secondly, the occurrence of constrictions (v. *infra*).

Now each of the cylindrical rods is the product of a linear contraction of the threads observed in early prophase. The extent of this contraction can be approximately estimated from measurements. The rather fragmentary estimates given are largely derived from observations of successive stages at meiosis, where the contraction at metaphase is greater than at mitosis (*cf. Bellevalia*), but they indicate the extent of the changes undergone during the prophases of both types of division (Table 3).

The degree of contraction at metaphase is, however, subject to genetical control, as will be seen later (v. Ch. III), and may therefore vary considerably in different races or species.

In this process of contraction we may probably distinguish two kinds of change in succession. First, there is the change which renders the chromosomes visible at the end of the resting stage. This may be regarded as a dehydration of previously more dispersed chromatin material, associated (according to Sakamura, 1927) with an increase in hydrogen-ion concentration in the nuclear sap.

Secondly, there is a linear contraction of the thread such as might be regarded merely as due to successive particles, the chromomeres observed at meiosis coming to lie closer together. But since the separation into chromomeres is probably to some extent an artefact (although characteristic and significant) such a simple contraction need not necessarily be distinguished from a spatial rearrangement of the thread. The rearrangement might be supposed *a priori* to be orderly or haphazard. Since each chromatid is uniformly cylindrical at metaphase, there seems no arrangement that can be assumed but an orderly one, and none of this kind but a spiral. Such an arrangement was first observed by Baranetsky in 1880, and his observations have since been generally confirmed whenever any definite structure has been observed in the chromosomes of plants or animals at meiosis.

Linear contraction of the chromosomes must depend on *spiralisation*. At the metaphase of mitosis it has been possible to see this spiral in special preparations (Geitler, 1935 a; Upcott, 1935 a). Comparison with the succeeding stages leaves no doubt that the same type of structure regularly obtains at mitosis. Moreover, it

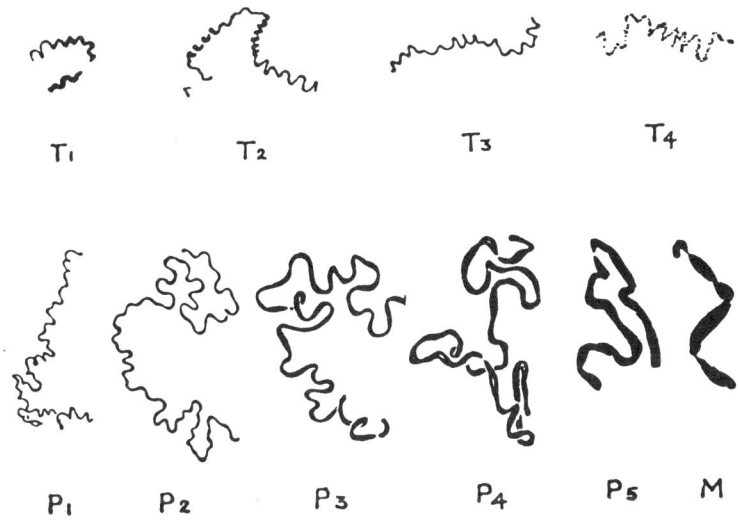

Fig. 6.—The uncoiling of relic spirals from telophase (T) to prophase (P) and metaphase (M) in *Fritillaria*. Super-spirals appear in P_2. × 1800. (D., 1935 a.)

shows a mechanical relationship between succeeding divisions (D., 1935 a). At the telophase a spiral becomes distinguishable, corresponding to the supposed structure at metaphase. This is the relic spiral. The same general structure is visible at the beginning of the following prophase, except that the arrangement is somewhat more contorted and the chromosomes have divided into chromatids which are coiled round one another in a way that presumably compensates for the relic spiral. As the chromatids contract and, we must suppose, form internally the spiral which they show at metaphase, they also gradually uncoil the relic spiral which in the longest chromosomes is still seen at the following metaphase. Whatever internal forces cause the coiling and uncoiling we see that there is a lag in the adjustment of the external form of the chromosomes to the changes in these forces: the coiling seen in a prophase is a relic of that at the preceding division, not the forerunner of that at the following division, as has been generally imagined. In some respects, to be sure, prophase reverses telophase. In other respects prophase continues telophase.

The changes inferred in the transition of the metaphase chromosome to the telophase condition have now been successfully imitated in experiment. Kuwada and Nakamura (1934 *a*) have changed the body of metaphase chromosomes in a pollen mother-cell into a resting nucleus by exposing them to ammonia vapour before fixation. The effect is to half-uncoil the spirals so that they form a compact mass having the optical properties of a resting nucleus that has been similarly fixed with acetic acid. The artificial differ from the natural nuclei merely in their irregularity of boundary and in their lack of nucleoli (*cf.* Ch. XII).

These observations show that the resting nucleus and metaphase chromosomes, which appear equally homogeneous in life and are equally susceptible to the production of non-characteristic artefacts through salting-out of their protein-water systems, are in fact, to be related in terms of a spiral structure which can be traced through successive stages in life and reproduced artificially in experiment. The change from metaphase to resting nucleus consists spatially in uncoiling and re-packing the chromosome thread.

(ii) **Constrictions, Centric and Nucleolar.** In the course of prophase another kind of discontinuity appears in the chromatin thread in many organisms. This is at first an unstained gap in the chromatid; afterwards it develops into the metaphase "constriction," at which the cylindrical body of each chromatid is drawn together as a sausage is drawn together with string or a blood vessel ligatured. All chromosomes have one such non-spiralising segment in their structure—the *centric constriction*. But there are also found in many chromosomes "secondary" constrictions which have no relationship with the centromere (*cf.* Plate II and Fig. 7).

CENTRIC CONSTRICTIONS

The centric constriction has the following three characteristics distinguishable at metaphase:—

(a) The chromosome is more easily bent and stretched at this point. This has been shown by micro-dissection (Chambers and Sands, 1923), and is equally evident from comparative observation of chromosome form in fixed material. The constriction is, as one might expect, a point of weakness in the chromosome. Secondary constrictions share this property. The chromosome, therefore, in the absence of secondary constrictions consists of two limbs separated by the intercalary centric constriction.

(b) During metaphase the two chromatids lie at this point in the axis of the spindle, although elsewhere lying at random. Evidently the centromere is strictly orientated. The chromosome seen in polar view therefore seems narrowest on either side of the attachment constriction.

(c) The two chromatids are held together tightly at this point, although elsewhere lying loosely parallel. Apparently the centromere is either single or its two halves are more closely paired than the rest of the chromatids. The ligature may be short, or may be extended so that a definite non-staining gap occurs in the chromosome. This also applies to secondary constrictions.

Constrictions give great diversity to the form of the metaphase chromosomes, and being constant in position and length they also give character to the individual chromosomes. Their study therefore has been of the greatest importance in investigating constancy and variation in mitotic chromosomes within and between species and between parents and their own gametes or gametophytes, particularly in the flowering plants (*e.g.*, in *Crepis, Vicia, Tradescantia, Muscari, Tulipa, Drosophila*). This advance we owe in the first instance largely to the work of S. G. Navashin and his school, and later to Sakamura (1915 *et sqq.*), Taylor (1924 *et sqq.*), Newton (1924, 1927), M. Navashin (1926, 1927, 1930), Delaunay (1926, 1929), Lewitsky (*loc. cit.*). *Cf.* also Belar (1926), Bridges (1927), Kachidze (1929), Hollingshead (1928).

The following are the main conclusions which are applicable to the higher plants in general (*cf.* Delaunay, 1929, and Lewitsky, 1931).

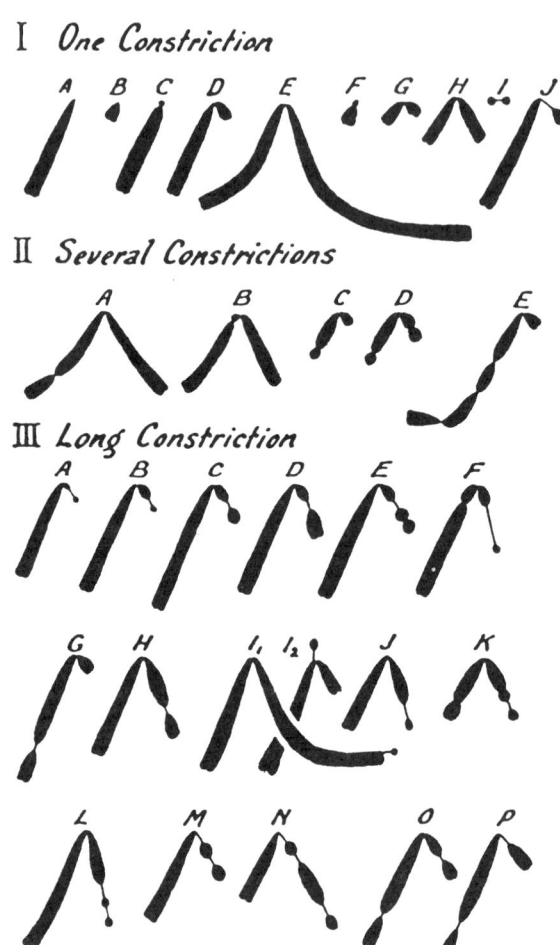

FIG. 7.—DIAGRAM OF CHROMOSOME FORMS AT MITOSIS.

Metaphase chromosomes of plants and animals represented uniformly with the same diameter and with the two chromatids in the same line of vision at the constrictions, which are therefore represented by single threads. This is only constantly the case with polar views of the centric constriction: in side views of this constriction the connection is double, the chromatids being seen side by side; in polar or side views of secondary constrictions the connection is double or single according to the chance arrangement.

I. Chromosomes with a single centric constriction.

A. " Terminal " (with one arm too small to be seen as a rule), *e.g., Stenobothrus lineatus* (Janssens, 1924; Belar, 1928), *Drosophila melanogaster*, X-chromosome (Stern, 1929), *Sphæromyxa sabrazesi* (Belar, 1926 *b*).

B. " Terminal," short chromosome, *e.g., Drosophila melanogaster* fourth chromosome (Stern, 1929), *Yucca filamentosa* (Taylor, 1925 *b*). Many

Lepidoptera. *Cambarus* (Fasten, 1914), *cf.* Wilson. Many fungi: *Camarophyllus* (Bauch, 1926), *Saprolegnia* (Mäckel, 1928).

C. Sub-terminal (i) minor element spherical. *Fritillaria* and *Crepis* species (very common).

D. Sub-terminal (ii), *e.g.*, *Fritillaria*, *Ornithogalum*, *Tulipa*, *Crepis* species (very common).

E. Median or sub-median, *e.g.*, *Tradescantia virginiana* and *Fritillaria ruthenica* (D., 1929 c), *Aulacantha scolymantha*, *Euglypha* sp., *Monocystis magna*, *Salamandra maculosa* (*cf.* Belar, 1926 b).

F. Sub-terminal, short chromosome, *e.g.*, *Tradescantia fluminensis*, fragments of *Fritillaria imperialis* and *Tradescantia virginiana* (D., 1929 c).

G. Sub-median, short chromosome, *e.g.*, *Ornithogalum narbonense* (Delaunay, 1925, 1929), *Veltheimia viridiflora* (Taylor, 1925 a), *Muscari tenuiflorum* (Delaunay, 1926), *Commelina bengalensis* (D., 1929 c), *Lachenalia* spp. (Moffett, unpublished).

H. Sub-median, medium length chromosome, *e.g.*, second and third chromosomes of *Drosophila melanogaster* (Stern, 1929), *Crepis* spp. (Hollingshead, 1928), *Prunus* spp. (D., 1928).

I. Median, both arms spherical, very short chromosome as compared with the rest of the complement. *Muscari latifolium* (Delaunay, 1929).

J. Sub-terminal long constriction, *e.g.*, *Aconitum* spp. (*cf.* Fig. 79.)

II. Chromosomes with secondary constrictions, all short.

A. Simple secondary constriction, *Hyacinthus orientalis* (D., 1926), *cf. Aucuba japonica* (Meurman, 1929), *Vicia Faba* (Sakamura, 1915).

B. " Intercalary trabant," *e.g.*, *Rhoeo discolor* (D., 1929 c). *Cf. Muscari polyanthum* and *M. pycnanthum* (Lewitsky and Tron, 1930) (?) *Cyrtanthus parviflorus* (Taylor, 1925 a).

C and D. *Muscari polyanthum* (Lewitsky and Tron, 1930), *Œnothera* spp. (D., 1931).

E. Multiple constriction, *e.g.*, *Fritillaria pudica* (D., unpublished).

III. Chromosomes with long secondary constrictions.

A. *Ranunculus acris* (Sorokin, 1927), *Crepis* spp. (Babcock and Navashin, 1930).

B. *Muscari longipes* (Delaunay, 1926), *Calochortus amabilis* (Newton, 1927), *Spironema fragrans* (D., 1929 c), *Crepis* spp. (Babcock and Navashin, 1930).

C. *Ornithogalum tenuifolium* (Delaunay, 1925), *Cyanotis zenonii* (D., unpublished) *cf. Lachenalia* spp. (Moffett, unpublished).

D. *Ornithogalum narbonense* (Delaunay, 1925), *Crepis setosa* (Taylor, 1925 b).

E. *Ornithogalum nanum* (Delaunay, 1925), *cf. Allium Moly*, (Levan, 1932).

F. *Bellevalia azurea*, *cf.* also *Muscari polyanthum* (Lewitsky and Tron, 1930).

G. *Fritillaria* spp. (Taylor, 1926; D., 1929 c, 1930 c).

H. *Crepis tectorum* (Hollingshead, 1928). *Fritillaria ruthenica* (D., 1929 c).

I_1. *Aloë saponaria* (Taylor, 1925 b), *Bellevalia romana* (D., 1926), *Tradescantia virginiana* and *Rhoeo discolor* (D., 1929 c), *Crepis pulchra* (Hollingshead and Babcock, 1930).

I_2. Mutant form with lateral trabant in *Tradescantia* (D., 1929 c, Fig. 77; *cf.* Levan, 1932).

J. *Ribes aureum* (Meurman, 1928; D., 1929 d). *Rubus* spp. (Crane and D., 1928).

K. *Dahlia Merckii* (Lawrence, 1929), *Œnothera* spp. (D., 1929).

L and M. " Tandem " trabants. L. *Allium Cepa* (Taylor, 1925 a). M. *Aucuba japonica* (Meurman, 1929).

N. Intercalary trabant. *Solanum Lycopersicum* (D., unpublished). (*Cf.* Lewitsky and Benetskaia, 1929 on *S. tuberosum*.) (*N.B.* This chromosome has led to difficulties in counting the chromosomes in *Solanum*, *cf.* Winkler, 1916).

O. *Cyrtanthus parviflorus* (Taylor, 1925 a).

P. *Aconitum* spp. (Fig. 79); long centric constriction.

38 MITOSIS: THE CONSTANCY OF THE CHROMOSOMES

(a) Each chromosome has centric constriction and perhaps one or more secondary constrictions which are constant in position. Both are usually intercalary, and it is doubtful whether they can ever be terminal (*cf.* Lewitsky, 1931; Kaufmann, 1934 on *Drosophila*; D., 1936 *b* on *Chorthippus*). The alleged terminal centric constrictions of grasshoppers and of the B chromosomes of *Zea Mays* have been shown to be near the end, but not at the end (D., 1936 and Fig. 102). A supposedly terminal nucleolar constriction gives the appearance of a " seta " (Kihara, unpublished).

(b) Constrictions vary in the length of the lacuna they create. This length is a characteristic property, although subject to variation as an artefact, and, in part, as a natural condition. We may therefore speak of a " long " constriction and a " short " constriction. The real variation in appearance is general amongst very long constrictions, especially those separating trabants (*v. infra*). Long constrictions may be two or three times as long as the chromatid is broad. At anaphase all constrictions are liable to be emphasised by the tension between the centromere and the parts of the chromosome distal to the constriction. Centric constrictions are nearly always short, but *Aucuba* and *Aconitum* are exceptional in this respect (*v.* Fig. 79).

(c) Prophase contraction of all sections of chromatids between two constrictions proceeds until the chromatid reaches the diameter characteristic of the race, unless it first reaches a spherical shape. When the whole chromatid, or any section of it, is less than a certain size, the characteristic breadth is not attained. In the first case we have the spherical type of chromosome found in many animals and in *Muscari latifolium* (Fig. 7 I B). In the second case, merely a part of the chromosome is narrower than the rest (Fig. 7 I C). But it often happens that such a part is terminal and is separated by a long constriction from the main chromosome, for long constrictions are more frequent near the ends. We then have what is called a " satellite " or *trabant*. Occasionally this condition is found in a small section intercalated in the middle of the chromosome. We then have an "intercalary trabant " (Fig. 7 III N).

This view that the trabant is distinguished from other parts of the chromosome merely by its small size (D., 1926) has been disputed by the Russian school (*cf.* Sorokin, 1927), who have endowed the trabant with special functions. The observation of intercalary as well as terminal trabants, however, leaves no doubt of their being determined merely by the position of constrictions.

(*d*) Where a number of constrictions occur in each chromosome the complement presents an appearance of some complexity. Extreme examples of this are seen in *Datura* and *Œnothera* (Lewitsky, 1931; D., 1931 *d*), and in the "tandem trabants" of *Allium* (Fig. 7 III L).

(*e*) The only restriction observable in the free combination of various forms of constriction in different chromosomes is indicated by the rarity of forms with long constrictions separating two large limbs of a chromosome. Further, the long constrictions are commonest where the distal element is smallest, *i.e.*, in a trabant, and they are not usually formed in the shortest chromosomes of a complement with a large size range. Selection has probably played some part in limiting the range of chromosome form, and it would work in two ways. First, a long constriction might not be strong enough to stand the strain of the anaphase separation with a large element distal to it. Secondly, a long constriction might interfere with the pairing of a short chromosome at meiosis on account of the properties we shall next consider.

All constrictions are regions of the chromosomes which fail to undergo spiralisation along with the rest. In the case of the centric constriction we can see that this special property is associated with the special functions of the centromere. The question arises as to what special functions are associated with the secondary constrictions. The answer was discovered by Heitz (1931), who found that nucleoli were developed at corresponding positions in daughter nuclei at telophase and, moreover, at positions and in numbers and sizes corresponding to the secondary constrictions of the chromosomes (*cf.* Geitler, 1935 *a*). It had been known for some time that nucleoli were associated with trabants at prophase (Navashin, 1926), that trabants were merely distinguished from other constricted bodies by their size (D., 1926), and that the number of nucleoli might vary with the number of chromosomes (De Mol, 1928). These observations were intelligible in view of the

40 MITOSIS: THE CONSTANCY OF THE CHROMOSOMES

origin of nucleoli from the particular unspiralised segments which we call secondary constrictions. At prophase such correspondences are difficult to trace, because the nucleoli often fuse at this stage, especially in meiosis, but the " nucleolar chromosomes " can still be recognised by their association with them. The nucleolus may lie round the chromosome, separating it into pieces (Kaufmann, 1933 on *Drosophila melanogaster*) or it may be merely attached to it by a thread (Dobzhansky, 1934, on *D. pseudo-obscura*). This difference is found between mid and late prophase in the meiosis of plants. As to whether there are non-spiralising segments giving

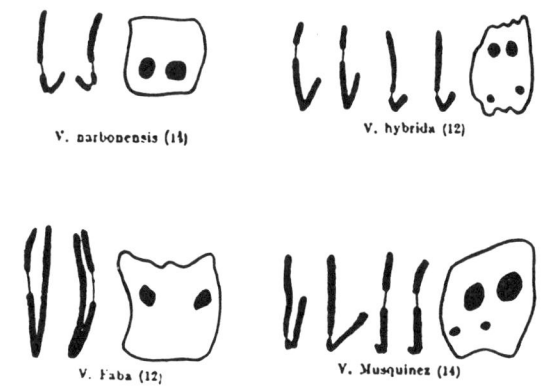

FIG. 8.—Anaphase chromosomes and telophase nuclei in four species of *Vicia*, showing the position relationship of secondary constrictions and nucleoli (after Heitz, 1931 *b*).

constrictions at metaphase apart from the segments adjoining nucleolar organisers, we have as yet no evidence. But it is certain on the other hand, that small nucleoli may leave no trace of constriction at metaphase.

(iii) **In the Resting Stage : the Theory of Permanence.** The numbers, sizes, shapes and constrictions of the chromosomes appearing at mitosis are constant. In the nuclei of an individual or of a group of individuals descended by mitosis from a single nucleus, apart from such special accidents as will be considered later, the same " complement " is found. With sexual reproduction other changes may occur, but these are not common enough to interfere with the generalisation that each species has a characteristic chromosome complement seen at mitosis (v. Plate I). This complement is a constant property of the nucleus, undergoing a constantly recurring series of changes in each cycle of mitosis. Hence each chromosome that appears at the prophase of one mitosis corresponds with a chromatid that entered into the nucleus at the preceding telophase in all its visible properties except that it is double : it has reproduced itself identically. It was on the assumption of this essential constancy of behaviour that Weisman correctly predicted the universal occurrence of a reduction that would compensate for the addition that occurs when nuclei conjugate in fertilisation.

The correspondence may be attributed to one of two conditions. Either it is due to conditions that are external to the chromosomes in the same sense as a correspondence in form between the walls of two cells is due to conditions external to the cell-walls, or it is due to conditions inherent in the chromosomes themselves. This would mean that the chromosomes, or essential parts of them, are, unlike the cell-wall, *permanent* structures, arising, like the nucleus as a whole, the cell as a whole, or the organism as a whole, only from pre-existing structures of the same kind. This hypothesis, which is expressed by saying that the chromosomes have the same permanence as nuclei, cells and organisms, was verified by the work of Boveri (1892–1909) on *Ascaris*. Boveri found embryos that developed from eggs in which only one polar body had been formed, owing to the failure of separation of two nuclei at the second meiotic division. Clearly the original failure of the separation would give egg nuclei with two chromosomes instead of one. Fertilised by a sperm with one chromosome the first segmentation nuclei should have three chromosomes. All the nuclei of embryos arising in this way (still recognisable from the single degenerated polar body) had three chromosomes at mitosis. Thus, if it is assumed that external conditions determine the number of chromosomes, it must also be assumed in this case that these conditions are permanently altered by an accident of separation which, on the hypothesis of permanence, would lead directly and simply to the observed results.

42 MITOSIS: THE CONSTANCY OF THE CHROMOSOMES

This type of observation has been multiplied in recent years, so that any other hypothesis than the simple one of individuality and permanence would now be preposterous. Abnormal chromosome numbers have been discovered in relation to so many investigations that they must be separately considered in their special categories. All these observations are compatible with the doctrine of permanence, for these reasons : first, such differences as are observed can be understood in every instance as the result of changes that have also been observed, or on other grounds inferred, in the individual chromosome or in the genetic properties of the nucleus as a whole. Secondly, the changes that take place are always irreversible in their effect on the chromosome complement and in the correlated effect on the organism as a whole, except through the reversal of the accident or change which is assumed to be directly responsible for the change in the complement on the hypothesis of permanence (*cf.* Buxton and D., 1932).

Now, if the constant structure of the chromosome complement is due to the inherent properties of the chromosomes, it may be supposed to be maintained in one of two ways. Each of the successive particles or chromomeres may have a specific affinity for the particles on either side of it, which causes it to enter into the same combination at successive mitoses. The changes in arrangement that will later be described show that this is not so. Treatment with X-rays will break up the chromosomes and recombine their parts more or less at random in new arrangements, most of which are as stable and permanent as the old ones (Ch. X). In fact the changes that take place in chromosomes under X-ray treatment are the strongest evidence of potential permanence without such treatment. Alternatively, the essential linear structure of the chromosome thread seen at prophase may be retained during the resting stage. This idea of the continuous existence of the chromosome we owe in the first instance to Rabl (1885). It was for long a matter of dispute, two opposed views being held with regard to the evidence.

Some consider that the chromosomes cannot be individually represented by any structure in the resting nucleus. Schaede (1925–30) found that the living nucleus, at least when " young,"

THE RESTING STAGE

is optically homogeneous or granular, and considers this to be inconsistent with the view that any threadlike structure is present. The granular structure that is sometimes found in fixed nuclei might be given by an optical section of a coil of threads, but it certainly does not always correspond to such a coil if the fixation is inefficient, because, as Belar points out (1928, Plate XIV, *cf.* Appendix I), the granulation is coarser the slower and more inefficient the fixation. Micro-dissection work has similarly suggested that the nucleus has no permanent structure and that its contents are "liquid" *in vivo* (Chambers, 1925). But failure to detect structure by these methods is not conclusive (*cf.* Heilbrunn, 1928). The structure that is sought is without analogy on a larger scale or in other materials.

What we have already seen, on the other hand, of artificial resting nuclei and of the continuance of the spiralisation cycle of the chromosomes from telophase in the next prophase can leave no doubt either of the structural relationship of the resting nucleus and the metaphase chromosomes or of the structural stability of the chromosomes in the nucleus. It is worth while, however, to consider some of the particular evidence of the permanence of the chromosomes during the resting stage since this principle is the foundation of chromosome genetics.

(i) The classical example was described by Boveri (1909), again in *Ascaris*. He showed that each of the several types of relative arrangement of the chromosomes seen at the beginning of a division in a young embryo corresponded exactly with one of the several types seen at the end of the previous division. Parts of the chromosomes, therefore, reappeared in just those parts of the nuclei where they had disappeared.

The same kind of evidence, but yet more striking, is afforded by the observations of Belar on *Aggregata* (1925, 1926). Here, owing to the rapid succession of divisions in the sporozoite, the longer chromosomes remain entangled at the ends at telophase (*v. supra*). They reappear entangled at the next prophase, and after two or three divisions the entanglement seen is precisely such as would have been expected had there been no resting stages. Similarly, Belar (1929 *b*) showed in living mitosis in the staminal hairs of

Tradescantia that the polarised arrangement of the chromosomes corresponded at prophase with that seen in the preceding telophase (*cf.* Manton, 1935). Finally the recognition of a spiral arrangement at prophase corresponding with that seen at telophase shows the persistence of the chromosome structure during the resting stage (Taylor, 1931; de Winiwarter, 1931). These observations show

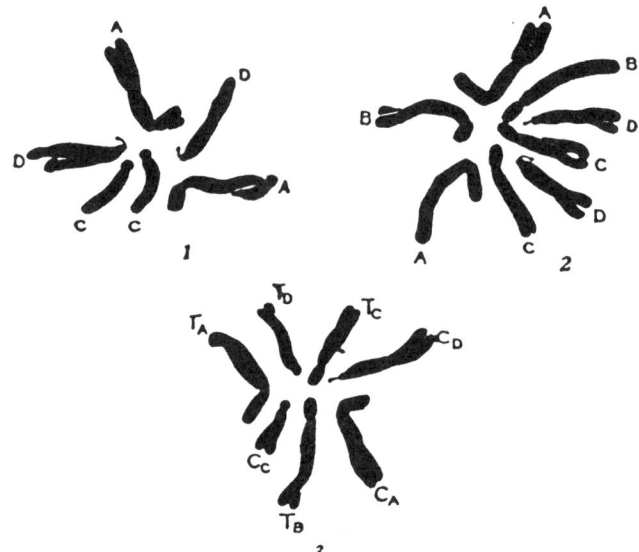

Fig. 9.—Mitotic metaphase: 1. in *Crepis capillaris* ($2n = 6$); 2. in *C. tectorum* ($2n = 8$); 3. in the hybrid between them ($2n = 7$). The chromosome types are marked with letters. Note that the D chromosome of *C. tectorum* shows no trabant in the hybrid. × 2300 (from Hollingshead, 1930; *cf.* Navashin, 1927).

that the resting stage does not interfere with the structure and relationship even of the constituent parts of chromosomes. The chromosomes are not merely permanent but immobile. The description as a "resting stage" is mechanically and morphologically correct.

(ii) Another kind of evidence consists in the demonstration of the separate existence of the chromosomes during the resting stage. This evidence is less decisive, because it may be held that such separation is only possible where the resting stage is short and therefore not a complete resting stage. For example, in the pre-meiotic resting stage of the spermatocytes (Wenrich, 1916, on *Phrynotettix*, Reuter, 1930, on *Alydus, et alii*) and in the early segmentation divisions of the fertilised egg (Heberer on *Cyclops*, 1925; Smith on *Cryptobranchus*, 1929) the chromosomes remain in distinct vesicles during the resting stage, from which they reappear at the ensuing prophase. In the spermatocytes of Orthoptera one chromosome (the sex chromosome) is more condensed than its neighbours and stains more deeply, so that it can be distinguished from the others throughout the resting stage. Similarly, in spermatozoa, even *in vivo*, it is sometimes possible to see the individual chromosomes and actually distinguish between those having different numbers (Mülsow, 1912, on *Ancyracanthus*; Meves, 1915, on *Filaria*; *v*. male gametes). Kater (1926) thinks it possible to distinguish the separate compartments in which the chromosomes lie during the resting stage in root tip nuclei.

(iii) Finally, a more elaborate kind of evidence is provided by the comparison of chromosome forms in hybrids with those found in the parents. Hybrids between species with clearly distinguishable complements have been studied in *Crepis*. It is found that each of the chromosome types of the parental species reappears, so that no two chromosomes are alike (Fig. 9, Hollingshead, 1930; Navashin, 1927). The constancy was not maintained in the second generation, for special reasons which do not concern the present problem (*v*. Ch. VII). This kind of evidence is afforded also by the observations of Avery (1929) on other hybrids of *Crepis*, Meurman on *Ribes* (1929), and by some older work.

(iv) The small resting nuclei of many flowering plants are to varying extents organised differently from the large nuclei on which accurate observation is easiest. They consequently show the permanence of the chromosomes in a different way. Small nuclei have been known to contain deeply-staining bodies, chromocentres or prochromosomes, in the resting stage which were distinguishable from nucleoli and corresponding in number with the chromosomes (*cf*. Rosenberg, 1909). These bodies have been shown by Doutreligne (1933) and others to consist of the proximal parts of the chromosomes. They correspond in number with the chromosomes in diploid and polyploid tissues of the same plant (Manton, 1935). According to Heitz (1929, 1932, 1935) this property is sometimes related with excessive prophase and telophase condensation of the affected parts, which, however, unlike the chromocentre, does not give the Feulgen reaction (*cf.* Ch. VIII).

Manton (1935) finds that small nuclei showing chromocentres differ from large nuclei of the ordinary type described in several correlated respects. The telophase nucleus is very much larger in relation to the metaphase chromosomes, as shown by root-tips (Table 4).

TABLE 4

Relative Sizes of "Solid" and "Vesicular" Nuclei

	Resting nuclei	Nucleoli	Metaphase chromosomes
Allium ursinum . . .	900 c.μ	70 c.μ	140 c.μ
Biscutella lævigata . .	120 c.μ	14 c.μ	4 c.μ

The telophase chromosomes consequently lie widely separated in the nuclear sap, as prophase chromosomes always do. They are therefore capable of movement and their visible parts, the chromocentres, move towards the surface of the nucleus at the end of telophase. Owing to the greater fluidity of the nucleus it changes its shape rapidly and the nucleoli are always coalesced into one body at this stage. For the same reason the nucleoli can be displaced by centrifuging in a way that is not possible in the compact nuclei formed in the same way as in Kuwada's experiment. These differences are undoubtedly significant and it is now necessary to find out how far they are determined by the size of the nucleus, which varies in different tissues, and how far by the size of the metaphase chromosomes.

CONCLUSION. We now see that the resting nucleus has a structure that may be identified both by comparison and experiment. The chromosomes do not appear like ghosts to flit for a transient moment

PLATE II

Smear preparation of pollen grains in *Paris quadrifolia* ($n = 10$), showing the first mitosis from prophase to telophase. Note the gradual disappearance of relic spirals during prophase. Fixation, La Cour's 2B.E.; staining, Newton's iodine-gentian-violet. × *ca.* 600. (La Cour, unpublished.)

PLATE II

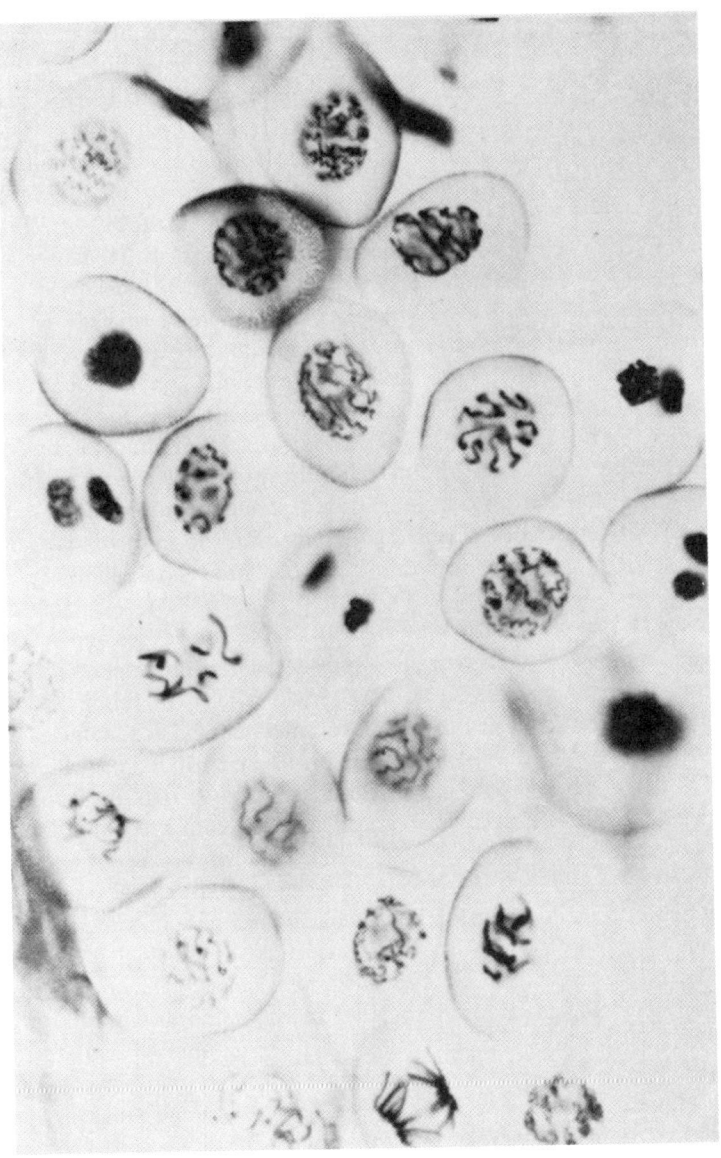

[To face p. 46.

across the mitotic stage, out of the shades of one resting nucleus and into the shades of another equally unknowable. They have a continuity which depends on the permanence of the *linear arrangement* of their characteristic constituents seen as a loosely coiled thread at early prophase of mitosis and as a closely coiled spiral thread at metaphase. This conclusion is of the first importance in genetics. The old dictum, *omnis cellula e cellula*, may be extended to nuclei, to chromosomes and to the unit-particles making up the chromosome, cytologically known as chromomeres, genetically as genes. It is even truer to say that all genes arise from pre-existing genes than that all cells arise from pre-existing cells, in this sense that the gene is an older structure than the cell.

The permanent thread is double at mitotic prophase. In the early prophase meiosis, as we shall see, it is single. We must therefore assume provisionally that it is always single at telophase. The one change that it undergoes during the resting stage is that of longitudinal division, presumably following the reproduction by each permanent particle or chromomere of one like itself, and the absence of this division distinguishes the resting stage before meiosis (*cf.* Chs. IV and X).

It seems that the centromere, on the other hand, does not divide at the same time as the thread itself, and has not the property, which the chromatids have, of associating in pairs. The special properties of the centromere will be clearer from a study of meiosis and of the results of irradiation (Chs. IV and XII).

4. ABNORMAL FORMS OF MITOSIS

Deviations from the described type of mitotic division are found in the higher plants and animals, but they are nowhere such as to render the process unrecognisable. In the lower organisms marked deviations from the described type occur (Fig. 10). These are described by Belar in a comprehensive survey of mitosis in the Protista (1926 *b*).

In many Fungi and Protista, an intranuclear spindle is found. In this type the centrosome, which is characterised by the presence of permanent granules colourable with nuclear stains, may lie outside the nucleus, as in some *Amœbæ*, or inside it, as in the

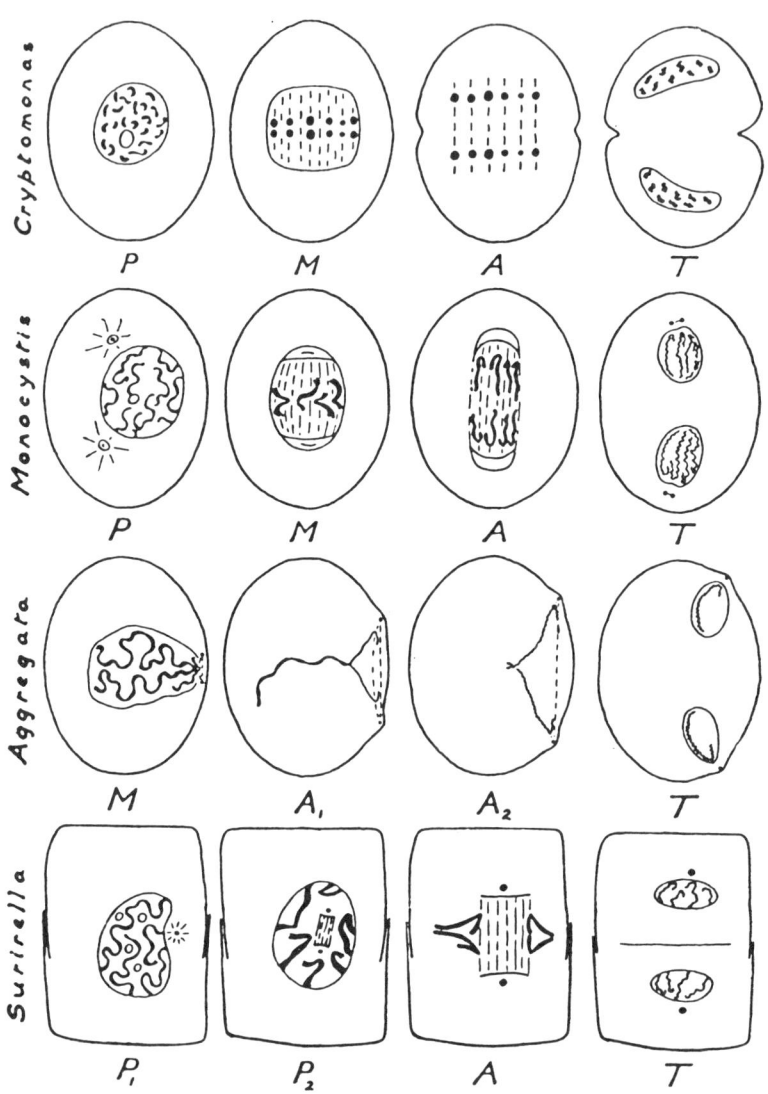

AMITOSIS

protozoan *Euglypha*. In the diatom *Surirella* it enters the nucleus during prophase. In the Fungi and many Protozoa the centrosome normally appears only at division.

Another type of abnormality is the extreme reduction in the size of the spindle relative to the nucleus. In this type, if the spindle is still distinguishable, it lies to one side of the dividing nucleus, as in *Aggregata eberthi* (Fig. 10). This organism demonstrates two other peculiarities of mitosis. It undergoes a process of "multiple nuclear division" during spore and germ-cell formation. In other organisms the relationship of this process to true mitosis is still obscure. Here the chromosomes do not show the high degree of

FIG. 10.—Diagram of mitosis in haploid nuclei of Protista. The spindle is shown by broken lines, the nucleoli in outline and the centrosomes by solid circles sometimes surrounded by "centrospheres." *Cryptomonas* (Cryptomonadina) : no visible centrosome, cylindrical spindle within the nucleus. *Monocystis* (Eugregarina, $n = 4$), progamic mitosis : centrosome plate-like at metaphase (M), invisible at anaphase (A). *Aggregata* (Coccidia, $n = 6$), sporogonial mitosis : two chromosomes shown at metaphase, one at later stages. Spindle outside nucleus. Greatest spiralisation at telophase (T). *Surirella* (Diatomeæ) : centrosome enters nucleus and spindle develops inside it during prophase. (After Belar, 1926.)

contraction characteristic of metaphase. They retain the length and appearance normally seen at early prophase, contracting further at anaphase. Divisions follow one another rapidly, however, and this further contraction is not always completed, so that the ends of two chromosomes which are longer than the rest do not separate. In this way daughter nuclei are constituted which run into one another and suggest the appearance of simple fission of a nucleus— what has always been called "amitosis." None the less the actual longitudinal splitting of the chromosomes can be seen with the utmost clearness, and in later divisions this separation is completed (*v. infra*).

These examples indicate the wide range of structure and

mechanism and the deceptive appearance that is found at mitosis in the Protista. They account for the difficulty in interpreting mitosis in these organisms and for its earlier description as " amitosis," that is, a type of division in which the chromatids of each chromosome were not separated to opposite daughter-nuclei. Extreme examples of this difficulty are seen in *Vahlkampfia* (*cf.* Belar, 1928 *b*) and in *Amœba* (Belar, 1926 *b*).

Improvement in technique in the Protista and elsewhere now makes it possible to show that all nuclear division is normally mitotic and that the apparent contradiction of genetic principles in the occurrence of amitosis need no longer be taken seriously. Amitosis, the division of a nucleus without the separation of chromatids, is found only in short-lived tissues, such as the endosperm of seed-plants and the deciduous embryonic membranes of animals when it is frequently unaccompanied by cell-division. It does not therefore give rise to independent and equivalent daughter-nuclei.

There is one apparent exception, the division of the macro-nucleus in the Infusoria. Here many forms of division are found which can scarcely be reconciled with the principles of mitosis (*e.g.*, *Spirostomum*, *cf.* Belar, 1926 *b*), and others which appear as very degenerate mitoses (*e.g.*, *Spirochona*). But the macronucleus of the infusorian is in a sense analogous with the body-cell nucleus in a multicellular organism, and like the body cells of *Ascaris* which lose part of their chromosomes, it need not carry the whole of the hereditary material, resulting from accurate mitotic division. It cannot give rise to a micronucleus, which alone is capable of reproduction by conjugation. It is off the germ-track. Its life is of restricted duration, for at intervals in the absence of sexual reproduction, when it always degenerates, it disappears and is replaced by a new macronucleus derived directly from its sister micronucleus by division and the growth of one of the products (Woodruff and Erdmann, 1914, on *Paramœcium*). Belar (1926 *b*) is inclined to believe that reported cases of a micronucleus arising from a macro-nucleus are due to the smaller nucleus being temporarily included in the larger and hidden by it. Races genuinely devoid of a micro-nucleus are apparently incapable of undergoing sexual reproduction (Woodruff and Spencer, 1924).

When the macronucleus undergoes an unequal division as in amitosis, we may therefore suppose that it suffers an irreversible change, as does the nucleus of *Ascaris* undergoing chromosome diminution (*q.v.*). And such a change cannot be undergone by a nucleus retaining the full genetic potentialities of the species (*cf.* Ch. VIII).

Summing up, recent work leaves no doubt that: whenever a nucleus divides to produce two daughter-nuclei physiologically or morphologically equivalent to it, it does so by mitosis. Further, the effective test of a nucleus is not primarily in its chemical or physical properties, but in its behaviour: *a nucleus is a cell-body which arises or reproduces by mitosis.*

CHAPTER III

MITOSIS : THE VARIATION OF THE CHROMOSOMES

The Subordination of the Chromosomes to the Genotype—Numerical Variation of the Chromosomes—Polyploidy—Structural Variation of the Chromosomes—Fragmentation, Translocation and Other Changes—Comparative Morphology of the Chromosome Complement—The Sources of Variation.

> Il n'y a réellement dans la nature que des individus.
> LAMARCK, *Discours*, 1804.

1. THE MITOTIC CONSTANTS

THE chromosome complement of an individual is constant not only in number but in certain other respects in which it is usually characteristic of its race or clone. These mitotic constants are three in number :—

1. **The Linear Constant.** Each chromosome consists of a linear arrangement of particles which is constant in order and normally has two ends, *i.e.*, it is not circular or branched. Two kinds of particles whose constant position is especially expressed at mitosis are the centromere and the nucleolar organisers which determine centric and nucleolar constrictions. Centromeres cannot arise *de novo* and chromosomes cannot divide without them. Whether they can exist without chromosomes is not known.

2. **The Volume Constant.** The bulk of the chromosome at metaphase is usually constant. It may, however, be affected by developmental conditions. Chromosomes are sometimes smaller in rapidly dividing nuclei.

3. **The Spiralisation Constant.** The degree of coiling of each of the chromosomes in the complement is equal and constant. This is shown by the equal diameter of their chromatids and by the constant length of each chromosome. Spiralisation does not proceed further than to reduce each part of a chromatid between constrictions to a sphere if this is of less than the standard diameter.

The developmental conditions of rapid mitosis reduce spiralisation below the constant value (as in *Aggregata*), while in pre-meiotic mitoses of animals it may be increased (as in *Chorthippus*, Fig. 11).

2. GENOTYPIC CONTROL

Considering these constants, we can see that there are some that can be taken as expressing the permanence of the thread which goes to make the metaphase chromosome. Thus the potentiality for a constriction, centromere or nucleolar, and the *relative* length of the chromosomes must be directly determined by the permanent character of the individual chromosomes. This character is, how-

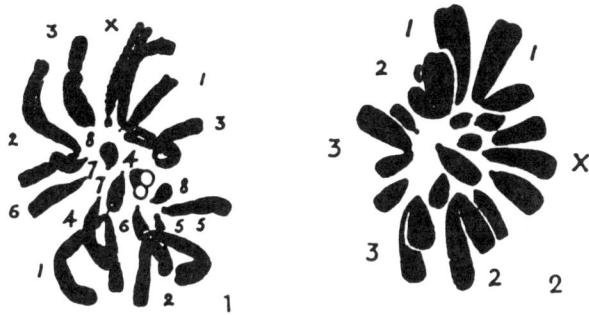

FIG. 11.—Metaphase in early and late spermatogonial mitoses of *Chorthippus parallelus* showing different degrees of spiralisation ($2n = 16 + X$). X understained, *cf.* Chs. VIII and IX. × 2500. (D., 1936 d.)

ever, consistent with variation in the appearance of the chromosomes in other respects. The constant diameter of the chromosomes making up one complement especially suggests that this diameter has nothing to do with the permanent structure of the individual chromosomes, but is a racial, *i.e.*, a genetic character, and therefore subject to variation through genetic change (*cf.* Ch. I).

This means that the genetic properties of the organism, which are controlled by certain permanent materials (the genes) in the chromosome thread, themselves control, *within certain limits*, the form that this thread takes up at metaphase of mitosis. It is of the utmost importance in interpreting observations of chromosomes in terms of genetics to know what these limits are. We must

54 MITOSIS : THE VARIATION OF THE CHROMOSOMES

therefore find out from a comparison of different species, races, mutants and hybrids what differences in the chromosomes can be attributed to changes in their permanent structure and what differences are due to a different joint reaction of the genotype controlling the chromosomes as a whole. Both these kinds of difference will, on the chromosome theory of heredity, be due to

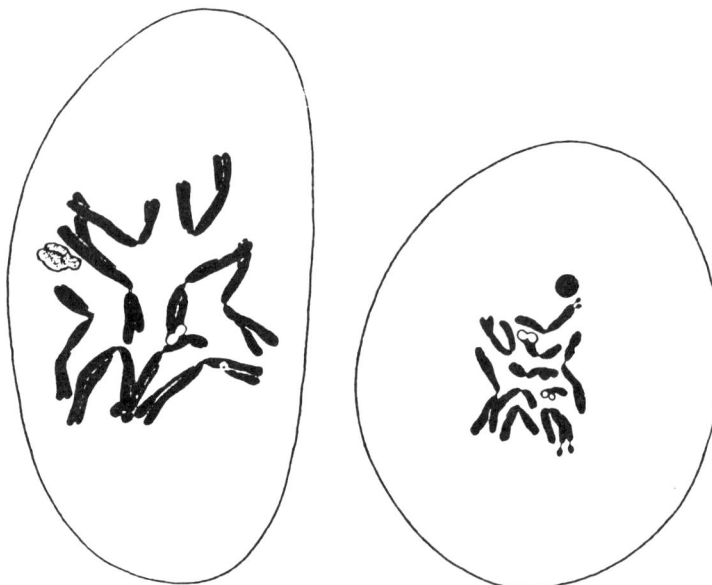

FIG. 12.—Mitotic metaphase in pollen grains from different flowers of *Tradescantia brevicaulis*, showing large and small chromosome types. A small chromosome is seen in the left-hand plate that has probably been lying in a separate nucleus, a persistent nucleolus in the right-hand one. × 1800. *Cf.* Table 47, p. 318 (after D., 1929 *c*).

changes in the arrangements or properties of genes, but the first will show in the part directly affected while the second, being due to gene mutation, will not be visible in direct effect at all, but will be seen by its altering the characteristic form or behaviour of the complement as a whole. Since both changes are genetic, the second kind will be distinguished as *genotypic*, for it takes effect through the action of the genotype as a whole (D., 1932 *a*).

CONTROL OF SIZE 55

The following examples show the range of genotypic control seen at mitosis.

(i) CHROMOSOME VOLUME. (*a*) *Tradescantia brevicaulis* is a triploid, apparently a hybrid between tetraploid *Tradescantia virginiana* with large chromosomes and a diploid relative with smaller chromosomes of proportionate diameter. It has chromosomes intermediate in length and width between those of its supposed parents. In the pollen grains of one bud of the plant studied the chromosomes were proportionately reduced to one-fifth and the resting nuclei to one-third of the normal size (Fig. 12). This was not associated with any change in the size of the cell (D., 1929 *c*, and unpublished; *cf.* Thomas, 1936, on *Lolium*, Fig. 122).

(*b*) The chromosomes of *Vicia sativa* and *Vicia angustifolia* are smaller in the hybrid than in the parental species, at mitosis and meiosis (Sveshnikova, 1929, *a* and *b*). Similarly F_2 hybrids from *Dianthus monspessulanus* × *D. plumarius* have chromosomes uniformly smaller than either of the parental species at mitosis (Rohweder, 1934).

(*c*) In *Sphærocarpus Donnellii* the chromosomes as a whole and the cells are 1·7 times larger in the female than in the male. This is largely accounted for by the presence in the female of the large X chromosome instead of the small Y chromosome, but partly also by the greater size of the other chromosomes. Apparently a genetic factor controlling chromosome size is correlated with sex in this species (Lorbeer, 1930). In man and the rat the chromosomes are longer (and probably proportionately broader) in the male than in the female (Evans and Swezy, 1928, *cf.* Table 12). Similarly the chromosomes (or at least the X chromosome) are larger in the male *Melandrium* than in the female (Belar, 1927). Witschi (1935) finds that the chromosomes are four times as large in the oöcytes as in the spermatocytes of hermaphrodite *Lepas anatifera* (Cirripedia). In this case differentiation has the same effect as has a genetic difference in other cases.

(*d*) A particularly significant change is found in haploid *Triton* embryos, where the chromosomes are larger than in diploids (Fankhauser, 1934 *b*).

These changes in bulk are paralleled by the effects of differences of hydrogen-ion concentration on living chromosomes (Kuwada and Sakamura, 1926), and are probably to be ascribed to differences of dispersion of the permanent materials (genes) in the resting nucleus and in the chromosomes rather than to a different degree of proliferation of these materials.

Cases which appeared to suggest the proliferation of permanent materials and to clash with the principle of genotypic control have been described by Tischler (1927) in *Ribes* and by Skovsted (1929, 1934) in *Æsculus* and *Gossypium*. A distinction between the large chromosomes of one species and the uniformly smaller chromosomes of another species (a distinction such as could hardly be due to

TABLE 5

Lengths of Chromosomes in Species and their Hybrids

	Species.	Hybrids.	
		capillaris-neglecta.	*capillaris-tectorum.*
C. capillaris	108	126	109
C. tectorum	156	—	148
C. neglecta	97	83	—

structural change) was found to be retained in the hybrid. These descriptions have not been confirmed (D., 1929 a ; Upcott, 1936 b).

(ii) CHROMOSOME SPIRALISATION. (a) *Phragmatobia fuliginosa* occurs in two races, one with constantly greater linear contraction, and greater width of chromatid than the other at mitosis, and at meiosis (Seiler, 1925).

(b) Two individuals of *Melandrium album* have been found to have chromosomes constantly about one-third the normal length. They therefore resembled the normal meiotic chromosomes in shape (Breslawetz, 1929).

(c) In the third generation from the cross *Viola tricolor* by *V. Orphanidis*, a segregate appeared having longer and thinner chromosomes than the parental species, or indeed any other species, of *Viola*. This reduced contraction is correspondingly effective at both meiotic divisions. It must be contrasted with the

Matthiola mutation in chromosome length (Ch. X), which affects only the first meiotic division (J. Clausen, 1931 *c*).

Table 5 shows the total relative lengths of the chromosomes at mitosis in the haploid sets of three *Crepis* species and their first generation hybrids (Navashin, 1931 *b*; *cf.* Fig. 7).

The chromosomes of *capillaris* are longer, and those of *neglecta* shorter, in the hybrid than in the parent. It is also found that their bulk is the same, greater length meaning less width. It is to be assumed that the species are genetically different in regard to the control of spiralisation, while the genotype of the hybrid determines

FIG. 13.—Mitotic metaphases in *Crepis capillaris*. A. Normal. B. In a root-tip that has been cooled before fixation. This change may be due to a difference in life or to a difference in the conditions of fixation. The difference is parallel to that found in different races of some species. × *ca.* 2000 (from Delaunay, 1930).

a uniform width in its chromosomes whatever their width in the parent from which they are derived (*cf.* Levan, 1936).

This difference has been imitated in *Crepis* by cold treatment before fixation (Delaunay, 1929, *cf.* Fig. 13; *cf.* Appendix II).

(*e*) In the neuropteran *Hemerobius pini* the chromosomes are longer and slenderer in the female than in the male (Naville and de Beaumont, 1933).

(iii) CHROMOSOME AGGREGATION AND DIMINUTION. In certain species of *Ascaris* the fertilised egg contains a small number of large chromosomes. At the second and later segmentation divisions in those cells which are not going to give rise to germ cells, these large chromosomes break up into numerous small ones which remain independent in all the nuclei derived from them. The change is thus irreversible. In eggs that have been fertilised by two sperms and have divided into three or four cells at the first division instead of two, the number of nuclei with unfragmented chromosomes at a later stage was correspondingly increased. Since it occurs in some species and not in others, this developmental property is evidently genotypically controlled, and the character may be described as the property of aggregation of the smaller chromosomes into larger ones, which is irreversible in the germ tract and reversible in the somatic nuclei, for the related species have the unaggregated condition in all their nuclei (Boveri, 1904).

(b) This requires that the centromeres of all the chromosomes in an aggregate, save one, shall be functionally suppressed, the breaking-up of the aggregate then being due to the revival of their activity. How this probably happens we shall see later.

In *Ascaris*, at the same time as fragmentation, and in *Miastor* and *Sciara* (Diptera), in the same circumstances, portions of the chromosome complement are left on the equatorial plate at anaphase. The portions lost are the distal ends of the fragmenting chromosomes in *Ascaris* and half the number of the chromosomes in *Miastor*. This change is necessarily irreversible (*Miastor*, Kahle, 1908; Hegner, 1914; *Ascaris lumbricoides*, Boveri, 1904; *Sciara*, Du Bois, 1933).

(c) In *Ascalaphus libelluloides* (Naville and de Beaumont, 1933) there occurs a fragmentation of two homologous chromosomes, the sex chromosomes. This change occurs gradually in mitosis and it is variable. It is also reversed entirely at the second meiotic division. In the hemipteran *Acholla* (Payne, 1910) the X chromosome seems to have broken up into five elements, but its breakage is permanent. Aggregation in *Phragmatobia* and *Philosamia* is probably also genotypically controlled, but diminution in *Drosophila* affecting particular chromosomes is more probably structurally controlled (*cf.* D., 1932).

(iii) **Conclusion.** These observations confirm the *a priori* views set out earlier. The relative form and behaviour of the chromosomes at mitosis in the same complement is determined by the length and linear structure of the chromosome thread which exists permanently at division and in the resting nucleus. Under different physiological conditions, however, the same thread will behave differently in regard to (i) the bulk it assumes at metaphase; (ii) the degree of spiralisation it undergoes during prophase; (iii) the aggregation

DUAL CONTROL

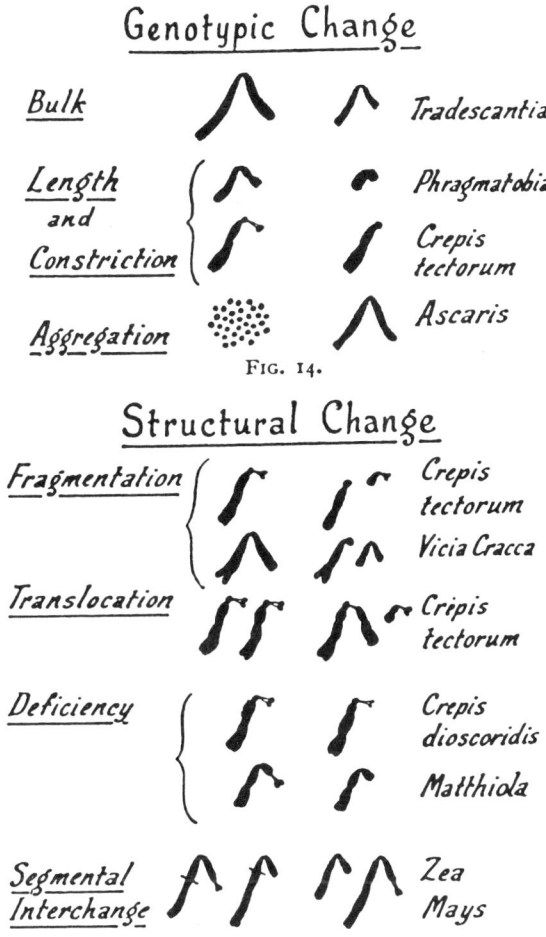

FIGS. 14 and 15.—Diagram to show the kinds of change produced genotypically and structurally in the chromosomes. In *Ascaris*, genotypic control determines a structural change.

of its parts at all stages; and (iv) the normal separation of its halves to opposite poles at anaphase. The last two are determined by the activity of the centromere which, having its own organisation, reacts in its own way to special conditions in the cell.

It follows that the chromosome differences observed between individuals and also between species must be interpreted in terms of physiological conditions. For example, a difference in the size of one chromosome may have much greater genetic and systematic importance than a difference in the size of all the chromosomes in the complement, for a general change may be determined by a change in a single gene in one chromosome, if this happens to alter the genetic properties of the organism in regard to chromosome behaviour. Further, the individuality of the chromosomes is a purely mechanical individuality. The individual chromosome has no independent physiological control over its external form any more than it has over anything else. The chromosome is a unit statically as a structure seen at mitosis, dynamically as an organ of segregation; physiologically, it is merely part of the genotype to which in many important properties it is subordinate. This subordination may be likened to the subordination of members of a legislature as individuals to the laws they enact as a body.

It is therefore necessary to distinguish in all chromosome behaviour between what is determined by the special properties of the individual particles and what is determined by the genetic reactions of the chromosomes as a whole, *i.e.*, by the genotype. This is particularly important for the analysis of the more complex phenomena of meiosis, where genotypic control plays an even more important part.

3. NUMERICAL VARIATION

(i) **Classification of Changes.** The characteristic group of chromosomes found at mitosis is spoken of as its complement. This complement in a diploid is made up of two haploid *sets*. Reduplication of some of the chromosomes of a set beyond the normal diploid number is called *polysomy*, reduplication of the whole set so that the nucleus contains three, four or more sets is called *polyploidy*. The two combined give *secondary polyploidy* (D. and Moffett, 1930). Any of these can result from abnormalities of (i) mitosis, (ii) meiosis, (iii) fertilisation.

Polyploidy in itself involves no change in the numerical proportions of the chromosomes in the set; it is said to be a

TABLE 6
Types of Chromosome Constitution

Deficient Gamete ($x - 1$).	Haploid (x).	Diploid ($2x$).	Triploid ($3x$).	Tetraploid ($4x$).
A B C — E *Uvularia*, Belling, 1925.	A B C D E Normal Gamete.	AA BB CC DD EE Normal zygote. " Unreduced " gamete (v. Ch. X).	AAA BBB CCC DDD EEE (v. Table 32).	AAAA BBBB CCCC DDDD EEEE (v. Table 32).

Monosomic Diploid ($2x - 1$).	Trisomic Diploid ($2x + 1$).	Tetrasomic Diploid ($2x + 2$).	Doubly Trisomic Diploid ($2x + 1 + 1$).
AA BB CC D— EE	AAA or AA BB BBB CC CC DD DD EE EE etc.	AAAA BB CC DD EE etc.	AAA BBB CC DD EE etc.
Zea Mays, McClintock, 1929 *c*; *Drosophila*, Bridges, 1916; Mohr, 1932.	*Œnothera* (no $2x + 2$), *Datura*, etc. (Tables 47–52).		*Œnothera*. *Datura*. *Solanum*.

Trisomic Tetraploid ($4x - 1$).	Pentasomic Tetraploid ($4x + 1$).	Doubly Hexasomic Tetraploid ($4x + 2 + 2$).
AAAA BBBB CCC DDDD EEEE etc.	AAAAA BBBB CCCC DDDD EEEE etc.	AAAA AA BBBB BB CCCC DDDD EEEE etc.
Datura, Belling and Blakeslee, 1924. *Primula kewensis*, Newton and Pellew, 1929. *Biscutella lævigata*, Manton, 1932. Cf. *Triticum vulgare* ($6x + 1$), Kihara, 1921; Huskins, 1928.		Secondary polyploids, *e.g.*, *Dahlia Merckii*. Pomoideæ (v. Ch. VI).

" balanced " variation. Polysomy, on the other hand, does involve such a change; it is said to be an " unbalanced variation," and

62 MITOSIS: THE VARIATION OF THE CHROMOSOMES

has on this account an entirely distinct physiological effect (v. Ch. VII).

The reciprocal results from the same series of errors as those which give polyploidy and polysomy are found in cells or in whole organisms which lack either individual chromosomes through an error of mitosis (*e.g.*, *Zea Mays*, McClintock, 1929 c) or meiosis (*e.g.*, Haplo-IV in *Drosophila*, Plate XII, Fig. 121) or whole chromosome sets (*e.g.*, haploid parthenogenesis, *q.v.*).

The table explains the nomenclature of polyploids by examples from a type with a haploid set of five chromosomes (*A* to *E*), and

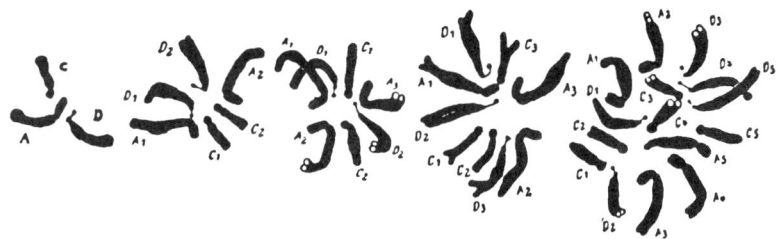

FIG. 16.—Mitotic metaphases in haploid (after Hollingshead, 1930), diploid, trisomic diploid, triploid and pentaploid (from Navashin, 1926) in *Crepis capillaris* ($x = 3$, chromosomes labelled A, C and D). × 1500.

shows the conventional formulæ in terms of the haploid number x. Thus x is 5 and a trisomic diploid has 11 chromosomes.

Since a zygote usually receives two similar sets of chromosomes from its two parental gametes, their number is conventionally referred to as $2n$; where the chromosomes pair regularly at meiosis they therefore form n pairs. Now in a particular individual these $2n$ chromosomes may consist of three sets or four sets of chromosomes relative to its own parents or ancestors. In the present work, therefore, the " basic number " of this ancestral set is distinguished by the sign x. Thus in *Triticum vulgare* $2n = 42$ and $x = 7$, the somatic chromosome number is therefore hexaploid ($6x$). Similarly in a trisomic tetraploid form of *Primula kewensis* (where $x = 9$) $2n = 35$, and this number will be represented as $4x - 1$. The endosperm, since it is a characteristically triploid phase of the life of the plant, may be referred to as $3n$.

POLYPLOID TISSUES

When polyploid and polysomic cells or whole individuals arise in experiment they can be identified by four criteria : (i) The physiological properties of the organism, depending on the balance of the chromosomes. (ii) The number of the chromosomes. (iii) The form of the chromosomes. This test is applied in the study of mitosis, and is completely satisfactory in favourable material, *e.g.*, in *Crepis*, where each type of chromosome can be recognised (Fig. 16). (iv) The pairing of the corresponding chromosomes at meiosis (*q.v.*).

(ii) **Polyploidy.** Tetraploid nuclei arise from diploid in the somatic tissue of plants and animals, probably through failure of the two bodies of chromosomes to separate at the anaphase of a mitosis (*v.* somatic pairing, Ch. VI). Such nuclei are normally found in certain specialised tissues of animals and very generally in some Hymenoptera. They have been observed amongst spermatocytes in the moth *Philosamia cynthia* (Dederer, 1928), and in *Drosophila* (Bridges, 1925 *a*). They are also usual in certain abnormal or degenerating tissues, such as the galls produced by parasites (*Beta maritima*, Nemec, 1926) the tapetal tissue surrounding the pollen mother-cells and in the embryonic membranes of animals. Here tetraploid and octoploid nuclei are regularly found. Such aberrations may also be induced by special treatment, as in Boveri's experiments with echinoderm eggs, where the division of the centrosome was inhibited and a single nucleus reconstituted at the end of mitosis (*cf.* Fankhauser, 1934).

Particular chromosomes, especially very small ones, sometimes fail to divide regularly at mitosis ; both halves pass to the same pole, so that nuclei with variable numbers arise (Mohr, 1932). This has been found with small chromosomes in the extreme case in *Tulipa galatica* (Upcott, unpub.) where the same plant may have some nuclei with four and others with as many as nineteen small chromosomes. Such a wide variation can come about only where the small chromosomes are inert and therefore do not affect the growth of the cells by the changes in their number.

The callus tissue developed round wounds by some of the higher plants is particularly liable to such abnormalities. Double nuclei have been found in the callus of a haploid tomato by Lindstrom and Koos (1931), and in this way tetraploid, octoploid and other abnormal nuclei may arise in the diploid. In the haploid *Triticum* multiple fusion occurs and in a root-tip of *Crepis tectorum* Navashin (1926) has found a dividing cell with 500 chromosomes instead of the normal 8 (*i.e.*, 125x). Tetraploid (4x) plants first arose in this way in grafting experiments in *Solanum* (Winkler, 1916) and Crane (*cf.* Jørgensen and Crane, 1927 ; Jørgensen, 1928 ; Sansome, 1931) developed the method to produce tetraploids systematically. Usually about 6 per cent. of adventitious callus shoots are tetraploid in *S. Lycopersicum*. Poisoning and ill-treatment of various kinds have been used to induce doubling of the chromosome number in root-tips (*cf.* De Mol, 1921 ; Nemec, 1929).

Doubled nuclei having arisen with normal frequency probably develop with exceptional vigour in haploids ; such plants have arisen by failure of fertilisation. They are weak in growth and are often distorted and modified by the occurrence of diploid tissues, with larger cells either as islands or as whole shoots (*cf.* Hollingshead 1930 ; *v.* Ch. VI). Thus amongst 110 root-tips of haploid *Crepis capillaris* there were 42 wholly diploid and 28 partly diploid, and one partly tetraploid. Amongst 82 root-tips of haploid *Nicotiana Tabacum*, 22 were wholly diploid and 8 partly diploid (Ruttle, 1928 ; *cf.* Webber, 1933). *Datura* and *Oryza* haploids, like those of *Crepis* (*cf.* Hollingshead, 1930 *a*) have yielded diploid shoots (*cf.* Meurman, 1933, on *Acer* ; Hruby, 1935 *b*, on *Salvia*).

In tissues in which polyploid nuclei have arisen, these nuclei have been observed to divide by multipolar mitoses. A spindle is then formed with three or four poles, as in a doubly fertilised echinoderm egg, and the chromosomes are distributed at random to these poles. This has been found both where the polyploid cells were developed in galls and following chloralisation. The result is the production of nuclei with chromosome numbers once more reduced (Nemec, 1926, 1929). This is perhaps the basis of rare variations such as the halving of the number in the tetraploid, observed by Winkler (1921) in *Solanum nigrum*. Randolph (1932) by treatment of the parents with high temperatures, has induced the formation of tetraploid cells (as well as other abnormalities) in young embryos from which

ORIGIN OF POLYPLOIDY

plants wholly or sectionally tetraploid have developed (*cf.* Dorsey, 1936).

The chief occasion of polyploidy in plants and in many animals is the reunion, or failure of separation, of daughter nuclei in the germinal cells at the last divisions before meiosis. This is known as syndiploidy (*v.* Table 7). In some cases multinucleate cells may be formed without fusion (*e.g.*, in *Zea*). In most cases the fusion takes place after the pachytene stage of meiosis (*q.v.*), for the nucleus, although tetraploid, does not produce quadrivalents at metaphase in *Datura, Zea, Prunus, Chorthippus*, or, as a rule, in *Brassica*. In *Prunus* and *Chrysanthemum* the abnormality is more far-reaching, for small supernumerary nuclei may be formed. This correlation of failure of formation of the cell-wall with failure of aggregation of the chromosomes into single nuclei is characteristic of extreme cases, for both processes depend upon the activity of the spindle.

The observations on animals and on *Brassica* and *Datura* show that the abnormality is often a racial, *i.e.*, a genetic, character. It has been more frequently discovered in hybrids, partly because they have been studied more closely, and partly because it is an undesirable character eliminated in species.

TABLE 7

Instances of Syndiploidy in Male Germinal Cells

A. ANIMALS

Lacertilia, various species, Painter, 1921.
Iguanidæ, all species, Painter, 1921.
Gomphocerus maculatus (Orthoptera), Eisentraut, 1926.
Chorthippus elegans D. 1932 *d* (*cf.* also Table 16B (p. 145)).

B. PLANTS

Œnothera	Geerts, 1909 ; Håkansson, 1927.
Lactuca	Gates and Rees, 1921.
Zea Mays	McClintock, 1929.
	Beadle, 1930 (asynaptic genotype).
Triticum compactum (*n*)	Gaines and Aase, 1926.
Raphanus × *Brassica*	Karpechenko, 1927.
Nicotiana tabacum (*n*)	Ruttle, 1928.
Pyrus Malus	Kobel, 1927.
	Nebel, 1929.
Triticum × *Ægilops* (5*x*)	Kagawa, 1929 *a* ; Kihara, Lilienfeld, 1934.

Prunus avium vars. . . D., 1930.
Brassica japonica (race) . . Fukushima, 1931.
Datura Stramonium (race) . Blakeslee, 1929.
Anthoxanthum odoratum (4x) . Kattermann, 1931.
Avena sativa (6x − 2) . . Nishiyama, 1931.
Chrysanthemum ornatum . . Shimotomai, 1931.
Taraxacum vulgare . . . Gustafsson, 1932.
Crepis capillaris (n) . . Hollingshead, 1930 a.

There is evidence for somatic doubling in other tissues being also determined by a special racial propensity in some cases. In *Kniphofia Nelsonii* one plant had independent and repeated doubling in root and stem, giving tetraploid and octoploid tissue (Moffett, 1932 ; *cf*. Meurman, 1933, on *Acer*). This is doubtless correlated with the spindle abnormalities found in the pollen mother-cells (Ch. X). Octoploid nuclei are also found in the abnormal race of *Datura*. In *Citrus limonum* tetraploid seedlings arise vegetatively in a regular proportion, probably through especially frequent doubling in nucellar tissue (Frost, 1925 ; *v*. nucellar embryony).

Tetraploid tissue in these cases and in others seems to arise regularly in certain structures as an adaptation of the species. For example, in *Cannabis*, the periblem cells are generally tetraploid, having 40 chromosomes instead of the 20 found in the plerome (de Litardière, 1925 ; Breslawetz, 1926, 1935).

Another source of polyploidy is from failure of reduction in the maturation in both male and female germ cells. This failure is particularly common at meiosis in hybrids, but may also be a racial character (*v*. Ch. X). The distinction between this method of origin and the first, by somatic doubling, can often be made in a particular case because a somatic change will give a whole tetraploid region with diploid germ-cells, while diploid germ-cells will only arise sporadically by the second method (Ch. VI).

Finally, tetraploidy may arise by natural apospory and artificial regeneration in ferns and mosses (Ch. XI).

Tetraploid animal forms have appeared in experiment only in *Drosophila melanogaster*. In the females of this species Bridges has found tetraploid (and hexaploid) tissue in the ovary. This, giving diploid eggs, yields triploid offspring which in the second generation has given a few tetraploids producing diploid eggs (Morgan *et alii*, 1925). Polyploids are only known naturally in a few animal species.

ORIGIN OF TRIPLOIDS

These are parthenogenetic in the tetraploid form when sexual in the diploid (v. Ch. XI). Polyploidy can be inferred from their chromosome numbers in the hermaphrodite Mollusca and Annelida.

This suggests that the rarity of tetraploid animal forms is due to the upset of the chromosome mechanism of sex differentiation, as has been found in *Empetrum* (Hagerup, 1927) where a diploid species is diœcious while its tetraploid relative is hermaphrodite. It may be thought, on the other hand, that the closed growth system of animals is responsible for the absence of polyploids, but, while this is a factor which will operate against their occurrence, it does not account for the complete absence of sexually reproducing polyploids (*cf.* Muller, 1925 *b*).

The nuclei of the endosperm in the flowering plants are triploid through the fusion of three nuclei, but such is not the origin of most triploid plants in experiment and in nature. One method of origin is by an abnormality in somatic mitosis in the haploid generation. Newton (1927) found nuclei at the third division of the embryo-sac with the diploid number of chromosomes, instead of the haploid, in many species of the Leiostemones section of *Tulipa*, of which this behaviour appears to be a characteristic. The doubling sometimes occurs regularly and presumably through the failure of anaphase separation at the division of one of the first two daughter nuclei in the embryo-sac (the second meiotic division). When it happens at the antipodal end it will result in the formation of endosperm with the tetraploid instead of the usual triploid number. When it happens, as it does exceptionally, at the egg-cell end, the embryo resulting from fertilisation will be triploid. This aberration therefore accounts for the occurrence of triploid forms and species in *Tulipa* and perhaps elsewhere in the Liliaceæ where non-hybrid triploids are common (*v.* Ch. V).

Triploids are probably derived more often through failure of reduction at meiosis (*q.v.*) on one side, and the consequent fusion of reduced (n) and unreduced ($2n$) gametes. In this way triploids have been produced in *Ascaris* (Boveri), *Drosophila* (Bridges, 1921 *a*), and in *Pygæra* hybrid back-crosses (Federley, 1912, 1931). This method of origin is particularly frequent in the dicotyledons (Ch. VII).

Finally triploids arise by hybridisation between diploids and tetraploids of the same species (*Œnothera, Primula, Zea, Datura*) or of another (*Rubus, Prunus, Fragaria, Avena, Triticum*) (*v.* Table 28D).

(iii) **Polysomy.** Polysomic forms arise in a diploid through two daughter chromosomes passing to the same pole at mitosis, or at meiosis. The first abnormality is doubtless stimulated by adverse external conditions, but disproportionately small chromosomes are especially liable to it (*v. infra*). The same abnormality affects the ordinary members of the chromosome complement less frequently. Navashin (1930) has observed the two halves of one chromosome passing to one pole in the plant of *Crepis tectorum* with the ring-chromosome. In *Drosophila*, irregular division in this way is the cause of gynandromorphism when a female cell with two X-chromosomes gives a daughter cell with one X which yields a male sector in an otherwise female animal, and of " diminished mosaics " when a fourth chromosome is lost (Mohr, 1932 ; *cf.* Crew on *Melopsittacus*, Fig. 121).

The second abnormality has been shown to follow in meiosis: (*a*) failure of pairing ; (*b*) " non-disjunction " of paired chromosomes which then pass to the same pole ; (*c*) " non-disjunction " or irregular distribution of the components of a multivalent configuration in a polyploid or of a ring in a structural hybrid. The first of these cannot strictly be called non-disjunction, although this term is used by geneticists to cover the results of all three types of abnormality, when they cannot be directly identified.

Trisomic plants regularly arise by crossing diploid with triploid in *Solanum, Crepis* and elsewhere (*v.* Ch. VI).

They also appear through two daughter chromosomes passing to the same pole in mitosis, either spontaneously, as has happened in *Datura* and *Scilla* (D., 1926) or following abnormal treatments which will be described later. How often trisomic nuclei arise in mitosis we do not know, because they are probably often eliminated in competition with the normal diploid nuclei. There is evidently a struggle for existence amongst meristematic cells, at least in the higher plants. This struggle favours the normal diploid at the expense of the abnormal haploid or trisomic or tetraploid, so that the abnormal types appear to be more changeable than the normal.

BREAKAGE AND REUNION

Thus we see frequently that trisomic plants give rise to diploid tissues or whole shoots (*Crepis*, Navashin, 1931 *a*; *Sphærocarpus*, Knapp, 1935 *b*). In high polyploid mosses artificially produced, Wettstein (1924) found that chromosomes were frequently lost. This " vegetative regulation " is probably the result of a similar competition of cells with different nuclear complements.

4. STRUCTURAL VARIATION

(i) **Classification.** The permanence in structure of the chromosomes is due to the permanence in linear order of a series of different particles, one of which must always be a centromere, the rest chromomeres. We can then represent two chromosomes by taking a letter to stand for each chromomere and an unspaced colon for the centromere, as *ab* : *cdef* and *ghj* : *klm*.

We may postulate several types of change in the structure of the chromosomes that are compatible with this linear arrangement, and with having one centromere in each chromosome. The following are the simplest :—

(i) *Inversion* of *de* segment :

$$ab : cdef \rightarrow ab : cedf.$$

(ii) *Translocation* of the intercalary *de* segment, either from one arm of a chromosome to the other, or to an arm of another chromosome, as :

$$ab : cdef + ghj : klm \rightarrow ab : cf + ghj : kdelm.$$

(iii) *Deficiency* or loss of part of a chromosome following its breakage.

(iv) *Interchange* between arms of different chromosomes :

$$ab : cdef + ghj : klm \qquad ab : cdlm + ghj : kef.$$

Interchange between two arms of one chromosome is the same thing as inversion of the *centric* or centromere-containing segment between them.

The detailed evidence of meiosis shows that these *a priori* types of structural change actually occur. The evidence of their occurrence that can be derived from mitosis, however, is incomplete, indirect and sometimes even misleading. The changes that are readily visible in mitotic chromosomes are necessarily restricted to four types ; first, independent changes of size that may be attributed to loss or gain following deficiency or translocation ; secondly, related changes of size in two chromosomes such as may be attributed to translocation or interchange between them ; thirdly, changes in the position of the centromere that may be attributed to inversion

TABLE 8

Analysis of Variation in Crepis

	Crepis tectorum ($n = 4$).	*Crepis capillaris* ($n = 3$).
Normal individuals	3,957	1,989
Numerical variants :		
$3x$	16	11
$4x$	5	—
$5x$	—	(1) *
$2x + 1$	10	
$2x + 2$	4	
$2x + 3$	4	
Total	39	11
Structural variants :		
Fragmentation	3 †	—
Translocation	1 ‡	—
Total	4	—
Total	4,000	2,000

* Different series of observations.
† One was trisomic.
‡ Also showed fragmentation.

of the segment containing it ; lastly, changes in number of chromosomes that may be attributed to fusion and fragmentation provisionally, but which must be due to secondary charges following interchange, since with normal monocentric chromosomes a change in the number of chromosomes means a change in the number of centromeres. Such a secondary change will usually have occurred as a result of the recombinations at meiosis which we shall consider later. Chromosomes with two centromeres and with none (dicentric and acentric) are found only immediately after they have arisen in special X-ray experiments, for they are unable to survive mitosis indefinitely. They must result from interchange that is asymmetrical with respect to the centromere (Ch. XII).

Thus the common observation of mitosis enables us to infer the occurrence of several different kinds of changes, but it does not alone enable us to define the nature of these changes. The observations of meiosis and of special types of mitosis is necessary for this. They reveal many changes that are necessarily invisible at mitosis such as equal interchanges, inversions not including the centromere or symmetrical with regard to it, and very small translocations. These show that all the conceivable structural transformations of chromosomes occur in nature and are responsible for the changes in number and size that are found in chromosome complements.

The relative importance of the different kinds of visible change occurring spontaneously has been shown in two species of *Crepis* from analysis of 6,000 plants by Navashin (1926) as in Table 8.

(ii) **Time and Place of Structural Changes.** From these observations several definite conclusions can be reached in regard to when, where, and how structural changes arise, as follows :—

1. Structural changes occur in chromosomes both in the mitotic and in the meiotic phase. They probably occur during the resting stage, or, at meiosis, during the prophase (*v*. Ch. VII, secondary changes and crossing over). The occurrence in mitotic nuclei is mostly clearly shown by Navashin's observations (1931 *a*) of chimæras in *Crepis* roots consisting of altered and unaltered cells and by four pollen mother-cells in *Secale* having an interchange while all their neighbours were normal, the interchange therefore having taken place two mitoses earlier. The occurrence at meiosis is most clearly shown by observations of new fragments both at the metaphase of the first meiotic division and at the first post-meiotic division in *Tradescantia*.

2. New structural types may fail to survive. Obviously a new acentric fragment will not survive mitosis unless it develops a new one, which it apparently cannot. Fragments of various sizes are found at the pollen grain mitosis in *Tradescantia virginiana*,

degenerating (Fig. 12). It seems probable that these are ones lacking a centromere. Others behave normally as new chromosomes. These are all the same size, about one-tenth the length of a normal chromosome in this and other genera (*e.g.*, *Fritillaria* and *Ranunculus*).

3. Except for such observations of the first post-meiotic division, it is only possible to see those new structural types which have survived many mitoses. These show a great range of form, but certain other restrictions are evident. Thus fusion only occurs

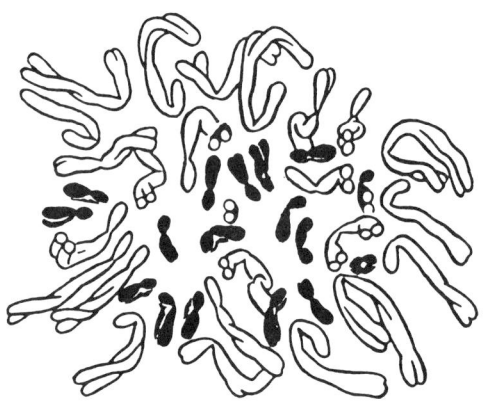

Fig. 17.—Metaphase of mitosis in the root-tip of a seedling of *Tulipa galatica*, $2n = 24 + 19ff$. Note the lack of synchronisation in the orientation and division of fragments (black). × 2,400 (Upcott, unpub).

between new fragments and old chromosomes (as in translocation) or between whole chromosomes which have a centromere close to the end. Fusion then occurs near the centromere (*e.g.*, in McClung's observations on the Acrididæ). Here again the fusions which occur may be at random and those that are seen may be the selected survivors. Such cases, as we have seen, are most easily explained as the result of unequal interchange.

4. Once the new structural type has survived the test of mitosis it is potentially as permanent as the chromosomes of the normal complement. But it often happens that in its size it is less well adapted than these to the conditions of mitosis or meiosis. The

problem of its adaptation to meiosis will be considered later (Ch. V). In some species an enormous range of size is found in the chromosomes of the normal complement (*v.* Section 4 (ii)), and great constancy is sometimes also found in the number of fragments in an old clone. Thus in a *Tradescantia* plant with 5 fragments, 4 and 6 were found exceptionally while in a *Fritillaria imperialis* plant with 6 fragments, 12 were found in an anther, evidently owing to both halves of each of the 6 going to the same pole in a mitosis, but otherwise the number was constant. In other species, behaviour is much less regular. In a group of *Ranunculus acris* plants, probably

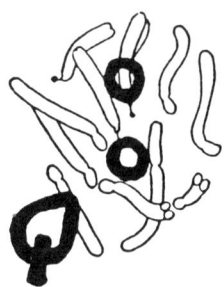

FIG. 18.—Metaphase of mitosis in root-tip of abnormal plant of *Crepis tectorum*, showing fifteen chromosomes instead of the normal eight. The four in black are the ring chromosomes, whose abnormal separation has led to their reduplication; two are interlocked. *Cf.* origin of polyploidy, Ch. III (from Navashin, 1930).

of one clone that must originally have had six fragments, Langlet found 252 divisions with the following fragment frequencies:—

No. of fragments	3	4	5	6	7	8	9	10
No. of mitoses	1	21	40	148	23	14	1	3

Similarly in *Nicotiana* the frequent loss of a fragment having an abnormal effect on flower colour gave patches of normal colour in the abnormal (*cf.* Blakeslee, 1928, on *Datura*).

Thus mitosis in some species seems to be better adapted to deal with a wide range of size of chromosomes than in others and a suitable mitotic mechanism is a second condition of the survival of new types of chromosomes.

5. It has usually been supposed that a simple linear arrangement is the only kind that can be produced by structural change, since that is the only kind found in the somatic chromosomes of mature organisms. Probably it is in fact the only one compatible with stable reproduction at meiosis and possibly also at mitosis. But in the translocation of a piece of one chromosome on to the side of another, *lateral translocation*, giving a branched structure, has been seen in three plants, *Tradescantia* (D., 1929 c), *Allium* (Levan, 1932) and *Alstrœmeria* (La Cour, unpub.). In all these cases the branch seems to occur at the centromere. Lateral trabants have also been produced by X-raying (Mather and Stone, 1933). In *Drosophila* the same structure has been inferred at some distance from the centromere (Bridges, 1923; Hamlett, 1926; Sturtevant, *cf.* Dobzhansky, 1931 a), but Bridges (1935) and Muller (1935) have been able to dismiss the possibility in particular cases.

On the other hand ring chromosomes are now known in five genera: *Crepis* (Navashin, 1930), *Drosophila* (L. V. Morgan, 1933), *Zea* (McClintock, 1932) and *Tulipa* and *Tradescantia* (Upcott, unpub.). Such rings may arise directly by interchange within a chromosome with loss of the small end segments which will have no centromere. Or they may arise by crossing-over between inverted segments in a way that will be considered later.

The behaviour of the ring chromosome is instructive in three respects: (i) In *Crepis* it varies in degree of linear contraction, sometimes appearing as a disc instead of a ring. Its structure therefore interferes with spiralisation, perhaps through interfering with the coiling of the chromatids which is characteristic of prophase chromosomes, but which is necessarily absent in these ring chromosomes (D., 1935 b). (ii) In *Zea* a segment of the ring chromosome is presumably lost during mitotic divisions, its broken ends joining up again since it is replaced by a smaller ring. Sometimes also the ring increases in size. These aberrations are perhaps due to the third peculiarity of the ring. (iii) Two ring chromosomes are sometimes found interlocked at metaphase in *Crepis* (Fig. 18) and the same thing no doubt occurs in *Zea*. This is presumably responsible for the irregular numbers in different cells both of these and of the normal chromosomes, for such interlocking interferes with

THE COMPLEMENT

the general course of anaphase separation as we shall see in dealing with meiosis. The fact that the interlocking is exceptional means that when the chromosome divides it does so along a cleavage surface which is helical in relation to the chromosome thread and, as a rule, exactly compensates for the relic spiral in which this thread is lying. The two halves will then separate freely at anaphase. Such a constancy of cleavage direction implies a constancy in the orientation of the particles making up the thread and a specificity in the direction of cleavage, or perhaps more properly the direction in which the new particles are laid down which go to make the sister chromatids (Ch. XII).

5. THE COMPARATIVE MORPHOLOGY OF THE CHROMOSOME COMPLEMENT

(i) Specific. In the previous sections comparison has been made of sister-cells, sister-individuals and sister-races. It may now be extended to related species and genera. The more distant the relatives compared, the less certain are the grounds of inference, but it will be seen that the relationships of species and genera can often be expressed conveniently in terms of the changes inferred from the study of smaller groups. It will also be seen later that more detailed study at meiosis bears out these conjectures in many instances.

In the first place our knowledge of polyploidy within the species suggests that when groups of species have chromosome numbers in a series of multiples of a common number and the chromosomes are of similar form, these species are probably derived from forms with this "basic number" by polyploidy (Winge, 1917). The application of this inference will be considered later in detail (Ch. VI).

Structural changes may be inferred with some certainty from a detailed comparison of chromosome form within genera and larger groups. Thus the chromosome complements within *Fritillaria* and the Acrididæ may be represented diagrammatically as follows (V, a long chromosome with median centromere ; I, a short chromosome with terminal or sub-terminal centromere).

76 MITOSIS: THE VARIATION OF THE CHROMOSOMES

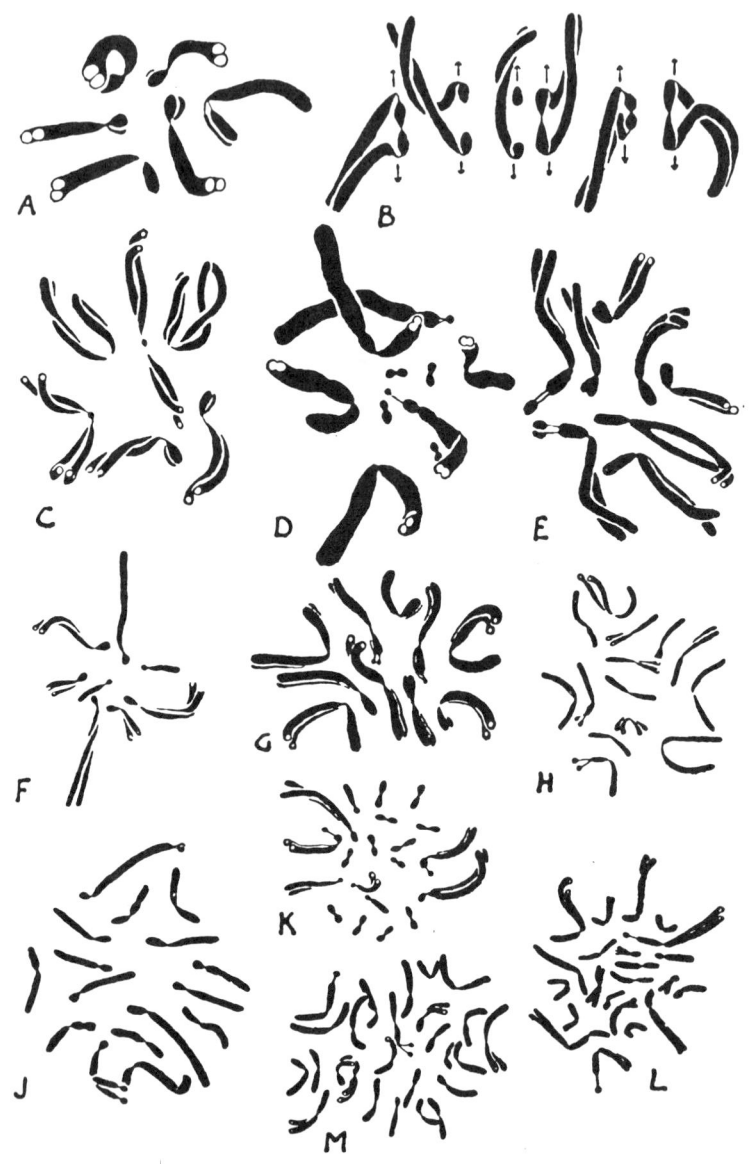

Fritillaria (Figs. 48, 95, and) :
 F. ruthenica ($n = 9$) . . V V V V V I I I I
 F. imperialis ($n = 12$) ⎫
 F. Meleagris ($n = 12$) ⎬ . V V II II II I I I I
 and six other species ⎭
 F. pudica ($n = 13$) . . V II II II II I I I I
Acrididæ (females) :
 Stenobothrus ⎫
 Chorthippus ⎬ ($n = 9$) . . V V V I I I I I I
 Chloëaltis ⎭ (Figs. 25–28)
Other genera ($n = 12$) . . II II II I I I I I

FIG. 19.—Mitotic chromosomes from root-tips of *Crocus* species. A. *C. Aucheri*, $2n = 6$. B, the same in side view of early anaphase, C, *C. graveolens*, $2n = 6$. D, *C. hyemalis*, $2n = 6 + 4ff$. E, *C. aureus*, $2n = 8$. F, *C. zonatus*, $2n = 8$. G, *C. stellaris*, $2n = 10$. H, *C. Hueffelianus*, $2n = 14$. J, *C. hadriaticus*, $2n = 16$. K, *C. karduchorum*, $2n = 20$. L, *C. corsicus*, $2n = 22$. M, *C. Imperati*, $2n = 26$. × 3,700 (Mather, 1932).

Note.—C is early metaphase with the limbs of the chromosomes separate ; D is full metaphase with the limbs joined throughout their length. In B, arrows indicate the position of the centromere at which the daughter-chromosomes begin to separate.

The direction of the changes suggested by this comparison necessarily cannot be settled by direct evidence. But the fact that the number 12 is constant in both the groups concerned, apart from the types given here with other numbers, suggests that these are derived from original forms with 12 chromosomes. Thus *F. ruthenica* and *Stenobothrus* are derived from ancestral forms by " fusion," *F. pudica* by " fragmentation."

A comparison of *Circotettix* with other Tettigidæ (Carothers, 1917, 1921 ; Helwig, 1929), of *Nicotiana alata* and *N. Langsdorffii* ($n = 9$,

Kostoff, 1929) with other species ($n = 12$), of *Cardamine pratensis* with other species (Lawrence, 1931 *d*), of *Orgyia thyellina* with *O. antiqua* (Cretschmar, 1928), and of various species of *Gryllus* (Ohmachi, 1929), indicates similar processes in the ancestry of these forms.

Within species "fusion" may be inferred in races of various species of *Hesperotettix, Mermiria* and *Jamaicana* (McClung, 1914, 1917, Woolsey, 1915) and *Crepis tectorum*. "Fragmentation" may be inferred in *Vicia Cracca* (Sweshnikova, 1928, and Fig. 14), *Phragmatobia fuliginosa* (Seiler and Haniel, 1921), *Gryllotalpa gryllotalpa* (de Winiwarter, 1927, Barrigozzi, 1933) and *Felis domesticus* (de Winiwarter, 1934).

More complex changes, such as inversion and interchange, necessarily cannot be inferred from a comparison of mitosis in different species. They will be considered in relation to meiosis and crossing over.

When larger groups are considered a more generalised effect is seen. Thus in the ninety-four species of the Lepidoptera examined the greatest number of species is found with thirty-one chromosomes, and the frequency with other numbers diminishes regularly with increasing remoteness from thirty-one. This indicates, as Beliajeff (1930) points out, that the group has arisen from ancestral forms with this number. The variation in number has arisen by random fragmentation and fusion. The fragmentation and fusion account for certain differences in size of the chromosomes also, but not for more than a small part.

Thus *Dasychira pudibunda* ($n = 87$), may be supposed to be derived by fragmentation from a *Lymantria*-like type with thirty-one chromosomes, for while most of the chromosomes are reduced in size, three pairs remain thirty or forty times the bulk of the rest.

On the other hand, where two species in the same family like *Spilosoma lubricipeda* and *Miltochrista miniata* have the same chromosome number and the chromosomes correspond in size-variation in each, but are throughout forty times larger in one species than in the other, probably genotypic control of chromosome size is responsible for the difference (*cf.* Beliajeff, 1930 ; D., 1932 *a*). *A fortiori* when between two species in the same family of plants

PLATE III

POLAR VIEWS OF METAPHASE OF MITOSIS IN ANIMALS AND PLANTS
(EXCEPT FIG. I WHICH IS LATE PROPHASE)

FIGS. 1 and 2.—*Chorthippus parallelus* (Orthoptera) ♂, $2n = 16 + x$. Smear of spermatogonia. *Cf.* Text-Fig. 11. (La Cour and Osterstock. Smears, 2B.D. — gentian-violet.)

FIG. 3.—*Locusta migratoria* (Orthoptera) ♂, $2n = 22 + x$. All except the five smallest chromosomes lie on the edge of the plate. All have subterminal centric constrictions. (Koller, unpublished.)

FIG. 4.—*Eremurus spectabilis*, root-tip, $2n = 14$. Medium Flemming — gentian-violet, × 2000. (Upcott, 1936.)

FIG. 5.—*Tradescantia bracteata*, pollen-grain, $n = 6$. FIG. 6.— *Brodiea uniflora*, pollen-grain, $n = 6$. (La Cour: 2B.E.—gentian-violet.)

FIG. 7.—*Puschkinia libanotica*, root-tip (Text-Fig. 5), $2n = 10 + 4\text{ff}$. × 1500 (D., 1936 *c.*: 2B.E. — gentian-violet.)

FIG. 8.—*Tricyrtis hista*, root-tip, $2n = 26$. (La Cour: 2B.E.—gentian-violet.)

PLATE III

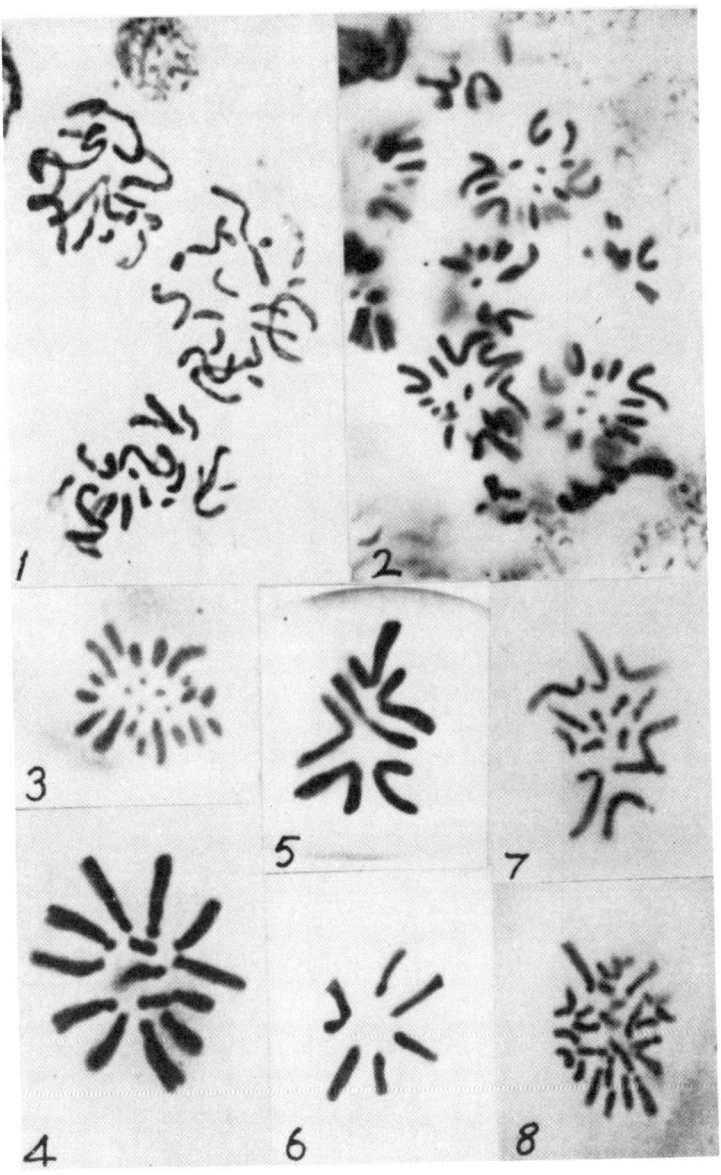

[To face p. 78.

(Droseraceæ), there is difference of size in the proportion of a thousand to one, genotypic control of the difference is to be inferred (Table 10).

Structural changes in themselves have no necessary genetic effect, but they may condition *genetic* isolation of stocks, and hence a later genetic differentiation between species. But this differentiation need not follow. We therefore find every relationship between structural and numerical and genotypic differences in the chromosomes on the one hand and systematic differences on the other, as follows :—

(i) The absence of readily detectable differences within large groups such as the gymnosperms—all of which, with a few exceptions, have twelve haploid chromosomes (*cf.* Sax, 1933)—and the Acrididæ (except three genera) ; or the absence of all differences except polyploidy (*e.g.*, the Pomoideæ).

(ii) The occurrence of differences between genera of an order while the genera are themselves fairly uniform, *e.g.*, *Antirrhinum*, *Orchis* (*cf.* Tischler, 1928 *b*), *Ribes* (Meurman, 1929), *Prunus* (apart from polyploidy, D., 1930 *a*).

(iii) The occurrence of differences between sections of a large genus, *e.g.*, *Primula* (Bruun, 1930).

(iv) The occurrence of differences between all the species of a genus, *e.g.*, *Drosophila* (Metz, 1916), *Fritillaria*, *Tulipa*, *Crepis*, *Tradescantia*, *Muscari*, *Vicia* (and polyploidy in nearly all large angiosperm genera except *Antirrhinum*, *Ribes* and *Orchis*).

(v) The occurrence of differences between geographical races or sub-species (*Nothoscordon* species, Matsuura and Suto, 1935 ; *Rumex acetosa*, Yamamoto, 1933 ; Ono, 1935 ; *Drosophila pseudo-obscura*, Dobzhansky, 1934 (*v.* Tables 8 and 37).

(vi) The occurrence of differences between forms not recognisably distinct (*v.* Table 37, structural differences in some Orthoptera and polyploidy in *Silene*, Blackburn, 1928 ; and numerical variation in *Viola canina*, Clausen, 1931).

A comparison of the mitotic chromosomes of different forms is therefore of little value in placing them systematically unless we know the type of variation that prevails in the groups in question. If the forms have different chromosome numbers they are probably intersterile or yield sterile hybrids, a fact which the systematist may or may not take into consideration.

(ii) **General.** We now see the whole capacity for variation in the

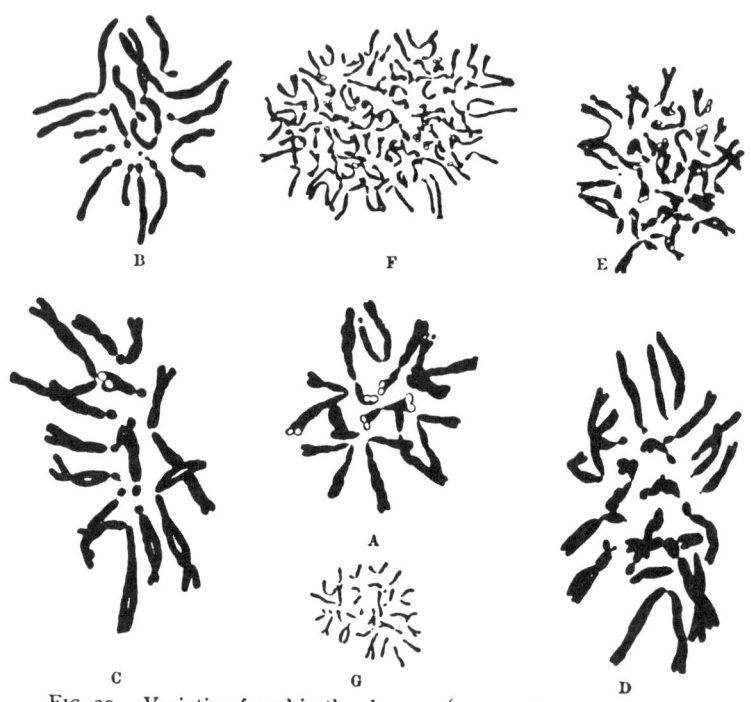

FIG. 20.—Variation found in the chromosome complement in species of a natural group, the Tradescantieæ. A. *Rhœo discolor*, $2n = 12$. B. *Cyanotis zenonii*, $2n = 16$ (by fragmentation). C. *Spironema fragrans*, $2n = 12$ (by loss of parts). D. *Zebrina pendula*, $2n = 24$ (by polyploidy). E. *Coleotrype natalensis*, $2n = 36$. F. *Commelina cœlestis*, $2n = 90$ (by polyploidy and fragmentation). G. *Cyanotis somaliensis*, $2n = 28$ (by fragmentation). F and G show reduced chromosome width. G shows reduced bulk, × 1,800 (from D.,1929 c, and unpublished; *cf.* Matsuura and Suto, 1935).

chromosome complement as it is shown to us at mitosis. The chromosomes are subject to two kinds of change: change governed directly by the activities of the part affected, and change governed by the reaction of the genotype as a whole, which usually has a uniform effect on the whole chromosome complement, although it is probably determined by a gene mutation which is itself invisible and strictly local in origin. The first is described as structural or numerical change according as it affects the internal arrangement, the structure, of a chromosome or the number of different chromosomes present. The second is described as genotypic change. When the chromosome complements of living organisms as a whole are compared, their variation is found to be such as can be described in terms of these two kinds of change.

The time has not yet come when their relative importance can be accurately assessed except in considering small groups of species. An internal and external comparison of the chromosome complements, however, indicates the part played by the two kinds of change. Differences within a chromosome complement indicate solely the range of structural change. Differences in the diameter of the chromosomes in different species indicate solely the range of genotypic change.

Most complements are uniform in the size of their members, which do not vary over a range of more than two to one. Exceptionally a much higher range of variation is found, reaching a maximum of 500 : 1 in the domestic turkey, *Meleagris* (Werner, 1931).

TABLE 9

Organisms with an Extreme Range of Size in the Chromosomes
(*cf.* Chiasma Frequencies, Ch. VII)

PLANTS (Monocotyledons)

Eucomis bicolor	Müller, 1912.
Yucca filamentosa	Taylor, 1925 *c*; Morinaga *et al.*, 1929.
Y. flaccida	O'Mara, 1931.
Agave sisalana	Suto, 1935; Doughty, unpub.
Muscari latifolium	Delaunay, 1926.
Uvularia grandiflora	Belling, 1926.
Fourcroya altissima et al.	Heitz, 1926.
Ornithogalum pyramidale et al.	Heitz, 1926; Matsuura and Suto, 1935.

ANIMALS

Aggregata eberthi (Sporozoa)	Dobell, 1925 ; Belar, 1926.
Drosophila melanogaster (Diptera)	Bridges, 1927 ; Morgan *et al.*, 1925. (See Fig. 121.)
Tettigonia albifrons (Locustidæ)	de Winiwarter, 1931.
Dasychira pudibunda (Lepidoptera)	Beliajeff, 1930.
Schistocerca paranensis (and other South American Acrididæ)	Saez, 1930 ; *cf.* White, 1933.
Vipera aspis (Reptilia)	Matthey, 1929, 1931, 1933.
Eumeces latiscutatus (Reptilia)	Nakamura, 1931.
Meleagris gallopavo (and other Galliformes and Anseriformes).	*Cf.* Werner, 1931 ; White, 1932 ; Koller, unpub.

In some of these species the small chromosomes in the complements are variable in number. They are perhaps unessential to the species, like the results of recent fragmentation already described.

To turn to the comparison of different species, the lowest chromosome numbers are found in the nematode *Ascaris megalocephala*, which has two races, one with one haploid chromosome, the other with two ; and in the Agaricaceæ (Wakayama, 1930), where species with two as their haploid number occur. In the flowering plants, nine species are known which have three as their haploid number, *viz.*, *Crepis capillaris* (*cf.* Babcock and Navashin, 1930), *Callitriche autumnalis* (Jørgensen, 1923), *Zacintha verrucosa* (Navashin, 1930), and six *Crocus* species (Mather, 1932, and Fig. 16). The Table (10) shows the range in size and numbers found in living organisms. It will be seen that the approximate ranges found in the various constants, represented as proportions, are as follows :—

Breadth, $1 : 10^{1 \cdot 2}$ . . (*Sphæromyxa : Cambarus*)
Length, $1 : 10^{1 \cdot 7}$ . . (*Saprolegnia : Hyacinthus*)
Bulk :
  Chromosome, $1 : 10^{3}$ . . (*Drosera : Drosophyllum*)
  Complement, $1 : 10^{4 \cdot 1}$ . . (*Saprolegnia : Aulacantha*)
Number, $1 : 10^{3}$ . . (*Ascaris : Aulacantha*)

It is clear that such great differences in structure and number demand the assumption of both structural and genotypic change.

RANGE OF SIZE

Through these agents the size and number of the chromosomes must be adapted to the needs of the organism. Later we shall see

TABLE 10

Range in Size and Number of Chromosomes in Animals and Plants [1]

	Diploid No. of Chromosomes ($2n$).	Approx. Breadth (2 Chromatids).	Approx. Length.	Author.
1. *Aulacantha scolymantha* (Radiolaria).	ca.1600	1·0μ	ca.20μ	*Cf.* Belar, 1926.
2. *Ascaris megalocephala univalens* (Nematoda):				
(i.) Germinal	2	0·7μ	20μ	Boveri, 1909.
(ii.) Somatic	ca.50	0·5μ	0·5μ	
3. *Cambarus virilis* (Crustacea)	200	3μ	3μ	Fasten, 1914.
4. *Drosophila melanogaster* (Diptera).	8	0·3μ	2·5μ	Bridges, 1927.
5. *Stenobothrus parallelus* ♂ (Orthoptera).	17	0·6μ	13μ	D. and Dark, 1932.
6. *Homo sapiens* (Mammalia)	48	0·6μ	5·4μ ♂ 4·1μ♀	Evans & Swezy 1928, 1929.
7. *Sphæromyxa sabrazesi* (Myxosporidia).	6	0·2μ	2μ	Belar, 1926 *b*.
8. *Saprolegnia mixta* (Phycomycetes).	22	0·4μ	0·4μ	Mäckel, 1928.
9. *Humaria rutilans* (Ascomycetes).	16	0·5μ	5μ	Fraser, 1908.
10. *Hyacinthus orientalis* (Monocotyledones).	16	2μ	21μ	D., 1926, Fig. 41.
11. *Prunus laurocerasus* (Dicotyledones-Rosaceæ).	176	0·3μ	2·5μ	Meurman, 1929.
12. *Œnothera biennis* (Dicotyledones-Onagraceæ)	14	0·3μ	2·5μ	D., 1931 *d*, Plate IV.
13. *Crepis capillaris* (Dicotyledones-Compositæ)	6	0·9μ	6μ	Navashin,1926, Fig. 9.
14A. *Drosera capensis* (Dicotyledones-Droseraceæ).	40	0·3μ	1·0μ	Behre, 1929.
14B. *Drosophyllum lusitanicum* (Dicotyledones-Droseraceæ).	12	2μ	25μ	*Ibid.*

[1] Illustrations showing the general variation of chromosome form in flowering plants are found in Lewitsky (1931) and Matsuura and Suto (1935).

in relation to meiosis and polyploidy what certain of these needs are. The most obvious, however, may be mentioned here. The minimum size of the cell in the life history clearly limits the size of the chromosomes, on the principle that you cannot put a quart into a pint pot. The effect of this is seen in various ways. In polyploids the area of the metaphase plate does not increase in proportion to the cell volume even where this increases in proportion to the chromosome number. Tetraploids do not occur in some genera with the largest chromosomes, such as *Lilium* and *Fritillaria*. In *Tulipa* a pentaploid is found in one species only, that with the smallest chromosomes (*T. Clusiana*) and tetraploids are not found in those with the largest chromosomes, the garden tulips. Polyploidy is commonest in the Dicotyledons where the chromosomes are smallest. In *Dianthus* the difficulty seems to have been overcome by the polyploid species having smaller chromosomes than their diploid relatives. They are genetically adapted. In considering limitation of cell-size we have to remember the whole life-history of the organism. In plants the most serious restriction occurs in the narrow cambium cells which produce secondary growth. It is therefore not without significance that secondary growth is not found in organisms with the largest chromosomes, the Gymnosperms being at the upper limit in this respect. Apart from sections of the Ranunculaceæ, Berberidaceæ and Droseraceæ, the Dicotyledons have small or medium-sized chromosomes. Those Monocotyledons with secondary growth have small chromosomes or very few large ones. Three evolutionary relationships seem to be explained by these considerations: First, the size of the cells conditions the size of the chromosomes. Secondly, the size of the chromosomes conditions the changes of shape of the cells at particular phases of growth, and so limits the internal habit of growth. Thirdly, the multiplication of the chromosomes by polyploidy is conditioned by the existence of a margin of space in the cell permitting an increase in the bulk of the chromosomes.

CHAPTER IV

MEIOSIS IN DIPLOIDS AND POLYPLOIDS

Introduction—Outline of Meiosis—Meiosis in Diploids—Meiosis in Polyploids—Forms of Bivalents and Multivalents at Metaphase—Separation at Anaphase—Conclusion—The Conditions of Chromosome Pairing.

1. INTRODUCTION

There must be a form of nuclear division in which the ancestral germ plasms contained in the nucleus are distributed to the daughter-nuclei in such a way that each of them receives only half the number contained in the original nucleus.
WEISMANN, 1887.

STUDIES of fertilisation between 1870 and 1875 showed that the process consisted essentially in the union of the nuclei of the fusing germ cells (O. Hertwig). Studies of mitosis during the succeeding years showed that at every observed division of the nucleus or mitosis there was a division of its constituent elements, the chromosomes, into equal halves. These elements were constant in appearance, and therefore they were, as later work has abundantly shown, " permanent " structures.

It follows that, in the history of the nucleus from one generation to the next, wherever fertilisation occurs, there must also be some compensating process. Addition must be set off by *reduction*. This conclusion was arrived at by Weismann in 1887, although at that time the nature of the reduction was only slightly indicated by observations of germ-cell formation on the female side. Weismann noted that in certain parthenogenetic organisms only one polar body was extruded instead of two as in sexual eggs. He thought, therefore, that the division which yielded the second polar body was a " reducing division "—a conclusion that was sound in a certain sense. He further concluded that a similar reduction would be found in male germ-cell formation.

Weismann's induction has been verified universally : wherever there is fertilisation there is also reduction, or, as we shall call it, *meiosis*. The course of this verification has been beset with many difficulties, which will later be considered in detail. They have been discussed at length by several authors (Wilson, 1925; Belar, 1928; Reuter, 1930; D., 1931; v. Appendix I).

We may now define meiosis by its superficial phenomena as *the occurrence of two divisions of a nucleus accompanied by one division of its chromosomes*. It results in the production of four nuclei, each of which has half the number of chromosomes of the mother-nucleus, provided that their distribution has been regular.

The two divisions of the nucleus are accompanied by the usual mitotic mechanism of spindle-formation. The single division of the chromosomes consists in the usual mitotic longitudinal reproduction by which each chromosome produces two chromatids. At once we see therefore that the difference between meiosis and mitosis consists in the external mechanism, which acts twice, getting *out of step* with the internal mechanism, which acts once. Precisely how the two systems get out of step we shall see from a detailed study.

From the principles set out in the first chapter, it will be clear that the occurrence of one form of nuclear division instead of another at a particular point in the development of the organism is an example of differentiation—in time, in the Protista, and in time and place, in the higher organisms. This differentiation being a racial characteristic must be genetically determined, although its origin is too remote for its exact conditions to be reconstructed. Organisms do occur, however, in which the reverse change has taken place, and meiosis has been more or less completely replaced by one or two mitoses. These cases show that the two forms of division are distinguished by a simple physiological difference which is controlled by the genetic properties of the organism. The nature of this difference and of this control will be considered later in relation to the theory of meiosis, of apomixis, and of cell-mechanics.

2. OUTLINE OF MEIOSIS

The following sketch shows the succession of events in meiosis, as they occur in the simplest clear examples, *e.g.*, pollen mother-cells

CHARACTER OF MEIOSIS 87

of diploid species of *Fritillaria*. Alternative nomenclature will be found in the glossary.

I. FIRST DIVISION. (*a*) *Leptotene*. The chromosomes appear in the nucleus in their diploid number. They are single threads, not double as in mitosis. Each has an uneven granular structure which gives it the appearance of a string of unequal beads, unequally strung together. The beads are the *chromomeres*. The threads are disposed evenly in the nucleus.

(*b*) *Zygotene*. The chromosomes come together in pairs, corresponding chromomeres lying side by side. They usually come into contact first near the centromere (and this body can be picked out by the greater staining capacity of the chromomeres on either side of it). But association often begins at other places independently. It extends along the chromosomes until the whole complement are present as double or bivalent threads in the haploid number.

(*c*) *Pachytene*. The paired threads of each bivalent coil round one another and their proximal chromomeres increase in size still further. The chromosomes thus show differential condensation in different parts.

(*d*) *Early Diplotene*. The chromosomes fall apart and at the same time each is seen to be double, consisting of two chromatids, which remain in close association like that of the chromosomes at pachytene. The chromosomes separate completely except at various points along their length, where their chromatids exchange partners. These cross-shaped exchanges are called *chiasmata* and number one, two, or more in each bivalent. They are due to *crossing-over* between two chromatids of the partner chromosomes, *i.e.*, the two chromatids have broken between corresponding pairs of chromomeres and rejoined across to give two new combinations. The chromosomes now begin their spiralisation.

(*e*) *Late Diplotene*. Spiralisation continues, the chromosomes straighten and successive loops between chiasmata come to lie at right angles.

(*f*) *Diakinesis*. Spiralisation reaches a maximum and the chromosomes are shorter than at a somatic mitosis. They continue to be evenly disposed throughout the nucleus. The nucleoli, which have been attached to their organisers on the chromosomes, disappear.

88 MEIOSIS IN DIPLOIDS AND POLYPLOIDS

(g) *(First) Metaphase.* The paired chromosomes come to lie with their pairs of centromeres evenly distributed on either side of the equatorial plane of the spindle, members of each pair of

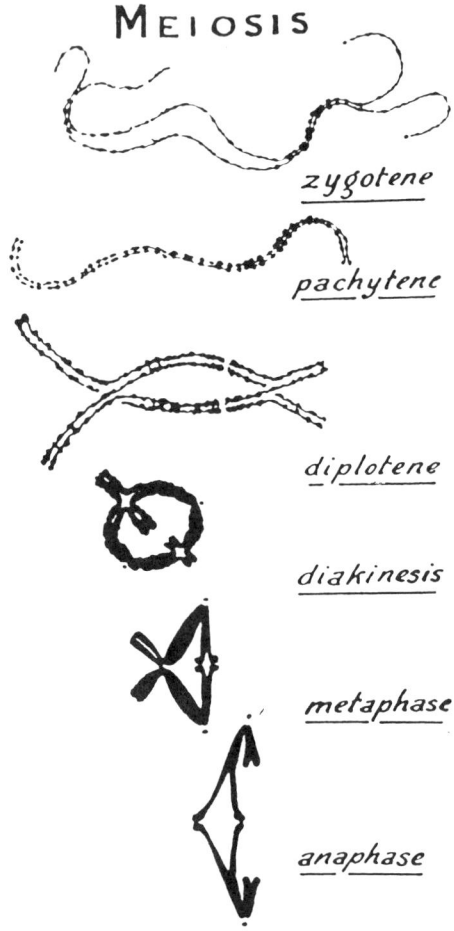

FIG. 21A.—Diagram showing the history of two pairing chromosomes at the first division of meiosis. The chromosomes reach the same degree of contraction between diplotene and diakinesis that they have at the mitotic metaphase. The centromere is not shown at diplotene.

centromeres orientated with regard to one another in the arc axes of the spindle.

(h) (*First*) *Anaphase*. The centromeres of each pair pass to opposite poles, each drawing after it the pair of attached chromatids so that in segments distal to a chiasma paired chromatids are separated. At the same time the attractions between paired chromatids usually lapse even where they are not forcibly drawn apart.

(i) (*First*) *Telophase*. The chromosomes form two daughter-nuclei at the poles, uncoiling and passing into a resting stage, the *interphase*, two separate cells being formed. The two daughter-nuclei each have the haploid number of chromosomes, but, since these are already divided, they have the diploid number of chromatids as in a mitotic telophase.

II. SECOND DIVISION. (*j*) (*Second*) *Metaphase*. At an early stage the two chromatids are widely separated—more so than in any somatic mitosis. They are then held together only at their centromeres. Later stages resemble mitosis, but metaphase is rapid and the chords never associate completely.

(k) (*Second*) *Anaphase*. The chromosomes are distributed as at an ordinary mitosis.

(l) (*Second*) *Telophase*. Four daughter-nuclei are formed, each receiving the haploid number of chromatids.

3. MEIOSIS IN DIPLOIDS

(i) **The Pairing of Chromosomes.** The prophase of meiosis begins after a resting stage which may be short or almost entirely omitted, so that the chromosomes which appear can be traced directly to telophase chromosomes whose individuality has never been even entirely lost. In the spermatogenesis of certain Orthoptera and Hemiptera, for example, the nucleus continues divided into as many compartments as there are somatic chromosomes until the prophase chromosomes appear (Wenrich, 1916, *Phrynotettix* ; Reuter, 1930, *Alydus* ; Chickering, 1927, *Belostoma*, etc. *Cf*. Ch. II). One chromosome comes from each compartment. It consists of a series of darkly staining granules, the chromomeres, of unequal sizes lying at unequal distances in a lightly staining thread (Plates V, XV, Fig. 29),

Both the granules and the thread are single so far as observation can show. In favourable material of some animals the threads can be counted from their centromeres and shown to be present in the diploid number, as in somatic mitoses (Gelei, 1921 ; Janssens, 1924). This stage with separate single undivided threads is called the *leptotene* stage.

The chromosome threads may remain freely distributed in the cell, as in most of the higher plants, or they may gradually orientate themselves so that one or both ends of each thread lie towards one side of the nucleus (near the centrosome where one is present). They are then said to be polarised, and as long as they remain so,

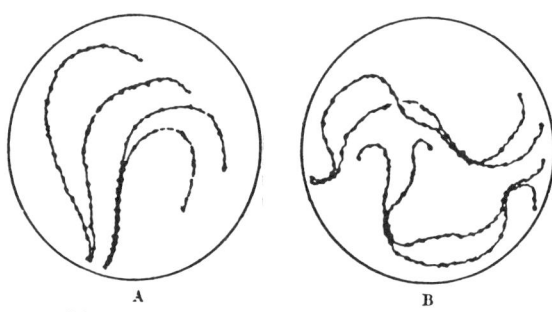

FIG. 21B.—Diagram to show the difference between the method of pairing at zygotene with and without polarisation. A, as in *Stenobothrus*; B, as in *Tulipa*.

that is, until pachytene, they present the appearance to which the name "bouquet" stage is given (Digby on *Osmunda*, 1919 ; Gelei on *Dendrocœlum*, 1921 ; Janssens on *Mecostethus* and *Stenobothrus*, 1924 ; Belar on *Actinophrys*, 1926 ; Kihara on *Rumex*, 1927 a ; Belar on *Tradescantia*, 1929 b).

The single threads soon begin to associate in pairs side by side. If they are freely distributed in the nucleus, their association sometimes takes place at random, *i.e.*, in different parts at the same time.

The bouquet stage has been regarded as a possible characteristic of chromosome pairing (Gelei, 1921), especially on account of its absence in certain hybrids without pairing (*e.g.*, *Pygæra*, Federley 1912), and in certain parthenogenetic organisms without pairing

THE BOUQUET STAGE

(*e.g.*, *Aphis*, v. Baehr, 1912). Certainly the failure to observe it in unfavourable material, such as many dicotyledons and Diptera, is inconclusive, but there can be no doubt of the absence of polarisation in many Lepidoptera (Seiler, 1914), and in some Liliaceæ except merely as a relic of the orientation of the chromosomes, with their centromeres towards the pole, at the preceding telophase (Newton, 1927 ; D., 1929 *b*, 1935 *a*). Polarisation may be regarded

FIG. 22.—Complete early pachytene nucleus in *Chorthippus parallelus*. Eight paired chromosomes shown as single lines, three long pairs broken or wavy. X chromosome ovoid. All ends but two polarised. Small nucleolus in outline. × 3,000 (D., 1936 *d*).

as an adaptation to secure regularity in pairing (*cf.* Gelei). It is bound to affect the distribution of pairing amongst the possible partners, where there is a choice, whether amongst segments in structural hybrids or whole chromosomes in polyploids.

With a bouquet arrangement pairing begins at the end or ends lying towards the surface of the nucleus, whether the centromere is situated there or not (Fig. 22 ; Wenrich, 1916 ; Gelei, 1921 ; Belar, 1928 *b*). In the absence of a bouquet it is more difficult to trace the course of pairing except where the chromosomes are differentially condensed. Thus in *Fritillaria* (Darlington, 1935 a) after pairing the chromosomes are more deeply staining in the neighbourhood of the centromere and pairing is variable in all species and incomplete in most. The defect of pairing is always

FIG. 23.—The chromosomes of *Bellevalia romana*. Top, at mitotic metaphase, $2n = 8$; four types, *A*, *B*, *C*, *D*. Middle, the four pairs of chromosomes at pachytene, drawn separately. Bottom, the four pairs at metaphase, each with two chiasmata except the rightmost, which has three. × 2,000 (Dark, 1934).

in regions remote from the centromere. Particularly frequent is a failure of pairing close to the nucleolus in those chromosomes which are attached to nucleoli. It seems likely that such chromosomes would be hampered and delayed in coming together by having to pull their nucleoli after them, when the nucleoli have not already fused.

PAIRING AND COILING

Probably, therefore, in *Fritillaria*, and many similar organisms that will be described later, pairing begins near the centromere and passes, sometimes regularly and sometimes intermittently, along the chromosomes until it is complete or until it is interrupted.

The unstable stage of pairing is called *zygotene*. The stable stage when pairing has ceased is *pachytene*. Threads that are still unpaired are then seen to be double, although the paired threads are still single (D., 1934 *a*, *Zea*, and 1935 *a*, *Fritillaria*). It is therefore possible that when delayed the chromosomes divide before they can pair, and their division inhibits their pairing.

At leptotene the chromosomes are contorted in a way that probably corresponds to the relic spiral of the prophase of mitosis. The chromomeres are seen to be distributed at unequal distances along the chromosome thread, and the chromosomes as they pair are seen to correspond in regard to this structure. The constant linear arrangement of the chromosomes already suggested by the constant position of centromeres and nucleolar organisers is now established by the characteristic size and position of every chromomere. At pachytene successive granules are more closely crowded along the thread. This may be due to their staining more deeply, since it cannot be due to a linear contraction. Sometimes the threads are longer at the end of pachytene than at the beginning (Belling, 1931, on *Lilium*; Koller, 1936, on various marsupials). In *Phrynotettix magnus* Wenrich (1916) was able to represent each of the paired chromosomes separately, showing that they were present in the haploid number and that corresponding types in regard to the granular structure appeared in different nuclei. In *Phrynotettix* the chromosomes were already contracted at this stage to not more than three times the length characteristic of the somatic metaphase. But in plants they are still five or six times their mitotic metaphase length in the few species in which the contents of the whole nucleus have been separately drawn (Belling on *Lilium* and *Aloë*, 1928 *d*; Dark on *Bellevalia*, 1934, and Moffett on *Anemone*, 1932, Fig. 23, *v.* Table 3).

During the pachytene stage, which may be almost indefinitely prolonged without perceptible change, the chromosomes become coiled round one another. This *relational coiling* is probably universal and takes place throughout the length where they are paired and in any intercalary unpaired segments between paired segments (D., 1935 a). Such a coiling must be due to a torsion in the individual threads which, as will be seen, is expressed in various other ways.

TABLE 11

Examples of a Diffuse Stage at Diplotene or Diakinesis

ANIMALS

Pristiurus (Pisces) ♀	Rückert, 1892.
Gryllus (Orthoptera) ♀	Buchner, 1909.
Cœlenterata, Annelida, Amphibia, etc., ♀	Jörgensen, 1913.
Phanæus (Coleoptera)	Hayden, 1925.
Alydus (Hemiptera) ♂	Reuter, 1930. Cf. Chickering, 1927 ; Wilson, 1928 et al.
Limnophilus decipiens (Trichoptera) ♀ and ♂	Klingstedt, 1931.
Dasyurus (Marsupalia) ♂	Koller, 1936.

PLANTS

Padina (Algæ, Dictyotaceæ) sporogenesis	Carter, 1927.
Hyacinthus, ♂	D., 1929 b.
Tradescantia, ♂	D., 1929 c.
Mitrastemon, ♂ and ♀	Matsuura, 1935.

Development is interrupted in some organisms during the pachytene stage, or (especially in the eggs of Trichoptera and Lepidoptera) later in the prophase, by the nucleus reverting partially or entirely to a resting stage condition. This complication of development, which has naturally hindered the interpretation of successive stages, has been found in spermatogenesis in some animals (*e.g.*, Hemiptera, Wilson, 1912 ; *cf.* Chickering, 1927), and in oögenesis in many more. In some species this change takes the form of a swelling of the chromosomes, which continue the normal course of development in their bloated condition (*Pristiurus*, Rückert, 1892). In other species the chromosomes disappear for a while and reappear again unchanged (*Gryllus, Alydus, Mitrastemon*). In others again (*Phanæus*) a part of the nucleus containing the chromosomes condenses to one side to form a " karyosphere," in which the condensed pachytene chromosomes can be separately distinguished (see Table 11). These changes are inessential to meiosis ; they appear to be determined by the special conditions of growth of the cell and the nucleus at this stage of meiosis. They are not due to the enormous increase in size of the egg cells in which they are most commonly found, for in the pollen mother-cells in which

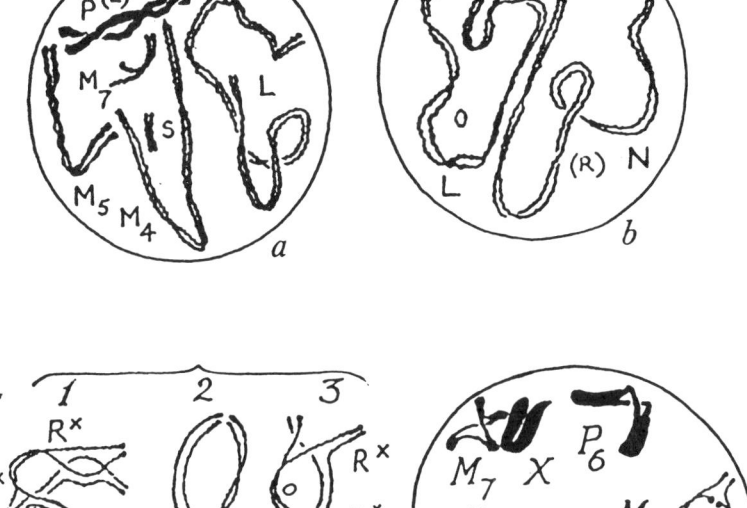

FIG. 24.—Above, separate drawings of late pachytene in *Chorthippus*. Diplotene separation has begun in the precocious P chromosome. X shown in each figure. (R), (L), direction of relational coiling. L, N, M, P, S, chromosome types.

Below, early diplotene nucleus with three L bivalents drawn separately with the numbers of total and terminal chiasmata under each and the directions of chromatid coiling shown at chiasmata, R^x and L^x. × 2,000 (D. 1936 d).

they are also found a contraction actually occurs at this period (*e.g.*, in *Mitrastemon* and other plants).

(ii) **The Separation of Chromosomes and the Origin of Chiasmata.**

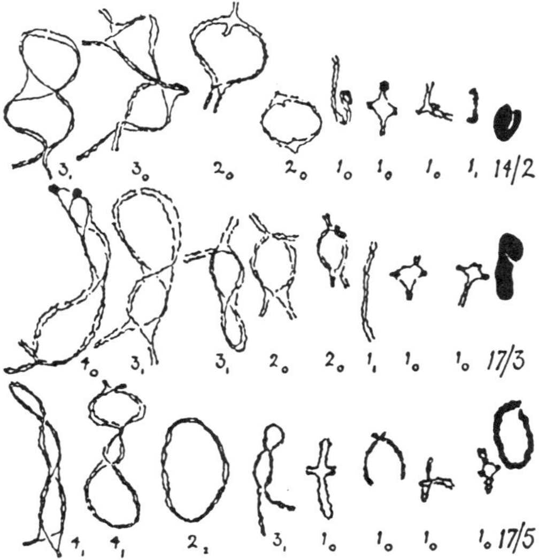

Fig. 25.

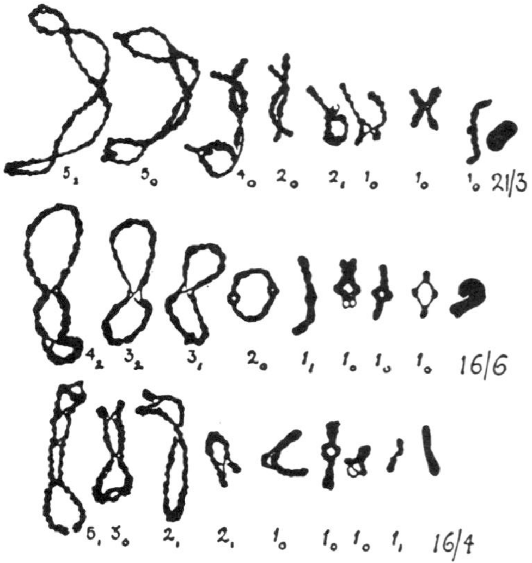

Fig. 26.

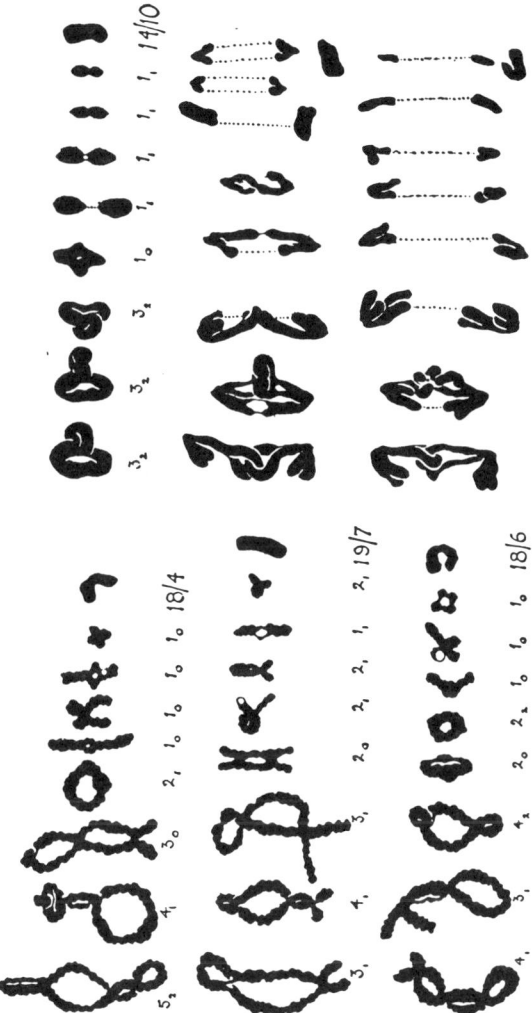

FIG. 28.

FIGS. 25–28.—The eight paired chromosomes of *Stenobothrus parallelus* and, rightmost, the unpaired sex chromosome, drawn separately. Three nuclei are illustrated at each stage to show the kind of natural variation that occurs in the form of the chromosomes, *i.e.*, in the number and position of the chiasmata. Under each bivalent are given the total number of chiasmata and the number that are terminal. Fig. 25.—Middle Diplotene. Fig. 26.—Late Diplotene. Fig. 27.—Diakinesis. Fig. 28.—Metaphase and Anaphase. Note the increase in the proportion of terminal chiasmata in the later stages and the expansion of the middle loop in the three leftmost bivalents. These have median centromeres while the shorter bivalents have centromeres close to the end. The two leftmost bivalents at anaphase correspond exactly with the metaphase figures above them. × 2400 (from D. and Dark, 1932.)

98 MEIOSIS IN DIPLOIDS AND POLYPLOIDS

The first evidence of further change is seen in the separation of the partner chromosomes making up the double thread. It is then found that each of the new threads is double. Each chromosome has divided into two chromatids. This stage is diplotene.

The evidence of the best material (Janssens, 1924; Newton, 1927; Belar, 1928 *b*) is conclusive that the separating threads are double. In *Fritillaria* where pairing begins near the centromere, separation begins in the same region. In *Stenobothrus* the splits appear in the paired threads at several points and extend until they

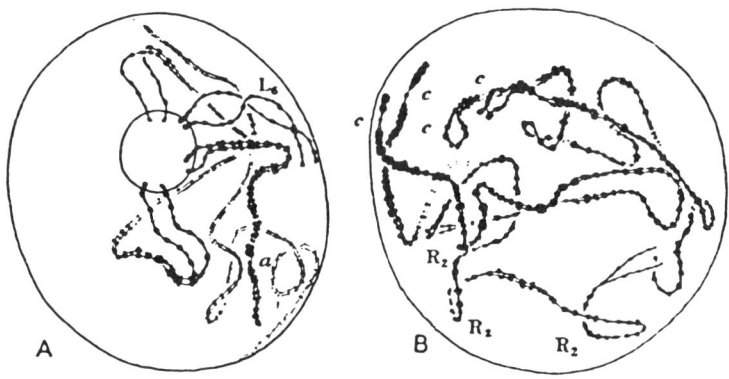

FIG. 29.—A. Pachytene in *Fritillaria imperialis*; pairing incomplete near the attachments of two pairs of chromosomes to the nucleolus (*cf*. Plate XV). B. Beginning of diplotene in *F. obliqua*, separation showing in two pairs at the centromere. L_6, relational coiling; R_2, relic coiling. × 2000 (D., 1935 *b*).

meet. It is then found that the splits separate different pairs of chromatids, for, at the point of meeting, the pairs change partners, giving, from one point of view, the appearance of a cross; hence the name *chiasma* is applied to the occurrence of this exchange (Janssens, 1909, 1924; Belling, 1928 *c*; D., 1930 *b*, 1931 *c*).

Chiasmata are formed as a rule in all paired chromosomes at this stage. But, as we shall see later, short chromosomes which have been paired at pachytene may in some cases be presumed to fall apart at this stage without the formation of chiasmata. They become " unpaired chromosomes."

Chiasmata probably always arise in the first instance interstitially, *i.e.*, at some point between the ends of the chromosomes, not at the ends (Figs. 25–28). They vary in number from one to as many as thirteen in the pair of chromosomes or bivalent, but their range of variation is characteristic for each bivalent of an individual under constant conditions.

This variation has certain characteristic properties in all organisms. Its curve is unimodal. In some organisms where the range of size is small, the mean number of chiasmata per bivalent is approximately proportional to the length of the paired chromosomes.

But in many species where there is a big discrepancy between the sizes of the chromosomes, the smaller ones have a higher proportional frequency (*e.g.*, *Stenobothrus parallelus*, D. and Dark, 1932; *Hyacinthus amethystinus*, D., 1932 *f*, *cf.* Sato, 1934, and Ch. VII).

This is not true of organisms in which new small chromosomes have suddenly arisen. Thus in *Fritillaria imperialis* and in *Tradescantia* new small fragments fail to be paired at metaphase in a proportion of cells owing to the failure of chiasmata to be formed. It seems therefore that the regular pairing of very small chromosomes that are part of the characteristic complement in *Stenobothrus* and elsewhere is a derived property permitting pairing of short chromosomes without an unduly high frequency of chiasma formation in the long ones.

(iii) **Localisation of Pairing.** In organisms with very large chromosomes, pairing at pachytene is sometimes incomplete, owing, as we have seen, to the time-limit for pairing. In *Fritillaria*, *Mecostethus*, and elsewhere, the pairing usually begins near the centromere, and it is in the distal parts that pairing fails. Chiasmata are formed in the paired parts and are thus localised near the centromere (D., 1935 *b*; White, 1936 *a*). In the extreme case one chiasma is found in every bivalent, long or short. The numbers, however, vary in different nuclei and a detailed study of the distribution of the chiasmata shows us indirectly how pairing varies in different chromosomes, in different nuclei and in different species. The chief conclusions to be drawn are that shorter chromosomes often have an advantage over long ones in the time of pairing and the ends of chromosomes have an advantage over the middles unless the centromere is median. Thus chiasmata are sometimes found at the distal ends and short chromosomes with subterminal centromeres have a higher chiasma frequency than longer ones with median centromeres. Both these results agree with the time-limit explanation of incomplete pairing, since small chromosomes and ends must be freer and quicker in movement than long chromosomes and middles.

In some species the pairing is interrupted later than in others, and all degrees of localisation occur on this account in *Fritillaria* (D., 1936 e). Probably elsewhere a time limit determines incomplete

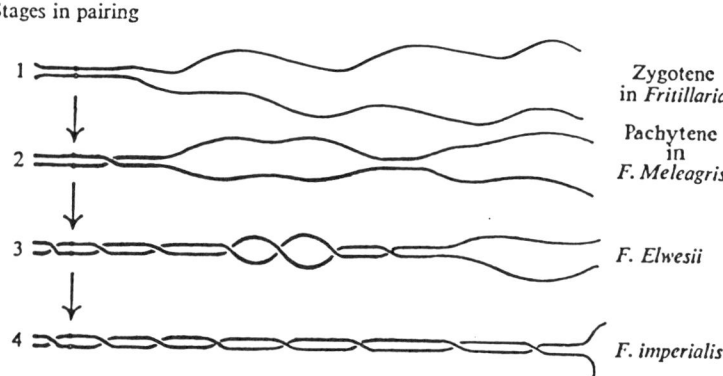

Fig. 30.—Stages in zygotene pairing in *Fritillaria*, showing how its interruption at various stages by the time-limit leads to localisation in some species. (D., 1935 b.)

pairing, but owing to a less regular beginning of pairing near the centromere the localisation of pairing and of chiasmata is less readily recognised (*e.g.*, *Lilium regale*, Mather, 1935).

(iv) **Relational Coiling.** At pachytene the chromosomes were coiled round one another. Coiling of three kinds is found at early diplotene (D., 1935 *a*, 1936 *b*) : (i) The loops on either side of a chiasma lie in the same plane and two chromatids therefore cross over one another in either a right- or left-hand spiral direction. Later, successive loops come to lie at right angles to one another, and thus abolish this coiling. (ii) In so doing they alter the relationships of the chromatids, which are also coiled round one another between chiasmata, probably in the opposite direction to the first coiling. (iii) The chromosomes are coiled round one another to a greater or less extent between chiasmata.

This last system is evidently the direct successor of the pachytene coiling of chromosomes, and its variation is particularly instructive. Where the pairing at pachytene is intermittent (as with partial localisation in some species of *Fritillaria*) the chromosomes are relationally coiled in the unpaired intercalary parts. This coiling must necessarily survive to diplotene. But in the paired proximal regions there is little coiling; the chromosomes are there associated by chiasmata. Now relational coiling of the chromosomes is a property shown by partner chromosomes before they divide into chromatids and whether they form chiasmata or not. Each pair of chromatids that is coiled round another pair at diplotene must be derived from a parental chromosome. And since such coiling can occur and does occur on both sides of chiasmata, chiasmata must represent and result from breakage and crossing-over between chromatids of the partner chromosomes. It seems probable that this breakage and re-union of chromatids replaces the relational coiling by permitting the chromosomes to uncoil after breaking before they reunite (D., 1935 *b*).

Between diplotene and diakinesis the chromosomes gradually uncoil so far as the chiasmata allow them to. Thus since the direction of coiling is always consistent in particular arms, so far as we know, their uncoiling is bound to coil their chromatids round one another, and this coiling cannot be undone in loops between chiasmata, although in free ends it can.

(v) **Diakinesis.** After chiasma-formation the chromosomes contract still further in length and begin to assume a more rounded outline until *diakinesis*, the stage of greatest linear contraction. The chromosomes are then shorter and broader than at metaphase of a somatic mitosis, except where, as in some Lepidoptera, the mitotic contraction is already a maximum, and the chromosomes are spherical at metaphase even at mitosis.

This contraction is achieved, at least as far as organisms with the largest chromosomes are concerned, by the assumption of a larger spiral over and above the small spiral developed at mitosis. These major and minor spirals were first discovered by Fujii in *Tradescantia*

102 MEIOSIS IN DIPLOIDS AND POLYPLOIDS

and their structural relationship has been proved by the experiments of Kuwada and his collaborators (1934, 1935; *cf.* D., 1935, and Ch. XII).

During diakinesis the nucleoli become detached from their

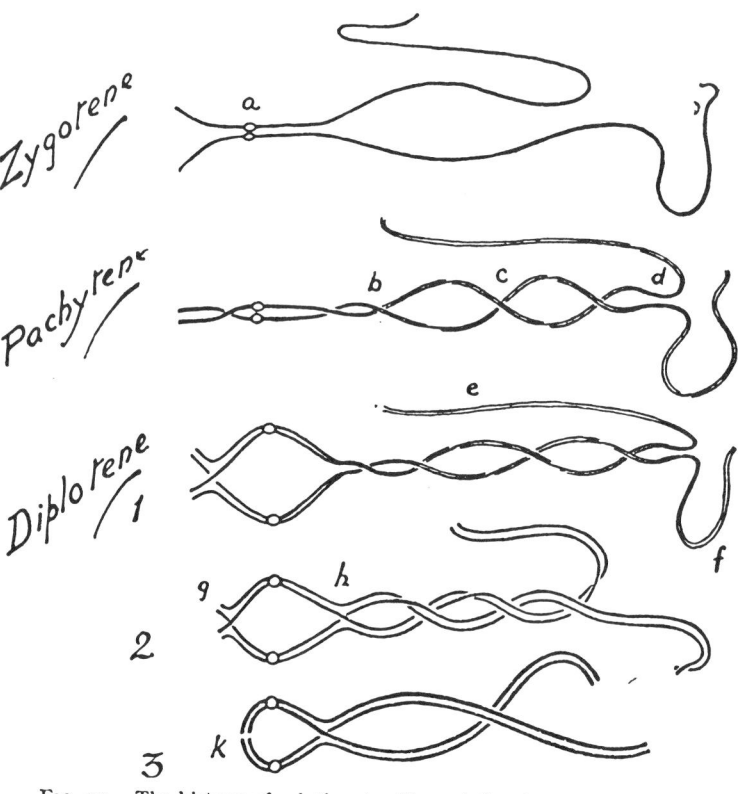

FIG. 31.—The history of relational coiling of the chromosomes with intermittent pairing. The coiling of unpaired segments persists at diplotene since they cannot form chiasmata. *a*, centromere; *b*, end of paired region; *c*, intercalary unpaired region; *d*, intermittently paired region; *e, f*, distal unpaired regions (*cf.* Fig. 29B); *g* and *h*, chiasmata; *k*, chiasma *g*, terminalised. (D., 1935 *c*.)

organisers on the chromosomes and, as a rule, gradually disappear. In some plants, however, they come to lie like a cap on the surface of the nucleus and at the end of prophase are extruded into the cytoplasm, where they degenerate (Frankel, 1936, on *Fritillaria*; *cf.* Catcheside, 1934, on *Brassica*, and 1933, on *Œnothera*).

Where, as in *Pristiurus* (Rückert, 1892) and *Scolopendra* (Bouin, 1925), considerable size variations occur in the nucleus during prophase, the chromosomes vary in size proportionately, probably through their retaining the same water-content relative to the nuclear sap. They occupy, whether in an organism with large or small chromosomes, less than one-hundredth of the space of the nucleus.

The relationship between the volume of the nucleus and of the chromosomes in different organisms is perhaps not so variable as would appear from a comparison of illustrations of this stage in organisms with chromosomes of different sizes.

But while this is true of sperm and pollen, and even of embryo-sac mother-cells, it is not true of the very large egg-cells of animals. In these the nuclei are ten or a hundred times as large as in the sperm mother-cells. We then find a very instructive difference in distribution. In the small nuclei the separate bivalents are evenly distributed in the diakinesis nucleus. Where they are few in number and much condensed this means that they lie chiefly on the periphery.

In the large nuclei, on the other hand, they are not evenly spaced. They lie in a haphazard group. The repulsions between them which are effective in giving an even distribution at short distances are ineffective at longer distances.

In some organisms the bivalents themselves continue to show no special internal changes. Successive loops between chiasmata lie at right angles to one another, but no general distinction can be made between them in regard to size (*e.g.*, *Vicia Faba*, *Fritillaria imperialis*). In others one loop is distinguished from the others by the sharp repulsion between its arms—so sharp that they are drawn out into fine threads as they are afterwards at late metaphase (Fig. 34). In these it is often found that the chiasmata are now no longer interstitial—they have been gradually pushed along to the ends of the chromosomes. All connection between the chromosomes may then seem to be broken by the strong repulsion, especially where the chromosomes are small, and this interpretation has often been put upon such configurations. Comparative study

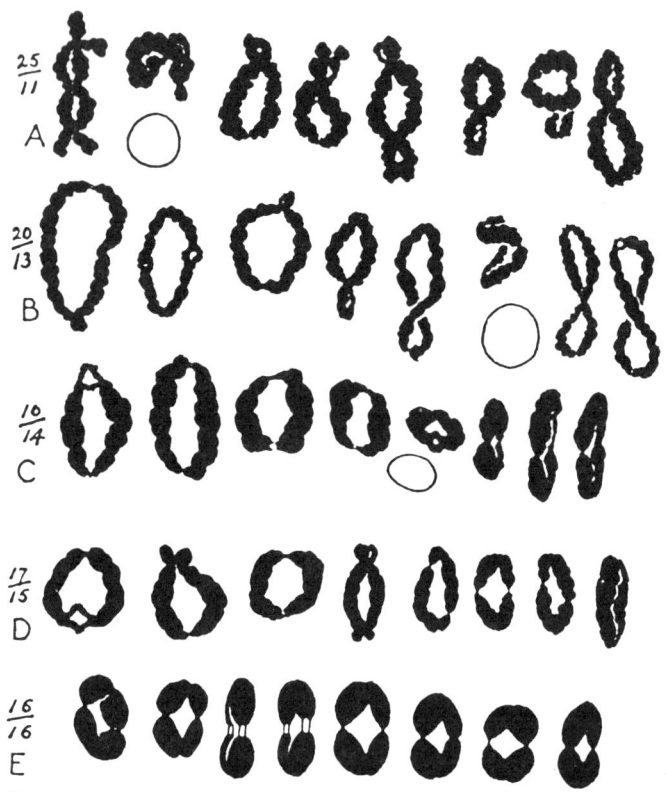

Fig. 32.—The eight paired chromosomes of *Campanula persicifolia* drawn separately, at successive stages between diplotene and metaphase to show the changes undergone with complete terminalisation. The numbers of chiasmata, total and terminal, in each nucleus are given at the side. D. is the same stage as C. The nucleolus is shown in outline. × 3000 (from Gairdner and D., 1931.)

shows that a different interpretation is possible. This will now be considered.

(vi) **The Movement of Chiasmata.** In *Fritillaria* the change from diplotene to diakinesis is unaccompanied by any marked change in the structure of the paired chromosomes. The chiasmata are almost stationary and are therefore to be found distributed at diakinesis as they arose—either localised or at random.

In most other organisms a change takes place. This is seen in its extreme form in *Primula* (D., 1931 *a*) and *Campanula* (Gairdner and D., 1931). It has three obvious characteristics :—

(i) The total number of chiasmata is reduced in each bivalent.

(ii) The chiasmata come to be concentrated nearer the ends.

(iii) These changes taking place *pari passu* eventually leave all the bivalents associated terminally and the number of chiasmata reduced to the number of ends associated. It still remains true that the connection is not between two whole chromosomes, but between their component chromatids. Each chromatid, which has been connected side-by-side with one partner, is connected end-to-end with another, and hence the terminal association means a change of partner after the lateral association. It therefore has the essential mechanical properties of a chiasma (Ch. XII).

A true terminal chiasma can be readily distinguished from a sub-terminal one both at diakinesis and at metaphase, when the limbs of each of the chromosomes connected by it are repelling one another ; this happens when they include the centromere. The connections between the chromatids are then reduced to the minimum and are drawn into fine threads by the tension they undergo. In the absence of this tension there is no sharp demarcation between the terminal chiasma and the interstitial chiasma that is nearly terminal.

There can be no doubt that this change of arrangement in the chromatids is essentially the same as that described in certain bivalents of *Phrynotettix* by Wenrich (1916). It was then said that the chiasma arose from the meeting of splits separating the four chromatids along two planes and that the movement was due to one split gaining at the expense of the other. It may equally be described as two types of association of chromatids replacing two others (a description that will be found most convenient for theoretical consideration) or as the movement of a chiasma along the chromosome towards the end or away from the centromere.

This last point of view has suggested the expression *terminalisation* to describe the change (D., 1929 c).

However many chiasmata are terminalised, the chromosomes remain associated by terminal chiasmata. It must therefore be assumed that they merge at the ends, for if they fused interstitially they would sometimes cancel one another out. Why they do not do so when they fuse at the ends is a problem that will be specially considered later (Ch. XII).

In many organisms, such as *Tulipa* and *Zea*, a slight movement of chiasmata takes place, but there is usually no fusion at the ends and interstitial chiasmata remain at metaphase. It is then found that when there is one chiasma in each arm this chiasma has been pushed to the end. When there are two, the loop between them is smaller than that containing the centromere. It seems that an equilibrium is reached, characteristic for a particular type of bivalent with a particular number of chiasmata, between two kinds of repulsion, the generalised *body repulsion* which effects the even distribution of chromosomes at all stages of mitosis and meiosis, and the localised *centromere repulsion*, which effects the separation of the chromosomes at anaphase in both mitosis and meiosis. The different degrees of terminalisation are then due to different degrees of centromere repulsion, in relation to the length of the chromosome. Means of testing these assumptions we shall discover later.

(vii) **Metaphase : the Structure of Bivalent Chromosomes.** After diakinesis the wall of the nucleus disappears and the spindle develops as in ordinary mitosis. The bivalent chromosomes arrange themselves on the equator with their pairs of centromeres orientated just as the two daughter centromeres of a mitotic chromosome are orientated in early anaphase. That is to say, they lie symmetrically on either side of the equatorial plane and in an axial direction from one another. Under favourable conditions of fixation in the largest chromosomes they appear as single spherical bodies a quarter of a micron in diameter (*v.* Ch. XII). Each chromosome, *i.e.*, pair of chromatids, now acts as though it had a single centromere. The two centromeres of the bivalent (whose repulsion has momentarily lapsed) begin to repel one another actively, so that they are separated by the two segments joining

POSITIONS OF CHIASMATA

them to the first chiasma. If these segments are short they are drawn out, as is the case with small chromosomes at diakinesis, into a fine thread.

The successive loops between chiasmata have, with the increasing thickness and, presumably, rigidity of the chromosomes, come to lie

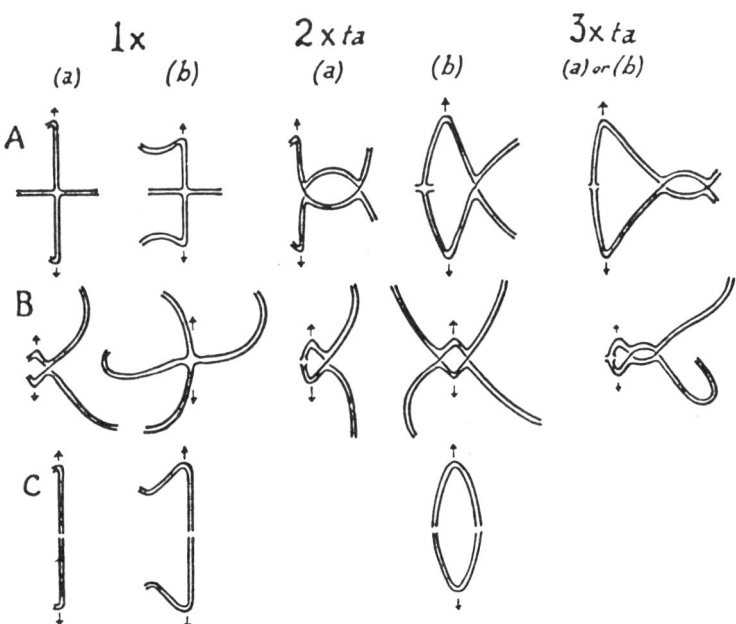

FIG. 33.—Diagram to show the chromatid structure of representative types of bivalents at metaphase. A, with unterminalised, random chiasmata. B, with unterminalised, localised chiasmata. C, with completely terminalised chiasmata. x, xta: chiasma and chiasmata. (a), with subterminal centromere (represented by arrow); (b), with submedian centromere. (From D., 1931 c.)

in planes at right angles to one another before diakinesis. This condition is still more strongly developed at metaphase, and is indeed the only direct criterion of the existence of chiasmata at this stage in ordinary preparations.

Metaphase bivalents are therefore found to fall into various types according to (i) size of the chromosomes; (ii) position of

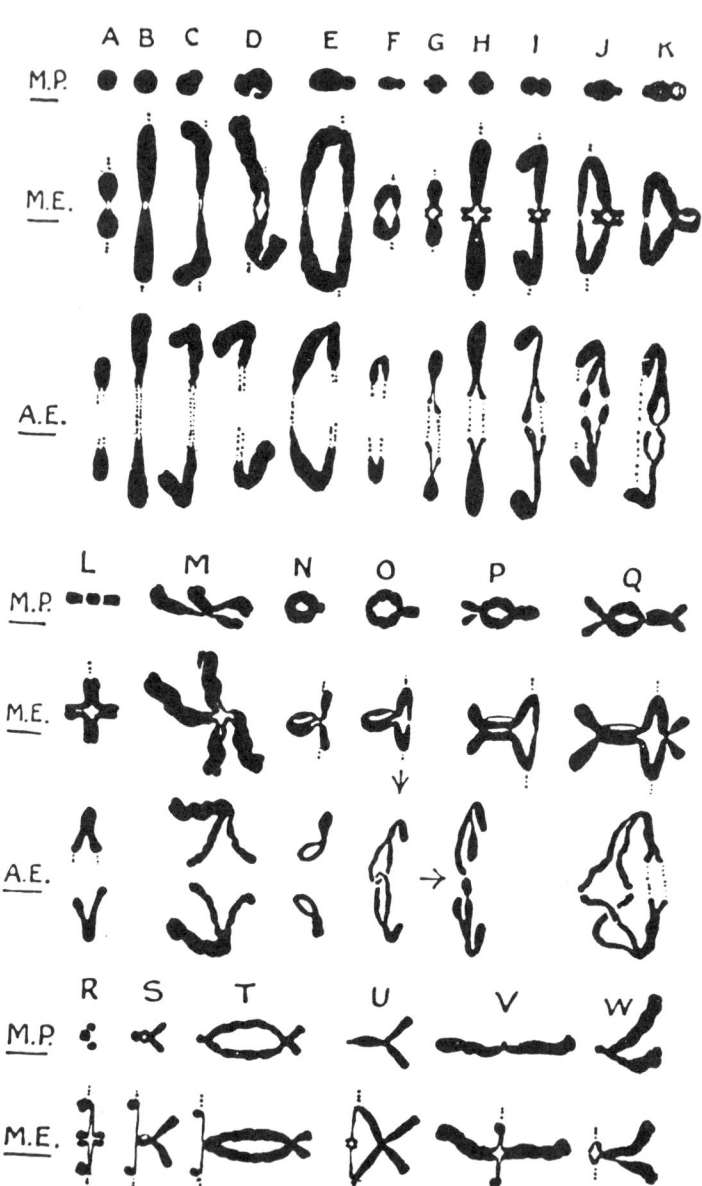

SHAPES OF BIVALENTS

their centromeres; (iii) the number and position of the chiasmata, and especially the distance of the nearest chiasma from the centromere. Hence all metaphase bivalents fall into classes of which types are illustrated in the diagram (Fig. 33). The three classes as determined by the position of the chiasmata are as follows: A, those with nearly random chiasmata; B, those with chiasmata

FIG. 34.—Diagram showing forms of bivalents at the first metaphase of meiosis. M.P., polar or top view of the metaphase plate. M.E., equatorial or side view of the same. A.E., equatorial view of anaphase showing daughter bivalents (not shown in types R to W, which correspond with earlier forms). Note that constrictions do not show at metaphase except by their position at the centromere. Compare with line diagrams of structure (Fig. 33) as follows:

Type A, 1 chiasma (a) :—	G, H, L
(b) :—	I, R
2 chiasmata (a) :—	N (one terminal), S, T
(b) :—	J, U
3 chiasmata :—	K, O, P (one terminal), Q
Type B, 1 chiasma (a) :—	V
(b) :—	M
2 chiasmata (a) :—	W
(b) :—	(Not shown)
3 chiasmata	(Not shown, *cf.* Q)
Type C, 1 chiasma (a) :—	A, B
(b) :—	C, D
2 chiasmata (b) :—	E, F

localised near the centromere; C, those with chiasmata confined to the ends or near the ends of the chromosomes, following terminalisation.

These classes are readily explained by reference to the observations of the origin of chiasmata at diplotene and their later movement in terminalisation. It is to be noticed especially that, at metaphase,

TABLE

Provisional Classification of Annuals and Plants according to the Dis

1. Localised Proximally: Slight Movement.		2. Distributed: Slight Movement.	
ANIMALS			
Bombinator ♀	Lebrun, 1902.	*Pristiurus*	Rückert, 1892.
Zoogonus minus	Grégoire, 1909.	*Tomopteris*	Schreiner, 1906.
Hippiscus	McClung, 1914.	*Pamphagus*	Granata, 1910.
Tropidolophus	,, ,,	*Dendrocoelum*	Gelei, 1913.
Mecostethus spp.	,, 1927.	*Marsupalia* (various).	Agar, 1923.
	Janssens, 1924.	*Actinophrys*	Koller, 1936.
	White, 1935 *a*.		Belar, 1922.
Aggregata	Belar, 1926 *b*.	*Stenobothrus* and *Chorthippus* spp.	Janssens, 1924.
Barrouxia	Wedekind, 1927.		D. and Dark, 1932; D., 1932 *c*, 1936 *d*.
Rattus	Minouchi, 1929 *a*.		
Chromacris miles	Saez, 1930.		
Tettigonia	de Winiwarter, 1931.	*Amœba*	Belar, 1926.
Reptilia (various)	K. Nakamura, 1931.	*Eumeces*	K. Nakamura, 1931.
	Matthey, 1933.		
Gampsocleis	Haseyama, 1932.	*Rattus*	Koller and D., 1934.
		Melanoplus	Hearne and Huskins, 1935.
		Culex	Moffett, 1936.
PLANTS			
Fritillaria Meleagris, etc. (15 spp.).	Newton and D., 1930	*Hyacinthus* spp.	Belling, 1925 *a*.
	D., 1935.		D., 1929 *b*, 1932 *d*.
Ranunculus acris	Larter, 1932.	*Uvularia*	Belling, 1925.
Anemone appenina	Bocher, 1932.		,, 1926.
Allium fistulosum	Levan, 1933.	*Ephedra*	Geitler, 1929.
A. Farreri and *A. Porrus* (4x).	,, 1935.	*Lathyrus odoratus*	Maeda, 1930.
		Vicia Faba	
Sagittaria Aginashi	Shinke, 1934.	*Fritillaria* (10 spp.)	D., 1936 *e*.
Trillium kamtschaticum	Matsuura, 1935.	*Taxus*	Dark, 1932.
Lilium regale	Mather, 1935.	*Kniphofia*	Moffett, 1932.
Tulipa linifolia	Upcott, unpub.	*Allium* spp.	Levan, 1932–35.
Paris quadrifolia	D., unpub.	*Crepis* spp.	Richardson, 1935.
		Lilium spp.	Mather, 1935.
			Richardson, 1936.
		Scilla	Dark, 1934.
			Sato, 1934 *b*.
		Podophyllum	D., 1936 *c*.

tribution of Chiasmata at First Metaphase in their largest Chromosomes

3. Equilibrated: Incomplete Terminalisation.		4. Fused: Complete Terminalisation.	
Leptynia	de Sinéty, 1901.	*Myxine and Spinax*	Schreiner, 1906.
Brachystola	Sutton, 1902.		Federley, 1914.
Allolobophora	Foot and Strobell, 1905.	*Lepidoptera* (various).	Seiler, 1923.
Salamandra	Schreiner, 1906.	*Forficula*	Payne, 1914.
Circotettix	Carothers, 1921.	*Acridium*	Robertson, 1915.
	Helwig, 1929.	*Rhodites* and *Cynips*	Hogben, 1920.
Phrynotettix	Wenrich, 1916.	*Rana*	Witschi, 1924.
Discoglossus	Champy, 1923.	*Belostoma*	Chickering, 1927.
Orgyia antiqua	Cretschmar, 1928.	*Bruchus*	Brauer, 1928.
Nyctereutes	Minouchi, 1929.	*Orgyia thyellina*	Cretschmar, 1928.
Homo	Evans and Swezy, 1929.	*Alydus*	Reuter, 1930.
Macronemurus ♀	Naville and de Beaumont, 1933.	*Macronemurus* ♂	Naville and de Beaumont, 1933.
		Odonata (various).	Asana and Makino, 1935.
		Bufo	Saez *et al.*, 1936.
Prunus spp.	D., 1930 *a*.		
Humulus spp.	Kihara, 1929 (v. text).	*Drosera*	Rosenberg, 1909.
Matthiola	Philp and Huskins, 1931.	*Rumex*	Kihara, 1927. Sato and Sinoto, 1935.
Cladophora	Higgins, 1930. Schussnig, 1931.	*Œnothera* *Rhoeo* and	Catcheside, 1931 *b*. D., 1929 *c*.
Rosa	Erlanson, 1931, 1933.	*Tradescantia. Aucuba japonica*	Meurman, 1929.
Pisum	E. R. Sansome, 1933.	*Campanula*	Gairdner and D., 1931.
Tulipa	D. and Janaki-Ammal, 1932.	*Primula sinensis*	D., 1931 *a*; Dark, 1931.
Festuca and *Lolium*	Peto, 1933.	*Dahlia* spp.	Lawrence, 1929.
Triticum	D., 1931 *d*; Mather, 1935.	*Anthoxanthum*	Kattermann, 1931.
Secale	D., 1933 *d*.	*Polemonium*	J. Clausen, 1931 *a*.
Avena	D., 1933 *e*.	*Dactylis glomerata*.	Müntzing, 1933.
Zea Mays	D., 1934 *b*.		
Agapanthus	D., 1933 *a*.	*Briza media*.	Kattermann, 1933.
Hevea	Ramaer, 1935.	*Solanum*	
Primula malacoides	Kattermann, 1934.	*Lycopersicum*.	Upcott, 1935.
Makinoa	Tatuno, 1933.	*Lathyrus odoratus*	
Fremurus	Upcott, 1936.	(mutant).	Upcott, 1936.
Setcreasia	Richardson, 1935.		

bivalent forms are to be found which correspond with different stages of terminalisation, and it is evident from observations on *Rosa* (Erlanson, 1931 c) that this is due to the process being interrupted before completion by the onset of metaphase, after which it proceeds no further.

We have as the extreme type of terminalisation organisms in which interstitial chiasmata have never been observed at metaphase, e.g., *Rhoeo discolor* (Fig. 53, D., 1929 c; Koller, 1932 c).

Secondly, we have organisms in which a proportion of interstitial chiasmata are found, e.g., *Circotettix*, *Matthiola incana* and *Rosa*.

Finally, we have organisms in which terminal chiasmata occur fairly frequently, but since there is no appreciable reduction in the number of chiasmata between diplotene and metaphase the occurrence of terminal chiasmata is presumably due to the terminalisation of only the distal chiasma, accompanied no doubt by slight movement of the proximal ones—e.g., *Stenobothrus*, *Mecostethus*, *Fritillaria imperialis* and *F. Meleagris* (Figs. 25–28).

All chromosome association at metaphase that is derived directly from pachytene association may therefore be regarded as derived merely by the formation and movement of chiasmata. Pairing that is not derived from pachytene association will receive special consideration (Chs. IX and XII).

The table classifying bivalent types is based on these considerations. The original distribution and later movement of the chiasmata is inferred where, as in so many organisms, it has not yet been observed, from the form of the metaphase chromosomes. This kind of inference has been made in *Œnothera* (D., 1929 a and c), where at the time interstitial chiasmata had never been observed. Later work has shown such inferences to have been correct.

The classification enables us to examine on a broader basis the conditions affecting terminalisation and localisation.

Thus it will be seen that the first two classes with minimum terminalisation have all large chromosomes. The fact that amongst these are also all those with localisation of chiasmata is not surprising, since this depends on delay in pairing of parts of chromosomes such as can only occur where a great length is to be paired. The fact that all types with the minimum terminalisation have

LOCALISATION AND TERMINALISATION

large chromosomes suggests that terminalisation depends on the size of chromosomes. When we recall that the effectiveness of the body repulsion depends on the size of the nucleus we see that this also is not surprising. Although similar or even greater forces may be engaged in moving these large chromosomes they will have less effect. This conclusion is borne out by the comparison of chromosomes of different sizes within the complement. In *Stenobothrus*, for example, we see that, while in the larger chromosomes there is a low degree of terminalisation, in the smallest one the process is always complete. The same is true in the small fragments of *Fritillaria imperialis* (D., 1930 d, cf. Ch. V).

The degree of terminalisation is probably correlated with the degree of longitudinal contraction. Thus in male-sterile *Lathyrus* as compared with the normal (Upcott, 1936), and in the male *Macronemurus* as compared with the female (Naville and de Beaumont, 1933) contraction is greater and terminalisation is more complete. It seems that the greater contraction may directly determine a greater repulsion and hence a greater movement, but the connection might be indirect in several other ways.

When we compare behaviour in related species we find that other variables are concerned: there is no direct relation between size and terminalisation, for example, amongst species of *Allium* (Levan, 1935). Species therefore differ in the degree of centromere repulsion during prophase, independently of the size of their chromosomes. Nevertheless those with the longest chromosomes never have a high degree of terminalisation.

The question then arises why all chiasmata should be terminal in one group of organisms with very large chromosomes, the tetraploid relatives of *Tradescantia virginiana* (D., 1929 c; Richardson, 1935). This species passes through a diffuse stage at diplotene, and the stages of movements of chiasmata have not been seen. It seems probable that pairing and chiasma-formation are localised near the ends, that this species in fact corresponds with *Fritillaria* in having interrupted pairing, but merely begins its pairing at the ends instead of near the centromere.

(viii) **Anaphase: the Separation of Chromatids.** When the bivalent chromosomes have been arranged on the metaphase plate

114 MEIOSIS IN DIPLOIDS AND POLYPLOIDS

so that their pairs of centromeres lie in the axis of the spindle they divide; the pairs of chromatids associated at the two centromeres move to opposite poles. This means that in segments distal to a chiasma pairs of chromatids pull apart. And it is evident that they resist this separation, for bivalents which have the greatest length of chromatids to pull apart lag behind the rest (Plate VI, and Figs.

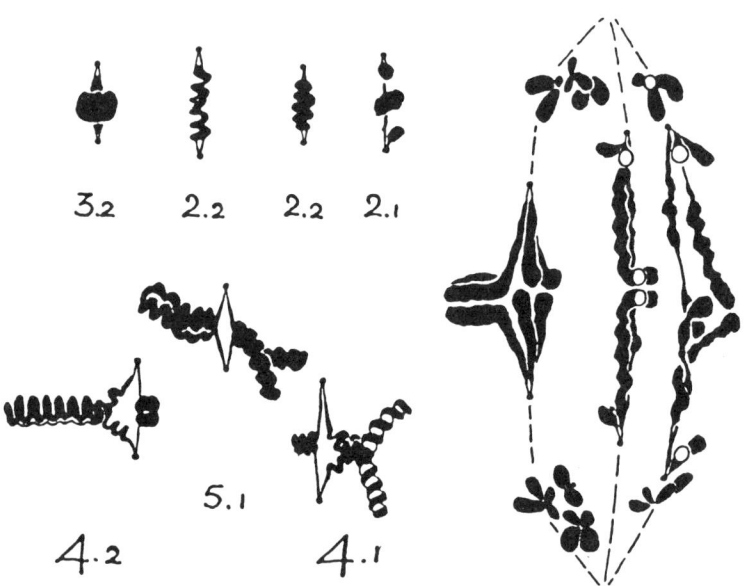

FIG. 35.—Left, metaphase, and right, anaphase, in *Uvularia perfoliata* with three long and four short pairs. Internal spirals shown by some chromosomes. × 2200 (D., unpub.).

28 and 35). Such chromosomes are seen at anaphase with their two chromatids separated throughout their length. Conversely, the simplest and quickest separation is found amongst bivalents with terminal chiasmata. These merely break their terminal connections and pass to the poles without any forced separation of their paired chromatids. Indeed, sometimes they give no evidence, except at their tips, of their double structure. The extremes are therefore found with proximal localisation (Fig. 34, M, V, W)

and with terminalisation (Fig. 34, A-E). For regularity in the separation of the paired chromosomes it is evident that uniformity in chiasma distribution is the essential condition, and the ideal is therefore complete terminalisation.

Terminalisation and likewise localisation of chiasmata are characteristic properties of species, and variation in the metaphase distribution of chiasmata must be regarded as another instance of the control exerted by the genotype over the behaviour of the chromosome complement as a whole (v. Ch. III). Hence we can see in the very general occurrence of terminalisation an adaptation to the conditions of regular anaphase separation. The opposite advantages of localisation will be considered in relation to the genetical theory (Ch. VII).

In organisms having complete terminalisation, as a rule, if an occasional bivalent has an interstitial chiasma, it is sharply distinguished at anaphase by its lagging. Such is the reason for the lagging of chromosomes described in the following species. (The list does not include cases where the cause of the appearance of the interstitial chiasma is probably arrest of terminalisation by change of homology in a structural hybrid, Ch. XII.)

TABLE 13

Lagging of Bivalents with Interstitial Chiasmata

Phragmatobia fuliginosa	Seiler, 1914.
Physaloptera sp.	Walton, 1924. (Large X chromosomes in homozygous sex.)
Stenobothrus spp.	Belar, 1929 *a* ; D. and Dark, 1932.
Cycas revoluta	Nakamura, 1929.
Spinacia oleracea	Maeda and Kato, 1929.
Viola striata	J. Clausen, 1929.
Prunus hybrids	D., 1930 *a*.
Yucca flaccida	O'Mara, 1932.
Lachenalia glaucina	Moffett, 1936.

The other side of the picture is shown by the *precocious* separation of bivalents in certain organisms. This is particularly noted where most of the bivalents have interstitial chiasmata. If then an exceptional one has a terminal chiasma, and especially if this is a short chromosome, it seems to divide precociously (*e.g.*, *Mus musculus*, Painter, 1927 ; *Ranunculus acris*, Larter, 1932).

Careful comparison of several such cases shows that the centromeres of the two " separating " chromosomes are no further apart than those of the larger chromosomes, but the parts of the chromo-

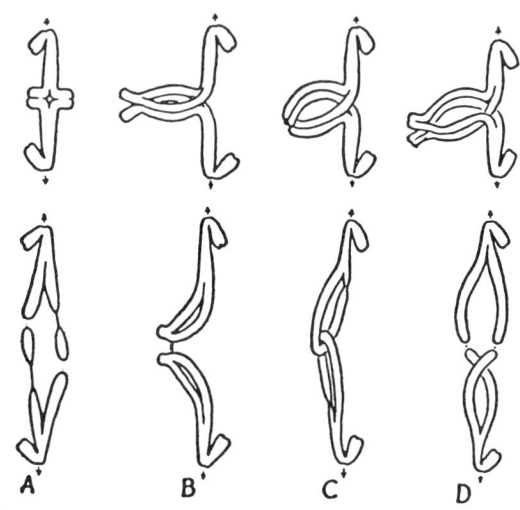

Fig. 36.—Diagram showing chromatid behaviour in anaphase separation of bivalents of different types. A. Single chiasma : one chromatid in each daughter chromosome shows a constriction. This constriction is momentary and has been best seen in living material (Belar, 1928, Fig. 266 ; *cf.* Janssens, 1924, Fig. 259 ; Belling, 1926 *b*, Fig. 6).

B and C. Two comparate chiasmata : the chromatids are associated distally to the second as they were proximally to the first (E. R. Sansome, 1932 ; *cf.* Janssens, 1924, Fig. 259 ; Belling, 1926 *b*, Fig. 6 ; Newton and D., 1930). The point at which the chromatids separate last appears to show a connection (D., 1929 *b*, Figs. 29 *a* and 56).

C. The loops formed by the two pairs of chromatids interlock (Janssens, 1924, Fig. 259). One loop then breaks its terminal association in separating (Belar, 1929 *a*, Fig. 9).

D. Two disparate chiasmata : the anaphase figures are asymmetrical in consequence (Janssens, 1924, Fig. 259 ; Newton and D., 1930).

N.B.—*Cf.* also Sakamura, 1915 ; Maeda, 1930, *a* and *b* ; Richardson, 1936, D., 1936 *d*.

somes between the centromeres and the terminal chiasma are so small as to be drawn into a fine thread which may be invisible (*Chorthippus*, D., 1936 *d*, and Fig. 49 c).

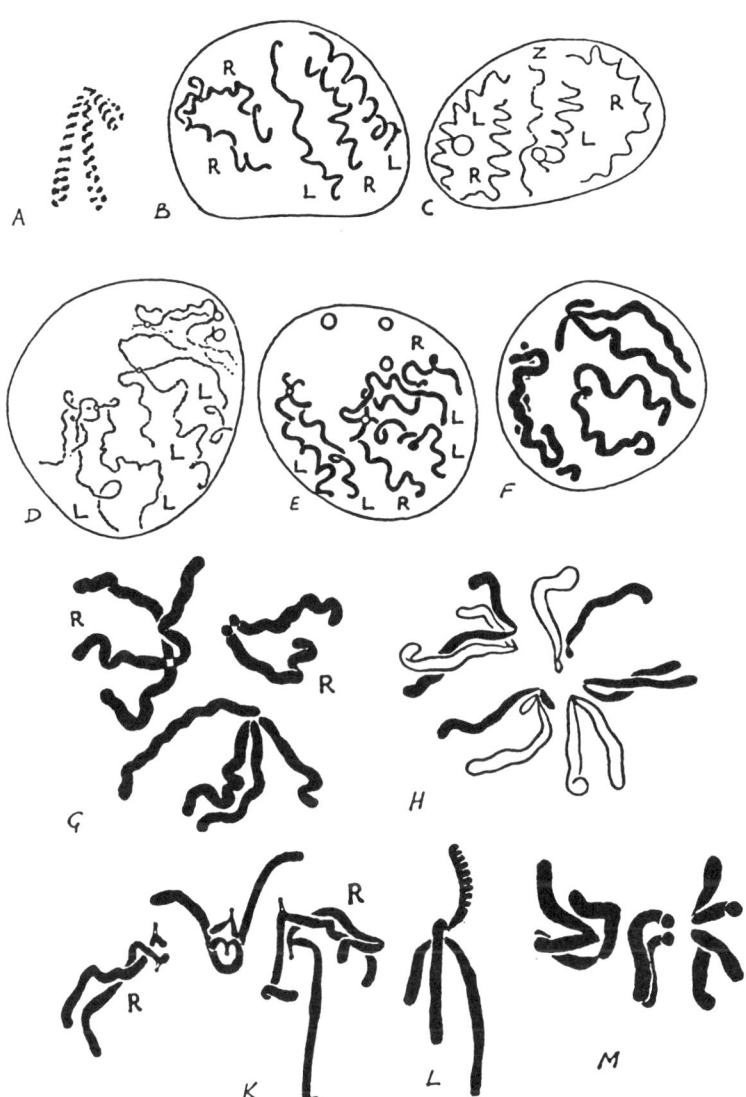

FIG. 37.—First anaphase to second anaphase changes in *Podophyllum versipelle*. A. First anaphase showing internal spiral. B. to F. Interphase showing the development of relic spirals (R. and L.) and sister chromatids held together at the centromere. Small nucleoli. G. Early second metaphase. H. and K. Polar and side views (with centromeres) of early anaphase. L. Chromosome showing internal spiral. M. Second metaphase with highly spiralised chromosomes. × 2000 (D., 1936 c).

The observation of the structure of anaphase chromosomes is of great importance. It shows, first, that what might have been regarded at metaphase as "points of contact" between the chromosomes really were chiasmata. This is particularly significant in organisms (such as *Pisum* or *Œnothera*) where it has not been possible to trace the development of the chiasmata in prophase. It shows, secondly, the relationship of the exchange undergone at one chiasma to that undergone at the next. Two types of relationship are to be expected (Newton and D., 1930). In one the second chiasma restores the association of chromatids that was lost at the first chiasma ; the two chiasmata may then be said to be compensating or *comparate* in a purely observational sense. In the other type the second chiasma does not restore the original association. The chiasmata are non-compensating or *disparate*. Both these types of relationship are found (Fig. 36) where anaphase chromosomes have been studied and are readily distinguished. They are derived from different kinds of relationship between the crossing-over at the two chiasmata which will be considered later (Ch. VII).

(ix) *Interphase*. The chromosomes pass to the pole in half the number characteristic of somatic mitoses. But since each consists of two chromatids instead of one, these are present in the normal number. We may say, therefore, the number of centromeres is reduced, the number of chromatids is unreduced.

When the chromosomes reach the pole they either reconstitute daughter-nuclei or they pass directly into a second division. In the first case the daughter-nuclei may pass into a fairly complete resting stage, with uncoiling of relic spirals and development of nucleoli. In these circumstances the chromosomes regain, or partly regain, their normal mitotic length. Otherwise, as in many animals and dicotyledons, they remain contracted to the same degree as at the first division.

The loss of contraction is shown to be due to the occurrence of an interphase by an abnormality in *Gasteria*. Under exceptional conditions, doubtless both genetic and environmental, shortening occurs in the length of the interphase with a corresponding lack of recovery to the length of a somatic mitosis found in the normal nuclei (Tuan, 1931 ; D., 1936 *d*).

PLATE IV

Fig. 1.—Part of a univalent chromosome at first metaphase of meiosis in *Hyacinthus orientalis*, pressed in fixation to show spiral structure of two chromatids side by side. × 5000. (D., unpublished.)

Figs. 2 and 3.—Chromosomes at first metaphase in *Tradescantia virginiana*, fixed in boiling water and unstained, to show spiral structure. (Sakamura, 1927.) × *ca.* 1000.

Fig. 4.—First metaphase in *Tradescantia rubra* ($4x = 24$) pretreated with nitric acid, fixed in medium Flemming, stained in gentian-violet. (La Cour, 1935. *cf.* Fig. 137.) × *ca.* 1800.

PLATE IV

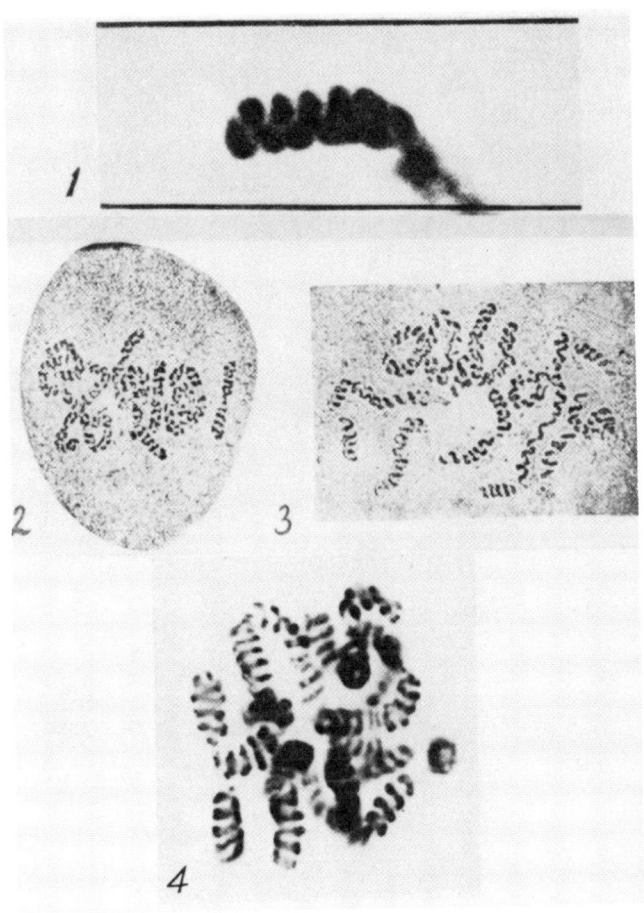

[To face p. 118.

COMPLETION OF MEIOSIS

One constant distinction is observed at the prophase from the normal behaviour at mitosis : the two chromatids are held together only at the centromere. Elsewhere they actually seem to repel one another until metaphase. This is perhaps to be associated with the unique property of the interphase period between the two meiotic divisions, *viz.*, that there is no division of the chromosomes, that the associated chromatids are derived partly from association with other chromatids, so that they have undergone spiralisation separately. The theoretical significance of this will be considered later (Ch. XII).

(x) **The Second Division.** The second division follows the course of an ordinary mitosis. The chromatids momentarily come together for part or all of their length. Where they are globular they lie in the axis of the spindle. They are pulled apart at their centromeres which here even more clearly than at mitosis are alone concerned in their separation. Four daughter-nuclei are formed, each with half the number of chromosomes of the mother-nucleus. Meiosis is complete.

4. MEIOSIS IN POLYPLOIDS

(i) **Prophase.** It is known from the fact that chromosomes may pair in parthenogenetic organisms (*q.v.*), that their origin, whether maternal or paternal, has no connection with pairing. Therefore in polyploids having several identical chromosomes of each type instead of two, we must expect pairing to occur indifferently amongst all the chromosomes of each type. But there are several different ways in which we might suppose this to come about. For example, at zygotene the threads might act as units and associate in pairs throughout their length, as they do in diploids. Or they might associate in threes or fours, according to the number present of each kind, to give a triple or quadruple thread at pachytene. Neither of these surmises holds good.

In both triploids (*Tulipa, Fritillaria, Zea Mays* and *Hyacinthus*), and tetraploids (*Hyacinthus*, D., 1929 *b*; *Allium*, Levan, 1935 *b*), the chromosomes come together in pairs. In a triploid, therefore, one thread is left out of association, while in a tetraploid two separate pairs are formed. But different pairs are formed at different points.

Each particle of the chromosome is potentially independent of its neighbours, so that exchanges of partner take place here and there along the paired chromosomes, and the paired threads at pachytene

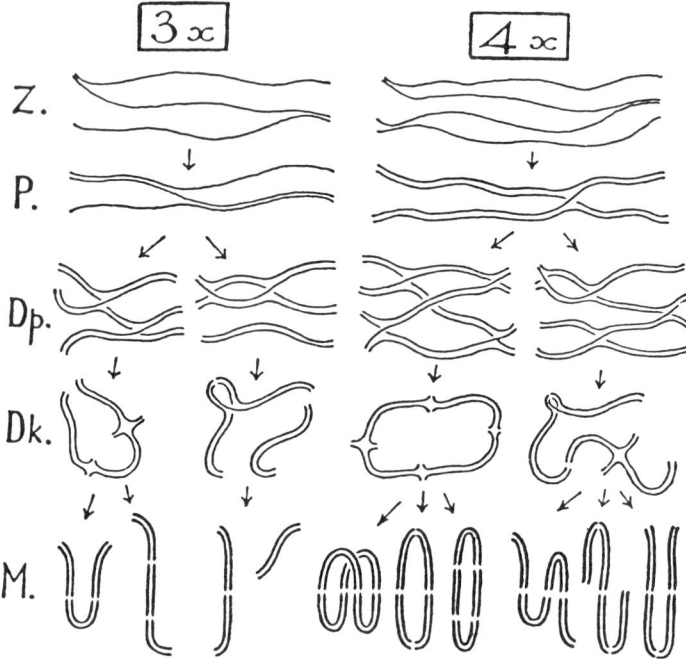

FIG. 38.—Diagram illustrating the structure of the pairing chromosomes during prophase and metaphase of the first division in triploids and tetraploids, with chromosomes having median centromeres and complete terminalisation. Z., zygotene; P., pachytene; Dp., diplotene; Dk., diakinesis (with partial terminalisation); M., metaphase (with completed terminalisation). The lines represent chromosomes in the first two stages, chromatids afterwards. There are three stages of development when behaviour varies: (i) choice of partner at zygotene; (ii) formation of chiasmata at diplotene; (iii) orientation in the spindle at metaphase. Variations through the last two chances are shown but the more complicated possibilities are omitted (v. Figs. 40, 42, 43, 45).

in the tetraploid resemble those at diplotene in the diploid. As many as six exchanges have been observed in one association (Newton and D., 1929; D., 1929 b).

Three conclusions may be drawn from this behaviour : (i) The pairing properties of the chromosomes are specific to their constituent particles (chromomeres), so that the whole chromosome does not act as a unit. (ii) The affinities of the chromosomes are satisfied by association in twos. (iii) Since the number of changes of partner is limited it is evident that, as one would expect, an unpaired particle is largely influenced in its choice of partner by the pairing of adjoining particles, the chromosomes pair in blocks. This, as we shall see later, must have an important bearing on the conditions of pairing in hybrids, which cannot be determined directly.

In the pachytene thread, chiasmata arise as in diploids approximately at random. And since the chromosomes were paired by blocks at random it follows that where enough chiasmata are formed every possible combination of them is found amongst the chromosomes associated. Amongst other combinations is that where no chiasmata are formed at all between two particular chromosomes where they have been paired, for sometimes with free exchanges two chromosomes may happen to be paired for a very short distance. Thus there arise, in triploids, single chromosomes unassociated with their two homologues, and, in tetraploids, pairs of chromosomes unassociated with their homologous pairs (*Hyacinthus, Primula sinensis*). There can be little doubt from the prophase observations that pairing has in these cases taken place at pachytene between chromosomes which are left unassociated at metaphase. It therefore appears that the association of chromosomes is preserved after pachytene and until metaphase solely by the occurrence of exchanges of partner, chiasmata, amongst their constituent chromatids (D., 1929 c). This conclusion is borne out by quantitative observations of various kinds (v. Ch. XII). Its significance will be considered later.

Terminalisation occurs in the trivalents and quadrivalents that arise from chiasma-formation at diplotene as in the corresponding diploids. The associations produced in *Primula sinensis* (D., 1931 a ; Dark, 1931) and *Solanum Lycopersicum* (Upcott, 1935) agree with the assumption that every pair of chromosomes that have been associated interstitially in one arm are at metaphase associated terminally in that arm. When chiasmata joining three or four chromosomes are terminalised to the same end a multiple chiasma is formed (see Figs. 40, 44 and 47). This consists in the terminal association of the two chromatids of each of the chromosomes with

FIG. 39.—Chromosomes of complete nuclei of diploid and triploid *Tulipa* at diakinesis. The numbers of chiasmata are given under each configuration. Top, twelve bivalents in diploid *T. Gesneriana*. Middle, nine trivalents, three bivalents and above, three corresponding univalents in a triploid variety (*cf.* Plate IV). Bottom, twelve trivalents in the same variety, and a chromatid diagram of these. (Note the triple chiasma in the fourth from the left.) × 2000 (from D. and Mather, 1932).

chromatids of two different chromosomes. Triple and quadruple chiasmata have been described in *Tradescantia*, *Aucuba*, *Primula*, *Œnothera* (Catcheside, 1931), *Rosa* (Erlanson, 1931), *Avena* (Kattermann, 1931), *Campanula* (Gairdner and D., 1931), and *Hemerocallis*

(Dark, 1932). They also occur in *Datura* (Belling, 1927) and *Triticum* (Kihara, 1929 et al.).

Thus each end of a chromosome in a triploid or in a tetraploid may be associated with all of the ends (two in the triploid and three in the tetraploid) that are homologous with it, *or* with some of them *or* with none at all. In other words, the kinds of terminal association at diakinesis correspond exactly with the kinds of interstitial association (by demonstrable chiasmata) observed earlier. Ten kinds of quadrivalents can be expected in this way, and they have all been found in *Datura* (Belling, 1927 *b*) or *Primula sinensis* (D., 1931 *a*). Clearly types where the chiasma is formed in each arm should be commoner, where the centromeres are approximately median, than those where two are formed in one arm and none in the other. It follows, therefore, that the simple ring or chain quadrivalents are the commonest types, in forms with median centromeres (*Primula, Datura, Campanula*).

In tetraploid *Primula* not only simple bivalents but very occasionally univalents (unpaired chromosomes) are found. Since the latter do not occur in the corresponding diploid it must be supposed that the exchanges of partner at pachytene are hindered by the presence of intervening chromosomes to a greater extent than in the diploid (*cf.* interlocking), and that pachytene association is therefore often incomplete. The time-limit comes into effect.

(ii) **Metaphase : the Limits of Chromosome Association.** The extent to which corresponding chromosomes associate at metaphase has been described in a number of polyploids in which these chromosomes are supposed to be identical or nearly identical. In several triploids and tetraploids the chromosomes have been said to associate *fully*, *i.e.*, in *x* trivalents or quadrivalents (*e.g.*, *Canna*, Belling, 1921 ; *Datura*, Belling, 1927 ; *Primula Sieboldii*, Ono, 1927 ; *Hemerocallis*, Sinoto, 1929 ; *Hyacinthus*, Belling, 1928). Recent work has shown 'that, while full association does occur in most such forms in a proportion of nuclei, in a proportion also it fails. Trivalents are replaced by bivalents and univalents in the triploids and by pairs of bivalents in the tetraploids. Association is therefore both incomplete and variable, as Table 15 shows.

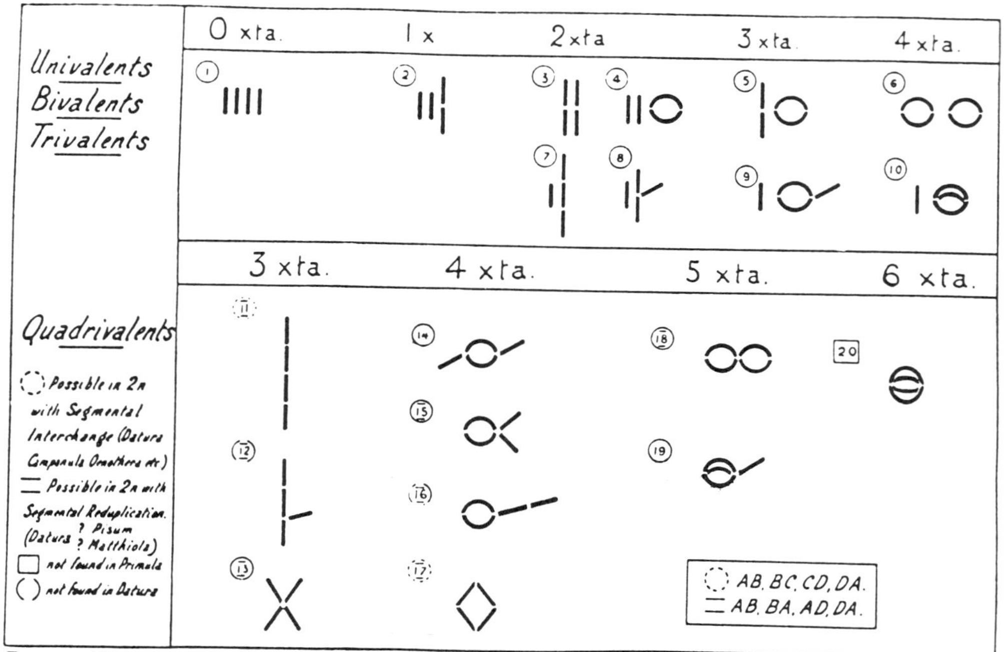

FIG. 40.—Diagram showing the multivalent and other configurations possible at diakinesis in a tetraploid with complete terminalisation. Certain of the associations of four are possible in diploid structural hybrids (v. Ch. V) xta. = chiasmata; the numbers given are those necessary for the formation of the different types of association. (From D., 1931 b.)

TABLE 14

Observations of Meiosis in Auto-Polyploids

1. (PLANTS ONLY) TRIPLOIDS

Hyacinthus orientalis	($x = 8$)	D., 1929 *b*; Stone and Mather, 1932.
Tulipa vars.	($x = 12$)	Newton and D., 1929; D. and Mather, 1932.
Lilium tigrinum	($x = 12$)	Takenaka and Nagamatsu, 1930; Mather, 1935.
Zea Mays	($x = 10$)	McClintock, 1929.
Solanum Lycopersicum	($x = 12$)	Lesley, 1929; Upcott, 1935.
Primula sinensis	($x = 12$)	Dark, 1931.
Allium spp.	($x = 8$)	Levan, 1931, 1933 *a*, 1935 *b*.
Hemerocallis fulva	($x = 11$)	Dark, 1932.
Oryza sativa	($x = 12$)	Morinaga and Fukushima, 1935.

A. PLANTS 2. TETRAPLOIDS

Solanum Lycopersicum	($x = 12$)	Jørgensen, 1928; Lesley and Lesley, 1930; Upcott, 1935.
Hyacinthus orientalis	($x = 8$)	D., 1929 *b* ($4x - 2$).
Primula sinensis	($x = 12$)	D., 1931 *a*.
Nicotiana suaveolens	($x = 16$)	Goodspeed and Avery, 1929 *a*.
Dactylis glomerata	($x = 7$)	Müntzing, 1936.
Allium schœnoprasum	($x = 8$)	Levan, 1935 *b*.
Petunia (hort.)	($x = 7$)	Matsuda, 1935.
Primula malacoides	($x = 9$)	Kattermann, 1935.

B. ANIMALS (tetraploid cells in diploid testes)—

Schistocerca gregaria	($x = 11$)	White, 1933.
Drosophila pseudo-obscura (hybrid)	($x = 5$)	Dobzhansky, 1934.
Culex pipiens	($x = 3$)	Moffett, 1936.
Chrysochraon dispar	($x = 9$,	Klingstedt, unpub.

Where only a single extra chromosome is present in trisomic plants it behaves in the same way as do each of the extra chromosomes in a triploid (*e.g.*, *Solanum*, Lesley, 1929; *Matthiola*, Philp and Huskins, 1931; *Datura*, Belling, 1927). Tetraploids with complete terminalisation show very clearly the further interesting property that although the chromosomes must be in

TABLE 15

Variation in Pairing of Chromosomes in Triploids and Tetraploids
(*Cf.* Table 14.)

Triploids $3x = 36$	Numbers of cells with different numbers of trivalents												
	0	1	2	3	4	5	6	7	8	9	10	11	12
Lilium tigrinum	—	—	—	—	—	—	2	4	9	25	14	12	4
Tulipa Gesneriana:													
Keiserskroon	—	—	—	—	—	2	—	1	4	8	5	—	1
Pink Beauty	—	—	—	—	—	2	1	3	2	6	5	2	3
Inglescombe Yellow	—	—	—	—	—	1	—	1	2	1	2	—	—
Solanum lycopersicum	—	—	—	5	13	17	10	5	—	—	—	—	—
Tetraploids $4x = 48$	Numbers of cells with different numbers of quadrivalents												
Solanum lycopersicum	—	3	12	10	2	15	6	—	2	—	—	—	—
Primula sinensis	—	—	—	—	—	—	—	—	—	1	11	9	—

approximately *symmetrical* associations at pachytene (every change of partner being reciprocal) their metaphase associations occur in all the possible ten types, of which only four are symmetrical (having both ends of all the chromosomes behaving similarly). Association is therefore variable, not only as affecting whole chromosomes but as affecting their parts.

Now it has been seen that the chromosomes of bivalents appear to be held together at diakinesis and metaphase merely by chiasmata, for the chromosomes themselves repel one another. And it has been seen also that chiasmata are usually formed in a number

PLATE V

Meiosis in Triploids

FIGS. 1 AND 2.—Zygotene, three chromosomes (running right and left in both nuclei) pairing, early and late stage. Note that the chromosomes are single threads and associate in pairs. *Tulipa* (D., 1931 *b*.) × *ca*. 1000.

FIG. 3.—*Fritillaria dasyphylla* ($3x = 36$), showing univalent and trivalent chromosomes from first metaphase to telophase. Flemming — gentian-violet smear taken with 4 mm. objective. (La Cour and Osterstock.) × *ca*. 500.

PLATE V

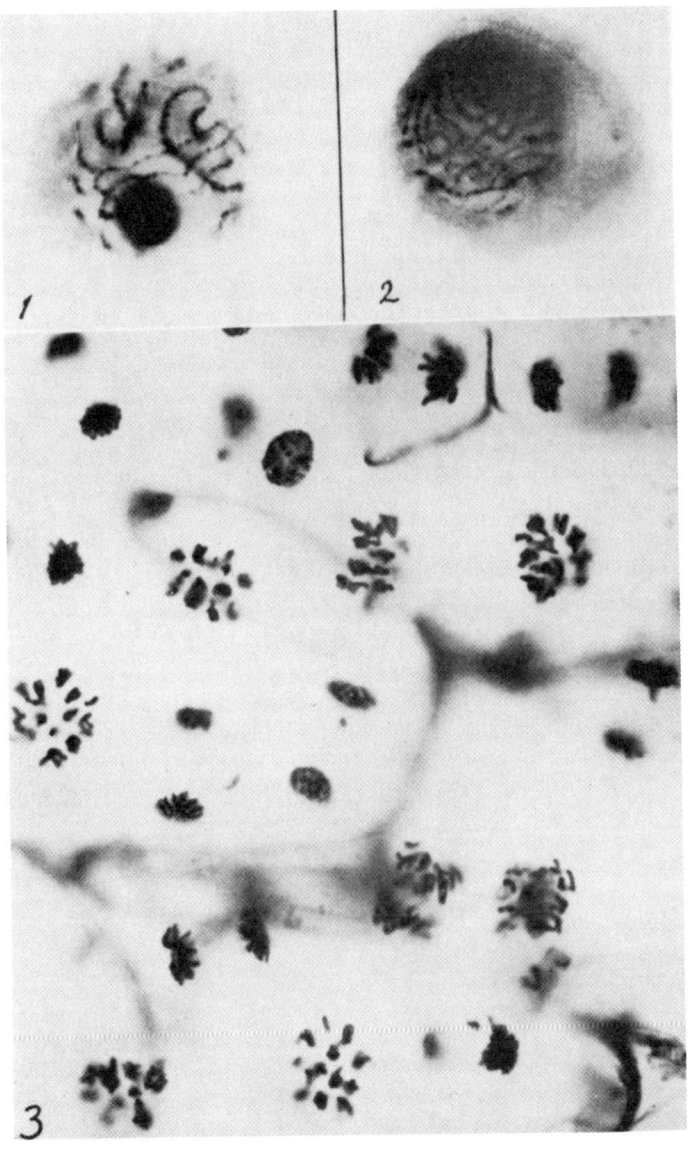

[To face p. 126.

proportionate to the length of the chromosome, or at least the length paired at pachytene, but varying in different nuclei both in number and position. The obvious conclusion is therefore that failure of chromosome association and its asymmetry is due to failure of chiasma formation owing to reduction in the length paired at pachytene, and hence that any two chromosomes are associated by virtue of a chiasma formed between them (D., 1929 b).

In order to test this hypothesis a closer analysis of chiasma formation in triploid *Tulipa* and *Hyacinthus* has been attempted

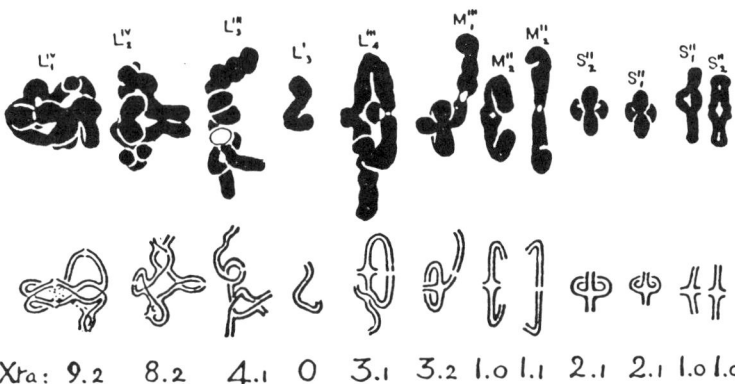

Xta: 9.2   8.2   4.1   0   3.1   3.2 1.0 1.1   2.1   2.1 1.0 1.0

FIG. 41.—Bivalents and multivalents in *Hyacinthus*, side view ($2n = 4x - 2 = 30$); four types of L, two of M and two of S. Four chromosomes of each type except L_4 and M. × 2000 (after D., 1929 b).

(D. and Mather, 1932 ; Stone and Mather, 1932). It is found that in *Tulipa* the proportion of univalents (with no chiasmata) as well as of chromosomes with different numbers of chiasmata can be approximately predicted on the assumptions :—(i) that the lengths of the chromosomes which pair at pachytene vary, as might be expected, from the randomness of association observed at this stage amongst the three chromosomes (only two of which can be paired at any one point), but the variation is very high, as though it were due to random association of a *small* number of " pairing blocks " ; (ii) that the numbers of chiasmata formed in these paired lengths vary in the usual way ; and (iii) that the association of the

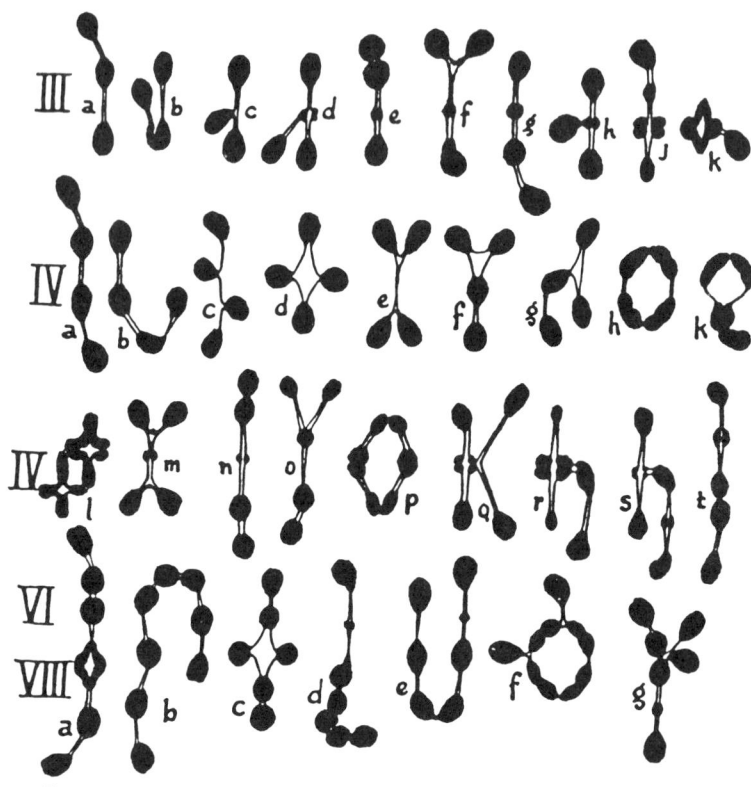

Fig. 42.—Multivalent configurations of small chromosomes, seen in side view of first metaphase. Roman numerals show the number of chromosomes in the configurations. III a–c, IV a–k, and VI b, c and f, have only terminal chiasmata. The rest have some interstitial chiasmata. IV c has an imperfect quadruple chiasma (*cf.* Ch. XII), IV d a perfect one. × *ca.* 6000.

These figures have been found in *Prunus* (D., 1928, 1930 *a*, Meurman, 1929), *Pyrus* (D. and Moffett, 1930; Moffett, 1931), *Solanum* (Jørgensen, 1928), *Primula* (D., 1931 *a*), *Galeopsis* (Müntzing, 1932), and will probably be found in many other genera where they have previously escaped notice.

chromosomes at metaphase is conditioned by the formation of the chiasmata.

In triploid *Hyacinthus* there are three types of chromosomes, long, medium and short, with average frequencies of chiasmata amongst the three chromosomes of each kind of 6·2, 3·6 and 2·0 respectively (*i.e.*, in proportion to their length). As expected, the three types have distinct frequencies of univalents—2·8, 0·1 and 11·9 per cent. respectively. The high frequency of univalents amongst the short chromosomes is explained by the low frequency of chiasmata. The fact that the long chromosomes have a slightly higher frequency than the medium ones is attributed to the interference with free assortment resulting from their extreme length, for they are the longest chromosomes known in any organism. This may be expressed by saying that they have fewer pairing blocks than the shorter chromosomes, two instead of four or five.

Thus quadrivalents are less frequent among the short chromosomes than amongst the long ones, in tetraploids not only of *Hyacinthus*, but also of *Schistocerca* (White, 1933) and of *Urginea* and *Scilla* (Sato, 1934). Similarly in polyploid forms with localised pairing and chiasmata we must expect fewer multivalents than in corresponding forms with complete pairing, since the effective length for pairing is shorter. This property is possibly shown by the tetraploid *Allium Porrum* (Levan, 1935 *a*).

More direct evidence that chromosomes are held together by chiasmata at metaphase will be considered in relation to pairing in hybrids. For the present it must be noted that this view conflicts diametrically with the old theory that chromosomes are held together at metaphase by an affinity between them, and adequately explains the present observations, while the old theory leads one to expect in pairing uniformity, symmetry, and completeness, instead of variation, asymmetry and incompleteness.

(iii) **The Two Divisions.** The arrangement of multiple bodies, with several centromeres, in a bipolar spindle is inherently irregular. The result seems to depend on three variable conditions : (i) the distribution of the chiasmata in the multivalent chromosome ; (ii) the spatial relationship of the centromeres of the component chromosomes, as determined by their position in the chromosomes, and by the position in which the chromosomes are held by the chiasmata ; (iii) the relationship of the size of the chromosomes to the area of the equatorial plate, *i.e.*, whether the chromosomes are crowded or not.

130 MEIOSIS IN DIPLOIDS AND POLYPLOIDS

The first condition determines the rigidity of the combination, for multivalents with interstitial chiasmata seem to be less pliable

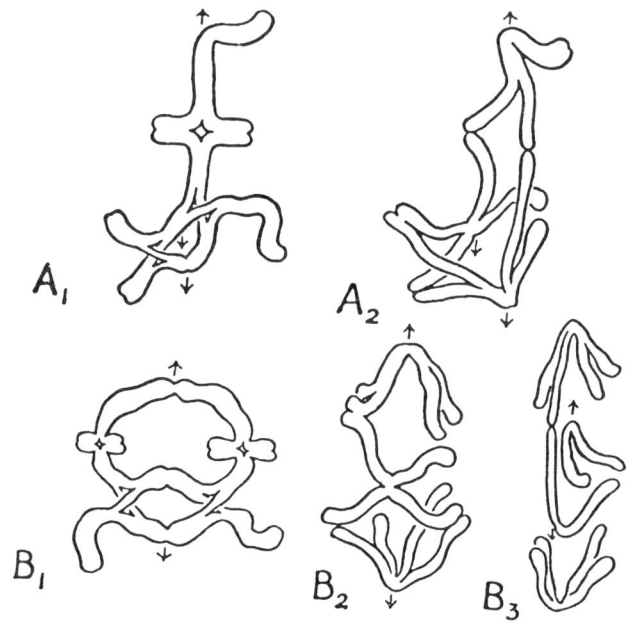

Fig. 43.—Diagram showing the division of trivalents in *Hyacinthus*. A_1 and B_1, metaphase; A_2, B_2 and B_3 the anaphase forms to which they give rise. A divides into $2 + 1$, B divides into $1\frac{1}{2} + 1\frac{1}{2}$. Arrows mark the positions of centromeres which are median. Note: (i) The chromatids are pulled apart distal to the chiasmata. (ii) Where, as in B_2, the chromatids of one chromosome are pulled in opposite directions, they are held together at the centromere. (iii) The centromere of a single chromatid is pulled back in B_3, where it is associated with a chromatid of a whole chromosome. (After D., 1929 b. A_1, Fig. 34 f; A_2, Fig. 36 b and 38; B_1, Fig. 34 r; B_2, Fig. 36 n; B_3, Fig. 39).

N.B. Fig. B_2 shows that the holding together of the chromatids at the centromere is something essentially different from their attraction to one another in pairs elsewhere. This difference is probably described by saying that the centromere has not yet divided and the separation of the chromatids waits upon its division.

than those with only terminal chiasmata, and for two reasons: the distance separating the centromeres of the associated chromosomes is shorter, and the cross-arms of an interstitial chiasma strengthen the connection while a terminal chiasma opposes the minimum resistance to any kind of torsion. The rigidity of a configuration with interstitial chiasmata is necessarily three-dimensional, that in one with only terminal chiasmata is two-dimensional (*v*. Ch. XII).

The second condition determines whether or not the multivalent is physically capable of lying in one axis, as does a bivalent. The third determines whether there is room for the chromosomes to lie transversely across the plate, as is necessary if each is to separate or

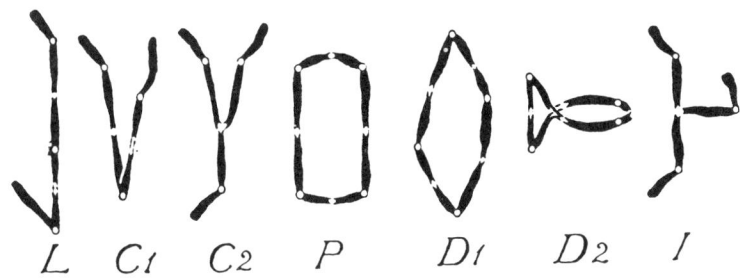

Fig. 44.—Different methods of co-orientation in trivalents (linear, L; convergent, C; and independent, I) and quadrivalents (parallel, P, and discordant, D). In D_2 the two loops are at right angles.

"disjoin" (to use a common phrase) from the one with which it is associated.

From the operation of these different factors the following variations may be observed :—

(i) The several centromeres may have one of four different relationships with one another (*cf.* D., 1936).

(*a*) *Linear:* all the centromeres lie in one spindle-axis. If this axis is near the side of the spindle it is arc-shaped, so that the centromeres do not lie in a straight line.

(*b*) *Parallel:* pairs of centromeres, where there are four or more, lie on parallel and independent axes.

(*c*) *Convergent:* two centromeres are lying axially with respect to a third and lie therefore on axes which converge.

(d) *Indifferent:* one centromere lies indifferently with respect to the others.

(ii) Each of the chromosomes may disjoin from those with which it is directly associated (provided it is not directly associated with more than two), or it may pass to the same pole with one or more of them.

(iii) Where an even number of chromosomes are associated the multivalents may divide evenly or unevenly; thus four chromosomes may divide into two and two or into three and one. In this way great differences may occur between the numbers of chromosomes in the two daughter-nuclei. *Prunus cerasus* ($4x$) gave exceptionally a distribution of $19 + 13$ (Plate V). Similarly, tetraploid *Datura* (Belling and Blakeslee, 1924 b) and *Primula sinensis* (D., 1931) give germ-cells with unequal numbers of chromosomes (*cf.* Levan, 1933 a).

(iv) Where an odd number of chromosomes are associated whole chromosomes may separate to the poles, or one of the component chromosomes may be left on the equator, and, after an interval, divide, the halves passing to opposite poles after the whole chromosomes. They then pass at random to one pole or the other in the second division or are lost. In triploids, therefore, germ-cells are formed with an approximately random distribution of the chromosomes in the third set. This has been achieved by the random distribution, either of univalents (or odd members of trivalents) at the first division, or of daughter univalents (or odd members of trivalents) at the second division (*v.* Ch. VIII). The detailed behaviour of unpaired chromosomes will be considered later.

Where associated chromosomes pass to the same pole, all trace of their association is lost at the second division if an interphase occurs (*e.g.*, in *Hyacinthus*, $3x$; D., 1929 b). The second division is therefore normal apart from such a difference as there may be between the numbers of chromosomes in the two daughter-nuclei.

5. THE CONDITIONS OF MEIOSIS

We are now in a position to ask ourselves what conditions distinguish meiosis from an ordinary mitotic division. Do new forces come into play to produce a pairing of homologous chromo-

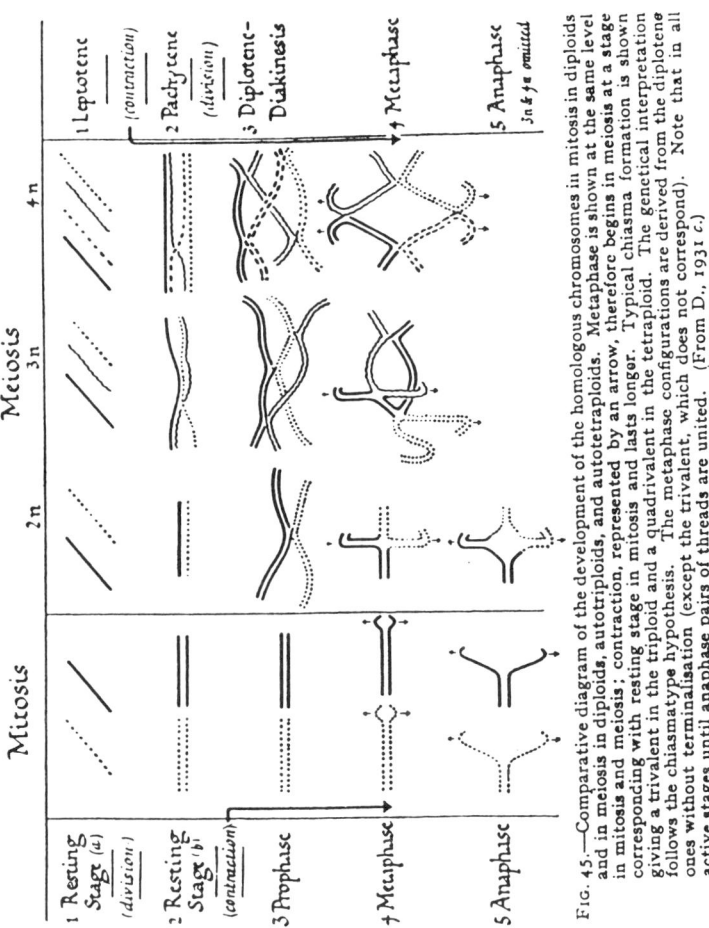

Fig. 45.—Comparative diagram of the development of the homologous chromosomes in mitosis in diploids and in meiosis in diploids, autotriploids, and autotetraploids. Metaphase is shown at the same level in mitosis and meiosis; contraction, represented by an arrow, therefore begins in meiosis at a stage corresponding with resting stage in mitosis and lasts longer. Typical chiasma formation is shown giving a trivalent in the triploid and a quadrivalent in the tetraploid. The genetical interpretation follows the chiasmatype hypothesis. The metaphase configurations are derived from the diplotene ones without terminalisation (except the trivalent, which does not correspond). Note that in all active stages until anaphase pairs of threads are united. (From D., 1931 c.)

somes instead of their simple division ? Or are the same forces working in a new combination or in a different sequence to produce this exceptional result?

Consider first mitosis. The chromosomes then consist throughout prophase of pairs of chromatids which appear to attract one another. Each chromosome has a centromere which is effectively single at metaphase and effectively double at anaphase. Its division seems to determine the separation of daughter chromatids.

At meiosis, on the other hand, the chromosomes are single at the earliest prophase. They unite in pairs, however many potential partners are present. This union ceases when they divide. Association afterwards is between pairs of chromatids, and chromosomes repel one another. They are held together merely in so far as their chromatids exchange partners at chiasmata. Moreover, if they happen to divide before pairing is complete, the process is interrupted. There is a time limit to pairing.

There is therefore potential association at all stages of meiosis of pairs of threads—chromosomes or chromatids. The same is true of mitosis. The difference is that the prophase begins earlier in meiosis. The chromosomes are still undivided. Homologous threads attract one another and pairing is possible for a short time before division takes place. The special properties of meiosis follow from the precocity of the prophase and not from the action of new forces. The external and internal processes of division are, as we saw earlier, out of step.

From this initial upset arise the later special properties of meiosis, the pairing of the chromosomes, their crossing-over, their specially strong contraction at metaphase, the postponement of the splitting of the centromere to the second division; and possibly the rapid sequence of the two divisions. A simple quantitative time-difference gives rise to the complex series of qualitative differences that distinguish meiosis from mitosis and have determined the origin of sexual reproduction.

CHAPTER V

STRUCTURAL HYBRIDS

Definition and Classification of Hybrids—Structural Hybrids—Fragments—Translocation and Interchange Heterozygotes—Unequal Chromosome Pairs.

> So wenig man eine scharfe Unterscheidungslinie zwischen Species und Varietäten zu ziehen vermag, ebenso wenig ist es bis jetzt gelungen einen gründlichen Unterschied zwischen den Hybriden der Species und Varietäten aufzustellen.
>
> MENDEL, 1866.

1. CLASSIFICATION BY FUNCTION

THE systematist recognises a hybrid as an individual or type intermediate between two species, and he assumes it to be either a first cross or a derivative of a first cross. The geneticist's definition of a hybrid, which he owes to Mendel, is entirely different. On the one hand, he cannot draw a sharp line between species-differences and varietal differences, because he finds they are of the same kinds and differ only in degree. Even in regard to inter-sterility, every gradation occurs between those pairs of species which are prevented from crossing in nature merely by incidental conditions and those which may be assumed to be intersterile under any conditions. For the geneticist, therefore, any cross between dissimilar races must be a hybrid, however slight their differences. On the other hand, the geneticist makes a sharp distinction between the first cross (F_1) and its derivatives, because the first cross is capable of showing segregation of the parental differences in its offspring, while its derivatives need no longer have these differences. An intermediate character is therefore for him no criterion of hybridity.

And finally the geneticist is not concerned with the difference between the parental zygotes but with that between the gametes which actually fused to make the hybrid or which are formed at meiosis in the hybrid. We define hybridity by *function* and not by *phylogeny*. Thus it may happen that a cross between two dissimilar zygotes, if these were themselves hybrids, is constituted of two gametes that are identical; speaking in mendelian terms, *Aabb* and *aaBb* differ, but may give homozygous offspring *aabb*. Hence it may happen that a first cross between two taxonomic species in *Œnothera* is less hybrid than its parents, and derivatives of such hybrids have been obtained which constitute pure lines in the strictest sense (Ch. IX).

A hybrid (or heterozygote) must therefore be defined as *a zygote which* either *arises from the union of dissimilar gametes*, or *gives rise to dissimilar gametes* (regularly, *i.e.*, apart from mutation). For most purposes it will be found that these alternatives lead to the same result, and the second can be ignored except in relation to polyploids, in which hybridity requires special consideration. Polyploids have chromosome sets corresponding with those of more than two ancestral gametes, and are capable of functional relations with one another, so that a simple definition of hybridity is therefore inapplicable to them.

Hybrids have very generally the property of hybrid vigour or " heterosis " which is sometimes due to the normal allelomorphs of depressive factors accumulated in their more homozygous parents and sometimes, perhaps, to an inherent virtue in the possession of a single dose of material in which the parents differ. Hybrids, both in plants and animals, are liable to abnormality, which is due to their having a new and untried hereditary complement. This liability is found in an exaggerated degree in one sex (the heterozygous sex) where the sexes are differentiated (Haldane, 1922) because the normal chromosome set of one species is not wholly represented in the hybrid and the deficiency may not be made up from the other. It may be said that the heterozygous sex in an animal hybrid is not strictly an F_1 hybrid at all, since it differs from the hybrid of the homozygous sex through the hybridity of one of its parents. Its properties are not therefore the characteristic properties of F_1 hybrids. They are incidental to special genetic conditions.

The characteristic properties of hybrids depend, therefore, not on the properties of the parents, but on the differences between these properties which are seen, first, in the relation of the corresponding genetic elements at meiosis and, secondly, in the differences between their offspring. The study of meiosis in non-hybrid forms and the variations already seen at mitosis enable one to classify hybrids and predict the kind of chromosome pairing that will be found in each class, as follows. (The term "hybrid" is used here for the broader classes and the term "heterozygote" for the individual categories discussed later).

(i) NUMERICAL HYBRIDS. The zygote is derived from the union of gametes dissimilar in regard to the number of their chromosomes. Trisomic and triploid organisms which owe their origin simply to the reduplication of one or more chromosomes in one of the gametes are the simplest of this kind. It is irrelevant to the consideration of their properties at meiosis whether they result from mitotic or meiotic irregularities. The behaviour of these hybrids at meiosis has already been considered.

(ii) STRUCTURAL HYBRIDS. The zygote is derived from the union of gametes dissimilar in regard to the structure of their chromosomes, $i.e.$, the linear arrangement of their chromomeres or genes. The change in structure may also involve a change in number, as in fragmentation. From what we know of the pairing of chromosomes in polyploids we can predict that the chromosomes of these hybrids will pair according to the relationships of their parts, subject to these parts being long enough to allow of their occasionally pairing at pachytene and forming a chiasma in the length paired.

(iii) UNDEFINED STRUCTURAL HYBRIDS. The zygote is derived from the union of gametes dissimilar as a result of changes which cannot be defined. In this class may be considered those whose parental gametes either had the same number of chromosomes or a different number, through a change that cannot be identified. These hybrids may be said to be "undefined" simply because the structural differences between their chromosomes are too slight or too numerous to be readily detected. We then find three types of behaviour in this kind of hybrid at the first metaphase of meiosis, according to the frequency of differences between the chromosomes:

(i) Normal behaviour owing to the differences being too slight to have a recognisable effect in reducing pachytene pairing and chiasma formation. (ii) A reduction in the frequency of chiasmata with or without occasional failure of pairing of the chromosomes. This failure always shows a curve of variation similar to that of chiasma-formation. (iii) A reduction of chiasma-frequency to such a point that pairing fails entirely or almost entirely.

(iv) COMPLEX HYBRIDS. Another class of hybrids occurs in which the differences can be partly defined. Amongst these are the complex ring-forming and sex heterozygotes which form permanent self-reproducing types. Interchange, translocation and deficiency can often be inferred in them, but together with such changes others occur that can be located but not defined (*v.* Ch. XI).

(v) POLYPLOID HYBRIDS (species or true-breeding varieties). The zygote is derived from the union of identical polyploid gametes, each of which is derived ultimately from the union of dissimilar haploid gametes in an ancestor. These hybrids are effectively true-breeding when they have an even number of sets owing to the pairing and segregation of identical chromosomes derived from opposite immediate parents. In this case they are functionally diploid. But they are also liable to yield segregates owing to the occasional pairing of dissimilar chromosomes derived from their ultimate diploid parents (*v.* Ch. VII).

(vi) NUMERICAL-STRUCTURAL HYBRIDS. All types of cytologically observable hybridity may occur in the same organism, *e.g.*, in triploid *Œnothera* (Catcheside, 1931 *a*), *Triticum* (Kihara and Nishiyama, 1930 ; Mather, 1935), and *Avena* (Nishiyama, 1929), and in trisomic *Datura* (Belling, 1927 *b*). These will be considered in relation to structural hybrids.

(vii) MENDELIAN HYBRIDS (*sensu stricto*). Where a difference, which might be empirically described as a mendelian difference or group of differences, can be described directly in chromosome terms (as in the case of the trabant of *Matthiola* or the properties of ring-formation in various genera), it is no longer convenient to speak of it in mendelian terms any more than it is convenient to speak of " mutation " where the change can be defined as translocation or inversion. It therefore follows that mendelian differences for the cytologist are those which show segregation

ANALYSIS OF HYBRIDS

(and, as we shall see later, linkage) but have no recognisable effect on chromosome behaviour. Probably most factor differences such as those found in *Drosophila* have a negligible effect on the pairing of the affected chromosomes. There is no evidence to the contrary, and in recent experiments in which wild type and mutant *Drosophila* have been compared genetically they show no difference in the frequency or distribution of crossing-over (Redfield, 1930) (*v.* Ch. VII). It will be seen, therefore, that cytological observations show many large differences, as in triploidy, which are incapable of mendelian analysis, while mendelian analysis shows many differences which are small in origin and incapable of cytological analysis. The visible spectra of variation shown by the two methods have different ranges which reveal two different aspects of the problem of variation.

2. MEIOSIS IN STRUCTURAL HYBRIDS

(i) **Method of Analysis.** The conclusion has been reached that only identical blocks or segments of chromomeres can pair at pachytene and form chiasmata which keep the chromosomes associated or "paired" at metaphase. On this assumption structural hybrids may be analysed. The differences between their chromosomes may be defined and hence the simple changes that have given rise to these differences specified.

Where terminalisation is complete in the organisms that have been studied from this point of view, the earlier association at pachytene, which is the characteristic of most importance, is only shown at metaphase by an association of the ends of the chromosomes. This association arises, however, in the ordinary way, as has been shown in particular cases (*v. infra*).

Where, in a diploid structural hybrid, several chromosomes are associated at meiosis it is better to speak of the configuration as a "group of three" or a "ring of four," reserving the terms trivalent, quadrivalent, and so on, for associations of homologous chromosomes in which every segment is represented as many times as there are chromosomes concerned.

(ii) **Fragmentation and Fusion Heterozygotes.** The simplest kind of structural hybrid is that resulting from the union of gametes which differ merely in one of them having two chromosomes corresponding to two parts of one chromosome in the other. The difference arises presumably from an unequal interchange followed by loss of the smaller product, having the effect of fusion or followed by reduplication of the smaller participant having the effect of fragmentation (Fig. 160). Such an organism may be described as a fragmentation heterozygote and has been found in nature, and produced by crossing different races, in *Phragmatobia fuliginosa* (Seiler, 1925 ; *cf.* Ch. XIII). The hybrid has regular pairing at metaphase of the two smaller chromosomes with ends of the large one. As a rule they pass to the opposite pole from it at the first division, but occasionally the smaller of the two fragments passes to the same pole with the large chromosome. The association is thus liable to the same kinds of variations of distribution resulting from linear and convergent arrangement as are trivalents formed from three identical chromosomes (Plate VI).

The hybrid *Vicia sativa* × *V. amphicarpa* (Sveshnikova, 1929, *a* and *b*) behaves similarly ; two chromosomes of the first species are evidently derived from one that is represented in the second. They sometimes both pair at metaphase with the corresponding parts of the larger homologue, and sometimes one of them fails to pair. It is apparently too short to establish a chiasma regularly.

The pairing in a cross between the domesticated silkworm (*Bombyx mori*, $n = 28$) and its wild relative (*B. mandarina*, $n = 27$) probably shows the same relationship, but one of the small chromosomes is sometimes paired by a lateral chiasma with the large one (Kawaguchi, 1928).

Fusion heterozygotes have been found by McClung (1905, 1917) in *Hesperotettix* and *Mermiria*, and by Woolsey (1915) and Robertson (1916) in *Jamaicana subguttata*. In the former a chromosome which is found unpaired in the males (the sex chromosome) is fused with one of the other chromosomes. This other chromosome pairs normally with its mate, while the fused element, still attached, is as usual unpaired (Fig. 46, *c*).

A hybrid between species of *Dicranura* (Lepidoptera) with 21 and

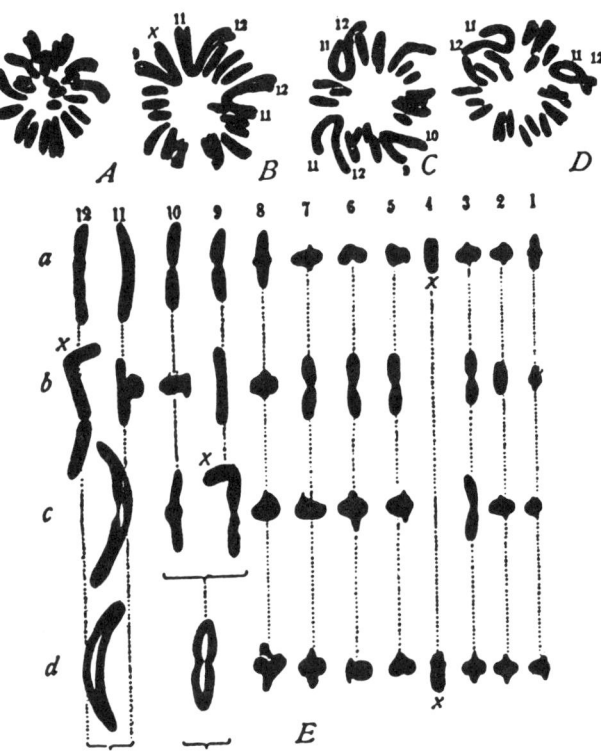

FIG. 46.—Permanent fusion of chromosomes shown by differences between *Hesperotettix brevipennis* (A) and different races of *H. viridis* (B–E). The complements shown are those of males and have a single unpaired sex chromosome (X, or No. 4, v. Ch. IX). A–D, mitosis, polar view.

    A, $2n = 23$; all chromosomes have terminal centromeres.

    B and C, $2n = 20$; three pairs of non-homologous chromosomes have fused and have median centromeres.

    (In B, X and 9, 11 and 12, both pairs; in C, 9 and 10, 11 and 12, both pairs.)

    D, $2n = 21$; two pairs have fused.

E. First metaphase of meiosis in side view; chromosomes drawn separately. *a*. The same type as A. *b*, $2n = 22$, X fused with 12. *c*. The same type as B. *d*, $2n = 19$. 9 fused with 10 and 11 with 12. All these are examples of fusion-homozygotes except in regard to the unpaired X-chromosome which necessarily gives heterozygotes in the male. (From McClung, 1917.)

29 chromosomes is probably a fusion or fragmentation heterozygote, and consequently has pairing of groups of chromosomes (Federley, 1915).

The most general type of fragmentation heterozygote is that in which one (usually the smaller) product of fragmentation is present as a supernumerary or reduplicated chromosome, so that the organism is polysomic in respect of the segment concerned. This

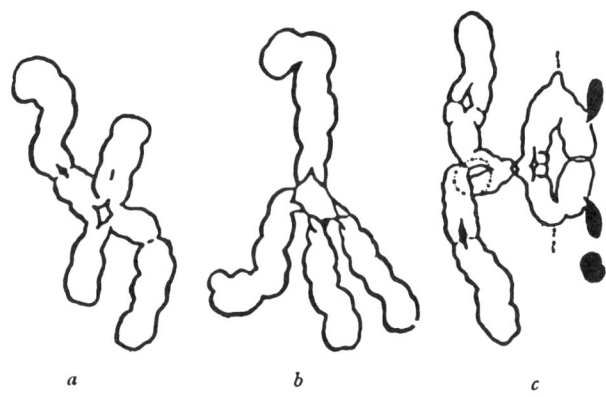

FIG. 47.—First metaphase in *Tradescantia virginiana* ($4x = 24$). *a*, an exceptional interstitial chiasma in a group of four chromosomes, due to arrest of terminalisation by change of homology, *cf.* Ch. XII. *b*, a normal quadrivalent held together by a quadruple chiasma. *c*, an association of six major chromosomes, and three homologous fragments (in black), two of them associated by lateral chiasmata at corresponding points in the major chromosomes. There is again an exceptional interstitial chiasma between two major chromosomes. × 2400. (After D., 1929 *c*.)

condition was first shown in *Metapodius, Diabrotica* and other Hemiptera (Wilson, 1905, *a* and *b*, 1909). In another hemipteran (*Alydus*, Reuter, 1930) supernumeraries occur, one of which is homologous with a part of a larger one, while both must be homologous with part of the sex chromosome, since all three may pair.

Recently large numbers of plants have been found with supernumerary fragments (*v.* Ch. III), and their pairing has been studied at meiosis in many of them (Table 17). It has the following characteristics.

(*a*) In an organism with negligible terminalisation in the major chromosomes, the fragments are always associated terminally (*e.g.*, *Fritillaria* and *Lilium*). Thus three homologous fragments may

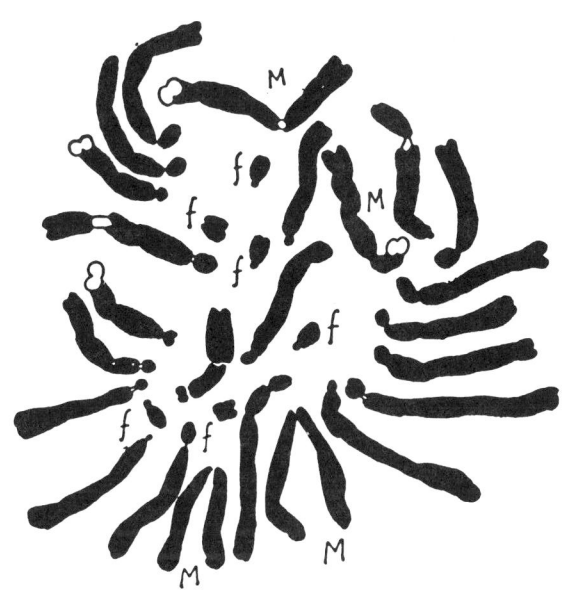

FIG. 48.—Mitotic metaphase in polar view in a clone of *Fritillaria imperialis* with six fragments ($2n = 24 + 6ff$). These are probably present along with the unfragmented sister chromosomes of those from which they are derived and are therefore reduplications. The chromosomes have been spread out in drawing; those with median centric constriction are marked M. *Cf.* Fig. 95. × 2200 (from D., 1930 *c*).

form a triple chiasma before diakinesis. This indicates that slight movement of chiasmata occurs, and is noticed in the larger chromosomes which are nine times as long only by their distal chiasmata becoming terminal (*cf.* Ch. XII).

(*b*) The fragment is usually associated at one end only, in organisms with complete terminalisation, although the major chromosomes are associated at both ends (*e.g.*, *Tradescantia*, D., 1929 *c* ; and *Datura*, Blakeslee, 1931). This is due to the fragment usually, on account of its shorter length, forming only one chiasma at prophase.

(*c*) Where the fragment corresponds to an intercalary part of a major chromosome, it forms a lateral chiasma, after terminalisation, with the part of the major chromosome with which it is homologous (*Tradescantia, Fritillaria, Aucuba, Lilium, cf.* Figs. 38, 39, 74 and 80, and Plate IV).

(*d*) Fragments which are a small fraction of the length of the large chromosomes fail to pair in a proportion of cases. This failure is expected in organisms with a chiasma frequency proportional to the lengths of the chromosomes, since the frequency of the chiasmata will be less in the short new chromosomes than in the old larger ones, and they will therefore fail to form chiasmata and associate at metaphase, although doubtless associated earlier like the unpaired third chromosomes of triploids.

The frequency of their pairing will be predictable within limits, on this assumption, when three conditions are determined : (i) the chiasma frequency of the major chromosomes ; (ii) the relative length of the fragments ; and (iii) the " condition " of the fragments, *i.e.*, whether the fragments correspond to parts of the major chromosomes and whether they are in themselves disomic, trisomic, and so on.

The limits to exact prediction are set, first, by our not knowing whether pachytene association is hindered, on the one hand, by the difficulty of exceptionally small chromosomes finding one another during pairing at zygotene and consequently being interrupted by the time limit, or facilitated, on the other hand, by their freer movement. In organisms with a wide size range in their normal complement and non-localised chiasmata, the evidence favours the second view, for although pairing is regular for long and short chromosomes alike, the shorter chromosomes usually have, as we saw earlier, a higher chiasma-frequency relative to their length. Secondly, limits to exactness are set by our not knowing the frequency with which two chiasmata will be formed in any given fragment. A high variance is taken to mean a low interference (*q.v.*) and will reduce the frequency of association (*cf.* D., 1933, on *Secale*).

TABLE 16

Examples of Supernumerary Fragment Chromosomes.

A. ANIMALS.
 (i) Apparently derived from Y chromosomes and inert.
  Metapodius (1–6 *ff*), *Diabrotica*, *Banasa*   Wilson, 1905, 1909, 1928.
  Alydus . . . . .   Reuter, 1930.
  Blaps lusitanica . . .   Nonidez, 1920.
 (ii) Apparently derived from autosomes.
  Tettigidea parvipennis . .   Robertson, 1916.
  Locusta danica (23 + 1–4 *ff*) (*ff* equal in size to the smallest members of the complement). . . .   Itoh, 1934.
  Aleurodes proletella . . .   Thomsen, 1927.
  Philosamia cynthia . . .   Dederer, 1928.

B. PLANTS.
 (i) *Dicotyledons*.
  Rumex acetosa (14 + 1–3 *ff*) . .   Ono, 1935 ; Yamamoto, 1933.
  Thalictrum aquilegifolium . .   Langlet, 1927.
  Ranunculus acris (14 + 3–10 *ff*) .   Langlet, 1927.
  R. Ficaria[4] (16 + 4 *ff*) . .   Larter, 1932.
  Rosa pyrifera (14 + 2 *ff*) . .   Erlanson, 1933.
  Primula spp. . . . .   Bruun, 1932.
  Alectorolophus (*Rhinanthus*) *major* (14 + 8 *ff*) . . . .   Fagerlind, 1936.
  Datura Stramonium (24 + 2 *ff*) .   Blakeslee, 1931.
  Solanum Lycopersicum (24 + 2 *ff*)   Lesleys, 1929.
  Crepis tectorum (8 + *f*) . .   Navashin, 1926.
  C. syriaca[2] (10 + 0–8 *ff*) . .   Cameron, 1934.
  Hieracium umbellatum (27 + *f*) .   Bergman, 1934.
  Leontodon hispidus (14 + *f*) .   Bergman, 1935.
 (ii) *Monocotyledons*.
  Lilium Henryi, etc. (24 + 2 *ff*) .   Mather, 1935.
  Fritillaria imperialis[1, 3, 4] (24+1–6 *ff*)   D., 1929 *c*, 1930 *c*.
  F. obliqua, etc. (24 + 2 *ff*) . .   D., 1936 *d*.
  Allium alleghaniense, etc. (14 + *f*)   Levan, 1932.
  Tulipa galatica and hybrids[1, 3, 4] (24 + 1–19 *ff*) . . .   Upcott, unpub.
  Naias marina . . . .   Winge, 1925.
  Crocus hyemalis (6 + 4 *ff*) .   Mather, 1932.
  Spironema fragrans (12 + *f*) .   Richardson, 1934.
  Tradescantia virginiana[1, 3, 4] (24 + 1–6 *ff*).
  T. bracteata (12 + *f*) . .   D., 1929 *c*.
  Secale cereale[1, 4] (14 + 1–4 *ff*) .   Gotoh, 1924 ; Belling, 1925 *a* ; D., 1933.
  Zea Mays[1, 3, 4] (20 + 1–25 *ff*) .   McClintock, 1933 ; Avdulov, 1933.
  Paspalum stoloniferum (20 + 3 *ff*)   Avdulov and Titova, 1933.
  Poa pratensis (67 + 2 *ff*) . .   Rancken, 1934.
  Alopecurus pratensis (28 + 2 *ff*)   Rancken, 1934.
  Festuca pratensis[4] (14 + 1–5 *ff*)   Rancken, 1934.

[1] Inert. [2] Active. [3] Varying in number somatically. [4] Not always pairing at meiosis.

Avena sativa $(41 + f)$ . . . Nishiyama, 1933.
Sorghum verticilliflorum $(20 + 2 ff)$. Huskins and Smith, 1934.
(iii) *Gymnosperms.*
Taxus canadensis $(24 + f)$ . . Dark, 1932.

TABLE 17.

Chiasma Frequencies of Fragments (calculated from metaphase pairing on the chiasma theory of pairing, and compared with those of major chromosomes, D., 1930 c; Philp and Huskins, 1931).

Species.	Mean Chiasma-frequency of Major Chromosomes.		Proportionate Length of Fragment.	Number of Fragments (even or odd multiple).	Chiasma-frequency of Fragment.	
	All stages.	Metaphase.			Calculated.	Observed.
Fritillaria imperialis						
(i) "Yellow"	2·58	—	1/9	6 ff.	0·29	0·22
(ii) "No. 10"	4·96	—	1/9	3 ff ($\times 2/3$)	0·37	0·35
(iii) "Crown upon Crown"	4·33	—	1/9	3 ff ($\times 2/3$)	0·31	0·23
Matthiola incana						
(i) "Smooth"	—	1·76*	3/4	1 f.	1·32	0·78
(ii) "Slender"	—	1·58*	1/2	1 f.	0·75	0·36

* This frequency has been calculated from the behaviour of the trivalent. No correction for reduction of chiasma-frequency in trivalents need therefore be made for the fragments.

These sources of error are not quantitatively definite and the observations lead to the following important conclusions:—

(a) Fragments fail to pair, as predicted, and this failure is evidently due to failure of chiasmata, since clones of *Fritillaria imperialis* with low chiasma frequencies have a lower frequency of fragment-pairing than those with a high chiasma frequency.

(b) Pachytene association of small fragments is probably liable to incompleteness, as expected, since the frequency is always below the expectation on the assumption of completeness. The pairing expectation is exactly fulfilled in *Secale*, where the fragments are larger. The major chromosomes have 2·4 chiasmata per bivalent, the fragments, one-third their length, 0·8. Their pairing frequency is slightly less owing to some fragments having two chiasmata (Fig. 49). In triploid *Hyacinthus* it was inferred that chromosomes

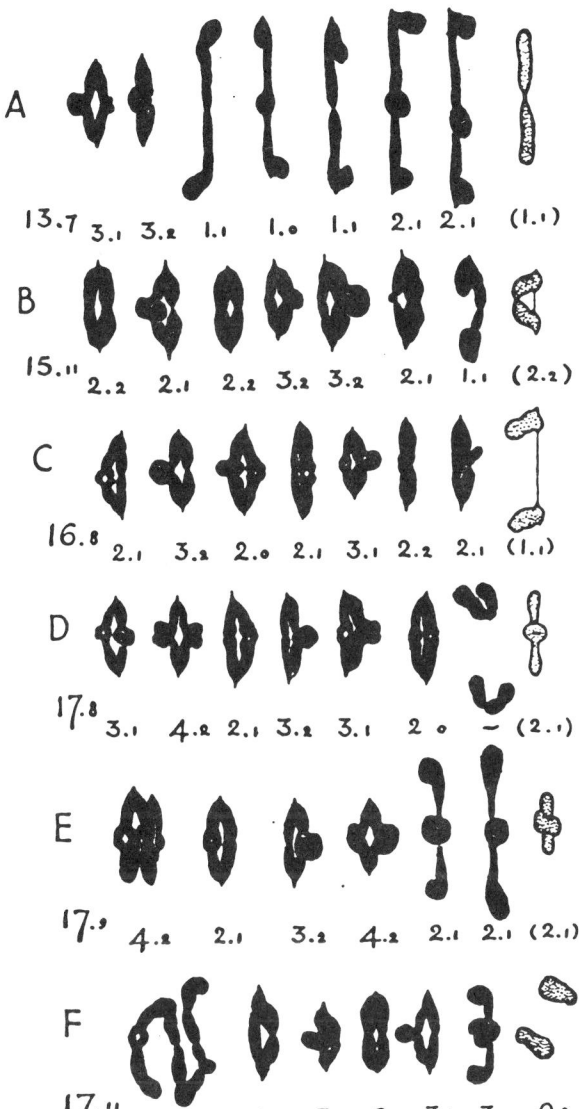

Fig. 49.—Side view of first metaphases in *Secale cereale* ($n = 7$) with one pair of supernumerary fragments. Six different forms of fragment—pair depending on the numbers and positions of its chiasmata. E and F, two mutant nuclei heterozygous for an interchange. × 1200 (D., 1933).

which were paired in part of their length at pachytene might be unpaired at metaphase owing to failure of any chiasmata to be formed. In the fragments of *Secale* as well as in the B chromosomes of *Zea* (Rhoades and McClintock, 1935) there can be no doubt that this is so (*cf.* Erlanson, 1933).

(c) Pairing of large chromosomes with large and fragments with fragments is commoner than cross-pairing, except in *Lilium*. Cross-pairing never occurs in *Solanum* (tetrasomic in respect of the fragment). In *Matthiola*, which is trisomic, the effect is still more drastic, since the odd fragment, deprived in competition of a partner amongst the large chromosomes, is evidently often left out of association and fails to reach two-thirds of the expected frequency. This indicates the importance of *competition* in pairing where there is a choice of partner: those which have the longest sequence of homologous particles will pair disproportionately often. This is particularly important in considering "differential affinity" in hybrid polyploids (Ch. VI) and *Œnothera* (Ch. IX) and pairing in undefined hybrids.

(iii) **Translocation Heterozygotes.** When a portion of one chromosome has been translocated to another the heterozygote shows pairing of the corresponding segments. If one is terminal and the other interstitial, the result is a lateral chiasma. This has been observed in *Tradescantia virginiana*, *Aucuba japonica* and a *Viola* hybrid (J. Clausen, 1931 c). The special behaviour found in the hybrid *Vicia sativa* × *V. angustifolia dolichosoma* (Sveshnikova, 1929, a and b) can also be taken as evidence of translocation. An open chain of four chromosomes is sometimes formed at meiosis, but more usually one or more, sometimes all, of these appear as univalents. The two chromosomes may be represented thus: *abcde* and *fg* in one species, and *abcd* and *efg* in the other. It is to be noted that failure of pairing is particularly frequent in the shortest chromosome (*cf.* Fig. 54).

(iv) **Reduplication Heterozygotes.** In certain trisomic individuals of *Datura*, the so-called "secondary trisomics" (Belling and Blakeslee, 1924 a), the extra chromosome was found to have the same pairing properties at its two ends. It was therefore supposed that this chromosome consisted of two identical segments united at identical ends, *i.e.*, if the parent chromosome was *abcdef* then the new chromosome was *abccba*. It gave a series of configurations in accordance with this assumption. The most interesting of these are the ones in which the two ends of the *same* chromosome are associated, by a terminal chiasma, for they show most clearly that the chromosome does not behave as a unit in pairing and that pairing is not determined by a difference in the origin of the mates. Similar configurations have been found in trisomic *Matthiola* (Philp and Huskins, 1931). This condition has

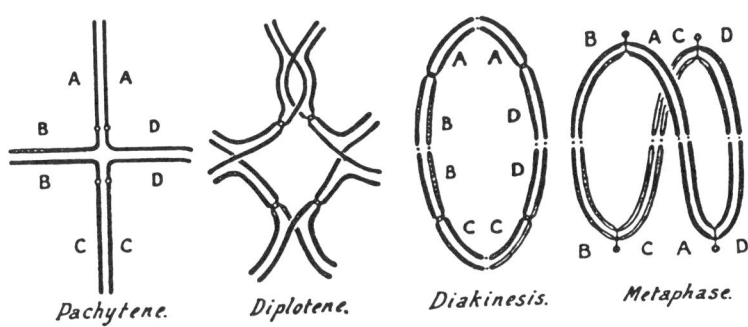

Pachytene.   *Diplotene.*   *Diakinesis.*   *Metaphase.*

FIG. 50.—Diagram showing the normal course of chromosome association in a simple interchange-heterozygote of the constitution *AB-BC-CD-DA* (type, *Campanula*). The A segments form two chiasmata at diplotene, the others one chiasma. Terminalisation is complete at diakinesis. The arrangement is "disjunctional" at metaphase; circles denote the centromeres. (From D., 1931 *c*.)

also been shown in regard to a small segment of the chromosome in tetraploid *Aucuba japonica* (Meurman, 1929 *b*) and *Tradescantia virginiana*.

(v) **Interchange Heterozygotes.** In many diploid plants, both natural races and artificial hybrids (*v.* Table 19) four, six or more chromosomes instead of forming two, three or more pairs at meiosis form a single association. This association, like those in polyploids, is not constant, but is liable to be replaced by smaller ones. Unlike those in polyploids, however, it is limited in its variation, as all associations must be in diploids, by the fact that each part of each chromosome has only one possible partner. Such variation as is observed is due solely to the failure of chiasma formation between some of the homologous pairs of segments.

The simplest condition which can lead to this behaviour is that of a heterozygote whose gametes contain chromosomes that have exchanged segments. Thus if such an exchange occurs in an organism of constitution *AB, AB, CD, CD* (each chromosome can be considered as made up of two segments when only a single structural change is involved) two new chromosome types will result : *BC* and *DA* (or *BD* and *AC*). The heterozygote derived from this interchange will have the constitution *AB, BC, CD, DA,*

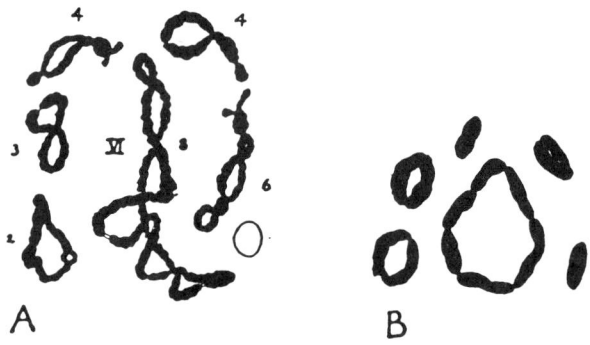

FIG. 51.—The development of a ring-of-six in a double interchange-heterozygote in *Campanula* (of the constitution, *AB-BC-CD-DE-EF-FA*). A, diplotene (with numbers of chiasmata given). B, diakinesis, (6) + 5 (2) = 16. × 2500. (From Gairdner and D., 1931.)

and pairing between all four pairs of homologous segments will give a ring of four when the chiasmata are terminal. Pairing between three of them will give a chain of four, and between two of them either two pairs or a chain of three and a univalent. In this way variations of chiasma formation will lead to variations in the size of the association, even though pairing is regular in the corresponding homozygotes, because the unit of pairing in this heterozygote is no longer a whole chromosome but a segment equal to about half a chromosome. The occurrence of an interchange of segments between non-homologous chromosomes was first inferred in trisomic forms of *Datura*, which will be considered later. In a number of

INTERCHANGE   151

diploids with rings or chains of chromosomes the whole course of meiosis and the special genetic properties of the heterozygote produced have now been determined. They may be summarised as follows :—

(i) Corresponding parts of the chromosomes can be seen to associate in pairs at pachytene, where all of them can be distinguished morphologically. There is an exchange of partner, where the homology changes, like that at pachytene in a tetraploid but obligatory (McClintock, 1931, *Zea Mays*; Levan, 1935, *Allium ammophilum*).

(ii) Chiasmata are formed at random in the paired segments at diplotene (Gairdner and D., 1931, *Campanula persicifolia*).

(iii) These may remain nearly stationary (Pellew and Sansome, 1931, *Pisum sativum*), or they may be terminalised (*Campanula, cf. Œnothera,* Ch. IX), in which case a simple ring or chain of chromosomes joined end to end appears at diakinesis (as found in *Datura, Œnothera,* etc.)

In *Campanula, Rhœo* and *Œnothera* unpaired chromosomes are frequently found owing to failure of association of both their ends. In *Campanula* even a ring of four may be replaced by a pair and two univalents, while in *Rhœo* and *Œnothera* a ring of twelve may break up in any of the conceivable ways—into $11 + 1$, $10 + 2$, $9 + 3$, etc., or $9 + 2 + 1$, $7 + 3 + 2$, etc.

(iv) The configuration arranges itself on the metaphase plate just as a corresponding configuration would in a polyploid. This means that the behaviour of the chromosomes at metaphase is directly determined by mechanical conditions and is not influenced by genetical conditions. Thus an association of four in *Pisum* is arranged " non-disjunctionally " in about half the divisions, *i.e.*, so that chromosomes with associated segments will pass to the same pole (Håkansson, 1931 a). This is just like a similar association of four with interstitial chiasmata in a tetraploid *Hyacinthus*.

A ring of four in *Campanula,* on the other hand, arranges itself " disjunctionally " in about two-thirds of the divisions, *i.e.*, so that pairs of chromosomes with associated segments pass to opposite poles (Fig. 52). The same holds good for a quadrivalent in tetraploid *Primula sinensis*. The reason for this difference of behaviour is fairly evident. An association with interstitial chiasmata is relatively rigid; its shape shows little variation, and is determined by the number and distribution of the chiasmata. It is a three-dimensional system. An association with terminal chiasmata is

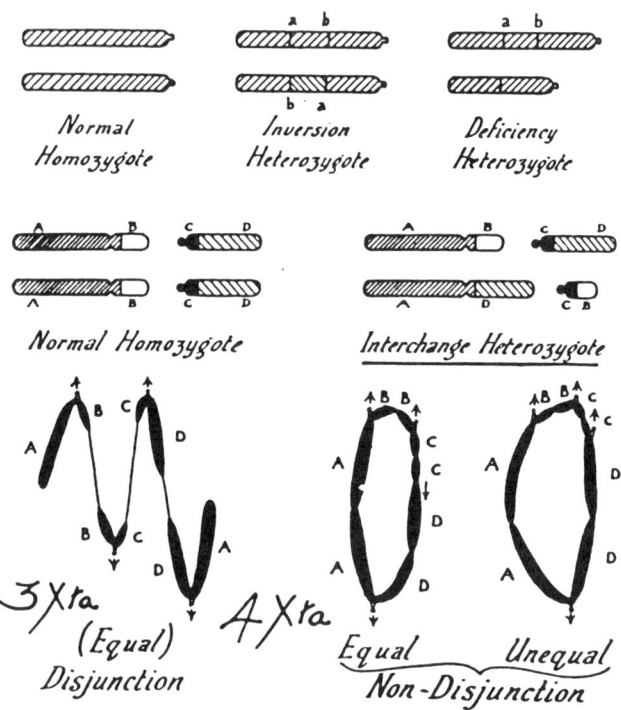

FIG. 52.—Above, structural relations of homozygotes and heterozygotes for different changes. Below, chain and ring formation with different numbers of chiasmata showing different kinds of disjunction.

extremely pliable; the variation found in its shape shows that (as might be expected) any two chromosomes move freely in relation to one another unless they are associated at *both* ends. It is a two-dimensional system. Forces of repulsion associated with the centromeres are therefore able to effect regular disjunction of a

PLATE VI

MEIOSIS IN INTERCHANGE HYBRIDS

FIGS. 1 AND 2.—First metaphase in *Campanula persicifolia* ($2n = 16$).
Fig. 1. Ring of four and six pairs. Fig. 2. Ring of six and five pairs. (Gairdner and D., 1931.)

FIGS. 3 AND 4.—Metaphase in *Rhœo discolor*, rings of twelve chromosomes, showing non-disjunction in side and polar view. (La Cour: medium Flemming — gentian-violet smear, × 2500.)

FIG. 5.—Normally segregating ring of chromosomes at beginning of first anaphase in *Œnothera muricata*, five chromosomes shown.

FIGS. 6–8.—The exceptional occurrence of interstitial chiasmata in *Œnothera biennis* at metaphase and anaphase.
Fig. 7. The critical "figure-of-eight" configuration. Fig. 8. The interstitial chiasma leads to lagging at anaphase (*v.* Chap. IX). (From D., 1931 *d.*) × 6000.

FIGS. 9–11.—The same in *Pisum sativum* with a ring of six. Note the critical chiasma in Fig. 10. (*Cf.* Text-Figure. From E. R. Sansome, 1932.) × *ca.* 3000.

PLATE VI

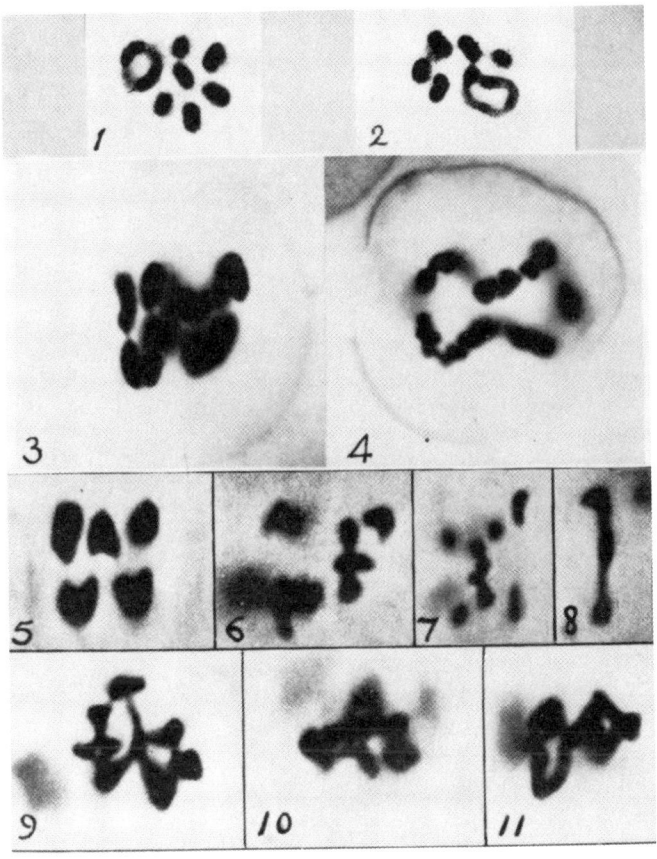

[To face p. 152.

ring with terminal chiasmata although they are powerless to distort a configuration held together by numerous interstitial chiasmata.

It appears that the proportion of non-disjunction is usually about 30 per cent. in rings of 12 or 14 chromosomes, just as it is in a ring of four or six (*cf.* Gairdner and D., 1931). This suggests that orientation of successive chromosomes in the ring usually begins at one part of the ring, and if begun disjunctionally will proceed regularly along the ring (*v.* Ch. XII).

(v) The genetical effects of non-disjunction are important. A ring of four segregates disjunctionally to give gametes *AB*, *CD* and *BC*, *DA*, non-disjunctionally to give *AB*, *BC* and *CD*, *DA*. Although in *Zea* and *Pisum* all the disjunctional products of segregation in the ring are viable, interchange heterozygotes are nevertheless semi-sterile. This is due to the non-viability of the 50 per cent. of non-disjunctional gametes, which lack one of the segments present in the parental complement. In *Campanula*, where there is less non-disjunction, partial sterility of the heterozygote is due to a second cause as well, *viz.*, the non-viability or lower viability of homozygous segregates. This is the probable explanation of the occurrence of the ring-forming heterozygote in nature, in *Campanula* as well as in *Œnothera* : the heterozygote breeds true because the homozygotes are not viable.

(vi) In all interchange heterozygotes so far found in plants the chiasmata are usually formed in both arms of the interchanged chromosomes. In *Trimerotropis citrina* (Carothers, 1931) a ring of four has been found at first metaphase, which is evidently due to the individual being an interchange heterozygote. But here the centromeres are near the ends, and when the ring of four is formed each chromosome has chiasmata with two other chromosomes on the *same* side of the centromere. The ring therefore lies in the equatorial plane and is never disjunctional. The significance of this will be considered later (Fig. 89, Ch. VII).

(vii) The fact that interchange heterozygotes, having four, six or more chromosomes in a ring at meiosis, occur in nature gives a great interest to the origin and inheritance of the condition. They are on our hypothesis hybrids, produced by the union of dissimilar gametes. This dissimilarity is due to interchange taking place in an ancestor. There is no direct evidence to show whether the interchange gave rise to the heterozygote at once in nature or whether it gave rise to a new homozygous race which afterwards by crossing gave rise to the hybrid. The second is obviously the source of those ring-forms that have appeared in experiments. There is some reason to suppose, on the other hand, that where they occur in nature their preservation is due to the elimination of the homozygous products of the original interchange hybrid (*v. Œnothera*), *i.e.*, that they arose directly from the change without crossing between different parental zygotes.

In experiments with *Datura*, *Pisum* and *Œnothera*, hybrids with a

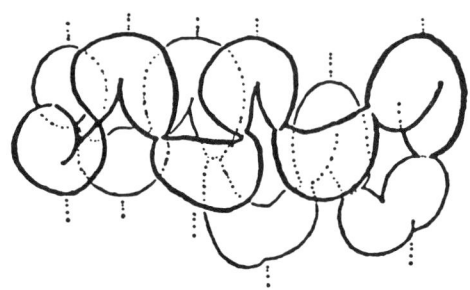

FIG. 53.—A ring of twelve arranged disjunctionally at first metaphase in *Rhœo discolor*. × 2800. (From D., 1929 *c*.)

ring of four have been obtained by crossing different homozygous races. The parents evidently differed in respect of one interchange, and the offspring (of the type $AB\text{-}BC\text{-}CD\text{-}DA$) may therefore be described as " single-interchange " heterozygotes. Hybrids with a ring of six or two rings of four are similarly " double-interchange " heterozygotes, and they have been produced either by crossing two parents which differed in two interchanges (AB, CD, EF and BC, DE, FA) as in *Datura*, or by crossing two single-interchange heterozygotes, which have one gametic type in common (as in *Pisum*, *Zea* and *Campanula*). From such a cross four kinds of progeny will result. Thus $(AB\text{-}BC\text{-}CD\text{-}DA, EF\text{-}EF) \times (AB\text{-}AB, CD\text{-}DE\text{-}EF\text{-}FC)$ will give homozygotes with simple pairing (*i.e.*, $AB\text{-}AB$, $CD\text{-}CD$, $EF\text{-}EF$), two kinds of " single " heterozygote

MULTIPLE HETEROZYGOTES

(*i.e.*, the parental types) and one kind of " double " heterozygote with a ring of six (*i.e.*, *AB-BC-CF-FE-ED-DA*). The proportions of the cytologically recognisable types will be 1 : 2 : 1, and the inheritance can be described in mendelian terms, if each interchange is regarded as making an allelomorphic difference. Thus the cross

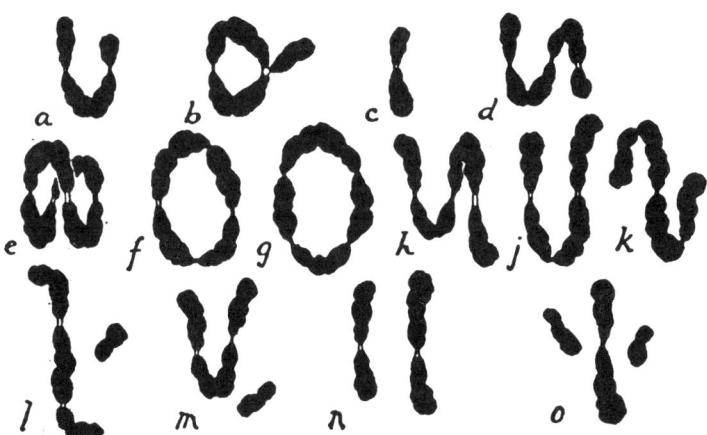

FIG. 54.—Configurations observed at first metaphase in diploid structural hybrids of the simpler kinds with complete terminalisation. *a*, fragmentation or fusion heterozygote (*e.g.*, *Vicia sativa* × *amphicarpa* or *Hesperotettix*). *b*, fragmentation-plus-reduplication heterozygote (*e.g.*, *Matthiola*); note triple chiasma. *c*, deficiency or reduplication heterozygote (*e.g.*, *Circotettix*). *d*, translocation heterozygote (*e.g.*, *Tradescantia bracteata* or *Vicia sativa* × *angustifolia*). *e—o*, simple interchange heterozygote (*e.g.*, *Datura*, *Campanula*) showing complete association in a ring of four and incomplete association in one or more chains. The chromosomes are represented as of three sizes—one long, two intermediate and one short in a constant order in the ring. *e* and *h* are disjunctional arrangements; *f*, *g*, *j*, *k* and *l* are non-disjunctional, the rest are doubtful.

is analogous to $AaBB \times AABb = 1\ AABB : 1\ AaBB : 1\ AABb : 1\ AaBb$ where $AaBB$ and $AABb$ are indistinguishable unless the chromosomes taking part in the rings can be distinguished (Fig. 52, Plate VI).

Where the interchanges in the two parents do not affect any chromosomes in common, the " double " heterozygote has two rings of four instead of one ring of six. Both these expected results have been found in *Campanula*, but in some crosses the homozygous forms, with simple pairing, are eliminated as mentioned above.

(vi) **Polyploid Interchange Heterozygotes.** A relatively interchanged condition was first found in the odd chromosome of a trisomic *Datura* (Belling and Blakeslee, 1924). This happens in two ways. First, the extra chromosome may contain two identical segments and may be supposed to be derived from *internal interchange*, *i.e.*, between non-homologous arms of the same chromosome or of homologous chromosomes. Thus two chromosomes *abc, def* and *abc, def* might give new types *abc, dcb* and *fec, def*, one of which in addition to the normal complement would make a " secondary trisomic " (Rhoades, 1933). Secondly, the extra chromosome may result from interchange of a segment of two complementary chromosomes. This type gives a " tertiary trisomic." Thus *AB* and *CD* will give *BC* and the trisomic will be made of *AB, AB, BC, CD, CD*, etc. Such a plant has various configurations, including a chain of five chromosomes.

In connection with these forms it is convenient to mention certain reduplication homozygotes, *i.e.*, organisms that are tetrasomic in respect of one segment, having four chromosomes that may be represented *abcd, abcd, efgd, efgd*. Since all four chromosomes can associate in the *d* segment, groups of four can be formed and, with terminalisation, a quadruple chiasma. Such configurations are found in *Datura* hybrids (Blakeslee, 1928) and probably in *Rosa* (Erlanson, 1931 *b*), *Matthiola* (Philp and Huskins, 1931), and *Tradescantia bracteata* (D., 1929 *c*). Probably configurations suggesting the same conclusions in *Œnothera biennis* (Cleland, 1926 *a*) are to be attributed to interlocking (*q.v.*).

Polyploid and polysomic interchange heterozygotes combine the large configuration made possible by their hybridity in structure with the variability in forming possible configurations which is characteristic of polyploids. Only the observation of " maximum association " (which is never found in complicated cases) or a knowledge of the ancestry can give the key to the constitution. The determination of their structure, therefore, requires extensive observation, and has only been conclusively carried out in a few forms.

A number of triploids have more than the haploid number of paired chromosomes at meiosis, although they are derived from diploids showing no evidence of a polyploid origin (*cf.* Ch. VII), *e.g.*, in *Festuca*, *Fragaria* and *Triticum* with basic numbers of seven.

This can be accounted for in one of two ways : the diploid from which they have been derived may have been either (i) an interchange heterozygote or (ii) a reduplication homozygote. The triploid would have a segmental constitution in the two cases as follows :—

(i) *AB CD*
    BC DA
    BC DA . . .

(ii) *AB BC*
    AB BC
    AB BC . . .

Either type can form three pairs instead of the expected two trivalents.

Since all the triploids in which this behaviour is found are interspecific hybrids of *ABC* or *AAB* types (*cf.* Ch. VI), or, in *Festuca*, *Œnothera* and *Campanula*, the offspring of interchange heterozygotes, the heterozygote rather than the homozygote explanation is favoured. Homozygous reduplication may also have contributed to the result in some cases.

A clearly defined heterozygote is the triploid derived from *Œnothera pycnocarpa*. The parent has a ring of 14 chromosomes which are presumably of the segmental constitution :

$$AB \quad CD \quad EF \quad GH \quad KL \quad MN \quad OP$$
$$BC \quad DE \quad FG \quad HK \quad LM \quad NO \quad PA$$

the upper row of chromosomes being derived from one gamete, the lower from the other.

The fertilisation of an unreduced gamete (with the constitution of the zygote) by a normal gamete has given a triploid of the constitution :—

        AB CD EF
        AB CD EF
           BC DE FG, etc.

This gives all the expected small configurations (like those in the tertiary trisomic *Datura*), but none with the maximum association (including all 21 chromosomes), since this would involve the apparently unlikely chance of 14 of the chromosomes forming two chiasmata in each arm (Catcheside, 1931 a).

Several tetraploid species show by their pairing that they are interchange heterozygotes, but for the reasons stated above an exact constitutional formula cannot be given with certainty. In these

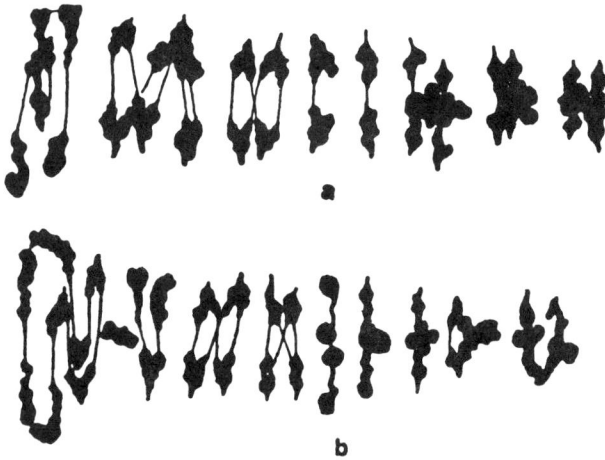

FIG. 55.—First metaphase in *Aucuba japonica* pollen mother-cells ($4x = 32$). *a*, two sixes, four fours, and two pairs. *b*, one eight, three fours, one three, four pairs, and one unpaired chromosome. × 2400 (from Meurman, 1929).

tetraploids an association of more than four chromosomes indicates that the plant is heterozygous for one interchange, and an association of more than eight, or two of more than four but not more than eight, indicates that the plant is heterozygous for two interchanges, and so on. The following table shows the percentages of configurations with different numbers of chromosomes in three such tetraploid species. It will be noticed that occasional unpaired chromosomes are characteristic of the polyploid as of the diploid interchange heterozygote.

TABLE 18
Pairing in Tetraploid Interchange Heterozygotes

Configurations.	I.	II.	III.	IV.	V.	VI.	VII.	VIII.	IX.	X.	XI.	XII.
Anthoxanthum odoratum ($x = 5$). Kattermann, 1931 (1120 chrs.).	5.8	59.9	2.4	20.4	0.7	4.8	0.0	3.8	0.0	1.4	0.0	1.4
Rosa relicta ($x = 7$). Erlanson, 1931 (380 chrs.).	1	62	3	23	2	2	2	5	0	0	0	0
Aucuba japonica ($x = 8$). Meurman, 1929 a (2496 chrs.).	0.3	34.9	0.8	52.7	0.1	5.2	0.1	4.8	0.1	0.8	0	0

Similar behaviour is found in *Tradescantia virginiana* (Stow, 1928; D., 1929 c) and *Campanula persicifolia* (Gairdner and D., 1931). In both of these tetraploid forms of various constitutions probably occur. The constitution of the tetraploid forms of *Œnothera* can be specified from knowledge of the behaviour of the diploids from which they are derived, and they show the types of configuration that are to be expected on these grounds (Håkansson, 1926; Catcheside, 1932 and 1935; Seitz, 1935).

TABLE 19
Classification of Interchange Heterozygotes

(a) DIPLOIDS.
  Artificial Hybrids.
    *Datura** ($n = 12$) . . . . Blakeslee, 1928, 1929.
    Pisum ($n = 7$) . . . . { Håkansson, 1929, 1931, 1934.
                                 E. R. Sansome (Richardson), 1929, 1932, 1933; Sutton, 1935. }

    Polemonium reptans (race cross)
    P. mexicanum × *P. pauciflorum* } J. Clausen, 1931 a.
    P. filicinum × *cæruleum*
    *Canna** . . . . . . Honing, 1928.
    Godetia amœna ($n = 7$) × *G. Whitneyi**
      ($n = 7$) . . . . Håkansson, 1925, 1931 b.
    *Œnothera** (4) and (6) . . . Catcheside, 1932.

  Spontaneous Forms.
    *Campanula persicifolia** ($n = 8$)
      (4) and (6) . . . . Gairdner and D., 1931.
    Œnothera spp.** ($n = 7$) (14) . Cleland, 1922–28; D., 1931 c.

* Chiasmata all terminal apart from those arrested by change of homology, *cf.* Ch. XII. Numbers in brackets are the numbers of chromosomes in a ring.

STRUCTURAL HYBRIDS

Rhœo discolor* ($n = 6$) (12)	Belling, 1927 ; D., 1929, a and c ; Nebel, 1932 ; Sax, 1931 ; Koller, 1932 c.
Humulus japonicus* ♂	Kihara, 1929 b (v. sex chromosomes).
H. lupulus* ♂	Sinoto, 1929 (v. sex chromosomes).
Briza media*	Kattermann, 1930, 1931.
Zea Mays (in experiment) (4) and (6)	Burnham, 1930 ; McClintock, 1931.
Eucharidium concinnum*	Schwemmle, 1926.
Clarkia elegans* ($n = 9$), 2 plants (4) + 7 (2)	Håkansson, 1931 b.
Polemonium cæruleum	J. Clausen, 1931 a.
Hypericum punctatum* ($n = 8$) (16)	Hoar, 1931.
Matthiola incana ($n = 7$)	Philp and Huskins, 1931.
Rosa forms ($n = 7$)	Erlanson, 1931 b.
Trimerotropis citrina ($n = 12$)	Carothers, 1931.
?Diaptomus castor	Heberer, 1924.
Secale cereale ($n = 7 + f$) 8 cells : (4) + 5 (2) + ff.	D, 1933
Allium ammophilum ($n = 8$), (4) + 6 (2)	Levan, 1935.
Rumex acetosa ($2n = 14 + XY_1Y_2$, (4) + 5 (2) + XY_1Y_2†	Yamamoto, 1935.
Crocus chrysanthus ($n = 4$), (4) + 2 (2)	D., unpub.
Festuca pratensis (4) + 5 (2)	Rancken, 1934.

Artificial Hybrids.

Vicia sativa (6) × V. angustifolia (6). (4) + 2 (2)	Sveshnikova, 1929 b.
Datura Stramonium (12) × (12). (10) + (2)	Bergner and Blakeslee, 1935 ; cf. Blakeslee, 1928.
Pisum humile × P. arvense (4) + 5 (2)	Håkansson, 1936.
Triticum ægilopoides (7) × Æ. squarrosa (7). (3) + 3^{II} + 5^I	Kihara and Lilienfeld, 1935.
Triticum ægilopoides baidaricum (7) × T. æ. stramineonigrum (7) (4) + 5 (2)	L. Smith, 1936.
Allium Cepa × A. fistulosum (4) + 6 (2)	Levan, 1936.

Progeny of X-rayed Stock (cf. Ch. X).

Drosophila melanogaster ($n = 4$)	Muller, 1930 ; Dobzhansky, 1931 ; Painter, 1934.
Circotettix verruculatus ($n = 12$)	Helwig, 1933.
Apotettix eurycephalus (X-autosome)	Robertson, 1936.
Zea Mays ($n = 10$)	McClintock, 1933 ; E. G. Anderson, 1935 ; D., 1934.
Triticum monococcum ($n = 7$)	Katayama, 1935 a.
Œnothera blandina ($n = 7$)	Catcheside, 1935.

* Chiasmata all terminal apart from those arrested by change of homology, cf. Ch. XII. Numbers in brackets are the numbers of chromosomes in a ring.
† Said to be reciprocal, but this cannot be proved.

(b) TRIPLOIDS.
  Zea Mays (10) × Euchlæna perennis (20), $2^{III} + 10^{II} + 8^{I}$. Longley, 1924.
  Artemisia nitida ($3x = 27$), 9-13^{II}. Chiarugi, 1926 ; cf. Rosenberg on Hieracium, 1917, and Ch. XI.
  Ægilops ovata (14) × Æ. caudata (7), 7-10^{II}. Bleier, 1928.
  Ægilops speltoides (7) × Triticum turgidum (14). Jenkins, 1929.
   Bivalents   .   .   .   4   5   6   7   8   9   10
   Frequency   .   .   .   1   5   9   71 15 1   1
  Avena barbata (14) × A. strigosa, $9^{II} + 3^{I}$. Nishiyama, 1929. (N.B. lateral chiasma).
  Triticum ægilopoides (7) × T. dicoccum (14), $1^{IV} + 1^{III} + 4^{II} + 6^{I}$. Kihara and Nishiyama, 1930.
  Fragaria bracteata vesca (14) × F. collina (7), 10^{II}, etc. Yarnell, 1931 b.
  Œnothera pycnocarpa ($3x = 21$) (see text). Catcheside, 1931.
  Campanula persicifolia ($3x = 24$), $1^{IV} + 4^{III} + 3^{II} + 2^{I}$. Gairdner and D., 1931.
  [Festuca pratensis (7) × Lolium perenne (7)] × L. perenne ($3x = 21$), 3^{III}, 6^{II}, etc. Peto, 1934.
  Potentilla chinensis (7) × P. nipponica (14). Araki, 1932.
  Triticum dicoccum (14) × T. monococcum (7). Mather, 1935.
  Narcissus Tazetta ($3x = 30$). Nagao, 1933.
  Petunia ($2x \times 4x$, $x = 7$). Matsuda, 1935.

(c) TETRAPLOIDS.
  Tradescantia virginiana ($x = 6$)   . D., 1929 c.
  Campanula persicifolia ($x = 8$)   . Gairdner and D., 1931.
  Aucuba japonica ($x = 8$)   . Meurman, 1929 b.
  Rosa relicta ($x = 7$)   . Erlanson, 1931 b.
  Œnothera Lamarckiana ($x = 7$)   . Håkansson, 1927.
  Anthoxanthum odoratum ($x = 5$)   . Kattermann, 1931 ; cf. Avdulov, 1931.
  Setcreasia brevifolia ($4x = 24$)   . Richardson, 1935.
  Œnothera biennis gigas ($x = 7$)   . Seitz, 1935.

(d) HEXAPLOID.
  Rumex acetosella ($x = 7$) (association of 8) . . . . . Kihara, 1927.

(vii) **Unequal Pairs of Chromosomes.** In accordance with the foregoing classification of structural hybrids it is possible to distinguish provisionally between four kinds of condition which give rise to the occurrence of unequal pairs of chromosomes at meiosis.

TABLE 20

Classification of Unequal Bivalents

1. DEFICIENCY, REDUPLICATION OR TRANSLOCATION HETEROZYGOTES.
  A. Animals.
   Forficula auricularia   . . { Payne, 1914.
                                 W. P. Morgan, 1928.
   Brachystola .   .   .   .   . ⎫
   Arphia   .   .   .   .   . ⎬ Carothers, 1913.
   Dissosteira .   .   .   .   . ⎭

Phrynotettix magnus * . . . Wenrich, 1916.
Acridium . . . . . Robertson, 1915.
Circotettix verruculatus . . . Carothers, 1931 ; Helwig, 1929.
Mus musculus (mutant) . . Painter, 1927 (deficiency).
Labidura bidens . . . . W. P. Morgan, 1928.
Mecostethus gracilis . . . McClung, 1928 b.
Stenobothrus lineatus . . . Belar, 1929 a.
Hemerobius stigma . . . Klingstedt, 1933.
Stauroderus bicolor . . . D., 1936 d.
Melanoplus femur-rubrum . . Hearne and Huskins, 1935.
B. Plants.
Ægilops ovata × Triticum polonicum (4n) . . . . Kagawa, 1929.
Æ. cylindrica × Æ. triuncialis (4x) Kagawa, 1931.
Bougainvillæa glabra . . . Cooper, 1931.
Zebrina pendula (4x) . . . D., 1929 c.
Crepis hybrids . . . . Babcock and J. Clausen, 1929.
Aloë purpurascens * . . . Belling, 1931.
Triticum vulgare . . . . Huskins and Spier, 1934.
Pæonia officinalis * . . . Dark, 1936.
Allium Cepa × A. fistulosum * . Levan, 1936.

2. FRAGMENTATION HETEROZYGOTES.
Vicia sativa × V. amphicarpa . Sveshnikova, 1929.
Crepis leontodontoides × C. marschalli (4x), etc. . . Avery, 1930 (Fig. 56).

3. INTERCHANGE HETEROZYGOTES.
Œnothera biennis . . . . D., 1931 d.
Tradescantia virginiana . . D., 1929 c.
Anthoxanthum odoratum . . Kattermann, 1931.
Orgyia thyellina × O. antiqua . Cretschmar, 1928.

4. COMPLEX AND SEX HETEROZYGOTES (v. Ch. IX).

The differences between these chromosomes can be accounted for by the assumption of simple structural changes in a way that will explain their behaviour at meiosis. Translocation, for example, affords a simple explanation where, as in *Circotettix*, there is a difference in position of the spindle attachment without difference of size. But it is probable, more especially in species hybrids, that numerous structural differences distinguish the chromosomes. This is clear in the hybrids of *Lycia zonaria* ($n = 56$) by *L. hirtaria* ($n = 14$) (Harrison and Doncaster, 1914), *Lycia pomonaria* ($n = 51$) by *L. hirtaria* ($n = 14$) (Malan, 1918), and *Orgyia thyellina* ($n = 11$) by *O. antiqua* ($n = 14$). In each cross the chromosomes of one species are smaller and more numerous than those of the other. Fragmentation or fusion is therefore to be inferred. But in the

* Having symmetrical lateral chiasmata.

CAUSES OF NON-PAIRING

Orgyia hybrid not only associations of three (as in *Phragmatobia*) but also of larger numbers are found. Their details cannot be made out, but interchange or translocation is to be inferred. In a *Dicranura* hybrid definite chains can be made out (Federley, 1915).

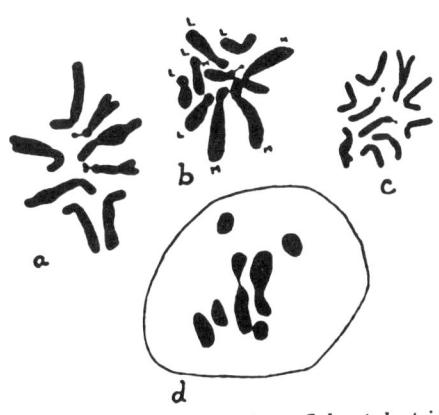

FIG. 56.—*a, Crepis marschalli*, $2n = 8$. *c, C. leontodontoides*, $2n = 10$. *b*, the hybrid at mitosis. *d*, the hybrid at first metaphase ; two unequal pairs and five unpaired chromosomes. × 1800. (*a* and *c* from Babcock and Navashin, 1929 ; *b* and *d* from Avery, 1930.) Note that the chromosomes of *C. leontodontoides* are larger in the hybrid owing to a change in genotypic control.

Closer study of many hybrids amongst those that will be classified as "undefined" would probably reveal analogous results and make it possible to specify some of the differences that constitute their hybridity.

3. THE CONDITIONS OF METAPHASE PAIRING

Before turning to more complicated kinds of hybrids we shall do well to consider how far observations of simpler kinds will help us. These show the following conditions of chromosome pairing.

(*a*) Where the chromosomes of a diploid organism can be seen at mitosis to consist of different pairs of identical chromosomes, the chromosomes associated in pairs at meiosis are always seen to be identical. Since the pairs segregate at random with regard to one another their members must be identical chromosomes and pairing must be a criterion of their common ancestry, that is, of their homology (Montgomery and Sutton). This conclusion is borne out by the usual absence of pairing in haploid plants (Ch. X).

(*b*) In tetraploids, as also in triploids, pairing takes place between corresponding parts of the pairing chromosomes at random, so that the chromosomes often change partner. Pairing is therefore not a property of the chromosome as a whole but of its constituent parts, and parts of chromosomes in structural hybrids pair according to their individual similarities with parts of other chromosomes.

(*c*) The pairing of chromosomes is subject to the conditions of meiosis being normal; and these conditions can be modified by genetic factors, the abnormal factor being recessive in the known cases (Ch. X). Meiosis, as we shall see later, is one of the most susceptible of all physiological processes to genetic abnormality (Ch. X). Many F_1 hybrids which betray no abnormality in any other respect (beyond hybrid vigour) go wrong in meiosis. Where this is due to errors in spindle behaviour we know it is not inherent in the structural differences of the chromosomes ; it must be genotypically controlled. But where it follows a failure of chromosome pairing this failure may be either structurally or genotypically controlled.

It follows that failure of pairing is not in itself evidence of dissimilarity of the chromosomes concerned, although pairing, which gives rise to chiasma-formation, is evidence of similarity.

(*d*) Likewise, genetic conditions may determine a failure to pair of *particular* members of the complement (as in the microchromosomes of Hemiptera, *v.* Ch. X), or of certain parts of the chromosomes (as in the distal parts of the chromosomes of *Mecostethus*). Such failure is certainly not the consequence of dissimilarity, because these chromosomes and parts of chromosomes are inherited in the same way as the rest of the complement.

(*e*) It has been seen that, while pachytene pairing is the first visible condition, chiasma formation is to be regarded as a second and immediate condition of the metaphase pairing of chromosomes. Therefore the interpretation of chromosome pairing in hybrids at metaphase must take into account the frequency and distribution of chiasmata in the parent. First, pairing in a hybrid of a type with localised chiasmata will be dependent on the identity of the parts in which the chiasmata are formed, and trivalents in a triploid will be less frequent if chiasmata are localised. This has not yet been proved. Secondly, proportionate dissimilarities will reduce pairing more considerably in a type with low than in one with high chiasma frequency, a respect in which clones of *Fritillaria imperialis* differ (cf. Ch. VII), and in short chromosomes more than in long ones. Thirdly, where pairing fails as a result of structural hybridity it will vary in the extent of its failure in different nuclei, following a unimodal curve such as that found in trivalent formation and in chiasma frequency itself. It will also show variation in different individuals from other conditions affecting chiasma frequency (cf. Ch. VII and Fig. 57).

(*f*) The occurrence of interlocking and its origin as shown by Gelei (v. Ch. VII) and the occurrence of exchanges of partner at zygotene in polyploids (D., 1929 *c*) indicate that the accidents of relative position may reduce the completeness of pairing. These observations apply equally to pairing in translocation-heterozygotes, for the same exchanges of partner occur optionally amongst the homologous chromatids of non-hybrid polyploids as occur necessarily amongst the interchanged chromosomes of the heterozygotes. It is therefore not surprising to find in certain polyploids a considerable reduction of chiasmata, and, with terminalisation, many more free ends than in the corresponding diploids. This is increasingly evident in the series : *Datura Stramonium* (Belling, 1927), *Primula sinensis* (D., 1931 *a*), *Tradescantia virginiana* (D., 1929 *c*), *Campanula persicifolia* (Gairdner and D., 1931).

Interference of one chromosome with another at zygotene is greater in its effect on the longer chromosomes. Thus in triploid *Hyacinthus* the long chromosome type has a much wider variation in the length paired at zygotene than the short and medium chromosomes. It pairs as though it consisted of only two blocks which pair, or do not pair, at random with regard to one another. It therefore has a higher proportion of unpaired chromosomes at metaphase than the medium type, in spite of its higher chiasma frequency (v. Ch. IV).

In hybrids with some polarisation of the nucleus at zygotene, interference with pairing from structural hybridity will give an advantage to the ends of the chromosomes (which pair first) in pairing, and chiasmata will be formed chiefly near the ends. This is probably responsible for an increase in the frequency of terminal association in *Triticum* hybrids in proportion to the reduction in frequency of pairing. The terminal association has been described as " loose " pairing in contradistinction to " close " pairing where the pairs have formed several interstitial chiasmata. It is directly determined by the formation of fewer chiasmata, and those nearer the ends than usual (Kihara and Nishiyama, 1930 ; *cf*. D., 1930 *c* and 1931 *d*, and Ch. VI).

We have seen that univalents, the result of incomplete pachytene pairing, often occur with permissive changes of pairing in a tetraploid. This is due to the time-limit in pairing which is responsible for certain parts of long chromosomes failing to get paired before they divide, and for the localisation of pairing characteristic of many species with long chromosomes. *A fortiori* we should expect that incomplete pachytene pairing would occur in any organism with frequent obligatory changes of partner resulting from a complicated structural hybridity. Evidently the reduction of metaphase pairing in undefined hybrids between species is in part attributable to the action of the time-limit.

There is another special way in which pairing is interfered with at pachytene in a structural hybrid. This has been made clear by the work of McClintock on *Zea Mays*. She discovered (1933) in defined interchange, inversion, and deletion hybrids a new principle in chromosome pairing. Chromosomes which have begun to pair at places where they are homologous often continue their pairing along their length irrespective of homology. They pair, it seems, like two pieces of twisted string set free together. The pairing begun by attraction continues by torsion, by the same strain, that is, which produces relational coiling in any case (D., 1935 *c*, 1936 *d*). This means that if we take the chromosomes of two different species which differ perhaps in numerous small details of arrangement, the greater part of their pairing in the hybrid will be between non-homologous segments. Such pairing does not permit of crossing-over and chiasma-formation so far as we know, and the chromosomes will therefore fall apart unpaired at diplotene. It does not therefore account for metaphase pairing in haploids. From these two circumstances the structurally determined failure of pairing in interspecific hybrids is to be derived.

Summing up: The foregoing observations show three general conditions of the pairing of chromosomes at metaphase of the first meiotic division in diploids, as follows :—

(i) The presence of similar pairs of chromosomes.

(ii) Non-division of the chromosomes permitting their pairing at pachytene.

(iii) The formation of chiasmata in the paired pachytene chromosomes.

In polyploids the pairing of a particular two similar chromosomes depends on a combination of the chance of assortment of them, determined by the number of pairing blocks of which they consist, and the chance of chiasma formation determined by the average chiasma frequency and the interference value.

Two questions then arise : (*a*) What kinds of dissimilarity inhibit or restrict chromosome pairing at pachytene ? (*b*) What relation has metaphase pairing (which can alone be considered in detail) to the degree of dissimilarity occurring ?

(*a*) There is evidence of two kinds of dissimilarity or differentiation in the chromatin material. First, there is a qualitative differentiation between particles or groups of particles. This is most simply supposed to be due to differences between their characteristic molecules. A second symptom of these differences is the capacity for specific variation, the property of particles at each locus in a chromosome giving rise to differences (mutations) of a specific character, which has led to such an identifiable particle being described as a " gene." It is probably through gene mutations that the qualitative (intramolecular) differentiation arises. Secondly, there are differences due to change in arrangement or structure of these specific particles ; these intermolecular changes determine genetic differences when they lead to changes in quantity, proportion and position.

Since intragenic changes affect single genes and intergenic changes affect larger or smaller groups of genes (perhaps many hundreds), though both may inhibit pairing, on the simple assumption that only identical genes may pair, the first will have a negligible effect as compared with the second. Therefore the assumption must be made that all changes effective in reducing pairing, apart from changes of number, are intergenic or structural.

Structural changes need not in themselves mean any change of genetical properties in the organism, except in a mechanical sense, since the same materials may be arranged in a different way without any phenotypic effect. But structural changes will nearly always lead to changes in proportion through recombination, and these have an important genetic significance. Their occurrence is therefore a measure of differentiation.

(b) Since organisms vary (within and between species) in the frequency and distribution of chiasmata in each unit of the length of paired chromosome, it follows that the pairing in different hybrids between species and between varieties of the same species cannot be compared as an indication of similarity without a knowledge of chiasma frequencies in the parents.

We are therefore led to expect that, in some hybrids between species, chromosomes will pair, in others they will fail to pair, and, in others again, their behaviour will be intermediate and will then vary in accordance with the variation in chiasma formation. The differences of behaviour between hybrids will have a very indirect relation to the differentiation in genetic properties of the parent species.

4. CHROMOSOME PAIRING IN UNDEFINED HYBRIDS

Undefined hybrids in which parental gametes have the same chromosome number, or a slightly different one, for no definable reason, are of three kinds: (i) those with little or no pairing at metaphase; (ii) those with partial, and, always, variable pairing; (iii) those with complete or almost complete pairing (*v.* Table 21). Prophase conditions have not been suitable for exact study in most of these hybrids, but failure of metaphase pairing has frequently been found to follow incomplete pachytene pairing, as would be expected (*cf.* Cretschmar, 1928). In other cases pairing appears to be complete at pachytene (Federley, 1915), but it may well be intermittent and appear complete as it sometimes does with localised pairing, owing to the relational coiling of the chromosomes in their intercalary unpaired parts.

The second class is characterised by a unimodal curve in the frequency of the pairing (v. Table 22). This curve shows some variation even between different preparations of the same individual, and *a fortiori* between sister individuals from the same cross. Such variation is paralleled by that found in chiasma frequencies shown

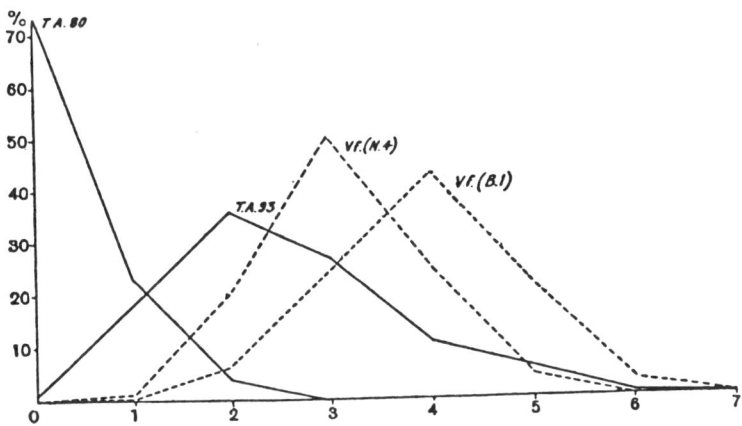

FIG. 57.—Percentage frequencies, $T.A.$, of pairs of chromosomes per nucleus in a *Triticum-Ægilops* hybrid (Kihara, 1929 c) from different preparations and $V.f.$ of chiasmata per bivalent in the *m* chromosomes of *Vicia Faba* (Maeda, 1930), also from different preparations showing the effect of different developmental conditions on these two variables. (From D., 1931 c.)

by the same chromosome pair in different preparations in non-hybrids (*cf.* D., 1931 c, on Maeda, 1930, and Fig. 57).

In the second and third classes it is also found that wherever the chiasma frequency of the parents is known or can be inferred, it is found to be lower in the hybrid. Thus in *Canna*, *Triticum* and *Papaver* two or three chiasmata are formed by the bivalents in the parental species (*cf.* Table 23, and Peto, 1930; Goodspeed, 1934). But in hybrids in these genera or between *Triticum* and *Ægilops*, where the pairing is small and variable, such pairing as occurs is

170   STRUCTURAL HYBRIDS

by single (usually terminal or sub-terminal) chiasmata (Figs. 58 and 72). The same is true in a less degree of a defined structural hybrid in *Lilium* (Richardson, 1936, and Ch. VII).

It will be observed that the difference between the two degrees

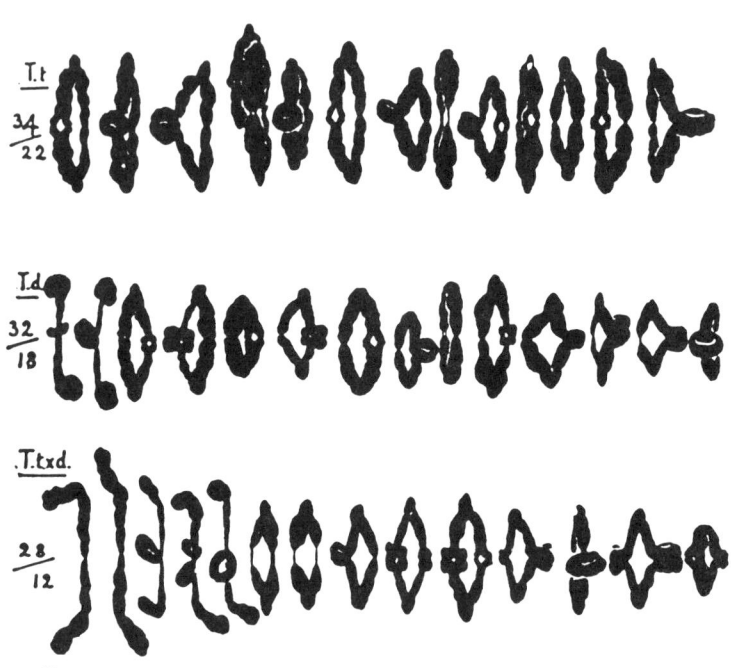

FIG. 58.—First metaphase of meiosis in *Triticum turgidum* ($2n = 28$), *T. durum* ($2n = 28$) and the hybrid between them. The total number of chiasmata and the number terminal is given at the side of each plate. *Cf.* Table 23, smear preparations: The top plate is fixed in aceto-carmine ($\times$ 1900), the rest in Flemming ($\times$ 2800). (From D., 1931 d. *Cf.* Fig. 57.)

of pairing is the same as that observed between the pairing of fragments which form only one chiasma and their large sister chromosomes which form two or more (*e.g.*, in *Fritillaria*, D., 1930 c, and *Datura*, Blakeslee, 1931). The pairing found in undefined hybrids is therefore in accordance with the assumption that a larger or smaller fraction of the total length of chromosome threads is paired at pachytene; a small reduction leading to rare failure of chiasma formation and metaphase pairing; a larger reduction, to more frequent but variable failure of pairing; and a very large reduction to almost invariable failure of pairing. The conditions that determine this reduction are clear from the genetical and cytological evidence of the conditions of pairing just considered. It is that corresponding chromosomes do not agree in structure.

If mere differences in the arrangement of the same materials are responsible for failure of pairing (and these differences do not in themselves imply differences in the genetic properties of the chromosome set) then frequency of pairing at pachytene is an uncertain indication of the genetic relationship of the parents. And since metaphase pairing is related to pachytene pairing through the formation of chiasmata, in the frequency of which species and varieties differ genetically, the indication becomes even more uncertain. This conclusion is verified in a general way by a consideration of chromosome pairing in hybrid Lepidoptera (*cf*. Cretschmar, 1927). Precise examples now show the effect genotypic control can have on the chromosome pairing of hybrids. Thus crosses of *Viola nana* and *V. arvensis* with a third species *V. tricolor* showed almost complete pairing, while when the first two species were crossed less than a quarter of the chromosomes paired (J. Clausen, 1931 c). In crosses of a particular decaploid species of *Chrysanthemum* with different diploid species different degrees of pairing occur amongst the chromosomes of the decaploid (Table 29, A and B). The same chromosomes pair or fail to pair under different genotypic conditions. Later we shall find cases where pairing may fail altogether in a homozygous organism (Ch. X).

But even where there is reason to suppose that genotypic conditions are unchanged we cannot infer the degree of differentiation of two species from the pairing in their hybrid without knowing the frequency of chiasmata in all three. This is most clearly shown by the *Anas* hybrid (*cf*. Ch. VII and Ch. X). Here the large chromosomes, with a high chiasma-frequency in the parents, pair regularly in the hybrid. The short chromosomes, with a low chiasma-frequency in the parents, fail to pair in the hybrid. Were all the chromosomes long, no failure in pairing would be noticed. Were they all short they would all fail to pair and meiosis would follow an entirely different course.

Finally when, as in polyploid hybrids, pairing is seen to be governed by "differential affinity" (Ch. VI), deductions from pairing must have an enormous margin of error. The extensive phylogenetic conclusions that have been based on observation of pairing in hybrids, especially in *Triticum* and *Nicotiana*, must therefore be regarded as little better founded than those based on chromosome number in *Crepis*.

The behaviour of unpaired chromosomes will be considered in relation to the breakdown of genetic systems (Ch. X).

TABLE 21

CLASSIFIED OBSERVATIONS OF PAIRING OF CHROMOSOMES IN UNDEFINED HYBRIDS

The haploid number is given after each species, and the number of pairs (x^{II}) formed by the hybrid where this is variable.

(1) **Potentially Complete Pairing.**

LEPIDOPTERA.

Oporabia filigrammaria (37) × *O. autumnata* (38), 34–37^{II}. Harrison, 1920 a.
Pergesa porcellus (29) × *P. elpenor* (29). Federley, cit. Cretschmar, 1927.
Amorpha populi austauti (28) × *A. p. populi* (28). Federley, cit. Cretschmar, 1927.
Celeris euphorbiæ (29) × *C. galii* (29). Bytinski Salz and Günther, 1930.

PISCES.

Platypoecilus variatus (25) × *P. maculatus* (24) 24^{II} + 1^{I} or 23^{II} + 1^{III}. Friedman and Gordon, 1934.

ANGIOSPERMS.

Primula floribunda (9) × *P. verticillata* (9) (= *P. kewensis*). Digby, 1912 ; cf. Newton and Pellew, 1929.
Digitalis lanata (28) × *D. micrantha* (28). Haase-Bessel, 1922.
Tragopogon pratense (6) × *T. porrifolius* (6) } Winge, cit. Babcock and J.
Geum rivale (21) × *G. urbanum* (21). } Clausen, 1929.
Prunus persica (8) × *P. Amygdalus* (8) }
P. triflora (8) × *P. persica* (8). } D., 1930 a.
P. triflora (8) × *P. cerasifera* (8) }
Fragaria vesca (7) × *F. Helleri* (7) }
F. bracteata (7) × *F. Helleri* (7) } Ichijima, 1926.
F. glauca (28) × *F. virginiana* (28) }
Sorbus Aria (17) × *Aronia melanocarpa* (17). Sax, 1929.
Ribes rubrum (8) × *R. petræum* (8) (= *R. holosericeum*). Meurman, 1928.
Viola tricolor (13) × *V. alpestris* (13) } J. Clausen, 1931 c.
V. cornuta (11) × *V. orthoceras* (11) }
Salix viminalis (19) × *S. Caprea* (19). Håkansson, 1929 b.
Crepis rubra (5) × *C. fœtida* (5). Poole, 1931.
C. leontodontoides (5) × *C. aurea* (5). Avery, 1930.
Triticum dicoccum (14) × *T. persicum* (14). Vakar, 1930.

UNDEFINED STRUCTURAL HYBRIDS 173

T. durum (14) × *T. turgidum* (14). D., 1931 *d* (v. Table 23).
Lolium perenne (7) × *Festuca elatior* (7). Peto, 1934 (*v*. Ch. X).
Dianthus neglectus (15) × *D. Carthusianorum* (15), 15^{II}. Rohweder, 1934.
(N.B. The larger chromosomes of the paternal parent are said to retain their size difference in the hybrid.)
Platanus orientalis (21) × *P. occidentalis* (21), 21^{II}. Sax, 1933.
Chrysanthemum Makinoi (9) × *C. lavandulæfolium* (9), 9^{II}. Shimotomai, 1933.
C. pacificum (45) × *C. yezoensis* (45), 45^{II}. Shimotomai, 1933.
Rumex acetosa (6 + X) × *R. montanus* (6 + YY), 6^{II} + XYY. Yamamoto, 1935.
Euchlæna mexicana (10) × *Zea Mays* (10), 9–10^{II}. Beadle, 1932.
Euchlæna mexicana (Durango × Florida), 6–10^{II}. Beadle, 1932.
Allium fistulosum (8) × *A. Cepa* (8), 8^{II}. Emsweller and Jones, 1935.
Crepidiastrum lanceolatum (5) × *Paraixeris denticulata* (5). Ono and Sato, 1935.
Gossypium herbaceum (13) × *G. arboreum* (13), 13^{II}. Skovsted, 1933.
Salvia grandiflora (8) × *S. officinalis* (7), 7^{II} + 1^{I}. Hruby, 1935.
S. nutans (11) × *S. Jurisicii* (11), 10^{II} + 2^{I}, etc. Hruby, 1935.
S. Jurisicii (11) × *S. nemorosa* (7), 7^{II} + 4^{I}. Hruby, 1935.
S. nemorosa (7) × *S. Baumgartenii* (8), 7^{II} + 1^{I}. Hruby, 1935.
S. nutans (11) × *S. Baumgartenii* (8), 9^{II} + 1^{I} (autosyndesis). Hruby, 1935.
Salix nigricans (57) × *S. phylicifolia* (57), 57^{II} (some IV). Håkansson, 1933.
Quamoclit angulata (14) × *Q. pinnata* (15), 14^{II} + 1^{I}. Kagawa and Nakajima, 1933.

(2) **Partial Pairing** (always variable).

LEPIDOPTERA.

Saturnia pyri (30) × *S. pavonia* (29), 5–6^{II}. Pariser, 1927.
Smerinthus ocellata ocellata (27) × *S. o. planus* (27), 5–18^{II}. Federley, 1914, 1915.

ANGIOSPERMS.

Crepis hybrids (4, etc.), 0–4^{II}. Babcock and J. Clausen, 1929 (v. Table 22); Avery, 1930.
Papaver somniferum glabrum (11) × *P. nudicaule* (7), 3–4^{II}. Yasui, 1927.
Ribes sanguineum (8) × *R. Grossularia* (8) (" *R. Gordonianum* "), 0–8^{II}. Meurman, 1928.
R. nigrum (8) × *R. Grossularia* (8) (" *R. Culverwellii* "), 0–8^{II}. Meurman, 1928.
Digitalis purpurea (28) × *D. ambigua* (28), 5–12^{II}. Buxton and Newton, 1928.
Viola arvensis (17) × *V. rothomagensis* (17), 14–15^{II}. J. Clausen, 1931 *c*.
V. nana (24) × *V. lutea* (24), 15–17^{II}. J. Clausen, 1931 *c*.
Ægilops cylindrica (14) × *Æ. ovata* (14), 7–13^{II}. Percival, 1930.
Æ. cylindrica (14) × *Æ. triuncialis* (14), 6–11^{II}. Kagawa, 1931.
Salvia Bulleyana (8) × *S. glutinosa* (8), ca. 5^{II} + 6^{I}. Hruby, 1935.
S. nemorosa (7) × *S. nutans* (11), 6^{II} + 6^{I}. Hruby, 1935.
Triticum ægilopoides (7) × *Ægilops squarrosa* (7), 0–6^{II}. Kihara and Lilienfeld, 1935.

(3) **Potentially Complete Failure of Pairing**

LEPIDOPTERA.

Pygæra anachoreta (30) × *P. curtula* (29), 0–2^{II} }
P. pigra (23) × *P. curtula* (29), 0–2^{II} } Federley, 1912.

Oporabia dilutata (30) × *O. autumnata* (38). Harrison, 1920 a.
Biston hirtaria (14, large) × *B. zonaria* (56, small). Harrison and Doncaster, 1914.

ANGIOSPERMS.

Raphanus sativus (9) × *Brassica oleracea* (9), $0-1^{II}$. Karpechenko, 1927, *a* and *b*.
Nicotiana Bigelovii (24) × *N. suaveolens* (16), $0-1^{II}$. Goodspeed and Clausen, 1927 a.
Populus Simonii (19) (presumed hybrid, male), $0-1^{II}$. Meurman, 1925.
Canna (9) (presumed hybrid), $0-3^{II}$. Belling, 1927.
Triticum dicoccoides (14) × *Ægilops ovata* (14), $0-5^{II}$. Kihara, 1929 c.
Narcissus intermedius, *N. Tazetta* (10) × *N. Jonquilla* (7), 17^{I}. Nagao, 1933.
Iris pallida (12) × *I. tectorum* (14), $1-2^{II}$. Simonet, 1931.
(The second species has smaller chromosomes which are probably recognisable in the hybrid.)
Nicotiana Rusbyi (12) × *N. silvestris* (12), 24^{I}. Brieger, 1934.

TABLE 22

The Fluctuation of Chromosome Pairing in Hybrids

	Percentage of Cells.							Total Number of Observations.
	0^{II}	1^{II}	2^{II}	3^{II}	4^{II}	5^{II}	6^{II}	
Crepis spp. ($n=4$). Babcock and J. Clausen, 1929.								
C. aspera × *C. bursifolia*	18	42	18	14	9	—	—	59
C. taraxacifolia × *C. tectorum*	30	27	37	6	0	—	—	30
C. aspera × *C. aculeata*	61	31	8	0	0	—	—	36
Triticum and *Ægilops* spp. ($n=14$).								
(i) Kagawa, 1929.								
Æ. cylindrica × *T. dicoccum*	76	22	2	1	0	0	0	118
(ii) Kihara, 1929 c.								
T. dicoccoides × *Ægilops ovata* (a)	2	24	28	31	15	0	0	100
(different florets) (b)	2	22	26	22	24	2	2	50
(c)	68	26	6	0	0	0	0	50
(sister seedlings) (d)	73	23	4	0	0	0	0	100
(e)	1	18	36	27	11	6	1	100
(iii) Kihara, 1931.								
T. Spelta (21) × *Æ. triuncialis* (14) (a)	7	28	41	13	8	3	0	100
(different florets) (b)	56	34	6	3	1	0	0	100

	0^{II}	1^{II}	2^{II}	3^{II}	4^{II}	5^{II}	6^{II}	7^{II}	8^{II}	
Ribes spp. ($n=8$). Meurman, 1928.										
R. sanguineum × *R. odoratum* (a)	0	0	0	2	2	28	36	22	10	91
(different anthers) (b)	0	2	2	3	5	16	25	32	15	141

	23^{II}	24^{II}	25^{II}	26^{II}	27^{II}	28^{II}	
Amorpha populi austauti ($n=28$) × *A. p. populi* ($n=28$). (Federley, 1915)	5	3	4	17	30	37	86

TABLE 23

Chiasma Frequency of Hybrids Compared with that of their Parents *

	Percentages of Bivalents with different Numbers of Chiasmata.				
	0	1	2	3	Mean.
Triticum, D., 1931 d.					
T. *turgidum*	—	—	57·1	42·9	2·43
T. *durum*	—	7·2	57·1	35·7	2·28
T. *turgidum* × T. *durum*	—	25·0	50·0	25·0	2·00
Kniphofia, Moffett, 1932.					
K. *Nelsonii*	0	30·8	56·7	12·5	1·8
K. *Burchellii*	0	42·4	54·6	3·0	1·6
K. *Uvaria* × K. *Macowanii*	2·6†	53·2	42·9	1·3	1·4

* *Cf.* Peto, 1934 ; Mather, 1935 ; and Ch. X.
† Unpaired chromosomes.

5. PERMANENT PROPHASE : THE SALIVARY GLANDS

The activity of secretory cells in glandular tissues of animals sometimes depends on mitosis and sometimes excludes it. In the latter case the nucleus loses its ordinary property of perpetuating a characteristic complement of chromosomes and may become more or less highly specialised for its immediate physiological functions. The most remarkable and important example of this specialisation is found in certain glandular cells of insects, particularly in the salivary glands of Diptera.

The nuclei in cells of the salivary glands in the fully developed larvæ of Diptera can be seen in life to be enormously enlarged and full of chromosomes which are correspondingly enlarged. In their proportions and marking they look like contorted earthworms. They are about a hundred times as long as metaphase chromosomes. In *Drosophila* the second and third chromosomes are a quarter of a millimetre long, and they show some hundreds of definite transverse striations. This structure was described by Balbiani in *Chironomus* in 1881, and by Carnoy in *Hymenoptera* and *Neuroptera*, as well as in *Diptera* in 1884. Alverdes, in 1912, considered that the striations were the result of spiralisation, but Kostoff, in 1930, pointed out that they corresponded rather with the lineally arranged genes. Heitz and Bauer (1933) finally put it beyond doubt that the bodies in the nucleus were separate and that each corresponded with a pair of chromosomes.

The structure of the salivary gland nucleus and its genetical interpretation have now been made clear by the work of Painter,

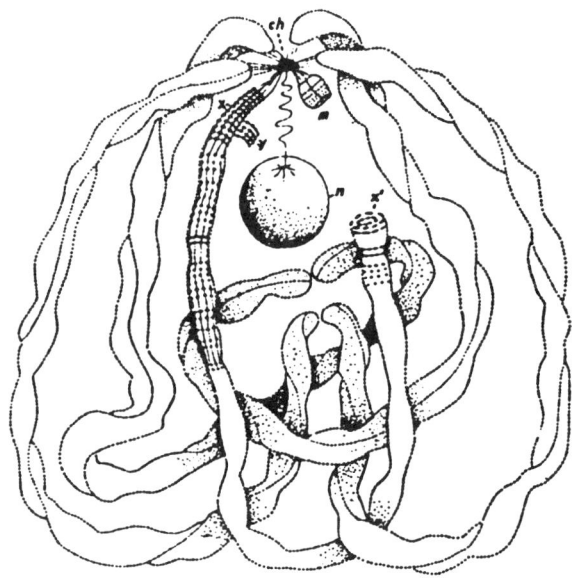

Fig. 59.—Complete nucleus in the salivary gland of ♂ *Drosophila funebris* after pairing is finished (diagrammatic). *ch*, fused centromeres; *n*, nucleolus; *x*, inert end of *X* chromosome; *x'*, distal end; *y*, visible segment of *Y*; *m*, the smallest chromosome pair (Frolowa, 1935).

Bridges, Koller, Bauer and others (Table 24). Each of the chromosomes before pairing is marked by transverse bands of characteristic size, number and order. The paired chromosomes in a homozygous larva result from the fusion side by side of pairs of chromosomes which correspond exactly in regard to these bands. Nucleoli are attached to the chromosomes in characteristic positions as in mitosis. In *Drosophila* all the chromosomes are fused at their centromeres into one central undifferentiated mass or magma, with which the separate arms of the large autosomes of D. *melanogaster* for example, appear to be separately connected. Certain regions of the X and Y chromosomes that are known to be inert seem to take part in this magma, so that the small active part of the Y has only recently been discovered (Prokofieva-Belgovskaya, 1935; Frolowa, 1936). The inert parts are either less hypertrophied or less closely paired than the active parts.

In different species the behaviour of the chromosomes differs. In *D. simulans* the chromosomes do not always pair, and the

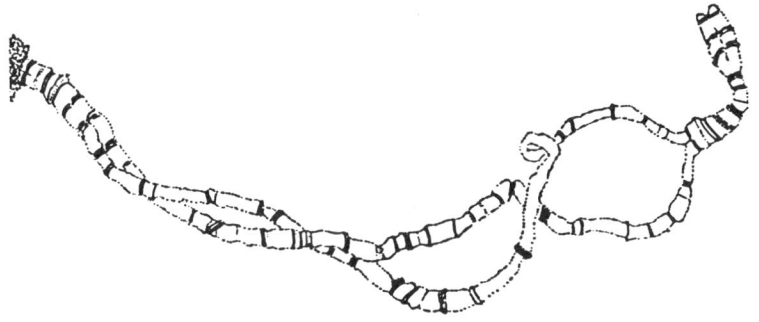

FIG. 60.—Two chromosomes before pairing is complete in *D. pseudoobscura*, showing left-hand relational and relic coiling in the unpaired segment. Only a few of the chromomere-bands are shown. × 1000 (Koller, 1935).

centromeres never fuse (Geitler, 1934; Bauer, 1935, *a* and *b*). It is particularly significant that pairing fails most often in the shortest chromosome to which a nucleolus is attached, since this indicates that the difference is due to delay in coming together. Delay in pairing means inhibition of pairing as we have seen in similar circumstances in meiosis, and it is possible, but unlikely, that this applies to the salivary gland nuclei.

While the thicker striations often appear as uniform bands, close comparison of a number of species has shown that each of the finer bands consists of a group of granules which resemble the chromomeres of meiotic chromosomes and may be properly so described (Koltzoff, 1934). A plate of thirty-two or sixty-four of these granules makes a band, and they are connected by delicate threads to successive granules along the chromosome. In other words each of the pairing chromosomes has divided four or five times, and the products of its division have remained in side-by-side association (Bridges, 1935; Metz, 1935; Koller, 1935). It is possible that they are associated closely in pairs resulting from primary attraction and that these pairs are more loosely held together by a secondary attraction corresponding to the secondary

TABLE 24

Observations of Salivary Glands in Diptera

Drosophila melanogaster.
    Kostoff, 1930.
    Painter, 1934, *a* and *b*.
    Bridges, 1935, 1936.
    Koltzoff, 1934.
    Muller and Prokofieva, 1935.
    Ellenhorn, Prokofieva and Muller, 1935 (*in vivo*).
    Koller, 1935.
D. simulans, and *D. melanogaster* × *D. simulans*.
    Pätau, 1935 (*in vivo*).
D. pseudo-obscura.
    Koller, 1935, 1936.
    Tan, 1935.
Chironomus spp.
    Balbiani, 1881.
    Bauer, 1935, *a* and *b* (*in vivo*).
    King and Beams, 1935.
    Koller, 1935.
Bibio hortulanus.
    Heitz and Bauer, 1933.
Simulium.
    Bauer, 1935.
Sciara spp.
    Metz, 1935.

attraction characteristic of somatic chromosomes in the *Diptera* (Ellenhorn, Prokofieva and Muller, 1935). The original plane of association of the paired multiple chromosomes is, however, soon lost, and the whole bundle of threads becomes a uniform cylinder, a *polytene* chromosome, as we might call it.

The knowledge that exactly corresponding parts of chromosomes pair in salivary gland nuclei has facilitated a closer analysis of the genetic structure of *Drosophila* in two ways. First it has been

PLATE VII

PAIRED CHROMOSOMES IN THE SALIVARY GLAND NUCLEI OF DIPTERA

FIG. 1.—Living nucleus of *Chironomus Thummi*, $n = 4$, in liquid paraffin. The small fourth chromosome has a nucleolus attached and lies in the middle. × 635. (Bauer, 1935.)

FIGS. 2 AND 3.—*Drosophila pseudo-obscura*, aceto-carmine smear preparations (*cf.* Text-Figs.).

Fig. 2. Complete nucleus of female showing inversion-pairing in third chromosome. Fig. 3. One pair of chromosomes in a heterozygote showing inversion and deletion pairing and relic coiling. (Koller, 1935, 1936.)

FIG. 4.—Part of chromosome of *Chironomus sp.* drawn out to show threads connecting chromomeres. (Koller, 1935.)

PLATE VII

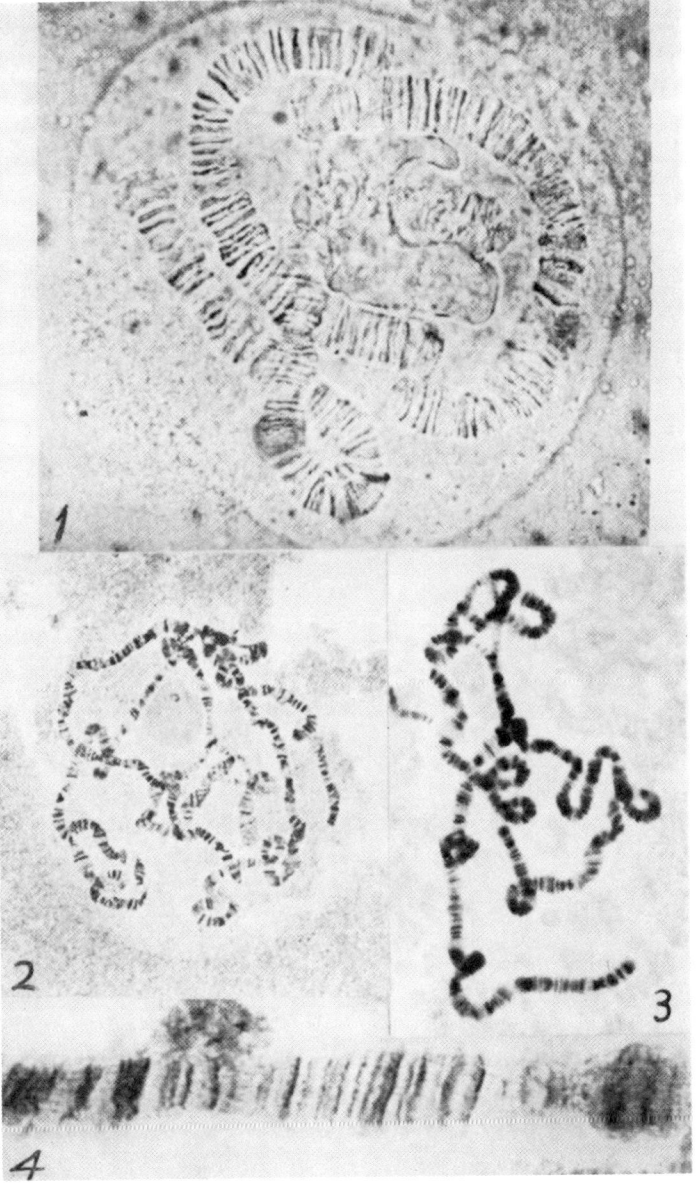

To face p. 178.

possible (Painter, Bridges and many others) to identify the structural differences such as in some cases exist within natural populations, or in other cases distinguish races and species, or in others again are produced by irradiation. Where a deletion has occurred in the middle of a chromosome, the chromomeres of the opposite chromosome having no partner are buckled. Where an inversion has occurred a reversed loop is formed. Where an interchange has occurred the chromosomes change partners. All these structures are strictly comparable with those observed at meiosis in a structural hybrid. The various kinds of differences have been referred to in the proper place. They have shown that the same kind of structural

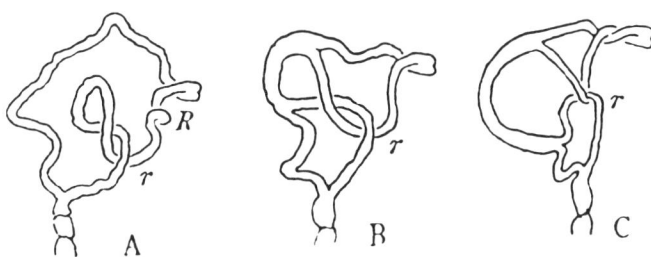

FIG. 61.—An error of pairing in a homozygous *D. pseudo-obscura* due to the torsion setting up non-homologous pairing within a chromosome. The result is internal interlocking. Right-hand relational and relic coiling. × 1000 (Koller, 1935).

differentiation has taken place in *Drosophila* as had been inferred from the study of meiosis in other organisms. In the hybrid *Drosophila melanogaster* ♀ × *D. simulans* ♂ the chromosomes fail to pair in certain places although they seem to be structurally similar. This might be due to invisible differences, but in view of the behaviour found in *D. simulans* it is more likely to be due to the pairing being genotypically controlled, and in the hybrid partially inhibited (Pätau, 1935).

Secondly, by comparing these differences with the crossing-over properties and phenotypic expression of the heterozygotes and homozygotes of the changed types it has been possible for Muller and his collaborators to introduce a new criterion of the gene: to find out that the unit of structural change, the smallest piece into which the chromosome can be broken, corresponds so far as can be seen with the unit of structure, the smallest chromomere visible in ultra-violet photographs, and both are smaller than the unit of

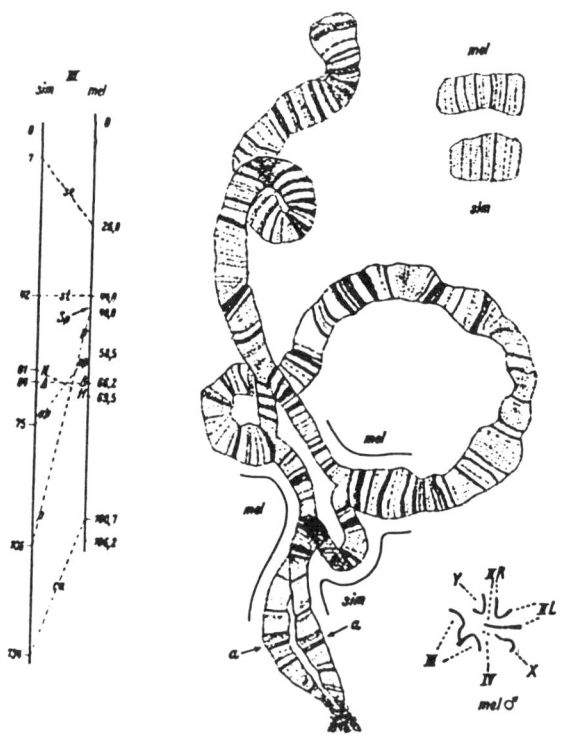

FIG. 62.—Left: comparative linkage maps of the right arms of the third chromosome in *D. melanogastes* and *D. simulans* (Sturtevant, 1929). Right: pairing of these chromosomes in the salivary gland of the hybrid between these species. Above: the two fourth chromosomes of the species. Below: mitotic chromosomes of *D. melanogaster*, ♂. (Patau, 1935.)

crossing-over (Ellenhorn, Prokofieva and Muller, 1935). The total complement of *Drosophila melanogaster*, which is less than a millimetre long, probably contains over 4,000 of these units. Furthermore Bridges (1935) has been able to show from a detailed description of the whole complement of normal flies that the band-pattern

INVERSION PAIRING

is often repeated; there has been a duplication of translocated segments in the history of the species.

The relationship of the enlarged chromosomes with those found in ordinary mitotic and meiotic nuclei has been traced by Koller (1935). As the nucleus grows the chromosome threads become

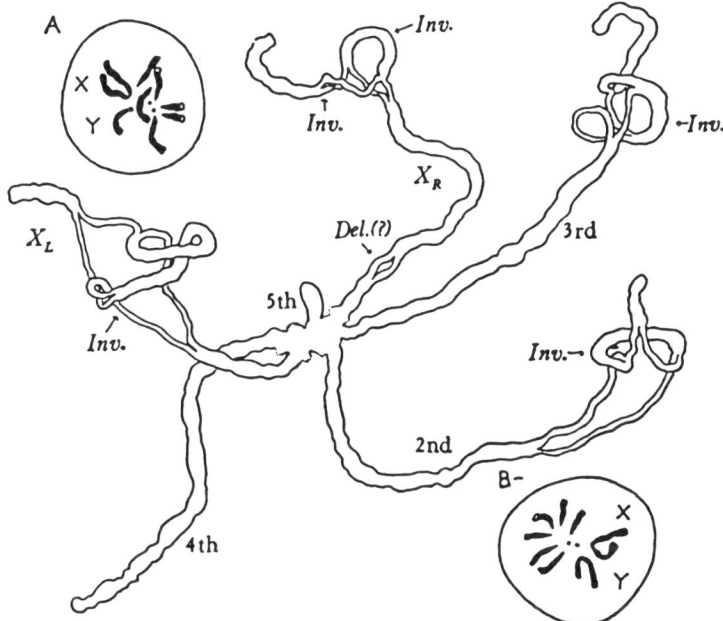

FIG. 63.—Complete nucleus in the ♀ hybrid, *D. pseudo-obscura*, Race A × Race B, showing the presence of five inversions and one possible deletion. × 750 (Koller, 1935). In opposite corners the mitotic chromosomes of these two races (♂) are shown. × 3000 (D., 1934).

thicker and it is presumably during this stage that they multiply by division. They show the relic coils characteristic of a mitotic prophase and consistent for given arms of a chromosome. In this condition they pair. They then show (but much more clearly on account of their great size) the relational coiling seen between meiotic partners at pachytene. The torsion that produces this coiling sometimes leads to abnormalities corresponding to the non-homologous pairing found at meiosis. Thus an intercalary unpaired piece of chromosome may coil round itself and prevent regular association. On the other hand, where an inversion is present the coiling seems to be a necessary preliminary to the formation of a reversed loop.

Mechanically therefore we can look upon the salivary gland nuclei as in a state of permanent prophase in which two special properties are developed to excess. These are the properties of gene-reproduction and gene-attraction. Since, as we shall see elsewhere, reproduction must itself be regarded as a function of attraction (for materials in the substrate) this is not surprising. The result is to produce an enormously greater bulk of the gene materials, or at least of those which are physiologically active, a bulk comparable with that found in glandular cells elsewhere which are polyploid, but continue to undergo mitosis.

Abnormalities of pairing arise in homozygous individuals of *Drosophila* in two ways. First the reduplication of segments leads to pairing within the chromosome as well as between chromosomes (Bridges, 1935). Secondly, owing to internal torsion an intercalary unpaired segment may become coiled round itself and then this coil breaks its normal pairing (Koller, 1935). These abnormalities, like the normal behaviour in structural hybrids, agree well with what we know of the pachytene stage of meiosis in similar organisms. They therefore extend our knowledge of structural change, of the hybrids resulting from it, in nature and in experiment, and of the effect of this hybridity on the processes of germ-cell formation.

CHAPTER VI

THE BEHAVIOUR OF POLYPLOIDS

Autopolyploids and Allopolyploids—Origin in Experiment—Meiosis in Allopolyploids—Differential Affinity—Autosyndesis—Classification of Polyploid Species.

The production of an indubitable sterile hybrid from completely fertile parents which have arisen under critical observation from a single common origin is the event for which we wait.

BATESON, 1922.

1. THE ORIGIN OF POLYPLOIDS IN EXPERIMENT

A POLYPLOID is an organism with more than two complete *sets* of chromosomes (Ch. III). When the chromosome number is doubled in a homozygous diploid such as *Datura Stramonium*, the new organism has four *identical* sets of chromosomes. Such a form is known as an AUTOPOLYPLOID (Kihara and Ono, 1926). Its chromosome complement can be represented as of the constitution $A_1A_1A_1A_1, B_1B_1B_1B_1$, and so on. Its chromosomes have a high frequency of quadrivalent formation, subject to random chiasma formation amongst the four homologues of each kind (Ch. IV). Where the chiasma frequency is low, as in the short chromosomes of *Hyacinthus* and in some chromosomes of *Solanum* and *Primula*, bivalents are formed instead in a proportion of cells (*v. infra*).

In contradistinction to the autopolyploid, the product of doubling in hybrids is known as an ALLOPOLYPLOID. The first of this type to be discovered and understood was in the genus *Pygæra*, but apart from this moth our knowledge of allopolyploids is derived exclusively from the flowering plants. If we take the diploid hybrid to be made up of pairs of chromosomes A_1A_2, B_1B_2—we may represent the tetraploid as $A_1A_1, A_2A_2, B_1B_1, B_2B_2$. Such an allotetraploid is homozygous inasmuch as it may be derived immediately from the union of two genetically identical gametes;

but it is also heterozygous inasmuch as its complement contains two dissimilar pairs of gametic sets. How this affects its behaviour we shall see later. But first we must make an arbitrary but convenient distinction between different kinds of allopolyploids. On the one hand, there are those derived from doubling in defined structural hybrids; where these are interchange heterozygotes pairing occurs between parts of structurally dissimilar chromosomes, and associations are formed correspondingly larger than those of the diploid (Ch. V). This group has been considered earlier because it agrees with autopolyploids in having no differentiation between the corresponding chromosomes such as seriously interferes with their pairing at meiosis.

On the other hand, the typical allopolyploids are derived from undefined hybrids; they show a gradation in behaviour from those with slight to those with strong differentiation between the homologues derived from opposite parents. At the extreme in one direction is such a hybrid as *Crepis rubra* × *C. fœtida* in which the difference between the parents is so slight that the diploid has complete pairing and the tetraploid derived from it behaves like an autopolyploid and occasionally forms five quadrivalents. In an intermediate position is the hybrid *Primula kewensis*: in the diploid pairing is complete, but in the tetraploid derived from it only a small proportion of quadrivalents are formed. At the extreme in the other direction is *Raphanus-Brassica*. In the diploid no pairing occurs; in the tetraploid there are no quadrivalents; pairing is simple, complete and regular. Both the *Primula* and *Raphanus-Brassica* types breed approximately true and are highly fertile. They are, therefore, with a slight qualification functionally diploid. Frequently from crosses between such allotetraploids, through doubling again, octoploids have been produced, and from crosses with diploids the triploids resulting have given rise to hexaploids. These polyploids must be considered along with the original tetraploids both in regard to origin and behaviour. They may be, in relation to their immediate allopolyploid parents, themselves autotetraploid or allotetraploid. Thus the allohexaploid species, *Solanum nigrum* ($2n = 72$), yields an autotetraploid sport by doubling ($2n = 144$), while its pentaploid hybrid with *S. luteum* ($2n = 48$) gives an allotetraploid sport ($2n = 120$), which is, in relation to the basic number of the genus, a decaploid.

TABLE 24A

Chromosome Association in Polyploids

SPECIES (giving auto-tetraploids)

Datura Stramonium (12). Belling, 1927; $2x : 12^{II}$; $4x : 12^{IV}$.
Primula sinensis (12). D., 1931 *a*; $2x : 12^{II}$; $4x : 9-11^{IV}$; $6 - 2^{II}$.
Solanum Lycopersicum (12). $2x : 12^{II}$; $4x : 7 - 11^{IV}, 10 - 2^{II}$ Lesley, 1930; $1 - 8^{IV}, 22 - 8^{II}$, Upcott, 1935.

UNDEFINED HYBRIDS (giving allo-tetraploids)

Crepis rubra × *C. fœtida* (5). Poole, 1931; $2x : 5^{II}$; $4x : 0 - 5^{IV}, 10 - 0^{II}$.
Primula kewensis (9). (*P. floribunda* × *P. verticillata*.) Newton and Pellew, 1929; $2x : 9^{II}$; $4x : 0 - 3^{IV}, 18 - 12^{II}$.
Triticum dicoccoides (14) × *Ægilops ovata* (14). Kihara and Nishiyama, 1930; $2x : 0 - 3^{II}, 28 - 22^{I}$; $4x : 0^{IV}, 28^{II}$.
Digitalis mertonensis (28). (*D. purpurea* × *D. ambigua*.) Buxton and Newton, 1928; $2x : 5 - 12^{II}, 46 - 32^{I}$; $4x : 0^{IV}, 56^{II}$.
Raphanus-Brassica (9). Karpechenko, 1927; $2x : 0^{II}, 18^{I}$; $4x : 0^{IV}, 18^{II}$.

The mechanically diploid behaviour of the allotetraploid means that the corresponding chromosomes that pair and pass to opposite poles to yield identical gametes, must always be those from the same ancestral diploid parent and not those from opposite parents. Thus in hybrids of the *Primula kewensis* type, the diploid has pairing of the homologues derived from opposite species, but its tetraploid descendants have not. Such a difference in behaviour may be described as due to a " differential affinity " (D., 1928) whose significance in relation to prophase behaviour will be considered later.

In *Primula kewensis* the corresponding chromosomes of the diploid parents, *P. floribunda* and *P. verticillata*, are not sufficiently different to inhibit their pairing in the diploid hybrid. Such a lack of differentiation is associated with two abnormalities in the tetraploid : (i) Chromosomes of opposite parents occasionally pair (in quadrivalents or bivalents) and pass to opposite poles. The progeny therefore differ in the proportions of the chromosomes of

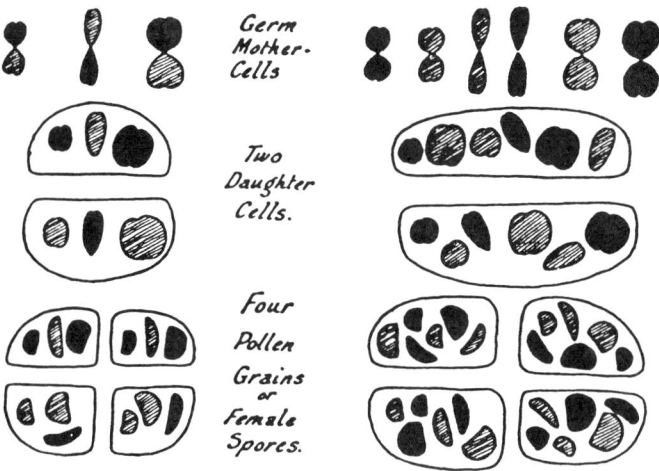

Fig. 64.—Diagram showing how chromosomes pair and segregate in a sterile diploid hybrid, such as *Primula kewensis*, by somatic doubling and in the fertile tetraploid derived from it (neglecting the consequences of crossing-over). The first gives various new (untested) combinations of chromosomes, the second gives uniformly the same combination that the parent had. (From D., 1932 b.)

the two species present. Moreover, owing presumably to this pairing being accompanied by crossing-over the normal type can never again be recovered from its aberrant offspring. (ii) Quadrivalents are formed which sometimes divide unequally (as in the auto-tetraploid *Datura*, *Primula sinensis* and *Dactylis*), giving, therefore, gametes with 17 and 19 chromosomes instead of 18, and progeny with 34, 35 and 37 chromosomes instead of 36. These differ from the normal tetraploid both owing to a change in proportion within the set of nine and owing to a change in the proportionate influence of the two species. Such plants, particularly those with 34 chromosomes, which have 2 chromosomes of one type and 4 of the other 8 types, are less fertile than the normal tetraploid (*cf.* Ch. VIII).

Plants derived from hybrids between species with different numbers (*e.g.*, diploid and tetraploid *Nicotiana*) may show a combination of the two extremes in the same nucleus. The *Nicotiana rustica-paniculata* hybrid is a triploid of this kind with 12 pairs and

TABLE 25

Fertility Showing the Origin of Polyploids after Hybridisation

1. Failure of Reduction (F_2 more fertile than F_1)

	Fertility of F_1.	Fertility of F_2.
Raphanus × *Brassica*, Karpechenko, 1927 a.	45 seeds per plant.	30 seeds per pod.
Digitalis purpurea × *D. ambigua*, Buxton and Newton, 1928.	200 seedlings in two plants.	400 seeds per capsule.
Phleum pratense × *P. alpinum*, Gregor and Sansome, 1930.	11·5 seeds per plant.	400 seeds per plant.
Ægilotricum, Kihara and Katayama, 1931	27 ears gave 5 seedlings.	61 per cent. fertile.
Crepis rubra × *C. fœtida*, Poole, 1931.	50 seeds from 42 heads. 9 seedlings (3, 2x; 5, 3x; 1, 4x).	55 seeds from 22 heads.

2. Somatic Doubling (F_2 of equal or reduced fertility).

Nicotiana glutinosa × *N. Tabacum*, Clausen and Goodspeed, 1925; Clausen, 1928.	155 seeds per fruit.	No greater.
Primula kewensis; *cf.* Sansome, 1931.	30, 90 and 180 seeds per capsule. 287 germinated.	122 seedlings per capsule.
Solanum nigrum × *S. luteum*, Jørgensen, 1928.	8-11 seeds per capsule (*i.e.*, 25 per cent.). ("Completely sterile" before doubling).	No greater.

12 univalents. Doubling, it gives rise to a hexaploid; this has two identical sets that pair with no other sets and therefore appear regularly as 12 bivalents, and it has four sets which form fluctuating numbers of quadrivalents and bivalents.

The sharp difference in chromosome behaviour between the allotetraploid and its hybrid parent causes a difference in fertility; this affords a means of determining the exact time and therefore mode of its origin, which would not otherwise be clear in all cases. The doubling of the chromosome number occurs in two ways: by failure of separation of two groups of chromosomes in a somatic mitosis and by failure of such separation at one of the two meiotic divisions, giving in consequence "unreduced" gametes. Where the tetraploid arises by doubling in the somatic tissue of the first generation (F_1) hybrid, the affected shoot of this plant is as fertile as the progeny (F_2) it gives rise to, or even more so. Where it arises from failure of reduction in both male and female germ-cells in the F_1 their successful union can only be occasional; the F_1 has therefore the characteristic sterility of hybrids and the F_2 is the first to regain full fertility (*cf.* Sansome, 1931). The contrast, it should be noted, is not so striking in the *Crepis* hybrid because it is not entirely sterile as a diploid; it gives diploid and triploid as well as tetraploid progeny. Further, it should be noted that frequent syndiploidy at divisions immediately before meiosis will be indistinguishable from failure of reduction except by direct observation.

Hybrid polyploids that have arisen in experiment may be provisionally classified in accordance with these considerations, and with direct chromosome counts, as in Table 26.

(*Note on Table* 26. The female parent is given first. The numbers in brackets are the gametic numbers of the parents. In the first column are the somatic numbers of the offspring before doubling (F_1) and in the second column the numbers after doubling (F_1 or F_2). Where the failure of reduction occurs in the parent the first column is therefore blank. The initials in the second column represent the sets of chromosomes of parental species. Classes A to D show doubling of the contributions of both parents, classes E to H, of that of only one parent).

These observations show that allopolyploids have arisen in five ways. In the first, type *A*, a fertile shoot usually appears which is readily distinguishable in the otherwise sterile hybrid, but the doubling may occur so early in the development that, as in certain cases of haploid parthenogenesis, the whole plant has the double number. In type *B*, the hybrid is usually highly sterile, but produces, here and there, viable seeds. In some hybrids a few of these seeds have the same chromosome number as their parent, or they result from doubling on one side only so that they are relatively triploid (*e.g.*, in *Crepis* and *Nicotiana rustica-paniculata*).

HYBRID ORIGIN

In these instances, the rest of the seedlings, and in other instances, all of them, have double the chromosome number of the parent; diploid gives tetraploid, triploid gives hexaploid, and so on. In type C, a triploid hybrid is produced through the fertilisation of an unreduced egg by a reduced pollen grain. This triploid, like most triploids, produces again an unreduced (triploid) egg which is fertilised by a reduced pollen grain of the second species again. In this way the doubling is achieved in two stages, which provide

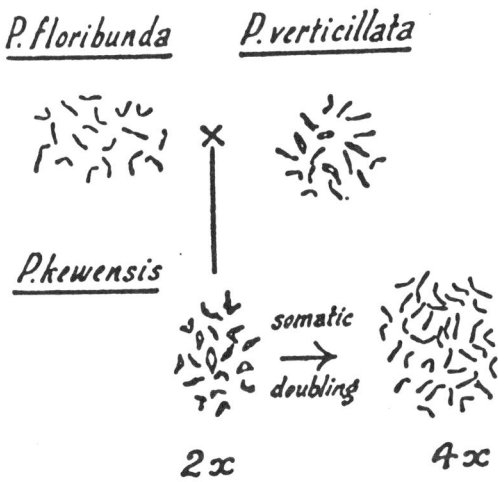

FIG. 65.—The origin of the tetraploid *Primula kewensis*, illustrated by drawings of mitosis; the diploid hybrid and *P. verticillata* are in early anaphase. × 2500, (after Newton and Pellew, 1929.)

excellent illustrations of the difference in genetic effect of one and two sets of one species combined with two of the other (*e.g.*, in *Raphanus-Brassica* and *Rubus*).

In type D, a triploid arises in the same way. It has three pairs of *Crepis capillaris* chromosomes and four unpaired *C. tectorum* chromosomes. When selfed it produces a tetraploid, together with other products of the random segregation of the extra set of unpaired chromosomes. The reason why a tetraploid has arisen in this way in *Crepis* and not elsewhere is evident; only where the chromosome

TABLE 26   THE ORIGIN OF ALLO POLYPLOIDS FROM HYBRIDS

A. SOMATIC DOUBLING IN F$_1$:			
1. *Nicotiana glutinosa* (12) × *N. Tabacum* (24) $(2x + 4x = 6x,$ "*N. digluta*").	36	72	Clausen and Goodspeed, 1925; R. E. Clausen, 1928.
2. *Primula floribunda* (9) × *P. verticillata* (9) $(2x + 2x = 4x,$ "*P. kewensis*," Fig. 65).	18	36	Newton and Pellew, 1929.
3. *Solanum nigrum* (36) × *S. luteum* (24)	60	120	Jørgensen, 1928.
4. *Vicia macrocarpa* (6) × *V. sativa* (6)	12	24	Sweschnikowa, 1929 *b*.
5. *Rubus strigosus* (7) × *R. rusticanus* (7)	14	28	Crane and D., 1927.
6. *Brassica Napus* (18) × *B. campestris* (10)	28	56	Frandsen and Winge, 1932.
7. *Prunus spinosa* (16) × *P. divaricata* (8)	24	48	Rybin, *cf.* Crane and Lawrence 1934.
8. *Aquilegia chrysantha* (7) × *A. flabellata nana* (7)	14	28	Skalinska, 1935.
B. NON-REDUCTION ON BOTH SIDES IN F$_1$:			
9. *Raphanus sativus* (9) × *Brassica oleracea* (9)	18	36	Karpechenko, 1927, *a*, *b*.
10. *Fragaria bracteata* (7) × *F. Helleri* (7)	14	28	Ichijima, 1926.
11. *Ægilops ovata* (14) × *Triticum durum* (14) $(4x + 4x = 8x,$ "*Ægilotricum*")	28	56	Tschermak and Bleier, 1926.
12. *Ægilops ovata* (14) × *Triticum dicoccoides* (14)	28	56	
13. *Triticum dicoccoides* (14) × *Ægilops ovata* (14)	28	56	
14. *Ægilops ovata* (14) × *Triticum turgidum* (14)	28	56	Kihara and Katayama, 1931; Katayama, 1935.
15. *Triticum turgidum* (14) × *T. villosum* (7)	21	42	Percival, 1930.
16. *Triticum vulgare* (21) × *Secale cereale* (7)	28	56	Tschermak, 1930, & Berg, 1934. Levitsky and Benetskaya, 1929, 1931; Lebedeff, 1932.
17. *Triticum durum* (14) × *T. monococcum* (7)	21	42	Thompson, 1931.
18. *Phleum pratense* (7) × *P. alpinum* (14)	21	32	Gregor and Sansome, 1930.
19. *Nicotiana Tabacum* (24) × *N. sylvestris* (12)	36	72	Rybin, 1929.
20. *Nicotiana paniculata* (12) × *N. rustica* (24)	36	72	Lammerts, 1929, 1931; Goodspeed *et al.*, 1926.
21. *Digitalis purpurea* (28) × *D. ambigua* (8) $(8x + 8x = 16x,$ "*D. merionensis*").	56	112	Buxton and Newton, 1928; *cf.* Buxton and D., 1932.
22. *Crepis rubra* (5) × *C. fœtida* (5)	10	20 (10 & 15 also)	Poole, 1931.
23. *Erophila verna* (15) × *Erophila verna* (32)	47?	94	Winge, 1933.
24. *Saxifraga adscendens* (11) × *S. tridactylites* (11) (and reciprocally).	22	44	Drygalski, 1935 (*cf.* Ch. VII).
C. NON-REDUCTION IN ONE PARENT AND IN F$_1$ (BACKCROSSED)			
25. [*Nicotiana Tabacum* (24) × *N. rustica* (24)] × *N. rustica*	—	(i) 72 (*TTR*) (ii) 96 (*TTRR*)	Rybin, 1927 *b*.
26. [*Galeopsis pubescens* (8) × *G. speciosa* (8)] × *G. pubescens* $(2x + 2x = 4x,$ "*G. pseudo-Tetrahit*").	—	(i) 24 (*SSP*) (ii) 32 (*SSPP*)	Müntzing, 1930 *b*.

HYBRID ORIGIN

ONE PARENT FOLLOWED BY	(i) 10 (CCT) (ii) 14 (CCTT)	
D. NON-REDUCTION IN SEGREGATION:		
27. *Crepis capillaris* (3) × *C. tectorum* (4)	—	Hollingshead, 1930 b.
E. NON-REDUCTION IN POLLEN ONLY.*		
28. *Chrysanthemum Makinoi* (9) × *C. japonense* (27)	—	Shimotomai, 1933.
29. *Euchlæna perennis* (20) × *Zea Mays* (10)	—	Emerson and Beadle, 1930.
30. *C. Makinoi* (9) × *C. Decaisneanum* (36)	—	Shimotomai, 1933.
31. *C. Makinoi* (9) × *C. pacificum* (45)	—	Shimotomai, 1933.
F. NON-REDUCTION IN EGG ONLY.†		
32. *Saccharum officinarum* (40) × *S. spontaneum* (56)	63	Bremer, 1923.
33. *Rubus rusticanus* (7) × *R. thyrsiger* (14)	40	Crane and D., 1927.
	72	
	81	
34. *Rubus* "Loganberry" (21) × *R. idæus* (7)	136	Crane and D., 1927.
35. *Nicotiana Tabacum* (24) × *N. sylvestris* (12)	28 (4x)	Webber, 1930.
	(also 21, 3x)	
36. [*Nicotiana Tabacum* (24) × *N. Rusbyi* (12)] × *N. sylvestris* (12).	49 (7x)	
	60 (5x)	
37. [*Nicotiana Tabacum* (24) × *N. Rusbyi* (12)] × *N. Tabacum* (24)	48 (4x)	Brieger, 1930 b.
38. [*Festuca pratensis* (7) × *Lolium perenne* (7)] × *Lolium perenne*	60 (5x)	Brieger, 1928.
39. *Pygæra pigra* (23) × *P. pigra* (23) × *P. curtula* (29)	21 (3x)	Peto, 1934.
40. *Dianthus sinensis* (30) × *D. Knappii* (15)	75 (3x)	Federley, 1912, 1931.
	60 (4x)	Andersson and Gairdner, 1931.
41. [*Triticum dincoccum* (14) × *T. monococcum* (7)] × *T. vulgare* (21)	42	Kostoff, 1932.
		(10[11] + 18[I]).
42. [*Zea Mays* (10) × *Tripsacum dactyloides* (10)] × *Euchlæna mexicana* (18)	38	Mangelsdorf and Reeves, 1935.
43. *Veronica orchidea* (17) × *V. longifolia* (34), and many others	68	Graze, 1935.
	52	Skovsted, 1934.
	32	Lawrence, 1936.
44. *Gossypium* hybrid (26) × *Gossypium* hybrid (52)	—	
45. *Delphinium nudicaule* (8) × *D. elatum* (16)	—	
G. NON-REDUCTION IN EGG OR POLLEN.		
46. *Salix viminalis* (19) × *S. Caprea* (19)	57 (3x)	Håkansson, 1929 b.
H. IRREGULAR NON-REDUCTION IN EGG OR POLLEN.		
47. *Saxifraga rosacea* (32) × *S. granulata* (24) (= *S.* "polternensis").	56	Philp, 1934.
48. *Salix viminalis* (19) × *S. Caprea* (19)	—	Håkansson, 1929 b.
49. *Saturnia pavonia* (29) × [*S. pyri* 23) × *S. pavonia* (29)]	84	Pariser, 1927.
	(3x)	

* Also in *Primula sinensis* (D., 1931 a) and *Solanum melongena* (Janaki Ammal, 1934).
† Also in *Campanula persicifolia* (Gairdner and D., 1931).

number is small will the triploid produce a moderate proportion of gametes having all the extra chromosomes, owing to their being distributed at random at meiosis (Ch. VIII). Further, of the diploid gametes most are evidently eliminated; since they should combine as frequently as haploids, yet only one tetraploid appeared in a family along with 59 diploids (Table 49).

In type E, tetraploid and diploid species are crossed, and, side by side with the expected triploid seedlings, tetraploids are found, differing from them in the preponderant influence of one parent and although they are not simple allotetraploids, in greater fertility. In the *Rubus* hybrid some quadrivalents are formed, as they were in its tetraploid parent, and genetic results show that the pairing in some types of chromosomes is between those derived from the same parent, and in others at random among the four homologues.

In crosses between races or species with different chromosome numbers we often find that exceptional unreduced gametes are favoured at the expense of reduced gametes. Thus in the following crosses the normally reduced gametes on the diploid side rarely function, and tetraploid seedlings may outnumber triploids:—

Primula sinensis (4x) × *P. sinensis* (2x). D., 1931 a.
Campanula persicifolia (2x) × *C. persicifolia* (4x). Gairdner and D., 1931).

Similarly the cross of diploid *Chrysanthemum Makinoi* with a high polyploid species succeeded only with the polyploids as male parents and in every case with tetraploid egg-cells produced by failure of both meiotic divisions in the diploid female parent (Table 26).

The greater success of the unreduced or even doubled gametes than the haploid in some circumstances has an important bearing on interspecific crossing. Diploid *Brassica oleracea* ($n = 9$) is useless in crossing with *B. chinensis* ($n = 10$) and *B. carinata*, but tetraploid *B. oleracea* gives highly satisfactory results. Thus doubling of the chromosome number makes possible interspecific crosses that would otherwise fail (Karpechenko, 1935).

In polyploid crosses a few cases of unexpected and so far inexplicable chromosome distribution have been found (Table 27).

TABLE 27

Polyploid Hybrids with Abnormalities of Chromosome Constitution

Newton and Pellew, 1929. *Primula kewensis* ($4x$) selfed. Seedlings, $2n = 20$ 26.
Steere, 1932. *Petunia axillaris* ($2x = 14$) × *P. hybrida* ($4x = 28$). F_1, all $2x$.
Shimotomai, 1933. *Chrysanthemum marginatum* ($10x = 90$) × *C. lavandulæ-folium* ($2x = 18$). One seedling $8x$.
Nishiyama, 1933. *Avena fatua* ($6x = 42$) × *A. barbata* ($4x = 28$). F_1, $5x$; F_2, $6x$; F_3 (one plant resembling *A. strigosa*) $2x$.
Darlington, 1933. *Prunus avium* ($2x = 16$) × *P. cerasus* ($4x = 32$). Three seedlings $2x$ (Knight's hybrids).
East, 1934. *Fragaria vesca* ($2x = 14$) × *F. virginiana* ($8x = 56$). One seedling $2x$.
Karpechenko, 1935. *Brassica oleracea*, green ($2x = 18$) × (*B. oleracea*, red × *Raphanus sativus*, $4x$). Progeny all $2x$ and heterozygous, red-green.
Brassica oleracea ($2x = 18$) × (*Raphanus sativus* × *B. oleracea*, $4x$). Progeny all $3x$.

Some of these abnormalities may be due to male parthenogenesis or to irregular fusion of nuclei in the egg (*cf.* Bergman, 1935, on *Leontodon*). Others may be related to unexplained abnormalities of mitosis that have been occasionally found in polyploids (*e.g.*, Huskins, 1934). An explanation of others again may be found perhaps in analogy with double fertilisation in echinoderms, which leads to the formation of spindles with three or four poles and hence to irregular reduction in the chromosome number (Boveri, 1907). In the *Brassica-Raphanus* hybrid, however, the nature of the abnormality has been accurately defined by using red cabbage for the parent of the tetraploid hybrid and green for the diploid. The diploid progeny were heterozygous and were therefore true crosses. The tetraploid hybrid which has *Brassica* as the female parent therefore produced haploid *Brassica* pollen and hence diploid offspring. On the other hand the tetraploid hybrids derived from *Raphanus* as the female parent produce normal diploid pollen and hence triploid offspring.

The progeny were abnormal in being diploid only when *Brassica* was the maternal parent of the first cross. Evidently the *Raphanus* chromosomes were all lost at meiosis in the F_1. Presumably at this stage there is an abnormal reaction between the *Raphanus* chromosomes or their centromeres and the *Brassica* cytoplasm or its spindle-determining agents.

2. PAIRING, SEGREGATION AND FERTILITY

When we compare polyploids with the diploids from which they have arisen we find a sharp change in the choice of partners and consequent segregation amongst homologous chromosomes and at the same time a sharp change in fertility. From this we can learn something important about the physiological action of the chromosomes. At present, however, we are concerned merely with inferring the constitution of polyploid species and other old polyploid forms from our knowledge of the constitution of new ones.

The increase in fertility in sterile diploids is in sharp contrast

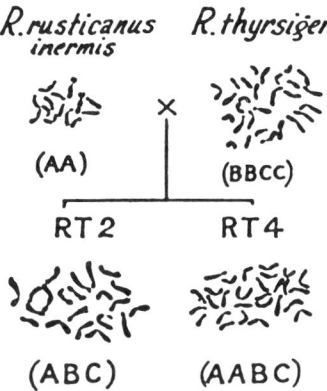

FIG. 66.—The origin of the tetraploid "RT4" and triploid "RT2" from a cross between two *Rubus* species, illustrated by mitotic metaphases. A, B and C represent sets of seven chromosomes. × 2000 (after Crane and D., 1927).

with the decrease which is always found in fertile diploids that double their chromosome number. In the following examples the tetraploid produced is less fertile than the diploid :—

TABLE 28

Autotetraploids with Reduced Fertility

Œ. Lamarckiana	.	. Lutz, 1907 *a* ; Gates, 1909.
Datura Stramonium	.	. Blakeslee *et al.*, 1923.
Primula sinensis	.	. *Cf.* D., 1931 *a*.

INFERENCE OF ORIGIN

Solanum nigrum . . ⎫ Jørgensen, 1928; *cf.* Upcott,
S. Lycopersicum . . ⎭ 1935.
Canna aureo-vittata . . Honing, 1928.
Campanula persicifolia . Gairdner and D., 1931.
Lathyrus odoratus . . Fabergé, 1935; Upcott, unpub.
Funaria hygrometrica . . v. Wettstein, 1924.

These are all autotetraploids except the structurally heterozygous Œnothera, which may be considered as an autopolyploid for this

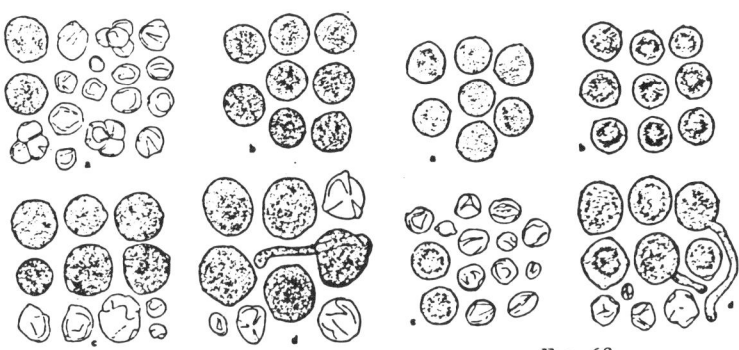

FIG. 67. FIG. 68.

FIGS. 67 AND 68.—Pollen grains of *Solanum*. Fig. 67.—*S. nigrum*: *a, x* (haploid); *b, 2x; c, 3x; d, 4x*. Fig. 68.—*a, S. nigrum* (*2x*). *b, S. luteum. c*, the diploid hybrid. *d*, the tetraploid hybrid. × 160 (from Jørgensen, 1928).

purpose, since, as has been seen, the differences between its homologues are so arranged as not to interfere with pairing. In *Funaria*, v. Wettstein found that auto-octoploids produced artificially without hybridisation were sterile. When octoploids were similarly raised from hybrids between different species and genera they were fertile, the diploid hybrids being sterile except in the production of unreduced gametes (*cf.* Burgeff, 1930; Haupt, 1932; Schweizer, 1932; Allen, 1935).

Further, diploid hybrids of intermediate fertility give tetraploids of intermediate fertility and with an intermediate frequency of quadrivalents, *e.g.*, in *Vicia* (Schweschnikova, 1929 *b*), *Crepis*

(Poole, 1931) and *Aquilegia* (Skalinska, 1935). It may therefore be said that there is negative correlation between the fertility of diploids and that of the tetraploids to which they give rise (D., 1928). The reason for this is evident. Two factors go to reduce generational fertility : (i) the segregation of dissimilar chromosomes (chiefly in the hybrid diploid) ; (ii) the unequal segregation of multivalents (chiefly in the autotetraploid). And the greater the dissimilarities in the diploid, the more regularly do the identical

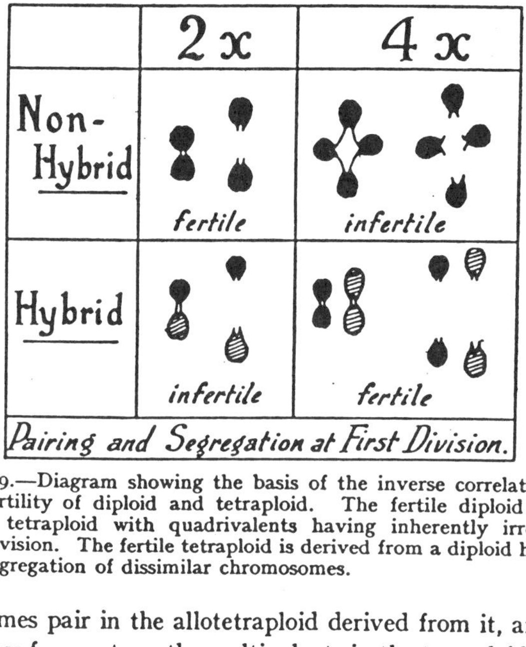

Fig. 69.—Diagram showing the basis of the inverse correlation of fertility of diploid and tetraploid. The fertile diploid gives a tetraploid with quadrivalents having inherently irregular division. The fertile tetraploid is derived from a diploid having segregation of dissimilar chromosomes.

chromosomes pair in the allotetraploid derived from it, and therefore the less frequent are the multivalents in the tetraploid. Multivalents are invariably liable to segregate unequally (*e.g., Datura,* Belling and Blakeslee, 1924 ; *Prunus,* D., 1928 ; *Primula,* D., 1931 *a*), and give rise to gametes and zygotes of reduced viability. The generalisation is therefore universally applicable subject to three qualifications :—

(*a*) The chromosomes must be capable of forming multivalents as they universally appear to be.

(b) The chromosomes of the haploid set must be numerous.

Crepis tectorum with 4 chromosomes gives a relatively fertile tetraploid because the chance of unbalance is reduced.

(c) There is a correlation between the genetic differentiation of the chromosomes of species and their structural differentiation. In other words, structural changes have always occurred along with other genetic changes and prevent the pairing of the chromosomes in hybridity. This follows from our earlier conclusion that all interspecific hybrids are also structural hybrids.

(d) The chromosomes of each set must be qualitatively differentiated in a fairly high degree. This third condition is not applicable to high polyploids (such as *Rumex Hydrolapathum*, $2n = 200$, Kihara and Ono, 1926; and *Prunus Laurocerasus*, $2n = 180$, Meurman, 1929 a), nor apparently to *Hyacinthus orientalis*, for this species combines high fertility of the triploid with viability of unbalanced forms and lack of differentiation amongst them (D., 1929 b). Its chromosomes are therefore not qualitatively differentiated within the set (Ch. VIII).

3. POLYPLOID SPECIES

In about half the species of the angiosperms the gametic chromosome number is a multiple of that found in some related species, the chromosomes being themselves comparable in the two forms. From this alone it is clear that they owe their origin to polyploidy, tetraploids being derived from crosses between diploids and diploids, hexaploids from crosses between tetraploids and diploids, and so on. They may therefore be expected to resemble the polyploids produced in experiment.

Winge (1917), in a remarkable survey of these conditions, first suggested that they were likely to be of the type derived from hybridisation (allopolyploids). He considered that the need for a partner with which a chromosome could pair would stimulate doubling in hybrids. This teleological argument led to a sound general conclusion, although the argument itself is admissible only as a figure of speech. Somatic doubling probably occurs no more freely in haploids (in which the chromosomes have no regular partner) and in hybrids (in which they have no effective partners) than in homozygous diploids; it is more frequently *noticed* in haploids and hybrids because of the striking change it produces in them. Gametic doubling or failure of reduction does occur more freely in haploids and hybrids, but for the demonstrable reason that meiosis has proved unworkable—and not in anticipation of its doing so (Ch. X).

The observations of the relative fertility of auto- and allo-tetraploids provide sufficient reason for supposing that polyploid species will be of the hybrid kind where they depend on their sexual fertility for their reproduction. The evidence of the relationship of the constituent gametic sets of polyploids (*i.e.*, of the extent to which they are allo- or auto-polyploid) is provided by a series of observations which, owing to their importance for the theory of chromosome pairing, may be conveniently considered in relation to " differential affinity."

4. DIFFERENTIAL AFFINITY : AUTOSYNDESIS

In the diploid *Primula kewensis* the nine chromosomes of *P. verticillata* pair with the nine chromosomes of *P. floribunda*, their presumed homologues. In the tetraploid, however, each chromosome pairs with its identical mate, *floribunda* with *floribunda* and *verticillata* with *verticillata*. Only exceptionally one, two or three quadrivalents are formed through all four homologues associating. The explanation of this difference lies in the method of pairing amongst the similar threads at zygotene and of chiasma formation in the paired threads at diplotene, described in polyploids. If *floribunda* and *verticillata* chromosomes can only pair in half their length at pachytene this will probably suffice to ensure regular metaphase pairing by the formation of one chiasma in the diploid hybrid. In the tetraploid their chance of forming a chiasma would be reduced by two-thirds *if the similar parts continued to pair at random*. But from what we now know of the structural differences between species we cannot assume that the differences in undefined hybrids are so simple as here suggested. Probably such portions of the corresponding chromosomes as can pair are distributed in several segments. There will then be several differences of linear sequence (or changes of homology) in the dissimilar chromosomes.

Since the chromosomes in polyploids behave at zygotene as though they consisted of a small number of " pairing blocks " each of these changes of homology will prevent pairing in the block affected, while pairing between identical chromosomes will proceed unobstructed. Thus the existence of limited numbers of effective pairing blocks means that similar parts of dissimilar chromosomes, if they are short, do *not* pair at random in a polyploid. There is *competition* in pairing : hence the differential affinity of dissimilar chromosomes in the presence and in the absence of identical partners (D., 1928).

An analogous observation has been made on a triploid hybrid containing two sets of *Crepis capillaris* and one of *Crepis tectorum* (Hollingshead, 1930 b). The *tectorum* chromosomes are regularly left unpaired, although they pair with the *capillaris* chromosomes in the diploid hybrid. Similarly the two Y chromosomes of *Rumex Acetosa* regularly pair with the corresponding parts of the X chromosome in the diploid, but in the triploid which contains $X\ X\ Y_1\ Y_2$ they are left unpaired (Ono, 1928). The two X's are completely identical and pair to the exclusion of the Y's, which may be inferred from this to be identical with them only in small segments (Ch. IX). The most striking examples of differential affinity are shown by the pairing of chromosomes in the haploid progeny of true diploids which have perfectly regular pairing. This will be described in another connection (Ch. VIII).

Differential affinity is characteristic of the behaviour of homologous chromosomes of different sets in allopolyploid species, as compared with that in their " haploid," " triploid," and hybrid derivatives. In these derivatives, chromosomes derived from the same parental gamete pair with one another, although such pairing rarely occurs in the normal " homozygous " polyploid species, where each has an identical mate. In these circumstances the change in conditions seen in the doubling of the diploid *Primula kewensis* is reversed, and it is possible to understand the relationship that the natural polyploid has to the product of experiment such as *Primula kewensis*.

This " internal pairing " is known as *autosyndesis* and is opposed to *allosyndesis*, the pairing of homologous chromosomes derived from opposite parents in a hybrid (Ljungdahl, 1924). Both consist in the pairing of dissimilar chromosomes, and the distinction does not arise in a true-breeding plant where identical chromosomes pair. The distinction between the two is merely relative to the immediate parents, but the occurrence of autosyndesis is significant in an allopolyploid's derivatives because it means that chromosomes pair in the absence of identical mates, although they rarely or never pair in the presence of such mates.

The distinction may be shown by symbols as follows (the hyphen joins the chromosomes that pass to opposite poles following pairing) :—

4x Parents		Allosyndesis	Hybrid	Autosyndesis
A_1–A_1	a_1–a_1	A_1–a_1 or	A_1–a_2	A_1–A_2
A_2–A_2	a_2–a_2	A_2–a_2	A_2–a_1	a_1–a_2
B_1–B_1	b_1–b_1	B_1–b_1	B_1–b_1	B_1–B_2
B_2–B_2	b_2–b_2	B_2–b_2	B_2–b_2	b_1–b_2
etc.	etc.	etc.	etc.	etc.

Autosyndesis occurs in allopolyploids under the following conditions :—

1. *In the normal polyploid species as an exceptional occurrence.* Thus quadrivalents are occasionally found in the tetraploid *Primula kewensis* and in many allopolyploid species. In the hexaploids *Triticum vulgare* and *Avena sativa* quadrivalents are found (Huskins, 1927, 1928). In all these instances aberrant offspring appear which evidently result from the segregation of dissimilar chromosomes (*i.e.*, through exceptional autosyndesis), since they have in a more marked degree the characters of one of the parental species (in *Primula kewensis*) or of an ancestral species (in *Triticum* and *Avena*). In aberrant plants of *Avena sativa* a genotypic abnormality prevents the chromosomes, or some of them, from pairing. These chromosomes or parts of chromosomes are therefore removed from competition. Autosyndesis occurs and multivalents are more frequently formed than in the regular and normal type (Ch. X).

2. *In the " haploid " derived parthenogenetically from a polyploid.* Thus if *Solanum nigrum* (6x) consists of six sets, *AABBCC*, its

SELF-PAIRING

" haploid " has the constitution ABC, and is, relative to its basic number, a triploid ($3x$). Pairing takes place between chromosomes of the different sets. From 5 to 12 bivalents in the pollen mother-cells and in the embryo-sac mother-cell even some trivalents are formed. Haploids derived from other polyploids have shown less autosyndesis.

TABLE 28A.

Chromosome Pairing in Polyhaploids

Triticum compactum ($6x = 42$), $0 - 3^{II}$. Gaines and Aase, 1926.
Nicotiana Tabacum ($4x = 48$), $0 - 3^{II}$. Chipman and Goodspeed, 1927 ; Lammerts, 1934.
Digitalis mertonensis ($16x = 112$), $5 - 12^{II}$. Buxton and D. 1932.
Ægilotricum ($8x = 56$), $0 - 3^{II}$. Katayama, 1935.
Solanum nigrum ($6x = 72$) $5 - 12^{II}$. Jørgensen, 1928

In these species the chromosomes are so far different that they pair very little even in the absence of competition. The species are approaching the condition of simple diploids. The variation found in pairing in all these " poly-haploids " as they are described by Katayama, is characteristic of that of the diploid, triploid and other hybrids with which they correspond. In *Digitalis* and *Ægilotricum* the process of doubling by which the new species arose from a diploid hybrid has been reversed in the origin of the haploid form from it by parthenogenesis.

3. *In "triploids" that have arisen through the functioning of an unreduced gamete.* These have each of the dissimilar sets of chromosomes reproduced identically three times. Thus a tetraploid $AABB$ gives a " triploid " $AAABBB$. Since only two chromosomes can pair at any one point at pachytene, the third set of A is left out of combination, and is therefore (at any one point) in the same relationship to the corresponding but dissimilar set of B as it would be in a polyhaploid AB. Chromosomes of different sets will therefore sometimes pair. But at other points such third-chromo-

somes will be free to pair with their identical homologues, so that sexivalents will occasionally be formed as well as bivalents derived from autosyndesis within the third complement. Such is the condition in " triploid " *Pyrus Malus* (Nebel, 1929 ; D. and Moffett, 1930), the diploid form of which is a secondary polyploid (*q.v.*). Seventeen bivalents are formed in the diploid ; quadrivalents are rare. But in the "triploid" more than seventeen bivalents are

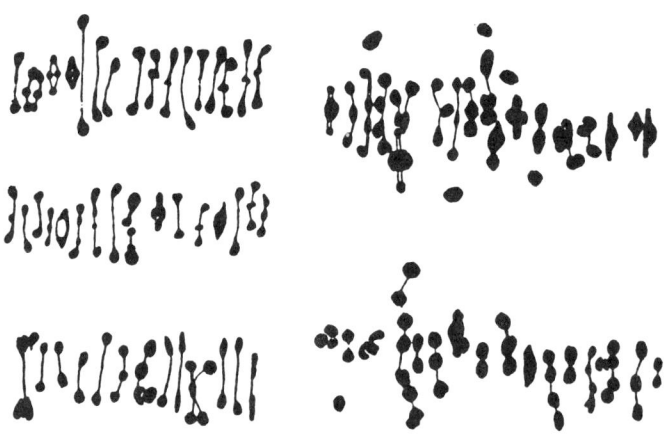

FIG. 70.—First metaphase in pollen mother-cells of *Pyrus* ($x = 17$). Left, diploid forms showing one or two exceptional quadrivalents. Right, triploids : below, 2^I, 5^{II}, 7^{III}, 1^{IV}, 1^{VI}, 1^{VII}; above, 5^I, 10^{II}, 6^{III}, 2^{IV}. × 2500 (from D. and Moffett, 1930).

usually formed and frequently associations of six, *i.e.*, the third set pair among themselves (Fig. 70).

4. *In hybrids of hexaploids, octoploids and higher polyploids with diploids.* Two, three or more sets of the polyploid are naturally left without a partner, as they are in a haploid and triploid. They are frequently found to pair by autosyndesis. The diploid-hexaploid hybrid not only has the chromosome number of a tetraploid but also behaves like one in pairing. In one such case (*Prunus*) quadrivalents are frequently found in the hybrid, so that pairing is taking place between all three sets of the polyploid parent (Fig. 71). In other cases (*Crepis* and *Papaver*) the hybrid can again be crossed

with a diploid species and again yield a hybrid with complete pairing, thus :—

$$10x \times 2x = 6x\ (3x^{\text{II}}),\ 6x \times 2x = 4x\ (2x^{\text{II}}).$$

Here also pairing is occurring between all the sets of the polyploid.

5. *In hybrids between two tetraploid species.* Most of the chromosomes may fail to pair altogether, but some quadrivalents are found as in the first and fourth cases. Higher configurations often occur in the hybrid than in its polyploid parent, although

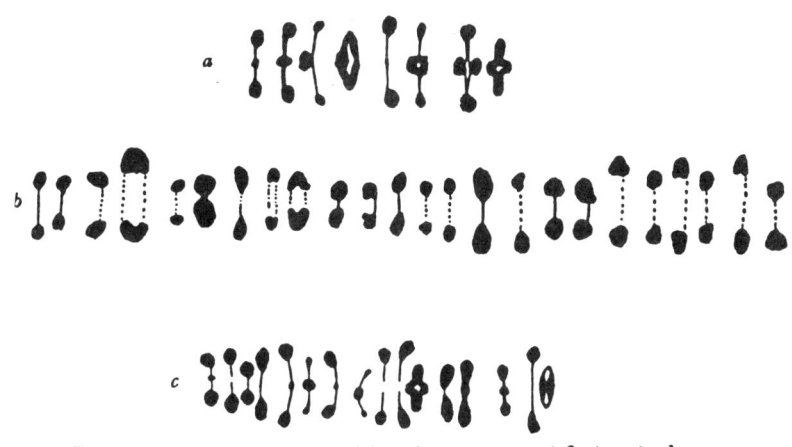

FIG. 71.—Side views of the pairing chromosomes at first metaphase : *a*, in *Prunus cerasifera*, $n = 8$; *b*, in *P. domestica*, $n = 24$; *c*, in the hybrid, *P. domestica* $\times$ *P. cerasifera*, $n = 16$. $\times$ 3000. (*b*, from Mather, unpublished ; *c*, from D., 1930 *a*.)

very little pairing occurs at all (Kihara and Nishiyama, 1930 ; *v.* Fig. 72 ; Kattermann, 1931). This condition is found in hybrids between two tetraploids (*Triticum-Ægilops*, Kihara, 1929) and in *Viola* hybrids (*e.g., V. nana* and *V. lutea*, J. Clausen, 1931 *c*). The formation of multivalents makes it possible to prove the occurrence of autosyndesis in a hybrid between two tetraploids. Thus in the cross *Ægilops cylindrica* ($4x = 28$) by *Æ. triuncialis* ($4x = 28$) a trivalent was sometimes found together with 6 to 11 bivalents (Kagawa, 1931). The trivalent could only arise through pairing of chromosomes derived from the same gamete, if not with one another at least with the same chromosome derived from the other gamete.

204   THE BEHAVIOUR OF POLYPLOIDS

Thus autosyndesis occurs in the hybrids when it does not occur in their parents.

The simplest case of this is the formation of occasional

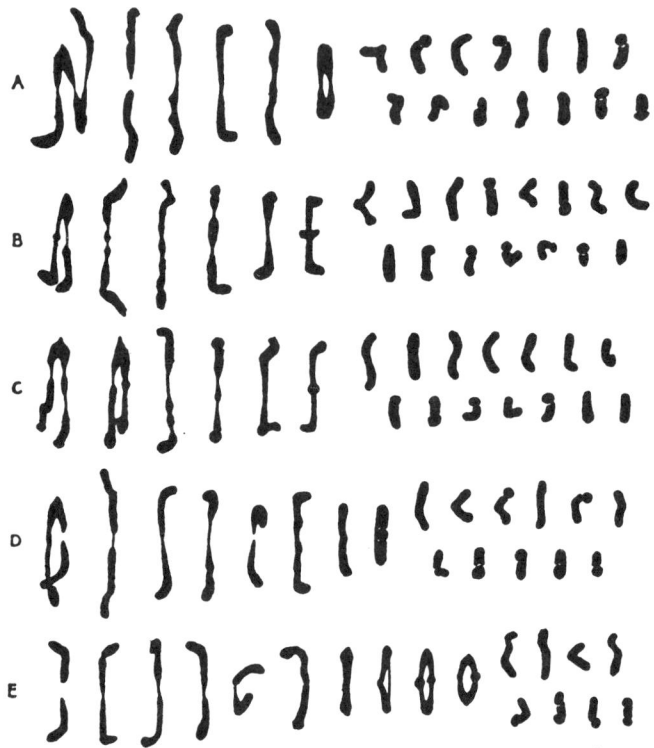

FIG. 72.—First metaphase in the tetraploid hybrid ($4x = 28$) of *Triticum Spelta* ($6x$) × *T. ægilopoides* ($2x$) showing the variation in pairing from nucleus to nucleus.   A, $1^{IV} + 5^{II} + 14^{I}$;   B, $1^{III} + 5^{II} + 15^{I}$;   C, $2^{III} + 4^{II} + 14^{I}$;   D, $1^{III} + 7^{II} + 11^{I}$;   E, $10^{II} + 8^{I}$.   All chiasmata are terminal or subterminal. × 1400 (from Kihara and Nishiyama, 1930. *Cf.* Fig. 58).

quadrivalents by the hybrid between the tetraploid species *Triticum turgidum* and *T. durum* which themselves have simple pairing as a rule (D., 1931 *d*). This indicates that among the pairs also in this hybrid there may be some consisting of chromosomes derived from the same parent. Thus segregation of the characters carried by the ultimate diploid parents may be expected. This "secondary segregation" has been found in the progeny of similar crosses in *Triticum* (*cf* D., 1928), and is probably characteristic of hybrids between allopolyploids (*cf.* Philp, 1934).

Finally there are hybrids (especially triploids) where the relationships of the pairing chromosomes cannot be stated with certainty. In the *Drosera* hybrid (Rosenberg, 1909) it was assumed that the unpaired chromosomes were those of the tetraploid parent, but they may (at least in part) be those of the diploid as a result of autosyndesis. Similarly, in the *Beta* hybrid it is not possible to know the provenance of the pairing chromosomes (*cf.* Table 28 D).

The different kinds of autosyndesis may be summarised, giving the same chromosome type the same capital letter, but a different number when dissimilar, as follows:—

1. Allotetraploid species ($4x$, $2x$ pairs): A_1–A_1, A_2–A_2, B_1–B_1, B_2–B_2, etc., shows exceptional autosyndesis, A_1–A_1–A_2–A_2, B_1–B_1, B_2–B_2, etc.

2. "Haploid" derived from it by parthenogenesis, or diploid hybrid giving rise to it by doubling ($2x$, x pairs) shows autosyndesis: A_1–A_2, B_1–B_2, etc.

3. "Triploid" derived from tetraploid by fusion of reduced and unreduced gametes ($6x$) shows autosyndesis:—

A_1–A_1, A_1–A_2, A_2–A_2, B_1–B_1, etc.
or A_1–A_1–A_1–A_2, A_2–A_2, B_1–B_1, etc.
or A_1–A_1–A_1–A_2–A_2–A_2, B_1–B_1, etc.

4. Hybrid with another tetraploid species (A_3–A_3, A_4–A_4, B_3–B_3, etc.) ($4x$) shows autosyndesis:—

A_1–A_2, A_3–A_4, B_1–B_2, B_3–B_4
or A_1–A_2–A_3–A_4, etc.
or A_1, A_2, A_3–A_4, etc.

(allosyndesis being A_1–A_3, A_2–A_4, B_1–B_3, etc.).

5. Hybrid of a hexaploid species ($6x$, $3x$ pairs): A_1–A_1, A_2–A_2, A_3–A_3, B_1–B_1, etc., and a diploid species ($2x$, x pairs), A_4–A_4, B_4–B_4, etc. ($4x$) shows autosyndesis:—

A_1–A_2, A_3–A_4, B_1–B_2, etc.
or A_1–A_2–A_3–A_4, etc.

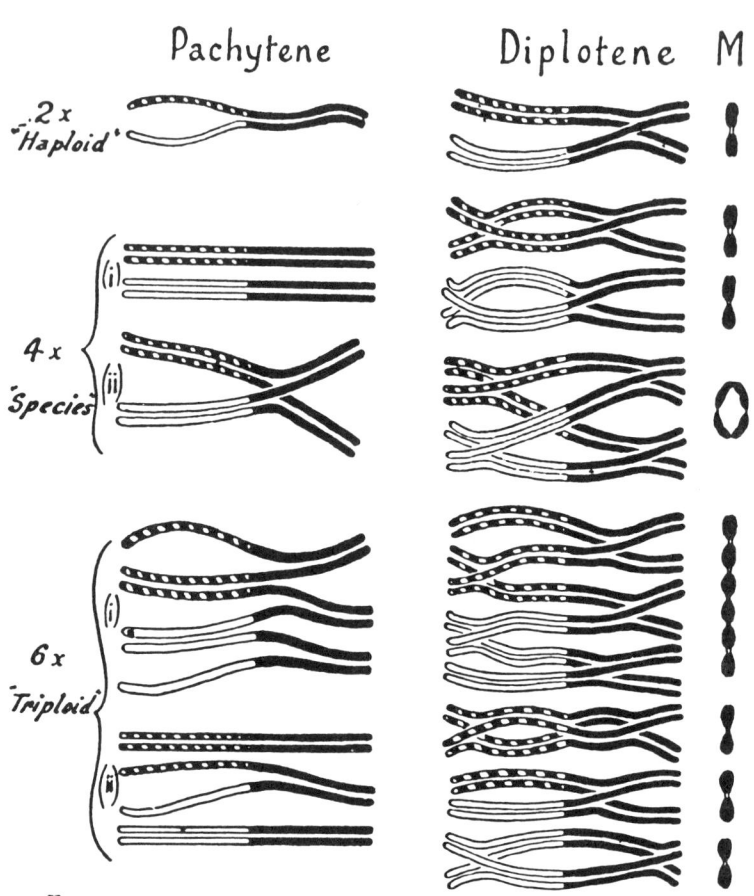

Fig. 73.—Diagram showing the method of pairing in an allotetraploid species and in the "haploid" and "triploid" derived from it on the assumption that the dissimilar chromosomes are similar and capable of pairing, in the absence of competition, in half their length. M, metaphase; black, similar segments; white and hatched, dissimilar ones. *Note*: pairing of dissimilar chromosomes is inevitable in "haploid" and "triploid" but only possible in the face of competition in the species, whence differential affinity. (i) and (ii) are alternative possibilities with and without autosyndesis.

The conditions determining these results may be explained on the assumptions (i) that the differences which reduce pairing at pachytene are structural; (ii) that pairing is limited amongst identical segments in a polyploid when they are not in the same sequence throughout the chromosomes; and (iii) that pairing at metaphase is determined by the formation of chiasmata in the pachytene thread proportionate in number to the length paired. The causal sequence is illustrated in the diagram (Fig. 73).

On such assumptions the corresponding chromosomes derived from opposite species might be represented as a series of gene-blocks or segments *abcdefg* and *adfgbe*. When two such chromosomes are free to pair together they may do so with great regularity, but in the presence of identical mates (in the normal allotetraploid of the constitution *abcdefg, abcdefg* and *adfgbe, adfgbe*) they will rarely if ever pair, on account of *d, b* and *e* being in a different sequence in the chromosomes of the two species. These observations of differential affinity are therefore in accordance with the theories of structural hybridity and chromosome pairing arrived at on other grounds.

All the different conditions of autosyndesis have not been obtained in one and the same polyploid, but it is possible to associate groups of species according to their behaviour in autosyndesis, since, whatever the conditions, this pairing indicates a certain degree of similarity between the pairing chromosomes. The classification is arbitrary, the more so since the observations on which it is founded have generally been defective in two ways: (i) they have taken little account of variations in the frequency of pairing from cell to cell; (ii) they have overlooked the formation of multivalents. Comparison, however, shows a close analogy between the results of the *synthesis* of new species by doubling and *analysis* of old species by parthenogenesis and hybridisation. It shows in allopolyploid species a range of variation from the *Primula kewensis* type to the *Raphanus-Brassica* type.

5. CLASSIFICATION OF POLYPLOID SPECIES

Observations of differential affinity show that every gradation occurs in the relationship of the homologous chromosomes in the gametic complement of polyploids. As a rule these chromosomes can be considered in sets, for the same conditions have operated

TABLE 29

Pairing of Chromosomes in Hybrids and Hybrid Derivatives of Polyploid Species

A. HYBRIDS BETWEEN HIGH AND LOW POLYPLOIDS SHOWING POTENTIALLY COMPLETE AUTOSYNDESIS

Authority.	Genus.	Basic Number.	Species Crossed (with numbers of bivalents).	Hybrid.	Pairing.
1. Haase-Bessel (1922) (cf. Buxton and Newton, 1928).	*Digitalis.*	7	*lutea* (28) × *micrantha* (14)	6x	21^{II}
2. Collins and Mann (1923).	*Crepis*	4 and 5	*setosa* (4) × *biennis* (20)	—	10^{II} (b) + 4^{I} (s)
3. Collins, Hollingshead and Avery (1929).	,,	,,	F_1 × *biennis*	6x	15^{II} (b)
4. ,, ,, ,,	,,	,,	F_3 ("*C. artificialis*," 10 (b) + 2 (s)) × *setosa*.	—	5^{II} (b) + 2^{II} (s) + 2^{I} (s)
5. Ljungdahl (1924)	*Papaver.*	7	*nudicaule* (7) × *radicatum* (35)	6x	21^{II}
6. ,, ,,	,,	7	*nudicaule* var. *striatocarpum* (35^{II}) × *nudicaule* (7).	6x	21^{II}
7. ,, ,,	,,	7	F_1 × *nudicaule*	4x	14^{II}
8. Helms and Jørgensen (1925).	*Betula.*	7	*verrucosa* (14) × *pubescens* (28)	6x	21^{II}
9. D. (1928, 1930 a)	*Prunus.*	8	*domestica* (24) × *cerasifera* (8)	4x	16^{II}
10. Crane and D. (1928)	*Rubus.*	7	*rusticanus* (7) × *thyrsiger* (14)	4x	14^{II} *
11. Tahara and Shimotomai (1927, 1931 & 1933).	*Chrysanthemum*	9	*marginatum* (45) × *lavandulafolium* (9)	8x	36^{II} †
12. Shimotomai (1933)	—	—	*Decaismeanum* (36) × *indicum* (18)	6x	27^{II}
13. —	—	—	*marginatum* (45) × *morifolium* (27)	8x	36^{II}
14. Anderson and Gairdner (1931).	*Dianthus*	15	*plumarius* (45) × *Knappii* (15)	4x	ca. 30^{II} (F_1 and F_2).
15. Mangelsdorf and Reeves (1931).	*Tripsacum* *Zea*	18, 10	Z. *Mays* (10) × *T. dactyloides* (36)	—	18^{II} + 10^{I}

* Occasional quadrivalents and tetraploid segregation.
† Suggested that each parental gamete had nine extra chromosomes, but more probably the diploid parent produced a triploid gamete.

B. Hybrids showing Partial and Variable Autosyndesis

1. Shimotomai (1931)	Chrysanthemum	9	marginatum (45) × indicum (18)	7x	$18^{II} + 27^{I}$ * variable $^{III, II, I}$
2. Shimotomai (1933)	"	9	Makinoi (9) × Decaisneanum (36)	5x	$14 - 15^{II}, 6 - 4^{I}$
3. J. Clausen (1931 c)	Viola	(?)	lutea (24) × elegantula (10)	—	$16 - 18^{II}, 5 - 1^{I}$
4. " "	"	"	tricolor alba (13) × nama (24)	—	ca. $1^{IV}, 2 - 3^{III}$, $3 - 6^{II}$, and bivalents.
5. " "	"	"	tricolor alba (13) × rothomagensis (17)	—	
6. Yarnell (1931 a)	Fragaria	7	vesca (7^{II}) × glauca (28^{II})	5x	$14^{II} + 7^{I}$ (and multivalents, v. Ch. V).
7. " "	"	7	vesca (7^{II}) × virginiana (28^{II})		

* Occasional quadrivalents.

C. Potentially Complete Absence of Autosyndesis

1. Melburn and Thompson (1927).	Triticum	7	spelta (21) × monococcum (7)	4x	$0 - 5^{II}$
2. Kihara and Nishiyama (1930).	"	7	spelta (21) × agilopoides (7)	4x	$6 - 10^{II}$ or ^{III}
3. " "	Triticum and Secale.	7	durum (14) × vulgare (21)	5x	$13 - 14^{II}$ or ^{III}
4. Kihara (1924), Kattermann (1934)	Fragaria	7	T. vulgare (21) × S. cereale (7)	4x	$0 - 6^{II}, 0 - 2^{III}$
5. Ichijima (1930)	"	7	bracteata (7) × virginiana (28)	5x	$7^{II} + 21^{I}$
6. " "	"	7	californica (7) × chiloensis (28)	5x	
7. " "	"	7	alba rosea (7) × chiloensis (28)	5x	
8. Morinaga (1929)	Brassica	?9	cernua (18) × Napellus (19)	$4x+1$	$10^{II} + 17^{I}$
9. Eleier, 1928	Beta	9	vulgaris (9) × trigyna (27)	4x	$9^{II} + 18^{I}$
10. Haase-Bessel (1916) (cf. Buxton and Newton).	Digitalis.	7	purpurea (28) × lutea (56)	12x	0^{II} (cf. Table A1)
11. Haase-Bessel (1922)	Viola		lanata (28) × lutea (56)	12x	0^{II} (cf. Table A1)
12. J. Clausen (1931 c)	"	7?	Kitaibeliana (7) × arvensis (17)		$2 - 6^{II}$
13. " "	"	"	nana (24) × Kitaibeliana (7)		$5 - 6^{II}$
14. " "	"	"	Kitaibeliana (i) (7) × Kitaibeliana (ii) (18).		ca. 6^{II}

Cf. Kihara and Lilienfeld (1935) and Hosono (1935) for further observations on Triticum-Ægilops hybrids.

D. PAIRING IN TRIPLOID HYBRIDS (cf. Table 26)

1. AAB type (crosses of diploids with autotetraploids or unreduced diploids) *

Authority.	Genus.	Basic Number.	Species Crossed.	Hybrid.	Pairing.
1. Håkansson (1929 b)	*Salix*	19	*viminalis* (19) × *caprea* (19)	$\left(vc + \dfrac{vc}{2}\right)$	$16 - 19^{III}$
2. Müntzing (1930 b)	*Galeopsis*	8	*pubescens* (8) × *speciosa* (8)	(pss)	$8^{II} + 8^{I}$
3. Poole (1931)	*Crepis*	5	*rubra* (5) × *foetida* (5)	$\left(rf + \dfrac{rf}{2}\right)$	—
4. Hollingshead (1930 b)	,,		*capillaris* (3) × *tectorum* (4)	(ccf)	$3^{II} + 4^{I}$ rarely 1^{III}
5. Karpechenko (1927, a and b).	*Raphanus* and *Brassica*.	9	[*R. sativus* (9) × *B. oleracea* (9)] × *R. sativus*.	(rrb)	$9^{II} + 9^{I}$
6. Pariser (1927)	*Saturnia*		*pavonia* (29) × [*pyri* (3) × *pavonia* (29)]	(pv, pv, pr)	—
7. Federley (1931)	*Pygaera*		*pigra* (23) × [*pigra* × *curtula* (29)]	(ppc)	$23^{II} + 29^{I}$
8. Steere (1932)	*Petunia*	7	*hybrida* (14) × *axillaris* (7)	(hha)	7^{III}

* The letters in brackets show the constitution of the triploid in regard to the sets of its two parents. The third set of the *Salix* and *Crepis* hybrids is mixed.

D. TRIPLOID HYBRIDS (continued)

2. *ABC* type (crosses between diploid and allotetraploid species). *Cf.* also structural hybrids.

Authority.	Genus.	Basic Number.	Species Crossed.	Pairing.
1. Rosenberg (1909)	*Drosera*	10	*longifolia* (10) × *rotundifolia* (20)	$10^{II} + 10^{I}$
2. ,, (1917)	*Hieracium*	9	*auricula* (9) × *aurantiacum* (18)	$9^{II} + 9^{I}$
3. Ljungdahl (1922)	*Papaver*	7	*atlanticum* (7) × *dubium* (14)	$0^{II} + 21^{I}$
4. Goodspeed, Clausen and Chipman (1926);	} *Nicotiana*	12	*sylvestris* (12) × *Tabacum* (24) *tomentosa* (12) × *Tabacum* (24) *paniculata* (12) × *rustica* (24)	$12^{II} + 12^{I}$
5, 6. Goodspeed and Clausen (1927 b).				
7. Goodspeed and Clausen (1927).	,,	12	*Bigelovii* (24) × *glutinosa* (12)	$0_1^{I} + 36^{I}$
8. Brieger (1928)	,,	12	*Tabacum* (24) × *Rusbyi* (12)	$12^{II} + 12^{I}$
9. Sax (1922),	*Triticum*	7	*monococcum* (7) × *turgidum* (14)	$7^{II} + 7^{I}$
10. Thompson (1926 b) Kihara (1924) *cf.* Mather, 1935	,,	7	*dicoccum* (14) × *monococcum* (7)	$4 - 7^{II} + 13 - 7^{I}$
11. Kihara and Nishiyama (1930).	,,	7	*agilopoides* (7) × *dicoccum* (14)	$0 - 3^{III} + 4 - 7^{II} + 6 - 7^{I}$
12. v. Berg (1931)	*Ægilops* and *Secale*.	7	*A. triuncialis* (14) × *S. cereale* (7)	$5 - 7^{II} + 11 - 7^{I}$
13. Jørgensen (1927)	*Lamium*.	9	*dissectum* (18) × *amplexicaule* (9)	$9^{II} + 9^{I}$
14. Fukushima (1929)	*Brassica* and *Raphanus*.	9	*B. cernua* (18) and *B. juncea* (18) × *R. sativus* (9).	$0^{II} + 27^{I}$
15. Wakakuwa (1931)	*Celosia*	18	*argentea* (36) × *cristata* (18)	$18^{II} + 18^{I}$
16. v. Berg (1934)	*Triticum*	7	*turgidum* (14) × *villosum* (7)	$0 - 4^{II}$
17. Nishiyama (1934)	*Avena*	7	*barbata* (14) × *strigosa* (7)	$0 - 5^{III}$

D. (continued)

3. Of Unknown Origin and Composition (Species, etc.).

Authority.	Genus.	Basic Number.	Species.	Somatic Number.	Pairing.
1. Rosenberg (1917, 1927)	*Hieracium*	9	*boreale, pseudoillyricum*, etc.	$3x=27$	few pairs
2. de Vilmorin (1929)	*Solanum*	12	*Commersonii*	$3x=36$	—
3. Rybin (1929 b)	,,	12	*coyacanum, medians, "tuberosum"*	$3x=36$	—
4. D. (1928)	*Prunus*	8	*avium nana* (? *avium* × *cerasus*)	$3x=24$	$1-8^{III}$
5. Moffett (1931)	*Pyrus*	17	*minima* (?*ABC* type)	$3x=51$	$0-4^{III}, 17-14^{II}, 13-16^{I}$

E. AUTOSYNDESIS IN "TRIPLOIDS" DERIVED FROM ALLOPOLYPLOIDS

Authority.	Species.	Somatic Number.	Basic Number.	Pairing.
1. Collins *et alii* (1929)	*Crepis artificialis*, mutant	$2n=36$	$x=5$	$17^{II} + 2^{I}$
2. Nebel (1929), D. and Moffett (1930).	*Pyrus Malus*, vars.	$2n=51$	$x=7$	} less than 17^{I}
3. Moffett (1931)	*Crataegus monogyna*, var.	$2n=51$	$x=7$	
4. Buxton and Newton (1928)	*Digitalis merionensis* × *D. ambigua*	$2n=84$	$x=7$	more than 28^{II}

F. Pairing in Multi-specific Hybrids

1. Pairing of chromosomes in the characteristic allopolyploid *Raphanus-Brassica* and its derivatives (trispecific hybrids). Karpechenko, 1927, *a* and *b*, 1929

Raphanus sativus	(Rs)	$n = 9 = x$	Rs.Bo	$(2x = 18)$	$9^{II} (18^I)$
R. raphanistrum	(Rr)	$n = 9 = x$	Rs.Rs.Bo.Bo	$(4x = 36)$	18^{II}
Brassica oleracea	(Bo)	$n = 9 = x$	Rs.Rs.Bo	$(3x = 27)$	$9^{II} (9^I)$
B. campestris	(Bcp)	$n = 10 = x$	Rs.Bo.Rr	$(3x = 27)$	$9^{II} (9^I)$
B. carinata	(Bcr)	$n = 17 = 2x?$	Rs.Bo.Bcp.	$(4x = 28)$	$5-6^{II} (15-17^I)$
			Rs.Bo.Bcr.	$(4x = 35)$	$9-17^{II} (17-1^I)$

2. Pairing in a hybrid of two new hybrid polyploids (quadrispecific hybrids) v. Berg, 1931 *b*. *Cf.* also Buxton & Dark, 1934, on *Digitalis mertonensis* and Kostoff, 1933, on *Nicotiana*.

Ægilotricum, $7x$ (= *Ægilops ovata*, 14, + *Triticum durum*, 14) × *Triticum turgidovillosum*, $6x$ (= *T. turgidum*, 14 + *T. villosum*, 7). $7x = 49$; $12-14^{II}$, $25-21^I$.

to secure the degree of differentiation found between pairs of homologues in any two. The different kinds of polyploid species may therefore be classified, arbitrarily but conveniently, in three groups.

The first class includes those in which autosyndesis never occurs in haploids, triploids or hybrids, or *a fortiori* in the normal species. This extreme of differentiation, found in *Nicotiana Tabacum* and in *Triticum compactum*, is analogous to that in the *Raphanus-Brassica* tetraploid. It is probably found most generally in old established sexually-propagated annual species.

The second class consists of those species with imperfect differentiation. The pairing observed in *Bromus erectus* may serve as a type of behaviour in this group. Table 31 shows pairing in one plant; in others a sexivalent and octavalent were found. The occurrence of univalents in 9 per cent. of the cells is probably due to mechanical interference with pairing at zygotene as in tetraploid *Primula sinensis*, and is therefore a symptom of autopolyploidy.

Following is a list of similar species whose pairing is normally and "legitimately," as we may say, that of diploids, but in which multivalents are occasionally and illegitimately formed. In these, more or less autosyndesis is found, or is to be expected, in haploids and hybrids. Experimental breeders are treading on dangerous ground when they treat such species as functional diploids.

TABLE 30

Polyploid Species of an Intermediate Type

Rumex Acetosella ($6x$, 42) . . . Kihara, 1927; Ono, 1930.
Triticum vulgare ($6x$, 42) . . . Winge, 1924; Huskins, 1928.
Avena sativa ($6x$, 42) . . . Huskins, 1927.
Bromus erectus eu-erectus ($8x$, 56) . . Kattermann, 1931.
Rubus thyrsiger ($4x$, 28) . . . Crane and D., 1927.
Prunus domestica ($6x$, 48) . . .
P. spinosa ($4x$, 32) . . . } D., 1930 a.
P. laurocerasus ($22x$, 176) . . . Meurman, 1929 a.
Pyrus Malus (4-$6x$, 34) . . . D. and Moffett, 1930 (v. p. 224).
Cardamine pratensis ($4x$, 30) . . Lawrence, 1931 b.
Lythrum Salicaria ($6x$, 30) . . Shinke, 1929.
Tulipa stellata ($4x$, 48) . . . D. and Janaki-Ammal, 1932.
Veronica spp. ($4x$, 68) . . . Graze, 1935.
Helianthus tuberosus ($6x$, 102) . . Kostoff, 1934.

TABLE 31

Pairing in an Intermediate Octoploid Plant, *Bromus erectus* var. *eu-erectus* ($2n = 56 = 8x$, Kattermann, 1931)

Type of pairing .	28^{II}	2^{I}	$1^{III} + 1^{I}$	2^{III}	1^{IV}	2^{IV}	Not Determined.	Total.
Number of cells .	1,006	99	5	2	11	2	26	1,151

The third class consists of species in which there is no clear differentiation of the corresponding chromosome sets. They may

FIG. 74.—First metaphase in *Tulipa Clusiana* ($5x = 60$) showing 8 univalents (hatched), 15 bivalents, 3 trivalents, 2 quadrivalents and one quinquevalent. × 4000 (from Newton and D., 1929).

even, in some forms, be considered identical. In others they may be equally dissimilar. They occur as triploids (the result of doubling in one parental gamete) tetraploids (the result of doubling in both parental gametes or somatically and pentaploids (the result of doubling on two occasions). The tetraploids are of somewhat reduced fertility, while the triploids and pentaploids are almost entirely sterile. In this they resemble polyploid species that are interchange heterozygotes, and must depend entirely on vegetative propagation for their continued existence.

With these species, polyploids having the behaviour of simple structural heterozygotes may conveniently be considered, for the reasons given earlier.

TABLE 32

Autopolyploid Species and Mutants (cf. Table 37)
(basic numbers in brackets)

A. TRIPLOIDS (cf. Table 19 (b))

1. Wild Species (i.e., clonal species asexually reproduced) :—

Lilium tigrinum (12)	Takenaka and Nagamatsu, 1930 (v. Table 15).
Tulipa præcox (12)	D., unpublished.
Tulipa saxatilis (12)	D., unpublished.
Hydrilla verticillata (8)	Sinoto, 1929.
Alnus cordata, etc. (7)	Jaretzky, 1930.
Lycoris radiata (11)	Nishiyama, 1928.
L. squamigera (9)	Takenaka, 1930.
Fritillaria camtschatcensis (12)	Matsuura, 1935.
Fritillaria spp. (12 and 13)	D., 1936 d.
Crocus sativus (8)	Karasawa, 1933.
Narcissus spp.	Nagao, 1933.
Allium carinatum (8)	Levan, 1933 a.
Tulipa lanata (12)	Upcott (unpublished).
Trichoniscus provisorius (8)	Vandel, 1927 (v. Ch. XI).

2. Mutants of Species (occasional progeny from unreduced gametes) :—

Drosophila melanogaster (4)	Bridges, 1921 a; Metz, 1926.
Crepis capillaris (3)	Navashin, 1926.
Rumex Acetosa (8) (discovered through examination of intersexes).	Ono, 1928, 1930, 1935; Ono and Shimotomai, 1928.
Ranunculus acris (6) (probably 3x, gynodimorphic)	Sorokin, 1927 a.
Zea Mays (10)	Randolph and McClintock, 1926; McClintock, 1929 (7–10^{III}).
Nicotiana alata (9)	Ruttle, 1928; Avery, 1929 (monosomic diploid selfed).
Funaria hygrometrica (7)	Wettstein, 1928.
Sphærocarpos Donnellii (8)	Allen, 1935, a and b.
Oryza sativa (10)	Morinaga and Fukushima, 1933.
Habrobracon juglandis (10)	Bostian, 1936.
Bombyx mori (29)	Kawaguchi, 1933.
Tulipa Clusiana (12)	Upcott and La Cour, 1936.

3. Cultivated Forms (vegetatively propagated) :—

Hyacinthus orientalis (8)	De Mol, 1923; Belling, 1925, 1927; D., 1926, 1929 (highly fertile).
Tulipa Gesneriana vars. (12)	De Mol, 1929; Newton and D., 1929; Upcott and La Cour, 1936.

NON-HYBRID POLYPLOIDS

Canna spp. (9)	Belling, 1921, 1925 d.
Fritillaria pudica (13)	D., 1936.
F. latifolia major (12), etc.	D., 1936.
Iris japonica (18)	Kazao, 1929.
I. florentina (16)	Kazao, 1929.
Narcissus poeticus var. (7)	Nagao, 1929.
Hemerocallis fulva (11)	Belling, 1925 d; Takenaka, 1929.
Morus spp. (14)	Osawa, 1921 (parthenocarpic).
Prunus Mume (8)	Okabe, 1928.
P. serrulata, etc. (8)	Okabe, 1928 (11 vars. $3x$; 30, $2x$).
Pyrus Malus vars. (17)	Rybin, 1927; D. and Moffett, 1930.
Cratægus monogyna var.	Moffett, 1931, a and b.
Primula Sieboldii (12)	Ono, 1927.
Thea sinensis (15)	Karasawa, 1932.
Agave cartula (30)	Doughty, unpub.
Nasturtium officinale (16)	Manton, 1932.

4. Experimental Hybrids ($2x \times 4x$):—

Vicia Cracca (7)	Sveshnikova, 1929.
Solanum Lycopersicum (12)	M. M. Lesley, 1929 (9–10^{III}).
Datura Stramonium (12)	Belling, 1927 (12^{III}).
Primula sinensis (12)	Dark, 1931 (8–9^{III}).
Campanula persicifolia (8)	Gairdner and D., 1931 (3–5^{III}).
Nicotiana Tabacum (24)	Goodspeed, 1930 (X-rayed).

B. TETRAPLOIDS (*cf.* Table 19 (*c*))

1. Wild Species:—

Avena elatior (7)	Kattermann, 1931.
Prunus Cerasus (certain forms) (8)	D., 1928.
Allium schœnoprasum var. *sibiricum* (8)	Levan, 1931.
Dactylis glomerata (7)	Müntzing, 1933.
Allium oleraceum (8)	Levan, 1933 a.
A. schœnoprasum (8)	Levan, 1935 b.
Tulipa turkestanica (12)	Upcott and La Cour, 1936.
T. Whittalli (12)	Upcott and La Cour, 1936.
Veronica Tournefortii (8)	Beatus, 1934.

2. Mutants of Species:—

Amblystegium serpens (11)	E. and E. Marchal, 1911, 1912.
Funaria hygrometrica (7)	v. Wettstein, 1924, 1928.
Primula sinensis (12)	*Cf.* D., 1931 a.
Datura Stramonium (12)	*Cf.* Blakeslee *et al.*, 1923.
Solanum Lycopersicum (12)	Jørgensen, 1928; Lesley, 1930; and Sansome, 1931.
S. nigrum (36)	Jørgensen, 1928.
Citrus Limonum, etc. (9)	Frost, 1925.
Crepis capillaris (3)	Navashin, 1926.
C. tectorum (4)	Navashin, 1926 (fertile).
Primula obconica (12)	Sansome and Philp, 1932.
P. malacoides (9)	Newton (*cf.* Tischler, 1931); Philp, unpublished; Kattermann, 1935.
Campanula persicifolia (8)	Gairdner, 1926; Gairdner and D., 1931.
Nothoscordon bivalve (8)	Beal, 1932.
N. fragrans (8)	Matsuura and Sutô, 1935.

C. PENTAPLOIDS

1. Wild Species :—

Tulipa Clusiana (12)	Newton and D., 1929 (Figs. 60–61).
Ochna serrulata (7)	Chiarugi and Francini, 1930.
Agave sisalana (30)	Doughty, unpub.

2. Mutant of Species :—

Crepis tectorum (5) . . . . Navashin, 1926 (Fig. 13).

D. HEXAPLOID

2. Mutant of Species :—

Rumex acetosa (8) . . . . Yamamoto, 1935.

E. OCTOPLOID

1. Wild Species :—

Crepis biennis (5) . . . . Collins, Hollingshead and Avery, 1929; *cf.* Babcock, and Cameron, 1934.

The triploids form a majority of trivalents, the tetraploids a majority of quadrivalents. The higher configurations possible in the pentaploids necessarily make the kind of association more variable, but it is characteristically different from that in allopentaploids. Thus in pentaploid *Triticum* and *Nicotiana* (Kihara and Nishiyama, 1930 ; and Webber, 1930) four chromosome sets are associated as pairs, and the fifth is chiefly unpaired, although they may take part in trivalents (Brieger, 1928 ; Table 29). A pentaploid hybrid species of *Agropyrum* is presumably of this type (Simonet, 1934). Chromosome behaviour in the pentaploid species *Ochna serrulata*, *Agave sisalana* and *Tulipa Clusiana* is entirely different. They both form a high proportion of multivalents (Fig. 74). A new mutant autopentaploid (such as that found by Navashin, 1926, in *Crepis capillaris*) has not yet been examined for comparison, but from what we know of the behaviour of auto-triploids and tetraploids, these pentaploids must be of this kind. It is, however, possible on cytological grounds that they consist of five sets, equally (and not strongly) differentiated. These species are sexually almost sterile, the one reproducing by apomixis, the other by suckers and by bulbs. They are clonal species.

This last class of polyploids includes high multiples, which probably always have different degrees of relationship between different sets, for two reasons : the complexity of their origin and the lower effectiveness of balance in acquiring constant allopolyploidy. The characteristic of these species is variability in the chromosome number. This variability arises in two ways. In sexually reproducing species the irregular segregation of the multivalents gives gametes with various numbers; owing to the high gametic number the unbalance of these is less serious than it would be in low polyploids, and they are not eliminated (v. Ch. VIII).

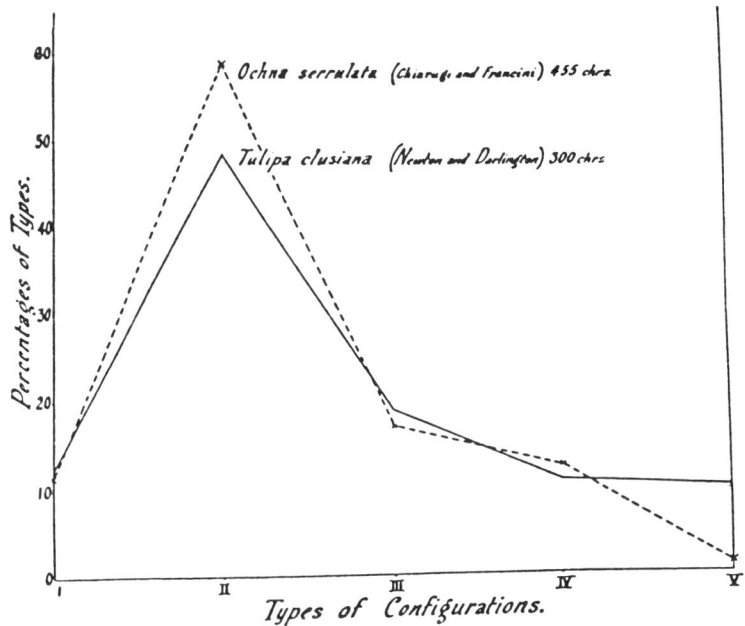

FIG. 75.—Graph showing the percentages of different types of configuration in two autopentaploid species at meiosis in the pollen mother-cells. The more frequent high configurations in *Tulipa Clusiana* are probably due to a higher chiasma frequency; *cf.* Yamamoto, 1935, on autohexaploid *Rumex*.

In the polyploid species of *Agave*, on the other hand, the smaller chromosomes are frequently distributed unevenly at mitosis, and owing to their presence being indifferent in a high polyploid they are often lost. The result is that the tetraploid and pentaploid species have the full complement of five long chromosomes regularly multiplied, while the short chromosomes are lacking to a variable extent.

Following is a list of species of this type with the numbers of somatic chromosomes found or inferred in each.

TABLE 33

Polyploid Species with Variable Numbers

Cochlearia anglica, 37–44, 49, 50	Crane and Gairdner, 1923.
Salix triandra, 38, 44	Blackburn and Harrison, 1924.
Prunus laurocerasus ($22x = 176$), ca. 176	Meurman, 1929.
Crepis biennis ($8x = 40$), 39–45	Hollingshead and Babcock,
C. ciliata, 40–42	1930 ; Babcock, 1935.
Viola canina, 40–47	J. Clausen, 1931 b.
V. Battaudieri, 52–60	J. Clausen, 1931 c.
Agave sisalana ($5x=150$), 135–145	Doughty, unpub.
A. Zapupe ($4x = 120$), ca. 110	
Chrysanthemum ornatum ($8x=72$), 72–74.	Shimotomai, 1933.

6. THE DEVELOPMENT OF THE POLYPLOID SPECIES

(i) **Introduction.** The increase of chromosome number by which a polyploid arises has been reversed in experiment, first, by crossing an autopolyploid with its diploid relatives and selfing the triploid progeny, or back-crossing them to the diploid (as in *Solanum, Campanula, Primula sinensis*) ; or, secondly, by parthenogenesis, in which an autotetraploid gives its fertile diploid parent (as in *Campanula*, Gairdner and D., 1931), and an allotetraploid gives its sterile diploid parent (as in *Digitalis*, Buxton and D., 1931). Thus when the allopolyploid reverts to the diploid, sterility is produced, which can only be got rid of by another doubling of the chromosome number. Chromosome doubling in a hybrid must therefore be an irreversible process in nature. On these and other grounds it is to be supposed that polyploid species are often of great antiquity, and this suggests two problems: How is a polyploid species to be recognised as such ? And how do polyploid species change after their origin ? The second problem will be considered first.

(ii) **Variation in Cell Size.** External and developmental factors

CELL-SIZE

influence the size of the cell and of its nucleus, but, these factors being equal, the size might be expected to be proportionate to the number of chromosomes of which the nucleus is constituted when the number is simply doubled or halved. This condition is realised immediately after the formation of a new nucleus in the pollen. Diploid pollen grains have 1·25 times the linear dimensions and twice the bulk of haploid pollen grains (Belling on *Uvularia*, 1925 ; *Solanum*, Fig. 67). The same applies with certain restrictions to the cells of all parts of the organism, and is equally true of the polyploid derivatives of hybrids and pure forms, as shown by v. Wettstein (1924) with diploid, tetraploid and octoploid mosses, and by the other authors cited.

TABLE 34
Observations of Cell Size in Haploids, Diploids and Polyploids

Animals :—
Drosophila melanogaster (x, $2x$, $3x$)	Bridges, 1925 b.
Artemia salina ($2x$, $4x$, wild races)	Artom, 1928.
Trichoniscus provisorius ($2x$, $3x$, wild races)	Vandel, 1927 b.

Plants :—
Datura x, $2x$, $4x$)	Blakeslee, 1922 ; Belling and Blakeslee, 1927.
Crepis (x, $2x$)	Hollingshead, 1928.
Crepis capillaris (x, $2x$, $3x$, $4x$)	Navashin, 1931 b.
Œnothera (x, $2x$)	Emerson, 1929 ; Gates and Goodwin, 1931.
Raphanus-Brassica ($2x$, $3x$, $4x$, $6x$)	Karpechenko, 1927, a and b.
Nicotiana digluta ($2x$, $4x$)	Clausen and Goodspeed, 1925.
Funaria, etc. ($2x$, $4x$, $8x$)	v. Wettstein, 1924.
Uvularia (x, $2x$)	Belling, 1925 (pollen grains).
Solanum (x, $2x$)	Jørgensen, 1928 (pollen grains).
Tradescantia (x, $x + 1$, $x + 2$, etc.)	D., 1929 c (pollen grains).

While the proportion between size and chromosome number is fairly closely maintained so far as cells are concerned, intracellular structures such as chloroplasts may be increased in number instead of in size, and multicellular structures are developed in changed proportions. The vigour of the organism as a whole is usually reduced by either reduction or multiplication of the chromosome number, and its size therefore does not change proportionally. The genotype of the new haploid or polyploid is still adapted to the cell size of the ancestral diploid and works less favourably under new conditions. A reduction in the number of cells in corresponding organs of the plant is found after doubling twice in *Funaria hygrometrica* (v. Wettstein, 1924), and after doubling only once in *Acrocladium cuspidatum*. The haploid is usually of a more immature stripling growth than the diploid, while the tetraploid is of a stouter growth (Jørgensen, 1928 ; Davis and Kulkarni, 1930 ; Clausen and Mann, 1924). It should be noted that this applies to the fruits for different reasons. The development of many-seeded fruits is often proportional to their seed content, which is lower in haploids, triploids and non-hybrid tetraploids than in the corresponding more fertile diploids. Hence the fruits are smaller in these, and often of different shapes (Blakeslee and Belling, Jørgensen).

Organisms are therefore modified in general character by a change in their size, *i.e.*, by a change in volume-surface relations. The same chromosome complement, in fact, always has a qualitatively as well as a quantitatively different phenotypic expression when represented different numbers of times, and changes from the condition to which it is adapted, whether haploid, diploid or tetraploid, are more or less deleterious, just as are changes of balance. This principle is of importance in considering the origin of the special differentiation found in species where the male is haploid and the female diploid. In these the complement is adapted to both the haploid and the diploid conditions. The differentiation in phenotypic expression is then carried to an extreme. It applies even to cell-size. Exceptional diploid males in *Habrobracon* are no larger than the normal haploids (Torvik, 1931). Special genes or gene-combinations can therefore be selected to modify the ordinary relationship of the size of the organism to that of the nucleus when the chromosome number is altered, just as they are evidently selected to modify the sexual character of the organism (Ch. IX). Simple doubling in species not specialised in this exceptional way is usually accompanied by more or less gigantism. Where the polyploid form shows no increase in size, this can be ascribed to segregation of dwarfing factors in the parent, *e.g.*, in *Crepis tectorum* (Navashin, 1926), where great size variation occurs in the diploid. It may be that in some species giant polyploids are not fitted to survive, and the failure of polyploidy to appear in many groups of flowering plants may be due to failure of such species to segregate or mutate to dwarfness.

On the other hand, in certain exceptional cases increase of size does not accompany the multiplication of the chromosome number. The only autohexaploid flowering plant known (*Rumex acetosa*, Yamamoto, 1935) is very much smaller than its triploid parent, and in *Anthoceros* doubling does not produce the same increase as in *Funaria* (Schwarzenbach, 1926). Similarly the autotetraploid *Lathyrus odoratus* (Fabergé, 1935) is no larger in stature than the diploid (*cf.* Dorsey, 1936). It seems as though doubling will increase the size of the normal type in most cases, but that some organisms are more strictly adapted to a specific size, so that particular genetic conditions influencing size, such as chromosome number, may have a deleterious effect. This condition is perhaps in part responsible for the general absence of tetraploidy amongst animals.

TABLE 35

Measurements of Cells in Haploids, Diploids, and Polyploids

I. (i) and (ii), *Funaria hygrometrica*, leaf gametophyte (v. Wettstein, 1924). (iii) *Crepis capillaris*, dermatogen (Navashin, 1931 *b*).

		x.	$2x$.	$3x$.	$4x$.
Cell number across leaf	(i.)	48	49	32	13
Cell volume	(ii.)	86·5	158·1	273·1	472·8
Cell volume (in 1,000 cu.μ)	(iii.)	1·8	4·0	6·0	9·0

II. *Raphanus-Brassica* hybrids (stomata) (Karpechenko, 1928).

	$\frac{R.}{2x}$	$\frac{B.}{2x}$	$\frac{RB.}{2x}$	$\frac{RRB.}{3x}$	$\frac{RRBB.}{4x}$	$\frac{RRRBB.}{5x}$	$\frac{RRRBBB.}{6x}$
Length	7·3	7·1	7·1	9·1	9·7	10·6	10·4
Breadth (in mm. × 1,350)	5·5	5·3	5·4	6·7	7·2	7·8	7·9
Cubical proportion	—	—	1·0	—	2·3	—	3·1

III. *Drosophila* (eye-facets) (Bridges, 1925 b). *N.B.*—The triploid is an unrelated individual.

	x.	$2x.$	$3x.$
Linear proportions	1	1·21	1·37
Cubical proportions	1	1·77	2·56

When these observations are compared with similar observations of diploid and polyploid forms of remote relationship, whether species or varieties, various kinds of difference are found, which may be classified under three heads.

1. Many polyploid wild species and cultivated forms have the same proportionate increase in size as is found in critical experiments, and it may then be inferred that little change has taken place in the polyploid since its origin. This conclusion is particularly obvious because wild species of this kind are found to be propagated vegetatively or " semi-vegetatively " (*e.g.*, *Vallisneria gigantea*, Jørgensen, 1927 *a* ; Caninæ roses), and cultivated forms have been subject to human selection in favour of size since their origin (*e.g.*, *Triticum* species, Sax, 1922 ; *Petunia*, Dermen, 1931).

2. Certain polyploid species and races are no bigger, or are even smaller, than their diploid relatives, yet in certain structures, such as pollen grains, they preserve what must be supposed to be their initial gigantism. This is true of the Cinnamomeæ roses, where the polyploid species are actually of smaller growth than the diploid

TABLE 36

Linear Measurements of Cells in Polyploid Derivatives (in microns)

	$2x.$	$3x.$	$4x.$	$6x.$
Petunia vars., pollen grains	35·3	37·6	41·7	—
Triticum species, pollen grains	44·0	—	51·1	55·3
Rosa (sect. Cinnamomeæ), pollen mother-cells	6·8	—	8·8	10·0
Primula obconica vars., pollen grains	29·4	—	32·9	—
Allium species, pollen grains	29·0	—	35·0	—

(Erlanson, 1933), and of *Primula obconica* (Philp, unpublished) and *Euphorbia granulata* (Hagerup, 1932), where there is no evidence of gigantism in the tetraploid forms except in the size of the cells (*cf. Allium*, Levan, 1932 ; *Narcissus*, Nagao, 1933).

3. Finally, the great body of polyploid species show a purely random size relationship with their diploid relatives, being on the average neither larger nor smaller as a whole or in any particular structure. Thus Kihara and Ono (1926) found that the size of pollen grains in *Rumex Hydrolapathum* ($2n = 200$) was no greater than in related diploid and tetraploid species ($2n = 20, 40$).

It is therefore necessary to explain why polyploid species have lost the initial size difference they must be supposed to have had, first in their general growth, and later in their detailed structure. Apart from changes in chromosome number, it is found that the progeny of new allopolyploids vary in fertility (*e.g.*, *Ægilotricum*, Kihara and Katayama, 1931 ; *Raphanobrassica*, Karpechenko, 1927, *a* and *b*), and in size (*Primula kewensis*, Newton and Pellew, 1929 ; *Nicotiana digluta*, R. E. Clausen, 1928). Such changes must be chiefly due to the pairing and crossing-over which must take place occasionally between chromosomes of opposite parents, even with the highest differentiation. Dwarfing, probably from such segregation, has been found in two original tetraploids (Kostoff, 1935 *a*).

In old-established polyploids such as *Triticum vulgare* and *Avena sativa* such abnormalities still occur, and when they occur give rise to variation. An occasional quadrivalent is formed owing to the pairing of dissimilar chromosomes from the ultimate diploid ancestors (autosyndesis) as well as pairing between identical chromosomes. When the dissimilar chromosomes pair, segregation of new types (actually resembling related species) results. This " illegitimate " pairing gives, as we have seen, " secondary segregation," and occurs more frequently in varieties newly derived from crossing different races.

In this way will arise variations on which natural selection can work to change the size of the new form. But such variations will not increase its fertility. Recombination of whole chromosomes or even of any large parts will mean a reduction in the differentiation of opposite sets, tending towards autopolyploidy and hence towards lower fertility. It is therefore probable that the occurrence of structural changes in the two separated sets in the normal course of variation is a more important source of differentiation between the chromosome sets, and thus of the evolution of allopolyploids in general. This will be shown more clearly by considering the occurrence of polyploidy within the species.

(iii) **Polyploidy within the Species.** Many species have been found to include a series of polyploid forms. In some cases these are indistinguishable from one another, except by distribution, while in

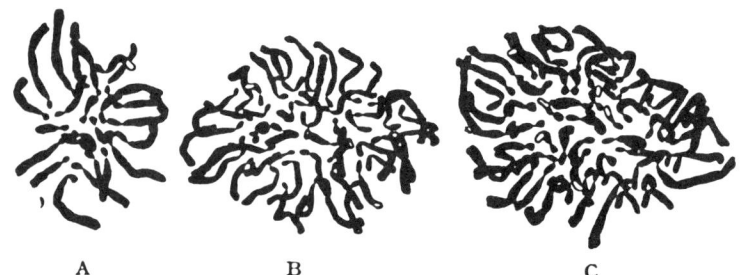

A B C

FIG. 76.—Mitotic metaphases from the three clonal types of *Tulipa Clusiana*. A. The diploid from Thibet. B. The tetraploid from Chitral. C. The pentaploid from the Mediterranean. *Cf.* Plate II. (D., unpublished.)

others they show size differences or slight differences of a general character.

In some species the multiple forms are autopolyploids, in others they must be allopolyploids, since they have high chromosome numbers and are seed-fertile. In one species, *Allium schœnoprasum* (Levan, 1935 *b*) chromosome behaviour shows that there are both kinds, one a giant auto-tetraploid, the other a smaller allo-tetraploid.

It might be assumed that such polyploid forms had arisen from their diploid relatives within the species, by hybridisation with a second species followed by selective elimination in the polyploid of all the characters of one parent which distinguished it from the other, *i.e.*, *convergence* of the two forms, parent and hybrid.

But it is more plausible to assume that these forms arose as autopolyploids, that is, as giants with free pairing amongst their homologous chromosomes. This condition is still found in certain forms which have presumably remained unaltered since their origin. Amongst these are the sterile polyploid clone-species in *Tulipa*, *Lilium* and *Fritillaria*, and probably the giant forms in *Allium* and *Portulaca*.

The rest, however, have changed in one or, as in the case of *Silene ciliata* or *Viola Kitaibeliana*, in both of two ways : they have lost their gigantism ; they have become allopolyploid. For example, the hybrid between the two forms of the *Viola* species shows no autosyndesis (Table 29). The loss of the gigantism may be attributed to mutation and segregation as already suggested. The loss of the autopolyploidy must be due to differentiation arising between the different pairs of haploid sets. Structural changes in the chromosomes might lead both to the differentiation and to the loss of gigantism ; it would be advantageous to any race to have pairs of chromosomes alike in regard to any new structural change, for this would prevent the formation of larger associations of chromosomes than twos. We shall see later, in the wide distribution of inversions, a means by which such differentiation could arise without a genetic change that is physiologically expressed. This method of differentiation is probably therefore the chief agent of change in polyploid forms after their origin. It is, in general, analogous in cause and effect to that which arises between particular pairs of chromosomes in sex heterozygotes and between the two whole sets in complex heterozygotes (Ch. IX).

TABLE 37

Examples of Polyploidy within the Species

A. *Not associated with systematic differences* (other than distribution

Rosa acicularis Lindl.	$n = (7), 21, 28$	Täckholm, 1922 ; Erlanson, 1929.
Potentilla opaca L.	$n = 7, 14$	Tischler, 1928.
Callitriche stagnalis	$n = 5, 10$	Jørgensen, 1923.
Salix aurita L.	$n = 19, 38$	Blackburn and Harrison, 1924 ; Håkansson, 1929 b.
Ranunculus acris	$n = (6), 7, 14$	Senjaninova, 1927 ; Sorokin, 1927 ; Larter, 1932.
Anemone montana	$n = 8, 16, 24$	Moffett, 1932 b.
Nasturtium officinale	$n = 16, 32$ (and 16×32).	Manton, 1932.
Draba magellanica Lam.	$n = 24, 32, 40$	Heilborn, 1927.
Crepis Bungei Ledeb.	$n = 4, 8$	Hollingshead and Babcock, 1930.

228   THE BEHAVIOUR OF POLYPLOIDS

Cassia mimosoides	$n = 8, 16$	Kawakami, 1930.
Silene ciliata	$n = 12, 24, 96$	Blackburn, 1928.
Dianthus carthusianorum	$n = 15, 30$	⎫
D. plumarius	$n = 15, 45$	⎬ Rohweder, 1929 ; Andersson and Gairdner, 1931.
D. arenarius	$n = 30, 60$	⎭
Festuca elatior	$n = 7, 21, 35$	Lewitsky and Kuzmina, 1927.
Phacellanthus tubiflorus	$n = 21, 35, 42$	Matsuura, 1935.
Pellia epiphylla	$n = 9, 18$	Heitz, 1928.

B. *Associated with slight systematic differences* (cf. Tables 33 and 70)

Plantago major L., s.l.	$n = 6, 12$	Miyaji (cf. Ishikawa, 1916).
Papaver nudicaule, s.l.	$n = 7, 35$	Ekstrand, 1918 ; Ljungdahl, 1924.
Erophila verna L., s.l.	$n = 7, 15, 32$	Winge, 1926.
Viola Patrini DC., s.l.	$n = 12, 24, 36$	Miyaji (1913, 1926), 1929.
V. biflora L., s.l.	$n = 6, 24$	Miyaji, 1929.
V. Kitaibeliana R. et S., s.l.	$n = 7, 8,$ ca. 12, 18, 24.	J. Clausen, 1927, 1929, 1931 c.
Valeriana officinalis L., s.l.	$n = 7, 14, 28$	Meurman, 1925 ; Senjaninova, 1927.
Ranunculus Ficaria	$n = 8, (9), 16$	Larter, 1932.
Vicia Cracca L.	$n = (6), 7, 14$	Sveshnikova, 1928.*
Prunus spinosa	$2n = 16, 24, 32, 40, 48$	⎫ D., 1928, 1930 a ; Mather, 1936.
P. domestica L., s.l.	$n = 8, 24$	⎭
Rosa blanda	$2n = 14$ (16), 45 ($= 6x +$).	Erlanson, 1929, 1931.
Phalaris arundinacea	$n = 7, 14$	Church, 1929.
Lythrum Salicaria†	$n = 15, 25$	Tischler, 1928, 1929 b ; Shinke, 1929.
Solanum nigrum, s.l.	$n = 12, 24, 36$	Vilmorin and Simonet, 1928.
Biscutella lævigata	$n = 9, 18 +, 27$	Manton, 1932.
Dianthus superbus	$n = 15, 30$	Rohweder, 1929 ; Andersson and Gairdner, 1931.
Portulaca oleracea	$n = 9, 27$	Hagerup, 1932.
Euphorbia granulata	$n = 10, 20$	Hagerup, 1932.
Tulipa Clusiana	$2n = 24, 48, 60$	D., 1932.
T. chrysantha	$2n = 24, 48$	Upcott and La Cour, 1936.
Crocus sativus	$2n = 14, 15$ and 24 ($=3x$, Karasawa, 1933).	Mather, 1932.
Allium nutans	$2n = 16, 24, 32, 40, 48, 56, 64, 108.$	⎬ Levan, 1935 b.
A. schœnoprasum	$2n = 16, 24, 32$	Levan, 1935 b.

* Three geographical races of *Vicia Cracca* are found (Sveshnikova, 1928) with 12, 14 and 28 chromosomes. The 12-chromosome race differs from the 14-chromosome race in having two short pairs in place of the one long one (v. Fig. 15). The 28-chromosome race has the complement of the 14-chromosome race represented twice except that one pair have lost their trabants, an indication that changes in nucleolar organisation have taken place in the polyploid since its origin. The same combined differences are found in *Ranunculus acris, R. Ficaria* and *Crocus vernus* (Mather, 1932).

† Since this species contains different forms, all polyploid, differences in the genetic basis of its heterostylism are understandable.

7. THE INFERENCE OF POLYPLOIDY

(i) **Statistical Evidence.** Winge (1917) was the first to point out that the high frequency of multiples and low frequency of primes among the gametic chromosome numbers of the flowering plants might be taken to show that polyploidy was a common source of new species amongst them. New data have been examined from this point of view by Fernandes (1931), who has shown that in the frequency graph of numbers between 3 and 100 (the highest number considered) not a single peak is found at a prime number. Further, the most important peaks are at the numbers with the lowest

TABLE 38

Chromosome Numbers in the Phanerogama

Chromosome number.	Number of species.	Chromosome number.	Number of species.
12	391	5	27
8	332	15	27
7	236	22	25
9	170	32	25
16	153	28	24
6	134	19	22
10	126	26	20
14	125	36	19
24	80	30	11
11	70	23	8
21	64	45	8
18	58	42	6
17	48	3	5
20	47	38	5
4	42	40	5
27	31	Rest.	39
13	30		

Totals: Below 12 . 1,242
12 and above . 1,171

Altogether . 2,413

230 THE BEHAVIOUR OF POLYPLOIDS

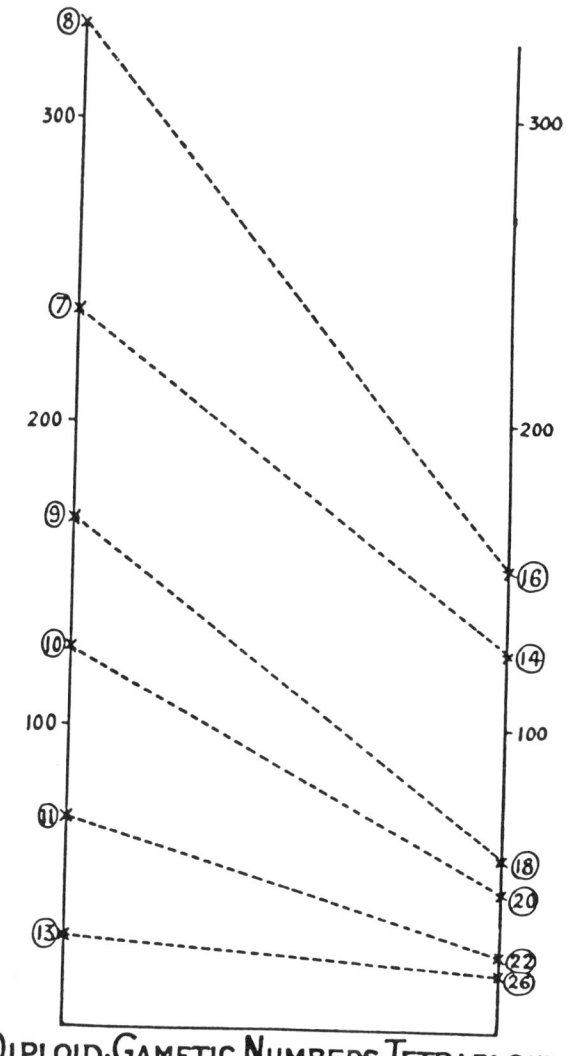

Fig. 77.—Graph showing the relationship between the numbers of species having certain low haploid numbers and the numbers of species having twice as many chromosomes. (From Table 38.)

factors, *viz.*, 8, 12, 16, 24, 27, 32 and 36. This means that these numbers are favoured by the conditions of origin or survival of new forms; such would be the case if they owed their origin largely to a process of reduplication which gave in the first instance doubling, in the second instance trebling, quadrupling, and so on; in a word, to polyploidy.

This conclusion is supported by the observation of polyploid behaviour of various kinds in nearly all species with these higher

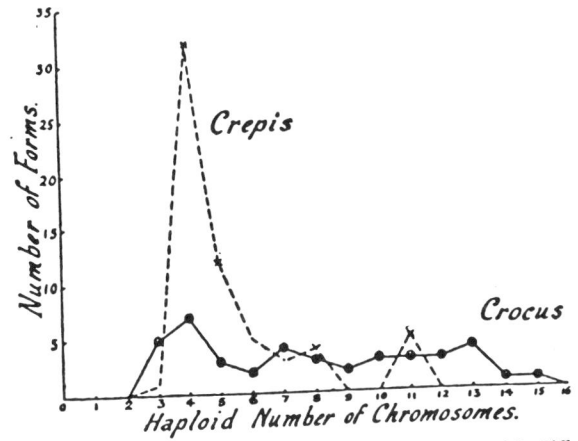

FIG. 78.—Graph showing the distribution of species with different numbers of chromosomes in *Crepis* and *Crocus* where polyploidy has played little part in the change in numbers. *Crocus* can be regarded as having the same type of differentiation as *Crepis*, but at a more advanced stage of development. (From Mather, 1932, *cf.* Fig. 16.)

numbers, but it is not supported by observations on species with 8 chromosomes or, for the most part, on species with 12 (*cf. Nicotiana*, Ch. VIII). The discrepancy is important, for it indicates that many of these were polyploid in origin but have since lost all trace of their origin, except in their number, through later differentiation of the kind already discussed.

Fernandes' enumeration is summarised in the list (Table 38 *cf.* Wanscher, 1934). It is chiefly derived from the lists compiled by Tischler (1936) and Gaiser (1933). It is a fairly random sample embracing records of 2,413 species. The most abundant contributors have not always been the most exact, but there are probably less than 5 per cent. of erroneous observations included.

In order to find out the relative importance of polyploidy and other changes in chromosome number, these records may be

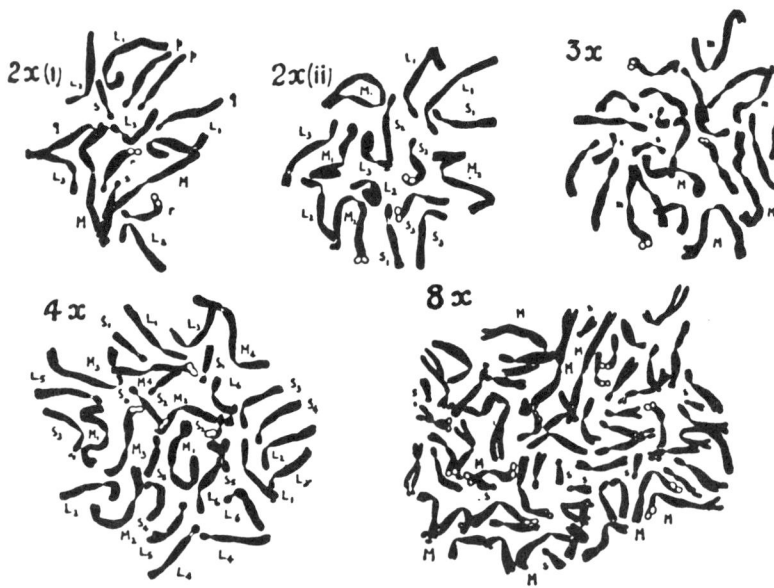

Fig. 79.—Mitotic metaphases in *Aconitum* ($x = 8$) diploid and tetraploid species, a triploid clone and an octoploid giant variety. The chromosome types are reduplicated in the forms with higher numbers. $2x$ (i) *A. barbata*. $2x$ (ii) *A. vulparia*. $3x$ *A. Napellus*, "Spark's Variety." $4x$ *A. anglicum*. $8x$ *A. Wilsoni*. × 2000 (D., unpublished; *cf.* Schafer and La Cour, 1934).

examined from a different point of view. There are certain numbers between 6 and 12 and the number 13 which can but rarely be regarded as having arisen from recent polyploidy. The number of species of each of these classes proves to have a definite relationship with the number of those having double the number. Fig. 77 represents graphically the results of the comparison. Three inferences can be drawn from this: first, that the double numbers owe their origin chiefly to polyploidy; secondly, that change in the basic number is, relative to change by doubling, somewhat rare; and thirdly, that no differential elimination of forms with certain numbers has occurred, as it might, owing to some numbers being more suitable mechanically than others. These conclusions are already probable on experimental grounds, for polyploidy is the most frequent of all changes observed in plant species (*cf. Crepis*, Table 8).

When the species of smaller systematic groups are considered, differences are found both in the amount and in the kind of variation in chromosome number. Some large groups are found to show no variation (Ch. III). Others show the influence of the polyploid series in various degrees. The graph (Fig. 64) of variation in number in *Crepis* (Babcock and Navashin, 1930), and *Crocus* (Mather, 1932) illustrates types where polyploidy has played little part. *Crepis* is more constant than *Crocus*, for it has a high proportion with the number four which must be considered the original number of the genus. In *Crocus* the distribution of numbers is almost at random. How far the changes responsible are merely numerical, *i.e.*, due to changes in the numbers of certain chromosomes, and how far they are structural, *i.e.*, due to fragmentation and fusion, can only be said after a detailed study of meiosis in hybrids. But there is clearly a similar variety of species in the two genera with less variation in chromosome number in *Crepis* than in *Crocus*. This is probably due to the greater importance of sexual reproduction (and therefore of regularity at meiosis) in the one than in the other.

Crepis and *Crocus* approach the distribution of numbers found in animals where there is no evidence of polyploidy. The opposite extreme is found in genera such as *Chrysanthemum, Triticum, Rosa, Prunus, Rubus, Solanum* and *Aconitum*, where all species have multiples of a common basic number (Fig. 79).

(ii) **New Polyploid Species.** Clearly it can be possible to reconstruct the method of origin of a species only when it has arisen recently. Suitable conditions could probably be found to allow of several new polyploids that have arisen in experiment becoming wild species. But amongst wild species there are a few the conditions of whose origin can be stated with precision. First there are those giant varieties found within species (Table 37) which can certainly be supposed to have arisen from simple doubling. Similarly, *Vallisneria spiralis* ($n = 10$) is diœcious, and appears to have given rise to *V. gigantea* ($n = 20$) of both sexes separately by somatic doubling (Jørgensen, 1927).

Secondly, there are allopolyploid species whose diploid parents are known with certainty (Table 39). Amongst cultivated plants there are also many forms of equivalent rank to new species, whose origin can be inferred with high certainty. Such are the autopolyploid *Petunia* varieties (Dermen, 1930) and certain allopolyploid *Rubi* (Crane and D., 1927).

In all these cases, therefore, there can be no doubt that a new species has been produced in one generation. This is possible with polyploidy and also with other changes which are in themselves of greater evolutionary significance. Some of these will be discussed in regard to mutation in *Œnothera* (Ch. IX).

TABLE 39

Allopolyploid Species whose Origin can be Determined

Genus	Species				Hybrid
1. *Æsculus* (10)	*Hippocastanum*	($4x$) × *pavia*	($4x$)		= *carnea* ($8x$), Hoar, 1927.
2. *Æsculus* (10)	*carnea*	($8x$) × *Hippocastanum*	($4x$)		= *planticrensis* ($6x$), Upcott, 1936.
3. *Galeopsis* (8)	*pubescens*	($2x$) × *speciosa*	($2x$)		= *Tetrahit* ($4x$), Müntzing, 1927, 1930 b.
4. *Phleum* (7)	*pratense*	($2x$) × *alpinum*	($4x$)		= *pratense* (America), ($6x$), Gregor and Sansome, 1930.
5. *Spartina* (7)	*alterniflora* ($10x$) × *stricta*			($8x$)	= *Townshendii* ($18x$), Huskins, 1931 a.
6. *Prunus* (8)	*divaricata*	($2x$) × *spinosa*	($4x$)		= *domestica* ($6x$), Rybin, 1936.
7. *Pentstemon* (8) *lætus*		($2x$) × *azureus*	($6x$)		= *neotericus* ($8x$), J. Clausen, 1933.
8. *Agropyron* (7) *junceum*		($4x$) × *littoreum*	($6x$)		= *acutum* ($5x$), Simonet, 1934.

(iii) **Somatic and Secondary Pairing.** At metaphase of mitosis the positions of the chromosomes seem to be governed by three rules: (i) the centromeres lie within the spindle and in its equatorial plane, which is the metaphase plate; (ii) they lie to a greater or less extent on the periphery of the plate according as the repulsion of the poles is greater or less; (iii) the bodies of the chromosomes lie as far apart from one another as their attachment to their centromeres allows. When the bodies of the chromosomes are long, those that are in the middle therefore lie on both sides of the plate and those that are on the edge lie off the plate and can be seen to move in the currents of the cytoplasm. These rules argue an equal repulsion of the two poles for the centromeres and a mutual repulsion of the bodies of all the chromosomes. The rules apply to meiosis when we make allowance for there being in each bivalent two centromeres which repel one another in the axes of the spindle and therefore like the anaphase daughter centromeres of mitosis lie on either side of the plate.

In some organisms the even distribution of the chromosomes on the metaphase plate is modified by another factor.

At mitosis in the diploid in many Diptera, pairs of chromosomes are seen to lie specially close together, although still never touching, and these are found to be similar pairs, with similar parts lying parallel. This is due to a specially exaggerated property of attraction in this group (*cf.* Metz, 1916, 1926 *et al.*; also Fig. 116). The same *somatic pairing* has been found between the chromosomes of polyploid plants at mitosis (*e.g.*, *Dahlia*, Lawrence, 1931). In this case more than two chromosomes being attracted to one another, the groups lie radially instead of parallel.

More striking, however, are the conditions in doubled nuclei which have arisen through the failure of the chromosomes to separate at a preceding division (*e.g.*, *Spinacia*, Stomps, 1911; *Sorghum*, Huskins and Smith, 1932; *Apotettix*, Robertson, 1930; *Iberis*, Manton, 1935). Probably in these the daughter chromosomes have remained together during the resting stage, and have therefore been in a suitable position to exercise their special attraction on one another during the prophase. At metaphase every pair is distinguishable by the chromosomes being of similar shape and lying parallel. This is evidence not only of the method of origin of polyploid nuclei, but also of the static condition of the chromosomes during the resting stage, and of their attraction at metaphase.

The special attraction of chromosomes seems to be specific to their parts, since it leads to the corresponding parts lying parallel ; and Sturtevant and Dobzhansky (1930) find that the four chromosomes in an interchange heterozygote in *Drosophila* lie in a ring at mitosis. Stern (1931) finds no evidence of attraction of the distal end of a normal X for the corresponding part of its homologue when this has been translocated to another chromosome, presumably owing to a conflict with the attractions of larger parts.

A similar juxtaposition of different bivalent chromosomes at meiosis has long been known in polyploid plants (*cf.* Kuwada, 1910 ; Ishikawa, 1911), but its relation to the homology of the associated bivalents and its independence of true meiotic pairing was not recognised till recently (D., 1928). Like somatic pairing, this *secondary pairing* is between chromosomes of similar size and shape and it appears first at metaphase, *i.e.*, it is not a continuance of a prophase association by chiasmata ; it consists in approximation, but never in " contact," *i.e.*, it reaches an equilibrium with forces of repulsion. It is too slight to determine or even modify the anaphase separation. It may be continued between the pairs of daughter bivalents during anaphase, and is often most marked at the second metaphase. Thus it may be supposed that whatever attraction is responsible for it it has a cumulative effect in sorting out the homologues during the two divisions. Again, like somatic pairing, it is variable in its occurrence from division to division. Cytologists have usually illustrated the nuclei that were freest of it, to avoid the suspicion of bad fixation, and in consequence its occurrence has been generally neglected, as Lawrence (1931) has pointed out in a general review of the problem. Secondary pairing, like many peculiarities of chromosome behaviour, is exaggerated in appearance by bad fixation, but its essentially differential character as between different chromosomes cannot be determined by an external agent—it is not an artefact.

Somatic pairing does not usually show at mitosis in plants with secondary pairing at meiosis. The reason for this is that the chromosomes in mitosis are always widely distributed, while at the end of diakinesis (" pro-metaphase " of meiosis), when secondary pairing begins, they are brought within close range of one another, and therefore on analogy should attract one another more strongly.

PLATE VIII

Secondary Pairing at Meiosis in Polyploids

Fig. 1.—*Cydonia cathayensis* (Pomoideæ), $2n = 17$. × 2500.

Fig. 2.—*Pentstemon lævigatus*, $2n = 96 = 12x$.

Fig. 3.—*Dahlia coronata* × *D. coccinea*, $2n = 32 = 4x$. × 2000.

Fig. 4.—Second metaphase in *Prunus cerasus*. $19 + 13$ chromosomes. $2n = 32 = 4x$ (*cf.* D., 1928).

Figs. 5–7.—*Æsculus carnea* and its parents. × ca. 3000.
Fig. 5. *Æ. Hippocastanum* ($4x = 40$). Fig. 6. *Æ. carnea* ($8x = 80$). Fig. 7. *Æ. Pavia* ($4x = 40$).

Figs. 8 and 9.—*Dahlia variabilis*, the same plate at two focuses. $2n = 64 = 8x$. × 2000.

Figures reproduced by kind permission of Mr. A. A. Moffett (Fig. 1), Mr. L. La Cour (Fig. 2), Miss M. B. Upcott (Figs. 5, 6 and 7), and Mr. W. J. C. Lawrence (Figs. 3, 8 and 9). All from medium Flemming — gentian-violet sections.

PLATE VIII

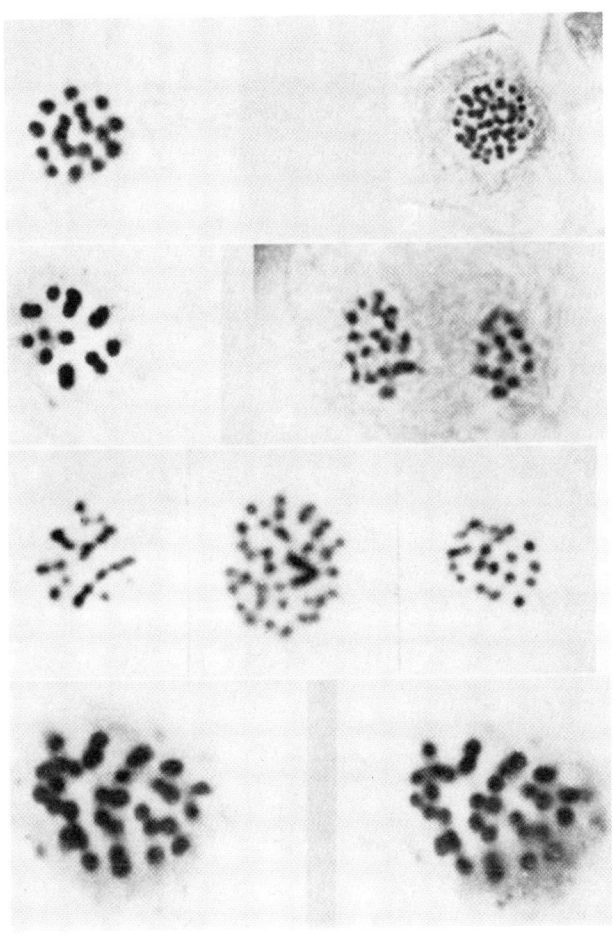

[To face p. 236.

Further, they are spherical at meiosis, and so should oppose less resistance to movement through a viscous medium than when they are long, as in mitosis (Lawrence, *l.c.*). Secondary pairing is not found at meiosis in plants whose chromosomes are still long at this stage (*e.g.*, in *Tulipa*). On the other hand, somatic pairing is shown by small chromosomes when the larger ones do not show it (Upcott, 1936).

Secondarily paired bivalents are distinguished from multivalents associated by chiasmata in four ways: (i) they lie evenly side by side, in twos, threes or larger groups according to whether the plant is a tetraploid, hexaploid or higher polyploid; (ii) they never "touch," except through collapse in fixation; (iii) they separate regularly into their daughter halves without interfering with one another; (iv) they show no association at the preceding diakinesis (D., 1928; Lawrence, 1929).

The distinction has a profound theoretical importance. It was formerly held that the characteristic pairing of chromosomes at meiosis was merely an exaggeration of the somatic pairing observed at mitosis. But chiasma pairing and "somatic" pairing are now seen side by side at meiosis, and the differences in their effects are as clear as the differences in their causes. These will be discussed later in more detail. In many polyploid species, where the two phenomena occur side by side, it is of great importance for genetic interpretation to tell one from the other, for chiasma pairing indicates a closer relationship than secondary pairing. It will have been seen that, while the separation of diploid and polyploid is a convenient one, every possible degree of relationship must be found within the gametic sets of different organisms. There must be polyploids so ancient that the original relationship between their chromosomes has been lost so far as pairing analysis can reveal it. There must also be diploids having parts of their chromosomes reduplicated, and therefore having internal relationships not unlike those of polyploids. In fact, the behaviour of haploids, with other considerations, makes it probable that this is true of all diploids in varying degrees (*v.* Ch. VIII). It follows, therefore, that all slight unevennesses of distribution of chromosomes on the metaphase plate might conceivably be interpreted in terms of genetical relationship if they could be measured sufficiently accurately. This is of practical importance in the analysis of polyploidy. Most ordinary cases are clear enough from purely numerical considerations, but where no species is known with a lower basic number or where the

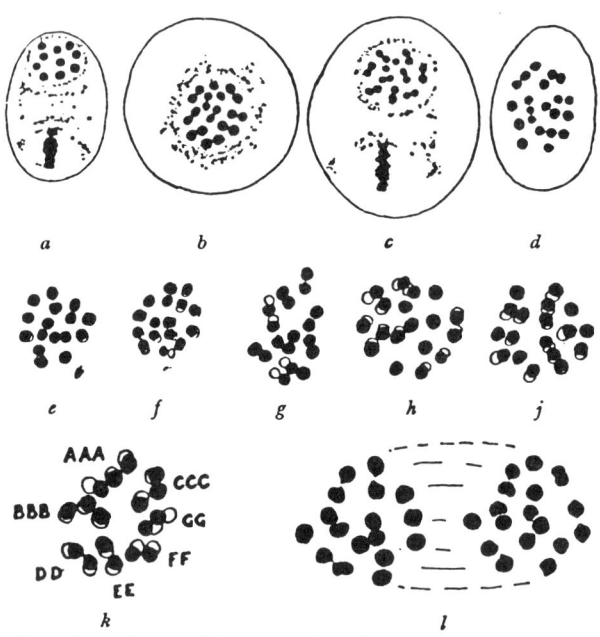

FIG. 80.—Secondary pairing at meiosis in the pollen mother-cells. *b* and *e* to *k*, first metaphase; *l*, first anaphase; *a*, *c* and *d*, second metaphase. *a*, *Euphorbia scordifolia* ($2x = 20$). *b*, *c*, *E. granula* ($4x = 40$). *d*, *Rubus* "Loganberry" ($6x = 42$). *e*, *f*, *Dahlia coccinea* ($4x = 32$). *g*, *h*, *j*, *D. Merckii* ($4x + 4 = 36$). *k*, *l*, *Pyrus Malus* ($4x + 6 = 34$). *a* to *c*, after Hagerup, 1932; *d*, from Crane and D., 1927. *e* to *j*, from Lawrence, 1929 and 1931. *k* and *l*, from D. and Moffett, 1930, and Moffett, 1931.

number is not a simple multiple, the evidence of secondary pairing becomes of use. The distribution of chromosomes on the equatorial plate may be used to show relationships between them, or, more exactly, between some of their constituent parts, which are beyond analysis by the two other methods—comparison of shape at mitosis

CHANGE OF BALANCE

or frequency of chiasma formation at meiosis. The application of this method will now be considered.

(iv) **Secondary Polyploidy.** Since new allo-polyploids already behave at meiosis to a great extent like diploids, it follows that polyploid species which have changed since their origin can be identified only by more or less circuitous inferences. And although such inferences can be drawn without much hesitation for the bulk of species in the angiosperms, there remains an interesting group of doubtful cases.

Where, as in one section of *Rumex*, species occur with " haploid " or gametic numbers of 10, 20, 30, 40, 50, 60 and 100, and the chromosomes are comparable in size, it is natural to assume without further evidence that the higher-numbered species arise through multiplication of a " basic " set of 10 chromosomes such as that found in the lowest-numbered ones. The inference is less direct in such genera as *Dahlia* ($n = 16$, 32) and *Digitalis* ($n = 28$, 56). These numbers are multiples respectively of 8 and 7 which are found in related genera. The conclusion that 8 and 7 are the basic numbers, however, requires the support of other evidence. Still less direct is the inference in those genera in which uneven multiples occur. Many of these, such as *Papaver* ($n = 7$, 11), are not yet elucidated, and in the hybrid *Nicotiana longiflora* (10) × *N. alata* (9) ($9^{II} + 1^{I}$, Goodspeed and Clausen, 1927 *b*), it is not clear whether the extra chromosome arises from fragmentation or reduplication. Others can be understood as the result of " fusion " between non-homologous chromosomes after the occurrence of polyploidy. Thus *Cardamine pratensis* (Lawrence, 1931, $n = 15$) is a tetraploid with an original basic number of 8, two chromosomes having fused, and *Nicotiana longiflora* ($n = 10$) is probably derived by fusion from a species with a haploid number of 12 (Ch. III). Others, again, show no evidence of fusion or other structural change, and it would appear that they result from the unbalanced multiplication of the basic set, two or three of the chromosomes occurring once more often than the rest. It is in relation to these supposed " secondary polyploids " that the criteria of polyploidy require to be strictly examined.

The evidence has been most fully adduced in the Pomoideæ

section of the Rosaceæ (D. and Moffett, 1930 ; Moffett, 1931, *a* and *b* ; 1934) in which it is assumed that a chromosome set of 17 (and its multiples) found throughout the section is derived from the set of seven found in other sections of the order (*e.g.*, in *Rosa* and *Rubus*) by unequal reduplication, so that the seven chromosomes *A* to *G* are present in the proportions

AAA, BBB, CCC, DD, EE, FF, GG,

making up 17 in all.

The evidence, chiefly from *Pyrus*, is of the following kinds :—

(i) *Occasional quadrivalents* are found in the diploid with normally 17 pairs. (These might also result from interchange hybridity.)

(ii) *Secondary pairing* reaches its maximum in the appearance of seven groups of bivalents, three being groups of three, four being groups of two (Fig. 80).

(iii) In auto-" triploids " (with 51 chromosomes), *autosyndesis* occurs in the third set of 17, so that more than 17 pairs may be formed (Nebel, 1929); and multivalents up to 9 chromosomes are formed (*v.* Fig. 70).

(iv) The progeny of triploids back-crossed with diploids, instead of showing the elimination of types with intermediate numbers so that most of the seedlings have approximately the diploid number, show the highest frequency with 41 chromosomes, *i.e.*, the secondary diploid number, 34, together with the primary basic number 7, Moffett, 1934. This indicates that the original *balance* of 7 still has a certain vestigial significance (*v.* Ch. VII).

(v) The somatic complement includes *four* chromosomes of one exceptional type, as is found in tetraploids.

(vi) Genetic evidence indicates more complex conditions than those found in strictly diploid organisms.

In the Pomoideæ the inference of polyploidy is necessarily the most indirect, since the original change from 7 to 17 by which such an important group arose must have occurred at an epoch more remote than any on which inference can usually be brought to bear. In another group, however, the case is simpler. *Dahlia Merckii* has 18 haploid chromosomes (Lawrence, 1929), while all other species of *Dahlia* have multiples of eight. The occurrence of secondary pairing agrees with the assumption that two of the eight chromosomes are present three times, while the rest are present twice (Fig. 80).

The same is true of *Solanum* and *Nicotiana*, which have an apparent basic number of 12, which might be derived from a lower basic number of 6. These assumptions have been verified by the study of secondary pairing (*cf.* Müntzing, 1933).

Chromosome pairing in certain hybrids suggests that reduplication of part of the set is responsible for the origin of new complements in *Brassica* and *Papaver*, and perhaps *Viola*. Observations of secondary pairing in *Brassica* confirm this conclusion (Catcheside, 1934).

Brassica cernua ($n = 18$, $4x$) $\times$ *B. chinensis* ($n = 10$) and other species: $10^{II} + 8^{I}$ (Morinaga, 1929).

Papaver somniferum ($n = 11$) $\times$ *P. orientale* ($n = 21, 6x$): $11^{II} + 10^{I}$ (Yasui, 1921; Ljungdahl, 1922).

Viola tricolor alba ($n = 13$) $\times$ *V. elegantula* ($n = 10$): $8-10^{II}$, $7-3^{I}$ (J. Clausen, 1931 c).

Brassica Napus ($n = 18$) $\times$ *B. rapa* ($n = 10$): $0-3^{III}$, $7-12^{II}$, $4-8^{I}$ (Catcheside, 1934).

The distinction between secondary polyploidy and structurally changed polyploidy (*i.e.*, polyploidy followed by fragmentation or fusion) is important, because the first means a great change in genetic balance, while the second means little or no change of this kind. On the analogy of most trisomic and tetrasomic forms a polyploid with a new balance will, if it breeds true, constitute a new species. The Pomoideæ are a sharply characterised group of the Rosaceæ and *Dahlia Merckii* is the most sharply characterised species of *Dahlia*.

The types so far elucidated are unbalanced modifications of tetraploids, not of diploids. It may be supposed on analogy that tetrasomic diploids would as a rule be sterile, the unbalance being too drastic. This is true also of artificial secondary polyploids.

Such new forms have been produced in experiment in the following ways. First, from a pentaploid having the complete complement of *Nicotiana Tabacum* ($4x = 48$) given by an unreduced gamete together with a haploid set of N. sylvestris ($2x = 24$), a plant was derived in F_6 having fifty chromosomes forming $2x + 1$ pairs (Webber, 1930). Secondly, from a cross with the diploid N. rustica a new race with 30 bivalents has been derived (Lammerts, 1932). Thirdly, from the cross between the probably autopolyploid *Crepis biennis* ($2n = 40 = 8x$) and the diploid *C. setosa* ($2n = 8$) it has been possible, owing to autosyndesis, to derive a new true-breeding

TABLE 40

Observations of Secondary Pairing at Meiosis

A. SIMPLE POLYPLOIDS.
Dahlia variabilis ($8x = 64$) . . Kuwada, 1910 ; Lawrence, 1929, 1931.
Prunus cerasus ($4x = 32$) . . . D., 1928.
P. domestica × *P. cerasifera* ($4x = 32$) D., 1930.
P. Laurocerasus ($22x = 176$) . . Meurman, 1929.
Primula kewensis ($4x = 36$) . . Newton and Pellew, 1929.
Digitalis mertonensis ($16x = 112$) . Buxton and Newton, 1928.
Veronica spp. ($4x = 68$) . . . Graze, 1935.
Solanum tuberosum ($2x = 24$) . . Müntzing, 1933.
B. SECONDARY POLYPLOIDS.
Pyrus Malus ($4x + 6 = 34$) . . D. and Moffett, 1930 ; Heilborn, 1935.
Cratægus, Mespilus, Cotoneaster, etc.
  ($4x + 6 = 34$) . . . . Moffett, 1931, 1934.
Dahlia Merckii ($4x + 4 = 36$) . . Lawrence, 1929, 1931.
Acer platanoides ($4x + 2 = 26$) . . Meurman, 1933.
Oryza sativa ($4x + 4 = 24$) . . Sakai, 1935.
Empetrum nigrum ($4x + ? = 26$) . Hagerup, 1927 ; *cf.* D., 1931.
E. hermaphroditum ($8x + ? = 52$) . Hagerup, 1927; *cf.* D., 1931; Wanscher, 1934.
Brassica Napus ($2n = 36$) . . Catcheside, 1934.
B. rapa ($2x + ? = 20$) . . . Catcheside, 1934.
B. oleracea ($2x + 6 = 18$) . . . Catcheside, unpub.
Dicentra spectabilis ($2x + 2 = 16$) . Matsuura, 1935.
C. BETWEEN UNIVALENTS.
Nicotiana . . . . . Lawrence, 1931.
Taraxacum . . . . . Gustafsson, 1934, 1935.

form, *Crepis artificialis*, having ten pairs of chromosomes from *C. biennis* and two from *C. setosa* (*cf.* Table 30). This new form is balanced in its *biennis* fraction, unbalanced in its *setosa* fraction (lacking 2 pairs of chromosomes). It therefore has the constitution of a secondary polyploid.

All these new forms are capable of breeding true, yet they differ in genetic balance and external form from their progenitors.

They therefore show how secondary polyploidy may be supposed to have arisen in *Dahlia* and in the Pomoideæ, viz., in the derivatives of crosses between original polyploid forms with different numbers of chromosomes. Thus, selfing a pentaploid, derived from crossing a tetraploid and hexaploid, would give, in the *Rosa* type with seven chromosomes, trisomic and tetrasomic seedlings with numbers between 28 and 42. The latter, if they bred true, would be secondary polyploids.

Further investigation will probably show that secondary polyploidy, and other variations due to change in balance (such as reduplication of segments), is an important source of species-formation in plants. Secondary pairing indicates such an explanation of the origin of the chromosome complements in many genera with basic numbers of 13, 17 and 19 (*e.g.*, *Empetrum*, *Gossypium*, *Salix*; *cf.* Lawrence, 1931; Skovsted, 1933). The observation of secondary pairing consistent with a particular assumption of primary or secondary polyploidy is not alone a sufficient basis of inference, for a secondary association can also presumably arise in a diploid from small reduplicated translocations which, as we shall see later, are very general in diploids. On the other hand, the evidence of secondary pairing has shown in certain instances that secondary polyploidy involving change of balance is an effective means of variation. This being established as an evolutionary principle the precise mechanism in particular and doubtful cases is not of immediate importance.

CHAPTER VII

THE CHROMOSOMES IN HEREDITY : MECHANICAL

The Theory of Heredity—Segregation—Crossing-Over—The Chiasmatype Theory—Structural Hybrids—Dyscentric Hybrids—Secondary Structural Change—Cytological Investigation of Crossing-Over—Its Universality and Biological Effect.

Fit quoque ut interdum similes existere avorum possint et referant proavorum saepe figuras propterea quia multa modis primordia multis mixta suo celant in corpore saepe parentis, quae patribus patres tradunt ab stirpe profecta.[1]

LUCRETIUS, *De Rerum Natura* IV.

1. THE CHROMOSOME THEORY OF HEREDITY

THE chromosome theory of heredity, which we owe chiefly to Weismann, is the hypothesis (and the corollaries of the hypothesis) that :—

the permanence of the physiological properties of organisms which is manifested in heredity is determined by the permanence in the structure of their chromosomes.

In order to examine the theory it is necessary first to recall the special properties associated with heredity and shown by genetical analysis, *i.e.*, by the comparison of parents and their offspring in their higher or *super-chromosomal* organisation.

They may be conveniently arranged under five heads, as follows :

1. *Potential Permanence.* Heredity consists in the same properties or characters being reproduced in successive generations ; a causal explanation, therefore, requires the assumption of permanence in the cause, although, since the character is only *liable* to be repro-

[1] Commonly also children may resemble their grandparents or even repeat the characters of remoter ancestors for this reason that the parents often conceal within their bodies many primordia combined in many ways which, derived from the stock, are handed down from generation to generation.

duced, the permanence is only potential. Potential permanence has already been shown to be a universal property of the chromosomes (Ch. II).

2. *Qualitative Differentiation.* Every organism has various and specific properties of variation (mutation) which are also permanent ; the permanent material must therefore be of various specific kinds, *i.e.*, must be qualitatively differentiated.

3. *Segregation.* All organisms with sexual reproduction have two stages in their life cycle, the one in which two corresponding elements or " factors " in each cell affect its breeding properties (one from the mother, the other from the father), and the other in which only one of these genes affects each character. There must therefore be a point in the life cycle at which corresponding elements separate or segregate to different daughter-cells having half the number present in the mother-cells, as well as a point at which the elements recombine.

A separation such as is required occurs at meiosis. Since the chromosome pairs which segregate at random at meiosis give corresponding products yet are qualitatively differentiated, it follows that the members of each pair correspond. The recombination is seen to take place at fertilisation (Ch. I).

4. *Linkage.* The elements or factors having different capacities for mutation are arranged in groups which segregate at random as between groups but with restricted freedom within the group. The restriction, known as linkage, is fixed in degree as between particular factors. It can be represented as fixed by their arrangement in linear order in a thread which has a fixed chance of exchanging segments (crossing-over) with the corresponding thread. The hereditary materials, as we have seen, give evidence of this character and behaviour. The linear character of the chromosomes has already been shown in considering the prophase of mitosis and meiosis.

It therefore becomes possible to refer the " factor " which determines by its change a mutation or hereditary difference to a particle lying in the thread. Such a particle is described by Johannsen's term as a *gene*. It is analogous in its behaviour to the chromomere seen in the cell. It might be thought that the observed chromomeres could at once be described as genes, but this is not so. A gene is the unit of crossing-over, and therefore the atom of inheritance. Until we consider more in detail therefore how crossing-over occurs, we cannot tell how big the gene may be; we cannot even tell whether the same particle always behaves as a unit.

We can, however, proceed to treat the chromosome as a string of particles, provisionally described as genes, whose changes are responsible for the differences in hereditary characters that we know as mutations.

5. *Mutation*. Experimental breeding has shown that the genes must be capable of at least two kinds of change: those affecting their individual properties (gene or point mutation) and those affecting their numerical proportions and arrangement (deficiency, translocation, etc.). Corresponding structural changes in the chromosomes have been shown (Chs. III, V). The existence of qualitative differences between genes in the same complement is evidence of gene-mutation on an evolutionary hypothesis, for all the different genes found to-day cannot have arisen independently at the beginning of life.

Some of the principal evidence for regarding the chromosomes as containing the hereditary materials has therefore already been given. The most unequivocal test of the chromosome theory, as indeed of any theory, is by the verification of prediction. It is possible to base prediction of chromosome form and behaviour on observations by genetical methods, and *vice versâ*, and these predictions can afterwards be tested. Thus it is possible in the case of certain flies mosaic for male and female characters to say that one X chromosome has been lost in the course of a somatic mitosis (Stern, 1927). It is possible to say in the case of seedlings such as those raised from the cross *Rubus rusticanus* by *R. thyrsiger* or from natural seed of the *Raphanus-Brassica* hybrid (Ch. VII) that some have one set of chromosomes more than others. These are predictions based on genetical evidence and they have been verified cytologically. On the other hand, it is known that meiosis occurs in the formation of spores of mosses and ferns, and that the gametophytes raised from them are haploid: they should therefore show the 1 : 1 segregation in haploid characters which Mendel supposed was the basis of the 3 : 1 segregation in the diploid generation. These are predictions based on cytological principles and they have been verified genetically (v. Wettstein, 1924 ; Andersson, 1927). Tests of this kind will be described where they affect debatable points, but in regard to the chromosome theory in general, they are too numerous to be recapitulated here. They have been dealt with elsewhere (Stern, 1928 ; Sansome and Philp, 1932).

More important, for our present purpose, is the showing of a parallelism between the rules of heredity and the rules of chromosome behaviour, for this parallelism is a help in directing enquiry. Bearing in mind this distinction, we will now consider the evidence as affecting (i) segregation of homologous chromosomes, (ii) crossing-over, and (iii) qualitative differentiation.

2. SEGREGATION

The pairing and separation of the chromosomes is clearly parallel to genetic segregation, as pointed out by Sutton in 1902. The assumption that the one determined the other made it possible to predict haploid or gametic segregation, which has since been abundantly demonstrated genetically. Special cytological observations have since made it possible to verify converse predictions in various ways, as follows :—

1. Carothers (1921) showed that when grasshoppers (*Circotettix*) with unequal pairs of chromosomes at meiosis were bred with related forms homozygous for chromosome structure, they produced offspring half with one type of chromosome and half with its mate. This was a combined demonstration of (*a*) chromosome permanence and heredity, (*b*) meiotic reduction and mendelian segregation, and (*c*) fertilisation and mendelian recombination.

2. Federley (1912, 1931) showed that *Pygæra* hybrids having no pairing and no reduction of chromosome number at meiosis (*v*. Ch. X) showed no segregation of parental characters.

3. Similarly in parthenogenetic organisms where pairing fails there is no segregation (*cf*. Ch. XI).

4. In all allotetraploids formed by doubling of the chromosome number in a hybrid there is a suppression of segregation. This suppression is partial or almost complete according as pairing is

partially or completely suppressed between chromosomes of opposite parental species. Thus in *Raphanus-Brassica*, where no segregation occurs, no pairing of *Raphanus* and *Brassica* chromosomes occurs—simple bivalents are always formed in the tetraploid (and no bivalents are formed in the diploid). In *Primula kewensis*, on the other hand, there is segregation in the tetraploid; quadrivalents are formed.

5. Where four, six or more chromosomes are associated in a ring in a diploid, each of these chromosomes must have a specifically different combination of materials from any of the others, and only two types of viable gametes should be produced—the opposite types arising from separation of each chromosome in the ring from the other two with which it is " paired " at opposite ends. This is verified both by genetical and by cytological tests of the progeny. It is found that two such gametic types are produced by wild species with chromosome rings (*v.* Ch. IX), and these differ genetically. Similarly the inheritance of the pairing properties of their chromosomes agrees with prediction. Each form with rings produces two kinds of gametes which give specific and different kinds of pairing or ring-formation with other kinds of gametes.

6. The direct effect of segregation in giving equal numbers of products bearing opposite allelomorphs can be seen in the haploid generation produced by a heterozygous diploid. This direct segregation is seen in organisms with an important haploid cycle, *e.g.*, Bryophyta (Wettstein, 1924; Allen, 1926, 1935 *a* and *b*); Ascomycetes (Dodge, 1936; Lindegren, 1936); Basidiomycetes (Buller, 1931); Pteridophyta (Andersson-Kottö, 1931).

7. It is even possible when the spores produced at meiosis remain together or when one of the divisions is suppressed, to show that segregation has occurred sometimes at the first and sometimes at the second division, and with special proportions for particular factors, as is the case with particular unequal bivalents (*v. infra*). Similarly in the flowering plants segregation of two types of pollen has been shown in *Œnothera* (Renner, 1919 *b*), *Oryza* (Parnell, 1921) and *Zea* (Demerec, 1924 *et alii*). The special type of reproduction of the Hymenoptera by which diploid females produce haploid male offspring without fertilisation has made it possible to study

direct segregation in these males, for as Dzierzon found in 1854, the opposite types of males arise in equal numbers from a hybrid queen (Newell, 1915; Whiting and Benkert, 1934; Whiting, 1935 *b*; *cf.* Ch. IX).

8. In every instance in which the sexes are visibly different in their chromosome complements (as they are in most animals and many plants) the segregation of dissimilar chromosomes can be seen in the sex heterozygotes at meiosis, and in some the whole history can be traced, from the random segregation to the development of sexual differences between the dissimilar individuals produced by this segregation (Ch. IX). In the higher plants the sexual differentiation does not express itself in external form until long after meiosis. This is true of the sporozoan *Aggregata* and has led Dobell (1925) to believe erroneously that the differentiated gametes might be derived from the same haploid cell and so the differentiation might not spring from segregation at meiosis (*cf.* Naville, 1931; Belar, 1926; M. Robertson, 1929).

9. Where segregation is occurring in factors carried by chromosomes represented more than twice (in polysomic and autopolyploid plants) the proportions of offspring bearing the opposite characters occur, not in the simple mendelian proportions but in the more complex proportions to be predicted from the random assortment of the larger number of chromosomes concerned (*cf.* Haldane, 1931 *a;* Crane and D., 1932; Sansome, 1933; Mather, 1935, 1936 *a*).

The parallelism between chromosome reduction and genetic segregation is therefore complete so far as whole chromosomes are concerned. The specific properties of segregation shown by parts of the same chromosome in the phenomenon of linkage must be considered next.

3. THE THEORY OF CROSSING-OVER

(i) **Introduction.** Crossing-over between chromosomes has been inferred from the structure of bivalents seen at meiosis (Chs. IV, V). It has also been inferred from the proportions of different types of progeny found in breeding experiments. It is now necessary to find out how far the two sets of observations agree, in order that we may combine them and use the conclusions in a joint attack on this problem. It is a fundamental problem, for on its solution depends our understanding equally of the mechanics of structural changes in the chromosomes and of the genetical consequences of these changes. These affect all sexually reproducing organisms, both in the pairing of their chromosomes and in the recombinations of their hereditary characters.

Let us consider first the evidence of experimental breeding.

(ii) **The Genetic Theory.** In all organisms in which the inheritance of a large number of mendelian factor-differences has been studied, certain of these have been found to be "linked," *i.e.*, when individuals differing in two such factors are crossed, the proportion of new combinations in the second generation is lower, and of old combinations higher, than would be expected from free assortment (according to Mendel's second law). The general conditions of this linkage are similar in most structurally homozygous diploids that have been studied, *e.g.*, in *Drosophila* species, *Zea Mays*, *Pharbitis*, *Primula sinensis*, *Pisum sativum*, etc., and observations of exceptionally high linkage in *Lebistes*, *Apotettix*, *Cepea*, *Funaria*, etc., are intelligible as the result of an irregularity in the spacing of genes in the chromosomes which is merely an exaggeration of that now known in *Drosophila*.

In 1909 Janssens suggested that the paired chromosomes broke and rejoined at meiosis and that the chiasmata, whose structure was then not clearly understood, were the result of this recombination. In the light of this *chiasmatype theory* Morgan in 1911 was able to put forward the explanation of linkage that is now accepted and has been the basis of genetical analysis since that time. Morgan assumed (i) that linkage of factors is due to the specific particles or genes that determine the characters concerned lying in the same chromosome in a linear order, and (ii) that the recombinations of factors are due to crossing-over or an exchange of homologous segments containing those genes between partner chromosomes.

Thus, if the factors are at opposite ends of chromosomes : $ABCDE$ and $abcde$, crossing-over might give $ABCde$ and $abcDE$. Crossing-over should be more frequent between A and C than between A and B, owing to its occurring at any point along the chromosome,

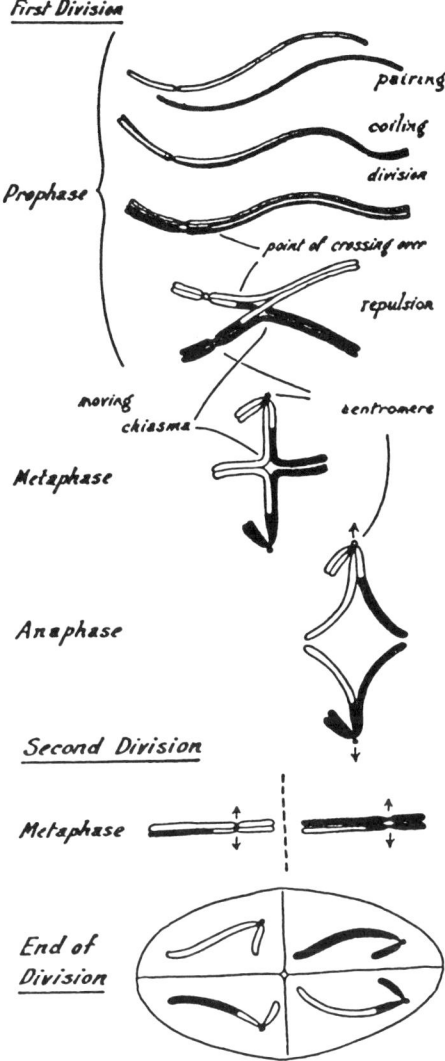

Fig. 81.—The history of two pairing chromosomes in meiosis, showing the time and place of crossing-over and its result in the production of daughter chromosomes in each nucleus of the tetrad, all different from one another in their origin.

and this can be verified where two factors are linked with a third, as Sturtevant showed in 1914. The amount of crossing-over between the two should, when small, approximate to either the sum or the difference between the amounts of their crossing-over with the third. This is always the case in organisms with simple pairing of chromosomes. But the exact relation is not a simple one as might be supposed, *i.e.*, the resultant crossing-over is not equal to the sum of the two smaller proportions *minus* twice their product (to allow for double crossing-over which would cancel itself out), but *minus* something less than this. It follows that double crossing-over between factors fairly close to one another is not so frequent as randomness requires; the occurrence of one cross-over may therefore be said to " interfere " with the occurrence of another in its neighbourhood, and the property is known as " interference " (Muller, 1916).

These principles must be understood in order to follow the cytological evidence. The conditions of crossing-over will be examined in detail at a later stage in relation to this evidence.

(iii) **The Cytological Theory.** The observation that the chromosomes paired at diplotene came together at various points along their length—points which we now recognise as chiasmata—suggested to Rückert in 1892 that the chromosomes exchanged material at these points. The view was ignored in the following years, being naturally overshadowed by the idea of permanence in the structure of the chromosomes. The four chromatids of the bivalent were usually supposed to be derived without change, two from one parental chromosome and two from its mate. The question then was merely whether the two chromatids which passed to the pole together at the first division were derived from the same parental chromosome or from the two partners. In the first case the first division would be " reductional " in respect of differences between the partners, in the second case it would be " equational."

The controversy was confused owing to a lack of distinction between numerical reduction and qualitative reduction. We now know that reduction in number does not occur at either division, but simply through the two divisions following one another so rapidly that no division of chromosomes takes place in the interphase

CHIASMATYPE THEORY

between them. We also know that reduction in quality, *i.e.*, the separation of unlike chromosomes, occurs both at the first and at the second division under conditions which will later be defined.

While this dispute was still undecided, Janssens (1909, 1924) put forward the objection that the formation of four spores or gametes at meiosis, the occurrence of two divisions, of pachytene pairing, and of the diplotene looping, would all lack *purpose* if reduction was achieved at a single division and without crossing-over. These arguments are of great value taken metaphorically; but Janssens also pointed out that without the assumption of an exchange of segments at the chiasmata the exchange of partner lacked any causal explanation. He pointed out that there was free association of larger numbers of factors in some organisms than there were chromosome pairs; this was genetic evidence of such exchanges. Applied to *Drosophila*, Janssens' theory became, as we have just seen, the basis of the successful study of the linear arrangement of genes.

There was, however, no unequivocal direct evidence either for or against the hypothesis. Two chromosomes went into pachytene association: four chromatids emerged at the diplotene stage. Whether an exchange had occurred no direct evidence could or can reveal. Moreover, at this time the regular structure of bivalent chromosomes, the uniform character of the chiasmata, and the relation of the two was not understood. Janssens therefore suggested (1924) several different mechanisms of crossing-over, and Belling, who accepted some of these suggestions (1928 *c*, 1931 *b*, 1933) left the relationship of the mechanism with the observed structures undefined.

In recent years three advances in interpretation have removed these difficulties. First, we understand that chiasmata are all of the same original structure (exchanges of partner among four chromatids) and differ only through movement to the ends. Secondly, we have evidence of special distribution of chiasmata in polyploids and structural hybrids which enables us to define the genetical change that determines them. Thirdly, we have a large body of comparative evidence from experimental breeding and chromosome observations which enables us to demonstrate the general validity of the hypothesis that is to be induced from these particular observations. The observations require that *all chiasmata result from crossing-over between two chromatids of the partner chromosomes*. This follows, in the light of our present knowledge of chiasmata, from Janssens' assumption that the diplotene split separates the partner chromosomes so that the pairs of chromatids that remain together after the formation of chiasmata are sister chromatids, derived from the same parent chromosome.

It now remains for us to describe the evidence for this simplified and defined chiasmatype theory and to show how it can be used in finding out genetical principles from chromosome behaviour.

4. CYTOLOGICAL EVIDENCE OF CROSSING-OVER

(i) **Kinds of Proof.** The proof that crossing-over has occurred between two chromatids of partner chromosomes follows on accepted assumptions from the demonstration that sister chromatids (from the same parent chromosome) are paired on both sides of the chiasma. This demonstration must depend on our ability to make a distinction between the two parent homologues. Such a distinction has been made in regard to three kinds of properties: of development, of function, and of form. The distinction of development was the first to be made. In auto-tetraploids the chromosomes pair at pachytene with exchanges of partner. If a chromosome could form a single chiasma between two such exchanges of partner, crossing-over must have occurred at the chiasma. Single "intercalary" chiasmata of this kind are formed in tetraploids and triploids (D., 1930 c, Figs. 39, 41). Similar developmental distinctions depend on interlocking and relational coiling. The distinction of function depends on the recognition of constant pairing properties in chromosomes. It enables us to infer crossing-over from special configurations in multiple interchange hybrids, and from the formation of chiasmata between the chromosomes of haploids (Catcheside, 1932). The distinction of form is naturally the easiest to make out. It is now regularly made in many structural hybrids which show not merely the occurrence of crossing-over, but also the occurrence of all the expected relationships between successive

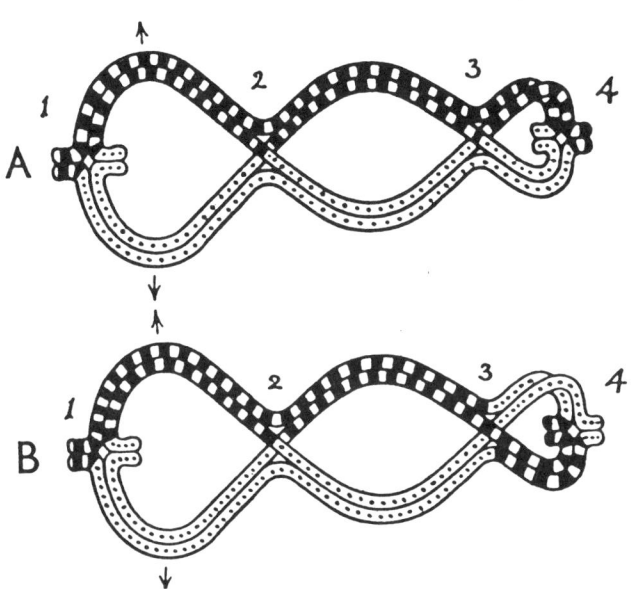

Fig. 82.—Diagram to show the distinction between different genetical interpretations of the same diplotene bivalent. This bivalent has four chiasmata which have different structural relationships with one another. Thus the second does not restore the association broken by the first; they are *disparate*; the second and third, the third and fourth, on the other hand, are *comparate* (v. Fig. 36, B and C). At anaphase there will be interlocking of chromatids in the loop between the second and third chiasmata (v. Fig. 36, C). A and B, alternative interpretations on Janssens' hypothesis of " partial chiasmatypy " as simplified by Belling, 1928 c, and Darlington, 1930 b. Disparate chiasmata (1 and 2) arise from crossing-over of one chromatid at both chiasmata and two different chromatids once at each chiasma. Comparate chiasmata arise either from both affected chromatids, being the same (2 and 3 in A, and 3 and 4 in B, " reciprocal crossing-over "), or from both being different (3 and 4 in A, 2 and 3 in B, " complementary cross-overs ").

cross-overs. These various kinds of evidence will now be considered from a more general point of view.

(ii) **Prophase Interlocking.** With few exceptions the chromosomes are fortuitously distributed in the nucleus before they pair at the prophase of meiosis, with regard to their homology. In the Diptera (cf. Metz and Nonidez, 1924) the homologous partners probably lie next to one another at the telophase of the preceding division owing to somatic pairing (q.v.) and persist in this relationship until they pair. Apart from such exceptions it would therefore be expected that the chromosomes in pairing would often interlock, a strange chromosome passing through a loop between associations of the two homologues. Necessarily, this would be detectable only in favourable material. Interlocking has been seen taking place at zygotene in *Dendrocœlum* (Gelei, 1921), *Viviparus* (Belar, 1928 b) and *Allium* (Levan, 1933 c) and maintained at metaphase, in organisms with and without polarisation of the zygotene nucleus

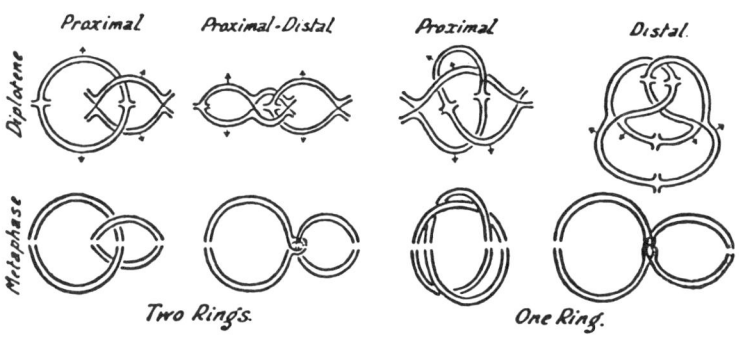

Fig. 83.—Chromatid diagram of diplotene and metaphase configurations showing interlocking of pairs of ring-bivalents (left) and within a ring-of-four (right) before and after terminalisation.

(*e.g.*, *Stenobothrus* and *Hyacinthus*), and in organisms with all degrees of terminalisation.

The detection and analysis of interlocking at metaphase is of importance, for the form it assumes will distinguish between various theories of chromatid relationship in organisms with and without terminalisation. On the hypothesis of chiasmatype crossing-over the chromatids derived from one parental chromosome fall apart from those of the other at diplotene so that an interlocked chromosome will lie between the separated chromosomes in the diplotene loops.

Simple interlocking is classifiable into three types at metaphase in an organism with terminalisation, according to whether the chiasmata, formed by the interlocked chromosomes, are moving

INTERLOCKING

away from the point of interlocking in both, one or neither of the pairs, thus (Fig. 83) :—

(i) *Proximal Interlocking.* The chiasmata move away from the point of interlocking in both pairs of chromosomes, owing to its being within the centromere loop.

(ii) *Distal Interlocking.* The chiasmata on either side of the interlocking move in the same direction, *i.e.*, towards one end, in

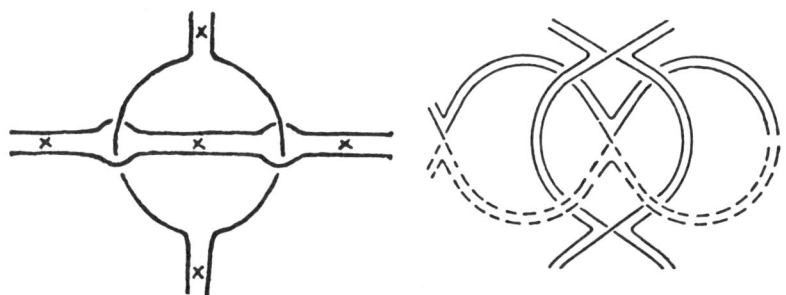

FIG. 84.—The genetical interpretation of double interlocking. Right, pachytene, and left, diplotene. Crossing-over must have occurred at the critical chiasma. (After Mather, 1932.)

both of the interlocked pairs, owing to the interlocking being distal to the most proximal chiasma.

(iii) *Proximal-distal Interlocking.* The interlocking is proximal as to one pair and distal as to the other.

The distinction between this type and the second cannot be expected to be made in practice. Both will appear, in an organism with complete terminalisation, indistinguishable from a multiple chiasma except in the most favourable material. They may both therefore be described as distal.

These kinds of interlocking should be found between separate pairs of chromosomes and between separate chromosomes of multiple rings, in interchange heterozygotes and polyploids.

A classification of the available observations based on these views shows that all the types that are expected on the chiasmatype hypothesis are found, and none that are not expected.

TABLE 41
Interlocking of Chromosomes at Meiosis

A. IN ORGANISMS WITHOUT COMPLETE TERMINALISATION (the distinction between proximal and distal being unimportant).

(i) Between single pairs :
Stenobothrus lineatus	Belar, 1928.
Hyacinthus orientalis	D., 1929 b.
Rosa blanda	Erlanson, 1931 b.
Pæonia spp.	Dark, 1936.
Allium spp.	Levan, 1933 c, 1935.

(ii) Within a multiple ring :
Pisum sativum	Pellew and E. R. Sansome, 1931.

(iii) Double interlocking of single pairs :
Lilium spp.	Mather, 1933, 1935.
	Beal, 1936.
Eremurus spectabilis	Upcott, 1936.

B. IN ORGANISMS WITH COMPLETE TERMINALISATION.

(i) Between single pairs or multiple rings :
 (a) Proximal :
Salamandra maculosa	Schreiner, 1906 b.
Œnothera spp.	Hoeppener and Renner, 1929.
	Håkansson, 1930, a and b.
	Catcheside, 1931 b.
Datura sp.	Blakeslee, 1929 (diagram).
Campanula persicifolia	Gairdner and D., 1931.

 (b) Distal :
Œnothera biennis	Cleland, 1926 a ; D., 1931 c.
Campanula persicifolia	l.c.

(ii) Within a multiple ring (only in interchange heterozygotes).
 (a) Proximal :
Œnothera biennis	D., 1931 c.
Campanula persicifolia	l.c.

 (b) Distal :
Œnothera and *Campanula*.	l.c.

PROOF BY COILING

The most significant type of interlocking is the *double interlocking* discovered by Mather where two successive loops of a bivalent have chromosomes interlocked with them. Each loop must then be separating the original parental chromosomes, and the chiasma between the two must be the result of crossing-over (Fig. 84).

(iii) Relational Coiling. Another kind of evidence, showing that

FIG. 85.—Diagram showing that crossing must have occurred before a chiasma could be formed where relational coiling is found on both sides of the chiasma. (D., 1935 d.)

the partner chromosomes always separate at diplotene, is found in the occurrence of relational coiling. This coiling develops, as we have seen, during pachytene and before the separation of the chromosomes and the appearance of chiasmata at diplotene. It must therefore be a coiling of the partner chromosomes around one another, yet it occurs on both sides of chiasmata (Fig. 85). If crossing-over has not preceded and determined such chiasmata, we must suppose (i) that the chromosomes were divided into chromatids before coiling began, (ii) that these chromatids were re-assorted into new pairs which acted as units in coiling, and (iii) that the chiasmata were thus pre-formed at the beginning of pachytene and before any stress has developed between the partner chromosomes such as could account for the special properties of distribution found in chiasmata. The observed development of relational coiling, such as is found at pachytene and diplotene in *Chorthippus* or *Fritillaria*, is evidently compatible with no other assumption than that of crossing-over at chiasmata (D., 1935 c, 1936 b).

5. CROSSING-OVER IN STRUCTURAL HYBRIDS

(i) **Deficiency Hybrids.** It is a matter of observation that the two chromatids between the centromere and the first chiasma pass to the same pole at the first anaphase. On the other hand, owing to the exchange of partners at the chiasma, paired chromatids are separated on the other side of the chiasma. Since apart from movement of the chiasma, the paired chromatids are sister chromatids from the same parent chromosome, the first division is reductional for the part of the chromosome proximal to the first chiasma and equational for the part immediately distal to it.

The results of this principle are seen in the different types of behaviour found in " unequal bivalents." These bivalents are usually the result of pairing of two chromosomes, one of which has lost an end segment (Table 20). If no chiasma and no crossing-over occur between the place where the difference shows itself and the centromere, then the first division is reductional in respect of the difference. If a single chiasma occurs in this segment, the first division is equational in respect of the difference. These types of behaviour are characteristic of particular unequal bivalents for the simple reason that they depend on the structure of the particular bivalents. The difference in some cases lies next to the centromere, and no crossing-over can ever occur between them; such bivalents always divide reductionally at the first division in respect of the difference. The difference in other cases lies at the opposite end of the chromosome from the centromere; they always form one chiasma, and the first division is always equational. It is naturally to be expected from what we know of the variable positions of chiasmata that certain unequal bivalents should be found sometimes with crossing-over between the centromere and the inequality

Equational   Reductional

Obligatory   Facultative

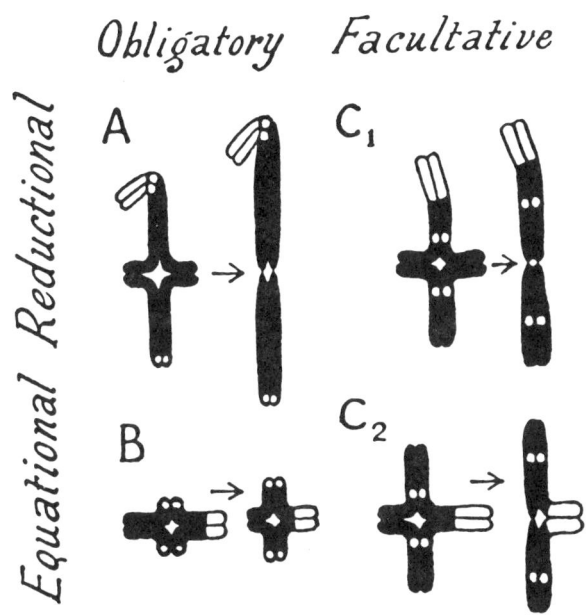

Fig. 86.—Diagram showing different types of division of unequal chromosomes owing, on the assumption of chiasmatypy, to the different relationships of the inequality, the centromere and the chiasma, which may or may not be formed between them. (Where several chiasmata are formed the possibilities are more numerous.) The arrows represent the direction of the change undergone by the bivalent between diplotene and metaphase with terminalisation (incomplete in B). The inequalities and the centromeres are shown blank. A, first division regularly reductional owing to the centromeres lying next to the inequality. B, first division regularly equational (second division reductional) owing to the centromeres lying at the opposite end from inequality and one chiasma being formed between them. C, first division reductional or equational according to whether a chiasma is formed on one side of the centromere or the other. C_2 shows a lateral chiasma. A, *Trimerotropis, Circotettix, Acridium, Stenobothrus*, most sex chromosomes (and autosomes fused with sex chromosomes). B, *Phrynotettix*, chromosome " B "; *Melanoplus* (Hearne and Huskins, 1935); C, *Phrynotettix*, chromosome " C "; *Mecostethus gracilis* and *Trimerotropis citrina* (Carothers, 1931); *Stauroderus* (D., 1936 *d*); *Peziza* (Matsuura and Gondo, 1935).

and sometimes not. Such bivalents are found frequently in Orthoptera, and we can now see why they behave in this way

262 CHROMOSOMES IN HEREDITY: MECHANICAL

(Fig. 86). They form chiasmata on either side of the centromere, *i.e.*, either between it and the difference, or not.

Terminalisation of a chiasma between two arms, one of which lacks an end, gives a symmetrical *lateral chiasma* (Fig. 86). Such chiasmata have been found in haploid *Œnothera* (Catcheside, 1932) and in *Pæonia* (Dark, 1936) as well as in Orthoptera.

In the light of these observations, we see how unfortunate was

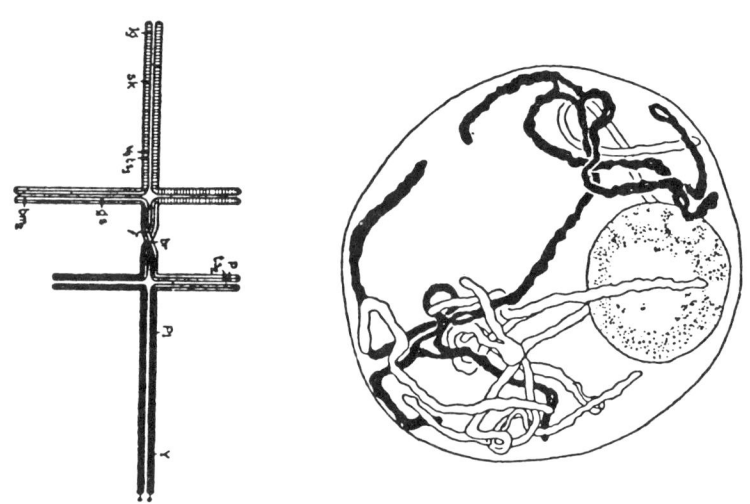

FIG. 87.—Pachytene pairing in a double interchange heterozygote in *Zea Mays*, diagram and observed arrangement, showing the differential segment where a chiasma will give a "figure-of-eight," *cf.* Ch. IX (Brink and Cooper, 1932).

the attempt made thirty and forty years ago to decide whether the first or the second division was the "reduction division." The behaviour in regard to reductional or equational division of a part of a bivalent depends on the number and relationships of the chiasmata that lie between it and the centromere. The behaviour of the whole of a univalent which forms no chiasmata depends directly on the division or lack of division of the centromere. The behaviour of the nucleus as a whole in having a reduction of chromosome number depends on the succession of two divisions so rapidly that only one division of the centromeres (and chromomeres) takes place. Two successive divisions, not one special division, are necessary therefore for reduction. It is no longer possible to imagine that by tacking on a third, as in " brachymeiosis," two reductions in the ordinary sense can be produced.

(ii) **Interchange Hybrids.** Interchange may occur at any point along a chromosome and the pachytene pairing of interchange heterozygotes therefore shows the centromere lying in one of the

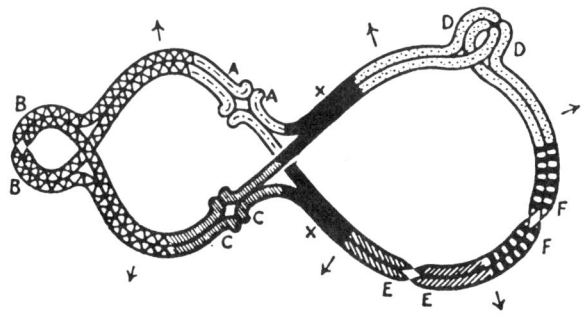

FIG. 88.—The " figure-of-eight " in *Pisum* at first metaphase. A chromatid diagram (*cf.* Plate VI). The six chromosomes before crossing-over had the constitution : AB-BC-CXD-DF-FE-EXA. Crossing-over has given two new chromatids (or " daughter-chromosomes ") : AXD and CXE (*cf.* Ch. IX, from E. R. Sansome, 1932.)

Note.—The chiasma formed in the X segment is represented as having been arrested in terminalisation by the change of homology.

two arms of the cross formed by the four chromosomes. If chiasmata are not formed between the centromere and the point of interchange a simple ring is formed at metaphase, which segregates, as we have seen. But if a chiasma is formed in one of these segments, then, owing to the crossing-over which has occurred, simple disjunction is rendered impossible (Fig. 88). There must always be " chromatid non-disjunction " (Sutton, 1935, *cf.* Sansome, 1933), whichever way the four chromosomes arrange themselves. The occurrence and length of such *interstitial segments* is therefore important for the behaviour and especially for the fertility of simple interchange heterozygotes (D., 1936 a). It should be noted that where interchange occurs between "rod-shaped" chromosomes having short arms with no chiasmata in them, a ring of four can be formed only when there is crossing-over in this interstitial segment (Fig. 89).

When we come to consider rings of six and larger numbers of chromosomes, a further complication arises. Owing to the negligible chance of two interchanges coinciding there will always be two or

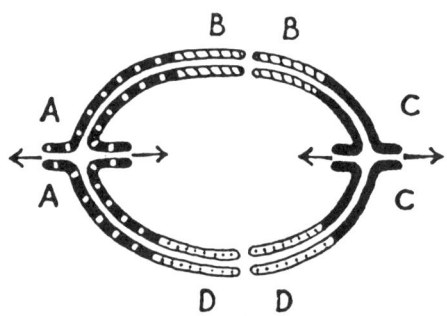

FIG. 89.—Diagram showing the genetic relationship to be inferred in the chromatids of the ring of four found in *Trimerotropis citrina* by Carothers (1931). The arrows indicate the spindle attachments, which are terminal. The pairs in the homologous segments (A and A, C and C) being closest together are lying axially in the spindle. The fact that the B and D segments are lying equatorially proves that there is a chiasma in A and in C. Since B is not homologous with D nor C with A, there must have been crossing-over between the centromeres and the points of interchange, as shown in the diagram, unless chromatids of opposite chromosomes are associated at the centromere. This is associated with the formation of a chiasma in each case.

Note.—This ring, unlike the normal type, will have the second division reductional and the two chromosomes dividing at random will give 50 per cent. of non-disjunctional progeny.

more chromosomes in such rings with interstitial segments which lie between end segments neither of which is homologous with the end segments of the homologous interstitial segment. Such segments are *differential segments* (D., 1936 a), and if they cross over they produce (secondarily) a reverse segmental interchange which will reduce the size of the ring in the progeny (Sansome, 1933, Brink and Cooper, 1935). In *Pisum* such crossing-over is frequent, in *Campanula* on the other hand, it is known only by its genetical results, and it is probably rare. This is no doubt due to the differential segments undergoing non-homologous pairing in continuation of the pairing of the ends, which alone can therefore cross over and form chiasmata. The genetical consequences of such behaviour will be considered in relation to *Œnothera* (Ch. IX).

(iii) **Inversion Hybrids** (*a*) *Structural Types.* We have so far considered organisms in which the changed and unchanged segments have the same linear order in relation to the centromere. These may be described as *eucentric* structural hybrids, and their behaviour at meiosis is remarkable merely in showing multiple associations, like those of polyploids, and unequal bivalents, sometimes with lateral chiasmata. Structural change may, however, reverse the linear order in relation to the centromere, and we then have a *dyscentric* structural hybrid. The simplest way in which this change can arise is by simple inversion of a segment. Such changes have long been recognised genetically in *Drosophila*, and have recently been identified in the salivary glands and at pachytene in maize. They may also arise with translocation of a segment from one chromosome to another or from one arm of a chromosome to another. Thus, if a comma represents the centromere, a chromosome *abcd,efgh* becomes *abcd,egfh* by simple inversion and *abfgcd,eh* by inversion of *fg* combined with internal translocation; this last by crossing-over with a normal chromosome will give *abfgcd,efgh*, in which the inverted segment is reduplicated. It will be noticed in this last chromosome that *fg* is not inverted in relation to the ends, but only in relation to the centromere, and inversion so defined is the essential property of a dyscentric hybrid. Inversion, including the centromere, it should be noted, is not inversion with respect to the centromere; thus *abcd,efgh* + *abce,dfgh* gives a eucentric hybrid (D., 1936 *d*).

(*b*) *The Results of Crossing-Over.* Inversion, whether including the centromere or not, results in the reverse pairing of a loop at pachytene (in *Zea*, *Tulipa* and *Chorthippus*), and is recognised in *Drosophila* by an exactly similar behaviour in the salivary glands (Ch. V). Sometimes, however, the inversions may pair straight at pachytene, non-homologous parts associating in a proportion of cells by torsion and not by attraction (McClintock, 1933, on *Zea*,

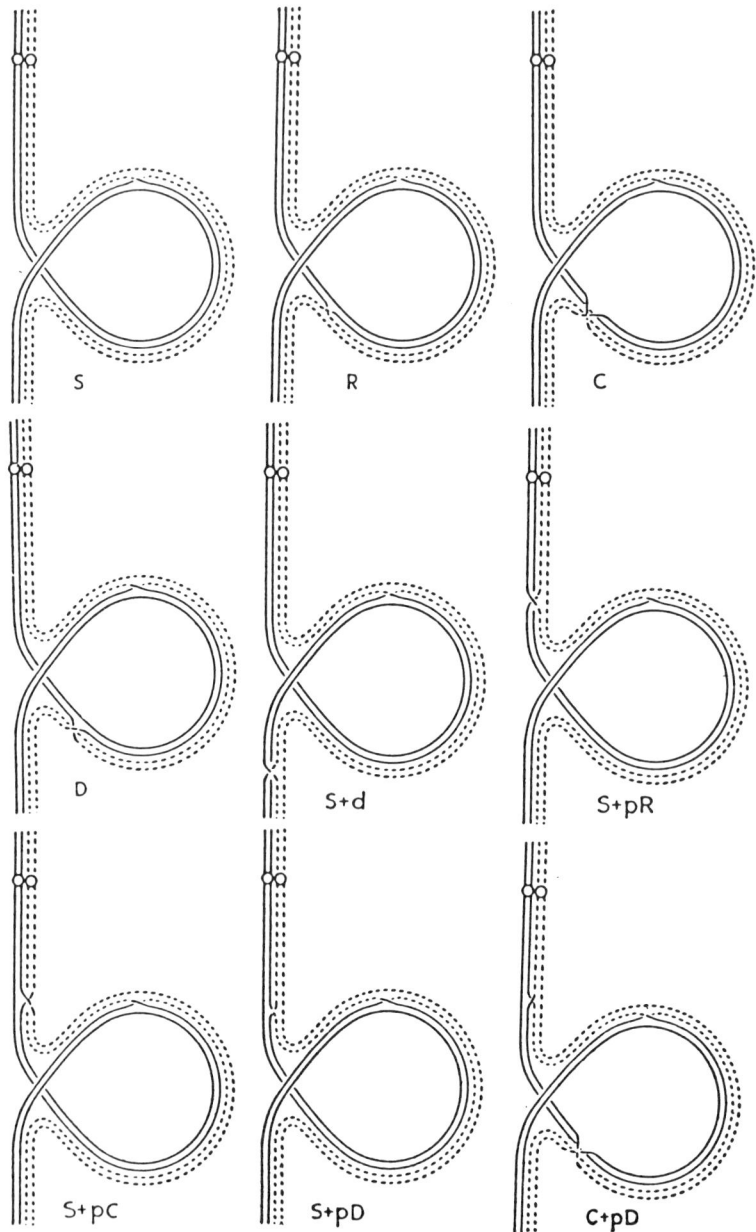

Fig. 90.

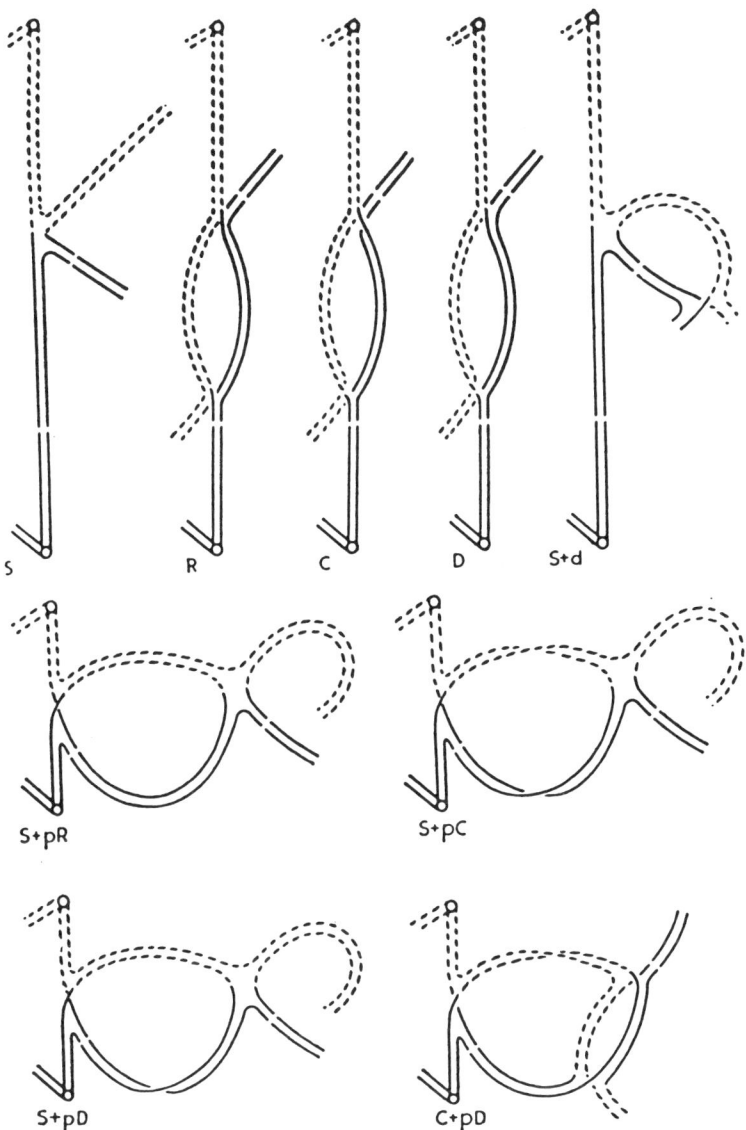

Fig. 91.

Figs. 90 and 91.—End of pachytene and metaphase configurations of a pair of chromosomes differing in respect of a single inversion. S, single crossing-over in the inversion; R, reciprocal, C, complementary and, D, disparate crossing-over in the inversion. *d* and *p*, crossing-over distal and proximal to the inversion. One chromosome broken, the other entire except for breaks at point of inversion.

D., 1936 d, on *Chorthippus*, cf. Ch. XII). Such association probably prevents crossing-over, and merely reduces the total chiasma-frequency of the hybrid.

The later behaviour which characterises a dyscentric hybrid appears only when crossing-over has taken place between the dislocated (*i.e.*, relatively inverted) segments. This behaviour depends on the number of crossings-over and their relationships *within* the dislocated segments and *proximal* to them, *i.e.*, between them and their centromeres (Richardson, 1936, D., 1936 d).

The result of crossing-over is to produce chiasmata which may be seen to be inverted, either directly (in *Chorthippus* and *Pæonia*) or by the inequality of the corresponding arms (as in *Tulipa*). Where two chiasmata are formed in the inversion they are no doubt liable to move apart, owing to greater repulsion within the loop than outside it.

At anaphase the behaviour depends on the fact that crossing-over between two dislocated chromatids gives one *dicentric* chromatid (with two centromeres) or loop chromatid, and one *acentric* chromatid (with no centromere). The dicentric chromatid makes a bridge at the first, and the loop chromatid at the second division; the acentric chromatid makes a passive fragment.

The five kinds of separation found at first anaphase in dyscentric hybrids without duplication within one chromosome are as follows (Fig. 91) :—

1. Normal separation, possibly delayed by exceptional tension: with *reciprocal* chiasmata within the inversion, whatever chiasmata are proximal to the inversion.

2. A chromatid bridge and a fragment: with a single chiasma in the inversion, or with two *disparate* chiasmata in the inversion, or with a chiasma proximal to the inversion, which is *comparate* (reciprocal or complementary) with regard to chiasmata in the inversion.

3. Two chromatid bridges and two fragments: with two complementary chiasmata in the inversion.

4. A loop chromatid, two normal chromatids and a fragment: with a chiasma proximal to the inversion which is disparate with respect to a chiasma, or chiasmata of Type 2, in the inversion.

5. Two loop chromatids and two fragments: with a chiasma proximal to the inversion, which is disparate with respect to complementary chiasmata in the inversion.

We thus see that pairs of crossings-over in the inversion may be in a complementary, reciprocal or disparate relationship with one another. On the other hand when crossing-over occurs proximal to the inversion it may be in a comparate (complementary or reciprocal) or disparate relationship with the crossing-over (single or double) in the inversion. The comparate relationship makes no observable difference to the effects of crossing-over in the inversion. The disparate relationship has the effect of replacing first division bridges by loops which will give second division bridges.

TABLE 41A.

Results of Dyscentric Crossing-over

Crossing-over in inversion	No crossing-over proximal to inversion or comparate crossing-over	Disparate crossing-over proximal to inversion
Single or disparate	$b^I + f$	$b^{II} + f^1$
Complementary	$2b^I + 2ff$	$2b^{II} + 2ff^2$
Reciprocal	Tension.	Tension.

(f, acentric chromatid; b^I first division bridge; b^{II} second division bridge from first division loop).

[1] Disparate crossing-over proximal to two disparate cross-overs in the inversion is that where a chromatid, which had a cross-over in the inversion, crosses over with one which had two, or none in the inversion. The other two kinds of crossing-over are comparate.

[2] Where two loops are formed they sometimes interlock at the first division. In this case the two second division bridges are interlocked unless they are broken (*Tulipa*).

It is difficult to observe the pachytene pairing of short inversions directly and to determine the exact length of long ones, owing, as we saw, to their pairing straight, wholly or in part, as though they were normal. It is therefore desirable to estimate the length of an inversion indirectly. For this purpose two assumptions can be made. (i) Its length is not less than the proportion indicated by the frequency of single chiasmata in it, as shown by the sum of first and second division bridges. (ii) Its length is less than that of

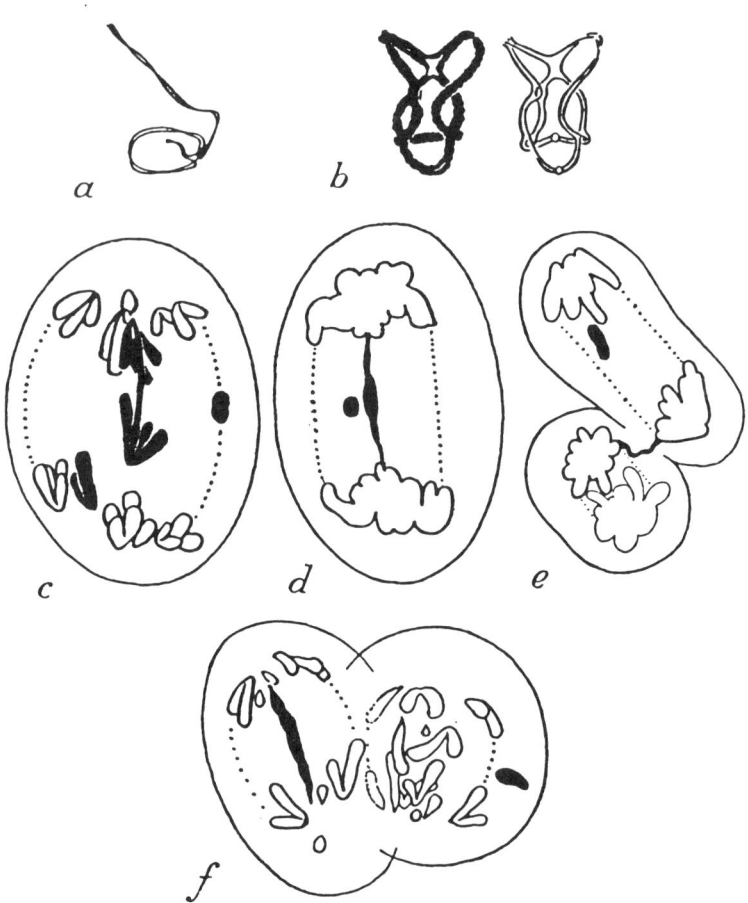

FIG. 92.—The results of crossing-over in inversions in *Chorthippus*. *a*, pachytene pairing. *b*, diplotene bivalent which has single crossing-over between two relatively inverted and translocated segments, giving an acentric and ring chromatid. *c*, *d*, first anaphase and telophase with bridge and fragment (*X* chromosome also solid). *e*, first division bridge surviving at second telophase. *f*, second division bridge (*M* chromosome) with fragment left in sister cell. × *ca*. 2000 (D., 1936 *d*).

the acentric fragment by the sum of the lengths of chromosome distal to it and less than that of the bridge by the sum of the lengths of chromosome proximal to it. Further the relative lengths of inversions are indicated by the relative frequency of double as well as single bridges, and in determining the position of an inversion the relative frequency of second to first division bridges should be correlated with the length of the bridges themselves. Where the fragment is a-shaped there has been a chiasma in the straight segment distal to the inversion.

In animals and plants with several inversions it is not always possible to define each one. But it is possible to obtain quantitative data showing what the comparative biological importance of inversion hybridity is in different forms. We can calculate the approximate frequency of crossing-over between relatively inverted segments by adding together the frequencies of bridges at the first and second division per mother cell (not per cell), counting at the same time all double first division bridges as six, instead of two, since these, resulting from complementary crossing-over, presumably represent equal numbers of reciprocal crossings-over which do not give bridges, and disparate crossings-over which give only one bridge for two. This calculation will not allow for the corresponding result in second division bridges, nor will it allow for triple crossing-over in inversions; these, however, will always be very rare.

The inversion crossing-over frequency obtained in this way is itself an index of the effect of the hybridity on the fertility of the organism. By dividing this value by the average chiasma-frequency per cell we can obtain a *coefficient of hybridity* in regard to inversions. With this datum comparison may be made of the hybridity equilibrium in different natural populations (D., 1936 d). The equilibrium is highest in plants that are largely propagated by asexual means, lowest in plants with self-fertilisation, and intermediate in the insect populations that have been examined.

(c) *Later Behaviour of Dyscentric Configurations.* With regular disjunction a short chromatid bridge is broken at the anaphase at which it appears, a long one may persist until it is broken by the separation of pollen grains or spermatids. The acentric fragment is entirely passive. It may either lie free on the plate or be strangled by the bridge chromatid coiling round it. Such a fragment may be carried into the nucleus, and has been seen lagging in the pollen-grain mitosis in *Podophyllum* (D., 1936 b).

So much applies to simple inversions or external inverted translocations (in which the two pairing segments are attached to two centromeres), but we may equally have the pairing of dislocated segments within one chromosome (inverted duplication), or as part of a trivalent combination. Crossing-over within one chromosome may be, so to speak, between a chromatid and itself or between two sister chromatids. The first will give a ring chromatid, which will perpetuate itself as a ring chromosome (as found in attached X's with an inversion in *Drosophila*, Tiniakov, 1935). The second will give a loop chromatid and a second division bridge.

Where three chromosomes are associated in an interchange hybrid or a polyploid, the two centromeres of the dicentric chromatid sometimes pass to the same pole. Such a dicentric chromatid may perpetuate itself normally, as in *Ascaris*, if the whole of each daughter chromatid goes to the same pole. But sooner or later they will go to opposite poles and a reciprocal bridge will be formed at the mitotic anaphase (*Tulipa*, Upcott, unpub., *cf.* Ch. X (iii)).

The consequences of crossing-over in a dyscentric hybrid are different in egg cells and male germ cells on account of the spatial arrangements resulting. This is important, because genetical observations on crossing-over in *Drosophila* are confined to the female. Sturtevant (1926 *b*) found that crossing-over was suppressed in inversion-heterozygotes between the dislocated segments apart from double crossing-over. Yet later workers (Stone and Thomas, 1935; Grüneberg, 1935) have found that the viability of eggs is not reduced by the proportion of the expected single crossovers. The reason for this is now clear. The four products of meiosis lie in a row in the embryo-sac or egg mother-cell (egg nucleus and polar bodies) and squarely or equidistant in the pollen or sperm mother cells. Beadle and Sturtevant (1935) have found that the cross-over chromatids taking part in the bridge never enter into the functional female nucleus, which is an end nucleus in the row. They conclude that the bridge holds together so that its parts never pass into this end nucleus. The deleterious effect of crossing-over in inversion hybrids will therefore be less serious in female than in male gamete formation.

(*d*) *Significance of Crossing-Over in Inversions.* The consideration of the crossing-over properties of dyscentric hybrids is important for several reasons. First, it enables us to distinguish the results of different crossing-over relationships between chiasmata—reciprocal, complementary and disparate. Secondly, it explains different kinds of chromosome behaviour found at meiosis in structural hybrids and not hitherto understood. It therefore enables us to detect the presence of these kinds of structural differences in nature. Thirdly, it shows how new chromosome types such as have been found in nature can arise from crossing-over in such hybrids, *i.e.*, by secondary structural change. Cases described earlier as the result of fragmentation are probably often due to this crossing-over, and spontaneous ring chromosomes such as have been found can undoubtedly arise from the crossing-over observed between inverted and translocated segments in *Chorthippus*. Fourthly, it provides us with a natural experiment in the behaviour of the centromere, showing its permanence and individuality, as well as what happens to chromatids with two centromeres and with none, and hence the vital importance of its position in the chromosome.

But crossing-over in inversions is perhaps of most immediate interest in showing the frequency with which inversions occur and give rise to inversion hybrids, not only when races and species are crossed, but also within a normally mating wild population. Their importance in a natural population depends to some extent on the abnormal results of crossing-over in giving new chromosome types, but to a much greater extent, at least in diploids, on their effect in suppressing effective crossing-over and therefore in holding groups of genes together in such a way that they will behave as units of crossing-over although they are not units of mutation (D., 1936 *a* and *d*).

TABLE 42

Dyscentric Structural Hybrids

1. Irradiated Stocks.
    Zea Mays (2x). McClintock, 1931. pach., ana. I.
    Drosophila melanogaster (2x). Painter, 1934 *b*; Painter and Stone, 1935; Grüneberg, 1935; Beadle and Sturtevant, 1935; Muller and Prokofieva, 1935 (salivary gland); Sturtevant and Plunkett, 1926; Sturtevant, 1926; Sturtevant and Dobzhansky, 1930 (breeding experiments).
    Vicia faba (2x). Mather, 1934. met. I.

2. Natural Populations.

PLANTS:
Trillium erectum (2*x*). S. G. Smith, 1935. ana. I.
Tulipa Orphanidea et al. (2*x*, 3*x* and 5*x*). Upcott, unpub. pach. dip. ana. I and II.
Fritillaria dasyphylla et al. (2*x* and 3*x*). Frankel, unpub. ana. I and II.
Agave sisalana. Doughty, unpub. ana. I.
Secale cereale (inbred). Lamm, 1936. ana. I.
Pæonia peregrina. Dark, 1936. ana. I.
Æsculus Hippocastanum. Upcott, 1936. ana. I.
Tradescantia gigantea (2*x*). Upcott, unpub. ana. I.
Pisum humile. Sutton, unpub. ana. I.

ANIMALS:
Chorthippus parallelus ♂
Stauroderus bicolor ♂ } D., 1936 *d*. pach., dip. ana. I and II.
Drosophila melanogaster. Painter, 1934 *a*, salivary gland.
D. funebris, etc. Dubinin, *et al.,* 1936, salivary gland.
Anas platyrrhynca ♂ Koller, unpub.
Chironomus dorsalis. Bauer, 1935 *a*, salivary gland.

3. Inter-racial Hybrids.

Avena sativa. Philp, unpub. ana. I and II.
Drosophila pseudo-obscura. Koller, 1935, 1936 ; Tan, 1935, salivary gland.

4. Interspecific Hybrids.

Triticum dicoccum × *T. monococcum* (3*x*). Mather, 1935. ana. I.
Lilium Martagon × *L. Hansoni* (2*x*). Richardson, 1936. ana. I and II.
Tradescantia (3*x*, probably 2*x* × 4*x*). D., and Upcott, unpub.
Primula kewensis (2*x*). Upcott, unpub. ana. I.
Æsculus carnea (8*x*). Upcott, 1936.
Pisum humile × *P. arvense.* Håkansson, 1936. ana. I.
Crepis divaricata × *C. dioscoridis* (2*x*). Müntzing, 1934. ana. I.
Drosophila melanogaster × *D. simulans* (2*x*). Pätau, 1935, salivary gland.
Anas platyrrhynca × *Cairina moscata* (2*x*). Crew and Koller, 1936. ana. I

5. Unanalysed Observations of Crossing-over in Inversions.

Lesley and Mann,	1924	*Matthiola incana.*
Belling,	1925 *c*	*Uvularia grandiflora* (Fig. 5, " fracture ").
Darlington,	1929 *b*	*Hyacinthus orientalis* (3*x*).
Meurman,	1929	*Aucuba japonica* (4*x*).
Belar,	1929 *a*	*Stenobothrus lineatus,* living cells (Fig. 68 *l.c.*).
Shinke,	1930	*Hosta* sp.
Tuan,	1930	*Gasteria* sp.
Lucas and Stark,	1931	*Melanoplus femur-rubrum,* living cells (Plate XVI).
Stebbins,	1932 *a*	*Antennaria solitaria.*
Nishiyama,	1932	*Avena sativa* (6*x*) × *A. strigosa* (2*x*).
Kattermann,	1933	*Briza media* (Figs. 3A, 16).
Nagao,	1933	*Narcissus tazetta* (Fig. 64).
Levan,	1935 *b*	*Allium schœnoprasum* (Fig. 36, 3*x*).
Levan,	1935 *b*	*A. nutans* (Fig. 52A, 3*x*).
Yasui,	1935	*Triticum monococcum* ($n = 7$, haploid).
Kato and Iwata,	1935	*Lilium longiflorum.*
Katayama,	1936	*Allium scordoprasum* (4*x*).

PLATE IX

THE RESULTS OF CROSSING-OVER IN INVERSION HETEROZYGOTES
(*cf.* Text-Fig. 92)

FIGS. 1–3.—*Chorthippus parallelus*, ♂. (D., 1936 *d*.)

FIG. 1.—Diplotene with chiasma between two relatively translocated and inverted segments within the same bivalent (Text-Fig. 92 *b*).

FIG. 2.—First anaphase, dicentric bridge and acentric fragment.

FIG. 3.—Second anaphase, second division bridge in *M* chromosome.

FIG. 4.—Beginning of first anaphase in *Pæonia*, showing inversion chiasma separating. (*Cf.* Dark, 1936.)

FIG. 5.—Bridge and fragment in diploid *Tulipa*, first anaphase × 2000. (Upcott, unpublished.)

FIG. 6.—Univalent and bivalent bridge formation in triploid *Tulipa*, three acentric fragments. × 2000. (Upcott, unpublished.)

FIG. 7.—Acentric ring chromatid produced by inversion crossing over at first anaphase in *Tulipa*. × 1400. (Upcott, unpublished.)

FIGS. 8 AND 9.—Acentric chromatid in pollen grain mitosis of *Podophyllum* at two focuses. (D., 1936 *b*.)

PLATE IX

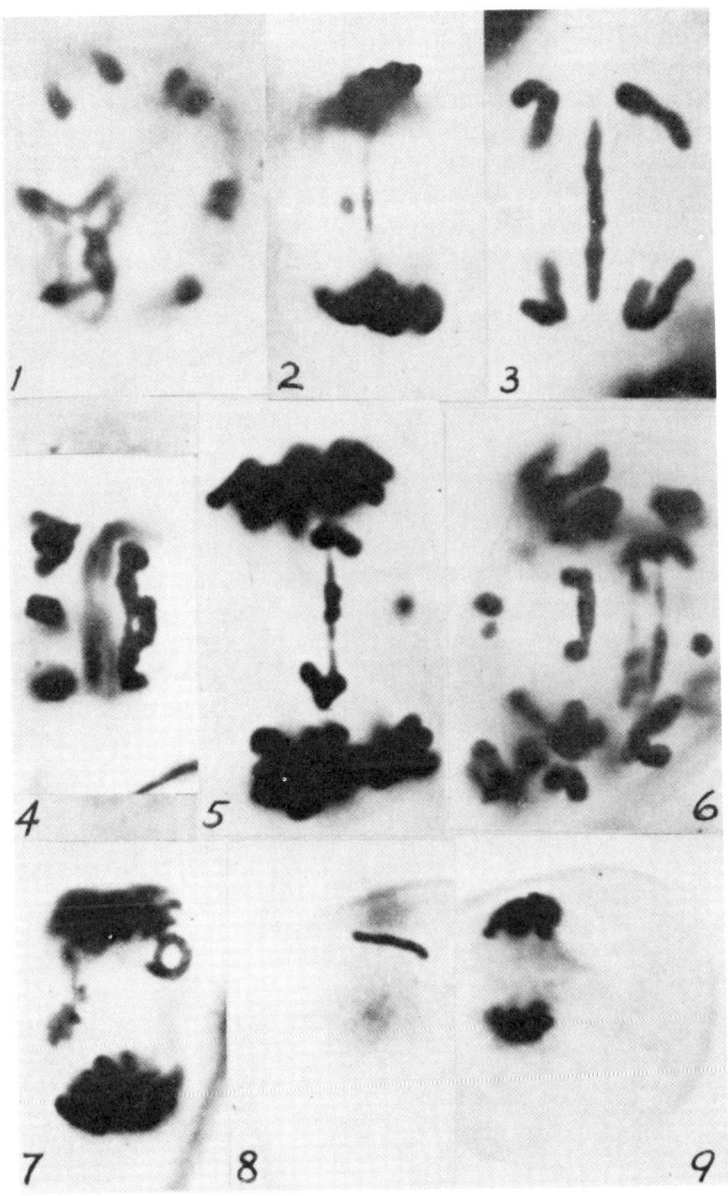

[*To face p.* 274.

ILLEGITIMATE CROSSING-OVER

(iv) **Secondary Structural Change.** Crossing-over in structural hybrids of the kind that has been discussed gives rise to recognisably new forms of chromosomes. There is an exchange of dissimilar segments between chromosomes with two such segments, as a result of crossing-over in the region between them. We have seen how it occurs in the differential segments of multiple interchange hybrids and also in dislocated segments in inversion hybrids. This crossing-over resembles interchange with the difference that it can occur only at meiosis and will occur only at particular places. Moreover, it will occur with a regular frequency. We can therefore distinguish by the later results between this secondary structural change and primary interchange even when its immediate consequence of chiasma formation is not seen.

The precise circumstances of secondary structural changes have been determined cytologically and genetically in two experiments.

(i) Females of *Drosophila* were taken having, in the first place, two *structural* differences between different parts of their X chromosomes, viz., attachment and non-attachment of Y' (Fig. 121); fragmentation into two parts, one of them attached to the fourth chromosome and non-fragmentation; and, in the second place, two *gene* differences lying between the two structural differences. They were crossed to males having a single normal (*i.e.*, non-fragmented and non-attached) X chromosome with the two corresponding genes recessive. The gene differences immediately adjoined the two structural differences, so that, in examination of the offspring, if crossing-over was seen to have taken place in an oöcyte by the genetic character of the individual derived from it, new chromosome types having the two kinds of structure at each end of the X chromosome should be found cytologically in that individual, and where no genetical evidence of crossing-over was seen, one of the two maternal types of X should be found. This was tested in four experiments. Crossing-over with formation of new chromosome types was found in 156 flies; no crossing-over was found with no new chromosome types in 203 flies; two unexpected types were found and three contrary to expectation; these are believed to be due to errors of genetical recording such as cannot be eliminated entirely (Stern, 1931).

(ii) An interchange and reduplication heterozygote in maize was recognised by its having a ring of four chromosomes, one of which had a small extra segment seen at prophase of meiosis. When this was crossed back to a form homozygous in respect of both the interchange and the reduplication the progeny could be classified cytologically according to the observed occurrence of crossing-over between the point of interchange and the point of reduplication. They could also be classified genetically according to the occurrence of crossing-over between factors known on previous evidence to occur in the segment between the interchange difference and the reduplication difference. The two classifications agree to the extent that would be predicted from the previously known linear order of the factors studied (Creighton and McClintock, 1931).

The characteristic results by which we recognise secondary structural change are the appearance of frequent new chromosomes of particular types, especially fragments, and genetically the appearance of new mutations. In *Œnothera* this method of mutation will be described in detail (Ch. IX).

The new fragments probably arise in one of two ways. Either an interchange results from crossing-over and is unequal, so that one very large and one very small new chromosome are produced. Or the crossing-over is between relatively inverted segments, giving a bridge which breaks and so leaves a deficient chromosome. This second method will probably be effective in polyploids.

The circumstances favouring secondary structural changes are of several kinds, which require separate consideration. First, we have the interspecific hybrids which are presumably structural hybrids such as those that have already been considered. Thus in derivatives from interspecific crosses with *Avena sativa* and of the hybrid *A. strigosa* × *A. barbata*, Nishiyama (1933, 1934) and Philp (1934 *b*) have found new types of small chromosome attributable to crossing-over (*cf.* also Buxton and D., 1932, in *Digitalis;* Katayama, 1935, in *Ægilotricum;* and Ono, 1935, in *Rumex*).

Hybrids that have been made between *Crepis* species having chromosomes of different shapes, reveal the occurrence of secondary structural changes. It would be expected that, crossing-over having taken place between them in the first generation, chromosomes of new shapes would appear in the second. Such new chromosomes have been found in the progenies of two triploid *Crepis* hybrids; *C. capillaris* × *C. aspera* (Nawaschin, 1927) and *C. capillaris* × *C. tectorum*, CCT (Hollingshead, 1930 *b*). In the first, a chromosome of *capillaris*, in the second, one of *tectorum*, had lost a portion of one arm. In both it is to be supposed that the conditions of zygotene pairing in the triploid, where one thread at any point is unpaired, had rendered possible crossing-over between relatively translocated segments. Evidence that such crossing-over occurs in the triploid with segregation of cross-overs and non-cross-overs is found in the fact that the derived tetraploid (*cf.* Table 26) is a weak plant unlike any tetraploid produced by simple doubling.

The attachment of the X and Y' chromosomes in *Drosophila* resembles a secondary structural change, occurs always in the same way and, like the half-mutants, with a definite frequency, once in 1,500 to 2,000 times (Stern, 1929 *a*). Since the association occurs at a point where the two chromosomes are probably homologous, it is possible that it arises through crossing-over (*cf.* D., 1931 *a*, and Fig. 121). This conclusion has now been confirmed by Kaufmann (1934, *cf.* D., 1935 *f*), who has found the complementary crossover types.

Analogous with this is the origin of new chromosome types following crossing-over in structural hybrid *Drosophila* whose hybridity is the result of X-ray changes, as in Stern's experiment described above (*cf.* Muller, 1930 *a*). Similarly, crossing-over within an inversion including the centromere will give the kind of chromosome that is found in secondary trisomics in *Datura* and *Zea*. Since it will survive only as an extra chromosome, its discovery indicates that the change determining it is not rare.

Secondly we have allopolyploids in which, as an exception, homologous but structurally differentiated chromosomes from different sets will pair. Two abnormalities of inheritance not found in diploids are conceivable in such allopolyploids, and both have been found. Consider three homologous but different pairs of chromosomes in a hexaploid, AA, BB, CC. One of these may be lost whether at mitosis or at meiosis to give ABC, AB–. In breeding this heterozygous type, AB– AB– will be produced, a homozygous form with more extreme abnormality than the heterozygote. Such a new form, however, usually results from pairing of B and C, and this pairing involves crossing-over between these dissimilar chromosomes. When the changed C chromosome is lost, this crossing-over makes no difference. But sometimes it is not, and we have a gamete of the make-up ABC^b (C^b being a C chromosome which contains a segment of B through crossing-over) and zygotes of the make-up ABC, ABC^b and ABC^b, ABC^b are produced. These are heterozygous and homozygous for the C^b chromosome, and appear as mutants from the original type. Such mutants must be taken to be the products of secondary segregation in the polyploid. They have been found in *Primula kewensis* and *Nicotiana Tabacum* (R. E. Clausen, 1931), but it is in the hexaploid cereals that their effect on inheritance has been studied in the greatest detail. Speltoid wheats and fatuoid oats both arise by these means (Winge, 1924; Huskins, 1927, 1928; Håkansson, 1931; Nishiyama, 1931, 1933; Philp, 1933). Moreover, it has been possible to identify the parts of the C chromosome whose loss is responsible for mutation in this way. Heterozygous fatuoids, ABC, AB—, produce some offspring in which the C chromosome has been replaced by a shorter chromosome which has lost either the shorter arm which suppresses the fatuoid character or the longer arm which conditions normal meiosis. (Nishiyama, 1935; Uchikawa, 1934). Both these losses are presumably due to crossing-over between the middle of the C chromosome and an homologous segment near the end of B or some other chromosome.

The third circumstance of secondary structural change is by intra-haploid pairing. This occurs in haploids, triploids and unbalanced forms such as trisomic diploids and trisomic tetraploids. In all these there are one or more unpaired chromosomes. If these contain amongst themselves reduplicated segments they will pair, crossover and give secondary structural changes. There is now evidence for every step in the process in different organisms. The reduplicated segments have been found in the haploid complement of *Drosophila* (Bridges, 1936; Muller, 1935). The pairing between different chromosomes in the haploid has been found in *Œnothera* (Catcheside,

1932), *Triticum monococcum* (Kihara and Katayama, 1932), *Pharbitis* (Katayama, 1935), and *Zea* (polymitotic pollen, Beadle, 1931, 1933). That this pairing is by chiasmata resulting from homologous association is indicated by the symmetrical lateral chiasmata in *Œnothera* and the inversion-chiasmata giving chromatid bridges in *Triticum* (Katayama, 1935), structures both characteristic of the pairing of homologous segments intercalated in non-homologous chromosomes. Quadruple chiasmata have been found in diploid *Pæonia* (Dark, 1936), and evidence of reduplication within particular chromosomes in *Tradescantia* (D., 1929 c). Particular fragments arise in trisomics in *Solanum* (Lesley and Lesley, 1929), *Matthiola* (Frost, 1927) and *Nicotiana* (R. E. Clausen, 1931). " Mutations " occur with specially high frequency in the progeny of haploid *Œnothera* (Davis and Kulkarni, 1930 ; Stomps, 1931).

The dyscentric crossing-over in *Secale* (Lamm, 1936), probably depends on intra-haploid pairing. It appeared in diploid plants produced by inbreeding and was associated with reduced precocity and some failure of pairing. It seems probable that the legitimate pairing was irregularly restricted on the time-limit principle, and segments deprived of their legitimate partner by its premature division have paired with reduplicated homologous segments elsewhere, and some of these have made dyscentric partners. The high frequency of intra-haploid crossing-over makes it clear that the common notion of a diploid derived from non-reduction in a haploid being necessarily homozygous is a delusion. Furthermore, a condition of intra-haploid reduplication is shown by these observations to be widespread if not universal. This conclusion is supported, as we have seen, by salivary gland structures in *Drosophila*. Its bearing on the theory of gene differentiation and haploidy will be considered later. Its effect on the ordinary mechanism may be summarised by saying that every organism with intercalary reduplications can, by crossing-over between them, give rise to new chromosome types and to new relatively unbalanced genotypes. Furthermore, if chromosomes containing these reduplications are unpaired, as in a triploid or trisomic, such changes will occur especially frequently. That is why crossing-over between relatively inverted segments seems to be commoner in triploids than in corresponding diploids, for in the triploid it results not only from structural hybridity, but also from intra-haploid reduplication.

We have seen that one of the essential properties of secondary, as opposed to primary structural change, is that it is exactly repeatable. All exactly repeated or exactly reversed changes that have been described as primary must therefore be under suspicion.

TABLE 43

Secondary Structural Changes due to Crossing-Over after Primary Changes

Primary Change.	Effect on Meiosis.	Secondary Change.
1. Dyscentric inversion.	Dicentric and acentric chromatids.	Terminal deficiency.
2. Dyscentric translocation to same arm.	Dicentric and acentric chromatids.	Terminal deficiency.
3. Dyscentric translocation to opposite arm of same chromosome.	Asymmetrical bivalent and as above.	Terminal deficiency and duplication.
4. Dyscentric interchange.	(At mitosis) dicentric and acentric chromatids.	Terminal deficiency and later breakage.
5. Simple external interchange.	Ring of four.	None.
6. Compound external interchange.	Ring of six, etc., and "figure-of-eight."	Reverse interchange.
7. External translocation.	Association of four.	External interchange.
8. External dyscentric translocation.	Association of four. Dicentric and acentric chromatids.	Dyscentric interchange, and hence deficiency.
9. Translocation to same arm.	Asymmetrical bivalent.	Intercalary reduplication and deficiency.
10. Internal interchange.	Asymmetrical bivalent.	Terminal reduplication and deficiency.
11. Translocation to opposite arm.	Asymmetrical bivalent.	Terminal reduplication and deficiency.

Important examples of these are the exact reversals of an inversion found in *Drosophila* to be accompanied by reversal of a correlated mutation believed to be due to position effect (Grüneberg, unpublished). For such reversals to occur by random structural change is out of the question. But if a small segment is translocated, reduplicated and inverted within the chromosome, as found by

CHIASMATA AND CROSSING-OVER

Bridges, crossing-over between the homologous dislocated segments will give a reversible inversion. If the segment is short enough for localised gene action to make a difference beyond it (as Muller found), position effect change will be possible at the same time (Ch. VIII). Thus mutation may be possible by secondary structural change in normal diploid *Drosophila*.

6. THE PARALLEL INVESTIGATION OF CROSSING-OVER

(i) **Method.** Having found an agreement in all particular cases that are available for study between chiasmata and crossing-over, we can now profitably consider how far their general properties show a parallelism. We can use the one method of study to eke out the other. This is the more important because, although the results are parallel, the techniques are complementary, which is just why it has taken so long to establish a connection between the two. Table 44 shows how the one reveals precisely what the other fails to reveal.

(ii) **Diploids (General).** In organisms with chromosomes of uniform size the average frequency of chiasmata in each diplotene bivalent is, so far as we know, usually between one and three. In *Primula sinensis* it must be greater than two, and is probably less than three. Where large size-variations occur the larger chromosomes have a higher average frequency or it ceases to be proportional to the length. In certain Liliaceæ and Leguminosæ the chiasma frequency is greater than 3; in the Tettigidæ it seems to be usually less than two. Further, where pairing is by chiasmata a single chiasma must always be formed to secure regular pairing.

These observations show definitely the amount of crossing-over to be expected in chromosomes in general and in particular instances, on the chiasmatype hypothesis. The average occurrence of one chiasma must correspond to 50 per cent. of cross-overs in the length of chromosome in which it occurs, *i.e.*, to 50 corrected units of crossing-over distance. Therefore the lengths of chromosomes completely measured in cross-over units should be greater than 50, and they should frequently be over 100.

No chromosome has been definitely measured and found to have a length of less than 50 units. If the pairing of the small fourth chromosome in *Drosophila melanogaster* is due to chiasma formation

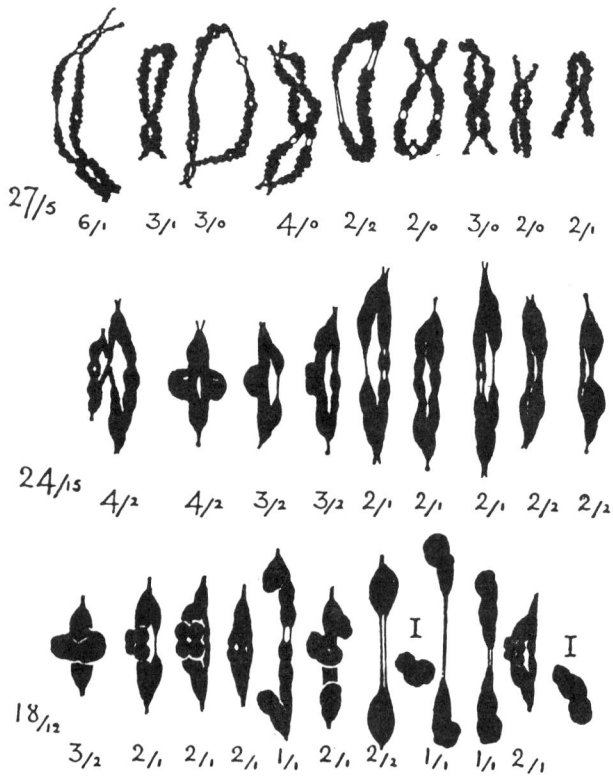

Fig. 93.—Chiasmata in *Zea Mays*. Above, late diplotene; 8 bivalents and one ring of four, 8–9 interchange. Note differential condensation. Middle, same plant at first metaphase, centromere obliterated by pressure in some bivalents. Below, double trisomic plant with reduced chiasma frequency. × 2000 (D., 1934, *cf.* Fig. 143).

the chiasmata would need to be strictly localised and map distances would not have the same relation to length as in the other chromosomes. Evidence of this condition has been found by Mather (1936 *b*). Two factors close together in this chromosome

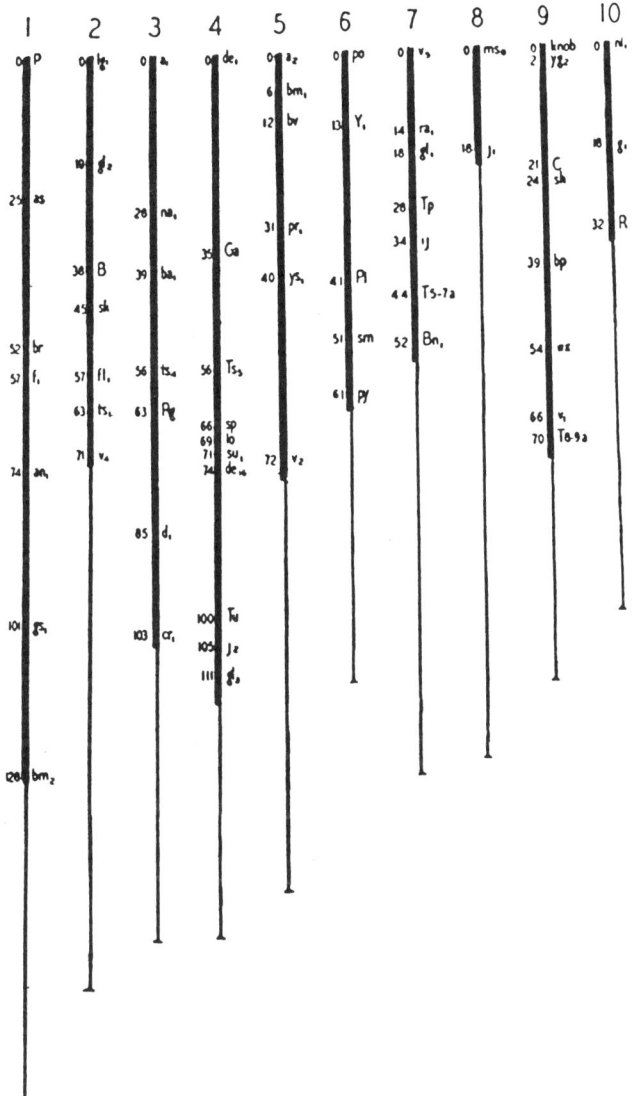

Fig. 94.—Lengths of the linkage groups of *Zea Mays* predicted from chiasma frequencies, compared with results already mapped from breeding experiments by Emerson, Beadle and Fraser, 1935 (after D., 1934).

would give 50 per cent. crossing-over. Its total genetical map is therefore still in doubt, and the possibility that it is really 50 units long cannot be dismissed.

In *Zea Mays* we can now compare the maximum possible crossing-over lengths for each chromosome with the lengths already observed in the study of linkage in this plant. We assume that every chiasma is equivalent to 50 per cent. of total or corrected crossing-over, since it means crossing-over between two of the four chromatids taking part in it (Fig. 142). The linkage studies can never provide a complete map of the chromosomes, because gene-differences will never be found between all the end particles of all the chromosomes. These studies are limited by the number of gene and structure differences available for experiment, and only a fraction of the 400 known genes have so far been mapped. The linkage limits should therefore always lie within the cytologically predicted limits. They show such a margin (Fig. 94): this margin will decrease with the increase of linkage data.

It is in a derivative of a *Zea-Euchlæna* hybrid that Beadle (1932) has succeeded in performing the *experimentum crucis* of comparing the chiasma-frequency with the crossing-over frequency in a pair of given recognisable segments. Between two such segments there was 12 per cent. of crossing-over while there was 20 per cent. of association at diakinesis and metaphase, which means this same frequency of chiasma-formation, and therefore agrees closely with the expectation of 24 per cent.

(iii) **Polyploids.** *Distribution among the Chromosomes.* In non-hybrid triploids and tetraploids the chromosomes pair at random and form chiasmata at random, *i.e.*, any one chromosome may pair at different points with any of the other chromosomes with which it is homologous and form chiasmata with them (Newton and D., 1927). Similarly, crossing-over is at random amongst corresponding chromosomes in triploid *Drosophila* (Bridges and Anderson, 1925), and tetraploid *Primula sinensis* (de Winton and Haldane, 1931).

Structural Units. Gametes produced by triploid *Drosophila* with three homologous chromosomes have one or two of these; and when they have two they are sometimes identical in one part, although the three homologues all differed in this part ; they are therefore derived from the chromatids of one chromosome. But such chromosomes are sometimes dissimilar in other parts ; they must then be the result of crossing-over between chromatids, not between chromosomes. In breeders' language, these " equational exceptions " are said to prove crossing-over in the " four-strand stage," *i.e.*, after division of the chromosomes into chromatids, and never between all four strands at the same point (Bridges and Anderson, 1925 ; Redfield, 1930 ; *cf.* Bridges, 1916). Such is the method of crossing-over demanded by the chiasmatype interpretation.

A structurally abnormal type of *Drosophila* yields similar evidence. *Drosophila* females with two X chromosomes " attached " to one centromere (*v.* Fig. 121) and differing in several factors have female progeny with both their X's derived from the same egg-cell of the mother, and these show crossing-over. It is therefore possible to see the constitution of two of the four X chromosomes derived from the four chromatids of one maternal cell and know whether the crossing-over took place before the division into four or after, and whether there is a rule as to the assortment of the four chromatids. Thus, if the two chromatids had factors $ABCDE$ and $abcde$, attached at Aa, it was found that flies appeared homozygous for recessive factors. Therefore, again, crossing-over must have taken place after division into chromatids, and between only two of the four, for otherwise crossing-over, being between whole chromosomes, would only yield complementary types, such as $ABCde$ and $abcDE$.

Segregation. In the attached-X experiment, it was further found that the proportion of recessives was lower the nearer the factor considered was to the centromere. The segregation of the chromatids is therefore determined at the centromere (Anderson, 1925 ; *cf.* Bridges and Anderson, 1925 ; and Rhoades, 1931). With this system of segregation, genes situated further from the centromere should show a higher proportion of homozygosis (and therefore a higher proportion of recessive segregates) approaching, as it were, asymptotically the proportion of one-sixth. The proportions in the X chromosome never reach this value, but agree very closely with the values expected in a chromosome with its known crossing-over frequency on the assumption of no interference between successive chiasmata in regard to the choice of chromatids which take part in them (Mather, 1935 a).

This nearly random segregation of chromatids where two chromosomes, that have undergone crossing-over, pass to the same pole at the first division, means that the segregation expected in tetraploids will not always be in the simple proportions given by Muller in 1914 for the assortment of chromosomes but will show a higher proportion of recessives. It will be possible for both divisions to be reductional in regard to a given part of a chromosome represented four times (Haldane, 1931 a ; Mather, 1935 a, 1936 a ; and cf. Fig. 146, factors $PPPp$). This random chromatid segregation will only be expected when factors are at the unattainable asymptotic distance where they show 50 per cent. of crossing-over with the centromere. It is therefore strictly in accordance with this theory that the chromatid type of segregation has been found in only a small proportion of the cases investigated (Crane and D., 1932 ; E. Sansome, 1933).

Frequency. The total frequency of crossing-over per chromosome in the triploid is the same as in the diploid, although the distribution is different (Redfield, 1930). This should mean that the total number of chiasmata per configuration is increased by half in the triploid. Exactly the same increase is found in triploid plants that have been compared with related diploids (*Tulipa*, D. and Mather, 1932 ; *Hyacinthus*, Stone and Mather, 1932).

(iv) **Structural Hybrids.** Crossing-over in dyscentric hybrids has already been considered. Observations of crossing-over in interchange hybrids agree with expectations based on cytological observation if crossing-over is supposed to be conditioned by pairing of identical segments of chromosome at pachytene, as are chiasmata.

On this assumption, Dobzhansky's results (1931) show that the amount of pairing is reduced in the neighbourhood of a break in *Drosophila* heterozygous for a translocation or interchange. Where the break has occurred near a median centromere of a chromosome it usually leads to a reduction of crossing-over in both arms. Where it occurs in one arm the reduction is stronger in the affected arm, while in the unaffected arm it is normal. Further, Dobzhansky has shown that this reduction is due to the interference of the newly associated segments with one another in pairing, for when a long segment of an autosome is translocated to the Y chromosome (which has no partner, at least for the greater proportion of

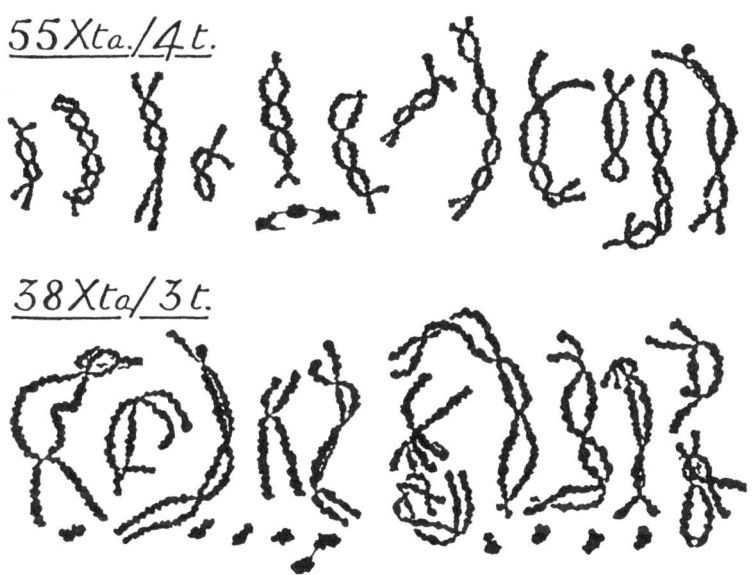

FIG. 95.—The twelve diplotene bivalents in two clones of *Fritillaria imperialis*, giving the total number of chiasmata and the number terminal in each. The upper one has three fragments united by a triple chiasma; the lower one has ten fragments, only two of which are paired. *Cf.* Table 17 and Fig. 48 (from D., 1930 *c*).

its length), its crossing-over is not sensibly reduced. There is no interference. When the translocated segment is short, its crossing-over is strongly reduced. Evidently, therefore, the chance of two homologous segments pairing is decreased when the segment is less than a certain length. Since the fourth chromosome is shorter than any segment concerned, and yet pairs fairly regularly, this decrease is clearly due to the change of association.

There is a reduction of crossing-over in *Drosophila* equally on both sides of the break, which indicates that pairing of the chromosomes may begin either distally or proximally to the break. This rule would probably not apply to organisms with polarised chromosomes at the zygotene stage.

In plants having more than two chromosomes associated in a ring, chiasmata are formed in such a way as can be derived only from a radial pachytene association instead of the linear type of association found in ordinary pairing both in diploids and polyploids (Fig. 106). The crossing-over map should therefore be radial or star-shaped (D., 1931 c). This prediction has been made independently and verified by observations on crossing-over between three factors in ring-chromosomes of *Œnothera* (Emerson, 1931 ; *cf.* Brink and Cooper, 1932).

In the progeny of interchange heterozygotes in *Drosophila*, Muller (1930 b) found that the interchanged segments had usually passed to opposite germ-cells at reduction as in a ring in *Œnothera*.

The reduction of crossing-over in interchange hybrids is paralleled by the reduction in the pachytene pairing, its replacement in part by non-homologous pairing (*cf.* Ch. V) and the reduction in the number of chiasmata formed (Gairdner and D., 1931) (Chs. V and VI).

(v) **Conditions of Variation.** The frequency and distribution of chiasmata and of crossing-over are liable to variation subject to environmental, genetic and developmental conditions.

(a) OBSERVATIONS OF CHIASMATA. The chiasmata in the *m* chromosomes of *Vicia Faba* (Maeda, 1930 b) show a different frequency variation at metaphase in different preparations. This is to be ascribed to an environmental or developmental difference (Fig. 96). The same conclusion can be derived on the chiasma theory of pairing from the variations of pairing in *Triticum* hybrids (Kihara, 1929 c), and the parallelism supports this theory (*cf.* Mather, 1935 b ; Oehlkers, 1935 ; Kattermann, 1933).

The chiasma frequency differs in different clones of *Fritillaria imperialis* (Figs. 95 and 96). The highest mean frequency at diplotene is 5·2, the lowest 3·0. This is a regular property of each clone, irrespective of the time of preparation, and is therefore assumed to be genetically determined. Different forms of *Tulipa australis* show a different degree of concentration in chiasma frequency, *i.e.*, different interference (*v. infra*). This modification of chiasma-frequency distribution is probably genetically determined (D. and Janaki-Ammal, 1932). Chiasma formation may also be genetically suppressed (Ch. X).

Similarly in crosses between *Allium* species, localisation of chiasmata has to be inherited as a factor difference showing segregation (Emsweller and Jones, 1935), and in a normal species, *Allium nutans*, localisation has appeared in inbred seedlings. Races of *Allium fistulosum* (Levan, 1933 *a*) and species of *Fritillaria* (D., 1936) differ in the degree of their localisation, presumably by genotypic control of this property.

(*b*) OBSERVATIONS OF CROSSING-OVER. The frequency of crossing-over is modified by changes : (*a*) in external conditions, *e.g.*, temperature (Plough, 1917 ; Kirssanow, 1931, on *Drosophila* ; *cf.* Stern, 1928) and irradiation (Muller, 1925) ; (*b*) in development, *e.g.*, crossing-over varies with the age of the female in *Drosophila* (Bridges, 1929) ; (*c*) in genotype, *e.g.*, factors occur reducing crossing-over and modifying its distribution (Detlefsen and Clemente, 1923 ; Bridges, 1929), or altogether suppressing it (M. S. and J. W. Gowen, 1922, 1928).

In testing the effect of differences of temperature on chiasma frequency a wider range can be used than in crossing-over experiments. Temperatures that are fatal to reproduction are not fatal to the cells in meiosis. Moreover the effect on all chromosomes can be studied at once. White (1934) has been able to show that the variation in chiasma frequencies agrees with that in crossing-over frequencies, but that this variation is shown by the longer chromosomes with the higher chiasma frequencies only. Whether this effect is a direct one or depends on variation in an interruption of pachytene pairing (as in *Fritillaria*) we do not know.

Most changes in crossing-over frequency are differential with regard to different parts of the chromosome, or, we may say, change in frequency is correlated with change in distribution. Thus reduction in crossing-over frequency is attended by an increase in the neighbourhood of the centromere.

290 CHROMOSOMES IN HEREDITY: MECHANICAL

Crossing-over does not occur at meiosis between the autosomes in male *Drosophila*, although it may occur at mitosis as a special form of primary structural change (Ch. XII). This difference is paralleled by the observed behaviour of the chromosomes (D., 1934; Dobzhansky, 1934, Fig. 116). The autosomes lie parallel to one another, in pairs, with all four chromatids equidistant and

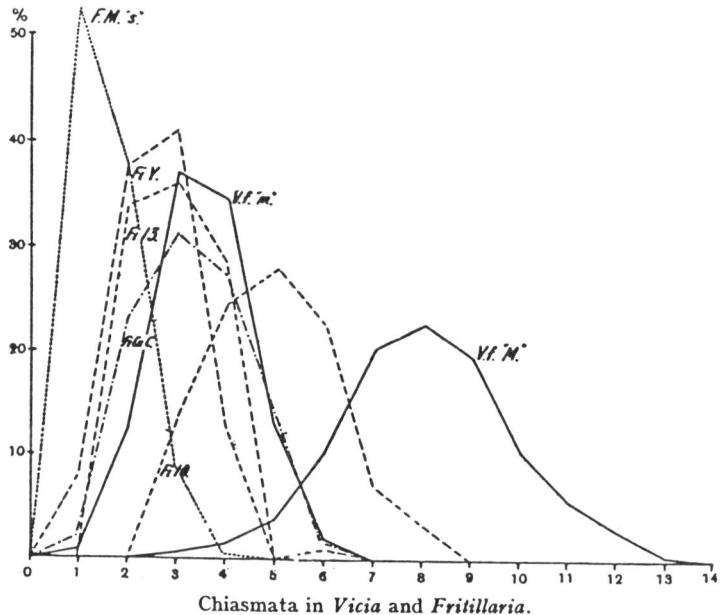

Chiasmata in *Vicia* and *Fritillaria*.

FIG. 96.—Chiasma frequencies in the bivalents of *Fritillaria Meleagris*, *F. imperialis* (various clones), and *Vicia Faba* (" m " and " M " types of chromosome), showing decreasing interference values with the higher mean values. (From D., 1931 c.)

associated at the centromeres only, until metaphase, when the centromeres separate. Pairing therefore takes place by an anomalous mechanism without the apparent formation of chiasmata. Of course crossing-over might have taken place without showing in the production of chiasmata since chiasmata require also a special association in pairs and a weaker secondary attraction. We can merely say that the cytological and genetical observations agree in showing no evidence of crossing-over. As to the sex chromosomes, the cytological and genetical observations agree in showing the occurrence of crossing-over, the method of which we shall consider later (Ch. IX).

The genetic suppression of crossing-over necessarily means the suppression of chiasma formation, pairing and fertile segregation.

Thus Gowen (1928 *et alibi*) and Bridges (1929) found that with suppression and reduction of crossing-over by autosomal factors in the female the fertility of the female was reduced, and that this was associated with irregularities in the pairing of the chromosomes. The factor had no effect on the male. This means that the male character of localisation or suppression of chiasmata inhibited the effect of this factor for crossing-over suppression.

Beadle (1933) has discovered that in " asynaptic " maize which has as a rule no pairing of chromosomes at meiosis, the progeny nevertheless show the results of normal crossing-over. This is because the progeny are derived from selected germ-cells resulting from regular segregation, and therefore from a complete pairing. Beadle's observations are particularly significant in showing that the minimum frequency of chiasmata and crossing-over compatible with this complete pairing is the same as the normal frequency of chiasmata and crossing-over.

The asynaptic maize and female *Drosophila* therefore show the necessity of crossing-over and chiasma-formation for pairing and segregation in these two widely separated organisms. The chiasma-type hypothesis and the chiasma theory of metaphase pairing are equally confirmed.

(vi) **Linear Distribution.** (a) IN RELATION TO ONE ANOTHER (INTERFERENCE). A statistical examination of chiasma frequencies in bivalent chromosomes of *Vicia, Fritillaria, Rosa* and *Tulipa* shows that their distribution is too great near the mode and too little near the extremes on an expectation of randomness (Haldane, 1931 *b*).

This means that the occurrence of one chiasma reduces the chance of occurrence of another in its neighbourhood. Chiasmata show " interference " like that shown by cross-overs. It should be noted that the distribution in position in the chromosome of chiasmata may be random (and apparently is in many organisms), although their distribution in frequency and in relation to one another is not so.

A second method of observing interference has been attempted by Belling in observing internode lengths. These distances, however, are likely to be determined more by the conditions of equilibrium in chiasma movement than by the original interference, for it is probable that movement occurs as soon as chiasmata are formed, in all organisms (Ch. XII). The frequency method is therefore the only one available for showing interference.

It might be supposed that interference between successive crossovers might work either between chiasmata of any kind or between chiasmata involving crossing-over between the same chromatids. It might be *chiasma* interference (as shown cytologically) or *chromatid* interference. Mather (1933, 1935 b) has shown that chiasma interference, and also the formation of at least one chiasma by every bivalent in the diploid, can be inferred from the breeding results in *Drosophila*, while on the other hand (1935 b) there is no evidence for chromatid interference. Whether chromatid interference also exists can be found cytologically in ordinary circumstances only from comparing the chromatid relationships at successive chiasmata in inversions, which has not yet been done. Hearne and Huskins (1935) conclude in *Melanoplus* that comparate chiasmata are more frequent than disparate chiasmata, but we still do not know whether the complementary or reciprocal types, or both, are responsible for this excess over expectation.

In the sex chromosomes of the male, unlike the autosomes of the female *Drosophila*, pairing and crossing-over show evidence of negative chromatid interference. This is absolute, giving pairs of reciprocal chiasmata and no others in the X-Y pair (Ch. IX). In female *Bombyx* there is a correlation in reductional as opposed to equational division of different parts of the sex chromosomes which themselves show no linkage (Mather, 1935 b). This seems to imply a similar negative interference.

(b) IN RELATION TO DIFFERENT CHROMOSOMES. Schultz found, from comparative studies of structurally homozygous and heterozygous *Drosophila*, that high crossing-over in one chromosome goes with lower crossing-over in others. Such a negative correlation between the frequency in different chromosomes of one nucleus has since been shown by chiasma studies in numerous plants. This can be done in two ways : either using a correlation table, where the frequencies of the different chromosome types, such as the large chromosomes and the fragments in *Secale*, can be recorded (D., 1933 ; Mather and Lamm, 1935) ; or by finding out whether the variance as between nuclei is less than the absence of interaction between bivalents would require (Mather and Lamm, 1935 ; Mather, 1936 ; Lamm, 1936). The first method requires dissimilar chromosomes within the nucleus, the second uniform chromosomes, at least as to part of the complement. Mather's extensive analysis shows first that negative correlation may occur in all or a part or none of the chromosomes of different species, and secondly that it may occur in some individuals of a species, those with a higher chiasma frequency, and not in others. Interference between crossing-over in different chromosomes begins therefore above a certain threshold frequency. How it works will be considered later (Ch. XII).

(c) IN RELATION TO POSITION ("MAP-DISTANCE"). In some organisms the number of chiasmata formed in a bivalent is a direct function of the length of the chromosomes paired. This is so in *Vicia Faba* (*v.* Fig. 96), where the M chromosome with a mean chiasma frequency of 7·1 is a little more than twice as long as the m chromosomes with a frequency of 3·4 (Mather, 1934). This is true of many organisms with complete pairing, and having a size range of less than 1 : 4 (*e.g.*, *Allium macranthum*, Levan, 1933 *c*). It is not, of course, true of those like *Fritillaria Meleagris* and *Mecostethus* which have localised pairing and in consequence the same or even a higher chiasma frequency in the shorter chromosomes. The proportionality with complete pairing may be shown in the many organisms where new very small fragments have appeared. These, as we have already seen, fail to pair in such a proportion of nuclei as may be predicted from the chiasma-frequency of the ordinary complement (Ch. V). Frequently, however, relatively small chromosomes form part of the normal complement. We then find that their regular pairing is not made possible by an increase of chiasma frequency all round, but by an indirect proportionality. We see this beginning in *Hyacinthus orientalis*, where with a size range of 1 : 4·2 there is a chiasma range of 1 : 3·1 (Stone and Mather, 1932), and also in *Spironema fragrans* (Richardson, 1934) and *Scilla peruviana* (Sato, 1934). Where the size disparity is greater, the lack

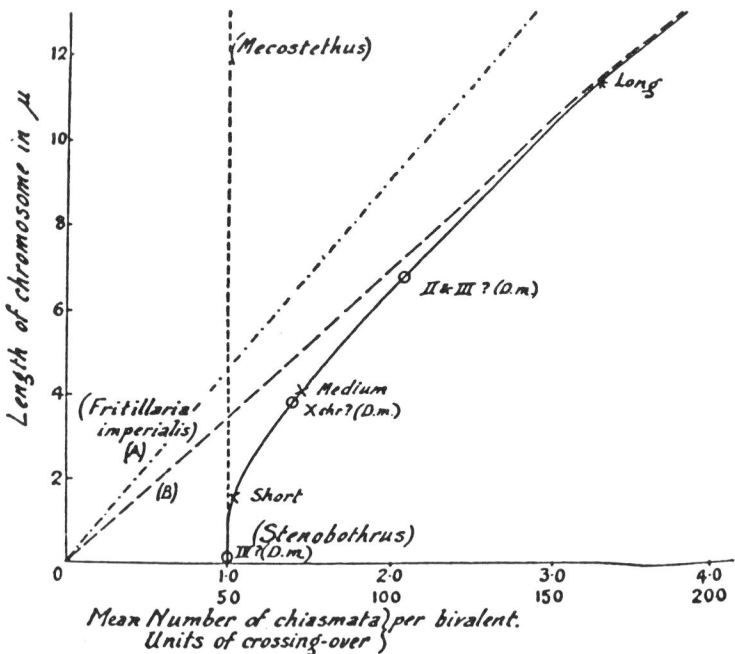

Fig. 97.—Graph showing three types of relationship possible between chiasmata or crossing-over and the length of the chromosomes: direct proportion in *Fritillaria imperialis* (two clones); fixed value with localisation in certain bivalents of *Mecostethus* and perhaps in some *Tettigidæ*; the compromise in *Stenobothrus* and perhaps in *Drosophila melanogaster* " D.m." (From D. and Dark, 1932; *cf.* White, 1933, 1936; Hearne and Huskins, 1935.)

of proportionality is more serious. In *Stenobothrus parallelus* (D. and Dark, 1932) the chromosomes with median centromeres are seven times the length of the shortest, yet they have only three times as many chiasmata. The same is true of *Syrbula acuticornis* (Robertson, 1916), *Hyacinthus amethystinus* (D., 1932 g), *Melanoplus femur-rubrum* (Hearne and Huskins, 1935), *Yucca flaccida* (O'Mara, 1932) and *Y. recurvifolia* (Sato, 1934) and in species of *Agave* (Doughty, unpublished). In none of these species is there evidence of localisation.

It would appear, therefore, that here chiasma frequency is no longer a direct function of length, and that this indirect proportionality and localisation are genetical adaptations necessary to the regular pairing of the short chromosomes. Such adaptations must be alternative to a general increase in chiasma frequency, and are to be expected wherever the chiasma frequency is low and considerable differences occur in length (Fig. 97).

The evidence of the behaviour of new fragments in always failing to show this adaptation is sufficient to show that proportionality is the simpler and original condition from which non-proportionality is derived. Further evidence is provided by White's observation in species of *Locusta*, *Schistocerca* and *Stenobothrus* that the non-proportionality which is characteristic of them at normal temperature is reduced or disappears when the chiasma-frequency is increased at higher temperatures. This suggests that the indirect proportionality is due to a partial failure of pairing of the longer chromosomes like that found with localisation, but near the centromere instead of away from it. Such a failure might be concealed in short intercalary segments by the development of relational coiling.

The possibility of testing an analogy to these observations in crossing-over frequencies is not yet available, but there are a few indications that similar departures from simple expectation occur. The X chromosome of *Drosophila melanogaster* is less than two-thirds the length of the longest chromosome, yet it has two-thirds of the map-distance length measured in crossing-over units. This is a departure in proportion in the direction expected from the analogy with *Stenobothrus*. A second indication is inconclusive ; the X chromosome in *Drosophila Willistoni* (Metz, 1916) is twice the length of that in *D. melanogaster*, but its map length is only one-fifth greater.

Finally, the group of observations already referred to show that wherever there are special conditions affecting crossing-over, the distribution of crossing-over is affected relative to the centromere.

All such conditions must affect the relation of length to map-distance. There is, therefore, no special reason for expecting that under any particular conditions the one should be a direct function of the other.

If the chiasma frequency in *Drosophila melanogaster* is such an indirect function of chromosome length as in *Stenobothrus*, then the fourth chromosome, although extremely short, should be of 50 units' length, a condition which would not be easy to verify. But if one of the fourth chromosomes were attached to a fragment of the X chromosome the chiasma frequency should be reduced in the new triple body, $IV + X + IV.X$, and the small unattached fourth should frequently fail to pair, and therefore sometimes pass to the same pole as its attached homologue. This is found by Stern (1931).

(vii) **Crossing-over with Ameiosis.** An important result of crossing-over between chromatids, pointed out by Janssens in 1909, is that the separation of dissimilar chromatids occurs at both divisions. Now one of these divisions may be suppressed, or the two nuclei resulting from it may fuse again into one (Ch. X). In these circumstances reduction of the chromosome number is suppressed ; but segregation should not be ; it can still occur at the other division if one of the pairs of chromatids that separate has crossed over with a dissimilar partner—if, that is, any of the chromosomes have paired by chiasmata.

It is now possible to consider, therefore (contrary to the earlier belief), that suppression of meiosis leading to failure of reduction will not lead to failure of segregation when any two chromosomes have paired. When the two chromosomes that have paired separate at the first division, but are included in a single restitution nucleus, their second division, which gives the two unreduced germ-cells, will, on the chiasmatype hypothesis, not be equational. Thus, if crossing-over must occur at the chiasma which conditions their pairing, the second division will always be reductional in regard to one of the cross-over segments. In either case the result will be equivalent to mendelian segregation of the differences in the pairs of chromosomes concerned. It follows, therefore, that a hybrid in which a pair of chromosomes is occasionally found at metaphase and in which failure of reduction is due to suppression of the first (or indeed the second) division should show a very limited amount of segregation. Suppression of the first division will give

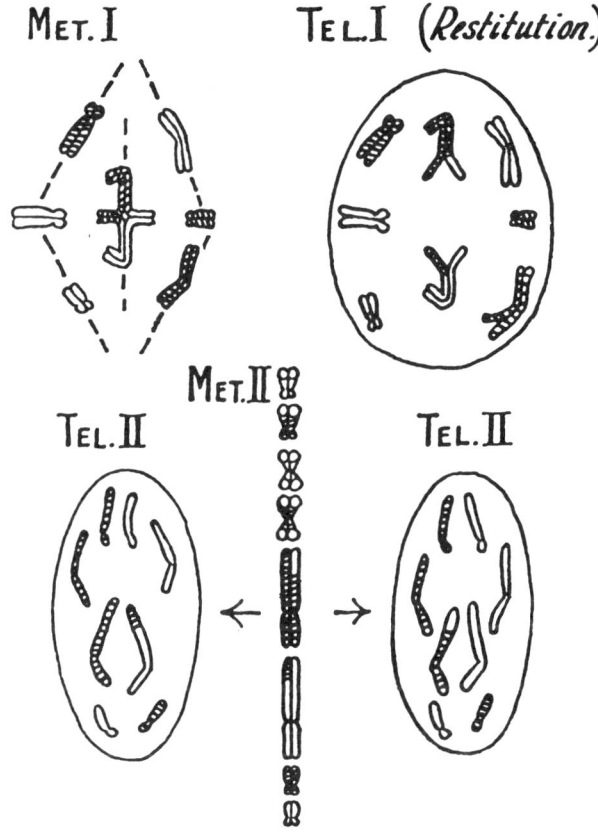

FIG. 98.—Scheme showing how a restitution nucleus will give numerically unreduced gametes showing segregation of differences so long as pairing of dissimilar chromosomes occurs.

diploid nuclei showing segregation, in respect of the differences between such chromosomes as paired, approaching the proportions 1 : 2 : 1 asymptotically. Thus, when hybridity reaches the point at which restitution nuclei are formed, segregation is appropriately toned down. This conclusion, on general grounds, is supported by v. Wettstein's observations (1924) of heterozygosity in diploid spores of *Funaria* when the second division was suppressed.

The consequences of segregation without reduction have now been shown in two ways. First, Kostoff (1934) has examined the progenies of several hybrids in *Nicotiana* whose unreduced gametes have functioned in crosses with a third species known itself to be homozygous. These progenies varied, and their variation must be due to crossing-over between unlike chromosomes in the hybrid parent. The same result is found in the tetraploid F_2 from a *Saxifraga* hybrid (Table 26B). Secondly, "mutation" in parthenogenetic plants whose unreduced egg cells develop directly may be ascribed to the same cause (Ch. XI).

TABLE 44

Comparative scope and validity of experimental and direct methods of studying crossing-over. (From D. 1935.)

From Proportions of Combinations in Progeny	From Chromatid Combinations in Germ Mother-cells
1. Event inferred from its genetical consequences on the assumption, justified by particular and statistical evidence, that the genes are located in structures having the observed properties of chromosomes. Fewer assumptions are made therefore in relating the genetical observations, such as frequency of crossing-over between two genes with the genetical consequences.	1. Event inferred from consequences on the assumption justified by particular and statistical evidence that chiasmata are determined by crossing-over. Fewer assumptions are made therefore in relating the cytological observations, such as frequency per chromosome, with the mechanical conditions (D., 1934 b).
2. Crossing-over data usually obtained after a lapse of one generation. There is an elimination of recessive lethal combinations in the interval, including those determining absence of crossing-over itself, since where crossing-over occurs it is a condition of segregation and fertility. Haploid generation can be studied only in lower plants and Hymenoptera.	2. Crossing-over data obtained after a lapse of a few minutes, during which no elimination can occur.

2A. Certain lethal effects, including segregation in many hybrids, therefore hinder or prevent study.

3. Crossing-over data usually refer to one of the four chromatids; in triploid and attached-X *Drosophila*, to two chromatids; in lower plants, especially in fungi, to all four. Except for the last type, therefore, no evidence is obtainable of the relationship of successive crossings-over in the chromosome.

3A. Interference determinable between crossings-over, and not between chiasmata except when related with pre- and post-reduction observations (*Bombyx* sex chromosomes; Mather, 1935).

4. Gene-positions of cross-overs determined from a limited number of genes serving as markers; certain double and terminal cross-overs are therefore missed (as in male *Drosophila*) and positions are always approximate.

5. Crossing-over determinable only in organisms heterozygous for a limited number of genes, and these must be situated in one chromosome. Possible therefore only in the analysis of selected and inbred populations, although equally in male and female cells.

2A. Effects of X-rays and extreme temperatures and hybridity, giving sterility, can be studied (D., 1932 b; White, 1934; Mather, 1934).

3. Crossing-over data refer to all four chromatids together. Which pair of the four chromatids have crossed over at a chiasma can be known only in specially favourable material and in certain structural hybrids (D. and Mather, 1932; D., 1936 d; Richardson, 1936).

3A. Interference only determinable between chiasmata, and not between crossings-over, except when all four ends of pairing chromosomes differ, as in male *Drosophila* sex chromosomes and dyscentric hybrids (Haldane, 1931; D., 1934 a, 1936 d).

4. Total number of cross-overs exactly ascertainable not only for given bivalents, but for the whole nucleus (D., 1933; Mather and Lamm, 1935). Movement of chiasmata may occur, so that positions are always approximate.

5. Crossing-over determinable in an unlimited number of species and hybrids, although only in male cells in practice. Equally possible (i) with the extremes of homozygosity and heterozygosity, (ii) with unselected natural populations, and (iii) irrespective of a lethal effect from an abnormal consequence.

(viii) **Conclusion.** The direct and parallel evidence now enables us to say that there is a one-to-one correspondence in all organisms studied between chiasmata and crossing-over. Further, we can say that chiasmata behave in all cases as though they were determined by crossing-over between two chromatids of the partner chromosome taking part in them. It now becomes possible, therefore, to investigate crossing-over cytologically. This has already been begun, as we have seen. It is important in bringing a different technique to bear on the problem and also in extending the observations of crossing-over to organisms that are not suitable for analysis by experimental breeding (Table 44). This not merely increases the material available for observation, it enables us for the first time to study crossing-over in homozygous organisms which have no variation in their progeny to show crossing-over, and in sterile hybrids which have no progeny to show variations.

The first consequence of applying this method is that it reveals for the first time the universal properties of crossing-over as an essential part of the mechanism of sexual reproduction.

We already know that chiasmata are a condition of metaphase pairing and segregation in all homozygous organisms, in fact in all organisms except a few sex-heterozygotes which will be considered later. It follows, therefore, that crossing-over itself is a condition of chromosome pairing and is approximately co-extensive with sexual reproduction, occurring even sometimes when reduction is suppressed. This leads to consequences of some importance. Thus the detailed properties of crossing-over determined in the female *Drosophila* and in *Zea* have an almost universal application, for they agree with the almost universally observed properties of chiasma formation.

This means that with sexual reproduction the unit of inheritance is not the whole chromosome but the smallest part into which a chromosome can be broken by crossing-over, that is, the gene. If this were not so the number of possible types of organism in a species would be an exponential function of the number of types of whole chromosomes in it, instead of being a slightly lower function of the number of types of genes. Every chromosome would have the properties of a clone. With mutation rates and breeding systems that are commonly known in the higher organisms there would be fewer types than individuals. Groups of individuals would be identical like the members of a clone and would be separated from other groups by sharp cleavages. So far as we know such an extreme system of reproduction occurs nowhere, but, as we shall see later, special modifications in this direction are found in permanent heterozygotes and in certain sexually differentiated organisms. For their analysis we need the knowledge we now have of the relationship of chromosome pairing to crossing-over, since they all depend on specially controlled conditions of crossing-over.

CHAPTER VIII

THE CHROMOSOMES IN HEREDITY : PHYSIOLOGICAL

Differentiation of Behaviour — Specificity — Nucleoli — Condensation — Differentiation of Activity—Balance—Inert Chromosomes—The Gene.

Besonders aussichtsreich scheint mir gerade bei diesem Problem, das wir ihm gewissermassen von zwei Seiten her beikommen können, einmal durch die Beobachtung der Vererbungs-Erscheinungen und dann durch die Beobachtung der (uns jetzt ja bekannten) Vererbungs-Substanz.

WEISMANN, 1892.

1. THE TWO METHODS OF ENQUIRY

THE modern study of heredity has been built up by the study of differences. The mendelian technique of experimental breeding depends on the classification and comparison of differences between parents and offspring and between brothers and sisters. We have now seen how closely this differential method is paralleled, in the deductions it makes possible, by cytological observation. The same modes of inheritance may be inferred from mendelian and from cytological analysis, although the special advantages of each method enable us to go further with it in certain directions than with the other. The direction in which the cytological method is particularly advantageous is (as pointed out in Table 44) that it can dispense with experimental differences. It enables us to study the hereditary properties not only of hybrids too sterile to have offspring, but also of homozygous organisms having no differences to segregate. We can therefore proceed to use the study of the chromosomes for finding out the hereditary properties of organisms, as Weismann, in some respects prematurely, attempted to do. We can proceed to deduce from the principles now induced.

One of the first problems to which this method can be applied is the problem of differences within the hereditary materials of the individual. With regard to this, alternative inheritance tells us the cardinal fact that different " loci " can be defined by their

QUALITATIVE DIFFERENTIATION

different mutation properties and must therefore be inherently different. We can now consider the cytological evidence.

2. DIFFERENTIATION OF BEHAVIOUR IN THE CELL

(i) Specificity. A qualitative differentiation of the chromosomes is shown by their specific pairing properties at all stages of meiosis. It was first found by Sutton and Montgomery that particular chromosomes paired at meiosis, and that these chromosomes were similar in size and shape. This meant that the different chromosomes showed specific and constant affinities at meiosis. Then it was found that in fusion-heterozygotes (McClung) and interchange-heterozygotes (Belling) each part of a compound chromosome similarly had specific properties. This rested on metaphase observations. Later the development of the structures seen at metaphase was traced back to the diplotene (Gairdner and D., 1931) and pachytene (McClintock, 1931) stages. The specificity was seen to apply to the smallest visible elements, the chromomeres, when zygotene pairing was studied in polyploids, for successive chromomeres, distinguishable in size, could be seen to pair independently of their neighbours when they had a choice of association (Newton and D., 1929).

This specificity applies equally, as we have seen, to secondary and somatic pairing. It is most strikingly demonstrated in the somatic pairing in the permanent prophases of the Diptera, where every chromomere associates with a corresponding one from the partner chromosome (Ch. V). Specificity therefore corresponds with the differentiation inferred genetically amongst genes by Muller (1916).

Three kinds of special differentiation in structure are visible in the chromosomes beyond the simple differentiation into chromomeres. First, differentiations in size that are produced by particular fixatives and are indications of specific reactions. Such are the knobs seen in the pachytene chromosomes of *Zea* when fixed with acetic acid (McClintock). Secondly, the special properties of the centromere. These will be considered in detail in relation to its mechanical properties (Ch. XII). Thirdly, the differentiation in nucleolar production associated with secondary constrictions (Ch. II). This now requires more detailed consideration.

(ii) **Organisation of Nucleoli.** McClintock (1934) has carried out

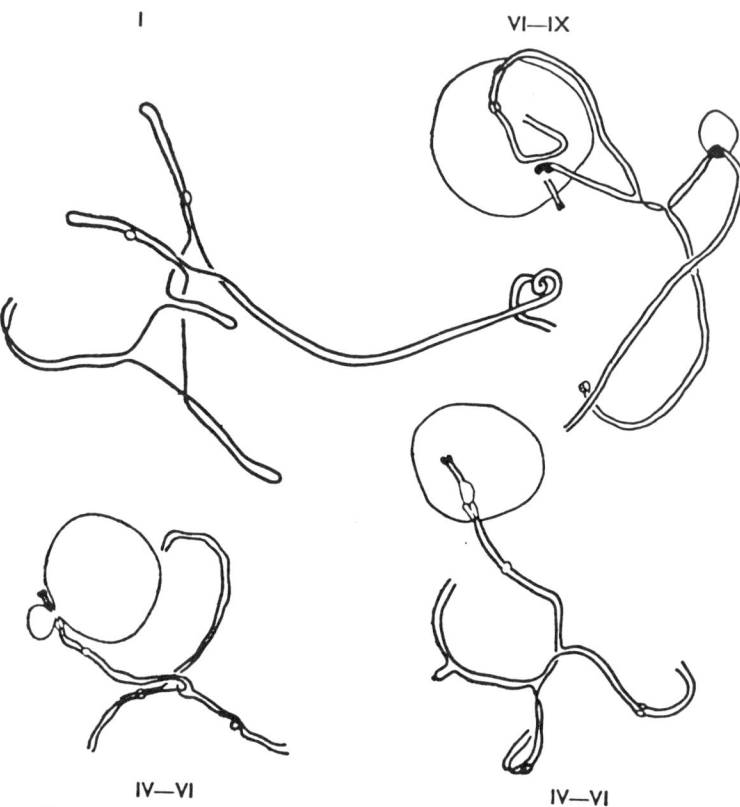

FIG. 99.—Diagrams of pachytene in *Zea Mays*. I, homozygous pair showing non-homologous torsion-pairing of chromosomes with themselves. VI—IX, interchange involving the nucleolar organiser at which the exchange of partner would occur but for non-homologous pairing. IV—VI, interchange without (left) and with (right) non-homologous pairing. The relational coiling associated with torsion-pairing is not shown. × 1200 (McClintock, 1933).

an analysis which makes it possible to define the factors controlling nucleolar production at the second telophase of meiosis in *Zea*.

A nucleolar organiser, staining deeply in aceto-carmine, lies next to the place where the nucleolus and the secondary constriction are developed. The organiser lies in chromosome No. 6 which underwent interchange with No. 9 as a result of X-raying. The interchange break in the chromosome occurred within the organiser and divided it into two parts that were now lying in different chromosomes. These parts proved to have a different degree of activity. The smaller distal part that was transferred to No. 9 gave a larger nucleolus than the proximal part left in No. 6, in normal circumstances (Fig. 99). When, however, this more active element was lying in an isolated piece of the nucleus at telophase, it gave a smaller nucleolus. Moreover, the weaker organiser developed a full-sized nucleolus if the stronger organiser in No. 9 had been left out of the nucleus (by non-disjunction at meiosis). It follows that the size of nucleolus developed depends upon the amount of material available in the chromosomes lying around it, and if there are two organisers, on their relative rate of action. When there are two organisers there is competition for the available materials. A further significant observation of McClintock is that the organising capacity can be suppressed by particular genotypic changes in the chromosomes, such as unbalance or loss of parts, even though these parts do not themselves contain specific organisers. The nucleoli then appear as droplets distributed over the chromosomes generally. These droplets may fuse and the chromosomes then appear to be embedded in the nucleolar material. The same result ensues when no organiser is present in a defective nucleus. Cases of loss or gain of a trabant (*Matthiola*, Philp and Huskins, 1931; *Rumex*, Yamamoto, 1933) require re-examination in the light of these results.

The competition amongst organisers is shown in another way in hybrids of *Crepis* (Navashin, 1934). Here a strong organiser from one parent seems to supplant a weaker one from the other. Since the nucleolus stretches the constriction, the weaker nucleolar chromosome fails to develop its characteristic long constriction in the hybrid. McClintock has established a scale of strength for *Crepis* species in accordance with this view. Competition is similarly shown by $\widehat{XY^s} + Y$ males in *Drosophila* where the attached arm of the Y does not develop its usual sized nucleolus owing to the proximity of the organiser in X (Kaufmann, 1934, *cf.* Ch. IX).

TABLE 45

Observations of Nucleolar Behaviour

1. **Attachment of Trabants to Nucleoli in Prophase of Mitosis.**
    S. Nawaschin, 1912 (1927). *Galtonia*, etc.
    M. Nawaschin, 1926. *Crepis* spp.
    Senjaninowa, 1927. *Ranunculus acer*.

2. **Specific Organisation of Nucleoli next to Secondary Constrictions at Telophase.**
    Heitz, 1931, 1932, many plants.
    Geitler, 1932. *Crepis*, 4*x*.
    McClintock, 1934. *Zea Mays*.
    Dearing, 1934. *Amblystoma*.
    Matsuura, 1935 h. *Polygonatum*, *Trillium*.

3. **Specific Attachment of Nucleoli at Prophase.**
    (*a*) To the Sex Chromosomes:
        Buchner, 1909. *Stenobothrus*, oögenesis.
        Metz, 1926; Bauer, 1932; various Diptera.
        Klingstedt, 1933; various Neuroptera.
        Kawaguchi, 1933; various Lepidoptera, oögenesis and spermatogenesis.
        Heitz, 1933; Kaufmann, 1934; Dobzhansky, 1934. *Drosophila*.
        Tatuno, 1933; some Hepaticæ.
    (*b*) Not to the Sex Chromosomes (plants only).
        McClintock, 1931; D., 1934. *Zea*.
        F. H. Smith, 1933. *Galtonia*.
        D., 1933, *Agapanthus*; 1935, *Fritillaria*.
        Richardson, 1934, *Spironema*; 1935, *Crepis*.
        Modilewski, 1934: Upcott, 1936. *Eremurus*.
        Levan, 1935 *a* and *b*. *Allium* spp.
        Ono, 1935. *Rumex acetosa*.

4. **Survival at Metaphase of Mitosis and Meiosis.**
    Modilewski, 1932. *Fourcræa*, root-tips.
    D., 1929 *c*. *Tradescantia*, pollen-grains.
    Klingstedt, 1931. *Trichoptera* ♀, meiosis (chromatin-diminution ?).
    Gwynne-Vaughan and Williamson, 1933 *et al.* Many Ascomycetes, meiosis.
    Webber, 1933. *Nicotiana glutinosa*, haploid, root-tips.
    D., unpublished. *Zea Mays*. Root-tip, plant with 3 B chromosomes.
    Geitler, 1935 *a* and *b*. *Spirogyra* spp., mitosis.
    Frankel, unpublished. *Fritillaria* spp., meiosis.

5. **Relationship with Chromosome Movements.**
    Manton, 1935. *Cruciferæ*.
    D., 1935 *a*. *Fritillaria* (Plate XV).
    Upcott, 1936 *b*. *Eremurus*.

6. Chemical Tests.

Modilewski, 1932. *Neottia, Galtonia, Allium* (two types of nucleolus).
Bauer, 1933. Oöcytes of insects.
Gardiner, 1935. *Tenebrio*.
Geitler, 1935 a and b. *Spirogyra* spp.

(iii) **Differential Cycles of Condensation.** Correlated with the cycle of coiling and uncoiling which the chromosomes undergo during mitosis and meiosis there is also a cycle of changes in fixation-staining reactions in the chromosomes. This cycle is not understood in physical terms, but it is recognised as being characteristic for particular nuclei at particular stages of the life-cycle and subjected to particular treatments. It consists in a change from deeply-staining metaphase chromosomes to lightly-staining or non-staining resting stage chromosomes. The behaviour of particular chromosomes or parts of chromosomes has long been known to be different in this respect from that of the rest of the complement, *e.g.*, Rosenberg's prochromosomes (1909) at mitosis in plants and the differentially condensed chromosomes at meiosis in many animals. Recently, owing largely to the work of Heitz, differences have been recognised at mitosis in both plants and animals to which a genetic foundation may be ascribed. The difficulty, however, in establishing this relationship lies in the great variability of the behaviour of differential parts of chromosomes in different organisms, even as seen under the same conditions of treatment, and the still greater variability with different treatment.

Broadly we may say that there are three types of differential behaviour which require the most careful comparison.

(A) First, there is the differential behaviour at telophase and prophase of mitosis. Certain parts of most of the chromosomes, usually proximal or distal, stain more deeply than the rest with aceto-carmine but not with Feulgen's reagent (Heitz, 1935). When, as sometimes happens, these are all proximal regions, they may remain condensed during the resting stage and reappear as the prochromosomes referred to earlier (Heitz, 1932, *cf.* Ch. II). The differential material or " heterochromatin " takes up a large part of the sex chromosomes in mosses and was earlier thought to be concerned with sex (Heitz, 1928). By contrast with ordinary mitosis, the prophase in salivary glands shows the heterochromatin understained or unrecognisable; the difference is there, but it is reversed (Ch. XI). An analogous reversal we shall see elsewhere.

(B) Secondly, there is a uniform " precocity " shown in some organisms by all the chromosomes in their proximal parts at the pachytene and diplotene stages of meiosis. This over-condensation on both sides of the centromere was first seen in *Agapanthus* (Belling,

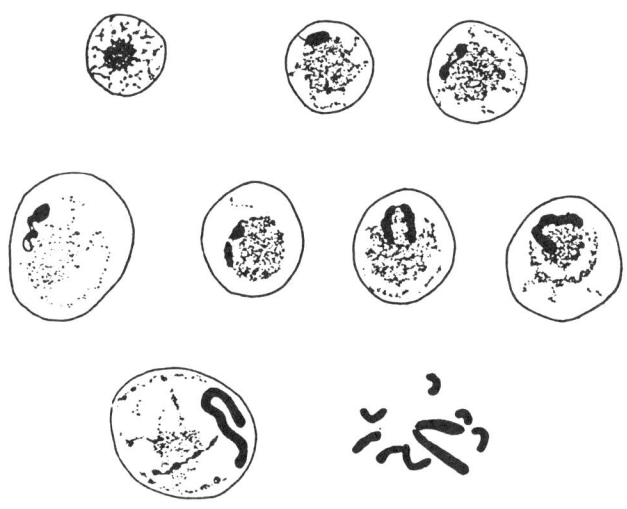

Fig. 100.—Mitotic nuclei of *Sphærocarpos texanus* ($n = 7 + X$). Above, resting nuclei with and without a condensed X chromosome. Below, successive stages of prophase showing the precocity of the X. × 3000 (Tinney, 1935).

1928; Geitler, 1933; D., 1933). It has so far been found only in plants, *e.g.*, in *Zea Mays* (D., 1934) with aceto-carmine treatment, and in *Œnothera* (Catcheside, 1931) and *Canna* (Offerijns, 1935) at diplotene. In *Fritillaria*, where it is accentuated by re-staining, it is associated with earlier pairing and earlier diplotene separation of the condensed part (D., 1935). It is not known to be correlated with different behaviour at other stages of the life cycle, except the meiotic interphase, and in *Pellia* it is clearly unrelated with the mitotic differentiation (Jachimsky, 1935). It can, therefore, be represented simply as a timing difference at meiosis, the proximal parts being in advance of the distal parts. Moreover, since it applies to all chromosomes regularly in exactly corresponding regions it can scarcely be due to a differentiation of the genetic properties of the chromomeres; it must be due to the position of the precocious parts in relation to the centromere. At the same time we must notice that its location is strictly comparable with that of prochromosomes in plants with small nuclei (*cf.* Ch. XII).

In *Mecostethus* a localised precocity of chromosomes is found

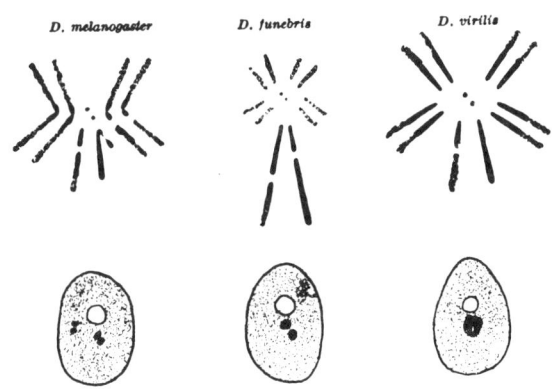

FIG. 101.—Diagram showing the parts of the chromosomes of three *Drosophila* species which are differentially condensed in prophase and, below, for comparison, their resting nuclei with dark prochromosomes. (After Heitz, 1933 *b*.)

at meiosis, but it is the distal parts of the chromosomes that are affected (Janssens, 1924; McClung, 1928 *b*; White, 1936).

(C) The third type of differential behaviour, like the second, is peculiar to meiosis or to cells that are about to undergo meiosis. Like the first, on the other hand, it is characteristic of particular parts of some of the chromosomes, and the parts may be proximal or distal or include the whole chromosome. This behaviour is the well-known "heteropycnosis" or differential condensation of the sex chromosomes in animals.

In order to understand the cause of this behaviour it is necessary to consider all cases of the kind together. These show that the property, as it is found in the sex chromosomes, is determined genetically and not by the pairs being structurally dissimilar, as a superficial study would suggest.

The whole of the sex chromosomes are affected in the heterozygous sex where they do not pair at pachytene, whether there is a single X as in some Orthoptera and Nematoda, or both an X and a Y as in the Hemiptera. In the Hemiptera certain fragment chromosomes

TABLE 46

Comparison of Cycles of Differential Condensation in Different Organisms

	Orthoptera			Marsupials $XY\male$	Rat $XY\male$	Hepaticae XY
	$XO\male$	$P\male$	$XX\female$			
Pre-meiotic mitosis.						
Prophase	+ (1)	O	O	O	O	+
Metaphase	—	O	O	O	—	O
Meiosis.						
Leptotene	+ +	O	O	+ +	O	+
Pachytene	+ +	+	O	+	+	+
Diffuse stage				+		
Diplotene	+	+	O	—	O	+
Metaphase I	O	O	O	—	—	O
Metaphase II	O	+	O (2)	—	O	O

Explanation : O, no differential condensation ; +, over-condensation ; —, under-condensation ; P, the precocious autosome of *Stenobothrus*, etc.

(1) Progressively increasing towards meiosis.

(2) No precocity, but according to Buchner there is a condensed body attached, which is probably a nucleolus.

Cf. Orthoptera : White, 1935 ; D., 1936 d.
  Marsupials : Koller, 1936.
  Rat : Koller and D., 1934.
  Hepaticae : Tatuno, 1933 ; Lorbeer, 1934.

and, in some Orthoptera, several smaller members of the normal complement, share the property. These are sometimes derived from the sex chromosomes, since they pair with the Y in *Metapodius* (Wilson, 1909), and with the X in *Alydus* (Reuter, 1930) (*cf.* Table 16). In certain species with an anomalous type of meiosis half the chromosome complement is affected (*e.g.*, in *Lecanium, Pseudococcus* and *Gossyparia, q.v.*).

The property may show itself long before meiosis in the failure of the affected chromosomes to enter into a complete resting stage between divisions. In the male of the coccid *Gossyparia spuria* half the autosomes already show differential condensation in the late blastula stage (Schrader, 1929). In some groups the unpaired X chromosome shows progressively increased condensation during the resting stages in the spermatogonia as meiosis approaches, and consequently shows very clearly at prophase the spiral form of the chromosome thread from the preceding division (*Tryxalis*, Brunelli, 1910, 1911; *Rhabditis*, Boveri, 1911; *Perla*, Junker, 1923; *Mecostethus*, Janssens, 1924; *Tettigonia*, Winiwarter, 1931). Frequently at the premeiotic metaphases the X chromosome is then understained (D., 1936), and in *Rattus* and *Mecostethus* it is understained at meiotic metaphase (Koller and D., 1934; White, 1936). We have the reversal that we noticed in the behaviour of heterochromatin. Generally, however, the differential condensation does not show itself before the prophase of meiosis, and is seen only in the more rapid condensation of the affected chromosomes after zygotene. From this is derived the name " chromatin-nucleoli " by which they were early described. The precocious chromosomes do not contract further than the rest at metaphase, but during the ensuing interphase and at second metaphase, even in *Chorthippus* they continue to show exceptional condensation (D., 1936 *d*). Correlated with the over-staining of the precocious chromosomes are two other anomalies of behaviour at pachytene. The X chromosome is bent back on itself and only gradually straightens out as it loses its over-condensation at diplotene. It seems to shed a coat of nucleolar substance and at the same time another change occurs. During pachytene the X chromosome seems frequently to attract to itself the other condensed chromosomes which stick to it by their ends. They separate from it at the same time that it straightens out (Janssens, 1924; D., 1936). In parenthesis we may notice that the temporary end to end contacts of chromosomes in the prophase of meiosis in female *Icerya* (Ch. XI) and haploid *Triticum monococcum* (Katayama, 1935) are perhaps analogous to this. The ordinary repulsions between chromosomes therefore fall into abeyance during the persistence of the differential staining properties. This is perhaps also true of the sex chromosomes of the rat at metaphase inasmuch as these chromosomes frequently lie off the plate or on the edge of it.

The possibility of an embedding of sex chromosomes in the nucleolus which is suggested by these observations in Orthoptera and Mammalia is in keeping with their observed association with nucleoli, of the kind found in plants, in meiosis of many animals which do not show the ordinary differential condensation of sex chromosomes. Thus the X chromosomes instead of being themselves condensed are associated with the nucleolus in oögenesis and spermatogenesis in Lepidoptera, in oögenesis in Orthoptera (Buchner, 1909) and in mitoses in *Drosophila* (Table 45). Against this view two objections must be taken into account. First, sex chromosomes that are equally precocious in mitosis and meiosis are attached to the nucleolus in some Hepaticæ, but not in others (Tatuno, 1933). Secondly, the X chromosome in *Stenobothrus* appears equally condensed at diplotene with the specific Feulgen treatment (Bauer, 1932). This stain is not, however, necessarily capable of distinguishing extraneous materials that are in close association with chromosomes. When we recall that the nucleolus can be condensed on the chromosomes under special physiological conditions in plants where it is normally organised at special centres, we see that precocious condensation is plausibly explained as the result of abnormal nucleolar formation in certain cases.

The general character of this behaviour makes it clear that we are dealing with a reaction of particular chromosomes to special genetic conditions rather than with the reaction of dissimilar homologues with one another in pairing, such as is characteristic of structural hybrids. Further critical evidence is afforded by the following observations.

(a) The two X chromosomes in the hermaphrodite generation of *Rhabditis nigrovenosa* condense precociously in oögenesis. In spermatogenesis in the same individual this precocity is perhaps exaggerated, since they fail to pair.

(b) The compound unpaired X chromosome of male *Perla marginata* (Plecoptera) behaves differentially in spermatogenesis, but not differentially in meiosis in an ovary-like organ produced by the male (Junker, 1923).

(c) Trisomic supernumerary fragments in *Fritillaria imperialis*, one of which must often be entirely unpaired, show no evidence of precocity (D., 1930 c), while supernumeraries in *Metapodius*, which may be present to the number of six and must be of common origin and homologous (Wilson, 1909), show strongly differential condensation.

(d) Precocious condensation is not usual in sex chromosomes in the homozygous sex, e.g., in Orthoptera (Mohr, 1915), but it has been found in male *Natrix* and *Gecko* (Reptilia) and female *Nyctereutes* (Mammalia) which are homozygous (Minouchi, 1929; Nakamura, 1931). It does not occur in *Enchenopa* (Homoptera), *Lebistes* (Pisces), or in angiosperms where the dissimilar sex chromosomes pair and cross over (Winge, 1923, 1930 et al.).

(e) Frequently the precocious autosomes in the Orthoptera form unequal bivalents, one of them being deficient (e.g., *Brachystola* and *Arphia*, Carothers, 1913; McClung, 1927. *Chorthippus parallelus*, Janssens, 1924; Belar, 1929; D., 1936). It has been thought by Belar and others that their precocity was due to their inequality. But the causation is, in fact, the other way round. They are precocious also when they are equal. It is their genetic property. It is also presumably their genetic property that their deficiencies do not matter because what is lost is inert. Hence their frequent inequality. The precocity of these autosomes does not show itself until pachytene and therefore does not interfere with pairing.

It is evident in *Mecostethus*, *Rhabditis* and *Alydus*, on the other hand, that the precocity of identical chromosomes prevents their pairing. That this should be so follows on the precocity theory, for the earlier development of the chromosomes will compensate for the earlier prophase and restore mitotic conditions. This has been shown in *Alydus* and *Mecostethus* where the chromosomes which do not pair are already divided at the time the other undivided chromosomes are pairing. Thus precocity of the chromosomes determines their failure of pairing, not the reverse. The importance of this failure of pairing in the evolution of sex chromosomes and the development of dissimilarities between them will be considered later.

Conclusion. While it is possible to classify the differential behaviour of chromosomes according to these three types, there

314 CHROMOSOMES IN HEREDITY: PHYSIOLOGICAL

still remain exceptions that do not agree with any of the rules described.

First, there are cases of precocious condensation of the third type that seem to depend on hybridity or the absence of a partner at zygotene. Such are the precocity of certain chromosomes in hybrid roses (Erlanson, 1929) asynaptic *Avena* (Huskins and Hearne, 1933) and *Saccharum* (Bremer, 1930), and in the fragments of *Camnula*, an orthopteran (Carroll, 1920). It may be that these represent the second type of differential condensation operating in hybrids.

Secondly, there are the B chromosomes of maize which are

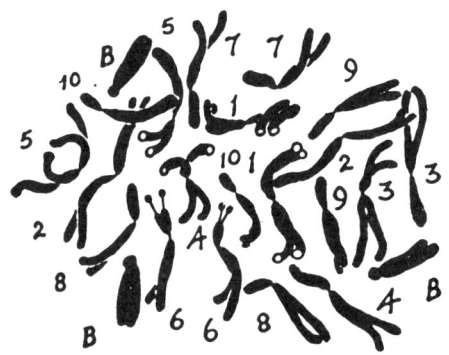

FIG. 102.—Metaphase plate of *Zea Mays* showing three condensed B chromosomes. No. 6 is the nucleolar chromosome. × 5000.

distinguishable by their behaviour at all stages of the life cycle. Thus they are over-stained at metaphase of meiosis (Randolph, 1928) and at metaphase of mitosis. They show reduced primary attraction at meiosis (McClintock, 1933) and probably reduced secondary attraction at mitosis, where they never show somatic pairing (Fig. 102). They are physiologically inert, and this property may be related with their reduced attraction or reduced specificity in attraction. It is difficult to know in what category to place them or indeed the condensed chromosomes of the hepatic *Frullania*, which show over-condensation equally at mitosis and meiosis.

Although it is not possible to put the whole of these heterogeneous observations into a consistent scheme we can see three principles connecting the differential staining capacity with other differential properties of the chromosomes. First, there is the inertness of the heterochromatic regions of the sex and other chromosomes in *Drosophila, Chorthippus* and *Zea*. Secondly, there is the reversal of staining properties in *Rattus, Chorthippus* and *Drosophila* at different stages of the nuclear cycle. Thirdly, there is the attraction or lapse of repulsion of heterochromatic chromosomes at one stage in *Rattus, Chorthippus, Zea* and elsewhere.

The first of these shows the working of a genetic property in the particles or genes concerned, and agrees with the particular observations of the last type of differential behaviour described above. The second shows that the behaviour depends essentially on a difference from the substrate, such, that if the usually over-condensed parts of the chromosomes have the difference the undercondensed parts have not, and *vice versâ*. What this difference is can be seen from the third principle: the absence of repulsion indicates a lower surface-charge and therefore probably a difference in iso-electric point of the chromosome materials. This agrees with the constant material basis in the genes arrived at earlier. But further it seems very probable that the staining capacity of the chromosomes with ordinary treatment should be related to the surface charge on them (although, as we find, not with the Feulgen reaction), since the deposition of the common nuclear dyes is a surface phenomenon.

We may conclude, therefore, provisionally, that the differential staining behaviour of chromosomes is conditioned most immediately by differential surface charge which in turn depends on the difference between the local iso-electric point and the pH of the medium, and also (in the case of proximal condensation) on more remote conditions such as the proximity of the heavily charged centromere (*cf.* Ch. XII).

3. BALANCE

Boveri showed from experiments on doubly fertilised sea-urchin eggs that all the chromosomes making up the complement of a gamete must be present for the normal development of the zygote.

He therefore properly inferred that the chromosomes of the complement were qualitatively different.

The same conclusion may be derived from recent experiments of various kinds, some of which carry us much further.

(a) MONOSOMICS. The first kind of evidence is derived from organisms which lack either a part or a whole of one chromosome of their diploid complement. The first are said to have a sectional deficiency; they are deficiency-heterozygotes. The second are said to be " monosomic; they are numerical hybrids."

Such organisms have arisen in all the conceivable ways, as follows :—

1. Through loss of one chromosome in a somatic mitosis of a diploid, *e.g.*, *Datura Stramonium*, Blakeslee and Belling, 1924; *Zea Mays*, McClintock, 1929 *b*; *Œnothera*, Michaelis, 1926.

2. Through parthenogenetic development of an " unreduced " egg (Ch. XI) which has lost one chromosome at meiosis, *e.g.*, *Œnothera Lamarckiana* in a cross with *Œ. longiflora*, Stomps, 1931.

3. Through functioning of a deficient egg cell $(x - 1)$ in an animal (*e.g.*, *Drosophila melanogaster* giving the haplo-IV type, Fig. 121) or in a plant (*e.g.*, *Nicotiana alata*, Avery, 1929; Goodspeed and Avery, 1929 *a*). This is also possible in regard to a segment, as in *Matthiola* heterozygous for lack of a trabant (Philp and Huskins, 1931) which transmits the deficient chromosome through the eggs but not through the pollen, and consequently, when selfed, yields 50 per cent. of offspring heterozygous for the deficiency and none homozygous.

4. Through functioning of a deficient male gamete $(x - 1)$ in an animal, *e.g.*, *Drosophila melanogaster*, whose sperm will function lacking an X chromosome (Kuhn, 1929 *b*). In special circumstances this can occur in plants. Thus Stadler (1931) obtained $2x - 1$ plants by X-raying pollen of *Zea Mays*. The $x - 1$ pollen functioned because no normal uninjured pollen grains were left to compete with it. Even pollen deficient for a small segment of a chromosome fails in competition with normal pollen in *Zea*, although embryo-sacs of this kind will develop (Stadler, 1932). As a rule, where it is known to be regularly formed at meiosis (as in ring-forming *Rhœo* and *Œnothera*; *cf.* D., 1929 *c*, 1931 *b*), $x - 1$ pollen never reaches

its first gametophytic mitosis. Sub-haploid pollen-grain divisions have, however, been found in *Uvularia* (Belling, 1925 c) and in an *Allium* hybrid (Levan, 1936).

These observations show that while embryo-sacs and animal gametes, lacking a part of the haploid complement, may function (on account of their very limited activities) the male gametophyte of plants is usually incapable of development in this condition. Further, the zygote of plant or animal can never develop when lacking either a section or a whole chromosome of one type entirely, *i.e.*, in both sets. The organism is slower and poorer in development where a section or a small chromosome is lacking in one set, and non-viable where a larger element is lacking, *e.g.*, one of the large autosomes in *Drosophila*.

The statement is satisfactory for general purposes, but it requires two qualifications, one of which is obvious from what has already been said. The haploid complement probably always contains reduplications. We might then expect that it would not all be necessary or equally necessary to the life of the organism. That this is so has been proved by Muller (1935). Certain deficiencies of small segments are not lethal in *Drosophila* in the homozygous condition. The segments lost are presumably present elsewhere. The organism is protected against their loss as a polyploid is. The second qualification depends on some parts of the chromosome being inert ; its consequences will be discussed later.

The fact that a certain group of materials is necessary for the successful working of the life processes enables us conveniently to define the haploid complement of an organism as the minimum combination of chromosomes or parts of chromosomes necessary in single or double number for the development of the organism to maturity.

(*b*) PROGENY OF TRIPLOIDS. The second kind of evidence is to be derived from the progeny of triploids. In these the extra chromosomes are distributed at random at the first and second divisions (with slight loss through lagging) shown by Mather (1935) with trivalents in *Triticum*, and therefore gametes are produced with an approximately binomial frequency distribution of chromosomes ranging from the haploid to the diploid number (Table 47).

It might be expected from these observations that the progeny of triploids would for the most part receive from their parents approximately $3x/2$ chromosomes.

TABLE 47

Distribution of Chromosomes at Meiosis in Triploids

Nos. of chromosomes	6	7	8	9	10	11	12	13	14	15	16
Hyacinthus (pollen grain) $x = 8$. Belling, 1925 ; D. 1926 . . .				4	28	61	91	88	69	33	5	2
Tradescantia (pollen grain) $x = 6$. D., 1929 *c* . .	—	4	13	30	21	4	—				
Petunia (second metaphase) $x = 7$. Dermen, 1931 .		1	12	32	40	41	20	5	1		

Note.—The slight excess of the numbers below $3x/2$ is due to loss of chromosomes at meiosis, *cf.* Levan, 1933 *a*, 1935 *b*. The chance is calculated as 10 per cent. in *Hyacinthus* (D., 1926), 11·5 per cent. in *Narcissus* (Nagao, 1935).

The zygote progeny, however, do not correspond at all with the gametic proportions. They show a different range according to whether the triploid is the male or the female parent, or both, and whether the opposite parent is diploid or tetraploid. In all these types of cross there is elimination of gametes and zygotes having extra chromosomes. In consequence triploids show a sterility which increases with the number of chromosomes in the haploid set, other things being equal. This elimination depends, first, on a failure of growth of abnormal pollen grains, which is sometimes complete and sometimes merely a failure to compete with normal ones—or with better suited ones, since the diploid pollen of a triploid *Œnothera* grows better on the tetraploid stigma and the haploid pollen on the diploid stigma (Table 49). In two cases in *Solanum* only the diploid gametes of triploids have functioned on selfing (Huskins, 1934 ; Janaki Ammal, 1934). Secondly, it depends on competition between zygotes in the embryonic stage ; this applies equally to the seed of *Solanum* and the larva of *Drosophila*.

UNBALANCE AND VIABILITY

Similar elimination is found in the breeding of plants and animals that are simply trisomic, either in a whole chromosome or in part of a chromosome (v. Table 50) or in the development of pollen of plants of these types (v. Table 51 and Fig. 103).

TABLE 48

A. Trisomic Organisms showing Differential Action of Chromosomes.

Œnothera Lamarckiana. de Vries. *Cf.* Lutz, 1907 b ; Catcheside, 1933, 1936 ; Emerson, 1935, 1936 (*cf.* Ch. XII).
Datura Stramonium. Blakeslee, 1922, 1928, 1931.
Matthiola incana. Frost and Mann, 1924 ; M. M. Lesley and Frost, 1928.
Crepis spp. Navashin, 1926, 1930 a ; Hollingshead, 1930 b.
Zea Mays. McClintock, 1929 a ; D., 1934 ; Rhoades and McClintock, 1935.
Solanum Lycopersicum. Lesley, 1928.
Avena. Nishiyama, 1934.
Antirrhinum. Propach, 1935.
Vicia macrocarpa (6) × *V. sativa* (6) F_2. Sweschnikowa, 1929 b.
Secale cereale. Takagi, 1935.
Drosophila melanogaster. Bridges, 1921 b.

B. Unbalanced Clonal Forms.

Hyacinthus ($2x + 1$–14). de Mol, 1923 : D., 1929 b.
Narcissus ($2x + 1, 2$: $3x + 1, 2$). Nagao, 1933.
Chrysanthemum ($6x + 3$–6). Shimotomai, 1933.
Crocus spp. Mather, 1932.
Iris spp. Simonet, 1934.
Poa alpina. Müntzing, 1932.

The relative viability of the different kinds of progeny of triploids and aneuploids may be tested. They differ from the normal in two ways :—

(i) in the size of the fraction of the whole complement which is present in excess, the fraction being in the following degrees : fragment or small chromosome, whole chromosome, two chromosomes, etc. ;

(ii) in the disproportion of the excessive fraction to the rest, the proportions being 3 to 2 (trisomic diploid) and 4 to 2 or 2 to 1 (tetrasomic diploid and disomic haploid).

They also differ at different stages of the life cycle owing to differences in the stringency of conditions of life.

Thus the observations on fertility of numerically abnormal zygotes may be used to measure elimination on three scales. Taking the last first, the stringency of elimination increases in the following series : Animal egg, animal sperm, embryo-sac, pollen, plant or animal zygote. Taking the size of the fraction : the higher the proportion of the haploid complement reproduced in excess (up to one-half) the greater the reduction of viability. Taking the disproportion of the excessive fraction : the higher the disproportion the more aggravated is the defect produced.

TABLE 49

Transmission of Extra Chromosomes of Triploids to their Progeny *

	$2x$	$2x+1$	$2x+2$	$2x+3$	$2x+4$	$2x+5$	$2x+6$	$2x+7$
$3x \times 2x$ (i.e., through eggs).								
Œnothera Lamarckiana. $x = 7$. van Overeem, 1921	1	9	4	2	2	—	1	1
Œ. biennis. $x = 7$. van Overeem, 1920	41	26	4	—	—	—	—	—
Datura Stramonium. $x = 12$. Belling & Blakeslee, 1922	215	381	101	—	—	—	—	—
Solanum Lycopersicum. $x = 12$. Lesley, 1928	10	38+	16+	2	—	—	—	—
Zea Mays. $x = 10$. McClintock, 1929	—	1	3	1	3	1	—	1
$2x \times 3x$ (i.e., through pollen).								
Œnothera Lamarckiana. $x = 7$	—	15	—	—	—	—	—	
Petunia. $x = 7$. Dermen, 1931	39	12	—	—	—	—	—	
Solanum Lycopersicum. $x = 12$	—	2	—	—	—	—	—	
Zea Mays. $x = 10$	37	27	6	2	—	—	—	

$3x$ selfed. Crepis capillaris ($x = 3$) × C. tectorum ($x = 4$). Hollingshead, 1930 b. CCT plant (from unreduced egg. $2x = CC = 6$; $3x = CCT = 10$; $4x = CCTT = 14$.

* Cf. Nishiyama, 1934, and Levan, 1935, for further data.

Chromosome number	6	7	8	9	10	11	12	13	14
Random expectation	1	8	28	56	70	56	28	8	1
Progeny	59	12	4	4	12*	3	3	3	1†

* All of the composition CCT ($3x$).
† Composition CCTT, i.e., true $4x$ (cf. Table 26).

Random expectation should be multiplied by 101/256.

UNBALANCE AND STERILITY

TABLE 50

Transmission of Extra Chromosomes of Trisomics to their Progeny *

		Through Pollen. Per cent.	Through Eggs. Per cent.
Blakeslee (1921) . .	*Datura* ("Globe")	3	26
Clausen and Goodspeed, 1924.	*Nicotiana Tabacum* † ("Enlarged," $4x + 1$).	3·4	35·4
R. E. Clausen, 1930 .	*Nicotiana Tabacum* ‡ ("Fluted," $4x - 1$).	2	ca. 60
Frost and Mann, 1924 .	*Matthiola* ("Slender") .	20	nil
Lesley, 1928 . .	*Solanum Lycopersicum.*		
	triplo "A" . .	0	24
	,, "C" . .	10	ca. 15
	,, "B" . .	23	27

* 50 per cent. expected : the reduction is chiefly due to elimination of abnormal gametes, but partly also to elimination of abnormal zygotes.
† Pentasomic allo-tetraploid.
‡ Trisomic tetraploid: the high transmission of $2x-1$ eggs is due to frequent loss of the unpaired chromosome at meiosis so that more $2x-1$ eggs than $2x$ are formed. No $4x-2$ seedlings appear.

TABLE 51

Pollen Abortion in Datura (after Blakeslee and Cartledge, 1926).

Type of plant . . .	x	$2x - 1$	$2x$	$2x + 1$	$3x$	$4x$
Percentage bad pollen .	88	ca. 60	1·2	ca. 5	44	5

Notes.—(i) There are no monosomics in the progeny of haploids, therefore $x - 1$ eggs and pollen or $2x - 1$ zygotes are non-viable.

(ii) The primary trisomics with the poorest pollen are not those with the poorest growth and lowest frequency of appearance in progeny of triploids. Therefore the disadvantageous effect of the extra chromosome is specific in regard to time of action.

(iii) The pollen abortion is higher in the tetraploid than in the diploid (*cf.* Table 28).

TABLE 52

Effects of Abnormal Chromosome Constitution on Viability

I. In *Drosophila melanogaster* (Fig. 12 ; *cf.* Li, 1927).

Deficiency.	Reduplication.
Gametes $(x-1)$. Nullo-X and -Y sperm function (Kuhn, 1929) and probably therefore all other aberrant types. All eggs probably fertilised.	$(x+1)$. All types of egg and sperm capable of fertilisation.
Zygotes $(2x-1)$. (i) Sections : some have no effect. Others are lethal to larva. (ii) Haplo-IV : development retarded, not lethal. (iii) Haplo-II and III : lethal in egg stage. (iv) Haplo-X : viable male (sterile).	$(2x+1)$. (i) Section : no deleterious effect. (ii) Triplo-IV : viability slightly reduced. (iii) Triplo-II and III : usually lethal to larva. (iv) Triplo-X : usually lethal to larva or pupa (" super-female ").
Zygotes $(2x-2)$. (i) Sections : homozygous deficiencies lethal in prelarval stages (including deficiencies in X of ♂). (ii) Nullo-IV : little or no development. (iii) Nullo-X : no development (similarly nullo-II or III).	$(2x+2)$. (i) Sections : " P III " lethal to larva. (ii) Tetra-IV : lethal to young larva. (iii) Tetra-X, II or III : cannot occur.

II. In various plant species referred to in earlier tables (for deficiencies, *cf.* Monosomics).

The following table summarises the effect of single-section and single-chromosome differences from the normal on various stages of the life cycle in certain animal and plant species.

Gametes, $x + 1$.

Pollen (i) Fragments : viability probably slightly reduced.
       (ii) Whole chromosomes : largely eliminated but to a varying extent according to the chromosome.
      (iii) Several chromosomes : all eliminated.

Eggs (i) Fragments   } Unaffected.
     (ii) Whole chromosomes

CHARACTERISTIC TRISOMICS

Zygotes.
- $2x + 1$ (i) Fragments: viability not clearly lowered; show specialised abnormality.
  - (ii) Whole chromosomes: viability slightly lowered. Show generalised abnormality.
- $2x + 2$ (i) Fragments: accentuation of $2x + 1$ type. Viability lowered.
  - (ii) Whole chromosomes: accentuation of $2x + 1$ type. Viability greatly lowered.

(c) TRISOMICS. Records of fertility and elimination showing the differentiation of the chromosomes are supported by a large series of qualitative observations. Thus in numerous species (Table 50), a whole series of trisomic phenotypes have been identified corresponding with the different members of the complement which are present in excess. In *Datura* and *Œnothera* there are not only the types of trisomic corresponding in number to the haploid complement, but there are the special types corresponding to viable interchange combinations and these are much more numerous (Ch. IX). The extra chromosomes are specific in the time of their action as well as in the kind of effect they produce. For example, the trisomic *Datura* type with the worst pollen at maturity is not the one with the lowest pollen effectiveness in fertilisation. Again, the extent of the disproportion always has a quantitative effect; the tetrasomic has the corresponding trisomic characters in an exaggerated degree, while the same number of chromosomes in a tetraploid or hexaploid have a reduced effect (*e.g.*, *Triticum vulgare*, $6x$, Kihara, 1919–24, Huskins, 1928; *Avena sativa*, $6x$, Huskins, 1927; *Primula kewensis*, $4x$, Newton and Pellew, 1929), whence the occurrence of only high polyploid species with variable numbers (*v.* Ch. VIII).

Simple trisomics do not normally breed true and multiple trisomics never do so. They are therefore found permanently only in clonal plants. Amongst these their special qualities have often led to their being selected for propagation (Table 48 B).

(d) BALANCE AND PROPORTION. Two principles may be inferred from these observations:—

324 CHROMOSOMES IN HEREDITY: PHYSIOLOGICAL

(i) The parts of the chromosome complement have specific and different properties; this is the principle of *differentiation*.

(ii) These parts co-operate in the developmental processes of the organism, so that every part of the chromosome complement is necessary for the normal development of every part of the organism; this is the principle of *co-operation*, which is an *a posteriori* statement of the *a priori* doctrine of epigenesis.

Now when these two principles are considered in relation to adaptation they give it a new character. The adaptation of an organism (gamete or zygote) must be held to consist in an

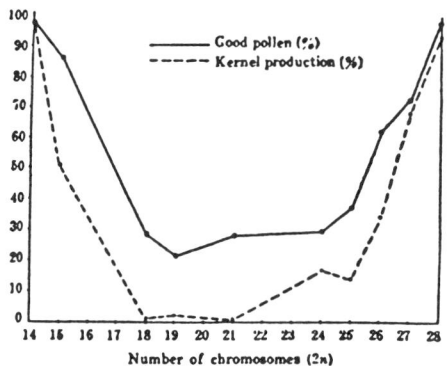

FIG. 103.—Relative fertility of plants with different chromosome members derived from the cross *Avena strigosa* × *A. barbata* ($x = 7$). (Nishiyama, 1934.)

adjustment of the proportions of the differentiated particles (genes) to co-operation with one another in joint reaction to external conditions. This adjustment has been described by Bridges (1922) as *balance*. The "normal" or successful type of gamete or zygote in a species is said to be "balanced" while the types with their constituent elements not so proportioned are said to be "unbalanced."

Consider a chromosome complement consisting of n qualitatively differentiated particles or genes. On Bridges' principle the properties of the organism depend on the proportions, *i.e.*, interactions, of these particles. There are $\dfrac{n(n-1)}{2}$ such specific interactions between pairs of genes. Where one particle is reduplicated n of these interactions are changed; where half the particles are reduplicated the maximum number of interactions are changed, *i.e.*, $\frac{n^2}{4}$. It follows that the maximum unbalance is attained where half the chromosome complement is reduplicated and the least unbalance where the least portion is reduplicated.

This rule applies to *Drosophila*, where the only viable autosome trisomics are the triplo-IV, and to *Solanum Lycopersicum*, where the $2x$, $2x+f$, $2x+2f$, $2x+1$ and $2x+2$ represent a descending series in relation to growth rate (Lesley and Lesley, 1929). It applies to all observed progenies of triploids, which show greater elimination of $2x+2$ and $2x+3$ forms than of $2x+1$. It applies to the size of unbalanced pollen grains in *Tradescantia* (D., 1929 c). The rule also means that the smaller the number of particles affected the more specific will be the effect on the organism. It is found that extra fragments have a more restricted effect than extra whole chromosomes in *Solanum*.

It also follows that where the change in proportion is from 1 : 1 to 2 : 1 the effect should be more drastic than where the change is from 2 : 1 to 3 : 2. Thus tetrasomics (whether of fragments or whole chromosomes) are more markedly abnormal (in viability as in other respects) than the corresponding trisomics in *Datura*, *Solanum* and *Matthiola*.

The principle of balance operates co-extensively with qualitative differentiation. Moreover, variation in the degree of differentiation between chromosomes is probably due to their being sometimes similarly differentiated—in plants, no doubt, owing to a remote polyploid ancestry. *Hyacinthus orientalis*, for example, shows reduced unbalance in trisomics and hence reduced sterility in triploids. It is possibly in its remote origin a tetraploid (*v.* Ch. VI).

This contrast between balance and unbalance conveys the impression of something denigratory about unbalance, and, as we have seen, unbalanced forms, like gene mutants, are usually maladjusted in proportion to the amount and frequency of the particles concerned. But particular exceptions show that this rule is no absolute one. Certain kinds of unbalance, like certain gene-mutants, are hardly inferior to the normal condition. The difference between balance and unbalance must therefore be expressed in relative terms thus : The balanced combination of genes has been exposed to natural selection. The unbalanced combination has not ; it is at random, and since the possibilities of random combination are vast (most organisms probably having some thousands of differentiated genes) and the number of successful combinations possible in any one environment are few, the chances of a new unbalanced type being as successful as the normal are remote.

These arguments may be inverted, and we then see natural selection from a different point of view, which is not without value. We may say that natural selection consists in the elimination of unbalanced forms from the body of proportion-mutants and gene-mutants in a species.

(*e*) SECONDARY BALANCE. From these considerations it follows that the possibility is always open of variation occurring by a change in proportion, either by reduplication of segments of chromosomes or of whole chromosomes. If such a new proportion were successful it would yield a new *secondary balance* having a different phenotypic expression from the primary, ancestral, type. Evidence of this probably universal type of variation is only beginning to appear, for the reason that the necessary inferences are elaborate ones.

Where the change of balance is concerned with whole chromosomes its occurrence seems to be confined to polyploids. Thus the 18-series arises in *Dahlia* secondarily to the 8-series and the 17-series in the Rosaceæ secondary to the 7-series (*v*. Ch. VI). The primary balance in the 17-series is shown by the critical test of triploid progenies. Thus in the progeny of triploid-diploid crosses the following distribution of chromosome number types is found in *Pyrus Malus* (Moffett, 1931 ; *cf*. Nebel, 1933 *b*) :—

Chromosomes in excess of 34 :	1	2	3	4	5	6	7	8	9	10	11	12	13	14	15	16	17
Nos. of seedlings	–	–	1	4	2	5	10	2	4	1	1	1	1	–	–	–	–

This distribution is different from that characteristically found in simple triploids already discussed where intermediate numbers are eliminated. The favouring of seedlings with seven extra chromosomes suggests that these are the seven that make up the primary set whose existence is inferred on other grounds.

In *Nicotiana* ($x = 12$) also the progeny of triploids having 12 bivalents and 12 univalents do not show the sharp elimination found in simple triploids. In these cases the middle number is favoured (Table 48, *cf.* East, 1933) :—

TABLE 53

Distribution of extra chromosomes of the triploid *Nicotiana paniculata* ($n = 12$) × *N. rustica* ($n = 24$) in the back-crosses to its parents (Lammerts, 1929, 1931).

No. of extra chromosomes	0	1	2	3	4	5	6	7	8	9	10	11	12	24*
1. 3x × *rustica* ♂ (78) $2n = 36+$	–	–	–	3	8	12	12	–	7	1	–	–	–	35
2. 3x × *paniculata* ♂ (117) $2n = 24+$	–	5	3	4	17	16	24	6	10	4	–	–	–	28
3. *rustica* × 3x ♂ (44) $2n = 36+$	–	–	–	–	–	–	4	3	2	2	–	3	30†	–

* Through failure of reduction. *Cf.* Ch. XI.
† Gametic selection on male side is determined by relationship to female parent as in *Œnothera*.

These results may be explained as due to a lack of differentiation in the chromosomes, or to the set of twelve being derived from an earlier set of six by polyploidy. The latter view is the more probable and is favoured by evidence of secondary pairing in *Nicotiana* and of the general evolutionary relationship of chromosome numbers (*v.* Ch. VI). Thus *Nicotiana* with an apparent basic number of 12 may be said to have a primary basic number of 6 from which the secondary basic number of 12 is in process of developing—without, however, any change in balance in the whole set.

The second kind of evidence of change of balance in evolution consists in the occurrence of intra-haploid homology, *i.e.*, the reduplication of materials (genes) within the haploid set. This may be inferred from both genetical and cytological observations, which indicate that pairing occurs exceptionally between chromosomes in the same haploid set in the absence of competition, as described in the previous chapter in relation to secondary structural change (*cf.* also p. 181).

It may be thought remarkable that pairing (and crossing-over) should occur between parts of chromosomes in the haploid, although these parts never seem to pair in the diploid, especially since in *Œnothera* this pairing is often terminal, proving that the ends of some of the different members of the haploid set are homologous. How does this agree with the notion that the parts of the chromosomes are qualitatively differentiated? A comparison with the observations of differential affinity in polyploids shows the explanation (Ch. VI). It is known that chromosomes which rarely or never pair, in an allotetraploid, may pair regularly in its diploid hybrid parent or its parthenogenetic diploid progeny. Thus it is not surprising that chromosomes which never pair in the diploid should occasionally pair in its parthenogenetic haploid progeny for the same reason. Two " non-homologous " chromosomes of the haploid set may consist of a series of particles (or genes) *abcdefghkl* . . . and *amnopqrs* . . . In the presence of identical mates, *i.e.*, with the same linear sequence of particles, the *a* particles of different chromosomes will never pair. In the haploid they will very often pair at pachytene and occasionally form a chiasma which will preserve their pairing at metaphase.

These observations are in agreement with the theory of mutation, by crossing over, outlined elsewhere (Chs. VII and IX), which demands that in *Œnothera*, as in other organisms, small sections should be homologous in different parts of the same haploid complement and of different complements. This would mean that there had been a considerable amount of reshuffling of small pieces of chromosome in the history of the species, the primary changes being inversion and interchange, and the secondary changes, reduplication and deficiency.

Every gene or series of genes must be supposed to have had a beginning at some time when it would have been singly represented in any chromosome complement. Reduplication of genes therefore means that there has been a change of proportion, which if successful becomes secondary balance. All chromosome complements are probably secondary in this sense, for such change is perhaps the most readily available source of successful variation in all organisms. Its further analysis is therefore a most pressing problem.

If single genes or blocks of genes are freely liable to reduplication within the haploid set, as these observations suggest, we must suppose that characteristic haploid sets will probably contain a proportion, perhaps a majority, of their genes represented two, three or more times. The gametic set of an autotetraploid will have all its genes present an even number of times and odd numbers will be restored to it by mutation, until it reaches an allopolyploid condition which will not be distinguishable from that of a diploid.

Haldane (1932 a) has suggested that chromatin-diminution in *Ascaris* requires localisation of certain genes; such a condition might be produced by the extensive reduplication of all or most of the genes and their later selective elimination.

This conception makes the occurrence of forms with different degrees of qualitative differentiation as between different chromosomes intelligible. If one species has very little reduplication, all its chromosomes must be differentiated in the highest degree compatible with random assortment of the genes among the chromosomes. If another form has a great deal of reduplication with most of its genes represented six or a dozen times it will have a very low differentiation with the same random assortment. Thus it is not necessary to assume that organisms with low differentiation have individual genes of any special kind but merely that they are reduplicated more frequently either at one stroke with polyploidy, or segment by segment with structural change.

4. INERT CHROMOSOMES AND GENES

Individuals of many species of plants and animals have chromosomes that produce no specific or general effect on their growth. Such chromosomes are said to be inert. They do nothing, and can therefore be lost entirely or reduplicated almost without limit and the organism will develop in just the normal way. They are never regularly members of the complement in the species, for since they can be lost without effect they frequently are lost.

Inert chromosomes arise, so far as we can judge, as small "fragments" derived from the breakage of larger chromosomes. Such are the supernumerary chromosomes of *Zea*, *Tradescantia*, *Fritillaria*, *Ranunculus*, *Tulipa* and *Paspalum* (Table 16). The large chromosomes from which these fragments have arisen must have been inert in the neighbourhood of the centromere, otherwise the new fragments will be active like those in *Matthiola* and *Solanum*. The small chromosomes of Orthoptera which are so frequently heterozygous for deficiencies are probably inert in part (D., 1936 *b*). How chromosomes that are partially inert could have developed is shown by the properties of Y chromosomes in certain animals with an advanced degree of differentiation of sex chromosomes (Ch. IX). This chromosome in *Drosophila* is almost entirely inert and has concurrently lost the property of mutation. The property of inertness is probably very general in Y chromosomes, since they vary in size most frequently both within species and between species. And again in many species they are lost altogether. The Y chromosome has become inert, according to Muller (1918) because crossing-over is effectively abolished in it. Probably, as we shall see later, the proximal parts of the other chromosomes in *Drosophila* are also inert, as they certainly are in the X. This may also be due to absence of crossing-over in them in the female (*cf.* Mather, 1936). Whatever the cause, their inertness is significant as a condition of the origin of inert fragments. It is apparently as fragments of the sex chromosomes that inert chromosomes have arisen in the Hemiptera, since such inert chromosomes sometimes pair with the X or Y in *Metapodius* (Wilson, 1909) and *Alydus* (Reuter, 1930).

The study of inert chromosomes has been greatly advanced by the hypothesis of Heitz (1929, *et seq.*) that the differential "heterochromatin" is always inert. Muller and Painter (1932) found that the proximal third of the X chromosome in *Drosophila melanogaster* included the loci of none of the genes which have mutated. Heitz (1933) was able to show that this region was differential in behaviour at mitosis, thereby substantiating his principle. Similarly he was able to show that the proximal parts of the autosomes were differentially condensed, and to recognise the sex chromosomes of *D. funebris* for the first time by their behaviour in this respect.

The application of the principle is, however, less simple than Heitz has supposed. The distinction between "heterochromatin" and "euchromatin" is not an absolute one. Lorbeer (1934) and Tinney (1935) disagree as to whether the Y chromosome of *Sphærocarpus* is inert, and as we have seen, chromosomes in different tissues of the same individual differ in their differential behaviour which is physiologically controlled. Probably in such an intermediate position is the X chromosome of *Sphærocarpus Donnellii*, which according to all observers is differential in its behaviour. Yet it is not by any means inert. Under X-ray treatment it produces mutations as frequently as all the autosomes together, and when differential parts are broken off and lost the plant is changed in different ways affecting both the sexual and vegetative characters (Knapp, 1935 *a* and *b*). These observations do not destroy Heitz's contention, but they show (what is equally clear from the cytological study of differential behaviour) that the differential properties of chromosomes, or even perhaps of separate chromomeres, cannot be described in terms of a simple all-or-nothing reaction.

Whether the distinction between active and inert genes is an absolute one we do not know. Inertness may itself be physiologically controlled, although we know of no example yet. But we know well enough that every transition occurs between an active chromosome and an inactive one. The change presumably takes place gradually by mutation to inertness of separate genes. If there is, therefore, a sharp distinction in behaviour of active and inert genes such a distinction will not always be found between chromosomes or parts of chromosomes at different stages of change, and we shall have such gradual variations of behaviour as make for the observed complexity.

Inert chromosomes are thus presumably made up of inert genes. Since the existence of a gene in the first instance is inferred from its action, "inert gene" may seem to be a contradiction in terms; nevertheless it represents a step forward; it enables us to see what we mean by a gene more clearly than before. Thus the active gene generally found in higher organisms must have two properties, (i) of reaction with its substrate, by which it creates a *milieu* permitting reproduction; and (ii) reproduction. The inert gene has lost the capacity for reaction, although it has retained the secondary capacity for reproduction which is conditioned by the reactions of the other genes associated with it in the complement of a higher organism. Now the capacity for reaction cannot have been born in the gene after the capacity for reproduction. It must rather have been born first. We can therefore take the two major steps in the evolution of the gene to be :—

reaction → reaction + reproduction → reproduction.

We also learn from inert genes that reaction and mutation are bound up with one another, for inert genes have lost these two properties together. Perhaps, therefore, natural mutation is an abnormality of reaction. This assumption agrees with the observations that rates of mutation, like rates of reaction, are both genotypically and environmentally controlled. They are conditioned by the substrate.

Muller and Gershenson (1935) take a different view of the change to inertness. But however the change takes place it is clearly irreversible. We might therefore expect that inert genes would be found very generally; the widespread occurrence of supernumerary fragments indicates that this is so, and that inert genes are usually concentrated near the centromere. In this position they cannot so easily be got rid of, and this is a possible reason why they are found here. We may say that they are clogging the genetic system with useless but inevitable material. On the other hand, it is possible that they are preserved near the centromere when crossing-over is reduced or absent in this part, as they are in the Y chromosome. If Heitz's method of identifying inert genes proves satisfactory as an approximation it will help in solving this problem.

5. THE GENE

We are now in a position to enquire more closely what we mean by the gene. The differences between organisms are found to behave in crossing-over as though they had a linear order. They are known to be determined by changes in the chromosomes. The chromosomes consist of particles, the chromomeres, shown by observation to be arranged in linear order. Therefore it is assumed that the units of crossing-over or genes are such particles. Moreover, the genes, like the chromomeres, are chemical individuals. The genes have characteristic properties of mutation, the chromomeres of attraction. We still do not know, however, what the relation may be between the differences from which the gene is inferred and the particles that are observed. We know that particles having a linear order may change in three very different ways to give hereditary differences.

First, they change externally by loss or reduplication (intergenic change) to give those differences in proportion that are known to have a physiological effect. By such a change the " bar " gene difference arises in *Drosophila* (Sturtevant, 1925 ; Bridges, 1936).

Secondly, they must be supposed to change internally (intragenic change) to give those differences that distinguish the different genes within the hereditary complement. Differentiation cannot have arisen merely by quantitative change.

Thirdly, it has now been shown beyond doubt that differences arise merely by changes in relative position of genes. This *position effect* must depend on a localisation of the products of gene action. It is responsible for two bar genes in one chromosome having a different effect from one bar gene in each of two chromosomes (Sturtevant, 1925). It is also responsible for changes occurring frequently when chromosomes are broken and their parts recombined in a new way (Muller and Prokofieva, 1934).

Now we see that when we speak of a gene as a particle whose changes are responsible for these differences we may not always mean the same thing. And this difficulty becomes more serious when we consider that the gene is a unit of crossing-over. As a rule, the unit of crossing-over agrees with the unit of mutation. But this is not always so. The reason is clear. A structural change such as inversion will suppress crossing-over at the same time that it determines a genetic difference by position effect. This is perhaps the explanation of *elongatus-vitellinus* crossing-over in *Lebistes* (Winge, 1934). The unit of crossing-over is therefore liable to be increased by the act of mutation. The old criteria of genetic structure therefore require re-examination.

It is in these circumstances that Muller and his collaborators have made use of new agents of analysis: the combination of X-ray breakage with the examination of the results both phenotypically and by salivary gland study. They have shown that the smallest unit of X-ray breakage corresponds with the smallest chromomeres detectable in ultra-violet light (*cf.* Ch. X). They have also shown that this unit is smaller than the unit of crossing-over and that changes inside the crossing-over unit are sometimes due to structural changes which have a position effect. What relation there may be between this classification of mutations by their chemical causes and that by their physiological effects (as classified by Muller, 1932) is not yet clear.

These conclusions, along with some others, show that a one-to-one correspondence of particles and differences can no longer be assumed. In particular they make it impossible to express all variation in terms of units directly derived from its material basis, the genes. Any advance in the theoretical treatment of variation must therefore wait on a more exact definition of the mutual relationships of particles and changes.

In spite of the complexity of these relationships we can assume without hesitation that an intra-molecular and therefore intra-genic change precedes and conditions all more complicated kinds of change. It follows that intra-genic change must inhibit pairing of the differing genes. Thus pairing of genes may be prevented at pachytene by either of the two changes which are ordinarily described genetically: intra-genic change will prevent pairing of two genes, quantitative change will prevent pairing of a larger or smaller group of genes according to the number shifted (which may amount to the whole chromosome). Therefore quantitative or structural change will be a vastly more potent means of preventing pairing, and the failure of pairing in hybrids may be taken as an indication of the part that structural dissimilarity between corresponding chromosomes has played in the differentiation of their parents. Further, it follows that allelomorphism essentially depends on a correspondence of position in the chromosome rather than on the chemical relationship which is usually associated with such a correspondence.

CHAPTER IX

PERMANENT HYBRIDS

Permanent Hybridity and Crossing-over—Complex Heterozygotes—Inheritance and Mutation in *Œnothera*—Its Chromosome Basis—The Differential Segments—Complex Trisomics—The Sex Heterozygote—Pairing and Segregation of Sex Chromosomes—The Haplo-Diploid System—Sex Determination—Evolutionary Changes.

> But in them Natures Coppie's not eterne.
> SHAKESPEARE, *Macbeth III.* (2).

1. PRIOR CONDITIONS

Two kinds of heterozygote exist in which the difference between the gametes is deeper in effect than that shown by mendelian and simple structural hybrids, but in which the segregation is a simple mendelian one: the difference behaves as a unit and only two types of gamete are normally produced. These properties are correlated with permanence; that is to say, the heterozygous condition is the property of the whole race or of a necessary part of it.

The first type is the "complex heterozygote," typically seen in many *Œnothera* species. These species breed true to their heterozygous condition owing to their producing only two kinds of gamete and owing to the homozygous form of either type being inviable.

The second is that found in organisms with hereditary sex determination (Ch. I). In these, where the differentiation of the sexes is in the diploid phase, one sex is homozygous for the sex determinants, the other heterozygous. The heterozygous sex can be shown genetically to produce two kinds of gametes, one the same as that of the homozygous sex, the other different; they therefore give the opposite sexes on fusing with the one kind of gamete produced by the homozygous sex. The heterozygous sex may be the

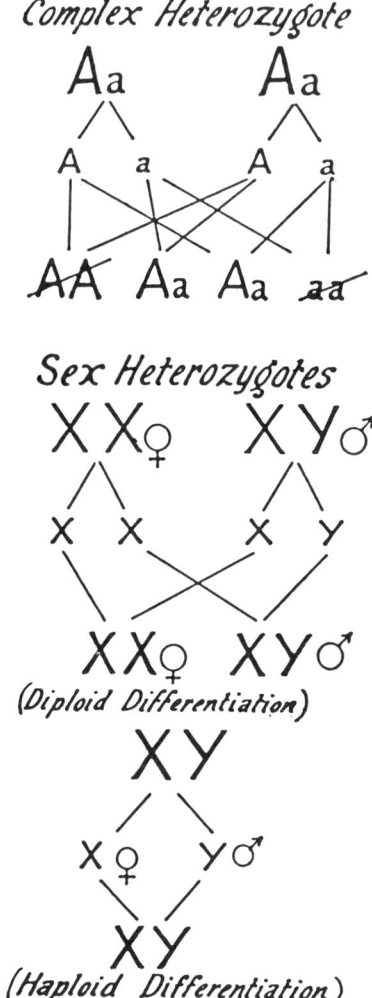

FIG. 104.—Scheme of inheritance in complex and sex heterozygotes showing that elimination of homozygotes in the one and the exigencies of reproduction in the other secure permanence in the system.

male or the female. Where, on the other hand, differentiation is in the haploid phase, the diploid undifferentiated generation is heterozygous in sex determinants and gives the two sexes in equal numbers. The heterozygous condition is therefore essential for differentiation by genetic segregation, but has no necessary connection with one sex or the other. Whether the heterozygote is male, female or asexual is indifferent to the origin and working of the chromosome mechanism, which depends on the genetic reaction of the original sex factors and the relative importance of the haploid and diploid generation. It is therefore convenient in considering the mechanism to speak of sex heterozygotes rather than of the heterozygous sex.

As in the complex heterozygote, so in the sex heterozygote, the differences (which determine sexual differentiation) behave as a unit in inheritance. The sex heterozygote differs in that one of its zygotic types can exist in a condition that is homozygous so far as the sex differences are concerned and show free segregation of all differences found in that condition. Permanence is secured by the complementary functions of the sexes in reproduction.

The two types of heterozygote have in common the two properties of permanence and unity in inheritance. Clearly the permanence could not be preserved without the unity, nor could differences more complex than any that have been known to arise singly have developed as a unit except through an accumulation of changes such as requires permanence in the system accumulating them. The one essential for these combined properties is, therefore, a suppression of crossing-over between the differences that are accumulated. Now crossing-over, as we have seen, is a normal condition of metaphase pairing and segregation in all organisms. Crossing-over is also a condition of preserving the similarity of homologous chromosomes in species, without which divergence sooner or later is bound to occur and pairing of any kind become impossible. How then can a suppression of crossing-over be brought about between chromosomes that continue to pair? Clearly there is only one means: a separation of the chromosomes concerned into two parts, one part which is similar to a corresponding part in the homologous chromosome and pairs with it, another part which does not pair or cross over and contains the differences that distinguish the two chromosomes. It is as we shall see on the evolutionary changes in the relative sizes and positions of these two parts, the *pairing segment* and the *differential segment*, that the different mechanisms of permanent hybrids depend.

2. COMPLEX HETEROZYGOTES : ŒNOTHERA

(i) Inheritance in Œnothera. A *complex heterozygote* is a diploid organism which produces two types of gametes differing profoundly in genetic properties and incapable of giving homozygous offspring. It breeds true primarily by virtue of the elimination of the homozygous embryos.

The determination of these properties by Renner depended on his genetic analysis of a group of *Œnothera* species. The notion of the complex heterozygote, however, is important not only in relation to the special genetic properties of these species and of other species and hybrids which are now being found to resemble them, but also in relation to their analogy with other permanent hybrids —sex heterozygotes and allopolyploids—and to the general problems of variation and hybridity.

Inheritance and chromosome behaviour are of such a special type in *Œnothera* and so closely related that, in order to show how they differ from the normal pattern, we must consider them together. The complex-heterozygote *Œnothera* species probably share the same hereditary properties with *Hypericum punctatum* and *Rhœo discolor* with which they also share the property of having permanent multiple rings (Ch. V). The following hereditary properties have been determined in *Œnothera* :—

1. They breed true in the main when self-fertilised, but throw a certain proportion (1 to 2 per cent. as a rule) of divergent forms. These were the mutants of De Vries. Most of the mutant forms have since been found to be trisomic or tetraploid (*cf*. Ch. III), and differ primarily from the same kinds of mutants in other species only in the high frequency with which they appear. Others, however, are diploid, and do not revert to the parental form in later generations.

2. They have a variable proportion (usually 30 to 60 per cent.) of bad seeds in which the embryo has developed little or not at all (Renner, 1916, *cf*. 1929).

3. When crossed together they usually produce more than one

type of hybrid, and reciprocal crosses may be of different types or the same types in different proportions.

4. Some of the diploid mutants (*e.g.*, *Œ. rubrinervis* from *Œ. Lamarckiana*, De Vries, 1918) and crosses (*e.g.*, *Œ. grandiflora* × *Œ. Hookeri*, Cleland and Oehlkers, 1929) yield on self-fertilisation a much higher proportion of new forms than their parents did; others resemble their parents in hereditary properties, and others again resemble the Californian species, *Œ. Hookeri*, in being absolutely true-breeding and fertile: they have entirely lost their parents' exceptional properties.

The special system of inheritance found in *Œnothera* can be expressed and understood in mendelian terms if we accept the complex-heterozygote hypothesis of Renner (1917, *cf.* 1925) and the balanced-lethal hypothesis of Muller (1918). The first assumes that each of the mutating species of *Œnothera* is a hybrid that produces two kinds of gametes. These gametes have genetic *complexes* which are distinguished by differences more profound that those that are inherited in a mendelian way in other organisms. They affect the whole habit and structure of the plant, although they behave as a unit in inheritance. In these respects they resemble the sex difference in animals.

The second hypothesis assumes that gametes of neither complex can yield viable homozygotes. Each complex is "lethal" in combination with an identical complex. In terms of Muller's observations on *Drosophila* (1917, *cf.* 1918), it may be said that each complex of factors is linked with a recessive lethal factor with no crossing-over between them, so that the heterozygote for two complexes is viable and breeds true. The whole system can then be described as depending on a "balanced lethal" mechanism (Fig. 104).

A third assumption can then be made, that dissociation through crossing-over occasionally takes place (*a*) between a complex and its lethal factors, and (*b*) between two parts of a complex. The first of these changes will permit the segregation of complete homozygotes; the second will result in the segregation of plants homozygous for part of the complex. The segregates will be mutants.

Further observations of Renner's (1919, *a* and *b*, 1921) explain differences in the behaviour of different species. He found that the two kinds of pollen grains produced by a heterozygote could often be distinguished by their size and the shape of their starch grains. Often one kind was entirely inactive, or, owing to its smallness, relatively feeble in growth. Again, in some species he found that, instead of the megaspore at the top (micropylar) end of the row of four, formed in maturation, always giving the functional embryo-sac as it does in a homozygous species such as *Œ. Hookeri*, the megaspore at the other end often pushed its way round the upper three and usurped the position of the micropylar megaspore. This competition is known as the " Renner effect." Now, if the first division of the mother-cell is the reduction division which separates the two complexes, it will be seen that this competition between top and bottom cells is in effect a struggle for predominance between the two complexes. Where half the embryo-sacs are formed from the micropylar cell and half from the chalazal cell it is natural to suppose that they are all of one complex type which has passed equally often to the top and bottom pole at the first division. The genetic results agree with these conclusions. *Œ. muricata*, which has half its pollen inactive and half its potential embryo-sacs overridden in competition, produces functional pollen of only one kind, with the *curvans* complex, and functional embryo-sacs almost all of the other kind, with the *rigens* complex. Hence, when this species is crossed with a homozygous species such as *Œ. Hookeri* reciprocal hybrids differ, one being a *curvans* hybrid, the other a *rigens* hybrid (Table 54). Such a species as this can breed true and with scarcely any loss of fertility, although it must be heterozygous in the same degree as many interspecific hybrids.

The assumption of the Renner complex, therefore, enables us to describe the phenomena of inheritance and variation in *Œnothera* in mendelian terms, and as an independent description it is unassailable. But now its mechanism has been analysed by the study of the behaviour of the chromosomes in the hybrid species. This shows its relationship with other systems and helps us in explaining them.

(ii) **Ring-formation in Œnothera.** Cleland (1922 *et sqq.*), first showed that the association of the chromosomes at meiosis has certain constant characters in each species. He found the 14 chromosomes in a diploid *Œnothera* were always associated at diakinesis and metaphase end-to-end in rings of two, four, six or other even numbers, up to fourteen. The size and number of the rings were constant for each species, except that occasionally a chain or two chains might replace a ring; just as with the simple pairs which usually form rings in other organisms with median centro-

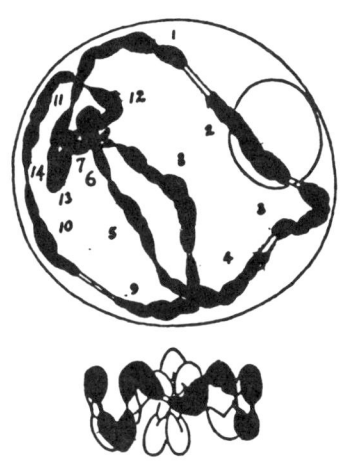

FIG. 105.—First meiotic division in *Œnothera muricata* with a ring of fourteen chromosomes. Above, at diakinesis (showing the self-interlocking of the ring, *cf.* Ch. VII). Below, at first metaphase (showing a disjunctional arrangement which gives regular segregation of the two complexes). × 4000 (from D., 1931 *c*).

meres and complete terminalisation (*cf.* Catcheside, 1933), a rod is formed owing to the failure of one of the two chiasmata.

The same constancy is found in the hybrids and mutants as well, and Tables 54 and 55 show some of the results that have been obtained.

Cleland found that the ring of chromosomes usually arranges itself so that adjoining chromosomes go to opposite poles (Fig. 101). He assumed that the chromosomes had a fixed position in the ring, so that in a ring of fourteen, each of the seven chromosomes of one parental gamete paired at opposite ends with ends of two chromosomes from the other gamete. In this way segregation would yield only two types of gamete, and these two types would be the same as those that had gone to make up the parent zygote. He also assumed that the type of configuration was a specific property of the two complexes of each species or hybrid *in relation to one another*.

This conclusion has been borne out by the extensive observations of recent years. All the complex heterozygote species have multiple chromosome rings. The homozygous forms, whether species or segregates, have seven pairs, or occasionally five pairs and a ring of four.

These observations showed the mechanism by which the segregation of complexes takes place. They showed the consequence of the chromosomes associating in rings, but owing to some confusion at this time in the interpretation of chromosome pairing and of meiosis in general, they did not at once reveal how the formation of rings and the inheritance of complexes came about and what it meant in genetic terms. In the light of later studies of the mechanism of chromosome pairing it became clear, however, that these large multiple rings were derived by normal processes from numerous interchanges, just as the smaller rings described in simpler structural hybrids are derived from one or two interchanges. As a result of six interchanges a ring of fourteen could be built up by association of all the interchanged segments, the formation of chiasmata in them and their later complete terminalisation (D., 1929, 1931 *d*).

Since the chromosomes are associated only at their ends, it is necessary, in considering their pairing properties, to regard them as made up merely of two " segments " which are effective in pairing. In this way a homozygous *Œnothera* with seven pairs of chromosomes at meiosis has a constitution as follows (homologous segments bearing the same letters) :

7(2)   (AB   (CD   (EF   (GH   (KL   (MN   (OP
        AB)   CD)   EF)   GH)   KL)   MN)   OP)

while a complex heterozygote with a ring of 14 chromosomes is made up as follows :

(14) (AB   CD   EF   GH   KL   MN   OP
      \ / \ / \ / \ / \ / \ / \
      BC   DE   FG   HK   LM   NO   PA)

It is supposed that this arrangement of segments arose as a result of successive interchange in the ancestry of the plant, as had been inferred in trisomics in *Datura* (Belling, 1927). Thus an ancestor with simple pairs, such as that given, would by interchange between B and D give new chromosomes BC and DA, which would form a ring of four with the parental type, as follows:

(4) + 5(2) (AB   CD   (EF   (GH   (KL   (MN   (OP
            \ / \
            BC   DA)   EF)   GH)   KL)   MN)   OP)

This hypothesis explains the constancy of the behaviour of the species, both in heredity and in chromosome association, for each of the complexes is made up in a specific and constant arrangement. The dissimilarity of the complexes is correlated with the dissimilarity in the arrangement of the segments in the two kinds of gametes produced by each species. Just as each plant is the result of the fusion of two complexes, so also is it the result of the fusion of gametes differing in respect of the arrangement of segments in such a way as to give the chromosome ring characteristic of the species. It is a structural hybrid.

The hypothesis also explains their behaviour on hybridisation, for Cleland and Blakeslee (1930) have been able to predict from the configurations seen in crosses of two species with a third what configurations would be found in a cross between the first two. Their predictions were verified.

Let us consider the way in which interchange will give rise to ring-formation and complex-inheritance in nature.

Interchange will always lead to the production of interchange heterozygotes in the first generation just as gene mutation leads to the production of mendelian heterozygotes. These will segregate homozygous new and old forms in the next generation unless the interchange is associated with the production of a lethal type. In the latter case a permanent interchange heterozygote will arise directly. In the former case the permanent heterozygote will appear only following hybridisation and after an interval during which changes have taken place that are necessary to make the segregation of homozygotes from the hybrid impossible or difficult.

The steps in this indirect process of origin are shown in *Campanula* (Gairdner and D., 1931 and unpub.) where crosses between structurally different forms give rings of six, eight and ten, which yield a high proportion of their like when selfed. Probably both the direct and the indirect method have played some part in the origin of the complex heterozygote species of *Œnothera*.

TABLE 54

Summary of Ring-forming Properties in Forms of *Œnothera* *

1. SPECIES.

Species	Complex constitution	Configuration
Heterozygotes :—		
Œ. grandiflora	acuens.truncans	(14)
Cockerelli	curtans.elongans	(14)
strigosa	deprimens.stringens	(14)
muricata	curvans.rigens	(14)
suaveolens	flavens.albicans	(12) + (2)
Lamarckiana	velans.gaudens	(12) + (2)
biennis	rubens.albicans	(8) + (6)
Homozygotes :		
Œ. franciscana	ʰfranciscana.ʰfranciscana	7 (2)
Hookeri	ʰHookeri.ʰHookeri	7 (2)
purpurata	ʰpurpurata.ʰpurpurata	7 (2)

2. HYBRIDS BETWEEN HETEROZYGOTES.

Heterozygotes :		
Œ. Lamarckiana × *strigosa*. (and reciprocally)	(i) velans.stringens	(6) + (4) + 2 (2)
	(ii) gaudens.stringens	(14)
	(iii) deprimens.velans	(10) + 2 (2)
	(iv) deprimens.gaudens	(10) + 2 (2)
Œ. Lamarckiana × *suaveolens*. (and reciprocally)	(i) velans.flavens	2 (4) + 3 (2)
	(ii) gaudens.flavens	(12) + (2)
	(iii) albicans.velans	(14)
	(iv) albicans.gaudens	(8) + (6)
Œ. suaveolens × *strigosa*.	(i) flavens.stringens	(4) + 5 (2)
	(ii) albicans.stringens	(12) + (2)
	(iii) deprimens.flavens	(12) + (2)
	(iv) albicans.deprimens	(not obtained)

* From D., 1931 *d*; Catcheside, 1933; Cleland, 1935.

HYBRIDS BETWEEN HYBRIDS

3. Hybrids between Homozygotes and Heterozygotes.

Species	Complex constitution	Configuration
Heterozygotes:		
Œ. muricata × Hookeri.	rigens.ᴴHookeri	$(6) + 4\,(2)$
Œ. Hookeri × muricata.	ᴴHookeri.curvans	$(6) + (8)$

4. Hybrids between Homozygotes.

Heterozygotes:		
Œ. deserens × blandina.	subvelans.ᴴblandina	$(6) + 4\,(2)$
Œ. deserens × purpurata.	subvelans.ᴴpurpurata	$(4) + 5\,(2)$
Œ. blandina × purpurata.	ᴴblandina.ᴴpurpurata	$(4) + 5\,(2)$

5. Derivatives of Hybrids.

Œ. lutescens (ex suaveolens × biennis).	ˡᵗflavens.ˣflavens	$7\,(2)$
Œ. suaveolens × pachycarpa. F_1	—	$(6) + (4) + 2\,(2)$
,, F_2	—	$(6) + 4\,(2)$

6. Mutants (v. infra).

Œ. rubrinervis (ex Lamarckiana).	subvelans.pænevelans	$(6) + 4\,(2)$
Œ. deserens (ex rubrinervis).	subvelans.subvelans	$7\,(2)$
Œ. ochracea (ex grandiflora).	acuens.acuens	$7\,(2)$

(iii) *Diploid Mutations.* Further, the hypothesis explains the origin of the diploid mutations. Most of these have changed chromosome configurations as the result of further segmental interchange. Two kinds of interchange are probably most important (D., 1931 *d*): first, interchange between chromosomes of opposite complexes and, secondly, interchange between chromosomes of the same complex (Fig. 108).

As an example of the first we can consider the so-called "half-mutants" of *Œ. Lamarckiana*. This species is a heterozygote made up of two complexes, *velans* and *gaudens*. It has a ring of twelve and one pair at meiosis. Take its composition to be as follows:

$$\text{velans} \quad (A_vB \quad C_vD \quad E_vF \quad G_vH \quad K_vL \quad M_vN \quad (OP$$
$$\text{gaudens} \quad B_gC \quad D_gE \quad F_gG \quad H_gK \quad L_gM \quad N_gA) \quad OP)$$

(Each chromosome is given the initial letter of its complex in the middle to distinguish it.

The exchange of segment A of A_vB with segment G of F_gG will

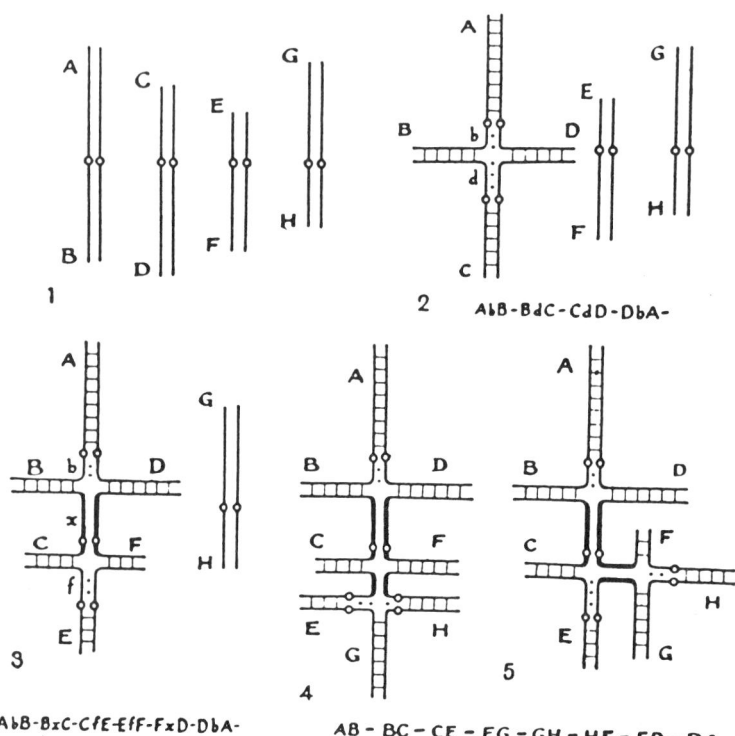

Fig. 106.—Pachytene diagrams of pairing in (1) a homozygote; (2) a simple interchange heterozygote with interstitial segments b and d; (3) a double interchange heterozygote with differential segment x; (4) and (5) triple interchange heterozygotes showing two ways in which the differential segments may develop. (D., 1936 a.)

give two new chromosome types G_vB and F_gA, and a new kind of segregation giving a gamete containing all the necessary 14 segments arranged as follows :

subvelans $B_gC, D_gE, F_gA, G_vH, K_vL, M_vN, OP.$

This will unite with an unchanged *velans* gamete to give a ring of six and four pairs, as follows:

velans $(A_vB$ C_vD E_vF $(G_vH$ $(K_vL$ $(M_vN$ $(OP$

subvelans B_gC D_gE $F_gA)$ $G_vH)$ $K_vL)$ $M_vN)$ $OP)$

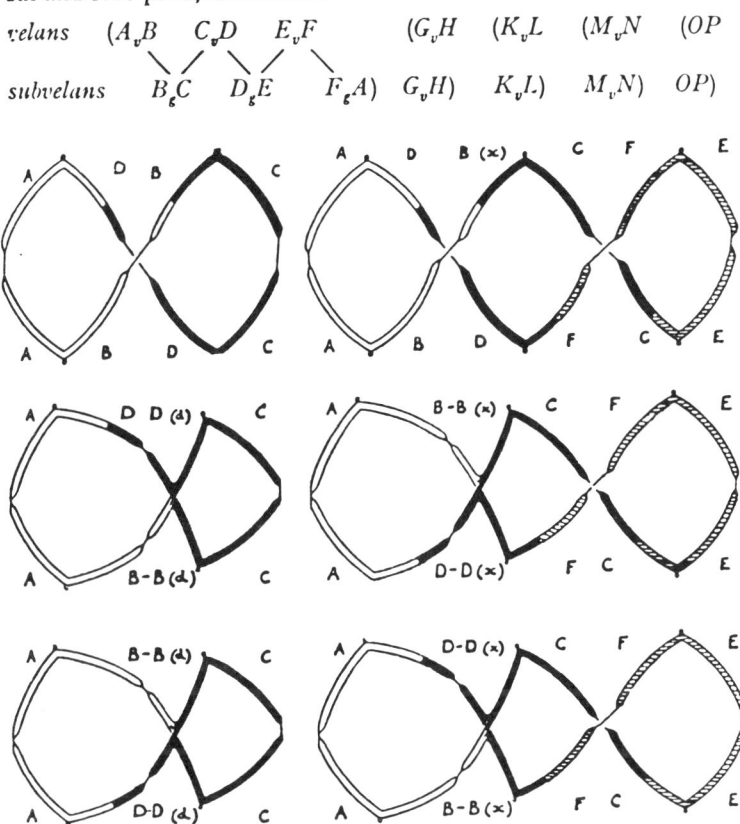

FIG. 107.—Left, three metaphase configurations derived from Fig. 106 (2); chiasma formation in the interstitial segment gives chromatid non-disjunction always. Right, metaphase configurations derived from Fig. 106 (3). Below, chiasma formation in the differential may give chromatid non-disjunction of the cross-over or of the old chromatids. (D., 1936 *a*.)

Such is the configuration found in the so-called "half-mutant" *rubrinervis*, which appears in the selfed progeny of *Œnothera Lamarckiana* once in a thousand seedlings.

Moreover, it will be seen that the mutant, on this hypothesis should consist of two complexes, one unchanged *velans*, the other a mixture of *velans* and *gaudens*. This is in agreement with the conclusions of the geneticist (*cf.* Hoeppener and Renner, 1929) who describes the two as "*pænevelans*" and "*subvelans.*" *Pænevelans* is the *velans* complex derived normally from *Lamarckiana*; the slight difference it shows in later generations from the *velans* of the parental form is probably due to crossing-over in the half-mutant with *subvelans*. This mutant yields one-quarter bad seeds, one-half seedlings like itself with the same chromosome configuration, and one-quarter *subvelans.subvelans*, which is the "full-mutant" *Œ. deserens*. This form, being homozygous, has seven pairs of chromosomes (7(2)) at meiosis. It arises, we may say, because *subvelans* no longer includes the chromosomes with the lethal factors of *velans*. Emerson (1936) has described genetical evidence of the same kind of process in *Œ. pratincola*.

Segmental interchange between chromosomes of the same complex will have a different result, for the seedlings, whose parental gametes have suffered the change, will not differ in appearance from their parents, but their chromosomes will pair differently. Where the parents had a ring of 14 chromosomes the seedlings will have two rings, (12) + (2) or (10) + (4) or (8) + (6). Now the rule that plants with the multiple ring breed true to this character need not apply equally to the small rings derived from it. *Œ. biennis* ((8) + (6)) breeds true because, presumably, each ring is preserved by its lethal system; but where one ring is broken into two by interchange each half will not necessarily have its lethal system. A new ring of four chromosomes in a plant from two gametes, one of them with chromosomes *AB* and *CD* the other with chromosomes *BC* and *DA*, may yield homozygous progeny either of the type *AB–AB*, *CD–CD* or the type *BC–BC*, *DA–DA* if either of these is viable (*i.e.*, has no factors lethal in the homozygous condition). Hence a plant whose parental gametes have undergone interchange between chromosomes of the same complex may be expected to yield progeny homozygous in regard to the members of one of the small derived rings in high proportions by a process which is essentially one of mendelian segregation. Such is a possible explanation of the origin of "mass-mutation" in *Œ. Reynoldsii*

(Kulkarni, 1929; Bartlett, 1915; D., 1931 c), but the chromosome constitution of the mutants has not yet been described.

TABLE 55

The Numbers of Species, Hybrids and Mutants in *Œnothera* having Different Types of Configuration compared with their Frequency if Segments were assorted at Random (after Haldane, in D., 1931 d).

Configuration type.	Relative random frequency (Total Possibilities, ×135,135).	Observed Frequency.				
		Species.	"Mutants" or segregates of species.	"Hybrids"	Derivatives or segregates of hybrids.	Totals.
1. (14)	23,040	12	—	15	—	27
2. (12) + (2)	13,040	5	—	16	—	21
3. (10) + (4)	8,064	—	—	5	—	5
4. (8) + (6)	6,720	2	—	2	—	4
5. (8) + (4)+(2)	5,040	—	—	1	—	1
6. (10) + 2(2)	4,032	—	—	11	—	11
7. 2(6) + (2)	2,240	—	—	—	—	0
8. (6) + 2(4)	1,680	—	—	1	—	1
9. (6) + (4)+2(2)	1,680	—	—	3	1	4
10. (8) + 3(2)	840	—	4	1	1	6
11. 3(4) + (2)	420	—	—	—	—	0
12. 2(4) + 3(2)	210	—	—	3	—	3
13. (6) + 4(2)	140	—	3	5	2	10
14. (4) + 5(2)	21	1	1	8	—	10
15. 7(2)	1	4	6	3	4	17
Totals	67,568	24	14	74	8	120

(iv) **The Differential Segments.** In order to examine these assumptions, it is necessary first to consider the conditions under which the chromosomes come to associate in rings during the prophase of meiosis in *Œnothera*. Recent evidence (Catcheside, 1931, 1932) shows that they are the same as in other interchange heterozygotes (Ch. V). Segments pair with their homologues so far as they correspond; where the correspondence ceases they change partners (McClintock in *Zea*, 1931; *cf*. D., 1931 d). Chiasmata are formed in the paired segments, and these are terminalised (*cf.*

Gairdner and D., 1931), in *Campanula persicifolia*. In polyploid *Œnothera* there are therefore multiple chiasmata, as in *Tradescantia*, *Rosa*, and *Primula* (Catcheside, 1933).

The question now arises, if chiasmata are formed, and crossing-over therefore occurs between each pair of segments : where are the differences situated which constitute the complexes and between which no crossing-over, and therefore no pairing, occurs ? Clearly they must lie in the neighbourhood of the centromere, for all association must take place distally to them if they are not first to arrest terminalisation and, secondly, to cross-over to opposite chromosomes, in a word there must be proximal *differential segments* (D., 1931).

It has now been found cytologically that whenever a pair of chromosomes suffers two interchanges these never coincide with one another or with the centromere (Ch. V). The first result of this property is that in a ring of four chromosomes there will necessarily be an *interstitial segment* lying between the point of interchange and the centromere in each chromosome. In such a ring crossing-over can occur between the interstitial segments (Fig. 107, Sutton, 1936). But when it occurs half the chromatids produced, with either of the possible segregations of the four chromatids, will be non-disjunctional. There will be what Sansome (1933) describes as "chromatid non-disjunction." This disadvantage, together with the failure of normal pairing seen near the change of partner at pachytene by McClintock (1933), will reduce the amount of crossing-over, *i.e.*, strengthen the linkage of genes in these interstitial segments with the centromeres which pass to opposite poles in the two pairs of chromosomes.

When rings larger than four arise from successive interchanges another special segmental property will appear : differential segments in two chromosomes will lie between terminal pairing segments which do not continue their homologies (D., 1931, 1933 ; Sansome, 1932). These segments may include the centromeres or not (Fig. 106). So long as the type is preserved, crossing-over is suppressed in these segments. Any differences arising in differential segments in opposite sets, whether by gene mutation or by further structural change, will therefore belong to a complex inherited as a unit

Any crossing-over between differential segments will result in *reverse interchange* of the kind found in Œnothera and *Pisum* and hence in the mutations described.

Cytological observations on *Œnothera* show that the segmental interchange which is responsible for regular mutation occurs by crossing-over in this way. Ring-forming plants occasionally have chiasmata formed interstitially between chromosomes which do not associate terminally in the ring. These chiasmata remain interstitial and are seen at metaphase. Evidently there are differential segments in the middle of the ring-forming chromosomes which do not follow the homologies of the ends (Fig. 108). Crossing-over between them at the chiasmata gives new segmental interchange as required by the theory of the origin of half-mutants set out above. Like all secondary structural changes, being due to crossing-over, it occurs at a specific place with a specific frequency and with specific results.

We now see that the existence of three kinds of segments, pairing segments, interstitial segments and differential segments in large rings expected *a priori* from simple interchange and established by direct cytological study accounts for (i) the existence of complexes located in the differential segments; (ii) the mutation resulting from exceptional crossing-over between the differential segments (D., 1936 *a*); (iii) the occurrence of rare crossing-over between genes characteristically associated with them (Renner, 1933) and lying in the interstitial segments or in the pairing segments very close to them. Diploid mutations not involving segmental interchange are due to this last type of crossing-over. This may take one of two forms. Either particular genes in respect of which the species has been heterozygous may cross over exceptionally and yield a homozygous segregate. Or the genes which cross over may be lethal genes and their allelomorphs. In this case one of the two complexes of a heterozygote emerges in a homozygous form. Thus the mutant *ochracea*, 7 (2), appears spontaneously from Œ. *grandiflora*, (14), (De Vries, 1918), and the mutant *lutescens* after hybridisation from Œ. *suaveolens* (Renner, 1927). Both can be understood as a result of exceptional crossing-over in the interstitial segments, so that lethal combinations are replaced by non-lethal.

Chain of Fourteen Chromosomes in Oenothera.

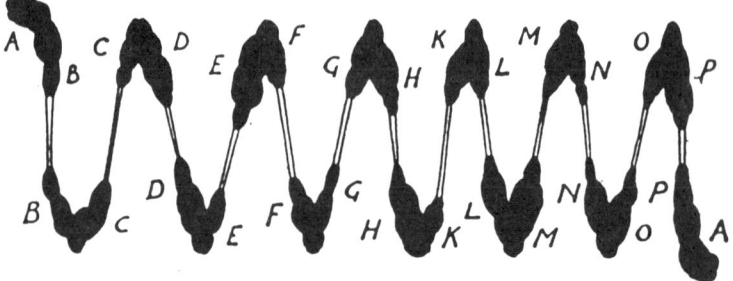

The Occurrence of Segmental Interchange between:—

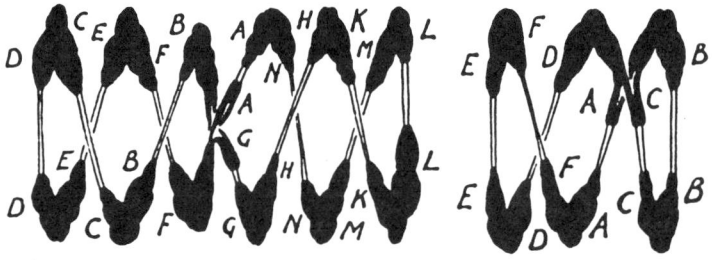

(i) *Non-Corresponding Segments of Opposite Complexes (to give half-mutants in Oe. Lamarckiana.)*

(ii) *Corresponding Segments of the Same Complex. (to give mass mutants in Oe. Reynoldsii)*

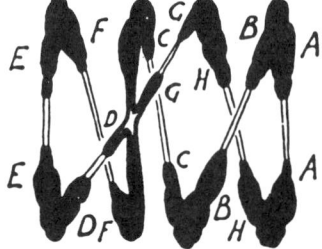
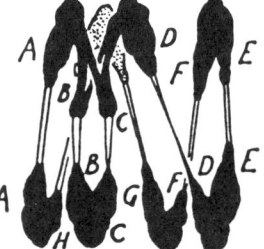

(iii) *Corresponding Segments of Opposite Complexes (new combination non-disjunctional.)*

(iv) *Non-Corresponding Segments of the Same Complex (to give change of pairing type in Oe. pratincola.)*

FIG. 108.—Diagram showing normal and anomalous pairing of chromosomes seen and inferred in complex-heterozygote Œnothera with rings of various sizes. Segmental interchange is due to crossing-over in differential segments. The first two types will yield mutant gametes. The third type will give no viable mutant and the fourth type will give a gamete having the same properties *vis-à-vis* its normal partner but changed properties *vis-à-vis* others. (*Cf.* D., 1931 *d.*)

This occurs in *Œ. grandiflora* normally ; in *Œ. suaveolens* only when the *flavens* complex is associated with a different complex in the hybrid with *Œ. biennis*. Such a difference in behaviour depending on a difference in partner is to be expected if crossing-over is the basis of these mutations, for crossing-over is conditioned by the presence of a partner for the region in which it occurs, and what are interstitial segments in one combination may be differential segments in another.

(v) **Trisomic Mutation.** The mutations involving changes in chromosome number are to be explained as the result of irregularities in chromosome distribution at meiosis, such as those occurring in other organisms. The fact that they occur in much greater numbers than in species with simple pairing is due to the mechanical conditions of separation in a ring. Normally a ring in *Œnothera* segregates " disjunctionally," *i.e.*, so that the chromosomes that are paired pass to opposite poles. It will be seen that if this orientation takes place almost simultaneously in different parts of the ring, sometimes chromosomes separated by an even number will come to lie to one side of the equator and two of the intervening chromosomes must then pass to the same pole, or one be left on the equator to divide equationally at the first division. These occurrences have been observed both in *Œnothera* (Cleland, 1926 *et al.* ; D., 1931 *d*) and in *Rhœo* (D., 1929 *c*) where a ring of 12 chromosomes is formed and conditions of segregation can be more easily studied than in *Œnothera*. Irregularities must necessarily occur in the intervening chromosomes in *both* directions round the ring, so that non-disjunction is always double. If the ring is replaced by a chain the non-disjunctional result occurs, although the two chromosomes which pass to the same pole may not have been joined, but only have been capable of having been joined. Where the non-disjunctions occur on the same side of the equator two gametes with eight and six chromosomes are formed instead of seven and seven. The high frequency of this occurrence is responsible for the high frequency of trisomics (1·5 per cent. in *Œ. Lamarckiana*). The lagging of chromosomes of undecided orientation is probably responsible for the failure of separation of two daughter-nuclei and the formation of unreduced gametes

(Schwemmle, 1928; D., 1931 d). From these arise occasional tetraploid and triploid seedlings, the *gigas* and *semi-gigas* mutants of De Vries.

It was formerly assumed that the trisomic mutations of *Œnothera* were similar to those of non-hybrid plants, and attempts were made by de Vries and others to reduce them to the seven expected types. This was not satisfactory, and their breeding behaviour was anomalous. Some, for example, bred true. The reasons have been shown by Catcheside (1933 b and 1936) and Emerson (1935, 1936 b). Thus in *Œ. Lamarckiana* an extra chromosome from the ring of twelve can provide a viable combination in company not merely with the six *gaudens* chromosomes or with the six *velans* chromosomes but also with a number of mixtures of the two sets. These mixtures arise from the different possible non-disjunctional arrangements. The arrangements giving rise to functional eight-chromosome gametes are of two main kinds ((1) and (2) below) ; the first gives whole complexes of the parental types, the second, new complexes formed by various mixtures of the old ones. One example of each will suffice (neglecting the OP pair) :—

1. $A_vB \quad C_vD-D_gE-E_vF \quad G_vH \quad K_vL \quad M_vN$ (8-gamete)
   $\quad B_gC \quad\quad\quad\quad\quad\quad F_gG \quad H_gK \quad L_gM \quad N_gA$ (6-gamete)

The 8-gamete will give with *gaudens* a zygote having the normal *Lamarckiana* complement together with an extra chromosome, D_gE :—

(i) $A_vB \quad C_vD \begin{pmatrix} D_gE \\ D_gE \end{pmatrix} E_vF \quad G_vH \quad K_vL \quad M_vN$
    $\quad B_gC \quad\quad\quad\quad\quad F_gG \quad H_gK \quad L_gM \quad N_gA$

With *velans* the 8-gamete will give a homozygote provided the D_gE chromosome covers the lethals of *velans* :—

(ii) $\begin{pmatrix} A_vB \\ A_vB \end{pmatrix} \begin{pmatrix} C_vD \\ C_vD \end{pmatrix} D_gE \begin{pmatrix} E_vF \\ E_vF \end{pmatrix} \begin{pmatrix} G_vH \\ G_vH \end{pmatrix} \begin{pmatrix} K_vL \\ K_vL \end{pmatrix} \begin{pmatrix} M_vN \\ M_vN \end{pmatrix}$

In the second type the non-disjunction occurs at two different places :—

2. $A_vB \quad C_vD-D_gE \quad F_gG \quad H_gK-K_vL \quad M_vN$ (8-gamete)
   $\quad B_gC \quad\quad\quad E_vF \quad G_vH \quad\quad\quad L_gM \quad N_gA$ (6-gamete)

The 8-gamete will give with *gaudens* a partly homozygous and partly heterozygous trisomic :—

(iii) $\begin{matrix} A_vB & & C_vD \\ \searrow & & \swarrow \\ & B_gC & \end{matrix}$ $\begin{pmatrix} D_gE \\ D_gE \end{pmatrix}$ $\begin{pmatrix} F_gG \\ F_gG \end{pmatrix}$ $\begin{pmatrix} H_gK \\ H_gK \end{pmatrix}$ $\begin{matrix} K_vL & & M_vN \\ \searrow & & \searrow \\ & L_gM & & N_gA \end{matrix}$

With *velans* it will give a different but analogous result :—

(iv) $\begin{pmatrix} A_vB \\ A_vB \end{pmatrix}$ $\begin{pmatrix} C_vD \\ C_vD \end{pmatrix}$ $\begin{matrix} D_gE & & F_gG & & H_gK \\ \searrow & \swarrow & & \searrow & \swarrow \\ & E_vF & & & G_vH \end{matrix}$ $\begin{pmatrix} K_vL \\ K_vL \end{pmatrix}$ $\begin{pmatrix} M_vN \\ M_vN \end{pmatrix}$

With the first type of 8-gamete any one of the twelve chromosomes in the ring can be present in addition to the six (and O P) of the opposite complex. In this way twelve kinds of 8-chromosome gamete should be formed and these might combine with either *velans* or *gaudens* normal gametes and thus give twenty-four potential types of trisomic. With the second type the number of gametic combinations is twenty-four, and these will give forty-eight potential trisomic types. Including the trisomic derived from the bivalent $(OP - OP)$, we thus have seventy-three types in all. Of these a large proportion will be homozygous for bivalent forming chromosomes which in the diploid parent species are heterozygous and within the ring.

A large number of these potential types of genetically different trisomics have actually been identified by Catcheside and Emerson in *Œ. Lamarckiana*; the proportion of the seventy-three which succeed depends of course on the viability of the combinations of the new types of gamete with normal *velans* and *gaudens*, *i.e.*, on the distribution of differential segments (or lethal genes) amongst the chromosomes.

Some of the trisomics, as we saw, have the remarkable property of being effectively true-breeding (Nos. (ii), (iii) and (iv)). Type (ii) is important. It will produce only two kinds of functional gametes, one with a normal complex which is non-viable in the homozygous state and one which has an extra chromosome and does not function in the pollen. This trisomic type is homozygous for a particular complex which exists only in the heterozygous state in the diploid species from which the trisomic is derived. On a specific lethal gene hypothesis it would be assumed that the particular extra chromosome from the opposite complex was covering, as it were, the recessive lethal of the homozygous complex. But this is not the case. Different chromosomes are equally effective. Catcheside, therefore, concludes that the lethal effect is not the property of specific genes, but is due to a general unbalance which may be rectified by different extra chromosomes. In considering the half-mutants of *Œ. Lamarckiana*, we must now say, therefore, not that *subvelans* has lost the lethal genes of *velans* and *gaudens*, but that it is a viable combination of parts of these two complexes.

The lethal properties of a complex are therefore due to unbalance, to an exchange of materials between the two complexes of such a kind that each becomes unsatisfactory by itself, while both together remain a working combination. Such a condition can arise by the differential segments becoming altered by the loss, gain or interchange of parts. Although other changes no doubt occur, interchange between chromosomes inside the ring will by itself lead to such a differentiation and will thus account for all the genetical properties of complex heterozygotes (D., 1933 a).

3. SEX HETEROZYGOTES

(i) **Types of Sex Chromosomes.** The difference between the sexes is inherited, as a rule, as a single alternative—male or female and no intermediate. When it is determined by genetic segregation it must therefore depend on a single gene difference or on a group of differences acting as a unit in inheritance. Such an inheritance of sex is often bound up with the inheritance of "sex-linked" characters, which also act as a unit in the heterozygote. It is then clear that the whole block must be determined by differences between one pair of chromosomes or between chromosomes in a configuration of the *Œnothera* type. This is found to be the case in all organisms with unitary genotypic sex determination. The chromosomes concerned are known as *sex chromosomes*. Other pairs of chromosomes whose differences bear no relation to the differentiation of sex are known as *autosomes*.

In the diœcious Bryophyta the diploid sporophyte produces two kinds of haploid spores, those giving gametophytes with only male organs and gametes, and those giving gametophytes with only female organs and gametes. The sporophyte, which itself has no sex differentiation, is therefore the sex heterozygote. In species of *Sphærocarpus* there are a large X chromosome and a very small Y chromosome which pair at meiosis in the diploid sporophytes by a terminal chiasma, and pass to opposite poles. The four spores of the tetrad formed stick together until germination. They produce two female gametophytes with X chromosomes and two male gametophytes with Y chromosomes. Sexual differentiation

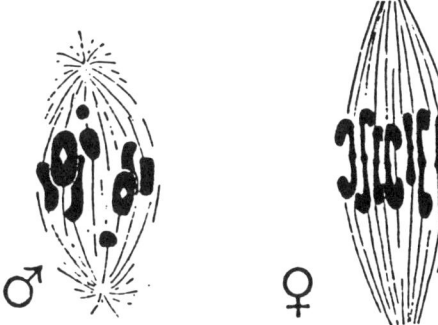

FIG. 109.—First metaphase in the two sexes of *Macronemurus appendiculatus*. XX not recognisable in the female, which shows lower spiralisation and terminalisation. X and Y lying unpaired on opposite sides of the plate in the male. × 1600. (Naville and de Beaumont, 1933.)

is therefore determined by the segregation of these X and Y chromosomes at meiosis (Allen, 1919, 1935 *b*).

In the higher animals and plants it is in the first instance the diploid generation which is sexually differentiated. We then find, as a rule, that one sex is heterozygous, with an XY pair of chromosomes, while the other is homozygous, having a pair of similar chromosomes XX. The convention, it will be noted, is necessarily different. In the Bryophyta, X and Y are associated with the female and male sexes respectively. They have exactly corresponding life cycles. In the higher plants and animals they have not. X meets an identical mate in a sex homozygote (which does not occur in the Bryophyta). Y, on the other hand, never pairs with any chromosome but X, and therefore resembles in its genetic history X or Y in the Bryophyta. The significance of this limitation we shall see later.

The sex heterozygote in the higher organisms is usually the male. Only in the Lepidoptera, Trichoptera, the birds and some fishes, amongst animals, and in *Fragaria* among plants is the female known to be the heterozygote.

The types of differences between X and Y chromosomes, as seen at mitosis, may be classified as follows:—

1. Organisms in which the X and Y chromosomes are not structurally distinguishable (*e.g.*, *Oncopeltus*, Wilson, 1912; *Drosophila Willistoni*, Metz, 1926) or scarcely so (*e.g.*, *Nezara*,

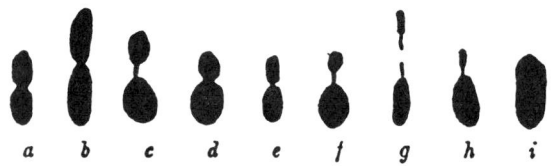

FIG. 110.—The X chromosome (below) and the Y (above) paired end-to-end at meiosis (first or second metaphase) in the heterozygous sex in various insects (after Wilson and Stevens) showing the transition from equality in size to the disappearance of the Y. *a*, *Oncopeltus fasciatus*. *b*, *Nezara hilaris*. *c*, *Lygæus bicrucis*. *d*, *Euschistus fisilis*. *e*, *Thyanta custator*. *f*, *Lygæus turcicus*. *g*, *Nezara viridula*. *h*, *Trirhabda*. *i*, *Protenor belfragei*. (From Witschi, 1929.)

Wilson, 1905, 1911). In *Drosophila funebris* two similar chromosomes are diagnosed as sex chromosomes by Heitz (1933) merely by one chromosome having a larger heterochromatic segment than its partner in the male.

2. Organisms in which the X and Y chromosomes differ more or less considerably. Every gradation is found between the slight difference in *Drosophila melanogaster* (Fig. 121) and the considerable difference in *Sphærocarpus Donellii* (Lorbeer, 1927, 1930), and in man (Evans and Swezy, 1929). The X chromosome may be the largest member of the complement (as in *Sphærocarpus Donellii*; *Leptophyes*, Mohr, 1915; *Phragmatobia fuliginosa*, *Gryllus campestris*, Ohmachi, 1929 *a*, and *Humulus japonicus*) or the smallest (as in *Asilus notatus*, Metz and Nonidez, 1923, and *Macropus*, Agar,

PLATE X

FIGS. 1–5.—Meiosis in *Phragmatobia fuliginosa*, Lepidoptera. (From Seiler, 1925.)
Fig. 1. Fragmented race, male, $n = 29$. Fig. 3. Fused race, male, $n = 28$. Figs. 2, 4 and 5. Hybrid between the two races, male, showing pairing of two small chromosomes with one large one.

FIG. 6.—First metaphase in *Melandrium album*, male: pairing of X and Y. (From Belar, 1925.) × *ca.* 1000.

FIG. 7.—First metaphase in *Zea Mays*, monosomic shoot of a diploid plant, showing 9 bivalents and 1 univalent. × 650. (From McClintock, 1929 *b*.)

FIGS. 8 AND 9.—Polymitosis in *Zea Mays* pollen grain.
Fig. 8. Late prophase. Fig. 9. Metaphase. Two chromosome pairs, probably with interstitial chiasmata in each nucleus. (From Beadle, 1931.) × 650.

PLATE X

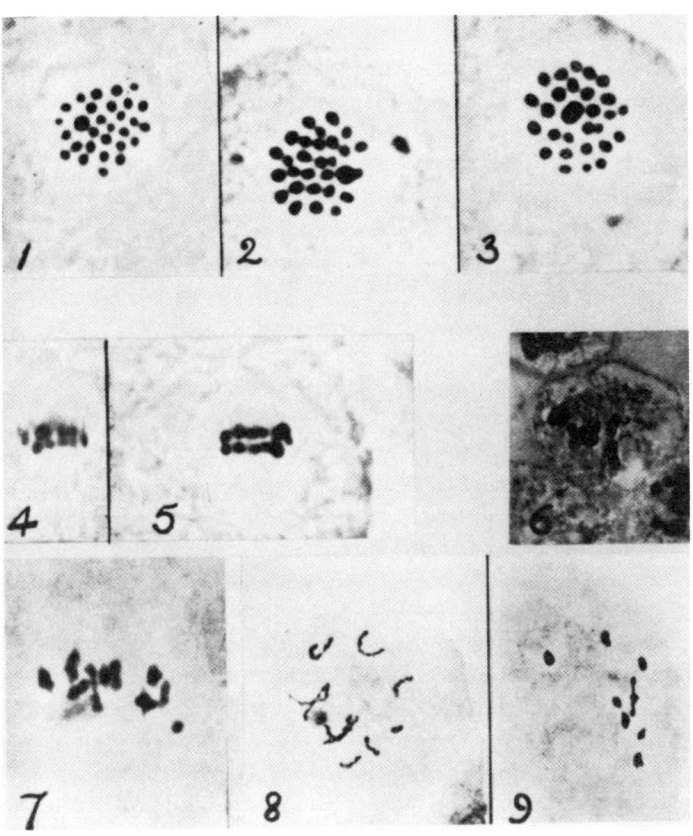

[To face p. 358

1923). The X chromosome is usually larger than the Y, but is occasionally smaller (as in *Drosophila melanogaster*).

3. Organisms in which the Y chromosome is represented by two

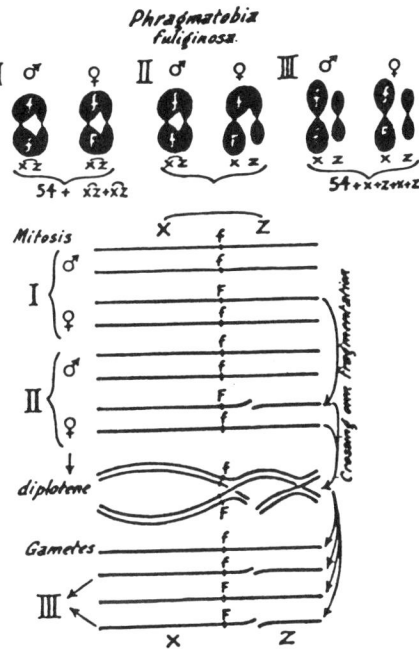

Fig. 111.—Diagram showing how fragmentation in a chromosome carrying sex factors makes the sex chromosomes distinguishable in the heterozygous sex and how crossing-over between the point of breakage and the sex-factors destroys the distinction. I, the unfragmented type with $54 + X + Y$ chromosomes in the female. II, breakage of the Y chromosome into two components, "X" and "Z," "X" containing the F factor or factors. Crossing-over in this female in the X and Y chromosomes between the breakage and the F factor will give the diplotene configuration shown and yield males with a fragmented X chromosome, *i.e.*, type III. The female is the sex heterozygote in the Lepidoptera. *N.B.* The change cannot be simple fragmentation, since it leads to reduplication of the centromere.

or more fragments. The mode of origin of this type from the first is made clear by Seiler's work on *Phragmatobia fuliginosa* (1925).

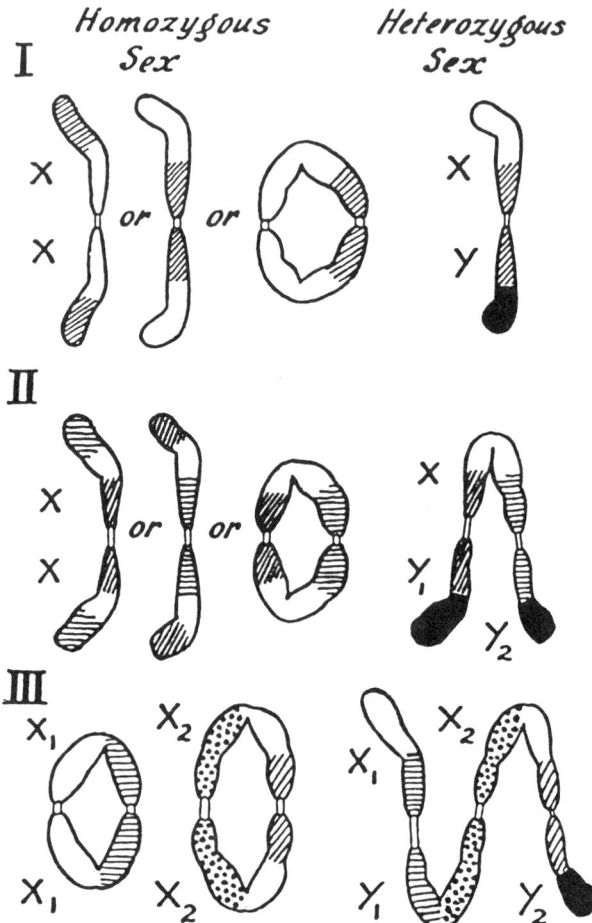

Fig. 112.—Pairing of chromosomes in the two sexes showing parts of X and Y that are homologous similarly hatched and stippled, parts of X that are not represented in Y blank, and parts of Y not in X black. I, *Humulus lupulus*, simple type. II, *H. japonicus*, simple type (fragmentation heterozygote). III, *H. lupulus*, with male as interchange heterozygote having X_1, Y_1, X_2 and Y_2 chromosomes, corresponding to segments BXA, AQ, QP, PY (cf. text). Note.—Types I and II are common in plants and animals. The centromere being median in the X_1 the two X chromosomes in the homozygous sex can form chiasmata in one or both arms to give a rod or a ring. From type II the chain of five is derived as the chain of four is derived from type I.

INTERCHANGE WITH SEX CHROMOSOMES

Two geographic races exist, one of which has two pairs of small chromosomes instead of one large pair found in the other. A third race is found in which the males have the single large pair while the females have one large chromosome associated with two small ones (v. Ch. V). Evidently therefore the sex factors are located in this group. Although the third race might have arisen as a cross from the first two it may equally be regarded as an intermediate stage in the change from the one type to the other, for, in the absence of crossing-over, fragmentation in the Y of the large chromosome type will lead to the intermediate type. Only if crossing-over occurs between the sex factors and the point of breakage will the intermediate type yield the small chromosome type. Evidently such crossing-over has taken place. This example therefore provides a diagrammatic illustration of the relationship between crossing-over and the preservation of differences between sex chromosomes (v. Fig. 111).

In a number of other animals and plants the sex chromosome pair is heterozygous for " fragmentation " of either the X or the Y, giving X_1X_2Y or XY_1Y_2 systems of segregation (Table 56 and Figs. 115 and 117).

4. In *Humulus japonicus* (Kihara, 1929 b) and *H. lupulus* (Sinoto, 1929) a further structural change has occurred in the extension of the differences in the sex-heterozygote to another pair of chromosomes, while the homozygous sex presumably continues to have simple pairs (Fig. 112). A race of *Humulus lupulus* has a chain of four instead of one unequal pair and an individual of *H. japonicus* has been found with a chain of five instead of a chain of three. Each of these modifications can be supposed to be derived from the simpler type by segmental interchange between sex chromosomes and autosomes in the same way as a ring of four is presumed to be derived from two pairs in ordinary interchange heterozygotes. Thus, if the three sex chromosomes of the simpler form of *Humulus japonicus* are composed of segments (homologues having the same letters):

$$YA-AXB-B$$

then segmental interchange between YA and an autosome PQ, giving two new chromosome types YP and QA, will form a chain of five in the interchange-cum-sex-heterozygote, thus:

$$YP\text{–}PQ\text{–}QA\text{–}AXB\text{–}B;$$

this will yield two types of gametes YP, QA, B and PQ, AXB. The

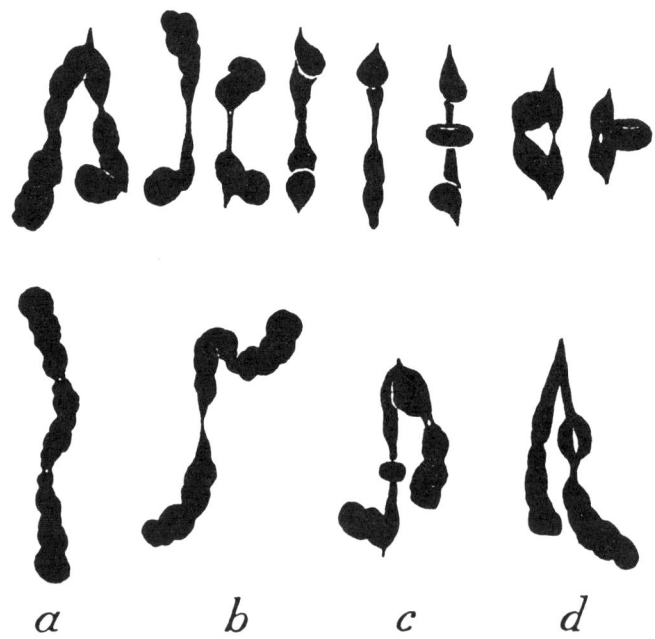

Fig. 113.—Above: first metaphase in ♂ *Humulus japonicus* XY_1Y_2 at left. Below: stages in co-orientation of XY_1Y_2, some showing exceptional interstitial chiasmata. × 5000.

homozygous sex will be unaltered by the change, still having the composition:

$$PQ\text{–}QP, AXB\text{–}BXA.$$

The chain of four in *H. lupulus* differs from this in their being no "B" chromosome.

The sex heterozygote is therefore, apart from other differences, an interchange heterozygote, and the mechanism combines the usual sex mechanism with that of the complex heterozygotes in *Œnothera*. This second instance of the preservation of a demonstrable form of structural hybridity by the sex heterozygote shows the essential similarity between the complex and sex heterozygote.

5. Even more complex conditions are found in *Blaps lusitanica* (Nonidez, 1921) and in various Hemiptera and Nematoda. In these the X or Y chromosome is represented by several separate bodies in somatic cells. They unite at meiosis, and pair with the opposite sex chromosomes. It would seem that these are apparently inert supernumerary chromosomes derived by fragmentation from X or Y and therefore continuing to pair with their homologues (Table 16).

6. The variation which occurs in the relative size of the X and Y chromosomes might be supposed to have two extremes, the complete disappearance of one or the other. But the X chromosome never disappears in organisms with the $XX - XY$ system. That is to say, the distinction between the sexes never consists in the presence of a single supernumerary chromosome in one sex which is not represented in the other. The opposite extreme, in which the Y chromosome has disappeared, is very common in animals, *e.g.*, Orthoptera, Hemiptera, Coleoptera and Nematoda. The sex chromosomes are said to be of the XO type. The random segregation of the single X chromosome gives two classes of gametes with and without X. These can be distinguished in the living sperm of the nematode *Ancyracanthus* (Mulsow, 1912), and the direct connection therefore established between segregation and sex determination. The X chromosome (as in the XY type) may have several independent components (*e.g., Perla marginata* and various Neuroptera). These may fuse temporarily at meiosis (*e.g.*, Nematoda and Aphides, Morgan, 1912). In *Perla* they fuse at diakinesis, but are independent at metaphase. They nevertheless pass to the same pole. The two poles must therefore be in some way asymmetrical in their development.

7. This last type may give a secondary XY mechanism owing to the fusion of the X chromosome with an autosome, giving an unequal pair at meiosis (de Sinéty, 1901 ; McClung, 1905, 1914, 1917 in Orthoptera). The origin of this type is betrayed by its

prophase behaviour. The attached X chromosome condenses in advance of the autosomes, and is therefore distinguishable from them until metaphase. Some species are uniform in having the attachment of the X (*e.g.*, in *Leptynia* and *Mermiria*) others have races with the free X and races with the attached X (*e.g.*, in *Menexenus* and *Hesperotettix*, *cf.* Fig. 46).

Such a fusion has probably taken place in *Ascaris megalocephala* (Edwards, 1910; Geinitz, 1915) for male individuals occur with a single X chromosome separate instead of included in the ordinary compound chromosome. In the male *Macropus* (Agar, 1923) the special behaviour of the two elements which condense precociously at meiosis shows them to be X and Y chromosomes. But the X chromosome is permanently fused with an autosome in both sexes. A configuration consisting of three unequal chromosomes is formed at meiosis and segregates like that in *Tenodera* and *Rumex*.

The apparent fusion is probably as elsewhere an interchange, in this case an interchange of the active part of an autosome for the inert part of the X chromosome; one product of interchange being entirely inert can then be lost. When it is not inert it is not lost and a chain of four, as in *Humulus Lupulus*, will arise, instead of a chain of three.

That changes of this type have been effective in *Drosophila* is indicated by a comparison of linkage maps and chromosomes in *D. melanogaster* and *D. Willistoni* (Lancefield and Metz, 1922). The sex chromosomes of the latter species are V-shaped, while one pair of autosomes are rod-shaped. The reverse is the case in *D. melanogaster* (Fig. 121). One arm of the X chromosome of *D. Willistoni* corresponds with the X chromosome of *D. melanogaster* in the linked factors located in it (*cf.* Koller, 1932 *b* on *D. "obscura,"* and Dobzhansky, 1934).

8. Hagerup's observation (1927) of two XY pairs in the tetraploid *Empetrum hermaphroditum* (corresponding with those in the diploid relative) is improbable on genetical grounds, while XX and YY pairs as found in trisomic *Rumex* (*v. infra*), are more probable on cytological grounds on account of differential affinity. Differentiation of sex chromosomes has not therefore been proved in any polyploid species but *Fragaria elatior*. Where it arises it will not

appear as a derivative of a diploid system, for a new polyploid will maintain itself sexually only if it is hermaphrodite; and it will be likely to breed true to hermaphroditism. There is nothing, however, to prevent differentiation arising *de novo* in an established polyploid, as it doubtless has arisen in *Fragaria*.

(ii) **The Differential Segment.** The dissimilar X and Y chromosomes pair at meiosis in the heterozygous sex. They diverge from

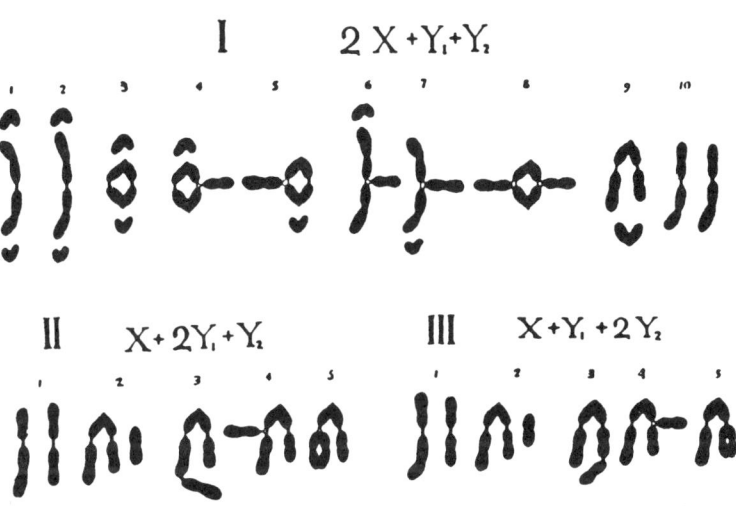

FIG. 114.—Methods of metaphase pairing in *Rumex acetosa* with extra sex chromosomes of each of the three types, X, Y_1 (equal in size) and Y_2 (the smallest). The limits of association define the homologies of the chromosomes and the extent of their differential segments. (Yamamoto, 1934.)

the autosomes in their behaviour at this time in varying degrees. This divergence can be shown to be related to the development of differential segments between which, as in *Œnothera*, pairing at pachytene, crossing-over and chiasma-formation are suppressed.

The evidence of differential segments is of three kinds. First the X and Y chromosomes may differ in size. It is then clear that the part of one which is in excess must be differential with respect to the other. The difference between the two may depend, as in the Mammalia, on no more than a gain or a loss. Secondly, the X and Y chromosomes may regularly have complete terminalisation, although the autosomes of comparable size retain interstitial chiasmata at metaphase. Interstitial chiasmata have been found in *Pellia Neesiana* (Tatuno, 1936) and in *Humulus* (Fig. 113). They are rare in *Rumex* where the process of terminalisation has been described (Sato and Sinoto, 1935). Its completeness shows that the pairing segments have been terminal and short, a part of the chromosomes being differential. Thirdly, the X and Y chromosomes may regularly fail to form a chiasma in one arm, which must then be differential, for when two X's are present in the female or exceptionally two Y's in the male, chiasmata are formed in both arms (Yamamoto, 1934, Fig. 114). If chiasmata fail in both arms the whole chromosome is differential and the pairing segments may be said to have disappeared. This extreme development is found in many Hemiptera, *e.g.*, *Lygæus*, *Oncopeltus*, etc. (Wilson, 1909, 1912), where the X and Y chromosomes may or may not meet at pachytene. In others (*Brochymene*, Wilson, 1905 a) they regularly meet. If they meet they fall apart again at diplotene (*i.e.*, they form no chiasma), divide equationally at the first division, and segregate after momentary touching of their daughter halves end-to-end at the second division. This extreme of differentiation is found in *Oncopeltus*, where no size difference is observed between the chromosomes, thus showing that mere loss and gain is only one of the kinds of structural change that may arise to distinguish the sex chromosomes.

TABLE 56

Representative Examples of Sex Chromosomes

PLANTS.

Hepaticæ (undifferentiated sporophyte).

XY	*Sphærocarpos Donnellii*	Allen, 1919 ; Lorbeer, 1927, 1930.
XY	*S. texanus*	Tinney, 1935.
XY	*Pogonatum inflexum*	Shimotomai and Koyama, 1932
XY	*Pallavicinia longispina*	Tatuno, 1933.
XY_1Y_2	*Frullania* spp.	Lorbeer, 1934 (*cf.* Tatuno, 1936).
XY	17 other species	Lorbeer, 1934.

TYPES OF SEX CHROMOSOMES

Dicotyledones.
$XY\male$	Melandrium spp.	.	Winge, 1923 b ; Belar, 1925 ; Blackburn, 1924, 1928.
$XY\male$	Humulus lupulus [3]	.	Winge, 1929.
$XYA_1A_2\male$	Humulus lupulus [2]		Sinoto, 1929.
$XY_1Y_2\male$	Humulus japonicus [3]	.	Kihara and Hirayoshi, 1932.
$XY_1Y_2A_1A_2\male$	Humulus japonicus	.	Kihara, 1929 a and b.
$XY_1Y_2\male$	Rumex acetosa [3] (and related species).		Kihara and Ono, 1925, 1928 ; Sato and Sinoto, 1935 ; Ono, 1935; Yamamoto, 1934.
$XY\male$	Sedum Rhodiola	.	Levan, 1933.
$XY\female$	Fragaria elatior [2]	.	Kihara, 1930.

ANIMALS.

Nematoda.
$XO\male$	Ancyracanthus	.	Mulsow, 1912.

Orthoptera.
$XY\male$	Œcanthus longicauda [3] [4]		Makino, 1932.
$X_1X_2Y\male$	Tenodera	. .	Oguma, 1921.
XY and $XO\male$	Gryllotalpa [3]	.	De Winiwarter, 1927 Barrigozzi, 1933.
$XO\male$	Phrynotettix magnus	.	Wenrich, 1916.
$XO\male$	Chorthippus and Stenobothrus.		Janssens, 1924 ; Belar, 1929 a ; D., 1932 e, 1936 b.
$XO\male$	Locusta spp.	.	Mohr, 1916 ; White, 1932.

Dermaptera.
$X_1X_2Y\male$	Forficula	. .	W. P. Morgan, 1928.

Odonata.
$XO\male$	Pantala et alii [4]	.	Asana and Makino, 1935.

Plecoptera.
$X_1X_2O\male$	Perla marginata	.	Junker, 1923.

Neuroptera.
$XY\male$	Chrysopa vulgaris [1] [3]	.	Naville and de Beaumont, 1933.

Mecoptera.
$XO\male$	Panorpa	. .	Naville and de Beaumont, 1934.

Coleoptera.
$XY\male$	Tenebrio	. .	Stevens, 1905.
$XO\male$	Phytodecta	. .	Galan, 1931.
$XY_1Y_2\male$	Blaps	. .	Nonidez, 1920.

Diptera.
?	Sciara	. .	Metz et alii, 1926 ; Du Bois, 1933 ; Metz, 1934.
$XY\male$	Drosophila spp.	.	Metz, 1916 ; Heitz, 1933.
$XY\male$	D. pseudo-obscura [5]	.	D., 1934 [2] ; Dobzhansky, [3] 1934.
$X_1X_2YO\male$	D. miranda	.	Dobzhansky, 1935.

[1] Distance-conjugation found.
[2] Occasionally failure of pairing, or non-disjunction resulting from it, observed at meiosis.
[3] Variation found in Y chromosome within the species (Fig. 63).
[4] Second division reduction.
[5] Proximal pairing segments.

Heteroptera.
$XY\male$	*Metapodius* [2], *Lygæus* [4]	Wilson, 1905, 1928.
$X_{1-5}Y\male$	*Acholla*	Payne, 1910.
$XO\male$	*Alydus, Protenor*	Wilson, 1928 ; Reuter, 1930.

Homoptera.
$XY\male$	*Enchenopa binotata*	Kornhauser, 1914.
$XO\male$	*Phylloxera*	Morgan, 1912, 1915.
$XO\male$	*Aphis*	v. Baehr, 1919.
$XO\male$	*Tetraneura*	Schwartz, 1932.

Lepidoptera.
$XO\female$	*Talæporia*	Seiler, 1920.
$XY\female$	*Phragmatobia* [3]	Seiler, 1925.

Trichoptera.
$XO\female$	*Limnophilus*	Klingstedt, 1931.

Aves.
$XY\female$	*Gallus* [6]	Crew, 1933 ; White, 1932.

Mammalia [5].
$XY\male$	*Mus musculus*	Painter, 1927.
$XY\male$	*Homo*	Painter, 1923 ; Evans and Swezy, 1929.
$XY\male$	*Rattus*	Koller and Darlington, 1934.
$?XY\male$	*Apodemus* [4]	Oguma, 1934.
$?XO\male$	*Evotomys*	Oguma, 1935.
$XY\male$	*Didelphus*	Koller, 1936.

Or in other words we may say that the pairing segments may be reduced to a minimum, while the differential segments, although genetically different, remain the same size in X and Y.

The pachytene observations seem to conflict with the assumption of differential and pairing segments, for in *Enchenopa* X and Y are paired throughout their length and in *Humulus* and *Rumex* a complete triradial association is often found, the two short arms of the Y's which never form chiasmata being associated (Kihara and Hirayoshi, 1932). Clearly in these cases we are dealing with non-homologous torsion pairing, which as we have seen earlier is probably characteristic of all structural hybrids, but which implies neither homology as a cause nor crossing-over as a consequence.

Where the differential segment is proximal, *i.e.*, where it includes the centromere, the simplest kind of metaphase pairing and segregation results. This is the condition found in all plants and most animals. It has usually been said in describing this type

---

[2] Occasionally failure of pairing, or non-disjunction resulting from it, observed at meiosis.
[3] Variation found in Y chromosome within the species.
[4] Second division reduction.
[6] The evidence for Y is genetical.

EVOLUTION OF X AND Y

that the first division is reductional for the XY pair. This means, as we have seen earlier, that the two chromatids with the differential segment of X pass to the opposite pole from those with the differential

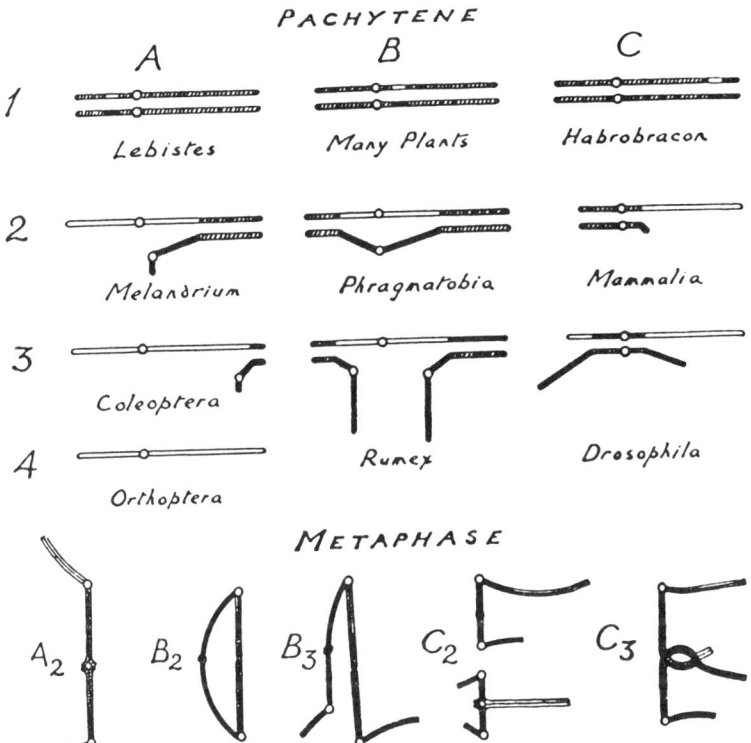

FIG. 115.—Diagram showing by the pachytene relationships of the sex heterozygote the development of the differential segments of X (above) and Y (below) leading to the disappearance of Y (4). The metaphase associations of certain types are given and depend on the position of the differential segments (black and white) in relation to the centromere.

segment of Y at the first division. Where, however, the pairing segment is proximal, crossing-over occurs in some cells between the centromere and the differential segment with the same results as have already been described in considering the general behaviour of unequal bivalents, viz., the differential segments divide equationally at the first division and reductionally in the second. This is what happens in *Rattus* in about one-twentieth of the spermatocytes. Genetic evidence indicates that it also happens in *Bombyx* (Goldschmidt and Katsuki, 1931 ; *cf.* Mather, 1935) and the observation of an equational first division of an " unpaired " X in *Apodemus* is most easily explained in this way.

Where the differential segment is proximal, no special property is required to permit of pairing segments at both ends of a sex chromosome, as with a chromosome in a ring in *Œnothera*. Such a system is actually developed in *Humulus* and elsewhere. But where the pairing segment is proximal a very special adaptation is required to permit of differential segments at both ends, for these reasons : the pairing segment is intercalary ; it cannot, therefore, pair by terminal affinity ; it can pair only by chiasmata ; but effective crossing-over giving a single chiasma cannot occur in segments lying between two pairs of differential segments without removing one of them. Nevertheless in several species of *Drosophila* the pairing segment includes or adjoins the centromere and both arms are differential. Thus in *D. melanogaster* the X and Y chromosomes have one gene in common (" bobbed."). It lies near the centromere, while both ends of the chromosomes are different.

To account for pairing without visible crossing-over in these species it was therefore necessary to suppose that crossing-over between them was regularly reciprocal and gave rise regularly to reciprocal chiasmata in a part of the chromosomes near the centromere where there were no differential genes (D., 1931 *a*). The Y chromosome was known to be almost entirely inert and such an inert region was afterwards found in the proximal part of the X chromosome (Muller and Painter, 1932). The view seemed to be supported by the frequencies of pairing of X and Y in trisomic forms, and also by various structural changes. For example, X becomes " attached " to an arm of Y in a regular way such as could result from exceptional single cross-overs near the centromere and with a regular frequency of once in 2,000 times. Such a change resembles the mutations which result from crossing-over in *Œnothera*. These expectations have been borne out in three ways. First, Kaufmann (1933) has found that the attached $\widehat{XX}$ become detached in $\widehat{XX}Y$ females by crossing-over between X and Y, such that two new chromosomes are produced, an X attached to a short arm of the Y, called $\widehat{XY^S}$, and an X attached to the long arm of the Y, called $\widehat{XY^L}$ (cf. D., 1935 e). Secondly, Philip (1935) has shown that the reciprocal crossing-over of normal

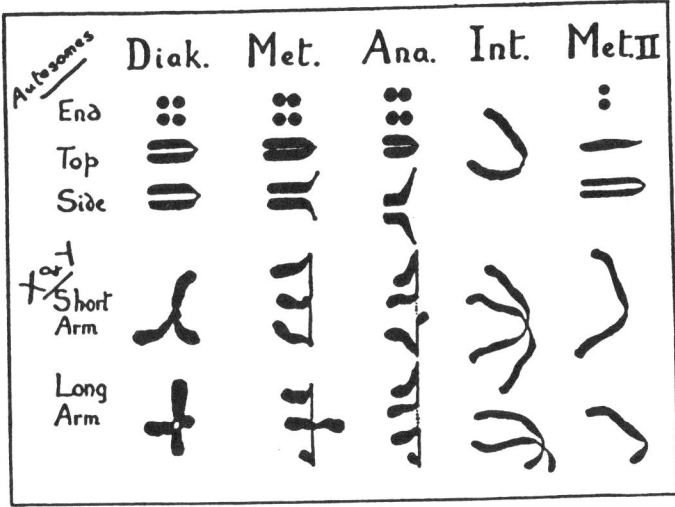

Fig. 116.—Behaviour of autosome and X–Y bivalents at meiosis in ♂ *Drosophila pseudo-obscura*. The autosomes are paired at the centromere in diakinesis, the sex chromosomes by reciprocal chiasmata either in the short arm or the long arm of Y. (D., 1934. cf. Fig. 63.)

males may sometimes take place on opposite sides of " bobbed," which adjoins the inert region, so that this gene may be transferred from X to Y and *vice versâ* without any other observable change.

Finally, the association of X and Y chromosomes has been seen in a pairing segment which lies on either side of the centromere in *D. pseudo-obscura* (D., 1934; Dobzhansky, 1934). That the association was in fact by reciprocal chiasmata could be shown as follows: The unchanged X and Y separate at anaphase, having been connected near the centromere at metaphase. Their connection may be on either side of the centromere, but is never on both sides. It therefore shows both the variation and the interference characteristic of chiasma-formation. Furthermore, the mechanical relationships of X and Y are entirely different from those of the autosomes. The former show the characteristic tension between the centromere and the nearest point of association, $i.e.$, the nearest chiasma. The autosome pairs, on the other hand, would be incapable of showing chiasma-formation, even if they had crossed over or not, because their four chromatids are equally attracted to one another, lying equally parallel at diakinesis. They are associated apparently at the centromere, but between diakinesis and metaphase suddenly turn away from one another at this point, very much as mitotic chromatids do at the beginning of anaphase. Thus in the relationships both of the centromeres and of the bodies of the chromosomes the autosomes are unique. The sex chromosomes, on the other hand, are normal in the formation of chiasmata except in their being regularly reciprocal.

These observations show that the properties of crossing-over that can be deduced from general observations of structural hybrids enable one to classify the types of sex-chromosome differentiation that is found, according to the arrangements of differential and pairing segments and to specify the kinds of crossing-over that will be necessary to preserve it. Thus the system of reciprocal chiasmata developed in *Drosophila* has been the condition of development of unique sex-chromosome differences in this genus. Furthermore, the pairing of sex chromosomes enables us to predict the occurrence of partial sex-linkage in all organisms where pachytene association and chiasma-formation occurs, in all, that is, except the most extreme types. Such linkage has now been found in *Drosophila* (Philip, 1935) and in man (Haldane, 1936).

(iii) **Segregation of Sex Chromosomes.** Paired sex chromosomes separate reductionally at the first or second divisions, according to the relative position of pairing and differential segments. Unpaired sex chromosomes are a permanent property of the race or species and behave regularly (unlike the unpaired chromosomes of adventitious hybrids) in one of the following ways :—

PAIRING OF X AND Y

(a) In the Orthoptera, the unpaired X chromosome (which is precocious during prophase) lies to one side of the equator and passes, without division, before the paired chromosomes to the nearer pole. It then divides at the second division.

(b) In the male Aphides (Morgan, 1912, 1915; Schwartz, 1932; Suomalainen, 1933) and Nematoda (Mulsow, 1912) it lags behind the bivalents, but otherwise behaves in the same way.

(c) In some Hemiptera (*Vanduzea arcuata*, and species of *Ceresa*, Boring, 1907), the unpaired X chromosome may pass to the pole either before or after the autosomes.

(d) An unpaired X chromosome is known to divide equationally at the first division in a few species and be included in two telophase

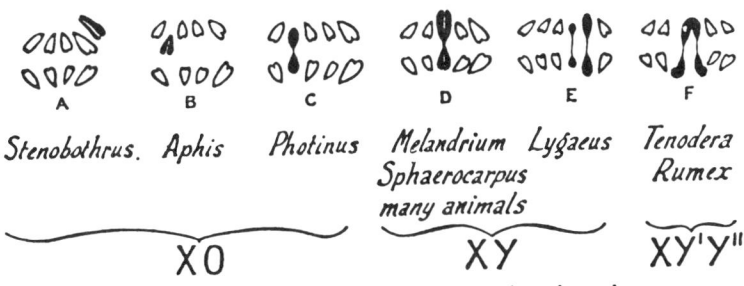

FIG. 117.—Diagram showing the methods of segregation of sex chromosomes at the first anaphase normally occurring in different species.

nuclei at the second division (*Photinus consanguineus* and *P. pennsylvanicus*, Coleoptera, Stevens, 1909 a; *Anasa tristis*, Wilson, 1905 b; *Alydus* Hemiptera, Reuter, 1930; *Protenor*, Schrader, 1935; various Odonata).

We see that the behaviour of unpaired sex chromosomes without partners does not differ in essentials from that of unpaired autosomes in plant hybrids, although animal cytologists have not usually been aware of this similarity. The distinction is that their behaviour is more regular, more exactly adjusted, by a means we shall consider later. The mechanism of the pairing of sex chromosomes with partners by chiasmata, on the other hand, is less well adjusted than that of the autosome bivalents.

Since only a part, and often only a small part, of the X chromosome constitutes the pairing segment, it is clear that the mechanism of chiasma formation, or whatever substitute for chiasma formation is effective in pairing will not be able to secure the same regularity in pairing as in the autosomes. Such is the case as shown both by direct observation and inference from the occurrence in genetical experiments of individuals with one or with three instead of with two sex chromosomes.

Where pairing fails, the unpaired chromosomes may divide at the first division (*e.g.*, in *Fragaria*), but will in any case be distributed amongst the four daughter-cells at random with regard to one

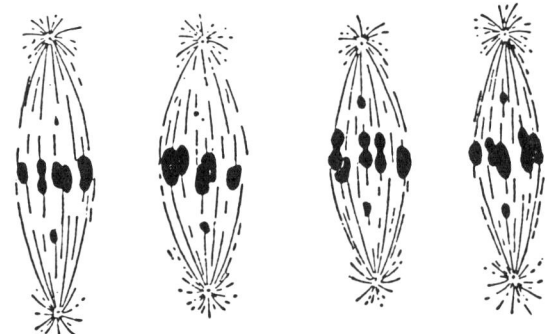

FIG. 118.—First metaphase in males of different races of *Chrysopa vulgaris*. *Y* varies in size and segregates from *X* without pairing ("distance-conjugation"). × 1600. (Naville and de Beaumont, 1933.)

another. Some will therefore have both the *X* and the *Y* chromosome. This result has been referred to in *Drosophila* (*cf.* Stern, 1929 *a*) as following non-disjunction, but is, as we now see, often more properly to be ascribed to non-conjunction, *i.e.*, failure of pairing.

True non-disjunction is, however, regularly found with multiple chromosomes, especially the chains of three and four chromosomes found in *Humulus*, *Rumex* and *Phragmatobia*. To give the normal sex segregation, successive chromosomes must pass to opposite poles, so that half the gametes have the *X* chromosome and half the *Y* chromosomes (Fig. 113). This segregation occurs in 85 per cent.

of mother-cells in *Humulus japonicus* (according to Sinoto). The non-disjunction is due to the same cause, irregular orientation of the chain at metaphase, and has the same effect, the formation of gametes with too many and too few chromosomes, as non-disjunction in other cases of diploid ring and chain formation in interchange heterozygotes (*cf.* Sinoto, 1929; Winge, 1929; Kihara, 1929 *b*).

Two kinds of segregation of sex chromosomes do not follow prophase pairing and chiasma formation. The first of these is found in certain Hemiptera (*Lygæus*, Wilson, 1905 *a*) where the X and Y chromosomes are unpaired at the first division but regularly lie on the equator and divide when the bivalents separate. Their halves then pair, end-to-end, at the second division and pass to opposite poles. This pairing is perhaps due to a property analogous to terminal affinity (*cf.* Ch. XII).

The second has been described as " distance conjugation." It is found in Neuroptera (Naville and de Beaumont, 1933; Klingstedt, 1933) and Hepaticæ (Lorbeer, 1934). In the Neuroptera the X and Y chromosomes, which have been lying on opposite sides of the diakinesis nucleus, pass to opposite ends of the spindle at metaphase. In the Hepaticæ, on the other hand, the distant conjugation described is spurious, being merely the result of one pair of chromosomes, usually the sex chromosomes or the smallest pair, having terminal chiasmata owing to their short pairing segments, while the rest have interstitial chiasmata (*cf. Secale*, Fig. 49).

Where pairing is by terminal affinity a minute pairing segment is presumably left. Where segregation follows distant conjugation there is no longer any need for a pairing segment. Since unpaired chromosomes in ordinary mutants and hybrids segregate irregularly, the regular segregation of unpaired chromosomes like the X in the Orthoptera and X and Y with distance conjugation in the Neuroptera requires a special explanation. The clue for this explanation is provided by three particular observations. The sex chromosome pair in the mammals has a weaker staining reaction than the autosomes at metaphase, and usually lies at the edge of the plate or to one side of it. It presumably has a lower surface charge (Koller and D., 1934). When the two are unpaired they can be seen lying on opposite sides of the plate (Koller, unpublished; Plate XI). Their behaviour is like that of paired chromosomes which are driven by overcrowding to form accessory plates (Ch. XII). The regularity in lying off the plate in the Orthoptera is due to the univalence of the X chromosome, and perhaps also to its lower surface charge. The regularity of X and Y in lying at opposite sides of the plate in the Neuroptera is due to mutual repulsion in a restricted space, following a similar distribution at diakinesis.

Other special properties of sex chromosomes seem to require no other explanation than precocious division of the centromeres where a univalent divides at the same time as the bivalents, and terminal affinity acting at a distance where X and Y pair momentarily at metaphase. The difference in time of division of unpaired X chromosomes in *Protenor*, *Chorthippus*, *Photinus* and *Vanduzea* is strictly comparable to that of other unpaired chromosomes in hybrids and mutants of plants. Univalents, as a rule, vary in their timing relationship, in some animals dividing only at the second division (probably *Chorthippus*, D., 1936 b) and in the extreme opposite case (in *Culex*, Moffett, 1936) scarcely lagging behind the bivalents at first anaphase (contrast Schrader, 1935).

(iv) **Haplo-Diploid Sex Determination.** (a) GENERAL. In several groups of Metazoa sexual females give female offspring when their eggs are fertilised, male when they are not fertilised, as first shown in the bee by Dzierzon. This is believed to be the method of sex differentiation in the Hymenoptera, Rotifera, and in some Thysanoptera, Acarina and Hemiptera. In the Rotifera, generations of females reproducing by diploid parthenogenesis can be intercalated between the sexual generations, as in the Aphides. In the hymenopteran *Neuroterus* certain females lay only fertilised eggs and others only unfertilised.

It has been shown in several of these groups that the females are, in fact, diploid and the males haploid. The evidence for this method of sexual differentiation, especially in Thysanoptera and Rotifera, is chiefly from breeding results. But in the following instances the assumption of male haploidy has been verified cytologically (*cf.* Schrader and Hughes-Schrader, 1931; Torvik-Greb, 1935).

TABLE 57

Examples of Haplo-Diploid Sex Differentiation

Acarina	*Tetranychus bimaculatus*	$n = 3$	Schrader, 1923.
Hemiptera			
(Aleurodidæ)	*Aleurodes proletella*	$n = 13$	Thomsen, 1927.
	Trialeurodes vaporariorum	$n = 11$	Hughes-Schrader, 1930.
(Coccidæ)	*Echinicerya anomala*	$n = 4$	,, ,,
	Icerya spp.		
	Icerya purchasi (☿ and ♂)	$n = 2$	,, 1927.
Hymenoptera	*Apis mellifica*	$n = 16$	Nachtsheim, 1913.
	Paracopidosomopsis floridanus.	$n = 8$	Patterson and Hamlett, 1925.
	Pteronidea ribesii	$n = 8$	Peacock and Sanderson, 1931.
	Habrobracon juglandis	$n = 10$	Torvik-Greb, 1935.
	(also probably in *Vespa, Neuroterus* and other genera).		
Rotifera	*Asplanchna* spp.		Whitney, 1929.

The mode of reproduction in *Icerya purchasi* is exceptional and particularly significant. Fertilised eggs give hermaphrodites. They produce 1 per cent. of eggs that go unfertilised, and give haploid males, and 99 per cent. that are fertilised either by the sperm of the hermaphrodite or by that of the males. The species therefore has an alternative system of reproduction between hermaphroditism and haplo-diploid sex differentiation.

Icerya purchasi also has a unique peculiarity in the development of the testes in the hermaphrodite. The haploid number of chromosomes is found in the spermatogonia, presumably through some process analogous to meiosis having occurred during their development. The testes are, therefore, genetically male, and the " hermaphrodite " diploid genetically female, since it produces sperm only in tissue developed by a special abnormality and genetically that of a different individual. The genetic differentiation of haploid and diploid is therefore as sharp as that in the Hymenoptera and the system of reproduction in *Icerya* is probably derived from the simple haplo-diploid type.

(*b*) HYMENOPTERA. By a prolonged series of experiments, Whiting and his collaborators have discovered the genetic mechanism underlying haplo-diploid sex-differentiation in *Habrobracon*. In this wasp diploid males are occasionally produced by fertilisation. They arise particularly in inbred stocks. They are accompanied by reduced, not increased, fertility of the parents. These diploid males show that diploidy alone does not determine femaleness; particular genes must also be concerned. From the inheritance of a character determined by genes linked with such hypothetical sex-determinants, Whiting (1935) has shown that the males are

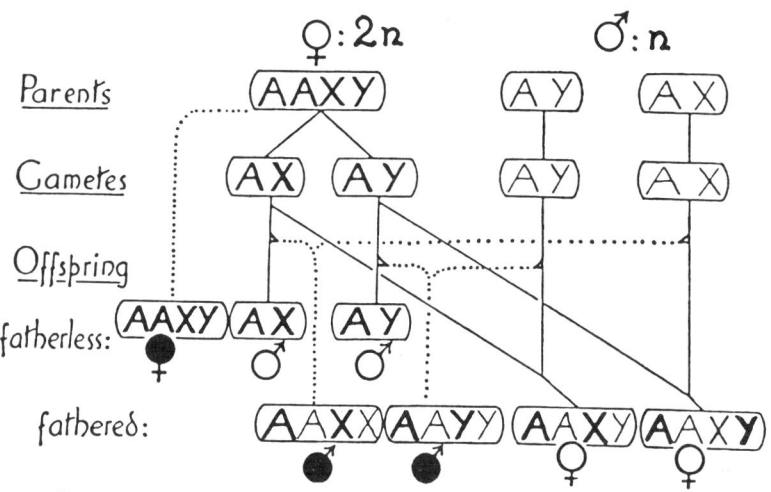

Fig. 119.—Normal and abnormal (giving black symbols) methods of sex-inheritance in *Habrobracon*. Non-reduction gives the fatherless females. Non-selective fertilisation gives the fathered males.

of two kinds, which we may call X and Y, while the females are produced by the combination $X - Y$. The regular femaleness of diploids is due to the complementary products of meiosis being present in the egg after fertilisation and the opposite type always fusing with the sperm nucleus, which will be either X or Y, according to the father. The biparental males are due to the breakdown of this inhibition; they are either XX or YY. Probably the products of the second meiotic division are the alternative mates of the sperm, and this would indicate that the second is regularly the reductional division for X and Y, as it sometimes is in mammals. Further, occasional parthenogenetic diploids arise (Speicher, 1934) presumably by fusion of the second division products and these are always female, that is XY, never XX or YY like the biparental males.

The ordinary haploid males have their first meiotic division suppressed as in *Apis*, the centrosome at one pole being separated in a bud. The diploid males undergo the same process (Torvik-Greb, 1935) and therefore produce triploid daughters (Bostian, 1936).

It is important to distinguish between the haplo-diploid system of sex differentiation found in so many different groups of organisms and the special XY mechanism of *Habrobracon*. Not only does the haplo-diploid system extend to a large group of animals; it also has important physiological implications. It demands the development of a special balance within all the chromosomes similar to that which Muller (1932) inferred in the X alone in *Drosophila*, viz., a balance such that particular genes have the same effect in double as in single dose. The development of this system to its present perfection indicates its antiquity. On the other hand, the particular XY mechanism shows some signs of newness. Whiting has represented the genic constitution of the differential segments of X and Y as $F.g$ and $f.G$ so that they would be complementary. Crossing-over still occurs between X and Y and gene differences are exchanged. The differential segments are not, therefore, developed to the extreme condition. Since neither in X nor Y are they ever homozygous (except in the diploid males), it seems likely that they never could develop except in the direction of inertness, and the fact that the two kinds of males do not differ in any observable respect bears out this expectation. It seems, therefore, that this system is probably short-lived, capable of being frequently replaced by similar systems, and probably secondary to the general haplo-diploid method of differentiation.

(c) COCCIDÆ. In the coccids *Pseudococcus* ($♀$, $2n = 10$) and *Protortonia* ($♀$, $2n = 6$) the haploid number of bivalents are formed in oögenesis and meiosis is normal (Schrader, 1923, 1931). In the male *Pseudococcus* ($2n = 10$) five of the chromosomes condense prematurely at the prophase of meiosis ; they do not pair with the other five and the first division is equational. At the second division the precocious chromosomes pass to one pole and the others then pass to the opposite pole. The two sides of the spindle are formed separately, as in the haploid bee, where only one side is effective. It appears that the two sets of five chromosomes are those derived from opposite parents, and that they are therefore permanently separated in phylogeny. The whole set is differentiated in regard to sex. The behaviour in another coccid, *Lecanium hemisphericum* and *Phenacoccus* (Hughes-Schrader, 1935), is similar (Thomsen, 1927).

In the male *Protortonia* ($2n = 5$), on the other hand, the condition is probably intermediate between that in *Pseudococcus*, and in the normal type of sex-determination in other Hemiptera. There is no distinction between the two sets, and pairing takes place between two chromosomes, the other three remaining unpaired. The four groups lie in separate vesicles at the first division and the chromosomes divide equationally. But at the second division the chromosomes arrange themselves in a row or chain which lies axially in the spindle and divides so that two chromosomes at one end go to one pole and the other three to the other. Their order is not constant, but one of the chromosomes which is definitely smaller than the rest lies at one end of the chain in three-quarters of the nuclei. This is probably the arrangement giving the effective segregation, which in this case is of one kind only—probably that giving the separation of unchanged maternal and paternal chromosomes. Diagrammatically, on this hypothesis, the chromosome relationship can be represented thus : ♀, $AABBCC$; eggs, ABC ; ♂, $AABB_1C$; sperm, ABC (♀-producing) ; and AB_1 (♂-producing). Since B and B_1 have no opportunity of crossing over, and B_1 is transmitted exclusively by the male sex, they will be permanently separated in phylogeny (as the whole complexes are in *Pseudococcus*). It may be noted parenthetically that it does not necessarily follow that they are not homologous or even identical because they fail to pair, for genetic factors must undoubtedly have been the original cause of this failure (as in facultative parthenogenesis, *q.v.*). But it does follow that since they do not

EVOLUTION OF SCALE-INSECTS

pair, but yet are segregated to opposite sexes they will vary independently in so far as they vary at all and therefore become less and less like one another as evolution proceeds.

In another Coccid, *Llaveia bouvari*, a condition is found more nearly approaching normal meiosis (Hughes-Schrader, 1931). The chromosomes as a rule pair normally, even forming chiasmata

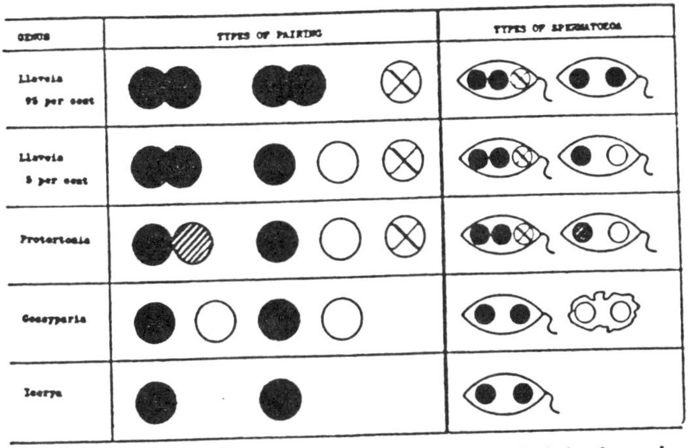

FIG. 120.—Diagram of chromosome behaviour at meiosis in the male in a series of forms showing a gradation from normal pairing to a complete failure. Circles represent chromosomes, white ones being those confined to the male sex (the "Y" set); the X chromosome is contributed to the female progeny. In *Llaveia* there is 95 per cent. of normal pairing, 5 per cent. of failure of one pair. There is probably no differentiation between the pairing chromosomes. In *Protortonia* only two chromosomes pair and these are perhaps differentiated. In *Gossyparia* there is no pairing and half the sperm degenerate. *Icerya* is haploid in the male tissue. (From F. and S. H. Schrader, 1931.)

apparently, though each pair remains in a separate vesicle. Occasionally, however, only one pair associates at metaphase, as in *Protortonia*.

These species are most easily considered as representing steps in the successive differentiation of the two sets of chromosomes in the male, so that the whole of a set develops the character of X or Y and the "Y" set is carried exclusively in the male line. Such a differentiation might be conditioned by, first, the replacement of ordinary chiasma-pairing by pairing as found in the sex chromosomes of other Hemiptera, which seems to depend on the property of terminal affinity, and, secondly, the development through genetic change of a special spindle mechanism. The progressive change in this series is represented in the diagram (Fig. 120). It is described here as differentiation, meaning the accumulation of structural differences. It has been described by the Schraders as " degeneration "—a term which should be avoided in speaking of chromosomes here and elsewhere, because it implies an analogy with evolutionary changes in gross structures that is somewhat misleading.

But there is a fourth genus, *Gossyparia*, whose behaviour seems to conflict with this or any other evolutionary hypothesis, although the Schraders have represented it as the last step in the process of differentiation. Here again meiosis is normal in the female: 14 bivalents are formed. In the male half the chromosomes are precocious and do not associate with the other half, but they do pair at metaphase amongst themselves. It is believed that this is not true pairing and does not lead to segregation, all 28 chromosomes dividing equationally at the first division. The precocious chromosomes separate from the rest at the second division, which is otherwise abortive. Thus four nuclei are formed with the haploid number of chromosomes, and if it were possible to consider these as of two alternative sex types, the precocious chromosomes representing those characteristic of the male, the " Y " set, and those that are homozygous in the female, the " X " set, it would be easy to see in *Gossyparia* an exaggeration of the differentiation of *Pseudococcus* and *Protortonia*. But Schrader (1929) finds that half the sperm die, and considers these to be of one type, the " Y " set. If this is so, obviously the chromosomes that are lost in the degenerating sperm of one male are precisely those that were not lost in its own male parent. Moreover, since the half that survive must contain both male-determining and female-determining chromosomes, these must have been derived in part from opposite parents. This upsets the notion of the constancy of behaviour being due to any properties inherent in the chromosomes concerned. The view that the sperm which die contain what they describe as the " degenerate " set of chromosomes is held by the Schraders

SEX AND PARTHENOGENESIS

(1931) to be the most decisive evidence in favour of the theory that this set is " progressively degenerating," but it seems rather to be the chief obstacle to regarding the series of steps as progressive.

If it were possible to assume that the sperm which survived carried both kinds of chromosomes the series could be taken to show a progressive change, and this would be the means by which the whole of one set might be lost in the same way and for the same reason as the Y chromosome has been lost in the Orthoptera and elsewhere. Hence the haplo-diploid system of sex-differentiation, which is found in yet another coccid, *Icerya purchasi*, might arise (v. diagram).

(v) **Meiosis in Cyclical Parthenogenesis.** Males arise in the cyclically parthenogenetic Rotifers, as they do in the Hymenoptera, presumably by haploid parthenogenesis. In the aphids and phylloxerans it has been established (*cf.* Morgan, 1912, 1915; Schwartz, 1932) that the development of males is due to the loss of one chromosome (an X chromosome) at maturation in the egg of the parthenogenetic mother. The $2x-1$ males might be expected to give offspring like themselves as well as females, but the true-breeding parthenogenetic line is restored by the functioning of only those sperms with the full haploid number.

The same aberration occurs less regularly in the Orthoptera. In *Apotettix* and *Paratettix*, Robertson (1930, 1931 c) found that the males exceptionally produced by parthenogenetic females lacked one of the X chromosomes of their parents, as do the normal males in the grasshoppers. This was supposed to be the result of non-disjunction, probably conditioned by a special property of the X chromosomes. Such a property as, for example, that of precocious development found in the single X chromosomes in the male might affect its pairing if equally present in the female, and hence lead to lagging and loss (v. Ch. IX).

(vi) **Genetics of Sex Determination.** In any species where the chromosomes of the two sexes differ in size the sex differentiation can be seen to follow from the segregation of dissimilar chromosomes at meiosis. In some species (as shown in *Ancyracanthus* by Mulsow) every step in development can be seen, from the act of unequal segregation to the differentiation of the sex types which is invariably correlated with it. This gives us the minimum information with the maximum certainty. It shows us the mechanism underlying sex determination, but it does not show us how the mechanism arose or how it operates. There are therefore an evolutionary and a physiological problem to be decided. Both have to be approached by the comparative method.

The evolutionary problem has been approached by attempting a unitary theory of sex determination. Correns (1928) considered that such a theory may rest on the assumption that homologous genes are present in all species for sex differentiation at each stage of the life cycle, and that allelomorphs of those affecting one stage are found in one group and affecting another stage in another group (*cf.* also Allen, 1932). This hypothesis is of little value. The assumption of genes in one organism affecting differences only found in an entirely different group of organisms depends on assumptions of the nature of variation for which there is no evidence. Sex differentiation has certainly arisen independently in different groups. The fact of its analogy of mechanism is due only to the inevitable limitations of its mechanism—segregation of alternatives at meiosis.

The second attempt is based on the study of sex conditions in hybrids. In these, sex seems to be determined by a balance of activities working in opposite directions—to maleness and to femaleness. They can be represented more or less quantitatively. The one activity is determined by the X chromosome, which is present in a single dose in one sex, in double dose in the other. The opposite activity is determined by the autosomes (or the cytoplasm or Y chromosome in *Lymantria*) and is in the pure species stronger than the single dose of X, but weaker than the double dose. When races or species are crossed this relationship does not hold, and intersexes result. This formula may be taken as an approximate description of the differences between races and between the effect of the X and of the other chromosomes. They may be expressed quantitatively and according to direction—male or female (Goldschmidt, 1934).

The formula does not, however, decide the question as to whether the effect is produced by differences in proportion or in summation, although we can assume proportion from what we know of unbalanced forms in plants. This is provided by work on forms of *Drosophila* with different numbers of chromosomes.

Bridges (1922) found that, in the progeny of triploid females of *Drosophila*, flies occurred having different numbers of X and Y chromosomes and autosomes (Fig. 121). Some of these were sterile, but the whole of the types found might be arranged in a more or less linear series from forms of male character to others of female character, through normal sexual types and intersexes. The extreme types at both ends were sterile and were described as super-males and super-females; those apparently normal males which had no Y chromosome were also sterile, because Y, although almost inert, still contains something necessary for fertility in the male (Stern, 1929). The succession in this series was found to be related to the proportions in which the autosomes and the X chromosome were present, and entirely unrelated to the presence of the Y. Thus if the autosome complement is denoted by A, the series (*cf.* Fig. 121):—

$$AAAX(Y), AAX(Y) \, \male, AAAXX(Y) \begin{cases} AAAXXX \, (3x) \, \female \\ AAXX \, (2x) \, \female \quad AAXXX \\ AX \, (x) \, \female \end{cases}$$

represents the transition from "super-male" to "super-female." Haploid patches of tissue (AX) in mosaics have since been found to be female in character as expected on this theory.

It is interesting to notice that types otherwise similar are more female with the higher number of fourth chromosomes. The fourth chromosome therefore acts in the same sense as an X chromosome, and in the opposite sense to the other autosomes. This departure from the rule can be simply understood by supposing that the small fourth chromosome has arisen by fragmentation of the X, a not improbable origin on general grounds.

When the principles of proportion-adjustment or balance to be derived from the simpler study of trisomic plants are applied to an organism with alternative X and Y inheritance it is seen that two systems of adjustment must work in such an organism. X and Y must both be adjusted in proportion to the autosomes—as we have

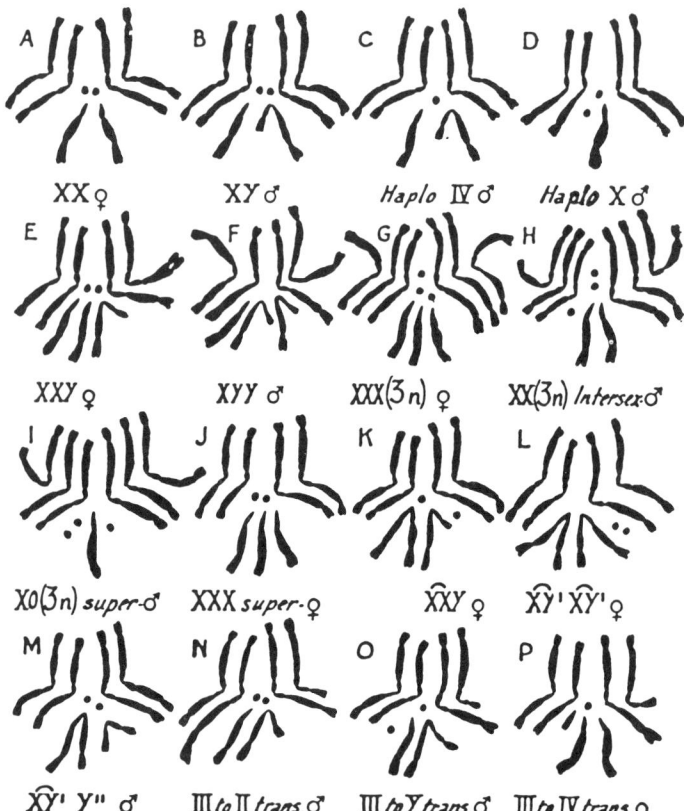

FIG. 121.—Diagram showing normal (A and B) and variant (C to P) chromosome complements of *Drosophila melanogaster*. Compiled from Bridges, 1916 (D, E, F), 1922 (G, H, I, J), 1927 (A, B), L. V. Morgan, 1922 (K), Stern, 1929 (L, M—alternative sex-mechanism), Painter and Muller, 1929 (N, O), and Dobzhansky, 1929 (P). Constrictions are emphasised. × *ca.* 9000.

D, E and K the products of primary non-disjunction. F of secondary non-disjunction. K of fusion. N, O, P of translocation. L and M of deletion with fusion. The structure of the chromosomes is otherwise constant. The XO male and the super-male and super-female forms are sterile, the first for mechanical reasons, the others for physiological reasons.

seen is the case, X having a resultant normal female effect, Y having a neutral effect. Further, both must be internally balanced. In the

Y neutrality implies balance. In the X recent work has indicated an internal differentiation. Certain fragments of the X which have a phenotypic effect on the males that bear them do not reduce their fertility (Muller, 1930 a), probably on account of a margin of safety, a threshold value, in the reaction of the normal type. When such fragments are added to intersexes they alter the sexual grade of the intersex in various degrees according to the fragment concerned (Dobzhansky and Schultz, 1931 ; cf. Patterson, Stone and Bedichek). These effects can be expressed more or less quantitatively as Goldschmidt has supposed. At the same time the X is shown to be differentiated and adjusted, i.e., balanced (cf. Muller, 1932 a; Kerkis, 1934).

Proportion can be shown to be important in sex determination in other organisms. Triploids and tetraploids in *Rumex acetosa* and *R. montanus* of the constitutions: $18 + 2X + Y_1Y_2$ and $24 + 3X + 2Y_1Y_2$ are intersexual (Ono and Shimotomai, 1928 ; Ono, 1930), but whether those lacking Y_1 and Y_2 are also intersexual is not yet known. Forms with an abnormal (probably deficient) autosome have, however, been found, and they, although diploid, are slightly intersexual (Ono, 1930, 1935 ; Yamamoto, 1934). In *Sphærocarpus Donnellii* (Knapp, 1935, a and b) gametophytes lacking a certain part of the X chromosome are male, although they have no Y chromosome at all. Intersexes in *Melandrium album* have the male complement (Belar, 1925), while those in *M. rubrum* have the female complement (Akerlund, 1927). These results indicate one essential conclusion (which is in accordance with genetic evidence) : all the chromosomes together establish the physiological system which permits of two sharply distinguished lines of development. The different proportions of the two sex chromosomes which result from this segregation and recombination are merely the trigger, as Muller puts it (1932), which sets development going in one or other of the two directions. How the trigger mechanism can be changed is shown by certain recent experiments which we shall now consider.

New mechanisms of sex segregation have arisen under two conditions, either in hermaphrodites which had no such segregation previously, or in sexually differentiated organisms where the new mechanism replaces an earlier one. No doubt also environmentally or developmentally determined systems can readily develop into genetic systems with new sex chromosomes. The beginnings of such systems are present everywhere.

In *Zea Mays*, which normally has unisexual flowers on different parts of the same plant, diœcious stocks have been produced in two ways, one giving a homozygous male, the other a homozygous female. In each case a combination of two mutant genes is necessary, one of them, however, being homozygous in both sexes (R. A. Emerson, 1932).

In *Lebistes* new gene combinations have been obtained in which the sex differentiation is determined by a single pair of allelomorphs in autosomes. Two such combinations have been selected, one giving males, the other females, homozygous. In both, the homozygous sex has been reversed to give the new heterozygous sex, while the old heterozygous sex type with its Y chromosome has been eliminated (Winge, 1934).

Hence, we see how the sex determining mechanism can be reversed in the course of evolution. We see also how nearly related fishes, as well as birds and mammals, come to have opposite systems. Similarly in *Drosophila simulans* (Sturtevant, 1929), *D. virilis* (Lebedeff, 1934) and *Lymantria dispar* (Goldschmidt, 1934) genes or gene-combinations or cytoplasmic variables are known which partially reverse one or both of the sexes. But in these organisms complete and fertile reversal of the homozygous sex seems to be impossible. Apparently the Y chromosome can be replaced only when it either differs from the X in the minimum degree or in the maximum degree when, as we shall see, it has become entirely inert, *i.e.*, in the simplest and in the most advanced types of sex chromosome differentiation. In the intermediate condition there is something essential in the Y that is not in the X; but at either extreme there is nothing.

Evolutionary changes of other kinds can be inferred from the comparison of related species. Of these the commonest is the complete disappearance of the Y chromosome in groups some species of which have an $XY - XX$ system (*e.g.*, in Hepaticæ and Orthoptera). The most remarkable of these is found in *Drosophila*.

PLATE XI

FIG. 1.—Diplotene in *Chorthippus* ♂, showing precocity of P chromosome which is attached to X. It has a single interstitial chiasma. (D., 1936 d.)

FIGS. 2-4.—First metaphase in *Hemerobius humulinus* (Neuroptera), showing non-pairing or " distance conjugation " of X and Y.

Fig. 2. Side view. Fig. 3. Polar view. Fig. 4. Two autosomes also unpaired. (Klingstedt, unpublished.)

FIG. 5.—Non-pairing of X and Y in the mammal *Putorius furo* (ferret), showing similar arrangement on opposite sides of the plate. (Koller, unpublished.)

FIG. 6.—First metaphase in man; X and Y paired terminally and lying off the plate. (Koller, unpublished.)

FIGS. 7 and 8.—First metaphase in the rat, showing pairing of X and Y, (Fig. 7) terminally to give reductional first division separation and (Fig. 8) interstitially to give a symmetrical bivalent having equational first division separation. (Koller and D., 1934.) × 2400.

PLATE XI

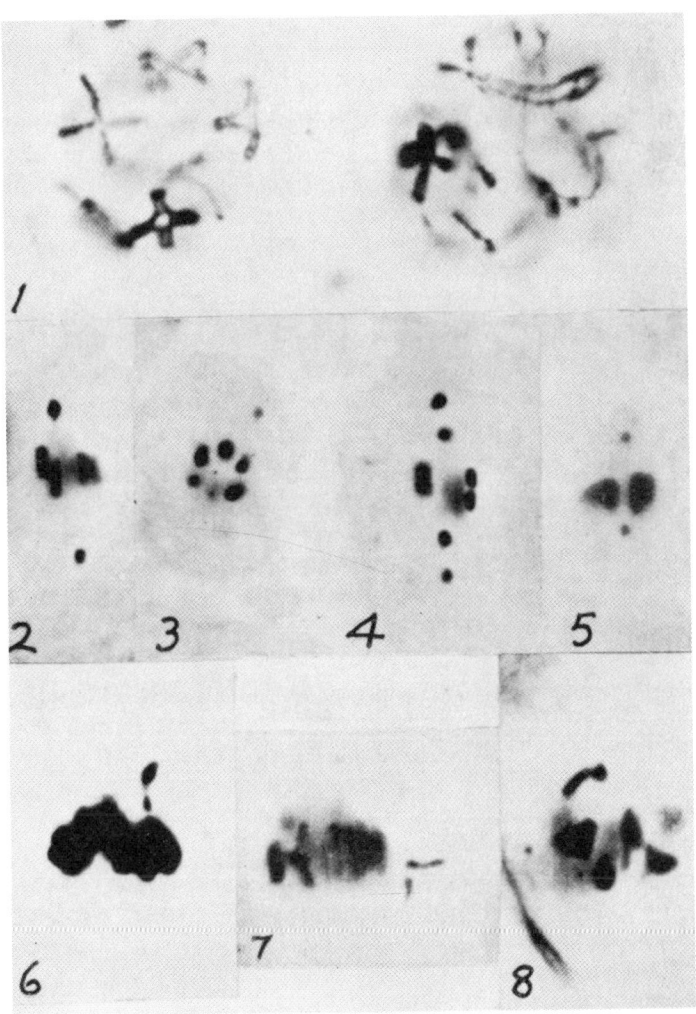

D. miranda will cross with *D. pseudo-obscura*, but has an unpaired chromosome in the male corresponding to two in the female. Together with this "X^2" chromosome there persist a pair of chromosomes which although equal, nevertheless associate by the reciprocal chiasmata characteristic of the sex chromosomes in the related species. Dobzhansky (1935 *b*) concludes that this is a second pair of sex-determining chromosomes (X^1Y) whose segregation is correlated by an unknown means with that of X^2, so that only two kinds of progeny are produced: $X^1X^1X^2X^2$, females, and X^1YX^2, males. It is, however, possible that X^1 and Y are the same chromosome, derived as to their pairing segments from X and Y of the other species and therefore continuing to pair by reciprocal chiasmata, but no longer taking part in sex-segregation (D., 1936 *a*). Whatever the precise mechanism, however, *D. miranda* shows the origin of a new system of sex-segregation in the Diptera on the foundations of the old one.

The experiments on *Zea* show how sex-differentiation can arise in a hermaphrodite. The experiment on *Lebistes* further shows how one gene may determine sex by its segregation, while others which formerly did so and are equally essential in establishing the system, being homozygous, no longer pull the trigger by their segregation. The first lies in an incipient sex chromosome, the rest lie in a former sex chromosome which is now an autosome. This relationship of the autosomes and sex chromosomes to sex-differentiation is true of every stage in their evolution. It follows that the autosomes govern the sex-determination from the beginning, although only special genetical experiments can show their influence.

(vii) **The Differentiation of the Sex Chromosomes.** The quantitative theory of sex, derived from the study of gene differences in hybrids, and the proportion theory derived from the study of proportion differences in polysomic forms answer different questions. The one shows how gene differences arise (and act), and the other shows how they interact. Both provide the basis on which a unitary theory may rest. The cytological evidence, on the other hand, reveals a third point of view—the mechanical one, which enables us to put the different types of sex determination in an evolutionary series. This can now be considered.

In the simplest cases of genotypic sex differentiation no differences can be seen to distinguish any sex chromosomes (*e.g.*, in the higher plants, *Spinacia*, Haga, 1935 ; *Taxus*, Dark, 1932 ; in the Polychæta, *Ophryotrocha*, Huth, 1933, and in the Amphibia ; in eleven out of thirty species of the Hepaticæ, Lorbeer, 1934). Genetically the same lack of differentiation is to be inferred in *Lebistes* and *Habrobracon*. But where complex genetic differences distinguish the two sexes we find complex differences between their chromosomes, either in their form or in their relative behaviour at meiosis. Many observations bear witness to the importance of structural change in the development of the sex chromosomes, *e.g.*, fusion in *Mermiria*, fragmentation in *Phragmatobia*, *Metapodius*, and *Alydus*, translocation in *Drosophila* species, interchange in *Humulus*.

The question then arises as to how a group of differences can act as a permanent unit in inheritance since crossing-over is a general property of pairing chromosomes, how, in other words, a differential segment can develop. The most obvious explanation is that the sex factors arise in association with structural changes. Small relatively translocated or inverted segments evidently have little chance of crossing-over. This hypothesis is an expression in cytological terms of Muller's suggestion (1918) that the differences between complexes in *Œnothera* and between X and Y chromosomes in the sex heterozygote were due to the association of lethal factors with crossing-over suppressors. We now know that structural changes may combine these two functions (of which the first is largely superfluous in sex differentiation). However, a variety of evidence indicates that structural changes have not been the primary agent in the suppression of crossing-over and the conditioning of differentiation.

Suppression of crossing-over might equally be supposed to be genotypically determined. This suppression would permit the development of structural differences which would themselves secondarily prevent crossing-over. The evidence that this is the origin of differentiation is both genetical and cytological, as follows :—

(i) Precocity of the sex chromosomes combined with pairing by terminal affinity, or precocity of parts of them, leading to localisation of chiasmata and crossing-over, is a reaction of the chromosomes to a genetic property of the organism. Localisation, for example, is found in *Fritillaria Meleagris*, but not in *F. imperialis*.

(ii) Abnormal forms of meiosis are characteristic of one sex in the Coccidæ and in the *Diptera*, and affect equally like and unlike chromosomes. They must therefore be determined by a genetic property and not by hybridity as such.

(iii) Crossing-over is generally reduced in sex heterozygotes relative to the homozygotes in the autosomes as well as in the sex chromosomes. This points to a unitary and therefore genotypic control of the difference (Haldane, 1922; *cf.* Huxley, 1928; Crew and Koller, 1932).

(iv) The mechanism of reciprocal crossing-over in *Drosophila* must be genotypically determined, since it does not occur in the female, and could not be due to a structural difference.

These remarks suggest the means by which crossing-over may be suppressed, *viz.*, non-pairing and localised pairing due to non-precocity or reduced precocity. The incidence of non-precocity and abnormal meiosis will be discussed later (Ch. X). Localisation is less easily detectable. It may be of two kinds. Most commonly sex chromosomes that appear to have paired normally during prophase are associated by terminal chiasmata at metaphase, as in all plants. In these the localisation must be near the distal ends of the chromosomes, and therefore terminalisation will occur rapidly and the original position of chiasmata will not be ascertainable. The differential segments will then be proximal and the pairing segments distal (*v.* Fig. 115). This condition is the same as that in the complex-heterozygote. A second possibility is that localisation occurs as in *Mecostethus* and *Fritillaria* in the neighbourhood of the centromere. When this was median and both arms were differentiated, chiasmata with reciprocal cross-overs would need to be regularly formed to maintain the differentiation of the X and Y chromosomes as in male *Drosophila* where visible crossing-over is consequently suppressed.

The best recognised change conditioned by the suppression of crossing-over between the differential segments is the change to inertness in the differential segment of the Y, a change evidently conditioned, as Muller pointed out (1918) by its crossing-over being completely suppressed. This inertness may be shown by the fact that the entire Y can be lost with no effect on the fly, apart from sterility, in *Drosophila melanogaster* (Bridges, 1916; Stern, 1929), and in *Banasa* and *Metapodius* (Wilson, 1905, 1928). It may also be shown by the inertness of fragments that are derived from the Y. As many as six may be found in *Metapodius* without apparent effect. Similar fragments are found in the Tettigidæ (Robertson, 1916) and probably in *Blaps lusitanica* (Nonidez, 1921).

In *Drosophila*, for a reason that is not apparent, the inertness extends to the pairing segments of X and Y (Muller and Painter, 1932). Possibly this change is general, for inert fragments probably derived from X are found in *Alydus* (Reuter, 1930) and elsewhere (*cf.* Table 16).

(viii) **The Evolution of Sex Chromosomes.** According to these different systems of crossing-over restriction, and the progressive changes that are conditioned by them, we may trace the development of the sex chromosomes, from their uniform beginning of minimum differentiation to their uniform end of the obliteration of the Y, by various different routes (*cf.* Diagram, Fig. 115). These steps represent the possible *direct* series of evolutionary changes. Such series may be interrupted at any point, as we have seen, by the establishment of a new mechanism in autosomes which will usually begin again at the first step.

The evidence shows that the differentiation of sex chromosomes evolves by the following steps:—

1. Origin of a gene-pair (the sex differentials) whose segregation determines the opposite sexes.

2. Genotypic suppression of crossing-over in the region of the sex differentials.

3. Structural and gene changes in this region, making it into a differential segment.

4. Special gene changes preserved which render the differential segment of Y inert and give that of X a new internal balance to compensate for the inertness of Y (Muller, 1932).

5. Further structural changes may, first, include autosomes in

EVOLUTION OF SEX

the sex-differential system by interchange, secondly, reduce the pairing segment to little or nothing, in which case a new method of pairing becomes necessary; and, thirdly, eliminate the Y, when an adapted segregation mechanism becomes necessary.

TABLE 58

Primitive and Advanced Types of Sex Determination

Rana-Lebistes-Melandrium-Rumex Type.	Drosophila-Protenor-Lygæus Type.
1. *Sex determination*: Partly by difference between XX and XY.	Purely by difference in balance between X/A and XX/A.
2. *Crossing-over*: Occurs between pairing segments of X and Y and hence partial sex linkage occurs.	Abolished between X and Y owing to the pairing segment being eliminated or inert.
3. *Differential segments of X and Y*: They are a small fraction and pairing is regular.	They are a large part, inert in Y, active in X.
4. *Recessive factors in X*: Show in XY individuals in reduced effect (Winge, 1931).	Show in XY individuals in unreduced effect, except those in pairing segment.
5. *Related forms*: They are hermaphrodite (*Melandrium*, *Rana*) or have an opposite system of sex-determination (*Lebistes* and *Platypœcilus*).	They have the same system of sex-determination, as a rule.
6. Intersexual forms appear in both types either through (i) gene mutation or hybridisation in the diploid or (ii) different proportions of the active chromosomes in polyploid and polysomic forms.	
7. *Variation in X and Y*: Still largely related on account of the small differential segment. Hence size difference not extreme.	Independent on account of the large differential segment. Hence size difference may be slight or extreme.
8. *Precocity of X and Y*: Absent. Pairing segments associate by chiasmata in prophase.	Frequent. Pairing may be postponed to metaphase of the first or second division.

Thus, if we assume that the differentiation of X and Y chromosomes has been accomplished by a combination of structural and intra-genic changes, and that this combination has been conditioned by a genotypic restriction of pairing and crossing-over, we can put the different types of sex-segregation in an evolutionary scale and thus express their mechanical and historical relationships in terms of the experimentally verified properties of chromosomes.

CHAPTER X

THE BREAKDOWN OF THE GENETIC SYSTEM: ADVENTITIOUS

Abnormal Mitosis—Polymitosis—Failure of Pairing at Meiosis—Genotypically and Structurally Determined—Failure of Chiasmata and of Spindle-formation—Non-Reduction and Diploid Gametes—Effects of Irradiation and their Use in Experiment.

> Without contraries is no progression.
> BLAKE, *Marriage of Heaven and Hell.*

1. THE BREAKDOWN OF MITOSIS

(i) **Introduction.** So far we have considered the origin, evolution and successful working of genetic systems of increasing complexity. In them sexual reproduction has been built up on a foundation of mitosis, meiosis and fertilisation. They show the operation of the laws of heredity which have been made out by experimental breeding in such systems.

Structural and numerical hybrids are an exception. In them the mechanism of reproduction and the laws of heredity partially break down. It remains now to consider the conditions and consequences of such a breakdown both in hybrids and elsewhere. We find in the first place that not only meiosis, but also mitosis may suffer abnormality. This is particularly important because here we know that the complex conditions of meiosis and particularly of hybridity as affecting chromosome pairing are not concerned. Genotypic and environmental factors are therefore readily distinguished from structural factors, and they will be specially considered later.

The occurrence of defective mitosis is, however, necessarily limited, either to special tissues or to slight and harmless irregularities, for this reason: any generally distributed and gross abnormality of mitosis will immediately stop the development of the organism, and ultimately kill it.

The commonest type of breakdown in mitosis is the failure of two nuclei to be formed at anaphase or their fusion at telophase to give a fusion nucleus with double the ordinary number of chromosomes. This abnormality has been described in relation to the origin of polyploidy (Ch. III). We have seen that it is conditioned by an abnormal environment, by injury, and so forth, as a rule. It is also often conditioned in its frequency by the genotype, some species and races being particularly liable to produce polyploid cells, either generally or in certain tissues. We have also seen that this " syndiploidy " is particularly frequent in the part of the life cycle approaching meiosis. It is at this stage also that its genotypic conditioning becomes most obvious (*e.g.*, in *Datura*). It is then often accompanied by other irregularities of mitosis. In *Drosophila simulans* one of these affecting the premeiotic divisions in the ovary as well as the cleavage divisions in the egg, has been referred to the action of a particular gene (Sturtevant, 1929).

The reverse effect of loss or breakage of particular chromosomes is determined by particular genes in *Drosophila*, but the mechanism seems to be structural, since only the chromosomes carrying the genes in question are affected (Bridges, 1925 *b* ; Stern, 1927).

(ii) **Pollen Grains and Spermatids.** Of the genotypically controlled abnormalities of mitosis, the most significant, the permanent and polyploid prophase in salivary glands, has already been described. Here, of course, we are dealing with nuclei which will never divide again. In pollen grains, on the other hand, abnormalities involve sterility and are therefore limited to aberrant conditions or genotypes. Abnormalities of development are frequent in them, such as the embryo-sac-like pollen grain in *Hyacinthus* (Stow, 1930 *et alii*), but these do not involve abnormalities of mitosis.

A minor abnormality of mitosis is found in species of *Fritillaria* and *Tulipa* (D., 1936 ; Upcott, 1936). Here in certain pollen-grains a group of correlated differences in behaviour is found. First, the chromosomes are more condensed than usual. Secondly, their centromeres do not synchronise in division, *i.e.*, in beginning the anaphase separation ; the chromosomes do not lie on a flat plate, but are irregular both in their distribution and orientation. Thirdly, their distribution is more confined to the periphery of the

spindle in the way that is characteristic of animals having the
" hollow spindle." These differences can probably be related best
by supposing that there is an irregularity in the timing relationship
of the spindle and of the chromosomes which as we shall see later
are independent but usually co-ordinated factors in metaphase and
anaphase movements.

(iii) **Polymitosis**. Meiosis may be more or less effectively
replaced by mitosis under special genetic conditions—less effectively
in adventitious mutants such as asynaptic maize, more effectively
in races with adaptive parthenogenesis (v. Ch. XI). It is equally
to be supposed that mitosis might under special genetic conditions
be replaced by meiosis. This is possibly so at an early stage in the
testis of *Icerya* and in the antheridia in the diploid gametophytes
of some aposporous ferns (v. Chs. IX and XI) where the normal
occurrence of meiosis is also omitted, in compensation. The
simple intercalation of meiosis is, of course, bound to be destructive
of the normal sexual cycle, and cannot therefore be expected save
in abnormal conditions or as an adventitious mutation. In this
way it seems to have been discovered in the form of supernumerary
divisions in the pollen grains of races of *Zea Mays* by Beadle (1931).
This property of having " polymitosis " is inherited as a mendelian
recessive character.

In an abnormal plant the four nuclei of the pollen tetrad formed
by a normal meiosis enter into a prophase condition soon after the
end of the second division. The chromosomes are still single at the
early stages but are certainly sometimes double at metaphase, when
it is found that occasional pairs are formed, characteristic bivalents
with single interstitial and terminal chiasmata (Plate X). At
metaphase the chromosomes have the extra contraction found at
meiosis; they sometimes form a metaphase plate, but they are
usually scattered over the spindle and are distributed at random to
the two poles at anaphase. The chromosomes have not been seen
to divide at their centromeres, but probably do so occasionally.
Their behaviour is therefore the same as that of univalents at
meiosis. This mitosis is followed immediately by two or three
more, and cell-walls are formed between the products. Thus 10
chromosomes or a few more are distributed to 8, 16 or 32 cells,

most of which presumably have only one chromosome. Fragments are occasionally found. The pollen is entirely sterile, but nearly 10 per cent. of the embryo-sacs are viable and haploid, owing to the failure to form separate daughter-nuclei after division or to less abnormality.

It will be observed that the first of these divisions has the property of beginning precociously, which in itself enables one to predict its later course on the precocity theory. This division corresponds in certain essentials to the first meiotic division in a haploid. The occurrence of pairing is analogous to that at meiosis in haploid Œnothera (v. Ch. VIII). It does not occur in pollen grains with reduplicated segments or whole chromosomes (Beadle, 1933). This is not surprising, since unbalance as well as particular genes found in inbred stocks will reduce chiasma-frequency in Zea (Randolph, 1928; McClintock, 1929 a; D., 1934; Beadle, 1933 a). The appearance of fragments is expe cted on the chiasmatype hypothesis, because crossing-over at chiasmata between relatively translocated segments will give new chromosome types large and small.

The abnormality differs from meiosis in that it is *reiterative*, so far as the spindle is concerned, while, on the other hand, the centromeres may never divide at all; the conditions that determine it persist as long as there are chromosomes to divide, while in meiosis it is assumed that the special conditions determining precocity act once for all at the prophase of the first division and determine the second division only through their effect on chromosome behaviour at the first.

Supernumerary divisions which are striking in their similarity to those in Zea have been found in a *Drosophila pseudo-obscura* race-hybrid with a genotypically abnormal meiosis (Dobzhansky, 1934). Here a new generation corresponding to the plant gametophyte is introduced into the animal's life-cycle after meiosis; this is associated with meiosis being pushed back two or three divisions in development. Unlike the polymitosis in Zea, the divisions are not precocious and are separated from one another by resting stages during which the chromosomes divide. Moreover, no separation of the daughter cells occurs. Consequently neighbouring mitoses fuse and multipolar spindles are formed. The poles are

GENOTYPE AND STRUCTURE

determined by centrosomes which thus keep pace in division with the chromosomes. The spermatids are originally tetraploid through double non-reduction and by three polymitotic divisions become $32x$.

The two abnormalities considered together show the independence of the spindle development from chromosome division on the one hand, and from cell division on the other. They indicate by their consistency that some body like the centrosome, visible or invisible, must always be responsible for its perpetual renewal.

2. THE BREAKDOWN OF MEIOSIS

(i) **Evidence of Genotypic Control.** We have seen how chromosome pairing may fail in genotypically and environmentally normal

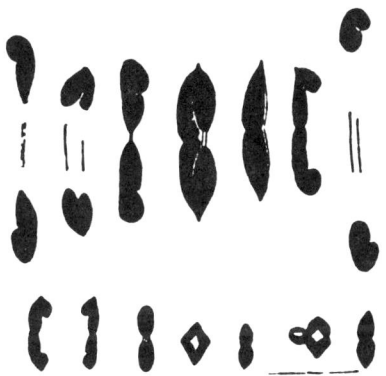

FIG. 122.—First metaphase and anaphase in two sister seedlings of *Lolium perenne* ($n = 7$) showing genotypic control of chromosome size. × 2000 (Thomas, 1936).

conditions as a result of structural or numerical abnormality, especially in hybrids. We shall later consider the results of this failure in detail. But we must first find out in what ways and to what extent the course of meiosis may be affected by genotypically abnormal conditions. The evidence by which we can decide that genotypic abnormality is concerned and structural abnormality is not, is of various kinds.

First, there are true breeding organisms whose abnormality is inherited on crossing as a mendelian character (*e.g.*, *Zea*, *Datura*,

Crepis, Matthiola) or is segregated by inbreeding (Zea, Randolph, 1928 ; Secale, Lamm, 1936). Secondly, there are organisms which differ from the normal merely in having one chromosome or part of a chromosome too few or too many (Avena, Allium, Primula kewensis). In these, abnormal meiosis is determined by unbalance. Thirdly, there are hybrids in which one sex (or kind of mother-cell) has normal pairing of chromosomes which do not pair in the other sex and do not pair even when the chromosome number is doubled

TABLE 59

Chromosome Pairing in Two Strains of *Crepis capillaris* and in Their Hybrids by *C. tectorum* (Hollingshead, 1930 b, cf. Tables 22 and 26).

	3^{II}	2^{II}	1^{II}	0^{II}
C. capillaris ($2n=6$). Normal strain .	219	5	—	—
X strain (different plants)	63	9	—	—
	84	62	4	—
	20	78	—	—
C. capillaris × *C. tectorum* ($2n = 7$).				
Normal strain (different plants)	47	16	—	—
	34	49	15	3
X strain (different plants)	32	44	44	45
	16	39	31	38
C. capillaris × *C. tectorum* ($3x = 10$, C C T), believed *X* strain derivative. (N.B.—Trivalents rare).	usual	rare	—	—

	7^{II}	6^{II}	5^{II}	4^{II}
($4x = 14$, C C T T, *X* strain derivative)	22	18	8	1

(e.g., *Viola Orphanidis*, *Drosophila pseudo-obscura*, *Smerinthus* and *Pygæra* hybrids). Similarly sister progeny in back-crosses or F_2s may show sharp differences in regard to chiasma-frequency and chromosome pairing which must be due to segregation of specific genetic factors and not to particular structural differences, since all the chromosomes are equally affected. Thus Peto (1934) found the following chiasma-frequencies : *Lolium perenne*, 1·81 ; *Festuca pratensis*, 1·88 ; F_1, 1·71 ; F_2, 5 plants 1·57 to 1·80 ; 2 plants with defective pairing 0·80 and 0·62. Again in a tetraploid backcross of derivatives of this hybrid to *F. arundinacea* ($6x = 42$) some seedlings had seven bivalents and fourteen univalents, while others had complete pairing.

Finally, there are organisms in which the later stages of meiosis are abnormal, although the earlier stages showed no defect in pairing.

The evidence of genotypic control may thus be direct or it may

TABLE 60

The Causal Relationships of Different Kinds of Non-pairing of Chromosomes at Meiosis.

General Character	Genotypic Abnormality ("Asynaptic" Mutants)	Lack of Genotypic Adaptation to Structural Conditions (New fragments)	Structural Differences (Structural hybrids)	Numerical Abnormality (Trisomics, Triploids, etc.)
Chromosomes affected.	All chromosomes.	Particular chromosomes.	Some or all chromosomes.	Some or all chromosomes.
Special condition.	Lack of precocity.	Chromosomes too short.	Frequent changes of homology in linear sequence.	Lack of partner for part of length owing to competition.
First result.	Reduced pachytene pairing (to a variable extent).	Length of pairing shorter than the average.	Reduced pachytene pairing (to a variable extent).	
Second result.	Numbers of chiasmata formed insufficient for regular metaphase pairing.			

merely exclude the possibility of structural control. The exclusion of environmental control is not important for an analysis of the mechanism, since in effect it is analogous and in function it is complementary to genotypic control. Thus in *Allium odorum* (Modilewski, 1930), which is true-breeding, embryo-sac mother-cells vary in the number of their bivalents from none to the full sixteen under environmental and genotypic control (*cf.* *Saccharum*, Bremer, 1930 ; *Kniphofia*, Moffett, 1932). The same is true of differences of behaviour in different cycles of cyclical parthenogenesis as in *Phylloxera* (Morgan, 1915, Ch. IX). Particular treatment with heat and other agents has produced effects on meiosis analogous to those resulting from abnormal genotypic conditions, but less precise and less significant (De Mol, 1928, 1933 ; Stow, 1930).

The occasional failure of pairing between partners that normally pair in approximately true-breeding organisms (Table 61) may arise from many causes. In some cases the shortest chromosome is affected, and we may then suppose that the margin of safety in chiasma-formation is too small and environmental conditions are directly responsible. The particular mechanism will probably be associated with the high variance in chiasma frequency between bivalents and the low variance between nuclei which Mather has ascribed to " competition " (Ch. VII). Where long chromosomes are affected, or the nucleus is polyploid, we may suppose that exceptionally awkward distribution in the zygotene nucleus has prevented the chromosomes from pairing in time.

TABLE 61

Exceptional Failure of Pairing not due to Hybridity

Diploids.
Nicotiana alata ($2n = 18$) . . Ruttle, 1927 ; Goodspeed and Avery, 1929 *a*.
Zea Mays ($2n = 20$) . . . Randolph, 1928.
Ranunculus acris ($2n = 14$) . Sorokin, 1927 *b* ; Larter, 1932.
Cycas revoluta ($2n = 22$) . Nakamura, 1929.
Dorstenia plumariæfolia ($2n = 26$) Krause, 1931 (? fragments).
Crepis capillaris ($2n = 6$) . . Hollingshead, 1930 *b* (*v. supra*).
Trimerotropis citrina ♂ ($2n = 23$) Carothers, 1931 (unequal bivalent).
Uvularia grandiflora ($2n = 14$) . Belling, 1925 *c*.
Secale cereale ($2n = 14$) . . Mather and Lamm, 1935.
Scilla italica ($2n = 16$) . . Dark (1932 *c*, Fig. 123)⎫
Fritillaria pontica ($2n = 24$) . D., 1935 *g*   ⎬ Chromosomes with lowest chiasma-frequency.
Chorthippus parallelus (S chromosome) . . . . D., 1936   ⎪
Mecostethus grossus ($2n = 23$) . White, 1936   ⎭
Stauroderus bicolor (L chromosome) D., 1936.
Triploid
Narcissus poetarum ($3x = 21$) . Nagao, 1929 ($6^{III} + 3^I$).

Tetraploids
  Nicotiana Tabacum ($4x = 48$) . . R. E. Clausen, 1930, 1931.
  Prunus Cerasus ($4x = 32$) . . D., 1928.
  Primula sinensis ($4x = 48$) . . D., 1931 a.
  Campanula persicifolia ($4x = 32$). Gairdner and D., 1931.
  Triticum durum ($4x = 28$) . . Thompson and Robertson, 1931.
  Tulipa stellata ($4x = 48$) . . D. and Janaki-Ammal, 1932.
  Kniphofia sp. ($4x = 24$) . . Moffett, 1932.
Hexaploids
  Triticum vulgare, etc. ($6x = 42$) Hollingshead, 1932.
Octoploid
  Bromus erectus, eu-erectus ($8x = 56$) Kattermann, 1931 (v. Table 31).

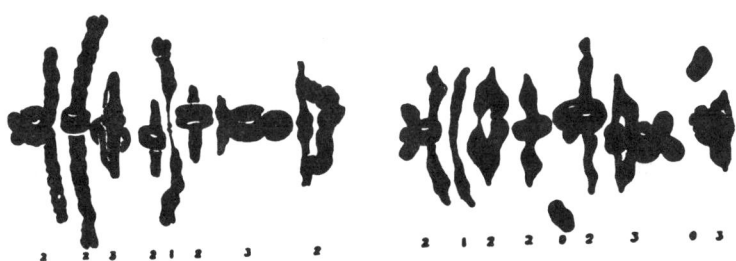

FIG. 123.—Side views of two first metaphases in *Scilla italica* ($n = 8$), the chromosomes drawn separately. The numbers of chiasmata are given below each bivalent. Right : the two shortest chromosomes are unpaired, having failed to form one chiasma. An example of lack of adaptation of chiasma frequency to the range of chromosome size as in organisms with new fragments. (After Dark, 1934.)

Genotypically controlled abnormalities, like those of structural origin, vary in degree from cell to cell and the consequences of their variations can be ascertained. But genotypic abnormalities are particularly important for the analysis of meiosis and of chromosome mechanics in general, because unlike those that are of structural origin they show themselves at different stages. The effects of divergence at each stage on the later stages can therefore be compared and the degree to which successive events are independent or otherwise can be determined. Sometimes this is not possible, because either the abnormality of behaviour is spread over a considerable space of time, or it is undefined in its nature (or in the description that is available). In other cases precise conclusions may be drawn.

TABLE 62

Classification of Genotypically Conditioned Abnormalities of Meiosis

1. FAILURE OF METAPHASE PAIRING DUE TO FAILURE OF CHIASMA FORMATION UNDER ASCERTAINED CONDITIONS.

(a) *Not Associated with Semi-mitotic Contraction.*
Zea Mays (asynaptic)	Beadle, 1930, 1933 b.
Datura Stramonium (bb)	Bergner *et alii*, 1934.
Crepis capillaris (" X-strain ")	Richardson, 1935 c.
Triticum vulgare *	Hollingshead, 1932.

(b) *Associated with Semi-mitotic Contraction.*
Matthiola incana (long-chromosome type)	Lesley and Frost, 1927; Philp and Huskins, 1931.
Chrysanthemum (small nuclei from premeiotic irregularity)	Shimotomai, 1931 b.
Secale cereale (inbred)	Lamm, 1936.
Ascaris megalocephala bivalens (one individual)	Geinitz, 1915.

(c) *Associated with Timing Irregularities.*
Ochna serrulata (facultative parthenogenetic embryo-sac mother-cell)	Chiarugi and Francini, 1930.
Eu-Hieracium spp. (parthenogenetic)	Rosenberg, 1927.
Nicotiana Tabacum (variety with unreduced gametes)	Rybin, 1927.
Aucuba japonica *	Meurman, 1929 a.
Prunus hybrids *	D., 1930 a.
Triticum vulgare *	Hollingshead, 1932.
Drosophila pseudo-obscura, inter-race hybrids	Dobzhansky, 1934.

(d) *Under Undefined Cytological Conditions.*
Nicotiana alata (2x + 1)	Avery, 1929.
Nicotiana Tabacum (" Pale sterile "), 4x + 1	R. E. Clausen, 1930, 1931.
Triticum vulgare × *T. durum*, 6x pure line derivative	L. A. Sapehin, 1931.
Triticum vulgare (6x − 2)	Huskins and Hearne, 1933.
Avena sativa (6x − 2s_2)	Huskins, 1927; Nishiyama, 1931, 1933, 1935.
Sorghum sp.	Huskins and Smith, 1932.
Primula kewensis (4x − 2)	Newton and Pellew, 1929.
Primula malacoides	Kattermann, 1934.
Hevea sp.	Ramaer, 1935.
Viola Orphanidis (male sterile)	J. Clausen, 1930.
Rumex Acetosa ♂	Yamamoto, 1934.
Angiostomum, *Rhabditis* and other Nematoda (X chromosomes in special circumstances, Ch. IX).	
Pygæra pigra × *P. curtula* ♂ (nearly complete pairing in female)	Federley, 1931.

* Irregularities in particular cells with reduced pairing or restitution nuclei resulting from reduced pairing.

2. Partial Failure of Spindle and Cell-Wall Formation at First or Second Division.

Kniphofia Nelsoni	Moffett, 1932.
Cannabis sativa (with environmentally controlled diœcism)	Breslawetz, 1935.
Impatiens pallida	Smith, 1935.
Zea Mays (variable sterile)	Beadle, 1933 (a).*
Triticum turgido-villosum	v. Berg, 1934.*
Drosophila pseudo-obscura	Dobzhansky, 1934.
Lathyrus odoratus (race with delayed first metaphase and complete terminalisation)	Fabergé 1936; Upcott, 1936.

2. (a) Total Failure of Second Division Spindle.

Datura Stramonium (stock derived from radium-treated plants)	Satina and Blakeslee, 1935.
Allium schœnoprasum ($3x + 3$)	} Levan, 1935 b.
A. nutans ($5x + 2$)	
Sphærocarpus Donnellii (possibly).	Allen, 1935.

1. or 2. Failure of Pairing at First or Second Division Induced by Treatment (cf. Table 65).

(ii) **Failure of Metaphase Pairing.** In all cases of the genotypic reduction or suppression of metaphase pairing there is some pachytene pairing, except in certain parthenogenetic organisms where, as we shall see, all relationship with meiosis, except position in the life-cycle, is lost. The circumstances in which pairing fails are clearly defined only in the "asynaptic" strains of *Crepis* and *Zea*, in both of which metaphase pairing is variable and the frequency of chiasmata varies from normal to none at all. In these plants pachytene pairing seems to be fairly complete as a rule, but at diplotene the chromosomes gradually fall apart. Their separation is delayed, especially in *Crepis*, by the untwisting of the relational coiling that has developed during pachytene. From this account it might appear that the abnormality began with the failure of chiasma-formation in a previously normal pachytene nucleus. But it is found that where failure of chiasmata is absolute, pachytene is also very defective, *e.g.*, in *Hevea*. Where it appears complete, chiasmata are always formed in a proportion of nuclei—a proportion subject to great local variation. Thus in *Crepis* one bivalent may have three chiasmata, while none of the other chromosomes in the same nucleus have any: they are unpaired as soon as they are uncoiled.

* A tripolar spindle is sometimes formed at the first metaphase.

406 BREAKDOWN OF GENETIC SYSTEMS

It is easier, therefore, to suppose that the failure of chiasmata springs from a failure of true pachytene pairing, such as has been determined with ordinary localisation, rather than as a new unexplained abnormality of chiasma-formation. Thus pachytene pairing may begin normally in these plants, but be interrupted as

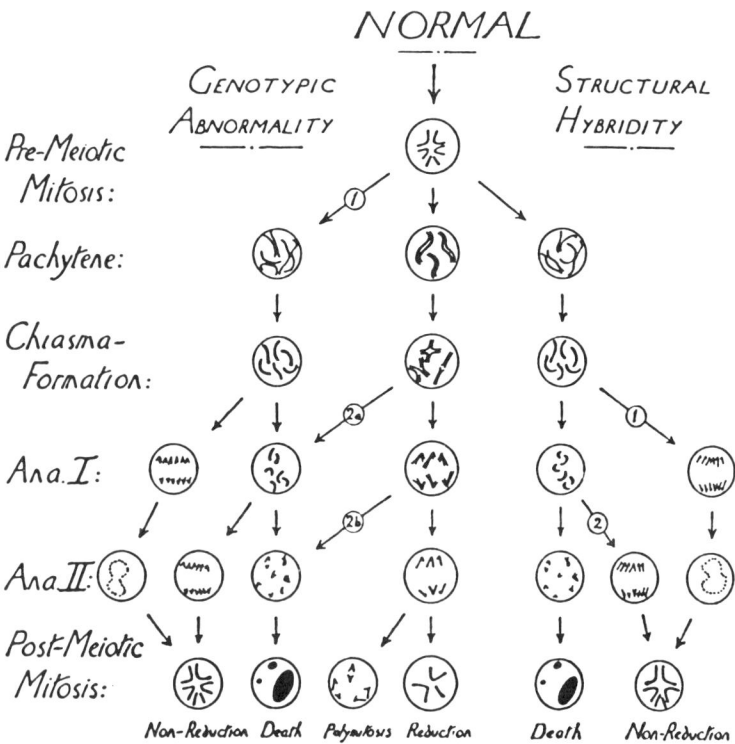

FIG. 124.—Diagram to show the differences between abnormalities of meiosis due to genotypic and structural causes (*cf.* Table 62).

it is with localisation by division of the chromosomes. It will then continue as torsion pairing in the intercalary unpaired regions but, owing to the chromosomes having already divided, will give no chiasmata in these parts. It will be effectively incomplete. As we shall see later, there is reason to suppose that paired chromosomes cross over and form chiasmata only at the moment of division.

On this view failure of metaphase pairing is, like localisation, a result of reduced precocity, but unlike localisation it is a disordered result; it depends on an incomplete pairing at pachytene, which is variable in its distribution, and gives rise to a reduced chiasma

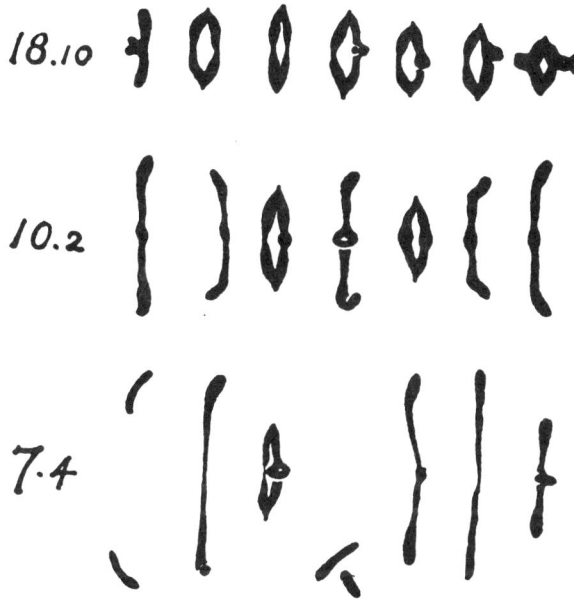

FIG. 125.—Different degrees of reduced precocity and reduced pairing in different inbred plants of *Secale cereale*. × 1000. (Lamm, 1936.)

formation, which is variable in its frequency. This assumption explains the enormous variation in chiasmata from nucleus to nucleus as the result of variable precocity and the enormous variation from bivalent to bivalent in the same nucleus as the result of chance variation in position of the homologous threads in the zygotene nucleus. We might then expect as a symptom of reduced precocity, a reduced contraction of the chromosomes at metaphase.

This is not by any means inevitable, since the first timing abnormality leads to others (Table 62, Fig. 124), and sometimes, as we shall see, upsets the development of the spindle, which itself determines metaphase. We find, however, that long " semi-mitotic " chromosomes are characteristic of several cases of reduced chiasma-formation (Table 62 (1) (a)).

In the absence of secondary abnormalities, the effect of genotypically controlled failure of pairing on the later course of division is in no way different from that of structurally controlled failure of pairing. They will be considered together later.

(iii) **Failure of the Spindle.** In particular races of a number of species pairing of the chromosomes at meiosis is complete and regular, but the chromosomes or daughter nuclei fail to separate at anaphase of the first or second division. Or if they separate, in plants, a cell-wall fails to develop between them. Now, as Belar and others have found, the spindle changes in shape in relation to both the anaphase movement of the chromosomes and the formation of the cell-wall. This relationship suggests that both depend on the activity of the spindle.

In *Triticum* the breakdown of the spindle leads to the formation of numerous small nuclei. In inter-race hybrids of *Drosophila pseudo-obscura* there is a genotypically controlled failure of pairing. Correlated with this abnormality is a premature stretching of the spindle which bends within the confined space of the cell. This is sometimes a consequence of lack of pairing, as we shall see later. But in this case it is not a consequence, for nuclei with complete pairing have bent spindles. It is correlated with it inasmuch as it occurs in the same organisms and therefore proceeds from the same prior conditions. Probably it is due to the spindle developing too early in relation to the chromosomes. It seems that to prevent the spindle stretching and bending it is necessary not only to have paired chromosomes, but also to have them there at the right time. On the other hand, stretching does not occur in another hybrid in this species when unpaired chromosomes are ready to divide and form a plate at the first division ; that is, when the spindle is less precocious in relation to the chromosomes.

The opposite case of the spindle failing before the chromosomes are separated at anaphase is perhaps due to its developing too soon. In *Kniphofia Nelsonii* (Moffett, 1932 *a*), a series of abnormalities occur during or after meiosis, apparently according to the time at which spindle degeneration sets in. This probably varies with the position of the flowers in the spike. In the extreme case the chromosomes form, instead of two nuclei at the end of the first division, a number of scattered nuclei which develop separately. If the first division proceeds regularly the same abnormality may befall the second division, and where this also is regular a wall may fail to form between two or all four of the daughter-nuclei. In consequence, these, or the numerous irregular nuclei, often fuse during the first mitosis of the single or double giant pollen grain. Side by side with these abnormalities normal tetrads are formed. Variation in the abnormality presumably depends on differences in developmental conditions.

In *Datura* and *Allium* the breakdown sets in at a definite period and inhibits the second division so that unreduced pollen grains are formed. Even here the result is not entirely regular. Normal second divisions may occur. The highest degree of precision is found as we shall see in parthenogenetic organisms in the suppression of either the first or the second division, and there is no reason to doubt that selected mutations have in these cases developed an analogous genotypic control.

Genotypic control of the spindle and of its relation with particular chromosomes has played an important part, not only in the evolution of parthenogenesis, but also in that of sex-chromosomes (*q.v.*). Probably there are no flowering plants which do not exceptionally fail to give four reduced nuclei at meiosis, under such conditions that they survive in nature. Levan (1935 *b*), for example, finds giant pollen-grains in nearly all species of *Allium*. Several workers have by special treatment succeeded in suppressing the first or second division regularly, even in some cases reducing the pairing of the chromosomes in otherwise normal organisms (Table 65). Whether pairing can be seriously reduced by external means without killing the cells before meiosis is complete is, however, doubtful.

These observations show that the mechanism of the spindle is, in its origin, genetically and physiologically independent of chromo-

410 BREAKDOWN OF GENETIC SYSTEMS

some behaviour, and therefore capable of independent adaptation. Such adaptation has led to the co-ordination which secures regular reduction in normal meiosis as well as to the special systems referred to above and discussed in Chapter XI.

(iv) **The Behaviour of Unpaired Chromosomes.** Unpaired chromosomes have the same general type of behaviour whether in numerical, structural or undefined hybrids or organisms with genotypically

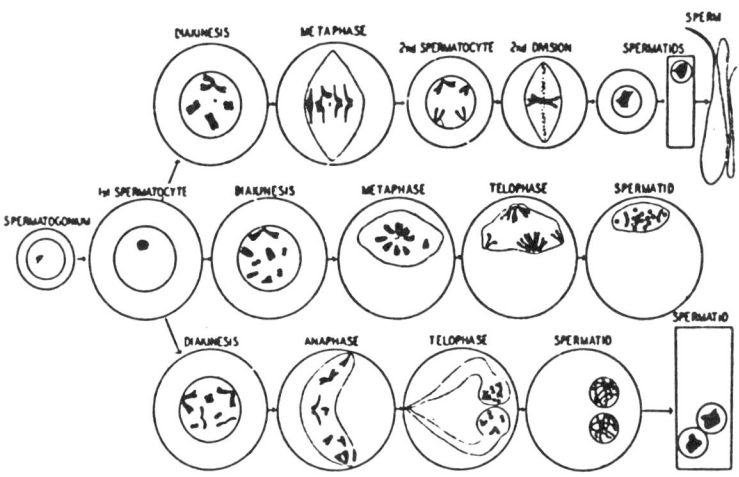

Fig. 126.—A scheme of the spermatogenesis in normal males (above), in B♀ × A♂ hybrids (middle), and in A♀ × B♂ hybrids (below) of *Drosophila pseudo-obscura*. (Dobzhansky, 1934.)

determined failure of pairing. One description will therefore apply to all these cases.

Unpaired chromosomes usually lie at random on the spindle at metaphase. They do not move towards the equator as early as the paired chromosomes. It is sometimes stated that unpaired chromosomes lying to one side of the plate are moving to the pole *in advance of* the bivalents at anaphase, but this conclusion is unjustifiable. Their position is due to their never having reached the plate, and they actually do not move until after the bivalents have divided.

When the paired chromosomes begin to separate at anaphase the unpaired chromosomes follow one of two courses: (i) those lying far away from the equator are included with the group of daughter bivalents passing to the nearest pole; (ii) those lying near the equator move on to the plate, orientate themselves axially, and divide after a short interval into their two chromatids, which then pass to opposite poles as in mitosis. The first gives *post-division* of the univalents, the second *pre-division*.

In *Triticum-Ægilops* hybrids Kihara (1931) in a detailed account has shown the probable history of the univalents to be chiefly of the second type, which is as follows: (i) The bivalents arrange themselves on the equatorial plate; the univalents remain distributed at random. (ii) The univalents arrange themselves on the edge of the equatorial plate. (iii) The bivalents divide; the univalents remain on the equator. (iv) The univalents divide and their halves follow the daughter bivalents to the poles. The variations commonly observed in univalent behaviour are probably due to various degrees of delay in the movement of the univalents relative to those of the bivalents.

In the extreme case they altogether fail to divide at the first division, some being included in the daughter nuclei and others lost in the cytoplasm.

Where multiple configurations are formed it often happens that one of the chromosomes is left on the plate at anaphase (*e.g.*, in *Hyacinthus*, Ch. IV). The same result may follow in a simple bivalent, dividing while lying on an accessory plate to one side of the main plate, owing to overcrowding (*e.g.*, *Fritillaria*, Ch. XII). The chromosome going to the further pole is stopped on the equator. Such a false univalent divides at late anaphase like a chromosome that has not been paired at all. But it may show two peculiarities in *Tulipa* (Upcott, unpub.). In the first place the two chromatids, owing to one or both of them having crossed over at chiasmata with the chromatids of a partner, do not lie close together, and may differ in structure and in length. In the second place, owing presumably to abnormal strain, the chromatids may be distributed unequally on either side of the centromere, two sister chromatids remaining attached to the same daughter centromere. This anomalous behaviour is presumably a means of structural change (*cf.* Nishiyama, 1931).

The daughter univalents usually lag at the second division or

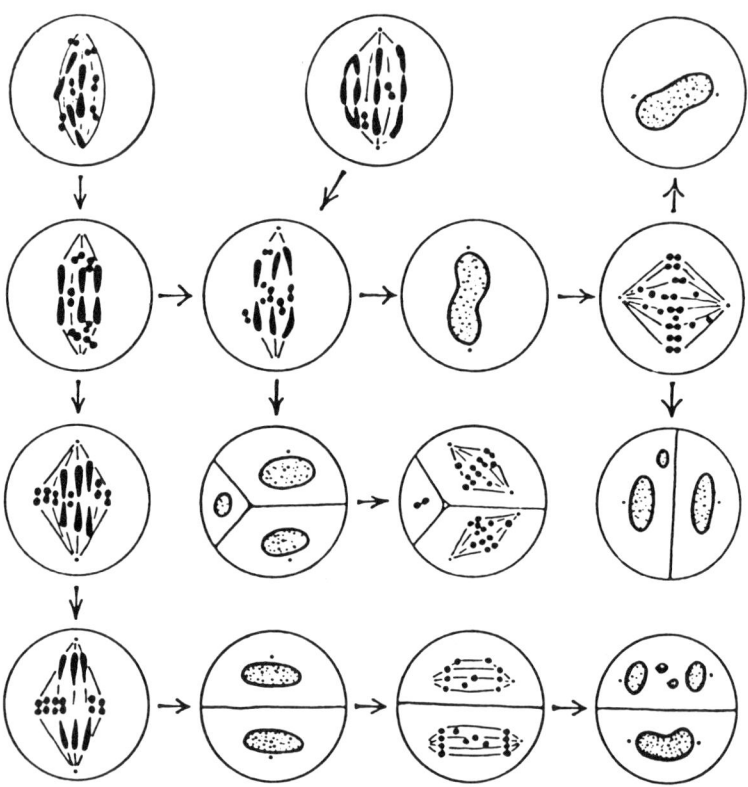

Fig. 127.—The various possible courses that meiosis may follow when several univalents or multivalents are formed, leading to (i) irregular reduction, (ii) non-reduction at either first or second division, (iii) double non-reduction.

pass without division to one pole or the other. With division at either the first or the second division, univalents are lost in the cytoplasm through failure to reach the poles.

Where all, or nearly all, the chromosomes are univalent their relationship to the spindle is often abnormal. Sometimes the nucleus relapses to a resting stage without any movement having taken place. This is a means by which unreduced gametes may arise (*Papaver*, Ljungdahl, 1922 ; *Hieracium*, Rosenberg, 1927).

With very few paired chromosomes and extreme delay in the unpaired chromosomes orientating themselves on the plate, a normally developed spindle increases in length, and, as we saw earlier, is bent round until the poles nearly touch (*Triticum*, Matsumoto, 1933 ; *Drosophila*, Dobzhansky, 1934 ; Koller, 1934 ; *Impatiens*, F. Smith, 1935 *et alii*).

Evidently the paired chromosomes prevent this lengthening and modify the shape of the spindle, just as convergent associations of three chromosomes modify the direction of repulsion between the centromeres of associated chromosomes in the spindle.

Reviewing these observations we see that the chromosomes which lag are univalents at the first division and daughter univalents at the second. These chromosomes evidently fail to move, although they resemble the other chromosomes in every respect save in not having partners. Movement, therefore, depends on a reaction—presumably repulsion—of pairs of elements. The presence of the spindle, and of the chromosome in the right position in relation to it, is not sufficient.

The behaviour of univalents thus shows that the ability to divide at anaphase is due to a change in the organ of division, the centromere. This change must be an assumption of bipolarity. It also shows that the time of this change in the centromeres of univalents is different in different organisms in relation to the metaphase, that is, in relation to the development of bivalents. But in all cases the univalents are later than this metaphase in becoming bipolar. The precocity of the metaphase in meiosis agrees well enough with the conditions of separation of bivalents, but it leaves the univalents more or less out of step. The degree of precocity of the metaphase, like that of the prophase, varies from one organism to another.

(v) **The Consequences of Non-pairing.** We have seen that chromosomes fail to pair either on structural or numerical grounds (in haploids, triploids and ordinary hybrids) or on genotypic grounds. Whatever the grounds the same kinds of results ensue, the particular kind depending on the proportion of chromosomes that are paired and are still able to behave normally, and on the time when the centromeres of the unpaired chromosomes divide in relation to the time of development of the spindle. Let us take first the extreme case of the haploids or hybrids or genotypic forms with no pairing at all. The univalent chromosomes come to lie on the spindle more or less at random. Some of the chromosomes may then move to the equator and begin to divide as at mitosis. In *Triticum turgido-villosum* ($4x$) all of the chromosomes may do so. In *T. monococcum-dicoccum* ($3x$) only a proportion do so (Mather, 1935).

The same is true of one of the *Drosophila pseudo-obscura* hybrids. In such cases the first division is normal. But since the chromosomes do not divide twice the second division is abortive. The chromosomes do not form a metaphase plate and a single nucleus is reconstituted. It may happen, on the other hand, that *none* of the chromosomes reach an equatorial position at the first division. A single *restitution* nucleus may then be re-formed and the second division is usually successful. The intermediate condition, in which *some* of the chromosomes make a plate at the first division, may lead to the formation of a restitution nucleus at one or both divisions. Thus one, two, three or four nuclei may result at second telophase.

In the *Pygæra* hybrids with complete failure of pairing, all the chromosomes divide at both divisions. Regular diploid gametes are then formed in fours. This regular double division of the chromosomes is not known in any plant.

When none of the univalents move to the equator at the first division, the spindle may stretch as it does between separating anaphase chromosomes. But, instead of separating, it then merely disperses the undivided chromosomes. Being scattered in this way they may form two daughter nuclei or several—the number being limited by the number of chromosomes and the size of the cell (*cf.* Fig. 127). Each nucleus then goes through a second division which is successful so far as the two main daughter nuclei are concerned.

When a moderate number of bivalents or multivalents are formed the case is different. The spindle does not stretch until the bivalents are divided and the regularity of their first and second division determines a regularity in the first and second division of the cell. Abnormality depends on the fact that, as a rule, the univalents can divide only once. They may lie on the spindle at first anaphase without dividing ; or having come on to the equator

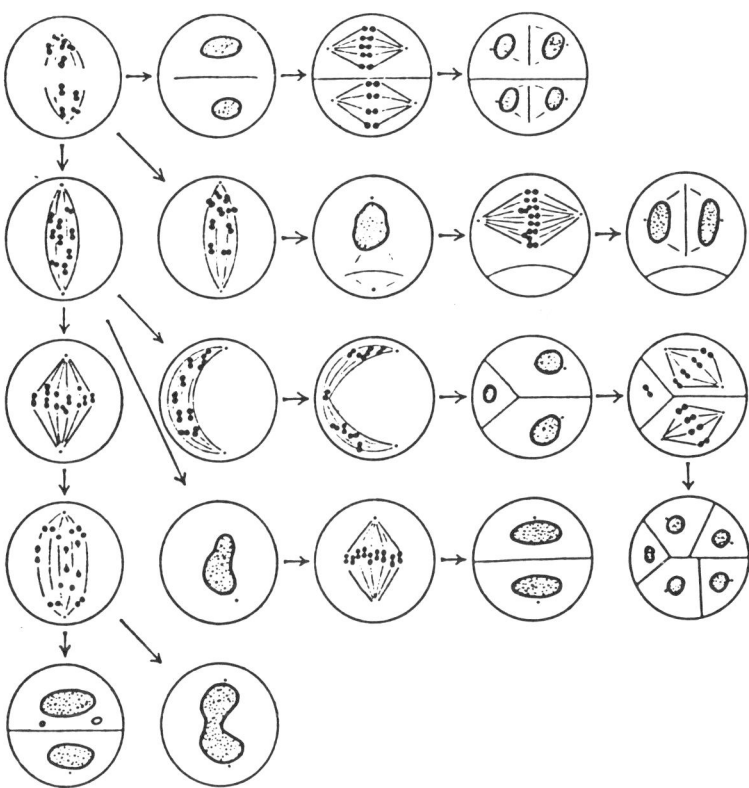

FIG. 128A.—Possible courses of meiosis in the total absence of pairing.

and divided at the first anaphase quickly enough to be included in the daughter nuclei, they may be unable to divide again, and consequently lag on the spindle at second anaphase. These lie cross-wise at first and are pulled lengthwise later by the stretching of the spindle (Mather, 1935). In either of these cases they may

416   BREAKDOWN OF GENETIC SYSTEMS

form a connection between the daughter nuclei, which will fuse to give a restitution nucleus either at first or second anaphase or both. In this last case only a single nucleus is produced with double the somatic number of chromosomes.

In all these circumstances we see that we are dealing with the varying relationships of the time of division of bivalents and univalents to one another and to the development of the spindle. We also see that while the position of the chromosomes depends in the first instance on the spindle, the changes of shape in the spindle depend on the separation of the paired chromosomes in bivalents and of the daughter chromatids in dividing univalents. The behaviour of dividing cells in which the normal relationships of the spindle to the chromosomes has been upset by failure of pairing is nearly always variable. No doubt this is due to a susceptibility to environmental changes from which the normal process is protected by a wide margin of safety.

(vi) **Non-reduction.** The results of a regular failure of the first or second division is the formation of unreduced or approximately unreduced gametes which function and yield polyploid progeny in the case of diploids, diploid progeny in the case of haploids. Sometimes failure of both divisions may yield gametes with double the parental number, so that diploids may yield hexaploid and octoploid progeny (*e.g.*, in *Raphanus-Brassica* and *Chrysanthemum*) and haploids yield triploid progeny.

The results of the functioning of polyploid gametes have already been shown in relation to the origin of polyploids. Their effect on the reproduction of haploids is shown in Table 64 to be analogous.

TABLE 63

Numbers of Cells in " Tetrads " of Haploid Plants

	Monads.	Dyads.	Triads.	Tetrads.	Pentads.	Hexads.
Datura (1867) $n = 12$	—	187	(a few)	1385	73	205
Matthiola (406) $n = 9$	3	285	99	19	—	—
Crepis (282) $n = 3$	12	195	57	19	—	—

UNREDUCED GAMETES

TABLE 64

Progeny of Haploids

Datura (Belling and Blakeslee, 1927). Selfed : 381 probable diploids, 12 trisomics.
Solanum Lycopersicum (Lindstrom, 1929). No selfed seed. 42 seedlings from 100 fruits with pollen of diploid.
Nicotiana glutinosa (Goodspeed and Avery, 1929). No seed when selfed or crossed reciprocally with diploid.
Nicotiana Langsdorffii (Kostoff, 1929). No selfed seed, few seeds when crossed with pollen of diploid.
Nicotiana Tabacum (Clausen and Mann, 1924). No seeds.
Œnothera franciscana (Davis and Kulkarni, 1930). Selfed : 694 diploid, 29 haploid, 17 mutant diploid.
Crossed with diploid : 531 diploid, 2 mutant diploid (*cf.* secondary structural change).
Œnothera franciscana (Stomps, 1930). Selfed : gave trisomics and triploids.
Triticum monococcum (Katayama, 1935). Open pollination : 3 haploid, 3 diploid.

TABLE 65

Cases of Non-Reduction

(*cf.* Table 62)

A. (i) Failure of Pairing due to hybridity (*v.* also Haploids and Parthenogenesis).

α. General Hybrids.

Drosera (3x)	Rosenberg, 1909.
Papaver	Yasui, 1921 ; Ljungdahl, 1922.
Saccharum	Bremer, 1923.
Triticum-Secale	Kihara, 1924.
Nicotiana (3x, 5x, etc.)	Goodspeed and Clausen, 1927 ; Rybin, 1927 ; Brieger, 1928.
Petunia	Matsuda, 1928.
Prunus (2x, 3x, 6x)	Kobel, 1927 ; D., 1930 a.
Œnothera (4x)	Schwemmle, 1928.
Ribes	Tischler, 1928.
Digitalis	Buxton and Newton, 1928.
Canna aureo-vittata gigas	Honing, 1928.
Triticum-Ægilops	Kagawa, 1929 ; Kihara, 1931 Kihara and Katayama, 1931.
Aucuba japonica (4x)	Meurman, 1929 a.
Triticum hybrid (3x)	Thompson, 1931.
Triticum turgidovillosum	v. Berg, 1934.
Brassica	Catcheside, 1934.

β. Purely Numerical Hybrids.

Datura (3x)	Belling and Blakeslee, 1923.
Solanum (3x)	Jørgensen, 1928.
Tulipa (3x)	Newton and D., 1929.
Tradescantia (3x)	D., 1929 c.
Salix (3x)	Håkansson, 1929.

418 BREAKDOWN OF GENETIC SYSTEMS

γ. Haploids.

Datura Stramonium	Belling and Blakeslee (1923, 1926).
Matthiola incana	Lesley and Frost, 1928.
Nicotiana glutinosa	Goodspeed and Avery, 1929
Nicotiana Langsdorffii	Kostoff, 1929.
Crepis capillaris	Hollingshead, 1930 a.
Œnothera franciscana	{ Emerson, 1929. Davis and Kulkarni, 1930. Stomps, 1930, a and b.
Aegilotricum (8x)	Katayama, 1935.
Triticum monococcum	Katayama, 1935.
Œnothera rubricalyx	Gates and Goodwin, 1930.
Œnothera blandina	Catcheside, 1932.

(ii) Due to a Genetic Property (cf. Section 4).

Perla marginata (♂)	Junker, 1923.
Avena sativa (6x − 2)	Huskins, 1928 ; Nishiyama, 1931.
Nicotiana alata (2x + 1)	Avery, 1929.
Zea Mays (asynaptic)	Beadle, 1930.
Drosophila pseudo-obscura hybrids	Dobzhansky, 1934.

All cases of facultative and cyclic parthenogenesis with suppression of the first division (v. Ch. XI).

B. Double division of univalents (partial except in the first instance).

Pygæra hybrids	Federley, 1912, 1931 ; Cretschmar, 1928.
Rosa canina	Täckholm, 1922.
Rosa subglauca	Erlanson, 1929.
Raphanus-Brassica	Karpechenko, 1927, a and b.
Prunus avium nana	D., 1930 a.
Saccharum officinarum	Bremer, 1930.
Ribes, hybrids	Meurman, 1928.
Brassica, hybrids	Morinaga, 1929.

C. Failure of Spindle ; due to genetic property.

Triticum vulgare var.	L. A. Sapehin, 1931.
Kniphofia	Moffett, 1932.
Hyacinthus orientalis	De Mol, 1923 ; Stow, 1930.
Zea Mays (var. sterile)	Beadle, 1932.

D. Due to External Conditions (effect shown by abnormal treatment).

Vicia	Sakamura, 1920.
Funaria	v. Wettstein, 1924.
Epilobium	Michaelis, 1926.
Gagea	Sakamura and Stow, 1926.
Liriope, Scilla	Shimotomai, 1927.
Lychnis	Takagi, 1928.
Triticum	Bleier, 1930.

(vii) Conclusion. These observations are important in four directions. They explain the origin of polyploids by failure of meiosis. They show the relationship of normal meiosis with its special modifications in apomictic organisms where, as we shall see, the sexual cycle is suppressed. They also show that the metaphase of meiosis, as well as the prophase, is precocious, since the univalent

BREAKAGE OF CHROMOSOMES

chromosomes are not then ready to divide. And finally, they show that the movements of the spindle and of the chromosomes which are independent in origin and independently controlled, mutually react as soon as they come together. How they come to react we shall consider in detail in relation to the special problems of chromosome mechanics.

3. THE EFFECTS OF STRUCTURAL CHANGE

(i) **Conditions of Change.** The chromosomes, as we have seen, are subject to changes which may be provisionally classified as intra-genic or molecular and inter-genic or structural. Apparently, as we have seen, all parts of the chromosome are subject to inter-genic, while only the physiologically active parts are subject to intra-genic, changes. Since, however, intra-genic changes are detectable only by their physiological effect we may suppose that inertness merely prohibits change back to an active form.

The rate at which these changes occur shows that the conditions determining them are various. Temperature, developmental conditions, genotype and the structure of particular genes, affect them. We have previously seen the physiological and structural conditions of chromosome behaviour in antithesis. They are alternative agents of change. We now see on a deeper analysis that they control one another. There is an inter-penetration of opposites.

Apart from indirect control of structural change certain special agents, short-wave irradiation and direct manipulation, such as centrifuging (Kostoff, 1935), determine its occurrence. There is no practical means as yet of making a distinction between " natural " changes such as may depend on inherent defects in the reaction of the gene with its substrate, and " induced " changes that depend on violent external interference. It is convenient, however, provisionally to separate natural from induced changes according to whether the method of control is internal or external to the organism.

(ii) **Natural Change.** Natural changes in the structure of chromosomes are known chiefly by their results at meiosis and in the ensuing generation. After this interval they can be described in kind. But during the interval we know that many of the nuclei with new chromosomes must have been eliminated and we cannot say how many or of what kinds these may be. Moreover, their occurrence is so rare that we can know nothing accurate about their frequency. The best indication we can get is from mitosis in exceptional organisms with a genotypically controlled high frequency of structural change.

Two stocks have been discovered with frequent structural changes. The more extreme was the one derived from a *Triticum-Secale* hybrid (Plotnikova, 1932) which showed as many as one-fifth of the mitoses with chromatid bridges and fragments. These are clearly the result of the formation of dicentric and acentric chromosomes by asymmetrical interchange, like those formed by crossing-over in dyscentric hybrids at meiosis. This type of interchange, let us remember, is precisely the type that will show itself at mitosis as no other will. And it is also precisely the type which will be eliminated during somatic life and will consequently never appear at meiosis or in the progeny (Plate XII).

A less extreme frequency was found by Beadle in the " sticky chromosome " mutant of *Zea Mays* (1932). A few changes in size and number at mitosis indicate that structural change is occurring with unusual frequency. At meiosis the result is a great but variable confusion in the association of the chromosomes. Sometimes a single chromosome group is formed as though a dozen translocations had taken place. At anaphase several bridges are found, showing that the changes include inversions. Beadle attributes this behaviour to " stickiness " of the chromosomes, but the spindle is normal and regular tetrads can be formed. The abnormality seems therefore to be sufficiently accounted for by a high rate of structural change.

If the natural change-rate is once in a million mitoses, the *Zea* rate is perhaps once in a thousand mitoses, and the *Triticum-Secale* rate once in a hundred. With X-ray treatment, as we shall see later, the rate may be increased to ten times in one mitosis ; the limit is merely the viability of the cell, of the tissue, or of the organism.

Normal changes continue or may even be hastened in resting seed, according to the work of Navashin (1933) on *Crepis tectorum* raised from six-year-old seed. As in the " sticky " plants in *Zea* there is a high cell mortality even amongst the 3 per cent. of seeds that germinate. Evidently many secondary changes take place during development through the breakage of dicentric chromatids at anaphase (*cf.* Cartledge and Blakeslee, 1934, on *Datura*). Peto (1935) and Schwarnikow and Navashin (1934) found that the rate might be greatly increased under special conditions of temperature and moisture.

In pollen grains of a triploid *Tradescantia* numerous structural changes, including interchange between chromatids, have been found by Upcott (unpublished). These give rise to mitotic pseudo-chiasmata between non-homologous chromosomes, and since the interchanges are sometimes asymmetrical, to anaphase bridges. Similar structures have been induced by Peto (1935) in *Hordeum vulgare* roots by heat treatment. Whether the chiasmata described by Kaufmann (1934) at mitosis in *Drosophila* are related to the somatic crossing-over that Stern (1934) and Friesen (1936) have found here, we cannot say.

Structural changes occurring within the organism occur independently in homologous chromosomes. They therefore develop a condition of hybridity within the organism in no way different from that produced by crossing different individuals. Structural hybridity therefore depends on the structural differentiation of homologues, whether they are derived from the same parental chromosomes, or from different chromosomes in the same parent, or again from widely separated parents. We may therefore say that the sticky-chromosome plant of *Zea* and the resting seeds of *Crepis* are generating hybridity within themselves, while in most organisms change is so slow that a high degree of hybridity arises only when different organisms are crossed or in permanent hybrids where, as we saw, hybridity is preserved with inbreeding or even self-fertilisation from generation to generation.

The internal generation of hybridity, by genetic changes within organisms in experiment, raises the question of how far such changes occur naturally. If they occur their results should appear in old clones of cultivated plants. The extreme hybridity of such clones has been shown in many species (*cf.* Crane and Lawrence, 1934). It has been attributed to crossing. But the occurrence of large incompatibility groups in clonal cultivated forms of *Prunus* and their non-occurrence in non-clonal populations shows that the clonal forms are, in fact, inbred (D., 1928). Their functional hybridity must therefore, in some measure, result from internal change and not from hybridisation. This question suggests another possibility of some interest in horticulture. Induced changes, of the kind that will be considered next, likewise determine structural hybridity and its concomitant sterility. Seed-sterility is an advantage for many ornamental plants. Triploids are often selected on this account. This property can now be produced artificially by irradiation.

(iii) Changes Induced by Irradiation. X-ray treatment was shown to determine gene-mutation in *Drosophila* by Muller in 1927. The natural rate of mutation was known accurately as a result of Muller's experiments and this rate could be increased 100 times or more, by appropriate doses of X-rays. Experiment with genemutation has now shown the ætiology of the changes induced with a high degree of accuracy (*cf.* Stadler, 1930, 1931 ; Muller, 1934, 1935 ; Timoféeff-Ressovsky, 1934, 1935). Mutation can probably be induced in all organisms, in all their tissues and at all stages of development. It is known that the frequency of change is independent of the wave-length used. It is merely proportionate to the energy that can be made available in the nucleus. Thus any short-wave radiation between ultra-violet of 2,500 A° and gamma rays of 0·001 A° will determine mutation, but the long wavelengths are usually impracticable because most of the radiation is absorbed before it penetrates to the nuclei and the absorption may be fatal (Altenburg, 1934 ; Stadler and Sprague, 1936 ; Brittingham, 1936).

The frequency of mutations induced in this way is, unlike that of natural mutations, independent of the genotype or of the temperature. The rate of induced mutation, however, depends on the gene concerned, although to a less extent than that of natural mutation. Thus some of the " mutable genes " in *Drosophila* (Demerec, 1934) and others of the least mutable genes in *Zea* according to Stadler (1932) are scarcely affected by treatment. The rate of induced mutation is influenced by anæsthesia and by the stage of development; seeds mutate less freely in dormancy than during germination. Organisms also vary greatly in their susceptibility, fungi being affected only by very high dosages (Dickson, 1933). It follows from these considerations that although the types of gene mutant induced are the same as the natural ones, the mechanism of change is different (*cf.* App. II).

Side by side with gene changes, structural changes are induced in the chromosomes by irradiation. Simple deficiencies perhaps occur like gene mutations in proportion to the dose (Stadler, 1932), but more complicated changes do not show these simple relationships (*cf.* Ch. XII).

In order to compare the effect of treatment on structural change with that on gene-change we must remember that the chromosomes may break (by definition) between any two genes. Muller and his collaborators (1935) have, in fact, shown in particular segments in *Drosophila* that any pair of adjacent chromomeres distinguishable in ultra-violet light may be separated by X-ray breakage. The natural frequency of breakage in any given interstitial segment is therefore probably lower than that of any genes whose mutation rate has yet been accurately measured. The frequency of breakage that may be induced is, on the other hand, perhaps as high as that of any genes. An exact comparison is impossible as we shall see later on account of differential mortality at different stages of development in different organisms. It seems, however, that in general the differences in the natural order of mutability; mutable gene—stable gene—gene thread are reduced by treatment.

The kinds and degrees of change seen in the chromosomes after treatment depend on three factors : (i) how far the general condition of the cell or the organism has been modified ; (ii) the stage of the nuclear cycle, whether meiosis or mitosis, that has been treated ; (iii) the extent of the nuclear cycle or the life-cycle which has elapsed between treatment and observation. We will first consider the effect of this interval.

When root-tips or spermatogonia are studied a few hours after treatment we can be sure that the nuclei are passing through their first mitosis. We then see that chromosomes of new shapes and sizes are produced. These have been described as the result of "translocation" (Navashin, 1931; Lewitsky *et alii*, 1931, 1934). Apart from these changed but characteristic chromosomes there occur others of a kind that are not normally found. These are commonly of four kinds: ring chromosomes, branched chromosomes (with lateral trabants) and the two types that are characteristically produced by crossing-over in inversions at meiosis, *viz.*, dicentric and acentric chromosomes. As we should expect, the acentric chromosomes are unaffected by the metaphase congression and anaphase separation, while the dicentric chromosomes divide, either regularly to give daughter chromosomes like themselves or criss-cross to give double bridges which break; the difference depends on whether the chromatids are parallel or have half a coil between the two centromeres (Mather and Stone, 1933). A third type of division is found where the centromeres lie farther apart, *viz.*, a complete coil between them. The two centromeres of each chromatid then pass to the same pole, but the chromatids, each M-shaped at anaphase, are interlocked in the middle loop of the M's (Koller, 1934, Plate XII).

The acentric chromosomes may be supposed to result from simple breakage, but some of them will arise from asymmetrical interchange, which must be responsible for the dicentric chromosomes and the ring chromosomes.

From the later results of these structural changes of whatever kinds, Navashin (1932) concluded that there was never any change in the number of centromeres and that centromeres were therefore self-perpetuating bodies. This conclusion is verified by the observations of the effects on the immediately following divisions by Mather and Stone; the two centromeres of one chromosome are independent of one another, while the chromosome without a centromere is deprived of the means of autonomous movement. It is usually lost in the cytoplasm at telophase.

We have seen from the comparison of mitosis and meiosis that the chromosomes must be supposed to divide after telophase always, and in mitosis, but not in meiosis, before the prophase. It seems probable, therefore, that their division occurs at the end of the resting stage in mitosis, possibly even determining the onset of prophase in certain circumstances (D., 1932). We cannot find out by direct observation whether this is so, since the finer details of structure are concealed in the resting nucleus. We can, however, resolve the question indirectly by X-ray treatment (Mather and Stone, 1933; cf. Lewitsky and Araratian, 1932, and Huskins and Hunter, 1935).

Thus when chromosomes are broken or interchanged under X-ray treatment before they have divided, all the changes produced will be exactly paired in the chromatids seen afterwards at metaphase. These changes in whole chromosomes may be described as due to chromosome-breaks. When, on the other hand, the cells are treated after their chromosomes have divided the changes they undergo will be changes in their chromatids; each chromatid will be broken or interchanged independently of its partner. These two conditions should be distinguishable at the following metaphase. Now in the higher plants, the resting stage between meiosis and the following mitosis in the pollen-grain continues for several days or weeks. In *Tradescantia* the interval is about one week. When, therefore, metaphases of mitosis are examined in the pollen-grain, three, two, or one day after X-raying, it is known that the nuclei giving rise to these metaphases have been treated at successively later stages of development of the resting nucleus.

Riley (1936) and Fabergé and Mather (unpublished) have confirmed these assumptions and shown that before a certain period all the changes are from chromosome breaks, and after this period they are all from chromatid breaks. The critical period is different in different species, twenty-four to thirty-six hours before mitosis in *Tradescantia*, earlier in *Allium*, always, however, some time before the end of the resting stage, which it cannot, therefore, be supposed to determine (Fig. 128 B). An exactly corresponding conclusion can be drawn from the mutations induced in ripe sperm in *Drosophila*. One-sixth of these produce mosaic offspring, one side mutant, the other non-mutant, owing to the genes concerned having divided before they were hit and the altered and unaltered daughter-genes having passed to opposite cells in the first cleavage division of the fertilised egg (Patterson, 1933). It might seem, therefore, that the genes, and the thread which connects them, divide at the same time. This does not follow without further evidence, however, since the gene changes in question may have depended on structural changes. Nor can we yet say whether the chromosomes divide in different parts at different times as they do at the end of pachytene.

New light is thrown on the time and control of division in the chromosome by the behaviour of an acentric fragment produced

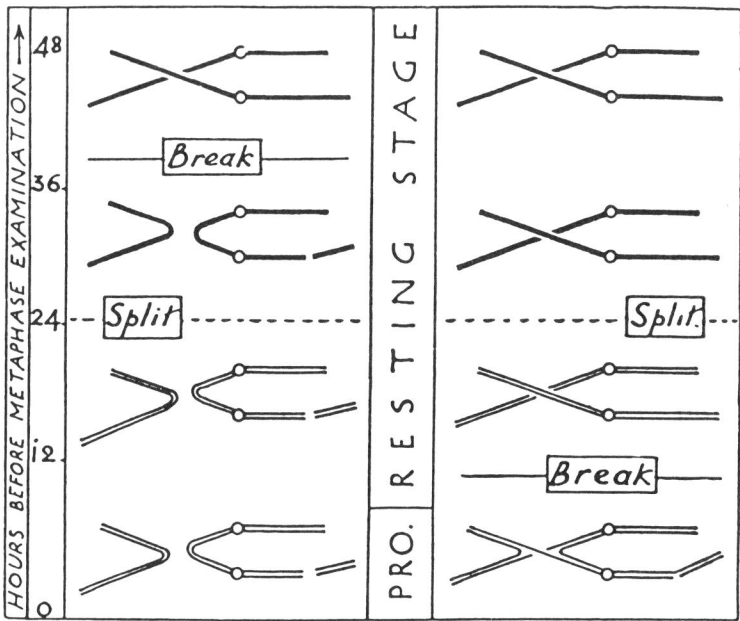

Fig. 128B.—Diagram showing how fragmentation and asymmetrical interchange will occur between chromosomes (left) and between chromatids (right), giving a pseudo-chiasma according to the time of treatment. (After Mather, 1933.)

at meiosis in *Podophyllum* (D., 1936). Such fragments produced by crossing-over in inversions are usually lost in the cytoplasm, but this very long fragment, having been included in the nucleus, is seen lying on the spindle at anaphase of the next mitosis. It is remarkable in that, unlike the normal chromosomes, it has failed to divide during the resting stage. It appears, when the pollen grain is rolled over, as a uniform cylinder of the same thickness as the anaphase chromatids. This shows that the division of the chromosome cannot have been already determined at the preceding meiosis and, further, that the division of a chromosome may be conditioned by its having a centromere at least at the beginning of the resting stage (Plate IX).

When structural changes occur between chromatids it is possible to define their character in a way that is not possible with chromosome changes, for the chromatids retain the association with their homologous partners and hence give pseudo-chiasmata. These chiasmata in some cases clearly result from asymmetrical interchange and show that this is how dicentric and acentric chromosomes arise when they are produced by change in the undivided chromosomes. Pairs of chromosomes with these chiasmata behave normally at metaphase and do not orientate themselves like meiotic bivalents (White, 1935).

Treatment has no effect on metaphase or anaphase chromosomes until half an hour has elapsed, but earlier stages of mitosis (Strangeways and Hopwood, 1927; Stone, 1933) are arrested for twenty-four hours or more in root-tips. This delay no doubt results from general damage to the cell such as is shown by the plastids (Hruby, 1935). Where the effects are more drastic they are also more complicated. White (1935) has distinguished three kinds of change, the probable causation of which may be inferred, in spermatogonia of *Locusta*. (i) A proportion of nuclei are killed, whether at a particular stage of mitosis or not cannot be said. (ii) All or many cells treated in the early prophase are delayed in coming to metaphase so that mitoses with chromatid breaks do not appear until twenty hours after the first mitoses with chromosome breaks. This suggests that during the prophase the normal development of the centrosomes or spindle-determinants is upset by the cytoplasmic abnormality resulting from treatment. (iii) Occasionally nuclei appear, sometimes in groups, whose chromosomes have an anomalous structure. They have four chromatids instead of two attached to each undivided centromere. These *diplochromosomes* differ from the attached X-chromosomes in *Drosophila*, in that the homologous arms are attached to the same side of the centromere and not to opposite sides. The origin of the diplochromosomes can be accounted for in terms of the conclusion already arrived at from general observations, that the chromosomes divide in the resting stage, while the centromeres do not divide until metaphase. The argument is as follows. We may suppose that mitosis is frustrated in these cases during prophase so that the nucleus relapses into the resting stage, the effect being merely an extreme example of the delay already noticed. Then before a new division can occur the chromosomes will divide again, while their centromeres, which divide only at metaphase, will be unaffected. Each centromere will then have four chromatids instead of two. Whether or not this explanation is sound we see that not only the independence of the two processes of division, but also the particular time-relationship assumed on general grounds is corroborated by the structure of the diplochromosomes.

The diplochromosomes show relational coiling, like the chromatids at an ordinary mitosis, but clearly visible owing to repulsion, like the chromosome coiling at diplotene. The special delay responsible for their occurrence might be due to direct damage to the centrosome, or to a secondary effect brought on by treatment at a particular stage of the centrosome-cycle. Since groups of cells are sometimes affected, and the nuclear cycles of groups are synchronised in the spermatogonia, we may infer the latter.

The immediate effects of irradiation on meiosis differ from those on mitosis in one respect, a disproportionately high production of acentric fragments (Stone, 1933; Mather, 1934). With a low dosage the number of bodies may be unaltered at mitosis owing to each breakage being followed by a reunion. With a similar dosage the number of bodies at meiosis is more than doubled in *Tradescantia* and *Vicia*. Further, *Tradescantia* is exceptional inasmuch as these fragments appear at metaphase; in *Vicia*, *Tulipa*, *Primula* and elsewhere, with and without terminalisation, the fragments do not appear until anaphase. Probably the explanation of this difference is that the fragments are held to their unbroken partners at pachytene in most organisms which have complete pairing, while in *Tradescantia* with incomplete pairing, they lie free. The explanation of lack of reunion of breakage-ends of chromosomes at meiosis is probably the pachytene pairing, which prevents the necessary movement of the broken parts in the nucleus (Ch. XII). Their high frequency indicates that a proportion of breakages are reunited, as they were before, in ordinary mitotic irradiation. Further quantitative study would be useful to find out what this proportion may be. Another and indirect effect is produced on meiosis after an interval, namely, an increase in chiasma-frequency (Mather, 1934; *cf.* Levan, 1935 *b*, on *Allium*). All the special properties of meiosis in relation to X-ray treatment depend, therefore, on the circumstances of pachytene pairing.

When mitosis is studied with a longer interval after irradiation, a considerable reduction in the frequency of changes is found. This is due to two kinds of selection. First, the acentric chromosomes are eliminated and the dicentric and ring chromosomes reduced in frequency by breakage and non-disjunction (*cf.* McClintock, 1933 *b*). Secondly, cells with severe abnormalities are eliminated in competition with less severely altered cells. Such competition is, of course, particularly severe in pollen. But where all the cells are damaged it becomes possible to effect fertilisation with such pollen grains as could never do so normally, and monosomic *Zea* and polysomic *Oryza* have been raised this way (Stadler, 1930; Ichijima, 1934). Damaged pollen has similarly been used to stimulate parthenogenesis in *Triticum* (Ch. XI).

The conditions of survival of gene or structural changes are very different in diploids and polyploids. Both are less severely selected in polyploids. Most induced gene changes, however, are invisible in the polyploids, *e.g.*, *Avena* (Stadler, 1930); while structural changes are visible, and, being less dangerous, are found as deficiencies such as would be eliminated in the diploid (*e.g.*, *Nicotiana*, Goodspeed, 1930).

Organisms are necessarily heterozygous in regard to the structural changes induced in them. Those such as interchanges, which survive to later generations can often but not always be obtained in the homozygous condition (McClintock, 1934; Catcheside, 1935; E. G. Anderson, 1935 *a*; Lewitsky *et alii*, 1934; Muller, 1932). In the heterozygous condition their behaviour resembles that of natural structural hybrids and has been considered in conjunction with them. After an interval of many mitotic cycles there is, as a rule, only the workable type of chromosome with two ends and one centromere. The changes in its size may have arisen in several ways, but losses will often have been eliminated with cell-selection by lower viability, and inversions, unless including the centromere, will not be visible. Observation of mitosis to show changes in cells derived from irradiation is therefore restricted to what have been called " translocations." These have been assumed by Lewitsky and Sizova (1934) always to be simple " donations " from one chromosome to another. Study of meiosis in the progeny of X-rayed plants, however, shows a great number of interchanges. Large numbers have been identified in *Zea*, occurring at random, but necessarily balanced since they have survived (Anderson and Clokey, 1934). In all the products of structural change in maize, there is no case of a simple fusion or translocation, or indeed of any change involving the attachment of unbroken ends of chromosomes, although there is no mechanical reason why such changes should not survive (McClintock, 1931 ; *cf*. Ch. XII). Changes of size must therefore often be due to the interchange of unequal segments, such as has been identified in *Zea* (Burnham, 1932).

On the assumption of a simple donation in *Crepis capillaris*, Lewitsky and Sizova have, however, concluded that since long arms " give " to short arms much more often than they receive from them, there must be a specificity in the direction of donation. But when we assume interchange we see that this apparent specificity is nothing more than the inevitable result of a size difference. When a large body exchanges parts at random with a small body it stands to lose by the transaction. There is therefore no specificity in the direction of transfer.

With regard to a possible specificity in the places of breakage, some workers have concluded on occasional evidence that breakage is more frequent near the centromere than elsewhere, but this is not generally so. It may be, however, that the X chromosome is more frequently broken than the autosomes in Orthoptera (White, 1935).

Not only may the genes and gene-string be broken by irradiation, but also, as we have seen, the centrosome may be prevented from dividing. We do not know whether this is a direct or indirect effect. It is possible, however, that a direct change may be induced in the

PLATE XII

The Formation of Acentric and Dicentric Chromatids and Chromosomes by Crossing-over (figs. 1 and 2) and Irradiation (figs. 3–8)

Figs. 1 and 2.—Criss-cross separation of dicentric chromosomes in a binucleate pollen grain of a triploid *Tulipa* following non-disjunction of dicentric chromatids at meiosis. × 1000. (Upcott, unpublished.)

Fig. 3.—Metaphase in root-tip of *Crocus Olivieri*. Six centric constrictions, two in one chromosome at 5 o'clock owing to chromosome breakage. (Mather and Stone, 1933.)

Fig. 4.—Pollen grain of *Allium Moly* with pseudo-chiasma due to chromatid breakage which follows irradiation at the end of the resting stage. (Mather and Fabergé, unpublished. 2B.E. — gentian-violet smear.)

Fig. 5.—*Tulipa* root-tip with dicentric chromosomes. (Mather and Stone, 1933.)

Fig. 6.—Anaphase of mitosis in a root-tip of *Tulipa*, showing criss-cross separation of dicentric chromatids. (Mather and Stone, 1933.)

Figs. 7 and 8.—The same in *Vicia Faba*, showing ring chromatid and *M*—interlocking of dicentric chromatids. (Koller, 1934.)

PLATE XII

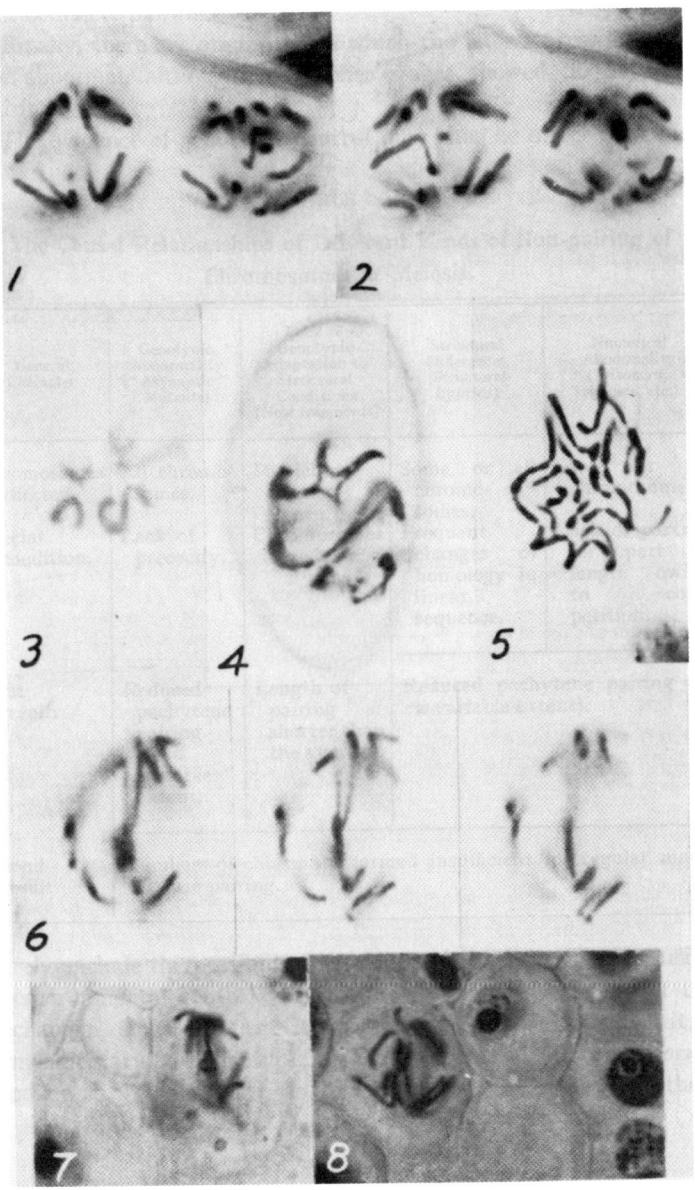

centromere by irradiation in two ways. First, areas of endosperm, after irradiation in *Zea*, may show the apparent results of "deficiency." This sometimes consists in the loss of a whole chromosome, since the recessive types of a group of linked factors may appear. Islands occur in which the dominants reappear. This may be explained as due to the failure of the chromosome carrying the dominants to divide, and its later recovery of this capacity (Stadler, 1930). Such an effect is most easily understood as the result of a single change, and it may be supposed that this change consists in injury to the centromere which temporarily prevents its division and consequently the separation and redivision of the two chromatids.

Secondly, McClintock found (1933) that on two occasions a ring and rod, each with a centromere, arose from one chromosome by X-raying. Whether this is due to breakage of the centromere or to the preservation of a ring from a multiple breakage and non-disjunction, it is difficult to say on account of the special conditions of survival of rings in diploids. Unless they are supernumerary the complement will be defective.

Evidence of a transverse doubleness of the centromere is uncertain in most organisms. Its appearance at pachytene and at first anaphase sometimes suggests transverse doubleness, but the interpretation is open to doubt (D., 1933 *a*). In *Ascaris*, however, the movements of the large multiple chromosomes indicate that there is a row of centromeres distributed along the chromosomes at intervals, for instead of the chromosome moving to the pole first at one point, the whole of the middle part moves together (Schrader, 1935; White, 1936). The existence of several centromeres would account for the chromosome breaking up into several independent bodies when "diminution" takes place. It is merely necessary to suppose that in the cells where no such breakage takes place the chromatids have lain parallel between the centromeres, while in those where breakage occurs there has been a persistence of chromatid coiling. The irregular numbers of chromosomes found after breakage may then be due to irregular coiling. In this way the structural change of diminution will be physiologically controlled. White has demonstrated the correctness of this view by irradiation.

432 BREAKDOWN OF GENETIC SYSTEMS

When the multiple chromosomes are broken in the middle, instead of giving monocentric and acentric fragments, they yield two or more functional chromosomes.

The theoretical consequences of the breakage of chromosomes

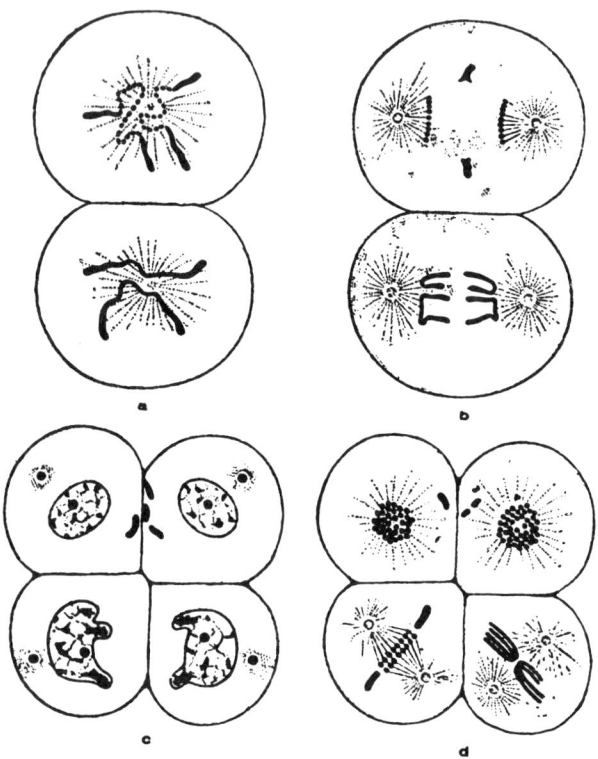

FIG. 129.—Metaphase, anaphase and telophase of the second cleavage division and metaphase of the third in *Ascaris* showing fragmentation and diminution of the polycentric chromosomes. (Boveri, 1904.)

by irradiation are manifold. Its bearing on the theory of the gene has already been discussed. Its bearing on the theory of structural change, especially the question of whether delayed reunion can take place, will be considered later. It is of immediate importance, however, to the theory of the centromere and of its time of division in relation to the chromosomes. The distinction between acentric fragments and functional monocentric chromosomes first became clear from the results of irradiation. They showed that the centromere was a self-perpetuating body and incapable of arising *de novo* or from other chromomeres. The inference had already been drawn both directly and on mechanical grounds that in mitosis the chromomeres divided during the resting stage and the centromeres divided at metaphase, while both these processes were relatively delayed owing to the precocity of meiosis (D., 1932). These conclusions are now borne out by both the genetical and cytological evidence of irradiated material. Chromatid breaks and chromosome breaks are found at the first mitosis after treatment of the resting nucleus at different stages. The evidence of the diplochromosomes is of especial significance in relation to the centromere, for it shows two of its properties; first, that its division is dependent on the action of the spindle or more properly of the centrosomes and is independent of the division of the chromomeres; secondly, that its division is later than that of the chromomeres. Finally, we see that mitotic " bivalents " united by pseudo-chiasmata which result from chromatid interchange do not orientate themselves on the spindle, and do not separate like meiotic bivalents owing to their centromeres being already in process of division and already having, therefore, an internal property such as polarity, which the centromeres of chromosomes at meiosis do not acquire until a later stage. These observations have an important bearing on the external mechanics of the chromosomes which will be considered later.

CHAPTER XI

THE BREAKDOWN OF THE GENETIC SYSTEM: CONTROLLED

Classification—Apomixis and the Life Cycle—Parthenogenesis—The Haplo-Diploid System—Apogamy—Apospory—Adventitious Apomixis—Meiosis in Functional Cells — In Non - Functional Cells — Pseudogamy — Conditions Determining Apomixis—Chromosome Numbers of Apomictic Plants.

And Earth . . . bare also the fruitless deep without the sweet union of love.

Hesiod, *Theogony*, 131–132.

1. DEFINITION AND CLASSIFICATION *

HAVING considered the breakdown of reproduction inductively from the great variety of unregulated examples arising chiefly in experiment, we can now turn to the natural and regulated types of breakdown which are included under the description of apomixis.

Apomixis may be defined (following Winkler, 1908) as a system of reproduction having the external character of sexual reproduction but omitting one or both of its essential cellular processes. These are meiosis and fertilisation. Clearly if only one of these is omitted the irregularity cannot become a habit, and examples of such non-recurrent abnormalities will be specially considered. The abnormality is recurrent where both processes are omitted from the life cycle, and this happens in species of all the principal groups except the Vertebrata, Echinoderma, Bryophyta and Gymnospermæ (*cf.* Ernst, 1918; Winkler, 1921; Prell, 1923; Ankel, 1929; Rosenberg, 1930).

Where the specialised reproductive bodies, spore and egg-cell, continue to function the abnormality is known as *parthenogenesis*. An important distinction can then be made between a type of

---

* The usage of terms in this chapter is that found most convenient in approaching the matter from a genetic point of view. Confusion in the terminology is due to writers approaching it from different points of view.

parthenogenesis in which fertilisation fails first, and a type in which meiosis fails first. The effect will be that, in the first, both generations (normally haploid and diploid) will be haploid, when the reproduction is called *haploid parthenogenesis*, and in the second, both generations will be diploid, when the reproduction is called *diploid parthenogenesis* (Fig. 104). It will be seen that this distinction can only be made with certainty where the origin of the new system is known or the normal type of reproduction can be compared with the aberrant ; this, however, is usually possible. Further, in such a case as that of apparent diploid parthenogenesis in *Chara crinita* (Ernst, 1918) it is possible, although not plausible, to assume that failure of a mitotic division, not failure of meiosis, was the starting-point of a parthenogenetic mode of reproduction (*cf.* Winkler, 1921).

In the higher animals the specialised germ-cell is always the immediate product of meiosis, or some modification of meiosis, in the specialised mother-cell, and where meiosis fails in parthenogenesis the specialised function of the mother-cell (oöcyte) and the gamete (egg-cell) are always retained. But in plants with an alternation of generations there are specialised spore mother-cells and specialised gametes developed at different parts of the life cycle, and unspecialised cells may be substituted for either or both of these when meiosis and fertilisation are omitted. Thus not only the characteristic processes but also the characteristic organs of sexuality are lost. Where the spore mother-cell either fails to develop or does not function in reproduction we speak of *apospory*, and where the normal gamete cell does not function we speak of *apogamy*. When both of these are combined only the external character of seed production distinguishes the development of the embryo from vegetative propagation.

Apomixis will first be considered in relation to the life cycle in the most important groups in which it occurs.

2. APOMIXIS IN RELATION TO THE LIFE CYCLE

(i) **Higher Animals (Metazoa).** (*a*) DIPLOID PARTHENOGENESIS. In many species of insects, crustaceans and nematodes this form of parthenogenesis is obligatory. Parthenogenetic females produce only their like, by a failure of meiosis at maturation in the oöcyte.

In others, parthenogenesis is facultative : parthenogenetic females may give rise to sexual females and sexual males. In some of these (*e.g.*, *Daphnia*, Banta *et alii*, 1926, and Rotifera) the succession of sexuality and parthenogenesis is irregular, in others they come in a definite cycle (*e.g.*, in Aphides). The conditions immediately determining this cycle will be seen later.

(*b*) HAPLOID PARTHENOGENESIS. With the haplo-diploid system of sex-differentiation found in many animals, haploid parthenogenesis gives male offspring (Ch. IX).

(ii) **Ferns (Pteridophyta)**. In the ferns the two generations have an independent existence, and it is therefore possible to distinguish the different abnormal sexual processes more readily than in the flowering plants. Nevertheless a great deal of the older work in this field is, as Manton (1932) says, ambiguous. All types naturally occurring involve the suppression of meiosis and fertilisation and may be classified as follows :—

(*a*) DIPLOID PARTHENOGENESIS, *i.e.*, without loss or substitution of the special organs—spores and gametes (*e.g.*, *Marsilia Drummondii*, Strasburger, 1907). In this type two diploid spores are formed as a result of the failure of reduction, instead of four haploid ones, and the egg-cells developed on the diploid prothallia (gametophytes) grow into sporophytes without requiring fertilisation (*cf.* Gustafsson, 1935).

(*b*) APOSPORY, *i.e.*, with substitution of unspecialised cells for the spores (*e.g.*, *Athyrium Filix fœmina* var. *clarissima* Bolton, and *Scolopendrium vulgare* var. *crispum Drummondæ*, Farmer and Digby, 1907). In this type the gametophyte does not develop from a spore but as a purely vegetative outgrowth of the sporangium or the tip of the frond. It is diploid, meiosis having been completely omitted, and it produces diploid egg-cells. These develop without fertilisation, but perhaps require the stimulation of the sperm as in pseudogamy (*cf.* Andersson-Kottö, 1931).

(*c*) APOGAMY, *i.e.*, with substitution of unspecialised cells for the gametes (*Nephrodium hirtipes*, Steil, 1919). In this type the sporophyte arises as a purely vegetative outgrowth from the gametophyte (prothallium), which is diploid, having arisen from unreduced spores. In the example, antheridia were produced although there were no archegonia. The failure of reduction was evidently due, as in *Marsilia*, to the formation of restitution nuclei following imperfect pairing of the chromosomes at the first meiotic division (*cf.* Rosenberg, 1927, 1930).

(*d*) APOSPORY WITH APOGAMY (*Athyrium Filix fœmina* var. *clarissima* Jones, Farmer and Digby, 1907 ; *Nephrodium pseudo-mas*, Rich. var. *cristata*, Digby, 1905). In this type both the alternating generations arise from unspecialised cells by purely vegetative processes.

These abnormalities may be variously combined with one another in the same individual (Digby, 1905).

(iii) **Flowering Plants (Angiospermae).** In the angiosperms the same types of abnormality occur as in the ferns. Relatively few of the great number of forms which are known to be apomictic from genetic experiment have yet been studied cytologically.

The following classification applies to the ways in which an embryo may develop asexually in the angiosperms; it is not a final classification of the habits of species, for it is often found that when sexuality fails it is replaced in several competing ways in the same species.

(*a*) DIPLOID PARTHENOGENESIS. In the sexual relatives of most angiosperms developing in this way the egg-cell with its eight nuclei arises from the end one of the four cells produced by meiosis. This is the "normal" type. A "*Scilla*" type, where it arises from two of the four, is found in the sexual relatives of the parthenogenetic *Erigeron*, and a "*Lilium*" type, where it arises from all four together, is found in the relatives of *Balanophora*. In many sexual forms behaviour is variable and in an aberrant sexual form of *Leontodon* the four cells sometimes separate and then later their wall breaks down and they join to form one embryo-sac (Bergman, 1935).

Evidently the cells that are going to contribute to the embryo-sac are distinguished from their sisters at the time when the first division occurs without forming cell-walls and this time is often variable.

With parthenogenesis it was formerly believed that an "*Alchemilla*" type corresponding to the normal sexual forms might still occur. If an unreduced embryo-sac were produced in this way there would need to have been a double division of the univalent chromosomes such as is never regularly found in plant hybrids. Gustafsson (1935) considers that unreduced embryo-sacs such as would be capable of development without fertilisation do not, in fact, arise in this way. Where "normal" embryo-sacs have been described in *Thismia* and *Artemia* they have probably arisen by reduction and will be incapable of parthenogenetic development (v. Table 68). In most parthenogenetic plants the normal meiosis is replaced by a single division, as it is in hybrids where pairing has failed. There are then two possibilities of development corresponding to the three of sexual plants. We should expect that those having a normal type in the sexual reproduction of their relatives would have the egg-cell developing from one of the two cells produced by the non-reducing meiosis. This is so, as a rule, in *Taraxacum* (Gustafsson, 1935) and *Chondrilla* (Poddubnaja-Arnoldi, 1933, 1934). But in many other groups the embryo-sac develops from both the daughter-cells, in other words the mother-cell develops directly into the embryo-sac (Fig. 130), e.g., in *Eu-Hieracium* (Rosenberg, 1917), *Antennaria* (Stebbins, 1932 b; Bergman, 1935 a), *Ochna* (Chiarugi and Francini, 1930), and *Ixeris* (Okabe, 1932). In these cases there has evidently been a shift in the development of the embryo-sac. Failure of cell-wall formation begins earlier than in the corresponding sexual species, and as is usual in such circumstances, the change is variable in its effect; most forms exceptionally have development from one instead of from both the products of meiosis. Such variations have been most accurately described in *Artemisia* (Table 68, Chiarugi, 1926).

The behaviour of the polar nuclei is variable in parthenogenesis. These nuclei in sexual reproduction fuse with a pollen nucleus to give the triple endosperm nucleus. Clearly in parthenogenesis the same three to two proportion in the nuclear content of endosperm and embryo cannot usually be established as in sexual reproduction. It has been found that in *Erigeron annuus* (Holmgren, 1919) the polar nuclei fuse, while in *Balanophora* (Ernst, 1914) they do not. *Zephyranthes* is remarkable in one generative nucleus of the pollen fusing with a polar fusion-nucleus, so that the endosperm is sexually produced, while the other generative nucleus does not fuse with the egg cell (Pace, 1913).

(b) APOGAMY, the development of cells other than the egg cell

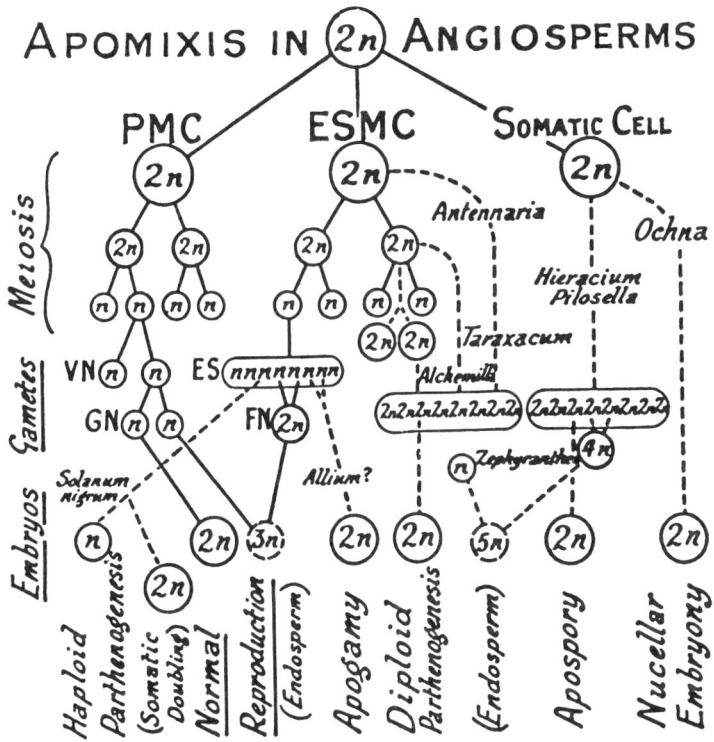

FIG. 130.—Diagram of different forms of apomixis found in the flowering plants. Unbroken lines show the course of sexual reproduction, broken lines, of apomixis. The species given are not all fixed in the particular kind of apomixis here assigned to them (cf. Table 68). N.B.—The normal embryo (2n) is derived from a nucleus of the embryo-sac (ES) and of the generative nucleus (GN).

into the embryo sporophyte, is probably not very important. The synergids and antipodal nuclei have been supposed to give rise to supernumerary embryos in the facultatively parthenogenetic *Allium odorum* (Haberlandt, 1923). In order to give diploid embryos they must first fuse in pairs or fail to separate at mitosis. Such a doubling in the antipodal nuclei is characteristic of a section of the related genus *Tulipa* (Ch. III).

(c) APOSPORY, the development of cells other than the embryo-sac cell into the gametophyte, is probably important in many species. In *Hieracium Pilosella* cells in the nucellus or integument may develop into embryo-sacs, which, of course, are diploid (Rosenberg, 1906, 1930). These compete successfully with the reduced embryo-sac in development and usurp its function. This is reminiscent of the Renner effect (Ch. IX). Apospory is also found in *Artemisia nitida* (Chiarugi, 1926) and in *Ochna serrulata* (Chiarugi and Francini, 1930).

In all these species the normal embryo-sac has a chance of developing, either by fertilisation, as in *Hieracium*, or parthenogenetically, as in *Artemisia* and *Ochna*. The first is conditioned by the formation of a viable gamete in a plant which is slightly sterile sexually owing to hybridity; the second is conditioned by the formation of a diploid embryo-sac cell by the successful suppression of meiosis in a plant in which this suppression has not become a constant process. Apospory may then be said to be characteristically developed in those plants which are too hybrid to be regularly sexual or not hybrid enough to be regularly parthenogenetic. It bridges the gap.

(d) APOSPORY WITH APOGAMY occurs in two ways. *Nucellar Embryony* is the purely vegetative development of cells in the nucellus or integument into the embryo sporophyte. It gives supernumerary embryos in sexually reproducing plants (*e.g., Citrus, Zygopetalum*), and occurs as with parthenogenesis in *Alchemilla, Nothoscordon fragrans*, and *Ochna " multiflora,"* which are sexually sterile.

Vivipary is the replacement of the flower by a purely vegetative outgrowth, which is capable of propagating the plant as a bulbil. It is principally developed facultatively and then shows considerable variation subject to environmental influence. It is found in *Allium, Poa*, and *Festuca*, especially in sexually sterile species, such as triploids and auto-tetraploids (Levan, 1933 *a*; Müntzing, 1933). The imposture is recognisable at once from the external morphology of the organism, so that this method of reproduction need not be described as apomixis.

For simplicity the different abnormalities of reproduction have so far been considered separately, but in their occurrence they are not easily separable. It is common for apomictically reproducing plants to show several abnormalities together. This is particularly well exemplified by *Artemisia nitida* (Chiarugi, 1926). This triploid species is sexually sterile and its normally produced megaspores die, their place being taken by aposporous embryo-sacs. Failure of reduction gives diploid parthenogenesis, so that seed may develop in two ways and in three other ways may fail to develop (Table 68). One ovule may contain several mother-cells and show all these variations side by side. In *Ochna serrulata*, a different complication arises: the differentiation of tissues in the ovule is not always clear, so that the distinction between parthenogenesis and apospory fades away.

Which of these different types of reproduction can be successful is probably decided by competition. Clearly the haploid or irregularly reduced embryo-sac of a hybrid will stand little chance against the somatic embryos. Where numerical reduction has failed the odds will be reversed. This competition has been well illustrated by Haberlandt (1921 b) in *Hieracium flagellare*. Such competition is perhaps characteristic of all stages of development of gametophytes and embryos in hybrids, as in *Œnothera*.

A different range of abnormality is found in *Allium odorum* according to Modilewski (1930). The egg-cell develops in one of four ways: (i) normally, after reduction and double fertilisation; (ii) after failure of reduction with fertilisation of the fusion nucleus only, so that the embryo is diploid, the endosperm pentaploid, as in *Zephyranthes*; (iii) after failure of reduction but without fertilisation of the fusion nucleus; (iv) by haploid parthenogenesis. The last two processes result in degeneration at an early stage: fertilisation, single or double, is indispensable.

(iv) **Lower Organisms (Thallophyta and Protista).** Amongst the lower organisms alone is found the type of reproduction in which meiosis immediately follows fertilisation, and is perhaps immediately conditioned by it (D., 1932). In these we find both diploid parthenogenesis (*e.g.*, *Chara crinita*, Ernst, 1918 ; *Cocconeis placentula*, Geitler, 1927) and haploid parthenogenesis (*e.g.*, many fungi, *Camarophyllus*, Bauch, 1926 ; *Schizophyllum*, Kniep, 1928 ; *Saprolegnia*, Mäckel, 1928).

3. NON-RECURRENT APOMIXIS

The omission of one of the two essential sexual processes (meiosis and fertilisation) without the other is necessarily a non-recurrent abnormality of the life cycle, for it would otherwise lead to indefinite reduction or indefinite multiplication of the numbers of the chromosomes. Such an omission can therefore be expected only as a result of exceptional conditions, whether genetic or external, whether natural or artificial—if the distinction can be made. These kinds of conditions have often been found. Of the three types of result, one, the formation of unreduced gametes, especially in hybrids, is of great importance in the origin of polyploids. It has been considered in relation to the breakdown of meiosis (Ch. X).

(i) **Parthenogenesis.** This abnormality, the omission of fertilisation without the omission of meiosis, has, with a few doubtful exceptions, been observed to give mature progeny only in certain flowering plants of various groups. The early stages of development of this kind have been found in some animals and in *Cutleria* (Phæophyceæ) by Yamanouchi (1912), where the female gametophyte was protected from fertilisation. The egg-cell develops without fertilisation into the embryo sporophyte. The development has been observed at all stages in *Solanum* (Jørgensen, 1928) ; in other cases it is inferred from the seedling being of a reduced size and having the reduced chromosome number and external character of one parent. Table 66 records the evidence of this kind in the best authenticated examples. In the last group, the seedlings resembled their male parent and had half its chromosome number, so that the inference was drawn that a male nucleus entered the egg-cell, whose nucleus died while the pollen nucleus gave rise to the embryo. This " male parthenogenesis "—or, it might be said, cuckoo-parthenogenesis—is analogous to that stimulated artificially in *Vaucheria* (v. Wettstein, 1920), and in echinoderms, where it is possible to " fertilise " an artificially enucleated egg fragment with a normal sperm (*cf.* Belar, 1933 ; Fankhauser, 1934).

Crane and Jørgensen have produced haploid plants of *Solanum nigrum* by inducing parthenogenesis in two ways : (i) a periclinal chimæra of one layer of *S. sisymbrifolium* over *S. nigrum* var. *gracile* was self-pollinated. Owing, it is believed, to the purely *nigrum* pollen grains growing less satisfactorily in the chimæra, few ovules were fertilised ; others, apparently stimulated by the pollination, developed parthenogenetically. One haploid plant was raised : it was a dwarf *S. nigrum* in character ; (ii) ninety flowers of *S. nigrum* were pollinated by *S. luteum* and yielded 43 fruits with 70 seeds, of which 35 germinated ; 7 were haploid, and the remaining 28 diploid. All were maternal in character. Genetical evidence later pointed to somatic doubling of the chromosome number at an early stage in the parthenogenetic embryos, and not to failure of reduction, as the explanation of the origin of the diploid progeny. Similar evidence shows that such haploid parthenogenesis followed by doubling is probably very common when interspecific crosses are made in *Fragaria* and *Nicotiana* (East, 1930), and it may reasonably be supposed that these observations are of general significance in the flowering plants.

When the embryo-sacs of *S. nigrum* were examined after cross-pollination with *S. luteum*, it was found that one or both of the generative nuclei of a pollen grain had often entered an egg-cell and later degenerated, as in artificial parthenogenesis, in the echinoderms. Probably in the same way the egg-cell nucleus degenerates in the occurrence of male parthenogenesis.

The *Triticum compactum* haploid arose from an exceptionally large seed—the opposite of what would be expected if the pollen played no part in its development. This suggested to Gaines and Aase that its origin was due to both the generative nuclei of the pollen fusing with the central nucleus to give a tetraploid instead of a triploid endosperm nucleus.

Triticum monococcum haploids have been produced by pollinating the plants with their own X-rayed pollen which presumably, like that of another species, will stimulate development without consummating fertilisation.

TABLE 66

Cases of Non-Recurrent Parthenogenesis in the Angiosperms

Species of True Parent.	Chromosome Numbers.		Origin or Cross.	Author.
	Parent.	Offspring.		
A. FEMALE PARTHENOGENESIS.				
1. *Datura Stramonium* (2x)	24	12	×Self or *D. ferox*.	Belling and Blakeslee, 1923.
2. *Nicotiana Tabacum* (4x − 1)	47	24 (2x)	×*Nicotiana sylvestris*	Clausen and Mann, 1924; McCray 1932 B.Z.
3. *Triticum compactum* (6x)	42	21 (3x)	×*Ægilops cylindrica*	Gaines and Aase, 1926.
4. *Campanula persicifolia* (4x)	32	16 (2x)	×*Campanula persicifolia* (2x)	Gairdner, 1926.
5. *Œnothera Lamarckiana gigas* (4x)	28	14 (2x)	Self	Håkansson, 1926.
6. *Crepis capillaris* (2x)	6	3	×*C. tectorum* (2n = 8)	Hollingshead, 1928 a.
7. *Solanum nigrum* (6x)	72	36 (3x)	×*S. luteum* (4x)	Jørgensen, 1928.
8. *Matthiola incana* (2x)	14+f	7+f	Self	Lesley and Frost, 1928.
9. *Solanum Lycopersicum* (2x)	24	12	Self	Lindstrom, 1929; Humphrey, 1934
10. *Œnothera franciscana* (2x)	14	7	Self	Davis and Kulkarni, 1930.
11. *Œnothera* sp. (2x)	14	7	×*Œ. franciscana sulfurea* (2x)	Emerson, 1929.
12. *Œnothera rubricalyx* (2x)	14	7	×*Œ. eriensis* (2)	Gates and Goodwin, 1930.
13. *Nicotiana glutinosa* (2x)	24	12	Self	Goodspeed and Avery, 1930.
14. *Digitalis purpurea* × *D. ambigua* (8x) (= *D. merionensis*)‡	112	56	×*D. ambigua* (8x)	Buxton and D., 1932.
15. *Œnothera franciscana* (2x)	14	7 and 14	×*Œ. longiflora* (2x)	Stomps, 1930 a.
16. *Œ. Hookeri* (2x)	14	7	×(i) *Œ. longiflora* (2x) ×(ii) *Œ. argillicola* (2x)	

17.	Œ. argillicola (2x)	14	7	× Œ. biennis (2x) (16 plants)	Stomps, 1930 b.
18.	Œ. biennis gigas (4x)‡	28	14	× Œ. biennis gigas cruciata	Stomps, 1931.
19.	Œ. franciscana lata (2x + 1, ex haploid).	15	7	Self (8 plants haploid in 178)	,, ,,
20.	Œ. Lamarckiana blandina (2x)	14	7*	× Œ. longiflora	,, ,,
21.	Œ. Lamarckiana (2x) (mutating to blandina).	14	7*	Self	,, ,,
22.	Fragaria vesca (2x)	14	14†	× F. chiloensis (8x = 56)	East, 1930.
23.	Oryza sativa (2x)	24	12 and 24	Dwarf variety	Morinaga and Fukushima, 1931, 1935.
24.	Nicotiana rustica × paniculata (6x)	72	36	Selfs (3)	Lammerts, 1932.
25.	N. tabacum (4x)	48	24	Species cross	Lammerts, 1934.
26.	N. glutinosa (2x)	24	12	Selfs (3)	Webber, 1933.
27.	Zea Mays (2x)	20	10	Heat treatment	Randolph, 1932.
28.	Phartitis Nil (2x)	30	15	Varietal cross (F_1)	(U., 1930, 1932); Katayama, 1935.
29.	Œnothera blandina (2x)	14	7	Species cross	Catcheside, 1932.
30.	Brassica Napella (4x)	38	19	Natural seedlings	Morinaga and Fukushima, 1933.
31.	Portulaca grandiflora (2x)	18	9	Selfs (3 in 268)	Okura, 1933.
32.	Gossypium hirsutum, etc. (4x)	52	26	Twin seedlings	Harland, 1936.
33.	Oryza sativa (2x)	24	12	Twin seedling	Ramiah et al., 1934.
34.	Triticum monococcum (2x)	14	7	Self with high tempre and with X-rayed pollen.	Kihara and Katayama, 1932; Katayama, 1935; Chizaki, 1934.
35.	Ægilotricum (8x)‡	56	28 & 27 +ff	X-rayed pollen	Kihara and Katayama, 1932; Katayama, 1935.
36.	Triticum vulgare (6x) × Secale cereale (2x).	28	53–56	× Secale (double non-reduction).‡	Lebedeff, 1934.

TABLE 66 (continued).

Species of True Parent.	Chromosome Numbers.		Ostensible Female Parent.	Author.
	Parent.	Offspring.		
B. MALE PARTHENOGENESIS.				
1. *Nicotiana Langsdorffii* (2x)	18	9	*N. Tabacum*	Kostoff, 1929.
2. *Nicotiana Tabacum* (4x)	48	24	*N. Tabacum* × *N. glutinosa*	Clausen and Lammerts, 1929.
3. *Fragaria virginiana* (4x)	28	28	*F. vesca* (*rosea* × *alba*) ($n = 7$).	Ichijima, 1930.
4. *Vicia sativa* (2x).	12	12*	*Lens esculenta* ($n = 7$).	Bleier, 1928.
5. *Euchlaena mexicana* (2x)	20	20*†	*Tripsacum dactyloides* ($n = 36$).	Collins and Kempton, 1916.

* Numbers presumed on good grounds.
† Evidently non-reduction or somatic doubling has occurred. In *Fragaria vesca* seedlings showed segregation of recessive characters.
‡ Haploid parthenogenesis in these cases has reversed an earlier doubling of the chromosome number in the ancestral hybrid.

N.B. Only approximately homozygous diploids, and those polyploids whose chromosomes have approximately identical mates give haploid parthenogenetic seedlings. No haploids are known in *Secale cereale*, or heterozygous species of *Œnothera*. Many are found in *Hordeum vulgare* (*cf.* Johansen, 1934).

FAILURE OF FERTILISATION

The fifty *Datura* haploids (*cf.* Blakeslee, Morrison and Avery, 1927) owed their origin to the stimulation of cold (*cf.* Haberlandt, 1927), or to pollination with *Datura ferox*.

Special conditions are responsible for the origin of haploids in *Linum* (Kappert, 1933). Here a single fertilised egg may divide to give twin or multiplet embryos. This polyembryony may take the form of a second haploid cell in the embryo sac growing up beside the normal diploid embryo to give a haploid-diploid twin. A similar process is found in *Gossypium* and *Oryza*.

We can therefore point to five conditions which lead to non-recurrent parthenogenesis in the flowering plants : (i) Exceptional external conditions (temperatures). (ii) Pollination without fertilisation or with fertilisation of neighbouring ovules only. (iii) Accidental fusion of both male nuclei with the fusion nucleus. (iv) The entry of an alien nucleus into the egg-cell without fusion. (v) A genetic propensity for polyembryony. Three of these conditions are often called forth by cross-pollination with another species, and it would appear that by this means parthenogenetic seedlings could be obtained in most groups of the dicotyledons, for they have been found wherever they have been thoroughly searched for. Therefore, although non-recurrent parthenogenesis has, for obvious reasons, been found only under experiment, there is no reason to doubt that it occurs in nature, at least in cross-pollinated plants.

Nevertheless, two genetical conditions appear to limit the occurrence of non-recurrent parthenogenesis. First, the plant must be sexually fertile and homozygous or nearly so, for the conditions which eliminate the segregates of highly heterozygous organisms (such as interspecific hybrids) and lead to their sterility are much more stringent in haploid parthenogenesis, since the single gametic genotype has to meet all the conditions of life, and not merely those normal to the gamete (*cf.* Ch. VIII). It is much more likely to do so if it has the same qualitative constitution as its parent, and this is possible only where the parent is homozygous. Attempts to obtain parthenogenetic seedlings of heterozygous species of *Œnothera* have therefore been unsuccessful (Davis, 1931 ; Stomps, 1931).

Secondly, a definite hereditary predisposition to produce

parthenogenetic offspring is indicated in a trisomic diploid *Œnothera franciscana* ($2n = 15$), itself derived from a haploid seedling (Stomps, 1931). Such a predisposition is equally to be regarded as a condition of diploid parthenogenesis, for failure of reduction in itself is no stimulus to parthenogenesis, it merely permits repetition of the lapse.

In connection with non-recurrent parthenogenesis two other errors of fertilisation, both occurring under normal conditions, may be mentioned. It has been possible to show in *Drosophila*, from genetical considerations, that a single, mosaic, zygote may develop from two separately fertilised eggs sticking together (Stern and Sekiguti, 1931) and from a binucleate egg, the two nuclei of which were separately fertilised (Stern, 1927; *cf.* Goldschmidt and Katsuki on *Bombyx*, 1931; and Whiting on *Habrobracon*, 1935).

(ii) **Regeneration and Apospory.** By injury or by vegetative propagation it is possible to " regenerate " the gametophyte of a liverwort (Marchals, *cf.* 1911) or a moss (v. Wettstein, 1924) from the diploid tissue of the sporophyte. The new gametophyte has escaped meiosis and is therefore diploid. This is artificial and necessarily non-recurrent apospory. From the diploid gametophytes tetraploid sporophytes are obtained, and by repeating this process v. Wettstein has succeeded in obtaining octoploid sporophytes. This is possible only in monœcious forms like *Amblystegium* (Marchals, 1911). In diœcious species the $2A + 2X$ and $2A + 2Y$ gametophytes do not give functional gametes. The same effect has been produced in the fern, *Osmunda* (Manton, 1932; *cf.* Schwarzenbach, 1926; Lawton, 1932).

A more complicated condition has been found by Andersson-Kottö (1931, 1932; Andersson-Kottö and Gairdner, 1936) as a result of mutation in the fern *Scolopendrium vulgare*, but being heritable the anomaly is recurrent. When the gametophytes from this plant were intercrossed (a process analogous to self-fertilising a flowering plant) one-quarter of the progeny developed the mutant character, which is therefore a mendelian recessive. The character consists in the development of gametophytic tissue on the fronds of the sporophyte. There is no spore formation or meiotic process of any kind at this stage. The gametophyte, therefore, has the

ALTERNATION OF GENERATIONS

diploid complement of the sporophyte. And, as in regeneration, it produces functional gametes which should yield tetraploid sporophytes in the next generation. This process is recurrent although indefinite repetition is scarcely conceivable.

When crossed with another race having the same chromosome number ($2n = 60$) the first generation was normal and its gametophytes all had 30 chromosomes. Yet these produced sporophytes all having about 45 chromosomes. By selfing the gametophytes from these plants and crossing them with normal types it was proved that the male gametes had often undergone a second reduction, to 15 chromosomes (cf. Manton, l.c.). One-quarter of the second generation were aposporous plants, and all these bred true to the character of having a redundant although apparently often imperfect meiosis. The combination of apospory with antheridial meiosis gave chromosome numbers increasing by about a half in succeeding generations, thus : F_2, ca. 45 ; F_3, ca. 65 ; F_4, ca. 95. These results point to the redundant meiosis being an independent hereditary character. It is analogous to the condition in the pseudo-hermaphrodite *Icerya purchasi*. The reduction to 15 indicates that the original plants were tetraploid.

In certain "normal" derivatives of the mutant the spores have been found to be replaced in part by spermatozoids (Gairdner, 1933). The traditional homologies which apply throughout the higher plants are thus overthrown at one step. Such revolutionary behaviour cannot perpetuate itself, since no corresponding archegonia are formed. The stability of the alternation of generations in the higher plants and the gradualness of its transformation is thus seen to depend, not on the absolute and mystical distinction which morphologists have made between the successive generations, but on the impossibility of making a decisive change in any one stage of development without making an exactly adapted change in others at the same time ; in other words without a coincidence in mutations such as does not occur in nature. The stability of the distinction between alternating generations depends on the complexity of their mutual adaptation and not on a profound genetic difference between them.

The unbalanced occurrence of apospory and parthenogenesis, as well as many observations on polyploids, show that the association of change of chromosome number with the alternation of form in succeeding generations is not due to a relationship of cause and effect, but that the alternation of forms is a genetic property of the race achieved by independent adaptation, immediately conditioned by the stage in the cycle of development, and therefore independent of the chromosome number.

TABLE 67

Methods of Suppressing Meiosis with Diploid Parthenogenesis

1. Little or no metaphase pairing. Restitution nucleus (as in *Raphanus - Brassica* hybrids, *cf.* Fig. 127). Usually alternative to type 2.	*Erigeron* spp.	Holmgren, 1919; Carano, 1921.
	Taraxacum spp.	Gustafsson, 1934.
	Chondrilla	Poddubnaja-Arnoldi, 1933.
	Ixeris	Okabe, 1932.
	Nephrodium	Steil, 1919.
	Balanophora	Kuwada, 1928.
2. After a more or less meiotic prophase the unpaired chromosomes divide once at metaphase (*cf.* Fig. 127).	*Artemia* (?)	Fries, 1910.
	Phylloxera	Morgan, 1915.
	Aphis palmæ	de Baehr, 1919.
	Cynips and *Rhodites*	Hogben, 1920.
	Rhabditis sp.	Belar, 1923.
	Daphnia	Schrader, 1925.
	Lecanium	Thomsen, 1927.
	Artemisia	Chiarugi, 1926.
3. After a prolonged resting stage a mitotic division with mitotically contracted or intermediate chromosomes replaces meiosis (alternative to type 2).	*Archieracium*	Rosenberg, 1917.
	Hieracium umbellatum.	Bergman, 1934.
	Antennaria	Stebbins, 1932 b; Bergman, 1935 a.
	Eupatorium	Holmgren, 1919.
	Poa serotina (*cf.* Leontodon, Bergman, 1935 a).	Kiellander, 1935.
4. Double division of unpaired chromosomes to give four nuclei (as in *Pygæra* hybrids).	*Bacillus rossii*	v. Baehr, 1907.
	Cocconeis *	Geitler, 1927.
	Thismia	Meyer, 1925.
	Artemisia	Chiarugi, 1926.
5. Normal pairing and first division with—		
(i) Failure of second anaphase.	*Cypris*	Woltereck, 1898.
	Rhabditis monohystera.	Belar, 1923.
	Solenobia pineti	Seiler, 1923.
(ii) Reunion of two daughter-nuclei at second telophase.	*Artemia* (i) or (ii)	Brauer, 1894; Fries, 1910; Gross, 1935.
	Apotettix (ii) or (iii)	Robertson, 1931 c.
(iii) Fusion of nuclei in early somatic division of embryo.	*Solenobia triquetrella.*	Seiler, 1923.
	Trialeurodes	Thomsen, 1927.

* Pachytene pairing seen.

4. MEIOSIS IN RELATION TO PARTHENOGENESIS

(i) **Meiosis in Functional Cells in Parthenogenesis.** Where the cell which would normally undergo meiosis still enters into the life cycle (as do the oöcyte, the embryo-sac mother-cell or the spore mother-cell in diploid parthenogenesis) it is clear that meiosis must be suppressed, so far as it effects a numerical reduction of the chromosomes. The ways in which this happens may for convenience be provisionally grouped in five classes, although gradations are commonly observed between some of them (*cf.* Ch. X).

The first two and the fourth of these types of behaviour correspond closely with the characteristic abnormalities of meiosis found in the absence of paired chromosomes (Fig. 127). The third and the fifth, on the other hand, are found only in conjunction with parthenogenesis and cannot be derived directly from the behaviour known in hybrids and autopolyploids. Their origin we shall consider later.

Side by side with these *regular* abnormalities others occur, particularly in the spore mother-cells of aposporous plants, but also sometimes in those reproducing facultatively by diploid parthenogenesis. The irregularities consist in the characteristic results of failure of pairing in hybrids—irregular distribution of the chromosomes at either the first or the second division, and formation of numerous daughter-nuclei (*Artemisia nitida, Nephrodium hirtipes*). Diploid parthenogenesis is out of the question in the cells affected, just as it would be following normal reduction, which is occasionally to be observed following more regular pairing in certain plants that are obligatorily apomictic (*Artemisia nitida ; Ochna serrulata*). Regular reduction with complete pairing is always to be found in those that are facultatively sexual (*e.g.*, *Elatostema sessile* and *Allium odorum*, Modilewski, 1931). Plants that are apogamous or aposporous and not parthenogenetic (*Hieracium pilosella, Citrus*, probably *Rosa canina*) are nearly always facultatively sexual and therefore have, as a rule, regular meiosis. The *Caninæ* roses, for example, have a meiosis which is not normal but which is regular in effect (*v. infra*).

The great range of behaviour that is possible in certain apomictic

452 BREAKDOWN OF GENETIC SYSTEMS

TABLE 68

Methods of Apomixis in *Artemisia nitida* (after Chiarugi, 1926)

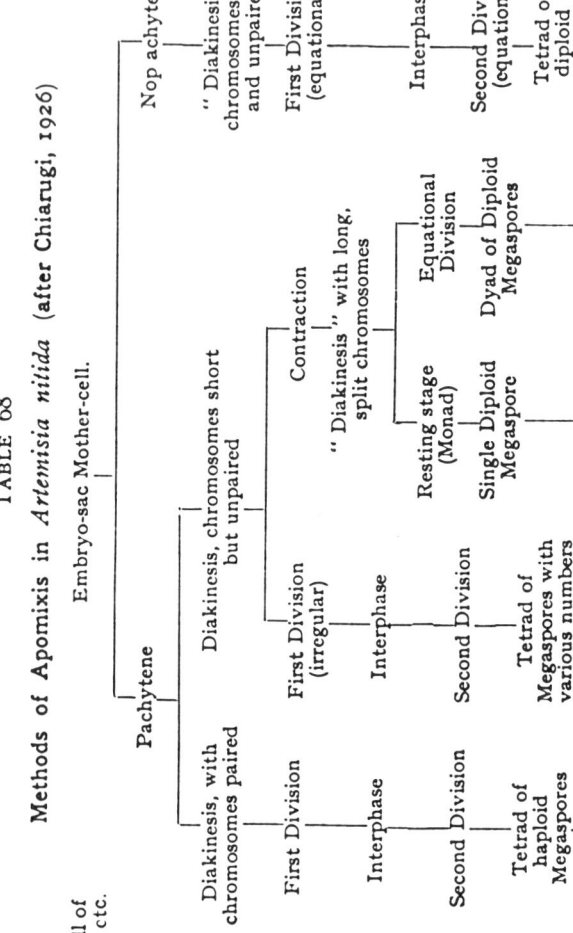

plants and its relation to the course of meiosis is shown by *Artemisia nitida* (Table 68).

(ii) **Meiosis in Non-Functional Cells.** The mother-cells of male spores and male gametes of apomictic plants and animals (where they are formed) show various kinds of behaviour which cannot be predicted merely from a knowledge of the behaviour in the functional cells on the female side. They have been studied more thoroughly than the functional cells and provide valuable evidence of the origin of apomixis.

In sexually reproducing organisms where a failure of pairing depends merely on mechanical chromosome relationships (*i.e.*, structural differences, numerical inequality among the pairing chromosomes, or their excessive number as in autopolyploids) we find a correlation between the conduct of meiosis in the male and female cells. The evidence of this correlation consists in the similar fertility of structural hybrids on male and female sides, and of the similar character of their offspring from male and female cells. The same correlation applies to many of the genotypically controlled abnormalities of meiosis, but it does not apply to all of them. There are many cases of such an abnormality specifically affecting either the male or the female cells of hermaphrodites. If we wish to know how far the conduct of parthenogenetic meiosis depends upon mechanical properties of chromosomes and how far on a genotypic property we must find out how far this correlation applies.

All facultatively apomictic plants have normal or fairly normal meiosis in the pollen mother-cells, *e.g.*, *Alchemilla speciosa* (Murbeck, 1902), *Thalictrum purpurascens* (Overton, 1904), *Hieracium aurantiacum* (Rosenberg, 1917), *Elatostema acuminatum* (Strasburger, 1910). In this species the pollen degenerates afterwards, as in so many male-sterile plants. Otherwise they are able to produce gametes with a normally reduced chromosome number.

In some organisms which are obligatorily parthenogenetic the male cells function in stimulating parthenogenesis; their meiosis is then fairly regular. In *Zephyranthes texana* it is entirely so, and the pollen actually functions in fertilising the endosperm nucleus. Moderate or even high pollen fertility is sometimes found in the pseudogamous *Potentilla* species. In spermatocytes produced in the ovary of a parthenogenetic *Rhabditis* (Belar, 1923) meiosis is also not completely irregular; some pairing of chromosomes is found at

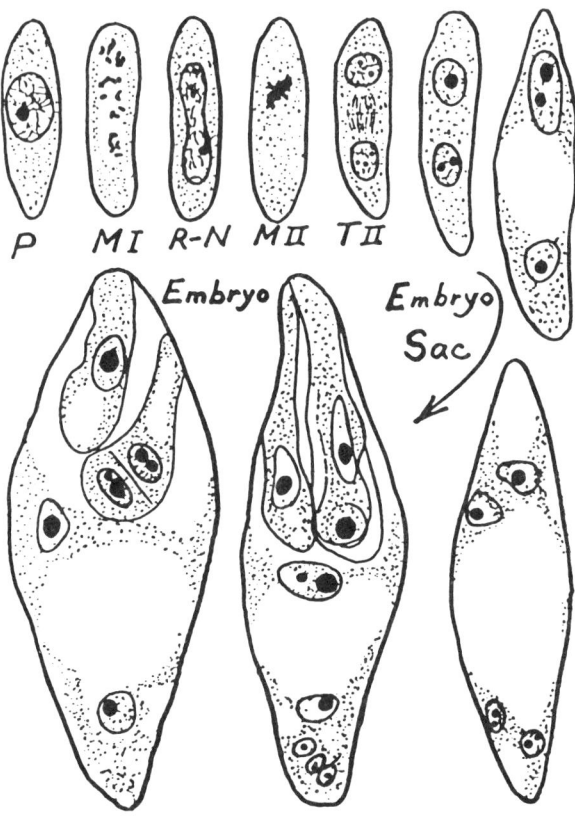

Fig. 131.—Development of the embryo-sac in *Ixeris dentata* by parthenogenesis: prophase—metaphase (21 univalents)—restitution-nucleus—second metaphase and telophase. Finally, the embryo in the two-cell stage (after Okabe, 1932).

the first metaphase, and the sperm formed actually functions in stimulating the egg to development.

Amongst plants with obligatory apomixis the pollen mother-cell behaviour is often closely parallel with that in the embryo-sac mother-cell, except that the purely mitotic *Antennaria* and *Euhieracium* type of development is missing. Two groups must be distinguished: (i) Species with apospory or nucellar embryony as well as parthenogenesis. (ii) Species that are solely parthenogenetic.

The first are naturally less dependent on the regular formation of diploid gametes (for the embryo-sac) and it is found that while meiosis is often suppressed owing to the reconstitution of a single nucleus at the first telophase (following the lagging of a variable number of unpaired chromosomes) yet often the results are entirely irregular or a tetrad is formed of nuclei with different numbers of chromosomes. This is the case in *Erigeron annuus*, *Artemisia nitida* and *Ochna serrulata*. In the last species pairing follows precisely the frequencies observed in the similarly autopolyploid *Tulipa Clusiana*, which is also a pentaploid and sexually fertile (Figs. 74, 75).

In purely parthenogenetic species (*e.g.*, *Eu-hieracium*) meiosis is rather more regularly suppressed. But several gradations have been described by Rosenberg in this apparent transition towards the regular formation of diploid gametes (if we may infer a progressive change). In all these the early prophase appears normal, and doubtless some association of chromosomes occurs at pachytene. The first stage of "degeneration" corresponds with that found in such hybrids as *Raphanus-Brassica*. Few (in *Hieracium boreale* and *Chondrilla juncea*, with embryo-sac of type 1) or no chromosomes (in *Hieracium lævigatum*, with embryo-sac of type 2) are paired and very often, probably as a result of the uneven distribution and lagging of the unpaired chromosomes on the spindle, a single resting interphase nucleus is reconstituted before the two groups are clearly separated. This "restitution nucleus" divides to give daughter-nuclei, each of which has the unreduced number of chromosomes. Various abnormalities occur, such as the formation of one or more, usually small, supernumerary nuclei.

The second stage of degeneration is found in *H. pseudoillyricum* and a few other species (with embryo-sac of type 3). The nucleus contracts after reaching a stage immediately before diakinesis, and on resuming its prophase condition develops directly into a somatic mitosis and gives a dyad. This type of division is often found together with the *lævigatum* type, for variations always occur in the same individual, as they do in all other irregular chromosome behaviour (Rosenberg, 1927). All such variation may be explained by the developmental variations having a certain threshold value

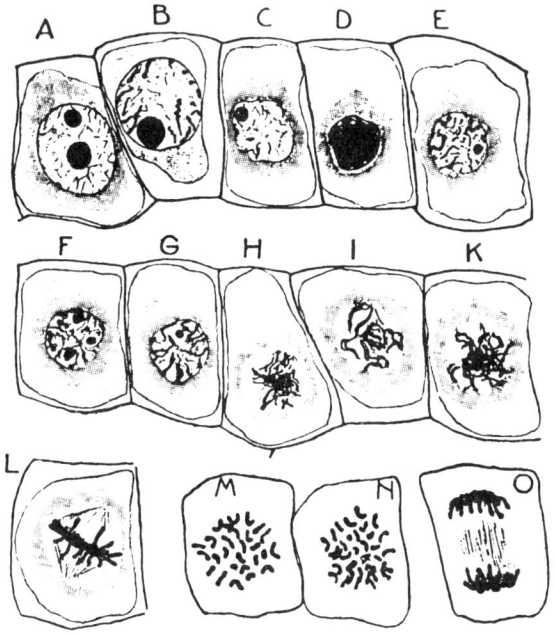

FIG. 132.—Meiosis in pollen mother-cells of *H. pseudoillyricum* ($3x = 27$). A–C, early prophase. D, contraction replacing first division. E–G, renewed prophase of second division. H–O, second division which gives dyad of pollen-grains with the unreduced chromosome number. (After Rosenberg, 1927.)

in the effect on the general course of meiosis in a sexually fertile organism (they can be detected only by their effect on chiasma frequency), while in hybrids and organisms with genetically abnormal meiosis there is no such threshold value.

A parallelism between pollen and embryo-sac mother-cell behaviour is by no means universal however. For example, *Antennaria*, *Poa serotina* and *Hieracium umbellatum*, all of which sometimes have a mitotic or semi-mitotic meiosis on the female side, have a high degree of pairing on the male side. In the *Hieracium*, which is triploid, five trivalents may be formed ; in the *Poa*, of which triploid and tetraploid forms occur, trivalents and quadrivalents are frequent ; in the *Antennaria* species, some of which are hexaploid or dodecaploid, nearly all the chromosomes are

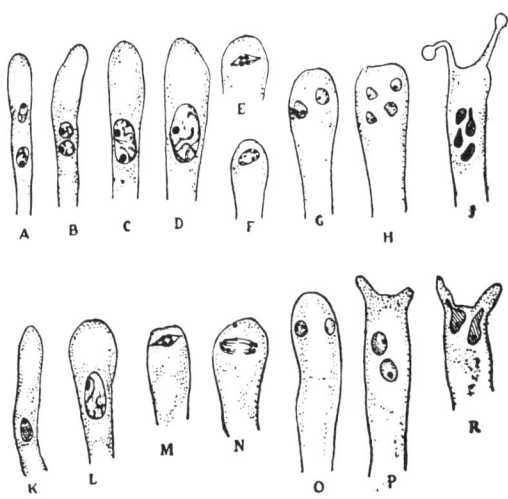

FIG. 133.—Spore-formation in the sexual four-spore (A–J) and the non-sexual two-spore (K–R) race of *Camarophyllus virgineus*. A, paired, diplophase nuclei. B, C, their fusion. D, zygotene. E, metaphase of the first division. F, anaphase. G, interphase. H, telophase of second division. J, migration of the four nuclei into their pouches. K, single nucleus. L, prophase. M, N, single division. O–P, spore nuclei formed. R, their migration. (From Rosenberg, 1930, after Bauch, 1926.)

associated in bivalents or multivalents. *Taraxacum Nordstedtii* similarly forms multivalents. These observations not only reveal the potential dissociation of behaviour on male and female sides. They also show that these species are autopolyploids and therefore, even where they have an even number of chromosome sets, and the pairing chromosomes are alike, they must have an irregular meiosis and be sexually infertile. In other words, although there is no evidence of their hybridity there is evidence that abnormality in their meiosis is conditioned by autopolyploidy which prevents regular reduction as effectively as does structural hybridity.

(iii) **Meiosis in Parthenogenetic Haploids.** The occurrence of meiosis in a mother-cell is affected in an entirely different way by haploidy according as the fertilisation which has failed is normally premeiotic or post-meiotic. The first type has been found in algæ and fungi, the second in the angiosperms (Ch. X), and in animals with haplo-diploid sex differentiation.

In *Vaucheria hamata* (v. Wettstein, 1920) it is possible to stimulate both male and female gametes to development without fertilisation or the meiosis which would normally follow it. Parthenogenesis can be similarly induced (by treatment with dilute solutions) in *Spirogyra*, *Ulothrix*, and *Saprolegnia mixta* (Mäckel, 1928).

In the sexual basidiomycetes the conjugating nuclei fuse in the basidial cell and immediately undergo meiosis. The four daughter-nuclei then pass into the four terminal sacs, where they develop into spores. But in some species only two sacs are produced, and in these it is believed that two nuclei pass into each (Sass, 1929). In parthenogenetic species the single nucleus of the basidial cell divides once where two sacs are developed (*Camarophyllus virgineus*, Bauch, 1926, v. Fig. 133), and twice where four sacs are developed (*Schizophyllum commune*, Kniep, 1928, p. 422). These observations are paralleled in the Uredineæ (*cf.* Rosenberg, 1930).

Thus, in forms where fertilisation immediately precedes meiosis, the omission of fertilisation means, sometimes, the suppression of meiosis, *e.g.*, in the parthenogenetic race of *Camarophyllus virgineus*, Bauch (1926) found early prophase stages reminiscent of meiosis, although the later course of the single division in the basidium was purely mitotic ($n = 4$). Elsewhere there appears to be a complete absence of any meiotic process, and the nucleus instead divides mitotically twice, once or not at all as purely vegetative conditions dictate (*Vaucheria*, v. Wettstein, 1920). Habitual parthenogenesis can therefore be induced (artificially or by mutation) at one step in such an organism as *Chara crinita*, while in diploid organisms the assumption of parthenogenetic habit must necessarily be more complicated, since it means a change in a specialised meiotic mechanism.

In haploids derived from organisms with a prolonged diploid generation two types of meiosis are found according as the haploidy is habitual (in certain male Hymenoptera and in coccids) or adventitious (in angiosperms arising by parthenogenesis from diploid parents). In both the chromosomes are contracted as in ordinary meiosis, but, of course, unpaired. In the drone bee (*Apis*, Meves, 1907) a spindle is formed and the chromosomes lie on a metaphase plate but do not divide; at one pole an enucleate bud is cut off, a second division spindle is formed and the chromosomes divide this time to give two daughter-nuclei, one of which is extruded as a polar body while the other gives a haploid (unreduced) spermatozoon.

In the haploid male coccids (*Icerya*, etc., $n = 2$; Hughes-Schrader, 1930) only one division occurs, and it is not clear whether this corresponds to the first or the second. In the haploid male rotifers (*Asplanchna*, $n = 13$; Whitney, 1929) it is the second division that is usually omitted; where an abortive second division occurs spermatozoa are formed with less than the haploid number, and these do not function. The mechanism is apparently imperfect.

This suppression of meiosis seems to be a sexual characteristic not directly determined by the haploid condition. Diploids sometimes show the male character in the hymenopteran *Habrobracon*, and these produce diploid sperm (Torvik, 1931; Torvik-Greb, 1935). They have the abortive meiosis characteristic of males. This shows that the mechanism of meiosis is here genetically adapted to give regular failure of reduction. The condition of the adventitious haploids in the angiosperms, on the other hand, is different. The ordinary meiotic mechanism has to deal with unpaired chromosomes, and the result is irregular (*v.* Ch. X).

Such observations also show that where meiosis takes place at the end of the diploid generation there is a mechanism of division external to the chromosomes and distinct from that of ordinary mitotic division (*cf.* Ch. XII). In the adventitious haploid angiosperms we see it acting on unpaired chromosomes, for whose division it is unadapted. In the haploid Hymenoptera we see it specially adapted to deal with the situation so as to produce regular results.

In the organisms in which meiosis immediately follows fertilisation, on the other hand, there is no evidence of any such mechanism. Since every cell which is diploid in these undergoes meiosis it seems possible to consider that meiosis, which in the higher organisms must be regarded as induced in specialised cells by their relationship with the rest of the organism, is in these simple forms determined directly by the condition of diploidy (*cf.* D., 1932, p. 458). In this connection the parthenogenetic *Chara crinita*, Ernst, 1918) is of interest. It is diploid, unlike its sexual relations ; but since, as Ernst supposes, it is probably a hybrid, the normal diploid condition, the presence of pairs of identical chromosomes, is lacking. Parthenogenesis will therefore follow the failure of meiosis in such an individual as a matter of course, if it happens to be female like the regenerated diploid gametophytes of *Sphærocarpus* (Lorbeer, 1927). Hybridity is then seen as a simple and immediate condition of parthenogenesis.

(iv) **Meiosis in Rosa canina.** The type of sexual reproduction found in *Rosa canina* and its relatives approaches closely to apomixis and can best be considered in relation to it. All species of *Rosa* have multiples of 7 chromosomes. Aneuploid forms reported by Täckholm (1922) and Erlanson (1929) are exceptional and probably sterile. Most diploid and even-multiple polyploid species have fairly regular pairing, although they are none the less to be regarded as hybrids (Blackburn and Harrison, 1921 ; Täckholm, 1922 ; Erlanson, 1929), and some have been shown to be interchange heterozygotes (Erlanson, 1931 *c*). They are sexually reproducing, with occasional apomixis in some species. All species of the section *Caninæ* (European and Asiatic) are facultatively apomictic. For example, in 38 seedlings of *R. coriifolia solanifolia* × *glauca concolor*, 37 were maternal in character, one was a hybrid (Täckholm, 1922). Gustafsson (1931) has shown that this apomixis is not spontaneous, but requires the stimulus of pollination, especially of cross-pollination, for self-fertilisation probably leads to successful sexual reproduction (*v. infra*).

All the *Caninæ* roses have, according to Täckholm and Hurst (1931), seven pairs of chromosomes and 21 univalents (or, in a few species, 14 or 28). The pairs are distinguished by the regularity of their occurrence and of their form. Erlanson (1931) has shown that metaphase pairing is of the partially terminalised type in *Rosa*. Uniformity, we now know, means that while in other groups of *Rosa* the chromosomes may be united by chiasmata at one or both ends, in the *Caninæ* they are always united at both ends, and this is an indication of regularly high chiasma frequency and homozygosity as to the pairs (Erlanson, 1933). Pairing behaviour is identical in embryo-sac and pollen mother-cells (Fig. 134), but the mechanism of distribution is entirely different. In the pollen mother-cells the bivalents and univalents lie in one equatorial plate. The bivalents divide first and then, as is usually the case, the univalents lying between the separating groups of daughter bivalents divide and follow them to the poles with very little irregularity. At the second division the same process is repeated; the daughter bivalents divide again and the daughter univalents also proceed to divide a second time (like the unpaired chromosomes in *Pygæra* hybrids). The delay, however, is greater than at the first division. Most of the univalents fail to reach the poles and many are left on the plate. In consequence many nuclei of irregular size are formed and the 4 nuclei which have received the 7 chromosomes derived from the bivalents have in addition a varying number of chromosomes derived from the univalents. Examination of the first mitosis in the pollen grains of *R. tomentella obtusifolia* and *R. seraphini* ($2n = 5x = 35$) showed the following chromosome numbers (*cf.* Table 47) :—

7	8	9	10	11	12	13	14	15	16	17	18	19	20	21	22
9	23	13	7	4	1	1	—	1	—	1	1	—	1	—	1

These do not necessarily correspond with the numbers originally received, but they indicate the irregularity of the process and the relatively high proportion that receive the simple haploid complement.

Several embryo-sac mother-cells are found in each ovule of these roses, and they are therefore exceptionally favourable for study. At metaphase the unpaired chromosomes, instead of lying in the same plate as the bivalents, are all, or almost all, grouped at the micropylar pole of the spindle, where they are joined at the first

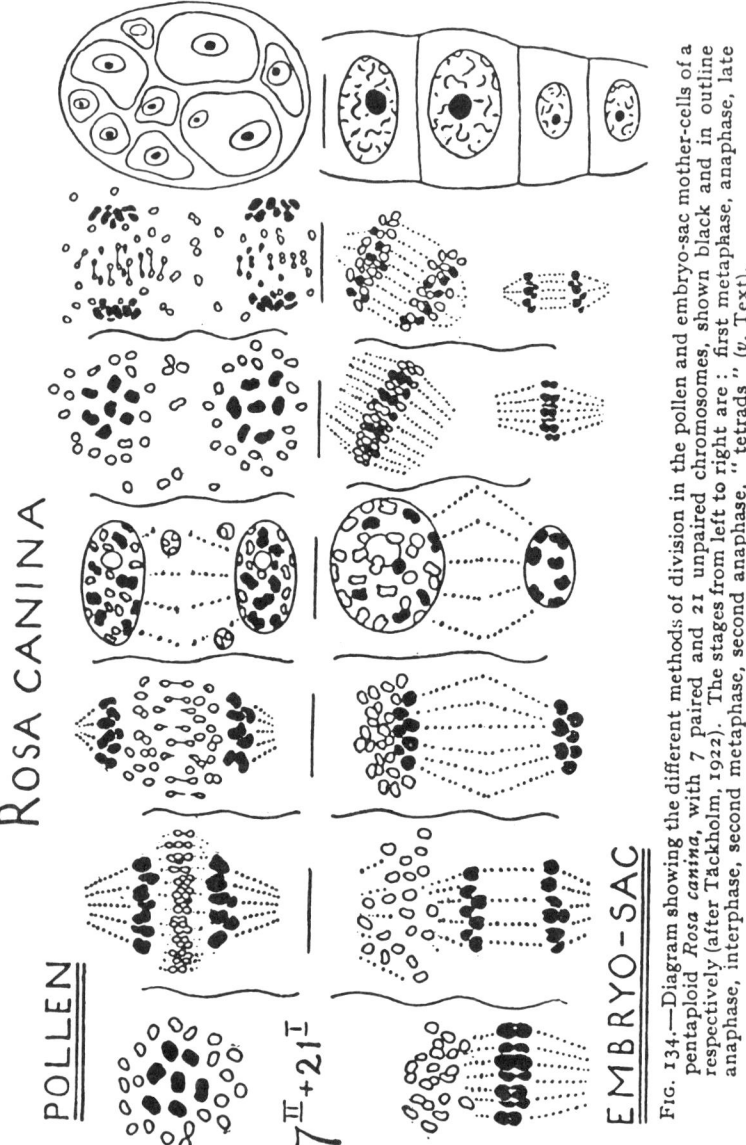

Fig. 134.—Diagram showing the different methods of division in the pollen and embryo-sac mother-cells of a pentaploid *Rosa canina*, with 7 paired and 21 unpaired chromosomes, shown black and in outline respectively (after Täckholm, 1922). The stages from left to right are : first metaphase, anaphase, late anaphase, interphase, second metaphase, second anaphase, "tetrads." (*v.* Text).

telophase by the seven daughter bivalents going to this pole. In the pentaploid forms there are therefore usually 28 chromosomes at the micropylar end and 7 at the chalazal end. The second division is regular and yields two small nuclei and two large ones ; one of the latter gives the functional embryo-sac, as is shown by the chromosome numbers of hybrid seedlings.

Self-fertilisation of these plants will therefore yield offspring like the parent having 7 pairs of chromosomes of identical origin and 21 chromosomes that are not directly related to them or to one another. A complement of a race of such a kind consists of 7 pairs of chromosomes that are transmitted by sexual reproduction (and no doubt inbred and homozygous) together with 21 chromosomes that are transmitted virtually by vegetative processes, for they are neither separated by meiosis nor brought together by fertilisation. Reproduction, in fact, may be described as *semi-clonal*. In regard to its genetic effect this mechanism is analogous with polyploidy and with the ring formation of complex heterozygotes, inasmuch as it enables a hybrid to breed true. The mechanism is responsible for the relative constancy of the innumerable species of Caninæ. It is also responsible for the ability of widely distinct forms to hybridise. The seven chromosomes of the pairs are probably highly constant throughout the group. Their pairing alone is necessary for fertility. The unpaired chromosomes are, on the other hand, the organs of variation. In this specialisation of function the two groups of chromosomes are analogous to the differential and the pairing segments of the chromosomes of the heterozygous *Œnothera* species.

The origin of the system does not present much difficulty. Repeated crosses amongst diploid species or a cross between two hexaploid species would give the condition with seven pairs of chromosomes and twenty-eight univalents, as in *Triticum-Ægilops* crosses. It is not now necessary to suppose, as Täckholm did, a cross between a diploid and an imaginary decaploid as the source of such a form. The hybrid would give at once a proportion of pollen grains with only the seven chromosomes derived from the pairs. The essential new property of these forms is therefore the polarisation of the embryo-sac mother-cell whereby it regularly gives embryo-sacs containing the full set of unpaired chromosomes. This property and the perfection of the mechanism of division in the pollen mother-cells must have been achieved by adaptation in the course of sexual reproduction. Täckholm's objection to this is also no longer valid. He considered that it would lead to frequent aneuploid forms in nature, but we now know that such forms would be largely eliminated (cf. Ch. VIII). The variation necessary for such adaptation will arise from the occasional pairing and crossing over of the chromosomes found by Erlanson (1933) in non-pairing chromosome sets. The present account, therefore, seems to provide a sufficient explanation of the origin and behaviour of the system.

5. THE ORIGIN OF APOMIXIS IN EXPERIMENT

The foregoing account has shown the characteristics of nuclear behaviour in apomixis. Apart from these immediate conditions certain events external to the nucleus can be related to the occurrence of apomixis as prior conditions. First we may note that the variations in conditions which are responsible for the change from sexual to non-sexual reproduction may be imposed artificially on organisms with facultative parthenogenesis (cf. Nabours, 1919 on *Paratettix* and *Apotettix*).

(i) **Pseudogamy.** In some species of plants and animals the male gamete functions in exciting the development of the unfertilised and unreduced egg in diploid parthenogenesis or of nucellar embryos in apospory. This is known as pseudogamy. We have already seen how non-recurrent parthenogenesis is stimulated in plants in this way.

In a nematode, *Rhabditis* "*XX*," Belar (1923) found that the sperm, although irregularly formed (in part of the ovarian tissue), entered the egg and stimulated its development, probably by the contribution of its centrosome—the middle-piece of the sperm—although no fusion of the nuclei occurred.

An analogous process is found in various monocotyledons, e.g., *Zephyranthes texana* and *Allium odorum* (Modilewski, 1930). Here the pollen is normally developed and one generative nucleus fuses with the central nucleus to give the endosperm; the other degenerates, but the first fusion stimulates the development of the unreduced egg-cell. Pollination is also necessary to stimulate apogamy and apospory in *Allium odorum* (Haberlandt, 1923) and *Zygopetalum* (Suessenguth, 1923).

In *Potentilla* (Müntzing, 1928) several species which fail to set seed without pollination produce entirely maternal offspring when cross-pollinated, as well as when self-pollinated. Some of these species have good pollen, others bad. In *Rubus* (Lidforss, 1914; Gustafsson, 1930) the position is more complicated. In many European species of blackberries (section *Eu-batus*) crosses with closely related forms yield a high proportion of hybrids, crosses with more distant species yield entirely maternal offspring—" false hybrids." These are identical with the female parent, yet when such plants are selfed they yield the heterogeneous progeny characteristic of a sexually reproducing hybrid. Evidently the proportion of sexual embryos diminishes with increasing remoteness of the male parent. Such a situation indicates competition between the sexual embryos and their substitutes, the asexual, probably aposporous, ones. The same conditions probably apply to the *Caninae* roses, although the evidence here is still scanty (Hurst, 1931), and to Ohmachi's orthopteran hybrids (1929 *b*).

Thus pseudogamy may stimulate either apospory (in *Allium*, probably in *Rosa* and *Potentilla*) or diploid parthenogenesis (*Rhabditis, Zephyranthes*).

In all these cases we may perhaps not inappropriately look upon pseudogamy as a physiological response that has been genetically conditioned. Originally development of the egg-cell was no doubt stimulated solely and directly by fertilisation. But fertilisation has been accompanied throughout a long period of evolution, that is of adaptive change, by certain invariable antecedents, such as the penetration of the egg by the sperm in animals and pollination in plants. Many organisms have therefore acquired by adaptation an equally satisfactory response by which the antecedents determine development as effectually as the act of fertilisation itself.

(ii) Injury. By electrical, chemical and mechanical stimuli the unfertilised egg has been induced to develop in many Echinoderma and Algæ, in the frog and elsewhere. Where the stimulus is applied after the two polar bodies have been extruded from the animal egg a haploid embryo develops. In the frog, where this has happened, somatic doubling has followed later as in the somatic origin of polyploid plants and in some kinds of diploid parthenogenesis. The general failure of a diploid organism produced in this way to reach maturity is probably due to the presence in most outbred organisms of factors lethal in the homozygous condition. The same result is obtained when a sperm fertilises an enucleated egg (" male parthenogenesis "). Where reduction has not taken place the stimulus may lead to the suppression of one of the meiotic divisions or to the fusion of the products of the second meiotic division (as in the production of diploid spores of *Funaria* by v. Wettstein, 1924). An embryo may then develop without fertilisation. Such individuals do not usually get beyond an early stage of development.

It is possible to induce the direct development of an unfertilised gamete in many Algæ where meiosis should immediately follow fertilisation, in various ways. Thus when the antheridia or oögonia of *Vaucheria hamata* (v. Wettstein, 1920) were pricked they developed directly into new thalli : both fertilisation and meiosis were omitted. Apospory was induced by layering the fronds in *Nephrodium* (Digby, 1905). This automatically led to apogamy.

The development of adventitious embryos from the nucellus or integument has been induced by wounding in *Œnothera Lamarckiana* by Haberlandt (1921 a). He also found that degeneration of the nucellar tissue and synergids was associated with parthenogenesis in *Taraxacum* and *Hieracium*, while in sexual species of *Hieracium* no such degeneration occurred (1921 b). These observations make it plausible that degeneration of neighbouring tissue may stimulate apospory and parthenogenesis, and thus be a prior, and genetically determined, condition of these phenomena.

(iii) **Mutation.** Paula Hertwig (1920) found a sexual strain of *Rhabditis pellis* which gave rise to a purely parthenogenetic strain breeding true to the new character. It had the same chromosome number as the parent race, but its egg-cells underwent a single division at meiosis and gave unreduced eggs. The entrance of the sperm was necessary to stimulate development without fertilisation, as in the related parthenogenetic species. In this case parthenogenesis must be regarded as a genetic character arising by mutation.

(iv) **Hybridisation.** Ostenfeld (1910), by crossing facultatively sexual species of the Pilosella section of *Hieracium*, with purely sexual species, *e.g.*, *H. aurantiacum* × *H. auricula*, produced a heterogeneous progeny of three types: (i) sterile; (ii) purely asexual, probably aposporous; and (iii) asexual when self-pollinated, sexual when crossed with good pollen. When we consider this in relation to Lidforss's observations on *Rubus* we see that with plants having the genetic faculty of producing aposporous offspring, the greater their hybridity the less the chance of their successfully producing sexual offspring—owing to the sterility inherited as a concomitant of high segregation—and therefore the greater the proportion of asexual seedlings produced on self-fertilisation. The extreme type, being sexually sterile, will be obligatorily apomictic. But in the same group we find plants not having the faculty of apomixis and sexually sterile as well. Thus the apomixis is conditioned both by the sterility of the hybrid and by its particular genotype, which is not inherent in its hybridity.

Harrison (1920) crossed two sexual species, *Tephrosia crepuscularia* and *T. bistortata* (Lepidoptera) and obtained four parthenogenetic females amongst normally sexual progeny. When this happened on a second occasion the offspring of the parthenogenetic female showed segregation of recessive characters (Peacock, 1925). This should happen with the kind of parthenogenesis found by Seiler in the Lepidoptera. Seiler (1927) crossed parthenogenetic and sexual races of *Solenobia triquetrella*, and the progeny showed variation in behaviour between the normal and that obtaining in the parthenogenetic race, where pairs of nuclei fused after the second segmentation division in the embryo. There were various other irregularities, with the result that different individuals and parts of individuals had different numbers of chromosomes.

(v) **The evidence of Chromosome Numbers.** It was early noticed that apomictic species of flowering plants had, as a rule, higher chromosome numbers than their sexual relatives. The following

BREAKDOWN OF GENETIC SYSTEMS

lists, giving the somatic chromosome numbers, show that this is still true of parthenogenetic as well as of aposporous species :—

TABLE 69

Chromosomes Numbers of Apomictic Species

DIPLOID SPECIES

Chondrilla juncea, etc. (15)	Rosenberg, 1912.
Zephyranthes texana (24)	Pace, 1913.
Allium odorum (16)	Haberlandt, 1923.
Potentilla argentea forma (14)	Müntzing, 1928.

TRIPLOID SPECIES

Hieracium boreale forms (27)	⎫
H. pseudoillyricum (27)	⎬ Rosenberg, 1917, 1927.
H. lævigatum (27)	⎭
Chondrilla juncea, etc. (15)	Poddubnaja-Arnoldi, 1933.
Erigeron annuus (27)	⎫
Eupatorium glandulosum (51)	⎬ Holmgren, 1919.
Artemisia nitida (37)	Chiarugi, 1926 (*cf.* Weinedel-Liebau, 1928).
Taraxacum officinale, etc. (24)	Gustafsson, 1932, 1934.
Nothoscordon fragrans (24)	Messeri, 1931.
Ixeris dentata (21)	Okabe, 1932.

TETRAPLOID SPECIES

Thalictrum purpurascens (48)	Overton, 1909.
Elatostema sessile (32)	Strasburger, 1910.
Hieracium boreale forms (36)	⎫
H. Pilosella (36)	⎬ Rosenberg, 1917, 1927.
Taraxacum croceum, etc. (32)	Gustafsson, 1932, 1934.
Poa serotina (28) [and 21]	Kiellander, 1935.

PENTAPLOID SPECIES

Rosa canina, etc. (35)	Täckholm, 1922.
Potentilla argentea forma (35)	⎫
P. Tabernæmontana (35)	⎬ Müntzing, 1928.
Ochna serrulata (35)	Chiarugi and Francini, 1930.
Taraxacum spp. (45)	Osawa, 1913 ; Gustafsson, 1932.

HEXAPLOID SPECIES

Potentilla collina (*ca.* 42)	Müntzing, 1928.
Allium roseum var. *bulbilliferum* (48)	Messeri, 1931.
Antennaria spp. (84)	Stebbins, 1932 *b* ; Bergman, 1935 *a*.
Taraxacum Nordstedtii (48)	Gustafsson, 1934.

UNSTABLE AND UNBALANCED POLYPLOID SPECIES

Poa alpina ($x = 7$, $2n = 22 - 38$)	Müntzing, 1932.
Poa pratensis ($x = 7$, $2n = 49 - 85$)	Müntzing, 1932.

APOMICTIC POLYPLOIDS

The chromosome numbers show that polyploidy and particularly odd numbered polyploidy favours apomixis. Two reasons suggest

TABLE 70

Cases of Polyploidy within the Species Associated with the Occurrence of Apomixis

	Sexual (haploid-diploid).	Apomictic (somatic number)	Authority.
CHAROPHYTA			
Chara crinita	12–24	24	Ernst, 1918.
ANGIOSPERMA			
Potentilla argentea, etc.	—	14, 42, 56	Müntzing, 1931.
Nigritella nigra	16–32	64	Afzelius, 1928.
			Chiarugi, 1929.
Hieracium umbellatum	9–18	27	Rosenberg, 1927; Bergman, 1934.
Crepis acuminata, etc.	—	33, 44, 55 ?	Hollingshead and Babcock, 1930.
Festuca ovina.	7–14 21–42	14, 21, 28, 35	Lewitsky and Kuzmina, 1927.
			Turesson, 1930, 1931.
Allium odorum	8–16	16, 24, 32	Haberlandt, 1923.
			Messeri, 1930.
			Modilewski, 1930.
A. roseum	8–16	48	Messeri, 1931 (viviparous).
INSECTA			
Solenobia pineti	30–60	120	Seiler, 1923.
S. triquetrella	30–60	120	,, 1923, 1927.
CRUSTACEA			
Trichoniscus provisorius	8–16	24, 336	Vandel, 1927.
Artemia salina	21–42	42, 84, 168	Artom, 1928; Gross, 1935; Barrigozzi, 1935.
Daphnia pulex	4–8	24	Schrader, 1925.

NOTE.— No other animal polyploids are known except in the hermaphrodite *Helix pomatia* ($n = 12$ and 24, *cf.* Wilson, 1928), and in experimental material of *Drosophila* and *Habrobracon*. Polyploidy may be inferred very generally in the Annelida and Mollusca from a comparison of chromosome numbers in different species.

themselves for this : (i) Meiosis is less regular in polyploids than in diploids, for the reason that a too great likeness of the chromosomes is here as serious a cause of irregularity as a too great dissimilarity (*cf.* D., 1928). (ii) All odd-numbered polyploids are numerical hybrids with inherent irregularity of meiosis. Nearly all triploids that have been examined, for example, form restitution nuclei whether they are the product of hybridisation of two species or of self-fertilising a diploid.

High chromosome number (polyploidy) therefore favours the conditions both of diploid parthenogenesis (by suppression of meiosis) and apospory (by irregularity of meiosis). Taking this into consideration, together with the fact that half the species of angiosperms are polyploids, the relative frequency is intelligible.

The reason for the association of a doubling of chromosome number with parthenogenesis in *Chara crinita* (Ernst, 1918), *Artemia salina* (Artom, 1928 *et al.*), *Solenobia triquetrella* (Seiler, 1927), and *Trichoniscus provisorius* (Vandel, 1927) is fairly clear. Polyploidy has prevented fertilisation, and therefore parthenogenesis has been a condition of their survival.

6. REPLACEMENT OF SEXUAL REPRODUCTION

(i) **The Development of Apomixis.** Experiment and comparison show that a failure of meiosis and the inheritance of special genetic properties may in certain instances independently determine the occurrence of apomixis. The problem of apomixis is the problem of the part these two properties play in its origin.

In the first place, it is clear that special genetic properties are necessary, permitting the development of unfertilised eggs, pseudogamously or otherwise, for parthenogenesis, and permitting the development of adventitious embryos in the absence of successful sexual embryos, for apospory and nucellar embryony. These properties are probably characteristic of certain genera such as *Hieracium* and *Aphis*. Secondly, it is necessary that unreduced eggs should be produced, for diploid parthenogenesis, and unsuccessfully reduced eggs, for apospory. It is this second property that needs special consideration.

For this purpose it is important to distinguish between the structural conditions of the breakdown of meiosis and a genetic property more clearly than has been done in the past.

The discussion of the causation of apomixis brings into the foreground the shortcomings of the notions of hybridity current amongst geneticists. There is no use in considering the effect of hybridity alone on sexual fertility because there is a group of conditions having the same effect on fertility as the structural hybridity characteristic of interspecific crosses, and this group of conditions must be considered with it. They are the numerical hybridity of triploids and pentaploids and the autopolyploidy which leads to multivalent formation and consequent irregular segregation even in tetraploids. The type of organisms in which the relationships of the chromosomes determine infertility, *i.e.*, hybrids and autopolyploids, must always be considered as a group in contrast with those in which irregular or suppressed meiosis is physiologically and not mechanically controlled.

In a hybrid their dissimilarities prevent the chromosomes coming together at pachytene, although we are not usually able to show this cytologically except by inference from the later behaviour. Now this property depends entirely on the relationship of the two chromosomes. But precisely the same result can be obtained, not through dissimilarity, but through the hereditary properties of the individual. In *Rhabditis* pairing of chromosomes fails, although they are known to be identical, and this has been shown to be genotypically controlled, *i.e.*, a property of the germ-plasm as a whole, and not of the relationships of its parts.

Viewed in this light, all apomixis can be seen to be primarily conditioned by the hereditary properties of the individual which makes development possible without fertilisation. Further, it can be seen that the development of this hereditary property is in many cases independent of any structurally controlled failure of meiosis. This is most obvious : (i) in organisms that are cyclically or facultatively parthenogenetic (Aphides, Rotifera, Cladocera, *Apotettix* and *Paratettix*, *Allium*, etc.) ; (ii) in others in which parthenogenesis is a means of sex determination (Hymenoptera, Rotifera) ; (iii) in many pseudogamous plants with good pollen (*Zephyranthes*, *Potentilla*) ; and (iv) in animals and plants in which the pairing of the chromosomes is perfect (*Rhabditis*, *Solenobia triquetrella*), especially those which are facultatively sexual and true breeding (*Elatostema sessile*, *Thalictrum purpurascens*, *Allium odorum*)

Thus, in all these examples, parthenogenesis is compatible with complete homozygosity.

There is a second class in which the decisive influence of the mechanical chromosome relationships can be seen: in those aposporous plants which are facultatively sexual but not true breeding. Apospory, in itself, is no more obscure than such teratological observations as those of Stow (1930), who induced the formation of a structure in the pollen grain of *Hyacinthus* resembling the female gametophyte. That such abnormal structures should be able to function is evidently conditioned by sexual failure. Lidforss's experiments show that the lack of viability of the sexual embryo stimulates apospory. In such forms apospory must be conditioned by partial sexual sterility which itself is sufficiently accounted for by either hybridity or autopolyploidy or cross-fertilisation.

It is probably of no account whether or not sexual failure is determined by interspecific hybridity. Irregularity of meiosis is as characteristic of non-hybrid tetraploids as of hybrid diploids, and tetraploidy of this kind is probably a determining factor in some species, *e.g.*, *Nigritella nigra*, which exists in a diploid sexual form ($n = 19$) and a tetraploid apomictic form, with nucellar embryony (Afzelius, 1928; Chiarugi, 1929). In triploids the conclusion to be drawn is unequivocal: the last sexual act in their history was an act of hybridisation, not necessarily between different species but between gametes with different chromosome numbers. If, in these, hybridisation determined the institution of apomixis, in many others where clear-cut evidence is lacking this kind of origin may be readily presumed from the irregularity of their meiosis.

In many plants with obligatory parthenogenesis we see that the institution of parthenogenesis was likewise determined by a mechanical defect in the chromosome organisation. Most of them are evidently triploids or autopolyploids of a kind that could not reproduce satisfactorily by way of meiosis. But since they began to be parthenogenetic they have evidently changed. They have acquired a genotypic property of regularly suppressing meiosis. This is shown in a variety of ways, as follows :—

First, we sometimes find that side by side with the regular *ameiosis*, as we may describe it, of the functional cells, there is still the original condition of pairing in the non-functional cells. Or again if this pairing is suppressed it is found to be equally suppressed when the chromosome number is doubled and every chromosome has an identical mate. Moreover, meiosis in the female cells is liable to much greater irregularity than in those of any hybrid, for occasionally, as in *Artemisia*, typical autopolyploid hybrid behaviour is replaced by the highly specialised ameiosis.

Secondly, in species of *Eu-hieracium* (Rosenberg, 1917, 1927) it has been possible to trace stages in the development of the suppression. Applying to these the results of our earlier analysis of aberrations of meiosis (Ch. X) it is possible to define the conditions of the development.

In the first stage of "degeneration of the reduction division," *i.e.*, abortion of the first meiotic division, the *Boreale* type, we see the ordinary symptoms of hybridity : pairing at pachytene but not at metaphase ; the failure of metaphase pairing is due to the same cause in these hybrids as in the fragments of non-hybrid species, an insufficient length of chromosome paired at pachytene to lead to the formation of a chiasma. But, in the last stage, *Pseudoillyricum*, with suppression even of pachytene pairing and the substitution of mitosis for meiosis, we have a phenomenon that is not characteristic of hybrids, as such. Clearly a primary condition of meiosis is wanting, and on the precocity theory this might happen in two ways : either (i) the division of the chromosomes is relatively advanced, or (ii) the beginning of prophase is relatively retarded, so that at leptotene the chromosomes are already doubled and therefore have no attraction for one another. That both these aberrations are indeed conditions of various types of abnormality found in meiosis is shown by the time discrepancies observed in several apomictic and sexual genera. The irregular nuclei are either advanced or retarded (Table 62).

Semi-mitotic meiosis occurs in *Hieracium* at an earlier stage in the development of the flower than the normal meiosis of sexual species (Rosenberg, 1927). This general timing abnormality is associated with a special one, characteristic of semi-mitotic meiosis in parthenogenetic plants: the prophase begins after a prolonged resting-stage and the chromosomes are contracted very little more than in mitosis (*e.g.*, in *Archieracium, Eupatorium, Antennaria,* Bergman, 1934, *cf.* Gustafsson, 1935; *Leontodon hispidus*, abnormal plant, Bergman, 1935).

We see then, on cytological grounds, that the suppression of meiosis in these plants is conditioned in its simplest form by the mechanical relationships of the chromosomes, in its more advanced forms, on the other hand, by a genotypical property. This property acts by removing the characteristic precocity in the prophase and with it the essential characteristic of meiosis, the pairing of chromosomes. It may have more or less of the adventitious characteristics of meiosis, which are secondary to this pairing and do not hinder the regularity of the substituted mitosis.

Thus three conditions appear to determine diploid parthenogenesis in its most advanced form in the *Eu-hieracia* and elsewhere: (i) A genetic property, particularly common in some genera, conditioning parthenogenetic development of the embryo-sac; (ii) hybridity or autopolyploidy conditioning the original formation of unreduced embryo-sacs, as found in the *Pilosella* section; and (iii) a genetic property acquired later by adaptation, perfecting the mechanism by which suppression takes place and probably consisting in a derangement of the ordinary time relationships of meiosis; the first stage of this is found in *Boreale*, last stage in *Pseudoillyricum*.

(ii) **Adaptation and the Mechanism of Variation with Parthenogenesis.** A difficulty arises when we come to consider how this third condition can be attained by adaptation in a parthenogenetic organism. This difficulty has been felt in another connection. Genera of plants in which apomixis is general, commonly show a remarkable polymorphism. This is particularly true of *Rubus* and *Eu-hieracium*. In *Eu-hieracium* Ostenfeld found 60 parthenogenetic species and only three sexual ones (*cf.* Rosenberg, 1917). How has such a remarkable polymorphism arisen in the absence of the two recognised means for the distribution and recombination of variations, viz., meiosis and fertilisation?

Ostenfeld and many others have sought to explain the phenomenon as the result of mutations (1921). As a rule the progeny of a parthenogenetic plant or animal are as like their parent as conditions permit ; in plants they constitute a " clone " of the same uniformity as one reproduced by cuttings or bulbs. Ostenfeld, however, showed that two " mutations " occurred in 1260 parthenogenetic seedlings of *H. tridentatum*.

Similarly, male plants appear from parthenogenetic *Antennaria*. The origin of these mutations becomes clear in the light of the present interpretation of chromosome pairing in terms of crossing-over. Both the general theory and the particular experiments show that where either the first or the second meiotic division is suppressed after certain chromosomes have paired and formed chiasmata there will be a segregation of any differences between these chromosomes (Ch. VII). Since, in all but the most extreme and stable forms of parthenogenesis, mother-cells occur with pairing of chromosomes, this pairing will account for the occasional " mutations " observed in parthenogenetic progenies of hybrid plants and animals (*cf.* Gustafsson, 1934 *a* ; Bergman, 1934).

In short, diploid parthenogenesis appears to provide a mechanism whereby an organism in which normal meiosis would give sterility through excessive segregation, may show limited segregation and be fertile. The mechanism is analogous in its effect to that whereby segregation is restricted in the tetraploid derivatives of diploid hybrids. Most segregates being more homozygous should be less asexually fertile than their parent, as these were. They will provide the basis of polymorphism. The mutations inferred in the relatively homozygous *Cladocera* (Banta and Wood, 1928) cannot, of course, be due to segregation of any kind. Nor is the segregation found in parthenogenetic *Apotettix* (*cf.* Robertson, 1931 *a*) of this kind, but merely first division mendelian segregation in an organism with perfect chromosome pairing.

A second ground for expecting mutations in the progenies of parthenogenetic plants is found in the imperfection of the failure of meiosis when sexual reproduction is first replaced. If, occasionally, univalents have divided before a restitution nucleus is formed in the first division, progeny will be produced with varying and unbalanced chromosome numbers. Such plants are produced and survive in Müntzing's *Poa* species which on this account must be regarded as representing the earliest and most unstable beginnings of parthenogenesis yet found in plants.

These mutations will also provide the basis of selection for adaptation of the meiotic mechanism to more perfect conditions for parthenogenesis, *i.e.*, for developing the genetic character which is the second step in the evolution of fertile parthenogenesis. Nor should this be difficult. The appearance of genotypes for the suppression or partial suppression of meiosis in mutants and segregates of sexual species and hybrids is common enough (Ch. X). And while in these any such mutation is a disadvantage, in incipiently parthenogenetic organisms, it will have the highest positive selection value.

Conclusion. The following are the chief independent conditions determining habitual apomixis :—

(A) A genetic property of allowing the development of the egg-cell without fertilisation and with or without other stimulus, such as false fertilisation (Lidforss, 1914) or degeneration of neighbouring tissues (Haberlandt, 1921, *a* and *b*). This is not inherent in all living organisms, but is widespread in certain groups of plants and animals, and is probably inherent in all those with normally zygotic meiosis.

(B) A genetic property determining the development of other cells than the spore in the embryo-sac or embryo (*cf.* Andersson, 1931) [condition of apospory and nucellar embryony] or of other cells than the egg-cell into the embryo [condition of apogamy]. This effect is usually conditional on failure of the facultatively sexual embryo.

(C) Functional hybridity, numerically or structurally incompatible with regular meiosis, leading to either (i) failure of numerical reduction and partial suppression of segregation (*cf.* D., 1930 *a*) which with (A) gives diploid parthenogenesis (*cf.* Harrison, 1920), or (ii) sterility through segregation, whether in a mendelian sense or as regards the distribution of whole chromosomes, which with (B) gives apospory (*cf.* Chiarugi and Francini, 1930).

(D) A genetic property determining : (i) fusion of a polar body with the egg nucleus, or of two segmentation nuclei ; or (ii) failure of numerical reduction (*cf.* P. Hertwig, 1920) owing to alteration of the time relationship of meiosis. With (A) this property gives diploid parthenogenesis in many plants and animals. It may be acquired as an adaptation by selection of segregates secondary to (C), which alone is an imperfect basis for diploid parthenogenesis.

(iii) **Subsexual Reproduction.** Apomixis has in the past been investigated, described and classified in terms of morphology and not in terms of genetics. This was at first inevitable since genetics is founded on the study of sexually reproducing organisms, and the rules it has established can be applied directly to them alone. Even for them we need the help of chromosome studies when we are dealing with hybrids of a complex character. These studies can be used as we have seen by extrapolating from the relationships between chromosome behaviour and breeding behaviour made out in specially simplified experiments. Such studies we now see can also be applied to the elucidation of apomixis from the genetical point of view.

The most obvious conclusions have already been drawn. They show the part taken by mechanical relationships of the chromosomes and by the adaptation of the genotype in the development of apomixis. But there is a further step to be taken. Sexual reproduction has always, since Weismann, been seen to consist in the two essential processes of meiosis and fertilisation. These processes are the means of recombining the parts of chromosomes. Such a recombination is possible, however, without either meiosis or fertilisation. The process of segregation which is responsible for the variation and development of parthenogenetic plants requires merely crossing-over. And crossing-over is as we have seen a universal concomitant of meiosis, an essential part of sexual reproduction, without which, in fact, sexual reproduction would become effectively clonal for each chromosome. The retention of crossing-over, even in the attenuated degree to which it is found in stable apomictics, is therefore genetically, although not morphologically a property of sexual reproduction. The parthenogenetic organism with crossing-over has thus, as we may say, a *subsexual* reproduction, the sexual character of which is morphologically concealed, but genetically effective, more effective indeed than in a regularly self-fertilised sexual diploid.

When we apply the genetical criterion of recombination to forms with sexual reproduction we see that many, if not all, have clonal chromosomes or parts of chromosomes. Thus all the unpaired chromosomes of *Rosa canina* are clonal, or if they occasionally pair they are subsexual. Similarly, the differential segments of *Œnothera* and of the *Y* chromosomes of higher organisms are clonal.

Such clonal parts of the hereditament in sexually reproducing organisms have always the capacity of returning to sexual life as a result of structural change. But if they have become inert before they recover in this way, their change is presumably irreversible. In obligatory apomictics the whole hereditament has made an irreversible step towards asexuality.

Another special modification of sexual reproduction results from all structural changes like short inversions which inhibit crossing-over in the hybrid, for such changes establish a condition of endogamy in the segment affected. All the genes within it are prevented from crossing-over effectively with the genes in the unchanged segment. It is therefore separated in evolution. Its descendants become an endogamous community corresponding genetically to an endogamous community of zygotes (D., 1936 *a* and *d*).

By pursuing investigation along these lines with the help of the new genetical interpretation of chromosome behaviour, it will be possible to determine how the genetic system works and changes over a wider range of organisms and a greater space of time than is accessible to experimental enquiry.

CHAPTER XII

CELL MECHANICS

Postulates—Unity of Mitosis and Meiosis—Internal Mechanics—The Spiralisation Cycle—The Molecular Spiral—External Mechanics—Specific Attractions—Gene Reproduction—Repulsions—Terminalisation—Centromere and Centrosomes—The Spindle—Congression and Orientation—Cell-Wall Formation—The Balance Theory of Mitosis—Ultra-Mechanics—Crossing-Over and Structural Change—Molecular Foundations.

Utinam cætera naturæ phenomena ex principiis mechanicis eodem argumentandi genere derivare liceret. Nam multa me movent, ut nonnihil suspicer ea omnia ex viribus quibusdam pendere posse, quibus corporum particulæ per causas nondum cognitas vel in se mutuo impelluntur et secundum figuras regulares cohærent, vel ab invicem fugantur et recedunt.*

Newton, *Principia Mathematica*, Pref. 1st Ed., 1686.

1. INTRODUCING AN AXIOMATIC

In studying movements in the cell, our object must be to find out how far they are consistent and hence to infer the principles governing their occurrence. Such mechanical principles must be related if possible to those inferred in other systems, especially non-living systems both of a molecular and of a macroscopic order of size and integregation. The most promising method at first sight would seem to be the method of externally controlled experiment which has proved most useful in the analysis of other physiological processes. This method has in fact yielded important, if somewhat isolated, results. But they have been limited in scope by difficulties of a similar kind to those that have limited the studies of electrons. It is difficult to observe the living chromosomes without altering them. It is usually impossible to alter them in one respect without altering them in several others and by means which cannot be exactly defined.

* I wish we might derive the rest of the phenomena of nature by the same kind of reasoning from mechanical principles, for I am led by many reasons to suspect that they may all depend upon certain forces by which in an unknown way the particles of bodies are either mutually attracted towards one another and cohere in regular figures, or are repelled and recede from one another.

In this case there is a way out : we can call in the aid of a different scientific method, hitherto successfully applied in geology and astronomy, to help out controlled experiment with natural experiment. In this method we compare conditions in nature which differ, so far as we can judge, in respect of the fewest variable factors. For such a purpose the study of the chromosomes is very well suited. When we take the whole of living organisms together, the chromosomes show a more uniform behaviour both mechanically and physiologically than any other structures. This is not surprising, since they lie at the root of all living processes, and on their accurate mechanical and physiological co-ordination the success of these processes depends. And, as we have seen in regard to the alternation of generations, co-ordination in function enforces stability in evolution. Moreover, the technique of genetics puts ready to our hands the means of analysing mutations which act physiologically, for the most part, in breaking down the genetic system. These mutations are not directly controlled by the observer, but are related in cause and effect to a large body of similar observations, similarly analysable.

In attempting to study cell mechanics we must take the further elementary (Baconian) precaution of beginning our induction from the most certain and most exact evidence and proceeding to the less certain and less exact. We must therefore begin with the chromosomes, the cell-components whose behaviour is most constant and most readily verifiable, and then on this foundation (and we are as yet little above the foundations) attempt to construct a general theory of cell mechanics.

The data for our study consist in observations of mitosis and meiosis in all living organisms. Our first task is to arrange the data in a suitable form for comparison, and to do this we must find out to what extent there exists a uniformity in each of the two important types of nuclear division, mitosis and meiosis. A uniformity in the general character of mitosis has long been understood. Its demonstration was completed with the masterly review of nuclear behaviour in the Protista by Belar (1926). The uniformity of meiosis has been more slowly realised. Its demonstration was completed by the evidence leading to the chiasma theory of meta-

ANALYSIS OF FORCES

phase pairing (D., 1929 b). We can now assume that with the same general character in all organisms each type of division works in the same way, on the same mechanical principles, in all organisms.

Our next task is to relate the two types of nuclear division. Classical cytology was unable to deal with this problem. It was baulked by a conclusion that seemed so obvious that it was never questioned: the homologous chromosomes which did not attract one another at mitosis, came together at meiosis by an attraction which consummated the hitherto incomplete sexual process. An insufficient analysis led to a teleological conclusion and the issue was closed.

The method that has been followed here is to assume every possible likeness between mitosis and meiosis and then by a process of exclusion to arrive at the essential difference. We have seen that in the attractions and repulsions, in the external and internal coiling relationships of the chromosomes, in the individuality and reproduction of the chromomeres and centromeres, and in the cyclical changes of the spindle and of the nucleus, there is indeed no essential difference between the two types of division. We have seen that on the other hand there is a difference in the timing relationship of the cyclical changes in the nucleus, spindle, centromeres and chromosomes. This difference can be expressed by saying that in meiosis the changes outside the chromosomes are advanced in relation to the changes inside the chromosomes. In meiosis, as compared with mitosis, the external changes are precocious. On this basis we can assume that the forces acting in the two types of division are the same, and the whole array of observations of mitosis and meiosis becomes available for the comparative analysis of chromosome movement.

Our third task is to separate this great mass of data into classes which are convenient for handling. Of course, every movement of a chromosome is related in some way with every other, but we have to separate them for individual consideration so far as possible, just as we have to separate the mechanical and physiological activities of the chromosome although these activities continually affect one another. The most important cleavage comes between the external and internal changes in the position of chromosomes.

The first consist in movements of individual chromosomes in relation to each other and to bodies outside them. These movements are the data of *external mechanics*. The second consist merely of changes in shape of the chromosomes. They are the data of *internal mechanics*.

A third class of data arises from the division into external and internal mechanics. This is a group of movements which depend equally on both and are necessarily more complex than either. Paired chromosomes during the pachytene stage express the laws of both internal and external mechanics, which are combined in the attainment of the special result of their relationship at this stage, crossing-over between them. This, and the mechanics of structural change of which it is a special form, may be dealt with in a special class of more speculative considerations as *ultra-mechanics*.

In order to analyse the movements assigned to these categories it is necessary to describe them according to a recognised canon of mechanical nomenclature. For the present elementary treatment it is easiest to follow the conventions of classical mechanics and attribute movements to *forces*, whether of attraction, repulsion or torsion. The student who finds this treatment objectionable is free to translate these shorthand expressions into terms of energy levels, but such a description is roundabout for our present purposes, except in regard to the especially complex spindle system. We have in the cell the same grounds for inferring forces that Newton had in the universe: movements and accelerations of bodies regularly occurring and accurately measurable. We use the same means of inference: comparison of the movements and equilibrium positions of bodies of different sizes and at different distances apart.

In attempting to establish a new and closed system of mechanics it is of the highest importance that we should begin our induction from the right group of data, taking by analogy the right assumptions. This is a matter of trial and error, and many false starts have been made in consequence. The sound basis of external mechanics lay in Lillie's observations in 1905, but this could not be used until meiosis and mitosis were correlated. The sound basis of internal mechanics was not revealed until Kuwada and Nakamura's experi-

PLATE XIII

FIGS. 1 AND 2.—Metaphase and telophase of mitosis of the generative nucleus in the pollen tube (*Tulipa*, $n = 12$), showing internal and relic spirals. × 2000. (Upcott, 1936, aceto-carmine.)

FIGS. 3, 4 AND 5.—Successive stages in prophase uncoiling from Plate II.

FIG. 6.—Major spirals of *Tradescantia virginiana* partly uncoiled by fixing in hot aceto-carmine. (Richardson, unpublished.) × *ca.* 1500.

FIGS. 7 AND 8.—*T. virginiana* fixed in aceto-carmine.
Fig. 7. Artificial resting nucleus produced by ammonia pretreatment of second metaphase chromosomes. Fig. 8. Natural resting nucleus in the pollen grain. (Kuwada and Nakamura, 1934 *a*.)

PLATE XIII

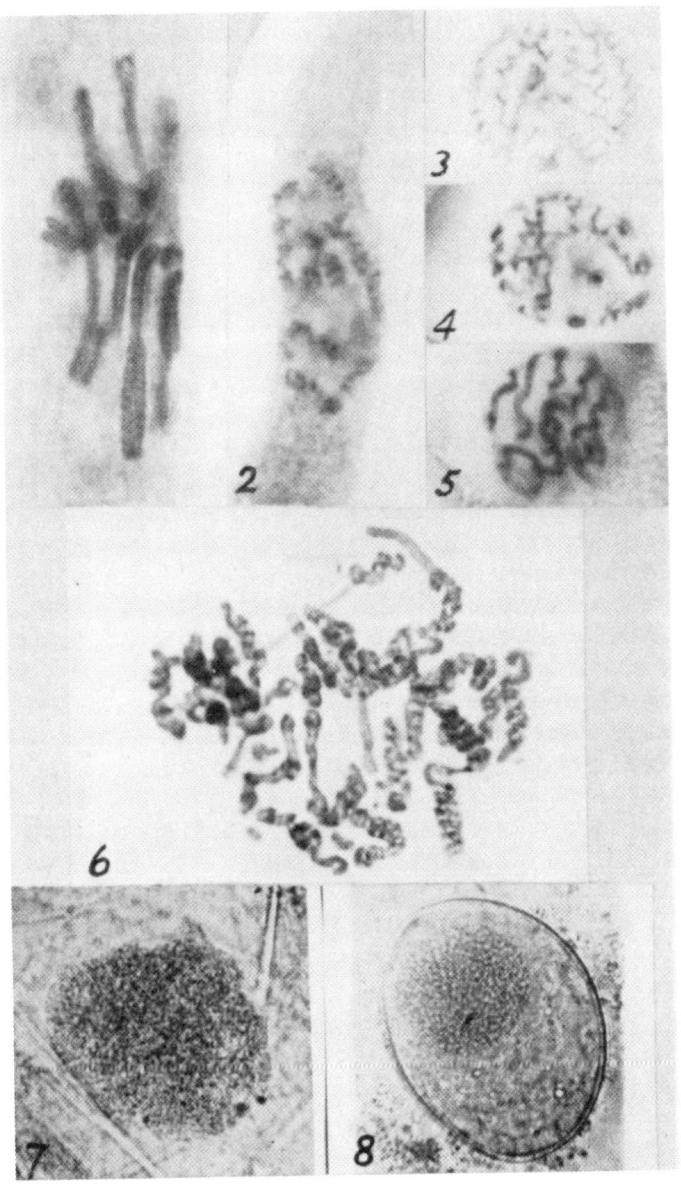

To face p. 482.

ments showed the way. The soundness of these beginnings has been shown only by their comparison with large bodies of related observations. How this has been done we shall now try to see.

2. INTERNAL MECHANICS OF THE CHROMOSOMES

(i) **The Problem.** The series of changes undergone by the chromosomes in the course of mitosis consist of a successive coiling and uncoiling. From the point of view of their internal mechanics, mitosis is therefore a spiralisation cycle. In the prophase each chromatid develops an *internal spiral* as it shortens and thickens. This is to be inferred from the structure observed at metaphase, for within the prophase nucleus it cannot yet be differentiated. At telophase in living cells the successive spirals can be seen to separate and in the following prophase their *relic spirals* gradually uncoil while the new internal spirals are being developed within each of the chromatids into which the chromosome has divided. These two chromatids themselves show a *relational spiral* which is, however, facultative, for the chromatids of ring chromosomes are usually able to separate and must therefore have been lying parallel.

From this series of events four mechanical principles may be inferred (D., 1935 *a*) :

First, the coiling and uncoiling of the internal spirals must be immediately determined by internal changes, for if they were externally determined, differences would always be found between chromosomes of different lengths (Kuwada, 1935). The agent must be torsion within the chromosome thread in the opposite direction to the internal spiral. This may be described most simply as due to the formation of a *molecular spiral*, a spiral torsion within the thread.

Secondly, the delay in uncoiling which leads to one cycle overlapping the next must be due to a failure of the external form of the chromosomes to respond immediately to the internal stresses due to changes in their molecular spiral. Such a *hysteresis* is clearly due to the chromosomes lying in a somewhat rigid medium and within a limited space, the resting nucleus.

Thirdly, the relational coiling of chromatids must depend on the *specific attractions* of their parts, for otherwise the chromatids would slip round one another freely as they shortened and straightened so as to lie parallel at metaphase. These attractions will be dealt with later.

Fourthly, the separation of daughter ring-chromosomes shows that a *cleavage surface* is predetermined within the coiled chromosome in the resting stage along a spiral compensating for the relic spiral.

Relationship	Structure	Type in order of increasing diameter	Duration					
			Mitosis			Meiosis		
			P	M-A	T	P	M-A	T
I Internal compensating spiral (1)	Single chromatid (1-4)	A—*Original spirals* (in metaphase of the mitosis of origin)—						
		1 Molecular spiral (R)	———————→			———————→		
		2 Minor spiral (L)	——————→			× ————→		
		3 Major spiral (L)	× × × × ×			× ————→		
II Individual spirals (2-4A) in opposite direction to (1)	Paired chromatids (3A-5)	B—*Derived or transferred spirals*—						
		4 Super-spiral (by transfer from 2 and 3 (L))	× × × × →			× × × × →		
		5 Relic spiral from 2 or 3 (L)	—→ × × ×			—→ × × ×		
III Relational spirals (5, 6)	Paired chromosome (6, 6A)	6 Inter-chromatid spiral (R) (opposite to II)	——→ × ×			× → × × ×		
		7 Inter-chromosome spiral (L)	× × × × ×			× → × × ×		

FIG. 135.—Relationships of different kinds of spiral structure, in space, time and (provisionally) in direction. (D., 1935 a.)

These conclusions, in answering some of the questions of internal mechanics, raise several new questions :—

(i) The internal spirals are evidently consistent for considerable parts of chromosomes. Are they consistent for whole chromosomes or for whole arms ?

(ii) If they are, and a molecular spiral determines coiling, what is it that co-ordinates the molecular spiral ?

(iii) Further, a molecular spiral must be supposed also to determine the relational coiling of chromosomes at pachytene by the homologues coiling in the same direction. Is this the same spiral as that determining the internal spiral ?

(iv) If so, why does it determine a relational spiral at one time and an internal spiral at another ?

(v) Finally, why do the chromatids not coil jointly instead of separately during the prophase, making use of the relic spirals from the preceding division instead of undoing them ?

New work on the coiling of the chromosomes at meiosis now enables us to answer some of these questions.

(ii) **The Meiotic Cycle.** An unpaired chromosome at the first metaphase of meiosis consists, like a mitotic chromosome, of two chromatids lying side by side with an undivided centromere. Two paired chromosomes differ merely in having an exchange of partner amongst their chromatids. Meiotic chromosomes, however, are distinguished from mitotic ones slightly but definitely in their superficial properties : on the one hand they are shorter and broader and on the other they are more regularly and closely associated. These differences depend on their internal structure.

The large meiotic chromosomes that have been accurately examined after appropriate treatment have been shown to consist of a minor spiral, coiled into a major spiral. The minor spiral corresponding in diameter to the mitotic internal spiral, was first suggested by Fujii and demonstrated experimentally by Kuwada and Nakamura. The secondary coiling accounts for the greater reduction in length of meiotic chromosomes. Moreover, the nucleolar constrictions are apparently constrictions of the minor spiral since they are invisible in a thread coiled into a major spiral as well. Kuwada (1935) supposes that this develops after the primary minor spiral. They may alternatively be supposed to develop side by side, since in this way variations in the degree of linear contraction would require the minimum of variation in the angle of the spiral.

This difference in spiralisation between meiosis and mitosis is generally correlated with precocity of the prophase. It is accompanied by a second difference. The two chromatids are sometimes jointly coiled both as to their major and perhaps their minor spirals. They then come apart at late metaphase or anaphase (Kuwada and Nakamura, 1933 ; Sax and Humphrey, 1934). Such seems to be

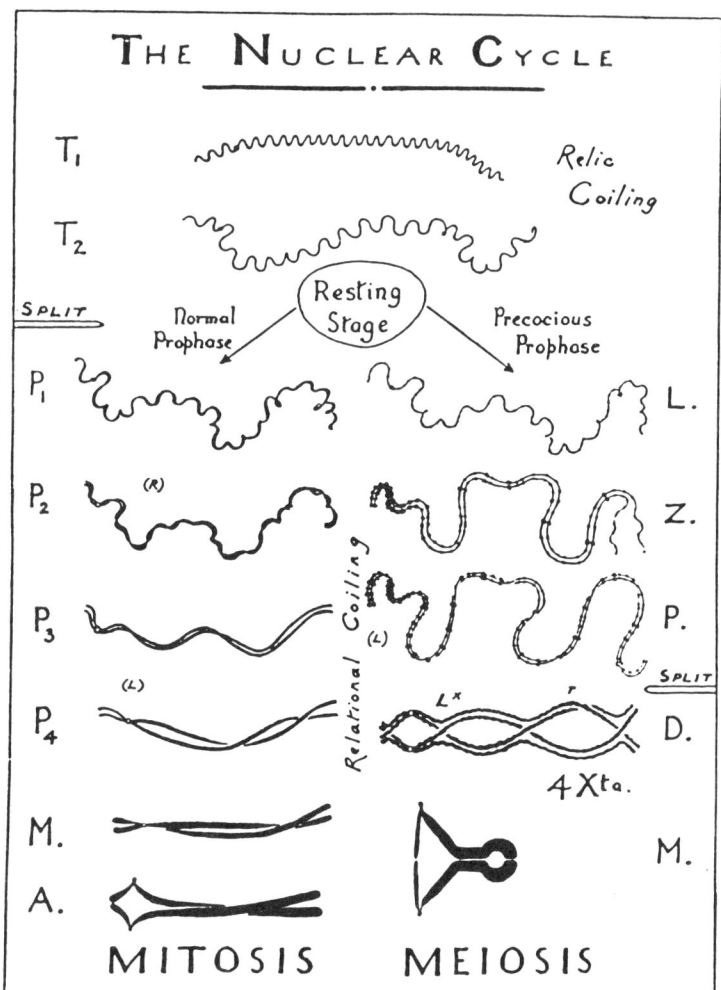

FIG. 136.—Diagram to show the history of spiral forms in the nuclear cycle, mitosis and meiosis, in relation to the division, represented by double lines in the side columns, and crossing-over of chromosomes, represented by chiasmata, Xta. With normal prophase the chromosomes divide during the resting stage, with the precocious prophase, at the end of pachytene. T, telophase; P_1 to P_4, prophase; M, metaphase; A, anaphase; L, leptotene; Z, zygotene; P, pachytene; D, diakinesis. (R) and (L), directions of individual and relational coiling of chromatids at mitosis and chromosomes at meiosis, numbered according to the seven categories of Fig. 135. A circle represents the centromere, which divides at the end of metaphase in mitosis and at the second division in meiosis.

the case in *Tradescantia*. In *Trillium* and *Fritillaria*, on the other hand, the coiling of the chromatids seems to be regularly separate after fixation (Huskins and Smith, 1935, D. 1935).

During the interphase between the two divisions we find relic spirals gradually uncoiling as at mitosis. Further, since the sister chromatids of each chromosome have separate coils we might suppose that these coils were embodied in the internal coils of the second metaphase. But this seems unlikely, for we find that owing to the shortness of the interval the relic coils may survive until metaphase, superimposed on the internal coils as they are in the prophase of mitosis (D. 1936 b).

The coiling system at second metaphase usually consists of a single internal spiral as at mitosis. The degree of contraction at this stage is, however, variable in many organisms (*e.g.*, *Gasteria*, *Podophyllum*, *Fritillaria*) and it is not surprising to find that major as well as minor spirals are found regularly in *Sagittaria* and occasionally in *Lilium longiflorum*.

A fixation that contracts the staining thread, obscuring the minor spiral, is especially favourable for showing the direction of the major spiral at first metaphase and anaphase. In bivalents of *Tradescantia* and *Rhœo* with only terminal chiasmata the simplest rules are found. The direction is constant for all chromosome-arms, and since the arms are jointly coiled the paired chromatids necessarily have the same direction of coiling. Pairing arms often have opposite directions of coiling however (Fig. 137). Where, therefore, interstitial chiasmata are found at metaphase in *Tradescantia* and *Trillium*, changes of direction occur not only at the centromere but in chromosome arms. These changes perhaps coincide with the interstitial chiasmata (Sax and Humphrey, 1934) and result from crossing-over—presumably between chromosomes with opposite directions of coiling (Huskins and Smith, 1935 ; Matsuura, 1935). Changes of direction are probably to be inferred from the observation of constrictions in two of the four chromatids at anaphase in *Stenobothrus* (Fig. 36A ; *cf*. Belar, 1928). They are not, however, inherent in crossing-over, for acentric chromatids, which must always result from crossing-over, have been found with no changes of direction (S. G. Smith, 1935 ; Upcott, unpub.).

Change of direction might be expected on the other hand following crossing-over between chromosomes with opposite coiling directions if the direction were a specific property of the arms of the parent chromosomes and were unaffected by crossing-over. Interpretations of three homologous chromatids having the same direction, while the fourth at a corresponding point has the opposite direction, contradict this conclusion but are perhaps mistaken. The constant direction in arms of *Rhœo* and *Tradescantia* with terminal chiasmata would require that chiasmata should have been formed originally

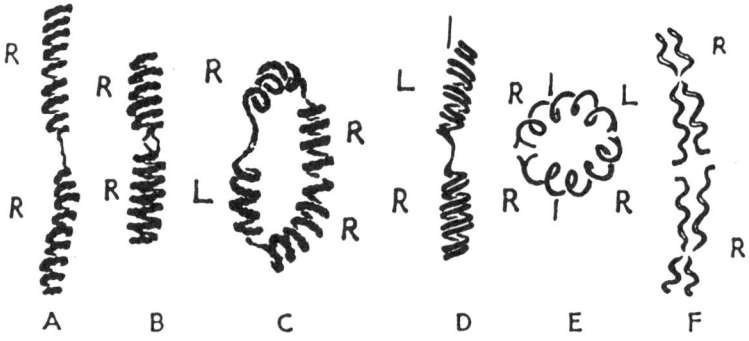

FIG. 137.—Major spirals at first metaphase in *Tradescantia* revealed by pre-treatment with hot water (*cf.* La Cour, 1935). A and B, normal and mutant type of spiralisation in diploid species; C, D, E, F, tetraploid species showing different degrees of chromatid association and contraction; C, ring of four; E, change of direction at the centromere. × 2,000.

very close to the end and that the amount of movement should not have been proportionate to that found in the terminalisation of chiasmata in small chromosomes. Such an abnormality of behaviour, a terminal localisation, has already been suggested on other grounds (Ch. IV).

In an organism with a considerable movement of chiasmata, a change of direction at the point of crossing-over will mean that chromatids with opposite directions of coiling will come to be paired. Such an organism cannot therefore be supposed to have a joint coiling of chromatids as in *Tradescantia*. It must have separate coiling as in *Fritillaria*.

The change of direction which results from crossing-over between two chromatids with an opposite direction of internal coiling is evidently lost at the next mitosis, since whole arms seem to give consistent results at each meiosis apart from crossing-over. Thus, although the internal spiral must be determined directly by the uncoiling of the molecular spiral, its direction must be facultative and determined by an action of, or at, the centromere in each division.

The consistent direction of molecular coiling of each arm is perhaps imposed by the centromere, through the coiling beginning near the centromere and being transmitted consistently from particle to particle along the chromosome. This view agrees with the observed priority of proximal parts of chromosomes at zygotene and diplotene in many plants and animals.

The conditions determining the direction of relational coiling of chromosomes in pachytene seem to be more complex than those determining internal coiling. Relational coiling seems to be universal both from direct evidence and from the general occurrence of non-homologous torsion pairing. To develop relational coiling the two chromosomes must suffer an internal torsion in the same direction as one another. We know in *Tradescantia* that their internal coiling may be in opposite directions. The direction of the spiral responsible for the two systems of coiling must therefore be separately controlled. Now the relational coiling, like the internal coiling, seems to be consistent throughout chromosome arms in *Fritillaria* and *Chorthippus* apart from any interruptions due to structural hybridity. But where such interruptions occur different chromosomes with opposite coiling properties must sometimes be brought together. These chromosomes must nevertheless be supposed to coil relationally, since crossing-over takes place between relatively translocated segments. Thus the torsion responsible for relational coiling must itself be not individually but relationally determined. We are thus brought face to face with a contradiction between the principles of internal and relational coiling and with a seemingly inexplicable property of the partner chromosomes influencing one another's coiling direction in relational coiling.

Relational coiling of chromatids must always be present in meiosis, since interlocking of chromatids may always occur at first anaphase where more than one chiasma is formed in a given pair of arms. It is also determinate in its direction (D. 1936 d). At mitosis, however, as we saw, relational coiling is facultative. It is also, so far as our present observations go, indeterminate in many chromosomes although determinate in some (D. 1936 c). We cannot yet say what the immediate cause of the mitotic coiling, still less of its variation, may be. Like the chromosome coiling at meiosis, it must depend ultimately on the specific attractions of the genes and on the molecular coiling of the threads. If the effective agent of association is the attraction between the particular pairs of genes nearest one

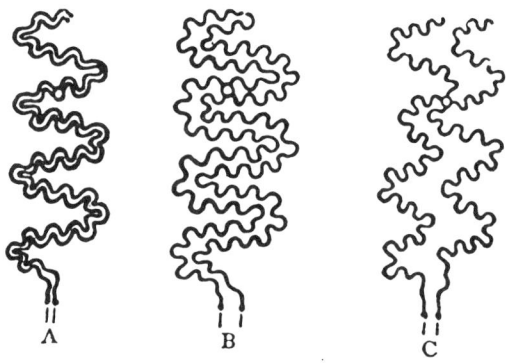

FIG. 138.—Three possible interpretations of a first metaphase chromosome in relation to its major and minor spirals. (D., 1935 a.)

another in the two internal spirals, then an inequality in the amplitude of the two spirals would give rise to a coiling of the chromatids —but consistently in the opposite direction to the internal coiling.

The reason why the relic spiral cannot be used in forming the new internal spiral is, on the other hand, fairly clear: it has already expanded to an amplitude far greater than an internal spiral ever assumes, and an internal spiral always develops by increasing its amplitude, not diminishing it.

There is no direct evidence to show why the chromosomes coil relationally at one stage and internally at others. The most obvious possibilities are that the difference depends either on a difference in the degree of attraction at work at different stages, or on the coiling

TYPES OF COILING

change being in opposite directions or in different degrees in the two cases (D. 1935 c).

Answering these questions of causation must needs wait upon a related study of all the spiral formations, internal and relational, in the same pair of chromosomes at different stages of the nuclear cycle. The direction of relational coiling of chromatids at meiosis may be recognised with difficulty in Orthoptera (D., 1936 b) or in anomalous nuclei in plants (D., 1936 e). The direction of the minor spiral has been said to be opposite to that of the major spiral (Iwata, 1935). These structures, however, are on the limit of visibility, hence it is necessary to test their observation by the most rigorous inferences at every step. We must not suppose, as some authors do, that the present apparent indeterminacy is final and irremediable. Meanwhile, it is important to recognise the contradictions between our present observations of relational and internal coiling in order to find out their cause. These contradictions do not affect our consideration of external mechanics.

TABLE 71

Observations of Chromosome Coiling

1. MITOSIS.
   (a) *Metaphase, Internal Spirals.*
   Allium, Gasteria. Geitler, 1935 b, c.
   Tulipa. Upcott, 1935.

   (b) *Telophase and Prophase, Relic Spirals.*
   Stenobothrus. Janssens, 1924.
   Tettigonia. de Winiwarter, 1931.
   Tradescantia. Kaufmann, 1926; Kuwada and Nakamura, 1934 b, Plate XIII.
   Tulipa. Upcott, 1936 b.
   Galtonia. Newton, 1924.
   Fritillaria. D., 1935.

   (c) *Prophase and Metaphase, Relational Chromatid Spirals.*
   Many plant species. Lewitsky, 1931 (presumed).
   Fritillaria, Puschkinia. D., 1935 a, 1936 c and d.
   Locusta (diplochromosomes). White, 1935.

2. MEIOSIS (1st Division).

(a) *Metaphase :*	*Major Spirals.*	*Major and Minor Spirals.*
Tradescantia	Baranetsky, 1880.	Fujii, 1926 cit. Kuwada.
	Sakamura, 1927.	Kuwada and Nakamura, 1933.
	Nebel, 1932 b.	Sax and Humphrey, 1934.
		Kato, 1935.
		La Cour, 1935.

Rhœo . . . Nebel, 1932 b. Sax, 1935.
Lilium . . Shinke, 1930. Iwata, 1935.
Trillium . . Huskins and Matsuura, 1935 a.
          Smith, 1935.
Gasteria . . Taylor, 1931.
          Tuan, 1931.
Lathyrus . . Mæda, 1928.
Sagittaria . . . . . Shinke, 1934.
Fritillaria . . . . . Darlington, 1935.
Podisma . . . . . Makino, 1936.
(Orthoptera).

(b) *Interphase, Relic Spirals.*
Tulipa. Newton, 1927.
Gasteria. Taylor, 1931.
Lilium. Kato and Iwata, 1935.
Tradescantia. Kato, 1935.
Fritillaria. D., 1935 a.
Podophyllum. D., 1936 e.

(c ¹) *Prophase, Relational Chromosome Spirals.*
Zea. McClintock, 1933 (presumed).
Fritillaria. D., 1935 a and b.
Chorthippus. D., 1936 d.

(c ²) *Prophase and Metaphase, Relational Chromatid Spirals.*
Chorthippus. D., 1936 b.
Fritillaria. D., 1936 c.

3. MEIOSIS (2nd Division).
(a ¹) *Metaphase and Anaphase, Minor Spirals.*
Tradescantia. Kuwada and Nakamura, 1935.
Lilium. Kato and Iwata, 1935.

(a ²) *Metaphase and Anaphase, Major Spirals.*
Sagittaria (sometimes). Shinke, 1934 ; Kato and Iwata, 1935.
Lilium longiflorum.

(b) *Metaphase, Relic Spirals.*
Fritillaria. D., 1935.
Podophyllum. D., 1936 e.

4. PERMANENT PROPHASE (Salivary Glands).
Relational and Relic Spirals.
Drosophila. Koller, 1935.

3. EXTERNAL MECHANICS OF THE CHROMOSOMES

(i) **Attractions.** (a) *Specific Attractions.* In homozygous diploid organisms an association occurs between corresponding chromomeres of homologous chromosomes in pairs at zygotene. In similar tetraploids a similar association occurs in pairs, but changes of partner take place along the chromosomes, each at which pairs evidently with the nearest of its possible partners at any particular place. This association evidently depends on a *primary attraction* between chromomeres which is qualitatively specific and quantitatively limited to pairs. There is no attraction between all four threads (chromosomes) of a tetraploid at zygotene, and there is no attraction between the four threads (chromatids) of a diploid at diplotene after the chromosomes have divided. This intermolecular force operates at a greater distance than intramolecular attraction, and although it constitutes one of the primary physical properties of living matter we can find a very ready inorganic analogy (cf. Hardy, 1919, on Static Friction cf. 1936).

In homozygous organisms there is, as a rule, no reason to doubt that the attraction which initiates pairing also carries it on throughout the chromosome. But in structural hybrids another force comes into play. Chromosomes are brought together in the parts where they are homologous by their specific attractions. But in linear sequence with these are non-homologous parts. These are under a torsion, as all chromosomes are at this stage, the torsion that determines relational coiling. They therefore continue this homologous pairing by a non-homologous torsion-pairing (McClintock, 1933, on *Zea*; Lammerts, 1934, on poly-haploid *Nicotiana*, D., 1936 *c* on *Chorthippus*). Sometimes a chromosome may " pair " with itself as a piece of twisted string does when released from longitudinal tension. The chromosomes that show this property most strongly are the inert B chromosomes of *Zea Mays*; they show a correlated lack of secondary attraction at mitosis and therefore perhaps have a weaker primary attraction (*v. infra*).

There are various other consequences of torsion pairing. First, in the triploids it is found that one of the three chromosomes is left out of association at zygotene, the three chromosomes changing partners intermittently. At a late pachytene it is found, however, that all three lie very close together and may appear to be equally associated in some small parts of their length (D., 1929 *b*, Olmo, 1934). This is evidently due to a chromosome which is unpaired in an intercalary segment being dragged round its partners and therefore closer to them as the relational coiling develops.

Further, we may recall that in ordinary homozygous diploids true pairing is often incomplete, especially where the chromosomes are very long. It is interrupted by the division of the unpaired parts

(Ch. IV). In such cases it does not seem impossible that pairing may sometimes be continued by a false association of divided parts of chromosomes, due to torsion and not to attraction, due in fact to torsion overcoming repulsion in short intercalary unpaired segments.

The evidence shows that none of these kinds of torsion pairing results in association that will permit of crossing-over. Many cases are inconclusive. Thus where there is non-homologous torsion pairing it is impossible to prove the complete absence of homologous attraction-pairing (Ch. VIII). Pairing and crossing-over within haploid sets, for example, may always be due to the duplications that are known to occur within them (Ch. XI). On the other hand, non-homologous parts pair at pachytene in a *Zea-Euchlæna* derivative (Beadle, 1932), and the same pairing is characteristic of the differential segments of some sex chromosomes (Ch. IX). But in no case do these segments form chiasmata and in no case does the progeny reveal crossing-over.

It follows from all these considerations that the analysis of the conditions of zygotene pairing is more complex than it was formerly believed to be. From time to time observers allege that they have discovered an unexpected peculiarity of the chromosomes in regard to their pairing (*e.g.*, Huskins and Smith, 1934; Lorbeer, 1934; Fig. 86). In order to arrive at such conclusions in the first place and *a fortiori* in order to assess their value for the general theory of chromosome mechanics it is necessary to see and understand the succession of changes before and after the supposed abnormality, a precaution that its discoverers have usually omitted to take.

The underlying principle in zygotene and pachytene association is therefore incontestable: an attraction between all homologous parts of threads in pairs. When we turn to the later stages of the prophase of meiosis we find that the same principle continues to apply in a system where the units have a different value. Instead of the single threads being chromosomes they are the chromatids produced by the division of the chromosomes at the end of the pachytene stage. Association between chromosomes continues by virtue of the changes of partner at chiasmata amongst the pairs of associated chromatids. This is shown by the failure of pairing after pachytene of chromosomes that have been paired at pachytene wherever crossing-over and chiasma-formation have failed. This happens in short chromosomes produced by fragmentation (Ch. V), in the odd chromosomes of triploids, and through partial failure of pachytene pairing in autotetraploids (Ch. IV). It happens through abnormal genotypes reducing the precocity of prophase (Ch. X) and through exceptional physiological conditions. The chiasmata which hold the chromosomes together are interstitial at first. They becomes terminal by movement. The evidence that terminal chiasmata correspond in structure, in development and in function to the interstitial chiasmata from which they are derived may be summarised as follows :—

1. The terminal connection between pairs of chromosomes can always be seen to be double in favourable material, *i.e.*, to be a connection between pairs of chromatids not associated proximally. It is therefore a change of partner amongst chromatids (*Tulipa*, Newton and D., 1929 ; *Tradescantia*, D., 1929 *c* ; *Prunus*, D., 1930 *a* ; *Primula*, D., 1931 *a* ; *Œnothera*, D., 1931 *d*, Catcheside, 1931 ; *Pyrus*, D. and Moffett, 1930).

2. The joint terminal connection between three and four chromosomes (triple and quadruple chiasma) can be seen similarly to be a change of partner amongst chromatids. The chromosomes are not associated as such, but by the property of pairing (and changing partner) possessed by their chromatids. By virtue of this, one chromosome is never directly connected at the ends with more than two others (D., 1929 *c*, 1931 *a* ; Meurman, 1929 *a*).

3. Where the association between one pair of chromatids fails, the " imperfect chiasma " resulting gives less resistance to anaphase separation (D., 1929 *c*) and in a quadrivalent with a quadruple chiasma gives rise to a special type of configuration where the chromosomes are arranged in a line instead of in the cross (Fig. 42, IV, *c* and *d*).

4. The movement towards the end may be shown by comparing the frequencies of chiasmata in different parts of the paired chromosomes at different stages (*Phrynotettix, Primula, Matthiola, Rosa*). They are more frequent near the ends at the later stages.

5. At metaphase, interstitial chiasmata may be occasionally

found instead of terminal associations in organisms with normally complete terminalisation (*Œnothera, Primula, Campanula*).

6. The frequency of terminal metaphase associations of fragments is in proportion to the frequency of interstitial chiasmata found in large chromosomes, where the fragments therefore must usually develop one chiasma (*Fritillaria imperialis*).

7. Further, these fragments are associated at only one end, as they should be, having only one chiasma formed, although the larger chromosomes are associated at both ends, having several chiasmata formed (*Tradescantia, Datura*, Blakeslee, 1931).

8. The terminal association of tetraploids and triploids does not consist in union of all homologous ends, but in a random assortment of possible associations of such homologous ends in every possible combination such as would be determined by chiasmata from randomly paired threads, as observed in a polyploid at zygotene (*Hyacinthus* and *Tulipa*), diplotene (*Primula, Campanula*) and metaphase (*Datura*).

9. Evidence from polyploids which agrees with the assumption that all chromosome association at meiosis is by chiasmata, necessarily implies that a terminal association is a chiasma. This evidence is the degree of, and variation in, the frequency of multivalent association of chromosomes (sometimes of different lengths) in triploids and tetraploids already described (*Hyacinthus, Lilium tigrinum, Primula sinensis*).

10. Similarly, the association of chromosomes in hybrids whether terminal or interstitial agrees in its frequency and in its variation between different nuclei of the same individual and between the nuclei of different individuals, as is expected from the conditions, frequency and variation of chiasma formation (*Triticum, Crepis,* etc., *cf.* Mather, 1935 *b*).

These considerations, as we have seen, are of primary importance in interpreting chromosome behaviour at meiosis. Chiasmata are a condition of metaphase pairing, and the understanding of their structural properties is a condition of understanding how this pairing comes about. They are also, as we have seen, of importance in interpreting meiosis itself. They show that specific attraction operates only between pairs of particles at all stages of meiosis.

The associations found at meiosis therefore require no assumption of forces of attraction that is not already justified by the study of mitosis.

We are not concerned here with the precocity theory of meiosis which is derived from this principle. We must consider how the specific attraction inferred is related to chromosome movements under special conditions and in exceptional organisms. When the chromosomes are brought close together at metaphase and anaphase of mitotic and meiotic divisions they show a secondary attraction (Ch. VI). This attraction is *residual* with respect to the primary attraction, and like it shows itself between similar parts of chromosomes. It is, as a rule, very slight, so that it only slightly affects the equilibrium position of the chromosomes on the metaphase plate. A number of variable conditions affect its expression. It shows itself more readily at meiosis than at mitosis because the chromosomes are then shorter and can move with less resistance. It shows itself more readily between smaller chromosomes, whether in organisms like most Dicotyledons, in which they are all small, or in the smaller members of a set. Thus fragment chromosomes reduplicated in large numbers usually lie in groups. On the other hand, it is genotypically and environmentally controlled. Its expression must depend upon the substrate. Thus pears show this association less than apples, although the relationships must be the same (Moffett, 1934). The chromosomes of the Diptera, on the other hand, have such a high residual attraction that it shows even in the prophase at mitosis where, according to the illustrations (Metz, 1916; Moffett, 1936) it is enough to permit the development of relational coiling. Probably the rare crossing-over in the male *Drosophila* occurs at somatic mitoses and is conditioned by this relational coiling (*cf.* Sect. 4).

At meiosis in the male this highly-developed residual attraction expresses itself in an entirely new type of segregation mechanism (Ch. IX) and at the permanent prophase of the salivary glands in a new type of chromosome relationship (Ch. VI). In meiosis the equilibrium position, with the four chromatids of the bivalent almost equally spaced, shows that the primary and residual attractions are almost equal and the distinction between them has therefore almost entirely lapsed. The contrary change may occur in the complete lapse of the residual attraction : this is probably, as we saw earlier, a corollary of a reduction in the primary attraction. The B chromosomes of *Zea Mays*, which are inert and over-condensed at metaphase, show by their frequent torsion pairing a reduced primary attraction and by their complete absence of juxtaposition at mitosis (Fig. 102) a reduced secondary attraction.

The association of centromeres in pachytene cells does not allow us to infer their own particular mechanical properties at this stage. It is clear, of course, that the centromere is not exerting the repulsion shown in anaphase movement and terminalisation (*q.v.*). But we cannot say whether it exerts an attraction on other centromeres in the pachytene nuclei because chromomeres adjoining it may be responsible for the fusion or pairing that it undergoes. This does not apply to the polytene nuclei where all the centromeres fuse, or to diakinesis in male *Drosophila*, for the sudden change that takes place between diakinesis and metaphase leaves no doubt that here an attraction has been followed by a repulsion.

There is, moreover, some evidence in *Agapanthus* and *Stauroderus* that at the pachytene stage an attraction exists between the centromeres of non-homologous chromosomes. This does not mean that the attraction is non-specific but rather that all centromeres are homologous, *i.e.*, their common forms and functions are inherited from a common progenitor, a single ancestral centromere (D., 1935, *b* and *g*).

These special properties express three important principles of chromosome mechanics. First, the anomalies of specific attraction at meiosis and in the permanent prophase of *Drosophila* are associated with other anomalies of movement. At diakinesis the centromeres of paired chromosomes, and in the permanent prophase the centromeres of all the chromosomes, are closely associated instead of repelling one another as the centromeres do at these stages in normal circumstances. This correlation of behaviour in different cells shows the importance of a *substrate effect* in chromosome movements. It shows that the attractions of chromomeres and centromeres are conditioned by a common substrate and therefore, like other properties of the chromosomes, subject to physiological control.

SUBSTRATE EFFECTS

Secondly, the expression of specific attraction seems to depend on the conditions of spiralisation in the chromosome threads. While on the one hand the molecular spiral may cause association in the absence of attraction by torsion pairing, on the other hand two separately spiralised threads show less attraction than two unspiralised threads. While the sudden lapse of attraction between homologous chromatids at first anaphase must be due to a change in the substrate, the absence of attraction between second prophase chromatids as compared with that shown at all other prophases seems more likely to be due to the spiralised threads having an attraction sufficient to maintain an association established at an unspiralised stage but

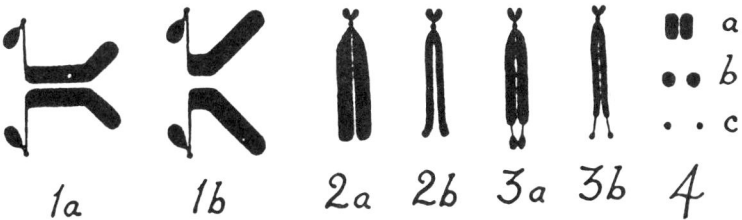

FIG. 139.—Equilibrium positions of the chromatids determined by different relative strengths of primary attraction and surface repulsion. (1) Meiotic bivalent, (2) mitotic chromosomes, (3) trabants, (4) acentric chromatids.

not sufficient to establish a new association in the face of the repulsion existing between all threads. The strength of the specific attraction between two spiralised threads that are jointly coiled is shown by the arrest of terminalisation, owing to change of homology, that will be described later.

Thirdly, the special properties of one chromosome (with differential condensation) as contrasted with the rest shows that these differences of behaviour are also conditioned directly by material differences within the chromosome.

Another source of misunderstanding lies in the distinction that may be drawn between the primary and residual attractions inasmuch as the second merely modify an equilibrium position determined by repulsion while the first seem to result in " contact." But the appearance of contact is not to be naïvely interpreted on a macroscopic analogy. It is itself always a position of equilibrium between attractions and repulsions. This is seen most clearly where two chromatids are paired at the first meiotic metaphase for a very short distance between two interstitial chiasmata (D., 1936 d). A gap may then be seen between them just as between the sister chromatids in *Drosophila*. The attraction is never, as on chemical analogy it never can be, absolute in its effect. It is always in equilibrium with the repulsions (which will be considered later) and whereas the repulsions should be a function of area, the attraction should be a function of volume. Smaller attracting parts will therefore lie further apart. Similarly, at particular stages, subject to particular substrate conditions the distinction between primary and secondary attractions may seem to lapse. The distinction is quantitative in cause, but it is, as we shall see, qualitative in effect under the conditions governing metaphase and anaphase movement in all organisms.

(b) *Non-Specific Attractions*. The only type of association between non-homologous parts of chromosomes in the nucleus consists of the association of ends. This is found in the prophase of meiosis in two ways. First, there is the special type of association found in the Coccidæ (F. and S. H. Schrader, 1923–32). This is perhaps analogous to the formation of a continuous thread in the prophase of meiosis in *Icerya* and in haploid *Triticum monococum*. Its causation is unknown. Secondly, there is the special type of arrangement of chromosomes at the zygotene stage described as polarisation. The chromosomes come to lie during leptotene with one or both ends turned towards a particular part of the surface of the nucleus. In some species this movement has nothing to do with the centromere. Thus it applies equally in *Chorthippus* to chromosomes with the centromere near the middle or the ends. In other species, the centromere ends of the chromosomes alone are polarised (*Mecostethus, Phrynotettix*). Two or even three centres of polarisation may occur in *Chorthippus*, one of them being the X chromosome which attracts certain other chromosomes specifically. It therefore seems that the attraction is between the chromosomes themselves rather than for something non-specific outside the chromosomes such as the centrosome, as Janssens (1924) and others have supposed (D., 1936 d).

The nature of this attraction is indicated, as we saw earlier (Ch. VIII), by the differential condensation of the parts of chromosomes concerned. It seems to be an attraction between nucleolar materials such as leads to the fusion of nucleoli at any stage in the life of the nucleus. It has nothing to do with the specific attraction of genes and chromomeres.

(c) *Reproduction.* We have seen earlier from direct observation that the chromosomes are undivided at the earliest prophase of meiosis and divided throughout the prophase of mitosis. The chromosomes must therefore divide during the resting stage in mitosis, while in meiosis the prophase must be supposed to anticipate this division. This conclusion is borne out by the evidence of abnormal precocious prophases in polymitotic and variable sterile maize (Beadle, 1932, 1933), where the chromosomes are still undivided as in meiosis. It has been confirmed by the results of X-ray analysis in showing that the genes and the thread divide at a particular period of the resting stage between meiosis and the following mitosis (Ch. X). The evidence of X-ray effects on the smallest visible objects must always have a higher validity than that of visible or ultra-violet light on account of the shorter wavelength, and it seems that these results must overbear any contradiction from direct observation. It is therefore unnecessary to account for the many records of a split in the metaphase and anaphase chromatids that have appeared in the past (Nebel, 1932; Huskins and Smith, 1935, *cf.* Lorbeer, 1934, for list). Comparison, in fact, shows that these records are internally as well as mutually inconsistent. The divided "leptotene" chromosomes have not been shown to pair. The supposed cleavages in the anaphase chromatids have cut across either the major spiral or the minor spiral, of whose existence the observers have been unaware. Those observations which have taken account of the internal structure have failed to reveal any split before the resting stage (Geitler, 1935; D., 1935 *a*). There are, however, two exceptions, observations made with a knowledge of structure, which suggest a split in the second anaphase chromatids in *Tradescantia* (Kuwada and Nakamura, 1935; Sax, 1935). They depend, however, on the detection of longitudinal doubleness in cylinders of less than one-half the wave-length of the light used. Experiments with infra-red light (Hruby, 1935) show that incorrect illumination or the use of light of a wave-length having a certain proportion to the diameter of the cylinder observed, may produce this result. This explanation seems to be all the more inevitable since the material and the stage are the very ones in which the chromosomes were tested by X-ray treatments and shown to be single.

The question now arises as to how the chromosomes " divide." The knowledge that primary attraction exists only between pairs of chromomeres at once suggests an analogy with the property of chromomeres reproducing only to give pairs. In reproduction a new chromomere or gene identical with the parental chromomere or gene is laid down beside it. This, at least, is the description necessary if we look upon the structures as having an organisation that is characteristic and unchanging like that of a molecule which retains characteristic and unchanging chemical properties. We may say, therefore, that the new particle that is laid down is attracted by the old one from the substrate. And the possibility of this attraction taking effect must depend on two general conditions: the degree to which the attractions of the particles are already satisfied, and the concentration of suitable materials in the substrate. A test of this view is provided by the behaviour of unpaired threads in pachytene nuclei as compared with that of paired threads. They divide earlier both in triploids and in organisms with incomplete pachytene pairing (D. 1935 b). Evidently, therefore, when the primary attraction is unsatisfied the concentration of substrate necessary for reproduction is reached earlier and presumably at a lower level than when this attraction is satisfied. Further evidence of a relationship between attraction and reproduction is shown by the salivary gland nuclei. An unlimited reproduction is associated with an unlimited attraction.

The question that next arises is as to whether the chromosome thread divides into two equivalent daughter-threads or as we have supposed gives a daughter-thread distinct in its properties from a parent thread. This question is answered by the observations of ring chromosomes (Ch. III). Such chromosomes divide so that either two free rings or two interlocked rings are formed. In the first case the threads have run parallel, in the second they have made a complete revolution round one another. In both cases they are distinct. A single continuous dicentric ring formed by their making half a revolution does not apparently arise. We are driven to conclude that there is an absolute distinction between a parent and a daughter-thread in the reproduction of the chromosome.

(ii) Simple Repulsions in the Nucleus. (a) *Between Chromosomes and Configurations*. During the prophase of mitosis and meiosis the chromosomes are limited in their movements by the nucleocytoplasmic surface. When the enclosed space is small, as it usually is, the separate chromosomes or configurations arrange themselves equally so far as their internal movements and occasional interlocking permit. Such an evenness of distribution at mitosis led Lillie (1903, *cf*. 1905) to suppose that they bore a surface charge and therefore repelled one another (*cf*. Hardy, 1911). This conclusion is supported at once by the arrangements found in the large prophase nuclei of egg-cells and embryo-sac mother-cells. Here an even distribution is no longer attained. The chromosomes are further apart than in corresponding small nuclei, but they are by no means equally distributed (*cf*. Gustafsson, 1935). Evidently the distances are too great for the repulsions to be effective. More critical evidence of the effect of surface charge, however, is provided by the movements occurring within configurations during the prophase of meiosis.

(b) *Within Configurations : Terminalisation*. The configurations of bivalent chromosomes found at diplotene present us with the largest field for accurate comparison of quantitative data in cytology. The configurations make up an isolated and protected system within the nuclear membrane. They vary in numbers of chiasmata that can be counted, in sizes of chromosomes that can be measured; they occur in nuclei of all sizes, in both sexes and in all organisms from *Amœba* to man. The observation of systems at this stage therefore constitutes a natural experiment in cell physiology capable of establishing the principles governing the external mechanics of the chromosomes under the simplest and most readily comparable conditions.

CELL MECHANICS

The configurations observed depend for their form, as we have seen, on the positions and numbers of their chiasmata. The forms of the chiasmata themselves depend on their positions. They may be classified in the first place according to whether they are *interstitial* or *terminal*. An interstitial chiasma is always of one kind. The only conceivable complication of this form is that where two such chiasmata involving the six chromatids of three chromosomes might coincide. This has not been observed. The terminal chiasma, on the other hand, has various complications. Several such chiasmata may coincide, and we then have a *multiple* chiasma. Again a chiasma which is terminal for one chromosome

FIG. 140.—Diagram of chromatids showing different types of chiasma observed at metaphase and early stages. A, interstitial chiasma. B, terminal chiasma. C, triple (terminal) chiasma. D, quadruple chiasma. E, imperfect terminal chiasma. F, imperfect lateral chiasma (asymmetrical type—for symmetrical *cf.* Fig. 86, c_1). G, lateral-terminal chiasma, one chromosome having a reduplicated segment. (From D., 1931 *c*.)

may be interstitial for its partner, with a part of which it is homologous, and then we have a *lateral* chiasma. Lateral chiasmata are of two kinds: those which must be supposed to arise from terminalisation of a single chiasma (*i.e.*, a single cross-over on the chiasmatype hypothesis) or several non-compensating chiasmata, and those which must be supposed to arise from two reciprocal chiasmata. The first gives a symmetrical configuration, the second an asymmetrical one. Multiple combinations of chiasmata, arising presumably through pairing of reduplicated segments in the same chromosome, may include lateral chiasmata.

When terminalisation is complete, every chromosome end that is associated by a chiasma with another end remains so associated through diakinesis and metaphase until anaphase, so that the same proportion of ends of chromosomes is found associated at metaphase by terminal chiasmata in any organism as there were arms earlier associated by interstitial chiasmata. One change only may take place, and this in special circumstances, the breakage of the attachment between one pair of chromatids. Where four chromosomes are associated by a quadruple chiasma the fusion of chiasmata necessary for its formation may break one of the four chromatid associations, so that the cross-quadrivalent becomes a special kind of chain-quadrivalent (Fig. 42, IV, c). This gives an *imperfect* chiasma (D., 1929 c).

Chiasmata are first seen through the opening of loops separating pairs of chromatids at diplotene. Their position cannot be accurately recorded at the very moment of origin, but nevertheless when the records of the earliest stages are compared with those of diakinesis and metaphase it is usually found that they have changed in position. The change has not been observed directly in living nuclei, but it is inferred from a comparison of records at the two stages in fixed preparations which show great constancy in its character. This comparison reveals the conditions summarised earlier (Ch. IV) for the purpose of understanding the form of the metaphase bivalents, namely, the occurrence of a range of types of behaviour from those in which the change is scarcely detectable to those in which the form of the bivalent is entirely altered owing to complete terminalisation of chiasmata. The proportion of the total chiasmata that are terminal in any given type of chromosome or at any given stage is conveniently used as a *terminalisation coefficient* (D., 1931 d). These different kinds of behaviour now call for exact analysis to show the changes in number and position of chiasmata, *i.e.*, in the association of the chromatids, for several reasons. Observations of meiosis are usually confined to metaphase. An exact understanding of the kinds of change in bivalent structure which lead up to metaphase is necessary in interpreting these observations. Such an analysis is also necessary in deciding what forces hold the paired chromosomes together during and after the observed changes in their relationship, and what forces determine these changes. Further, the changes in number and position of chiasmata during prophase had to be followed in order to know whether there was any possibility of exchanges in linear connection amongst the threads such as might determine genetic crossing-over

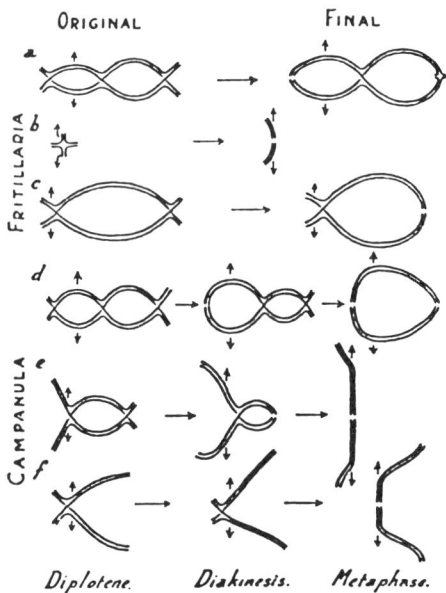

FIG. 141.—Diagram showing different types of terminalisation.
 a, b, c : *Fritillaria* type with slight change of position and little or no reduction in number of chiasmata through fusion. d, e, f : *Campanula* type in which the movement is completed and results in terminalisation of all chiasmata. *Tulipa, Lilium* and *Rosa* are intermediate between the two. b shows that the same amount of movement as gives little change of appearance in the large chromcsomes, in a fragment (of *Fritillaria imperialis*) gives complete terminalisation. c_1 a characteristic bivalent in *F. Meleagris* with localised chiasmata, shows that movement occurs in a distal chiasma although scarcely evident in the proximal one. e, f, show alternative behaviour to d where chiasmata are formed only on one side of the centromere, but conditions are otherwise the same. This explains the origin of the two types of bivalent form in " type C " in *Phrynotettix* (Wenrich, 1916, cf. Fig. 86), and " type 7 " in *Circotettix* (Helwig, 1929).

(Ch. VII). With the same end in view a record of the development of chiasmata was necessary in order that their number and position at the moment of origin (before they can be recorded) might be inferred. Finally, a comparison of analysis of hybrids with that of pure forms must throw light on the differences between chromosomes which constitute hybridity and would show the principles on which unsupported metaphase observations of hybrids should be interpreted. These expectations have now been fulfilled.

Chiasmata must be supposed to arise interstitially, as exchanges of partner amongst chromatids, on four grounds : (i) a strictly terminal origin excludes the possibility that chiasmata arise always as exchanges of partner, and demands therefore a dual explanation of their origin ; (ii) a terminal origin is incompatible with the chiasmatype theory, which supposes that chiasmata arise through crossing-over, and therefore interstitially ; (iii) terminal chiasmata are found at late diplotene more frequently in small chromosomes (*e.g.*, fragments in *Fritillaria imperialis*) than in large ones in the same nucleus, and while it is intelligible that the conditions of terminalisation are different in small chromosomes it is contrary to observations of frequency to suppose that the conditions of origin are different (Ch. V) ; (iv) the terminalisation coefficient of bivalents having different numbers of chiasmata at diplotene in *Tulipa* shows differences such as are accentuated later in the course of development, and may therefore be attributed to terminalisation (*v.* Fig. 142). Furthermore, the terminalisation coefficient is higher at the earliest recordable stage in organisms with complete terminalisation like *Campanula* than in those with slight terminalisation like *Tulipa*.

Where several chiasmata are concerned the process has been most clearly shown in *Campanula* (D. and Gairdner, 1931) where terminalisation is complete, in *Tulipa* (D. and Janaki-Ammal, 1932), *Stenobothrus* (D. and Dark, 1932), *Zea* (D. 1934), *Spironema* (Richardson, 1934), *Culex* (Moffett, 1936) and *Melanoplus* (Hearne and Huskins, 1935) where it is incomplete. The observations show the following rules of behaviour :—

(*a*) The movement is a movement away from the centromere because when terminalisation is complete a proportion of metaphase chromosomes are rod-shaped having a terminal chiasma at one end, the rest ring-shaped having terminal chiasmata at both ends. The first type arises from diplotene bivalents with chiasmata all on one side of the centromere. The second type arises from those having chiasmata on both sides. Metaphase configurations of chromosomes with unequal arms agree with this assumption.

This conclusion is favoured by the behaviour found in many organisms, especially those with small chromosomes. The paired

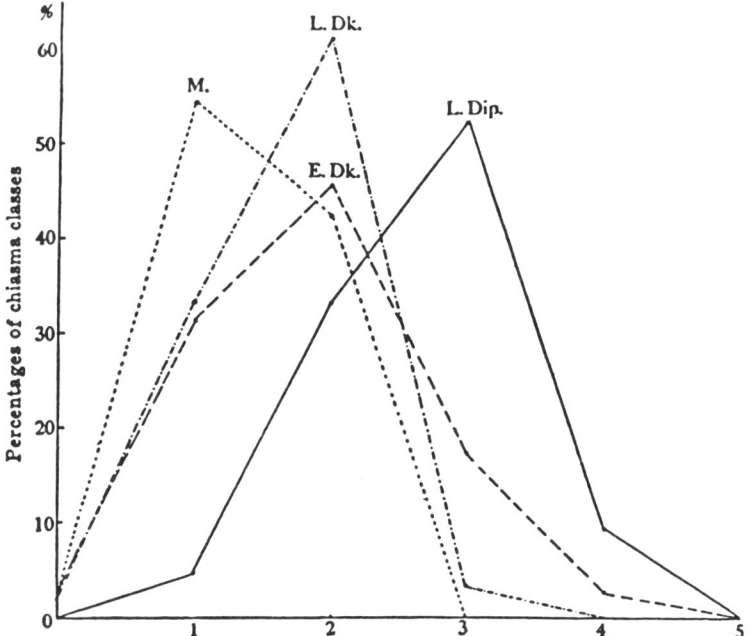

Fig. 142.—Percentage frequency polygons of numbers of chiasmata (abscissæ) in bivalents of *Rosa* " Orleans " at late diplotene, early diakinesis, late diakinesis, and metaphase of the first division in pollen mother-cells. (From Erlanson, 1931 *c*.)

chromosomes repel one another at diakinesis to such an extent that all the chromosomes, paired and unpaired, appear to be distributed evenly in the nucleus. Where interstitial chiasmata are still present (as in *Prunus*) it is then found that the repulsion is effectively between the centromeres and the parts of the chromosomes between these and the first chiasma are therefore drawn out into a fine thread while the other parts show no tension.

CONDITIONS OF MOVEMENT

(b) All the chiasmata, both near the centromere and further away, move at the same time towards the ends. The nearer ones do not catch up and meet the further ones. Thus the first change observed in terminalisation is that the distal chiasmata become terminal—without any reduction in the total number through a fusion taking place (*e.g.*, in *Fritillaria*, *Tulipa*, *Stenobothrus*). Moreover, if the proximal chiasmata caught up the distal ones they would in a proportion of cases cancel one another out, for the changes of partner at adjacent chiasmata can often be seen at anaphase to be reciprocal. The movement is not, therefore, merely an opening out of the proximal or centric loop at the expense of the distal loops, but a simultaneous movement of all chiasmata towards the ends and leading to fusion only at the ends.

(c) In *Fritillaria* and *Stenobothrus* there are long chromosomes with several chiasmata having incomplete terminalisation and very short ones with single chiasmata having complete terminalisation. In neither is there any reduction in the total number of chiasmata. There is merely a change in the distribution, the interstitial chiasmata nearest to the ends becoming terminal. In *Campanula*, where several chiasmata are completely terminalised, they all move towards the ends before they fuse. They must therefore fuse at the ends. Where there are two interstitial chiasmata fusing with one terminal one they presumably do so simultaneously, since in the intermediate stages the diminishing distal loops are always of the same size in any one arm.

We now have information which shows in a general way the changes that may take place in the course of terminalisation. Numerous conditions have been found to influence these changes in each paired arm of a bivalent, of which the following are the chief :—

(a) Length of the arm.
(b) Original frequency and distribution of chiasmata.
(c) Agreement in the homologous linear sequence of the pair.
(d) Rate of movement of the chiasmata.

(a) *Length of the Arm*. In *Nicandra* (Janaki-Ammal, 1932) terminalisation is complete in the shortest chromosome pair at late diplotene, and in the other short pairs at diakinesis, while in the longest chromosomes it is never complete. In *Fritillaria* (D., 1930 c) terminalisation is complete in the small fragments at late diplotene although it never has any effect in the long chromosomes, beyond leading to the movement of the distal interstitial chiasmata to the ends. This difference is due in part to the chiasmata being formed nearer the ends in short chromosomes and in part to their being nearer the centromere.

(b) *Number of Chiasmata.* Where the degree of terminalisation is low, as in *Lilium* (Belling, 1931) and *Tulipa* (D. and Janaki-Ammal, 1932), the number of chiasmata terminal, and even the

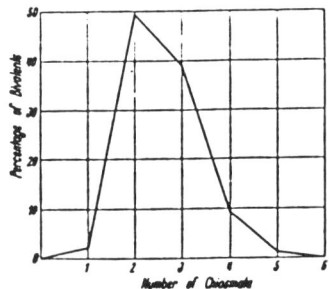 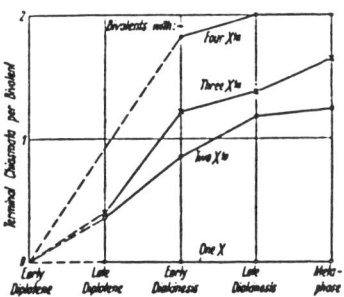

FIG. 143.—(Left) Chiasma frequency polygon of *Zea Mays* where no fusion of chiasmata occurs. (Right) Terminalisation in bivalents with different total numbers of chiasmata, *cf.* Fig. 93. (D., 1934.)

proportion terminal, is higher the greater the number of chiasmata in the bivalent (Fig. 142). The *Tulipa* observations show that this is due to failure of chiasmata lying in one arm only to be terminalised. It appears therefore that, other things being equal, movement is increased by the centromere lying in a closed loop.

(c) *Chromosome Homology.* In certain organisms with more or less complete terminalisation interstitial chiasmata are occasionally found, sometimes very near the centromere. When these were observed in a known structural hybrid (*Tradescantia*, D., 1929 c) the explanation suggested was that terminalisation was arrested by a change in the homology of the chromosomes. Thus, if two dissimilar chromosomes with a linear sequence *abcdef* and *abcxyz* with

ARREST OF MOVEMENT

centromeres at *a* form a chiasma at *c*, terminalisation will be impossible without the particles *xyz* coming to pair with the particles *def* and leaving their identical partners. If we suppose that identical partners exert a certain attraction on one another and that dissimilar potential partners exert none (*i.e.*, that the specificity of chromosome attraction observed at zygotene still continues at diakinesis and metaphase), then it is possible that whatever forces determine terminalisation are unable to overcome the resistance to change from identical to non-identical partners, or rather to no partner at all, and in consequence movement of the chiasma is arrested. This conclusion has been borne out by the observation of exceptional interstitial chiasmata in *Œnothera* (D., 1931 *c*) between parts of chromosomes whose distal segments were dissimilar and were therefore, on this hypothesis, bound to prevent terminalisation (Ch. VII).

TABLE 72

Examples of Arrest of Terminalisation by Change of Homology

Established cases:
- *Œnothera biennis* . . . D., 1931 *d*.
- *Pisum sativum* . . . E. R. Sansome, 1932.
- *Œnothera blandina* . . Catcheside, 1932; (haploid) 1935, Fig. 4*b* (X-rayed).
- *Rosa blanda* . . . Erlanson, 1931 *c*.

Possible cases:
- *Tradescantia virginiana* . . D., 1929 *c*; Koller, 1932 *c*.
- *Aucuba japonica* . . . Meurman, 1929 *a*.
- *Forficula* sp. . . . Payne, 1914.
- *Datura* hybrid . . . Bergner and Blakeslee, 1932.
- *Œnothera* (triploids) . . Håkansson, 1930.
- *Campanula persicifolia* . . Gairdner and D., 1931.
- *Dahlia* hybrid (6*x*) . . Lawrence, 1931 *a*.
- *Cardamine pratensis* . . Lawrence, 1931 *d*.
- *Zea Mays*, haploid gametophyte Beadle, 1931 (*cf.* Plate X).
- *Anthoxanthum odoratum*. . Katterman, 1931.
- *Zea-Euchlæna* derivative . Beadle, 1932.

512 CELL MECHANICS

(d) *Rate of Movement.* The differences found in degree of terminalisation in organisms with chromosomes of similar sizes such as *Lathyrus* and *Campanula* may be said to be immediately due to differences in the rate of movement relative to external development. Such differences are found between male and female mother-cells in the neuropteran *Macronemurus* (Naville and de Beaumont, 1933). They are also found between pollen mother-cells of normal and male-sterile *Lathyrus* (Fabergé and Upcott, unpub.). In the

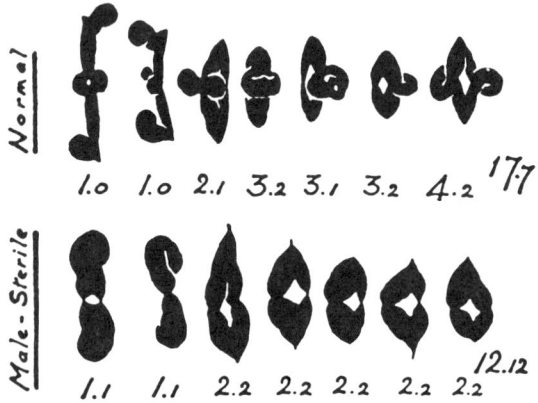

FIG. 144.—First metaphase in normal and male-sterile *Lathyrus odoratus*, showing correlated difference in terminalisation and spiralisation, *cf.* Fig. 109. × 4,000 (Upcott, 1936).

normal the chiasmata reach the intermediate equilibrium position with no fusion of chiasmata. In the male-sterile terminalisation is complete. Here it is possible to show that the abnormality is not due to quicker movement but to slower external development, for the cells do not reach first metaphase until the anther has reached the size at which the normal plant has mature pollen-grains. Correlated with the greater movement is a greater spiralisation in both cases (Fig. 144). Similar changes have been produced by abnormal temperatures (Straub, 1936).

Let us now consider the forces at work. The shapes of bivalent chromosomes at all stages of terminalisation show that a repulsion is acting between all parts of all chromosomes just as it does at other stages of meiosis and mitosis.

TABLE 73

The Change in Chiasma Frequency from Diplotene to Metaphase with Partial and with Complete Terminalisation

	Mid-diplotene.	Late Diplotene.	Diakinesis.	Metaphase.
1. *Stenobothrus parallelus* *				
Total number of chiasmata per bivalent	2·05	2·07	2·12	2·09
Number of terminal chiasmata per bivalent	0·40	0·55	0·57	1·04
Terminalisation coefficient	0·19	0·26	0·27	0·50
2. *Tulipa persica*				
Total number of chiasmata per bivalent	2·13	2·04	2·00	1·95
Number of terminal chiasmata per bivalent	0·23	0·40	0·74	0·91
Terminalisation coefficient	0·11	0·20	0·37	0·47
3. *Rosa* " Orleans "				
Total number of chiasmata per bivalent	2·66	1·80	1·65	1·53
Number of terminal chiasmata per bivalent	0·65 †	0·46	0·77	1·16
Terminalisation coefficient	0·23	0·25	0·47	0·76
4. *Campanula persicifolia*				
Total number of chiasmata per bivalent	2·67	2·29	1·96	1·90
Number of terminal chiasmata per bivalent	1·60	1·70	1·93	1·90
Terminalisation coefficient	0.60	0·75	0·98	1·00

* Great size variation among the chromosomes.
† Observations of position at this stage uncertain.

Terminalisation must then be regarded as the effect of unequal forces of repulsion acting between the pairs of paired chromatids on opposite sides of the chiasmata so that the association of the pair that repel one another more gains at the expense of the other and the chiasma moves along.

There are seven kinds of conditions on the two sides of a chiasma, five of them differential, according to whether the chromosomes are in closed loops or lie free (with open arms), and according to whether they have or have not the centromere in the loops or arms. These show which differences are associated with movement and may therefore be attributed to differential repulsion, as follows (*cf.* Fig. 145):—

(i) Closed loops repel one another more than open arms: for distal chiasmata move towards the ends even though no other movement is detectable. And chiasmata in *Tulipa* move away from the centromere when it is in a closed loop but not when it is an open arm (*i.e.*, when only single chiasmata are present).

(ii) Loops or arms, including the centromere, repel one another more than others: for the proximal loop is always the largest, where considerable movement takes place. But if the centromere is in an open arm and on the other side of the chiasma is a closed loop the extra repulsion between the centromere may not give equilibrium until the centromeres are drawn closer by the chiasma moving nearer them.

From these two series of observations, it may be inferred that two kinds of repulsion occur between the chromosomes (paired chromatids): First, a general repulsion between all parts of the chromosomes such as might be determined by a surface electrical charge; this would be stronger between loops than between free arms, because their parts would be held closer together. Secondly, a localised repulsion between the centromeres of the chromosomes. The first assumption is justified by the staining properties of the chromosomes, by the fact that unpaired chromosomes never touch one another in the nucleus, and by the known behaviour of ampholytes in solution. The second assumption is justified by the observed special state of tension found at diakinesis and metaphase at all terminal chiasmata and between the centromere and the first interstitial chiasma wherever these are close together, as they regularly are in small chromosomes and in large chromosomes where the chiasmata are localised (Fig. 145).

These two assumptions explain the movements of chiasmata so far as they are yet known (D. and Dark, 1932). In organisms with the least degree of terminalisation, such as *Lilium*, *Hyacinthus*, *Fritillaria* and *Vicia*, the generalised repulsion alone is effective and no fusion of chiasmata seems to occur. On the contrary, an equilibrium position is reached in which all the chiasmata in each arm are equidistant (D. 1933 on *Agapanthus*). In organisms with slightly more terminalisation such as *Stenobothrus* and *Tulipa* the centromeres are seen to repel one another by their loops becoming larger than others, but again little or no fusion of chiasmata occurs, and this repulsion even is scarcely effective to move chiasmata when the centromere lies in an open arm. In organisms with still more terminalisation such as *Rosa* and *Matthiola* the proximal loops expand considerably at the expense of the others, and most of the chiasmata are forced to the ends, where the supernumerary ones fuse. Finally, in organisms with complete terminalisation such as *Primula* and *Campanula* these changes are already well advanced at the stages of diplotene observed, and all the chiasmata, with rare exceptions, have reached the ends and fused

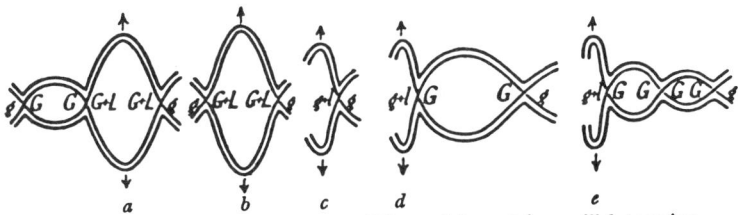

FIG. 145.—Diagram showing how differential repulsions will determine the movement of chiasmata. G, g, the generalised repulsions between all parts of all chromosomes. L, l, the localised repulsions between centromeres (marked by arrows). Capitals represent the forces within a closed loop, small letters between open arms. In all types G overcomes g. In *Tulipa* $g + l$ does not overcome g, but in *Campanula* it overcomes even G. In *Fritillaria* G may be greater than $g + l$.

before diakinesis. It can therefore be supposed that the generalised repulsion is a common property of all chromosomes after they have divided into chromatids and that the differences in degree of terminalisation observed in different organisms are due to differences in the degree of repulsion between centromeres.

This difference seems to depend to a less extent on differences in surface charge on the centromere than on differences in the size of chromosomes. Where the chromosomes are small the centromeres are held closer together; the repulsion between them and the resulting movement is therefore greater in relation to the size of the chromosomes. In other words, the repulsions between centromeres are related less to the sizes of the chromosomes than to their own functions in cell-division, which, as we shall see later, depend on the size of the cell.

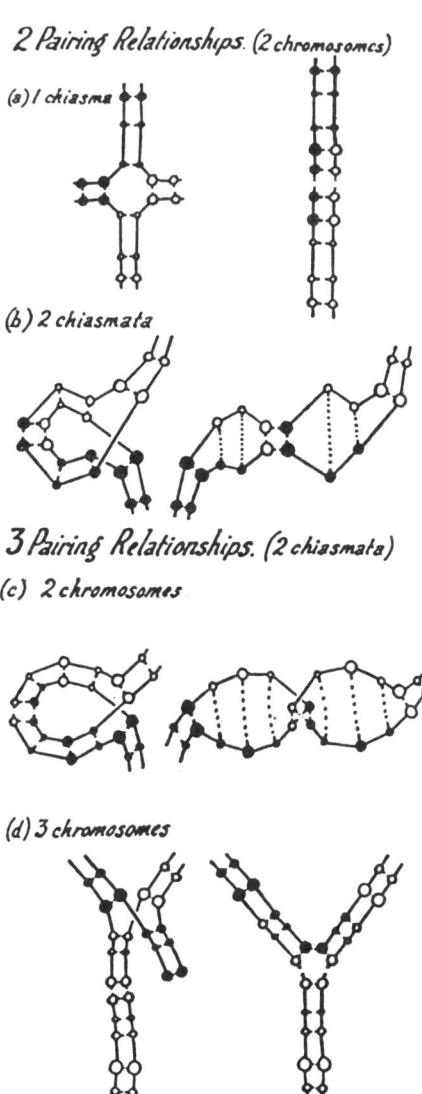

Fig. 146.—Diagram to show how terminal affinity will preserve terminal association with terminalisation of chiasmata having different relationships. The genetic interpretation follows the chiasmatype hypothesis. Left, before fusion of chiasmata; right, after fusion. If the end particles had not both lateral and terminal affinity the second type would give failure of pairing owing to its chiasmata being compensating. The dotted lines indicate the affinities that are satisfied after terminalisation.

TERMINAL AFFINITY

(iii) **The Theory of the Terminal Chiasma.** The conditions under which interstitial chiasmata are converted into terminal chiasmata are of three kinds, of which four examples may be given (illustrated diagrammatically in Fig. 146) : (*a*) The chiasma is single, and there are two associations of chromatids. (*b*) There are two chiasmata, and therefore three associations of chromatids, but the third is a continuation of the first. The chiasmata are then comparate. They result from reciprocal or complementary crossing over. (*c*) There are two disparate chiasmata : the second exchange does not restore the relationship found before. One chromatid which has crossed over at the first also crosses over at the second chiasma. The chiasmata are also bound to be disparate where (as in (*d*)) a third chromosome associates at the second chiasma with one of the other two.

The maintenance of terminal association following the movement to the ends by chiasmata having all these different relationships leads to the following conclusions : (i) The proximal or penultimate association replaces the distal or ultimate one in terminalisation (Fig. 146 (*a*)) in every particle save the terminal one. (ii) Comparate chiasmata (Fig. 146 (*b*)) do not cancel one another out, but the penultimate association replaces the ultimate one. (iii) Disparate chiasmata terminalise without replacement of the ultimate association. The penultimate association simply disappears (Fig. 146 (*c*)).

These three unexpected properties can all be explained on a single assumption : the terminal particles have, unlike the intercalary particles, a double affinity, lateral and terminal ; the latter is satisfied only on terminalisation. That terminal particles should inherently have a special affinity not found in intercalary ones does not seem improbable on analogy with attraction based on chemical, electro-magnetic or surface tension phenomena, though we need not inquire for the moment which kind of attraction is here concerned.

The result of *terminal affinity* will be that when one thread displaces another from association with a third, *either* the end of the laterally displaced thread will retain its lateral association as a terminal one (as in case (*a*) with a single chiasma, or in case (*b*) where the displacing association is the same as the terminal one owing to the chiasmata being comparate) *or* the lateral association at the end of the displaced thread will be lost (as in cases (*c*) and (*d*) where

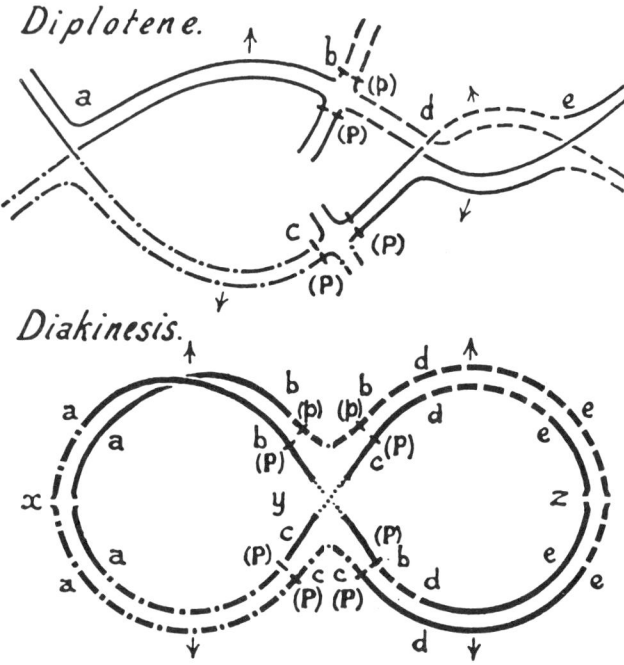

FIG. 147.—Diagram to show terminalisation in a quadrivalent giving a quadruple chiasma and the possibility of "double reduction" (or "equational exceptions") following crossing-over between chromatids (on the chiasmatype theory). a, b, c, d and e, diplotene chiasmata and points of crossing-over. x, y and z, terminal chiasmata. P and p, alternative factors in respect of which the organism is triplex (PPPp). Segregation will give double reduction in a proportion of cases yielding reduced nuclei of the constitution pp. Arrows show the positions of the centromere.

the three associations are different owing to the chiasmata being disparate). The terminal association, therefore, cannot be eliminated by cancellation (as suggested earlier by the present writer, 1929 *c*), it can only be changed (as in (*b*)) to become the lateral one.

It remains to be said that the property of terminal affinity can be supposed to develop only *after* the beginning of diplotene, for, were the ends already attractive at this stage, terminal chiasmata would arise at every chromosome end immediately and as a matter of course. Actually, it is improbable that any chiasmata are terminal at the moment of origin, for the reasons given above.

Terminal affinity is exerted as a rule only when one lateral association is displaced by another in terminalisation. But if terminal affinity develops after diplotene the possibility arises of its being a secondary cause of pairing at diakinesis or metaphase. Exceptional cases of this have been found in the sex chromosomes of *Lygæus* (Wilson, 1905), the microchromosomes of *Anasa* (Wilson, 1905), and *Alydus* (Reuter, 1930). These chromosomes do not as a rule pair at the prophase of meiosis. In *Alydus* this is apparently because they have already divided at the pachytene stage when the paired chromosomes have not yet divided. The microchromosomes pair momentarily at metaphase and separate at anaphase. The sex chromosomes divide equationally at the first division and their daughter chromatids pair at the second prophase. In the anomalous meiosis of *Llaveia* a pair of autosomes behaves in the same way (Hughes-Schrader, 1931). In either case the pairing is *end-to-end*, and not side by side. It indicates a special capacity of attraction in the ends of the chromosomes such as that which is assumed to maintain the terminal chiasma. Probably the sex chromosomes which pair in this way are homologous for too short a length to permit the formation of a chiasma if they were to pair at pachytene, but retain minute homologous segments that are sufficient to exert a specific terminal affinity.

Conclusion. Terminalisation has now enabled us to distinguish three forces acting within the resting nucleus. The two repulsions are non-specific. One is generalised and probably due to a surface charge evenly distributed over all the parts of the chromosomes. The other is localised and is probably due to a specially high charge, concentrated on the centromere and developing after chiasma formation. The different balance between these two forces of body repulsion and centromere repulsion is responsible for the forms of bivalents with different degrees of terminalisation.

The third force is an attraction which is specific to the parts of the chromosomes and is shown only by their ends. We now have to consider how these forces, inferred from the relatively simple system of the prophase nucleus, act when the chromosomes are brought into a specially differentiated substrate, during the extra-nuclear stages of their life-cycle.

(iv) **Balanced Repulsions in the Spindle.** (*a*) *Introduction.* The metaphase and anaphase movements of the chromosomes are more complicated than the prophase movements, for an obvious reason. During the prophase the chromosomes are free from outside interference. Their movements show what they can do by themselves.

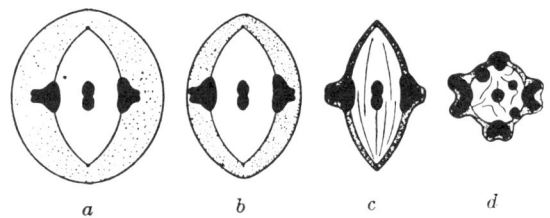

a b c d

FIG. 148.—Diagram to show the effect of dehydration on the different cell constituents at metaphase of mitosis or meiosis. Chromosomes black, cytoplasm stippled, spindle clear. *a* In isotomic solution. *b, c, d* In hypertonic solution, successive stages of dehydration. *a–c* axial section ; *d* equatorial section. (From Belar, 1927.)

At metaphase they are brought into relation with another set of variable conditions, conditions bound up with the activity of the spindle. It is only therefore, with a clear understanding of the simpler movements that take place in the prophase nucleus that we can hope to consider their later activities profitably. This we are now in a position to do.

(*b*) *The Structure of the Spindle.* The special properties of the spindle as distinct from the cytoplasm may be inferred in three ways (Belar, 1929, *a* and *b*). When the cytoplasm is shrunk by dehydration of living cells in a hypertonic solution, as in the course of fixation or in special experiments, the spindle shrinks less than the rest. When living cells are smeared on a slide in a single layer, as is readily done when the cells are large, they lie with their spindles flat; since the cells are globular the spindles must have opposed special resistance to pressure. Finally the spindle is seen to be more refractive to light than the rest of the cytoplasm.

The same tests show the relative rigidity of different cell organs. The nuclear ground substance at late prophase is less rigid than the spindle; the chromosomes at late prophase and metaphase are more rigid.

The special rigidity of the spindle is of a peculiar kind. The shrinkage when the spindle loses water is greater crosswise than axially. It may thus be caused to split axially either in numerous small fissures or in a few broad clefts. These, especially when stained, give the spindle a striated appearance, and they have been described as the "spindle-fibres." What is their significance? Evidently the water content of the spindle is not evenly distributed but lies in axial channels between the more rigid portions of the spindle. This is shown equally by the special axial expansion of the spindle at a later stage of mitosis. There are therefore in the spindle axial structures of high rigidity which may be described as fibres, although the "fibres" that are seen are merely the channels between them. Their structure is not inherent in their materials like that of the fibres of connective tissue; it must be determined by special conditions external to them, as is the structure of fibrillæ or cilia in the Protozoa or the arrangement of iron filings in a magnetic field. This has been shown most clearly by Chambers (1925); a needle passed through the spindle at metaphase does not disturb the chromosomes to which *connective* fibres might be supposed to be attached.

Another possible source of misunderstanding about spindle fibres must be removed. The threads connecting the centromeres to the bodies of the chromosomes are often stained at the first metaphase of meiosis although the centromeres themselves are not. These threads have often been illustrated and described as spindle fibres. We now see that these "fibres" are an integral part of the chromosomes and, although they depend for their arrangement on the structure of the spindle, they are not themselves a part of that structure.

Experiments showing the absence of connective fibres do not solve the problem of the molecular structure underlying the spindle. That it has a special molecular structure is shown not only by the distribution of its contained water and its reaction to fixatives but also by special experiments. Centrifuged cells show a distortion of the spindle after fixation of the kind that is to be expected of a body that is both fluid and orientated (Schrader 1934 on *Cyclops*, etc.). Moreover these observations leave no doubt on another point of some importance. The fibres or channels that are visible in certain organisms and indeed with special fixations in all organisms are not uniformly distributed in the spindle but are of two localised kinds at metaphase, *viz.* those " interzonal fibres " joining, or rather lying between, the two poles, and those joining each pole with the several centromeres of the chromosomes (F. Schrader, 1932 ; Upcott, 1936 a). It seems likely that the differentiation is not strictly confined to these types, but is rather locally exaggerated by the poles and centromeres to give the differentiation we see.

We may conclude therefore that the differentiation of the spindle is a differentiation of its own water content. Such a differentiation could be brought about by the orientation of molecules of suitable shape such as chain molecules with which water is laterally associated. Further, this differentiation takes place under the action of centrosomes in the first place and is influenced also by the centromeres.

We next have to consider how the centrosomes and centromeres can determine the orientation of the spindle. Hardy (1899) pointed out that an internal heterogeneity such as that shown by the structure of the spindle could arise from a stress such as stretching. We shall see, however, from the repulsions inferred in the spindle that it lies in an electrical field. We know also that its heterogeneity depends on a special distribution of molecules of a high dielectric constant, molecules of water. Finally, we know that an orientation, to give a structure of a liquid crystal such as that of the spindle, is affected by an electrical field (Bragg, 1933). It seems more likely therefore from our present knowledge that the differentiation of the spindle depends on the orientation of particles having an anisotropic dielectric constant in an electric field and that this orientation takes place in such a way that the repulsions are most efficiently trans-

PLATE XIV

FIGS. 1–3.—Meiosis in the oöcytes of the earthworm *Allolobophora*. (From Foot and Strobell, 1905, smears slightly pressed before fixation.)
Fig. 1. Diakinesis.   Fig. 2. Early metaphase.   Fig. 3. Full metaphase. Note terminal and interstitial chiasmata.

FIGS. 4 AND 5.—Anaphase of the first division in *Kniphofia* × 2000. (From Moffett, 1932 *a*.)
Fig. 4. Separation of bivalent with single interstitial chiasma.
Fig. 5. Separation of bivalent with two compensating chiasmata leading to interlocking.

FIG. 6.—First metaphase in *Mecostethus grossus*, ♂. 11 bivalents visible, X chromosome off the plate. 5 long bivalents have one proximal chiasma only, 3 more have also a distal chiasma to give a ring.   × *ca.* 1200. (La Cour: 2B.D. − gentian-violet, section 30μ thick.)

FIG. 7.—Metaphase, anaphase and telophase of mitosis in the morula of a white fish (*Coregonus clupeoides*), showing the changing shape of the spindle. (By kind permission of Dr. P. C. Koller.)

PLATE XIV

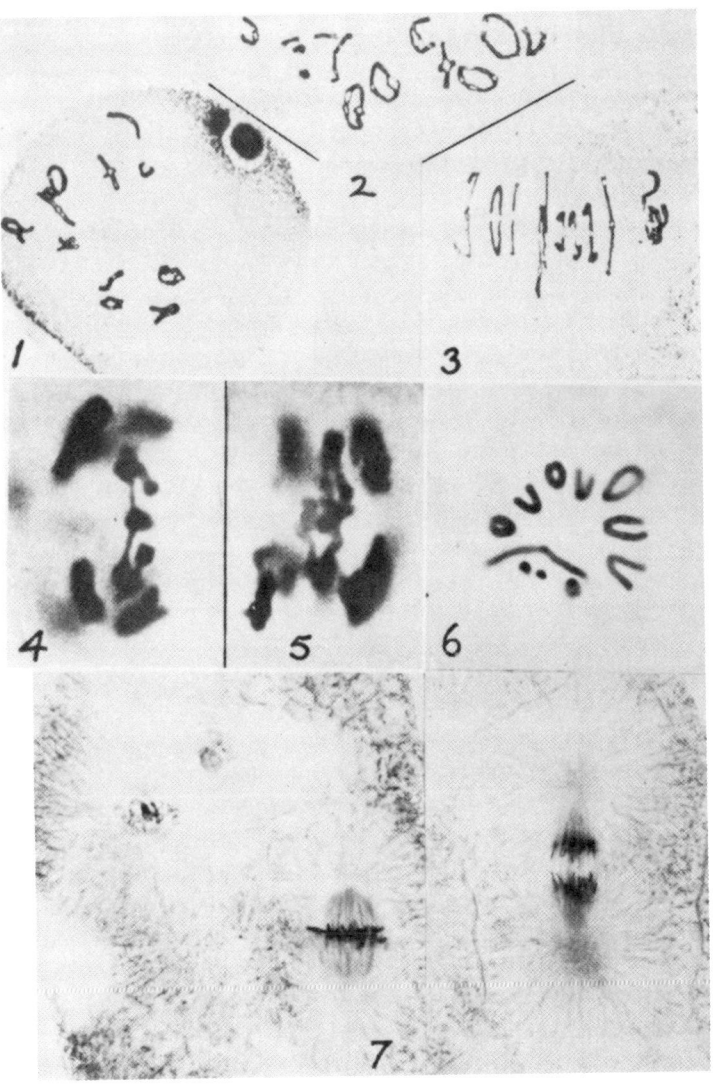

[*To face p.* 522.

mitted. The stretching of the spindle would then be the result of an internal orientation and not the cause of it.

(c) *Mitotic Metaphase.* At metaphase the chromosomes form a plate equidistant between the two poles. When, owing to the presence of extra centrosomes, there are three or four poles, a three-armed or cross-shaped plate is formed equidistant between them. When, however, two of these have arisen by division at the beginning of metaphase no plate is formed between them, showing that the development of the plate to some extent depends on the chromosomes (Kuhn, 1920; Koller, unpublished). Spindles can, however,

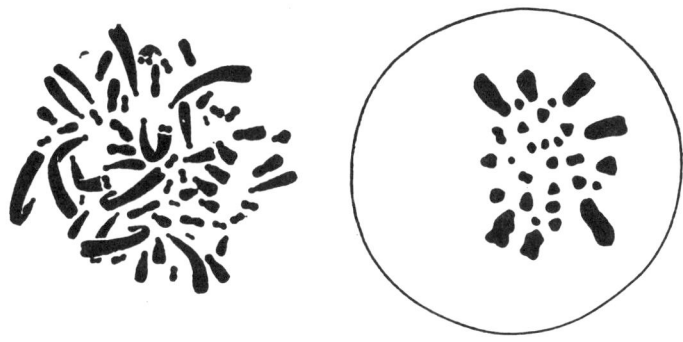

FIG. 149.—Comparison of mitotic and meiotic plates in *Hosta* ($n = 30$). Large chromosomes being rigid in meiosis are peripheral; being flexible in mitosis lie at random on the plate. (Akemine, 1935.)

be formed in fertilised egg-fragments of *Triton* without any chromosomes (Fankhauser, 1934, *a* and *b*). This indicates that the lack of spindle in the other case is due to competition for spindle-forming components, a competition in which time of action is all-important. How this comes about will be seen later.

The formation of the metaphase plate at mitosis takes place in three stages. The first is *congression*. The centromeres come to lie in a plane equidistant between the two poles of the spindle. The second is *orientation*. The centromeres come to lie so that the chromatids on either side of them lie in the axis of the spindle; the centromeres themselves must therefore be polarised and orientated.

That it is the centromere which is concerned in these movements has long been clear from the fact that it is only at this point that the chromosome lies on the spindle or is " attached " to it, as the phrase used to be. This view has now been verified by the study of acentric chromatids produced by crossing-over in inversions and by X-ray breakage. Such chromatids are entirely passive. Congression and orientation, together with the anaphase separation of chromatids and the movement of terminalisation, may therefore be defined as *centric reactions* of the chromosomes.

The third process is *distribution*. The centromeres come to lie so that the bodies of their chromosomes are more or less evenly distributed on the plate. It consists therefore in a modification of the centric reaction by the body repulsion of the chromosomes.

These three processes take place as a rule at the same time. The reason for distinguishing between them is that each may fail or vary independently of the other two and for special reasons. There is no evidence of congression and orientation varying at mitosis except for small chromosomes, which may conduct their movements out of step with the large ones of the same complement—either sooner or later (Upcott, unpublished).

Distribution, on the other hand, is subject to three natural sources of variation at mitosis. The first is the existence of the secondary attractions which modify the simple equilibrium position produced by equal repulsion. The second is the generalised body-repulsion considered at prophase. This leads to a characteristic difference in the distribution of large and small chromosomes. Necessarily, since only the centromeres regularly lie on the spindle, the widest distribution of the chromosomes is attained when the longer chromosomes lie on the periphery. This always happens when the bodies of the chromosomes are held in the equatorial plane by the orientation of their centromeres and by their own rigidity. This condition applies to meiotic bivalents with several chiasmata. It does not apply however to chromosomes that bend freely and may turn perpendicular to the plate, as long thin chromosomes always do at mitosis. Hence the contrast in the arrangement of the chromosomes of *Hosta* at mitosis and meiosis (Akemine, 1935; Fig. 149, *cf.* Fyfe, 1936).

The third source of variation is that which determines the hollow spindle in many animals, where all the chromosomes lie round the edge of the spindle instead of being evenly distributed on the plate. This difference has hitherto been represented as associated with characteristic developmental properties, such as the enfolding of the spindle by the prophase nucleus (*cf.* Belar, 1926). But the arrangement on the metaphase plate must be due to repulsion from the poles acting on the centromeres; it follows that a distribution round the edge of the plate will result from a stronger polar repulsion. This view is borne out by the discovery that a more or less peripheral

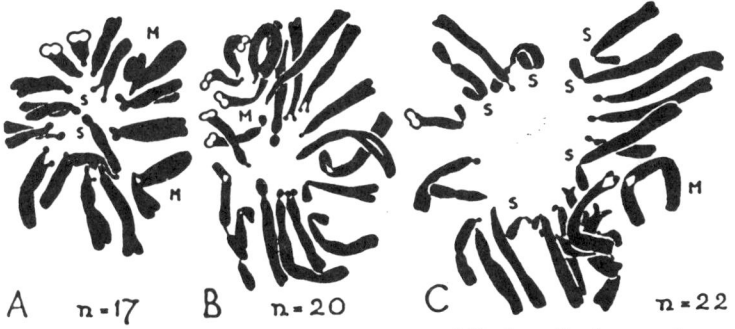

FIG. 150.—Three pollen grain mitoses in *Fritillaria pudica* ($3x = 39$). The increasingly hollow spindle is correlated with decreasingly spiralised chromosomes and decreasingly efficient orientation of centromeres. × 1,500 (D., 1935).

distribution may occur in certain pollen-grain mitoses of plants associated with other abnormalities of spindle-relationship (D., 1936, *Fritillaria*, Fig. 150; Upcott, unpub., *Tulipa*). Furthermore, both in normal and exceptional circumstances every gradation is found between the two extremes.

(*d*) *Meiotic Metaphase.* At the first metaphase of meiosis the behaviour of the chromosomes is like that at mitosis in some ways, very unlike it in others. Congression, orientation and distribution follow similar rules, but the agent of these changes is different. It is not a polarised, potentially double centromere, but two separate centromeres, which determine the movements of each bivalent. These lie on either side of the equatorial plane, somewhat further apart than in the prophase nucleus as a rule. Congression may be hindered by the body repulsions of the other chromosomes on a crowded plate, and in this case one bivalent may form an *accessory plate* half-way between the primary plates and a pole (D., 1936 c). The sex chromosomes are particularly liable to behave in this way (Koller and D., 1934). Orientation, as well as even distribution, may fail with interlocking (Dark, 1936, *Pæonia*).

Orientation is also liable to fail where the centromeres are exceptionally far apart. For example, in a hybrid where two long chromosomes have formed a single chiasma, instead of three or four, the

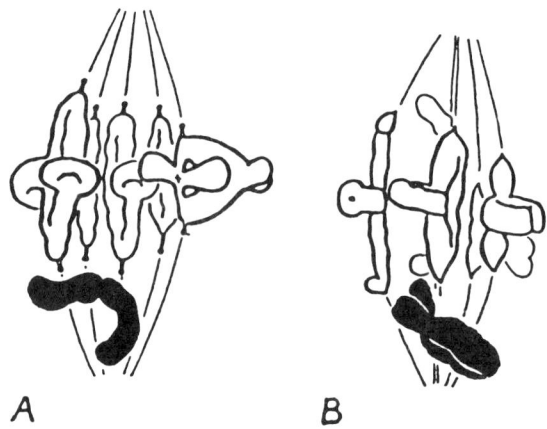

FIG. 151.—Non-congression and non-orientation of bivalents in *Podophyllum versipelle* ($n = 6$). × 1,800 (D., 1936 b).

centromeres may be five times their usual distance apart, further apart indeed than the centromeres of different bivalents. They then fail to orientate and also fail to show any effective repulsion even such as bivalents show at diakinesis (Richardson, 1936, *Lilium*; D., 1936 b, *Podophyllum*). This type of observation shows that repulsion is the agent of orientation. The converse is also true; bivalents that have not become orientated and continue to lie crosswise in the spindle fail to show tension between their centromeres.

Such repulsion as occurs crosswise in the spindle is less than that shown at diakinesis. This is true equally of the centromeres

ORIENTATION

of unorientated bivalents and of those of different bivalents. It indicates that the structure of the spindle which enhances repulsions in one direction diminishes them in another. Hence the lengthwise orientation of the contained water which is associated with the differential lengthwise repulsion must be responsible for this differential repulsion. But it is significant in two other respects. It enables us to understand why the chromosomes are bunched together in the short interval between diakinesis and the metaphase orientation (" pro-metaphase," Lawrence, 1931). Evidently the invasion of the nucleus by the spindle reduces repulsions until orientation has occurred. The second respect is more important. The reduction of repulsions does not imply a reduction in specific attractions, since the two are physically unrelated. The equilibrium position between repulsions and secondary attractions will therefore be influenced more strongly by the attractions crosswise in the spindle than at any time, such as diakinesis, where there is no spindle. Hence secondary pairing of chromosomes is found at metaphase in organisms which do not show it at prophase (*cf.* Matsuura, 1935).

Orientation must consist in the centromeres lying in such a direction as to exercise maximal repulsion on one another. This repulsion is shown by the pairs of centromeres only in meiotic chromosomes, for where mitotic chromosomes are united by pseudochiasmata as a result of X-ray breakage their centromeres show no relative orientation (White, 1935).

As we saw, owing to their body-repulsions, the bodies of long chromosomes whose centromeres are lying on the periphery of the spindle themselves lie outside it in the cytoplasm. Further, where the parts of the chromosomes between the centromeres are under very little tension, the pairs of centromeres lie equidistant, in two parallel planes (Fig. 152, Plate XV). This represents the simplest equilibrium position between the repulsions of the two poles and the two centromeres. When, however, the parts of the chromosomes between the centromeres are very short they are drawn out under tension. The equilibrium position is modified and the pair of centromeres lie a little closer together. This happens necessarily whenever a chiasma is formed in a very short arm of a chromosome and the connection between the two chromosomes may become so fine as to be invisible. Such cases have sometimes been misinterpreted as due to failure of pairing or precocious separation of a particular pair (D., 1936 d, on *Chorthippus*, cf. Belar, 1929 a, on *Stenobothrus*, Larter, 1932, on *Ranunculus*, D., 1933, on *Secale*; cf. also " distance conjugation " in the Hepaticæ).

The property of the centromeres of bivalents which leads to their relative orientation is revealed by the behaviour of univalents. It might be supposed that univalents at meiosis would behave like

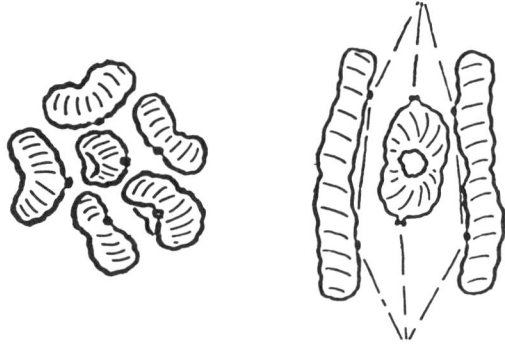

FIG. 152.—Positions of the centromere in *Tradescantia* (2x) in relation to the spindle. Rod chromosomes are on the periphery and lie outside their centromeres owing to body-repulsions. All pairs of centromeres are equidistant. × 2,400.

chromosomes at mitosis, but they do not do so until late metaphase or the beginning of anaphase. They are delayed in their reaction to the spindle, as we should expect from the precocity of the prophase of meiosis. And, as we should expect also, the delay affects all their centric reactions—congression, orientation and division. The question, then, is what change the centromeres undergo between the time when they can be orientated in pairs but not singly (as in bivalents and, at early meiotic metaphase, in univalents), and the time when they can be orientated in pairs but cannot be orientated singly (as in pairs of chromosomes with pseudo-chiasmata at mitotic metaphases and in univalents at early first meiotic anaphase). Clearly it is a change which permits one body to behave like two, and we must suppose it is the initiation of division, the mode of which will be considered later.

If, then, the centromeres of a bivalent at metaphase are in the same condition as those of a meiotic chromosome at anaphase why, it may be asked, do they linger in equilibrium on either side of the plate instead of passing at once to the poles? This question can be answered from the arrangements of multiple configurations on the plate. It is a matter of indifference for their movements on the spindle whether such configurations are the result of polyploidy

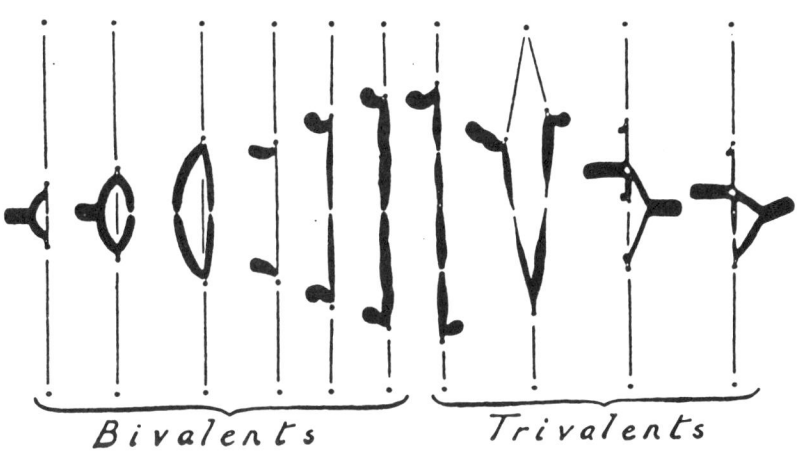

FIG. 153.—Equilibrium positions of centromeres of bivalents and trivalents with different chiasma positions in relation to the two poles. (D., 1935.)

or structural hybridity. They arrange themselves according to five different systems: (i) linear, (ii) convergent, (iii) parallel, (iv) discordant, (v) indifferent (Fig. 153).

The linear arrangement is usual where the centromeres are close together as in small chromosomes or with chiasmata very close to the centromere. Again we see that there is a limit to the distance apart at which orientated centromeres can lie, and this distance is less than the length of the spindle.

The convergent system is the characteristic " disjunctional " arrangement in rings of numerous chromosomes, but it is also the most frequent in many trivalents and quadrivalents. Now, if we consider the two centromeres of a bivalent as lying in a more or less predetermined axis, the three centromeres of a convergent, and therefore biaxial, trivalent must be regarded as having modified one or both of its axes. In other words, the centromeres modify the action of the spindle, and since this action depends on structure, they must be supposed to modify its structure. The same modification is sometimes seen with interlocked bivalents. But this modification is limited to a twist of 10° or 15°, as is shown by the discordant type of arrangement. Here the orientation of two members of a quadrivalent succeeds at the expense of the other two, which owing to the rigidity of the interstitial chiasmata in the configuration are prevented from entering into any axial relationship. The same rigidity is probably responsible for the indifferent type, where an odd member of a trivalent lies indifferently with regard to its partners.

The effect of the centromeres on the spindle may also be shown by what happens in their absence. Where spindles are formed without any chromosomes they are narrower than the normal (Fankhauser, 1934, *a* and *b*). Where all the chromosomes are univalent and orientation and congression fail, the spindle gradually stretches and becomes longer than it would be at any stage in normal development. The cumulative effect of the polar repulsions in orientating the spindle, which itself increases the effectiveness of the repulsions, lengthens the spindle so that it is bent round on itself in a confined cell (Matsumoto, 1933, and Katayama, 1935, on *Triticum*; Dobzhansky, 1934 on *Drosophila*; Morinaga and Fukushima, 1935, on *Oryza*; Brieger, 1934, on *Nicotiana*; Bergner *et al.*, 1934, on *Datura*).

The effect of the chromosomes, therefore, is to broaden the spindle when they come on to the metaphase plate. This effect may be inferred in another way, from the behaviour of univalents in *Triticum* and *Æsculus* (Kihara, 1929; Upcott, 1936 *b*). They come on to the plate after the bivalents and form a ring outside them. If they had not widened the plate they could not have found room on it. While the spindle is therefore entirely independent of the chromosomes in its origin it depends on a centric reaction for the changes in shape and structure that it undergoes after it comes into relation with them.

(e) *Anaphase*. The similarity of the centromeres and the centrosomes or spindle poles in their action on the spindle and their reaction in orientation suggests that in the metaphase of meiosis we are dealing with an equilibrium in repulsions between four bodies, two poles and two centromeres, instead of between three as at mitosis. As we have seen, this equilibrium position is such that bivalents with a single terminal chiasma in a short arm appear already separated at metaphase. The change from metaphase to anaphase should then be different in meiosis from that in mitosis.

In mitosis it seems to depend directly on the division of the centromere. That this division does indeed take place may be inferred, as we have seen, from the behaviour of diplochromosomes and from the lack of synchronisation in division of unorientated small chromosomes in *Tulipa galatica* (Upcott, unpublished). The two daughter centromeres at mitosis repel one another like the two partner centromeres at meiosis during terminalisation. In meiosis, on the other hand, no change such as division occurs at the beginning of anaphase, and indeed the distinction between metaphase and anaphase is more difficult to make when terminalisation is complete and the chromosomes are short (as in *Œnothera*). The centromeres merely move nearer to their poles. The equilibrium position shifts, but there is no sudden change of character. Where numerous interstitial chiasmata remain to be disentangled, there is a lapse of attraction between chromatids, and it seems likely that this plays some part in the change of equilibrium. The important agent in anaphase movement both at meiosis (and one which no doubt plays a part also at mitosis) seems to be the waning of the spindle pole repulsion, for at late anaphase the chromosomes always approach close to a pole from which they were repelled at metaphase.

A secondary agent of movement depends on the centromeres having already moved. The part of the spindle through which they have passed changes its shape. It narrows and stretches, and in doing so may be seen in living cells to push the chromosomes further apart (Belar, 1929 *a* and *b*). The stretching of the spindle was first clearly seen by Kuhn (1920) in *Vahlkampfia*, where T-shaped tripolar spindles became arrow-shaped at anaphase. It may be inferred regularly from the behaviour of anaphase chromosomes in pollen-grain mitoses (Geitler, 1935; D., 1929 c, 1936 b). The cell-wall stops one group of chromosomes from moving further in mid-anaphase, but the stretching of the spindle continues to push the other group, which is still seen in the characteristic attitude of movement. This anaphase change in the spindle is not surprising in view of the modifying action of the centromeres on the spindle at

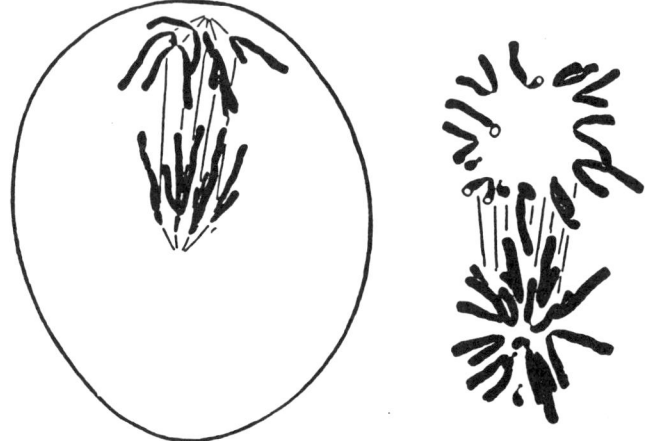

FIG. 154.—Pollen grain divisions showing asymmetrical anaphases caused by stretching of the spindle, one end of which abuts on the wall. (Left) *Podophyllum* (D., 1936 b). (Right) *Tradescantia*, $n = 11 + 2\,ff$. (D., 1929 c.) × 1,600.

metaphase, for they have converted the *centrosome-spindle* between two points into the *centromere-spindle* between two plates (Fig. 154).

The stretching of the spindle between the separating chromosomes at anaphase is clearly analogous to its stretching at metaphase in the absence of orientated centromeres. It seems that cumulative stretching in the axis of repulsion is inherent in the orientation of spindle particles under the influence of that repulsion. If the orientation depends on the gradual concatenation of long molecules, this property is inevitable, for the chains can grow only in length.

It has been supposed since the early hypotheses of van Beneden that a sudden change of charge at the poles determined a sudden change from repulsion at metaphase to attraction at anaphase (Kuwada, 1928). Such an assumption is difficult *a priori* when applied to an ideal meiosis, but is altogether impossible when applied to cases of non-congression, non-pairing and non-orientation at meiosis, and its application to mitosis has never even been attempted.

The fact that the later anaphase movement is not due to the action of forces at a distance, but to the spindle stretching, has some bearing on this old theory that the chromosomes move primarily by attraction to the poles and not by repulsion from one another. If this theory were valid the forces should, on all analogy, increase in effect as the chromosomes approach the poles; evidently they diminish, as they should if they are repulsions between the chromosomes which are moving apart.

With the present hypothesis the continuous action of the centres of repulsion on one another and on the spindle, having a cumulative effect in increasing the efficiency of those repulsions, is capable of explaining the series of movements that make up metaphase and anaphase in terms of changing equilibria. In order to see exactly how these equilibria will change we must examine in more detail the organs of movement, the centrosomes and centromeres.

(v) **The Organs of Movement.** (*a*) *The Centrosomes.* The centrosome is a body associated in its cycle of division and in its apparent function with the nucleus. During the resting stage it lies inside the micro-nuclei of ciliates, but elsewhere as a rule just outside the nucleus. In *Aggregata* it probably enters the nucleus at telophase (Belar, 1926). It can be found in every stage of the life-history in many protozoa and higher animals; but even in these it fluctuates in different tissues and different stages of its cycle both in its staining capacity and in its apparent influence on the cytoplasm around it (*e.g., Piscicola*, Geitler, 1934). In the resting stage it is most difficult to see, and its continuity, like that of the chromosomes themselves, is best shown in rapid divisions such as the cleavage mitoses in Echinoderms. In many organisms it is visible only at certain stages of development; in mosses, ferns and Cycads, for example, it is found only in the cells forming spermatozoids (*cf.*

Sharp, 1920). In *Drosophila* it is probably, as a rule, too small to see (Dobzhansky, 1934). It may be represented by a diffuse " centrosphere " or by a compact granule at the limit of visibility (0.2 μ in diameter). In the polychæte *Ophryotrocha* the daughter centrosome on the periphery of the cell disappears in the first division of the oöcyte, while the other remains visible and active. In the Protista the centrosome is often associated with other permanent cell-organs, whose exact relationship with those found in the higher organisms is difficult to make out (Cleveland, 1934 and 1935). Wherever centrosomes are found, with rare exceptions, they control the formation of the spindle. They divide and during the prophase the halves separate to opposite sides of the nucleus; they develop a radial structure in the cytoplasm around them and this structure extends to form a spindle between them.

Continuity in the inheritance and function of the centrosome has been shown experimentally in many ways. It seems to be brought into the egg by the sperm in animals, the egg losing its own centrosome. In the parthenogenetic *Rhabditis* the sperm introduces into the egg its own centrosome, which is invisible at every other stage, although no fusion of nuclei, no true fertilisation, takes place. Where several sperms can be induced to enter an echinoderm egg their several centrosomes develop multipolar spindles (Boveri, 1907 *et al.*). Similar spindles arise in cells whose centrosomes have divided but whose nuclear division has been suppressed by etherisation (Wilson, 1928, p. 175), in fertilised egg-fragments of *Triton*, and in anomalous mitoses in *Vahlkampfia* (*cf.* Table 74). In these organisms, therefore, the centrosome is a self-propagating body controlling the formation of the spindles.

In the higher plants it is doubtful whether centrosomes ever occur. They have been described by Guignard, Feng (1934) and others in the germ cells, but Geitler (1934) has failed to confirm their presence. The failure to find them may, as the studies elsewhere show, merely indicate their smaller size or failure to react to the system of fixation and staining used. Moreover, the more cylindrical shape of the spindle in higher plants does not argue against single pole-determinants, for such spindles are found with centrosomes in Gregarina and Oligochæta (Belar, 1928, p. 255). On the other hand,

TABLE 74

Abnormal Spindles

Organism	Author	Stage	Treatment or Condition
TRIPOLAR.			
Vicia Faba	Sakamura, 1920.	p.m.c.	Chloral hydrate.
Epilobium	Michaelis, 1926.	p.m.c.	Cold.
Melandrium	Heitz, 1925.	p.m.c.	Natural.
Psilotum nudum	Okabe, 1929.	Spore m.c.	Usual habit of $4x$ race.
TRI- AND QUADRIPOLAR.			
Echinoderms	Boveri, 1907 et al.	Cleavage mitosis.	Induced dispermy.
Triton	Fankhauser, 1934 a.	Cleavage mitosis.	Fertilised egg fragments (centrosomes divide out of step).
Anas-Cairina	Crew and Koller, 1936.	Exceptional double nuclei at meiosis.	Abnormal hybrid.
Drosophila pseudo-obscura.	Dobzhansky, 1934.	Polymitosis.	Abnormal hybrid.
UNIPOLAR.			
Sciara	Metz, 1933.	M I spermatocyte.	Normal habit (no pairing).
Urechis	Belar and Huth, 1933.	First cleavage.	Not stated.
Anas-Cairina hybrid.	Crew and Koller, 1936.	Meiosis.	Exceptional.
SPINDLES WITHOUT NUCLEI (and vice versâ).			
Triton	Fankhauser, 1934 a and b.	Cleavage divisions.	Fertilised egg fragments.

in the protozoan *Acanthocystis* bodies are present, having the appearance of centrosomes, but no evident connection with spindle formation (Belar, 1926 b). It therefore seems that the function of the centrosomes in controlling spindle-formation may be transferred to some other cell-organ, perhaps to an organ which lacks the property of self-propagation that is characteristic of the centrosome. How this may happen will be seen later.

(b) *The Centromeres.* The existence of a characteristic body forming part of the chromosome at its " spindle attachment " has long been recognised. Owing to its variable appearance and interpretation, different names have been given to it by different workers (Table 75). It has been identified at metaphase of mitosis in plants, at pachytene in animals and later in plants, at first metaphase of meiosis in plants (Minouchi cit., Kuwada, 1928 ; D., loc. cit., and 1936 d). Special conditions of fixation and staining are necessary to show it, and probably it is beyond the limit of visibility in small chromosomes. In the largest chromosomes it appears to be 0·2 μ in diameter with a chrom-acetic fixation, and larger with acetocarmine. With the first method its differential staining is

TABLE 75

Names Applied to the Centromere and the Centric Constriction

Name.	Stage.	Organism.	Author.
Substantial.			
Leitkörperchen	Anaphase of mitosis.	*Salamandra.* *Galtonia.* *Crepis.*	Metzner, 1894. S. Nawaschin, 1912. Trankowsky, 1931.
Polar Granule	Pachytene.	*Phrynotettix.*	Pinney, 1908. Wenrich, 1916.
	Metaphase I.	*Pungitius.*	Makino, 1934.
Granule proximale	Pachytene, anaphase I.	*Chorthippus* *Mecostethus*	Janssens, 1924 (*cf.* Belar, 1928).
Attachment Chromomere.	Pachytene, anaphase I.	*Zea Mays* *Agapanthus*	Darlington, 1933 *b*, 1934 *b*.
Kinetochore	Anaphase of mitosis.	*Trillium.*	Sharp, 1934.
Centromere	Mitosis and anaphase II.	*Chorthippus.*	Darlington, 1936 *b*, *d, e*.
Spatial.			
Point of Fiber Attachment.	Metaphase.	*Orthoptera.*	McClung, 1914.
Insertions-stelle der Zugfaseror.	Metaphase I.	*Stenobothrus.*	Belar, 1929 *a*.
Insertionslücke			
Spindle Fiber Insertion.	Moment of segregation (genetical).	*Drosophila.*	Bridges, 1927. Kaufmann, 1933. Dobzhansky, 1934.
Insertion region	Pachytene.	*Zea Mays.*	McClintock, 1933.
Spindle-fiber Attachment Constriction.	Mitosis.	*Gasteria.*	Taylor, 1924.

somewhat variable, the whole adjoining region of the chromosome may share its properties, and in some cells in *Tradescantia*, following acid vapour pre-treatment, it has appeared as a double particle transversely orientated at first metaphase (D., unpub.). This doubleness, however, being inconsistent, is probably a non-characteristic artefact.

Since the centromeres that have been seen are located at characteristic and permanent positions in the chromosome we may expect them to have the permanence of other chromosome parts. This has been shown most simply by the results of crossing-over in inversions, and of the breakages of chromosomes by X-rays already described (Ch. X). They make it clear that each centromere continues to function even when more than one lies in a particular chromatid, and that when a chromatid arises with no centromere it ceases to make any movement in response to the spindle. Centromeres do not arise *de novo* (Navashin, 1932 ; Mather and Stone, 1933). Wherever they are found they control the spindle movements of the chromosomes and the special prophase movement of terminalisation. We may therefore infer that bodies analogous to the centromeres are present wherever their functions are fulfilled and the centric constrictions that they determine are observed. The failure to see them in the smallest chromosomes is to be expected if they are proportionate in size to the bodies of the chromosomes, for in the largest they are near to the limit of visibility.

When we apply this conclusion generally we find that the chromosomes of some organisms have a less clearly defined relation to the spindle, which may imply an anomalous character in their centromeres. In *Ascaris*, as we have seen, the apparently diffuse character of the anaphase repulsion is due to the chromosome being *polycentric*. The possibility of doubleness in the centromere suggested by direct observation (D., 1933) and inferred from the results of X-ray breakage (McClintock, 1933 *b*), seems to be incompatible with its mechanical individuality. In other organisms the anomaly is not clear (*e.g.*, in the Coccidœ, F. Schrader, 1932).

The question of the genetic relationship of centromeres is important in considering their mechanical functions. It must not be supposed that the centromeres, since they belong to individual chromosomes, share their differentiation and hence their homology. There is no reason to doubt that they are all equivalent in structure where they are equivalent in behaviour and that they have a common origin as well as a common function (Muller and Painter, 1932, D.,

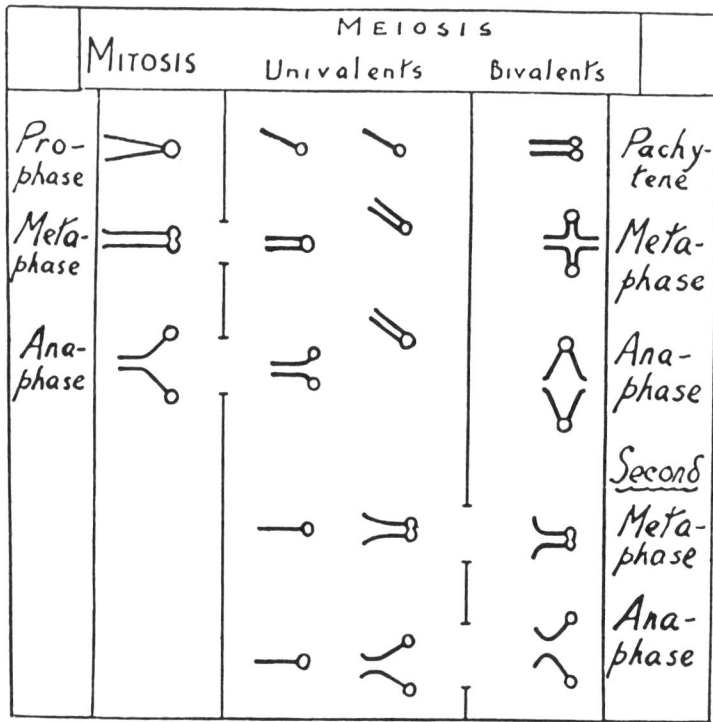

FIG. 155.—The division cycle of the centromere at mitosis and meiosis. Univalents at meiosis vary between the mitotic and meiotic cycles. (D., 1935 g.)

1933, *Agapanthus*). The differences in their movement at anaphase and in the condensation of special chromosomes at prophase may, however, depend on differences in their structure (*v. infra*).

The centrosomes and chromosomes both divide between mitoses. We have already noted some evidence of the time of division of the centromere. Undoubtedly it has divided at the mitotic anaphase. Most observations agree that it is undivided at the first meiotic anaphase. Moreover, the chromatids, which are free elsewhere, are always united at the centromere, and therefore presumably by the centromere, until the second metaphase. This is clear in *Drosophila* even at the first metaphase (Dobzhansky, 1934; D., 1934). The centromere, therefore, divides at metaphase in mitosis but (owing to the precocity of the first metaphase) not until second metaphase at meiosis (D., 1932). The best test of this view is functional. If we follow its behaviour at mitosis and meiosis (*cf.* Table 76) we see that its orientation on the spindle depends, either on its relationship with another centromere, giving *co-orientation*, or on an internal change which the centromeres of univalents undergo to permit their *auto-orientation* between metaphase and anaphase. Moreover, the mitotic chromosomes at metaphase are no longer capable of co-orientation. When they are united by pseudo-chiasmata they continue to show auto-orientation instead of co-orientation (White, 1935).

Since co-orientation depends on non-division and pairing, and auto-orientation always goes with division, the conclusion follows, as we have seen, that the internal change permitting orientation is a preparation for division which is functionally equivalent to it. This pre-division may be described as polarisation. The centromere presumably becomes hour-glass shaped. The fact that orientation does not occur until anaphase in univalents at meiosis shows that polarisation itself does not occur before. Nor can it be supposed that the separation of centromeres is determined by the association of the chromatids in pairs, for, as we have seen in the diplochromosomes, produced by arrest of mitosis, the chromatids can be associated in fours by the still undivided centromeres. On the other hand, the normally inert centromere of a daughter univalent may behave in an exceptional and remarkable way: it may divide transversely to give two fragments at second anaphase (Nishiyama, 1933 *b*; Philp, unpub., on *Avena*). This is significant in indicating different internal organisations and different methods of reproduction in the centromere and in the rest of the chromosome thread. The centromere, therefore, is not a gene, *sensu stricto*.

These observations go to show that the centromeres resemble the centrosomes in their permanence, in their dimensions (so far as these are constantly recognisable), in their cyclical staining properties, and in their correlated division and repulsion cycles. They differ from the centrosomes, however, in the timing of their division and in its effects. How this comes about we can discover from their relations with the spindle.

TABLE 76

Relationship of the State of the Centromere to the Movements of the Chromosomes

State.	Occurrence.	Behaviour.
1. Unpolarised. *(a) Inert* (no centric reaction).	Univalent at first metaphase. Daughter univalent at second metaphase.	Lies on the spindle, but has no other relationship with it. No division. No movement except from body repulsions. May pass to polarised state at first anaphase.
(b) Co-orientated.	Configurations of chromosomes united by chiasmata at first metaphase.	Two or more are orientated with respect to one another and to the spindle and move in accordance with this orientation at anaphase. No division.
2. Polarised and auto-orientated.	Mitotic or second metaphase chromosome or univalent at first anaphase.	Characteristic orientation and congression on equatorial plate. Always precedes division. Incapable of co-orientation, even when chromosomes are united by pseudo-chiasmata at mitosis.

We have already seen how they modify the shape of the spindle at metaphase and anaphase, acting on it in a similar way to the centrosomes. Indirect evidence of this action may be obtained from irregularities of meiosis. In many plants the spindle has been found to control in some way the formation of the wall between the daughter-cells formed at mitosis. Its changes of shape can be seen in the living cell to follow the growth of the partition (Belar, 1929 *b*). It is found that in such plants when pairing has entirely failed a cell-wall or cell-partition may be formed between a pole of the spindle on one side and the whole body of the chromosomes on

PLATE XV

THE CENTROMERE AT MEIOSIS

FIG. 1.—Pachytene : *Agapanthus umbellatus* (9 and 2 o'clock). Navashin-brazilin preparation given to the author by the late Dr. Belling. × 2700.

FIG. 2.—Diplotene : *Fritillaria Elwesii*. Chiasma on each side of the centromeres of an *M* bivalent (9 o'clock). Flemming — gentian-violet. × 1600.

FIG. 3.—First anaphase in *Alstrœmeria* dividing univalent lagging on the equator shows its two centromeres. (La Cour, unpublished, aceto-carmine fixation.)

FIG. 4.—First metaphase bivalent of *Galtonia candicans* with two terminal chiasmata. (La Cour : 2B.E. and gentian-violet.)

FIGS. 5 AND 6.—First metaphase in *Tradescantia bracteata* pre-treated with nitric acid. Flemming — gentian-violet. Polar view shows five of the six centromeres at one level, side view two pairs and one of a third pair. (La Cour, unpublished, Fig. 152.)

PLATE XV

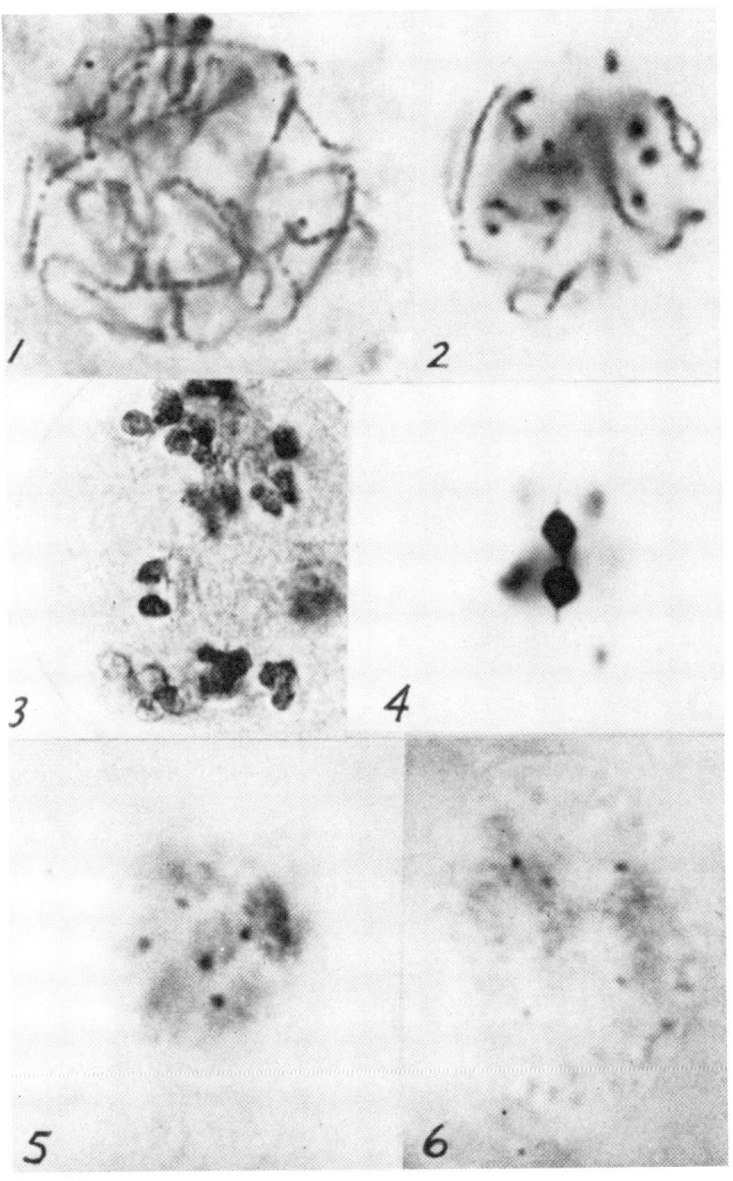

[To face p. 540.

the other, as in male Hymenoptera (Katayama, 1935, in haploid *Triticum monococcum*, and Fig. 127). Similarly it often happens that a lagging chromosome at first anaphase has a cell-wall formed around it within which it forms a nucleus (D., 1929 b).

When, however, an acentric fragment is left on the plate it never forms a cell-wall. Nor do chromatid bridges affect the formation of a cell-wall, which merely cuts through them (D., 1929 b). It therefore seems that, while the bodies of the chromosomes are relatively negligible, centrosomes, spindle poles and centromeres are of equivalent importance in determining cell-wall formation and hence, we must infer, in their action on the spindle.

(c) *The Spindle in Movement*. The uniformity that we have seen in the structure and function of the spindle in no way corresponds to a uniformity in its method of origin. In this respect it is extremely variable and we have to find out how far its variations are mechanically significant. In the first place there is the difference between those spindles which develop between two daughter centrosomes while they are still close together and those which arise from two poles on opposite sides of the nucleus. The first characteristically gives a central spindle, as found in *Salamandra* and *Aggregata*, where the long chromosomes lie with their centromeres on the edge of the plate and their bodies in the cytoplasm (Belar, 1926). The other method of origin characteristically gives a spindle with an even distribution of the chromosomes on the plate, as in the *Allium* type of the higher plants. Variations in this behaviour are found in organisms of both types. The early cleavage divisions in *Amblystoma* are of the *Allium* type, and sometimes, as we saw in pollen grains of *Fritillaria* and *Tulipa*, the chromosomes lie entirely on the edge of the plate. Here the abnormality goes with an abnormality in the degree of contraction of the chromosomes and in the timing of the division of the centromeres. The peripheral distribution is what would be expected with a higher repulsion from the poles, due to a higher charge or a smaller spindle, and variation in this direction is the commonest departure from the even distribution given by Meyer's floating magnets (*cf*. Kuwada, 1928). The significance of a relationship of distribution and timing abnormalities will be seen later.

A second important difference depends on the presence or absence of centrosomes. This problem has always seemed to be a difficult one, but its difficulty does not lie so much in the observation of the centrosomes as in the inference of their action. In the higher plants no centrosomes are seen, but the behaviour of the spindle is not sharply distinguishable from that in animals, in which centrosomes are seen. Abnormal treatment produces multipolar spindles and accessory nuclei, formed by lagging chromosomes at meiosis inde-

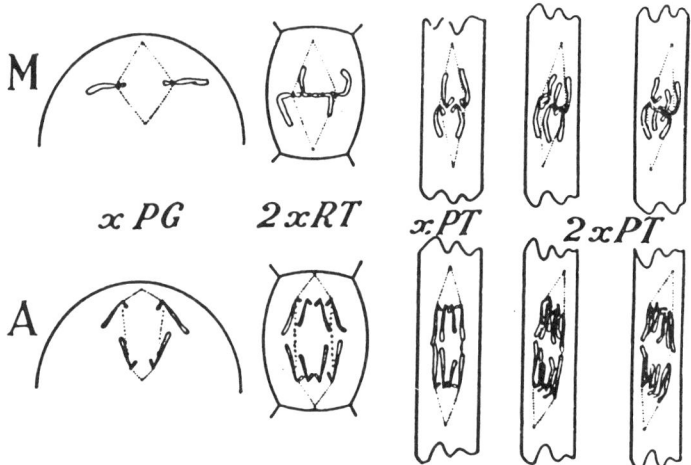

Fig. 156.—Comparison of metaphase and anaphase arrangements in pollen grains and pollen tubes of *Tulipa*, showing the effects of chromosome number, cell size and spindle size on plate formation. (Upcott, 1936.)

pendently of the spindle-poles, fail to develop a normal spindle at their second division.

The extreme opposite of the central spindle type of mitosis is found in the pollen-tubes of plants (Upcott, 1936 *b*). Here the nuclear membrane breaks down before the chromosomes are fully contracted, possibly by the premature action of the spindle. There is a considerable delay in congression on the plate, owing presumably to the confined space of deployment for the chromosomes. As in other cases of delay in congression, the spindle lengthens to an exceptional extent. The plate is restricted in width, and in polyploid species it is forced to lie oblique or it may even be buckled. The variations in position of the anaphase chromosomes correspond with those of the metaphase plate. Here we see a combination of spatial and timing abnormalities, the second required by the first if a regular division is to ensue.

The distinction, therefore, between these three types of spindle depends on timing differences and is co-ordinated with spatial limitations. The distinction, on the other hand, between these types and those which develop before any poles are established is very important. At meiosis in the Coccidæ and in some other animals each bivalent forms its own spindle orientated independently of the others, and only later is a uniform orientation produced, presumably by an external agency (F. Schrader, 1932). Already we have seen how the centromeres influence the shape of a spindle which has been developed by the centrosomes or pole-determinants. Presumably, independent spindles are determined by the centromeres coming into action before the centrosomes. In these cases and in others, in which the poles begin to act as soon as the centromeres, the pole-determinants are not located in any organised bodies, but consist presumably in a diffuse charge on opposite faces of the cell. Such a diffuse charge may act as the opposite pole in an abnormal unipolar (or monaster) spindle (*e.g.*, *Urechis*, Belar, 1933; *Triton*, Fankhauser, 1934 *a* and *b*; and *Sciara*, Metz *et al.*, 1926). Here the chromosomes range themselves in a spherical plate around the defined pole and their daughter halves are divided between it and the undefined polar surface.

In order to see the significance of the centromere spindle we must consider a fourth type of spindle formation, *viz.* the intra-nuclear spindle. In the Protista generally, in Fungi, and occasionally in the higher organisms the spindle is formed inside the nucleus. It may be controlled by definite centrosome-like bodies, as in *Monocystis*, or it may be determined by a less localised charge on opposite sides of the nucleus interacting with the centromeres as in *Cryptomonas* or *Pamphagus* (*cf.* Belar, 1926). The question then arises as to why the spindles can be formed inside the nucleus and by the centromeres in some conditions, but are never formed in this way in the greater numbers of higher organisms where the activity of the centromeres has been made clear by their effect on terminalisation at meiosis. The answer to this question seems to be provided by observations on different kinds of spindle formation in *Artemia salina* (Gross, 1935). In the cleavage divisions of the egg, spindles are developed by the centrosomes and by the centromeres, independently and simultaneously. These spindles combine and their combination produces metaphase. At meiosis on the other hand the centrosomes play no part at all in the formation of the spindle. It arises entirely within the nucleus, presumably as a centromere spindle. But before this spindle develops the nucleus contracts to a small proportion of its previous size. Now, so long as a nucleo-cytoplasmic surface persists, the difference of structure and behaviour on the two sides of it shows it to be acting as a type of semi-permeable membrane. The sudden contraction of the nucleus, therefore, probably means a loss of water and an increase in the concentration of whatever large molecules are present in the nuclear sap from which the spindle develops.

The effect of water-content on spindle-formation has also been shown in two ways experimentally. When the spindle is dehydrated it lengthens as at meiosis with non-pairing, presumably owing to an exaggeration of the polar repulsions (Belar, 1929 a). When it is hydrated the orientation of the chromosomes is upset, presumably owing to a decline in all repulsions (Wada, 1935). Thus the formation of the spindle must be supposed to depend not only on the action of repulsion centres such as centrosomes and centromeres but also on the presence of certain materials in the substrate and in a certain concentration. Also, since the changes in the repulsion centres (centrosomes and centromeres) are cyclical and independent, and their effects upon the spindle are cumulative and related, they must be co-ordinated or balanced in their time of action. *Artemia* shows that this *balance* may be achieved in different ways.

(vi) **The Balance Theory of Mitosis.** The present account has shown that mitosis takes many different forms so far as its external mechanics is concerned. The special mechanical systems at work in many of the aberrant forms, particularly of the Protista, cannot yet be analysed in detail, although they provide useful materials for comparison. The commoner systems, found in the higher

BALANCE THEORY OF MITOSIS

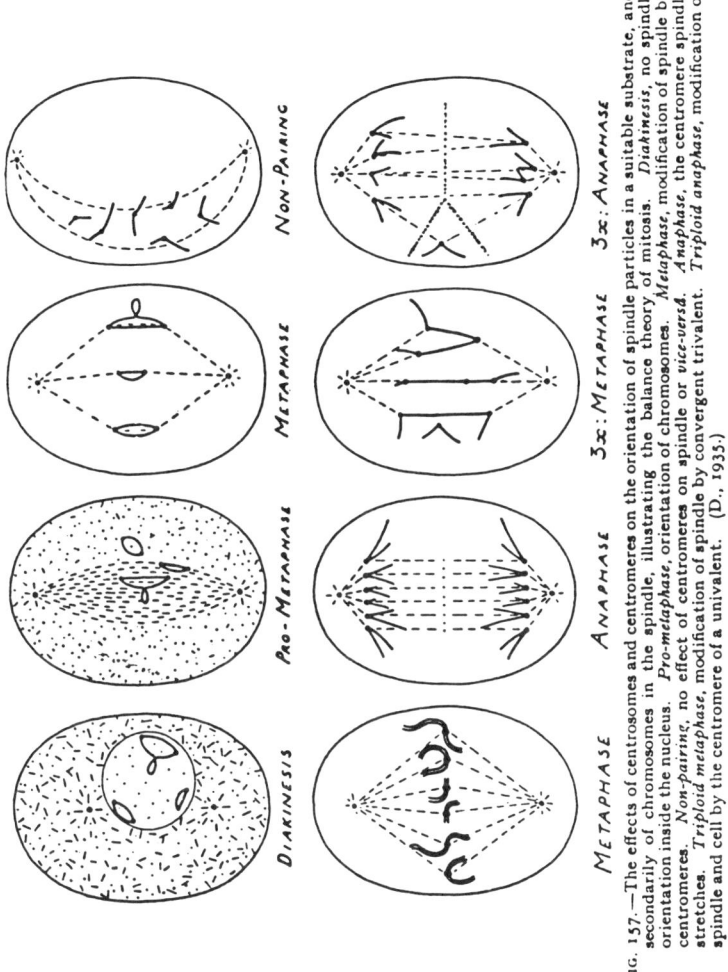

FIG. 157.—The effects of centrosomes and centromeres on the orientation of spindle particles in a suitable substrate, and secondarily of chromosomes in the spindle, illustrating the balance theory of mitosis. *Diakinesis*, no spindle orientation inside the nucleus. *Pro-metaphase*, orientation of chromosomes. *Metaphase*, modification of spindle by centromeres. *Non-pairing*, no effect of centromeres on spindle or *vice-versâ*. *Anaphase*, the centromere spindle stretches. *Triploid metaphase*, modification of spindle by convergent trivalent. *Triploid anaphase*, modification of spindle and cell by the centromere of a univalent. (D., 1935.)

organisms having large chromosomes, are, however, capable of exact comparison and accessible to experiment. They allow of a general account of the sequence of events in mitosis, which shows how the several agents concerned combine to produce a regulated or balanced result (D., 1935 g).

1. Two charged centres of repulsion or centrosomes, lying in the cytoplasm, orientate the substrate in their neighbourhood in such a way that it transmits their repulsions most efficiently. This change is produced by the orientation of long chain molecules with water associated laterally to give indirectly an orientation of water which could not be produced directly. An effective orientation therefore depends on the relative concentration of water and chain molecules. The orientated region of the cytoplasm invades the nucleus, when its surface breaks down, and constitutes the spindle of which the centrosomes are the poles.

2. The result of the orientation is to give a medium of the type of a liquid crystal, with a directionally differential dielectric constant. Hence, secondarily, the polarised centromeres of mitotic chromosomes and the paired centromeres of mitotic bivalents are held on the spindle and are orientated along the spindle-arcs where their repulsion is most efficient.

3. The centromeres bearing a similar charge to the centrosomes are forced to lie midway between them in an equatorial plate, which is an equilibrium position for two poles and one centromere at mitosis and for two poles and two centromeres at meiosis. The repulsion of the poles does not push the centromeres off the spindle, since this, in reducing the repulsion between paired or within the polarised centromeres, would increase the potential energy of the system.

4. Being themselves charged, the centromeres secondarily react with the centrosomes to modify the shape of the spindle, so that from having been a centrosome-spindle it becomes a joint centrosome-centromere spindle. The different distributions on the plate (solid or hollow) are equilibrium positions determined by the body-repulsions of different numbers of chromosomes and modified by different pole-repulsions.

5. This repulsion equilibrium is upset by the decline of the polar

repulsion, accompanied at mitosis by the division of the centromeres into daughter bodies repelling one another like those of bivalents at meiosis and therefore leading to their anaphase separation.

6. The separation of the chromosomes at anaphase is assisted by the change of shape which the part of the spindle between them undergoes owing to its lying between and being determined by, two plates instead of two poles and one plate. In this third stage it is a centromere spindle.

7. The formation of the spindle and the successful orientation and division of the chromosomes therefore depends on a balance in the strength and timing of the cycles of division and repulsion of the centrosomes and centromeres. This balance is determined in different ways in different types of mitosis. In higher organisms the most important condition is in the centromeres lying inside the nucleus while the centrosomes lie outside in a substrate with a presumably different cycle of changes in hydrogen-ion concentration. Most changes in the balance give unregulated results, such as polymitosis and non-pairing at meiosis. The one regulated change is that produced by the degree of precocity of the external cycle which has turned mitosis into meiosis.

4. ULTRA-MECHANICS

(a) *Introduction*. The mechanical hypotheses that have been discussed so far provide the clue to certain problems of chromosome movement, especially to those concerned with visible movements. In doing so they should also present us with a new means of approaching other problems of chromosome movement, those concerned with movements that are inferred from their later consequences. And these invisible movements in their turn should tell us something about the physico-chemical conditions underlying the mechanical properties of the chromosomes and other cell-components.

The first of these problems that we have to consider are the problems of chromosome breakage. They are concerned with two not unrelated aspects of breakage, viz., crossing-over and structural change. These properties, as we shall see, being universal and fundamental, are presumably inherent in the chromosome thread.

(b) *Crossing-over*. Many of those who have put forward hypotheses connecting observed structures of the chromosomes with the occurrence of crossing-over between them, have supported their views with hypotheses explaining what forces determined the crossing-over. In some cases this mechanical proposal merely added an ornament that was lacking in Janssen's original teleological construction. In others it was made the sole foundation of the theory. In all it depended on the assumption that the coiling of the chromosomes round one another, at the various stages at which it was thought that crossing-over might occur, determined its occurrence.

Now, however, we are in a different position. On the one hand we know the exact time and place of crossing-over. And on the other hand the forces acting on the chromosomes at different times can be distinguished and related according to their origins as well as according to their effects. The paired chromosomes are relationally coiled, and crossing-over between two of their four chromatids at the moment of their origin removes this coiling and determines a chiasma (D., 1935 c and d).

Crossing-over consists in a sequence of events, presumably a rapid sequence, at the end of the pachytene stage of meiosis. Two chromatids of partner chromosomes break, and they break at exactly opposite points. The breakage of one must therefore determine that of the other and precede it in time. The four ends must then move in a way that allows them to come together in a new combination. They may then rejoin to form two new chromatids. This rejoining presents no new problem, for as we have seen, ends of different chromosomes, free and unpaired, are capable of rejoining after X-ray breakage. The problem consists in the original breakages and the movement.

The position of the paired chromosomes at the end of pachytene can be exactly defined. Each particle is associated with a corresponding particle of the partner; it is undergoing division; the two threads are twisted round one another by relational coiling. The forces at work can be similarly defined. The lateral association is by a specific attraction between successive particles. The relational coiling must depend on the action of a second force apart from the specific attraction, a longitudinal cohesion. These two forces are in equilibrium, for pachytene may be indefinitely prolonged without change. Further, the attraction must be positionally specific, not permitting the threads to slip round one another. Such a specificity implies an asymmetrical molecular pattern such as might (since the time of Pasteur) be expected in the gene. The mechanical position may be imitated well enough by releasing two woollen threads that have been placed alongside one another in a state of torsion. Their internal molecular relations

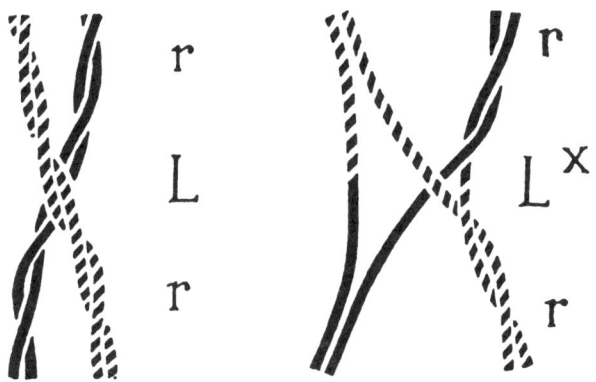

Fig. 158.—The coiling relationships of two chromosomes before (left) and after (right) crossing-over has occurred. Right-hand chromatid coiling (r) and left-hand chromosome coiling (L) and chiasma coiling (L^x). (D., 1936 d.)

N.B.—The production of the equilibrium of pachytene is the object of all spinning operations.

then determine a longitudinal cohesion. The entanglement of their fibres represents a positional specificity of attraction which prevents slipping. The two threads take up a position such that the torsion of their internal coiling is in equilibrium with the opposite torsion of their relational coiling (Fig. 158).

Such is the position when the chromosomes begin to divide. The daughter chromatids will be relationally coiled in the opposite direction to that in which the chromosomes are relationally coiled. Division is bound to upset the equilibrium in two ways. First, it abolishes the specific attraction which works only between pairs of threads and therefore only between the new chromatids. The chromosomes begin to repel one another. Secondly, it reduces the longitudinal cohesion, for the new half-threads at the moment of their origin must needs be weaker than the parent threads. Now the first change will reduce the torsion of the threads by uncoiling, but as observation has shown, uncoiling is very slow (D., 1935 c). It lags behind the operation of the forces which determine it. The second change on the other hand is immediate. The weakened chromatids are therefore exposed to an unreduced torsion and one of the four breaks under the strain.

The break of one chromatid causes another upset of equilibrium which is slight but sudden. The two broken ends will twist round their unbroken sister chromatids, thus releasing the coiling of the two which determined the breakage. But the relational coiling of the chromosomes has been in equilibrium with that of the chromatids. The release of that of one pair of chromatids therefore imposes an extra strain on the unbroken pair. Since the first pair yielded under the original strain the second pair is bound to yield under the increased strain. One of its chromatids will break at a point exactly opposite to the first break, since the strain will be greatest at this point.

The breakage of the chromatids will permit the uncoiling of their relational coiling in both chromosomes. The broken chromatids will reunite when, in the course of uncoiling, one of their ends first meets another. This will always be the end of a chromatid of a partner chromosome. Crossing-over will have occurred, and when the lapse of attractions leads to separation, a chiasma will appear.

On this view crossing-over depends on the action of forces that may be deduced from the behaviour of chromosomes at earlier and later stages of their history. The reason for its occurrence at meiosis is the division of the chromosomes at a time when they are paired and coiled in such a way that when their equilibrium is upset exceptional strains are imposed that can arise in no other way.

The cross-over chromatids cannot be distinguished from the non-cross-overs at a late diplotene stage because the chiasma has by this time become symmetrical. But immediately after the chiasma appears, the four chromatids can often be seen still lying in one plane on both sides of it. Two chromatids then appear to cross one another at the chiasma, and it has usually been thought that these middle threads were the cross-over chromatids. On the present view the reverse is the case. The middle threads still preserve the relational coiling of the chromosomes which is lost in the outer threads, and which they themselves lose just afterwards. Experiments with wool models confirm the conclusion that the outer chromatids are the cross-overs. They also show that a variable amount of coiling may be lost by each crossing-over, so that an estimate of the proportion lost in this way is not possible. It would appear that the special systems of crossing-over relationships found with and without chromatid interference may depend on the regulation of the relative transverse positions at which breakage may occur (D., 1936 *d*).

Various corollaries of this hypothesis follow : (i) Since crossing-over releases, as any movement must do, the strain that determines it, interference is inherent in the assumption of a longitudinal cohesion which will transmit this strain ; (ii) the fact that crossing-over is not reduced in proportion to the length of chromosome in triploids, although the length paired is reduced, is presumably due to the total amount of coiling tension being proportionate to the total length of the chromosomes and not to the length paired ; (iii) the occurrence of interference in crossing-over between chromosomes on the other hand is probably not due to conditions at the time at all but rather to an interference in pairing at zygotene, *i.e.*, to an average capacity for the nucleus to permit pairing before division of the chromosomes, the amount of true pairing (as opposed to post-division torsion pairing) being the variable factor limiting the frequency of crossing-over in individual chromosomes.

It is now possible to consider the special relationship of the centromere with crossing-over in the light of a new analysis of *Drosophila*. Mather (1936) finds first that there is no evidence of interference between cross-overs on opposite sides of the centromere. The centromere is therefore an interference-inhibitor. Secondly, he finds that the crossing-over in a given arm is specially distributed in relation to the centromere. The distribution is of the same kind as if the centromere were a point of constant crossing-over. The centromere is therefore an interference centre. Thirdly, he finds that the mean distance of the proximal chiasma from the centromere is a function of the length of the arm. There is therefore a proximal region with no crossing-over which no doubt when long established has become inert.

The double and contradictory action of the centromere as an interference centre and an interference-inhibitor can be regarded in the light of the present mechanical hypothesis as due to a single property, viz., that the centromere reduces or even abolishes the torsion on the adjacent chromatids at the moment of crossing-over owing to a lack of longitudinal cohesion and a free rotation of bonds. This property is implied by the failure of spiralisation at the centromere. Hence it interferes with crossing-over in its neighbourhood and at the same time prevents interference acting through it, since the strain on either side of it can be reduced no further. The action of the centromere is probably greater than that of a crossover in reducing strain, since it seems (at least in *Drosophila*) to interfere with crossing-over for a greater distance than crossing-over itself does. This means that crossing-over leaves an average unreleased strain greater than zero, a conclusion which can be deduced directly on the present hypothesis from the necessary discontinuity in uncoiling, one revolution for a pair of chromatids being the indivisible unit.

The action of the centromere implies two underlying properties. First, as Mather points out, that crossing-over should begin in its neighbourhood. Such a proximal priority has been observed cytologically in *Fritillaria* (D., 1935 *b*); it is conceivable that in other organisms the ends might enjoy priority, and crossing-over show a similar end-relationship. Secondly, the torsion-strain between the chromosomes and chromatids must be released at the centromere by the absence of longitudinal cohesion, while separation is beginning. Since there is no evidence of any plane of symmetry, or at this stage of any preparation for division, in the centromere, positional specificity must certainly be lacking.

The special length-frequency relationship found in *Stenobothrus* (Ch. VII) can be accounted for on the same assumption that the average distance of the first chiasma from the centromere is a function of the length of the chromosome-arm. We must therefore suppose that in *Drosophila* and in *Stenobothrus* the average strain is also a function of the length. This can be understood in the following way : if pairing begins near the centromere or near an end and passes along the chromosome, the strain developed will be a function of the time that the chromosomes have been paired before equilibrium is reached (*cf.* Fig. 30) because segments that pair late will partly uncoil first. Some equalisation of strain will then take place after pairing between the parts of the arm. The average strain in a long arm will thus be less than in a short arm which completed its pairing earlier.

The extent to which the torsion is released at the centromere can probably be measured more accurately from frequencies of chiasmata than of crossing-over in the progeny. Two-armed chromosomes should show a greater variance than one-armed chromosomes of the same average frequency. There is no evidence of this as yet.

There remains to be considered the effect of exceptional conditions, especially X-ray treatment, on crossing-over. Its normal occurrence in *Drosophila* is confined to the prophase of meiosis, where it affects all pairs in the female and the sex chromosomes only in the male. Exceptionally, however, it occurs somatically (Stern, 1934) or in gonial nuclei before meiosis, *e.g.*, crossing-over in the male (Muller, 1916, p. 304). This may also be inferred where groups of gametes are formed having the same exceptional crossing-over, *e.g.*, detachment of XX in $\widehat{XX}$Y females (Kaufmann, 1933; Fig. 121). Crossing-over is increased by X-ray treatment. In the male, where natural crossing-over is too rare to be traced, the induced crossing-over has been shown to take place somatically (Friesen, 1934, 1936; *cf.* Kikkawa, 1935 *b*; Patterson and Suche, 1934).

Somatic crossing-over is probably peculiar to the Diptera, being conditioned by the exceptionally strong somatic pairing of the chromosomes which even leads to relational coiling. It seems that this will not itself determine crossing-over, since the chromosomes do not divide while coiled, but if one of their chromatids is broken by X-rays at the earliest prophase the result should be the same as where a breakage occurs in meiosis ; it should impose an increased strain on the partner chromosome exactly opposite and thus lead to double breakage and crossing-over. Rare natural crossing-over would be due to rare natural breakage, which should lead to the same regular result. Where crossing-over follows natural breakage the frequency of inversion, translocation and other structural changes resulting from such breakage should be correspondingly reduced for the reasons we shall next consider.

(c) *Structural Change*. Structural changes can be classified from three points of view, according to their modes of origin, spatial relationships, and mechanical and genetical results. These have so far been treated separately. We may now see how they have to do with one another. The chromosomes with which we begin are the normal efficient chromosomes with one centromere and two ends. From these we obtain new types with no centromere or with two, and with no ends or with three or four. According to these mechanical results we can therefore classify chromosomes in the following way, giving the first seven types initial symbols (Table 77).

TABLE 77

Mechanical Types of Chromosomes

Ends	Centromeres			
	0	1	2	Many
0	Acentric Ring (a)	Monocentric Ring (c)	Dicentric Ring (d)	—
2	Acentric " Fragment " (A)	Efficient Chromosome (C)	Dicentric Chromosome (D)	Polycentric Chromosome (P)
3	—	Branched Chromosome	—	—
4	—	Diplo-Chromosome	—	—

To consider next the genetic relationships of structural change, some of these are simple and others very complex. *Deficiency*, the loss of a terminal segment, is the simplest, since it can affect only one chromosome, dividing it into a monocentric and an acentric fragment. *Deletion*, the loss of an intercalary segment of a chromosome, requires that the two breaks shall be *intra-radial*, in the same arm of the chromosome. *Inversion*, which is always so far as we know intercalary, may include the centromere or not. When it includes the centromere it is the same as internal interchange, that is, interchange within the chromosome. *Translocation* consists in the removal of an intercalary segment from one position to another. It may be intra-radial or *extra-radial*; it may be internal to the chromosome or *fraternal*, that is, between two homologues, or *external*, that is between two unrelated chromosomes; it may be eucentric or dyscentric. Further, translocation may be *symmetrical* with respect to the centromere or not, in which case the piece removed contains the centromere and so gives rise to a dicentric and an acentric product. *Interchange* is internal, fraternal or external, but it may be balanced or unbalanced and more or less unequal with regard to the sizes of segments exchanged.

These differences of spatial relationship have important genetical results, as we saw in considering structural hybridity and secondary structural change. By crossing-over all intercalary changes give reduplications and deficiencies, and all dyscentric changes give dicentric and acentric products, and these, whether arising in this way or directly, cause breakage and loss.

Thus both the spatial and genetic properties of a structural change condition its survival. Acentric, dicentric and ring chromosomes suffer *mitotic elimination*; dyscentric changes, whether inversion or translocation, suffer *meiotic elimination*; duplications and deficiencies suffer *cellular elimination*, that is, death, either immediate or delayed, of the whole cells containing the changed chromosomes. These last two do not depend on the purely spatial properties of the change but on its genetic consequences. In considering any analysis of structural changes we must therefore bear in mind the degree, kind and time of elimination that the changes may have suffered before they were identified.

The spatial relationships of structural change are interesting, however, from another point of view. They can be considered in relation to the number of breakages and number of reunions of *breakage-ends* required to produce the observed result (Haldane, unpub.). Stadler (1932) has pointed out that all viable X-ray changes are apparently due to the reunion of breakage-ends. This seems to be true except in the formation of branched chromosomes.

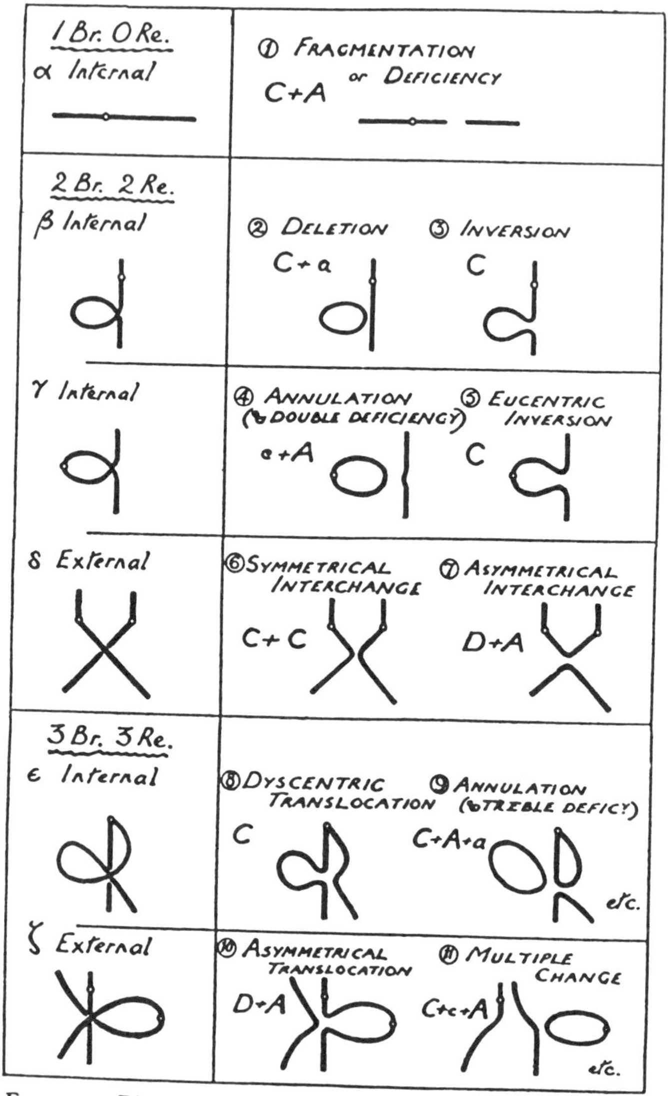

556 CELL MECHANICS

1 Br. 0 Re. α Internal	① FRAGMENTATION or DEFICIENCY C+A	
2 Br. 2 Re. β Internal	② DELETION C+a	③ INVERSION C
γ Internal	④ ANNULATION (≈DOUBLE DEFICIENCY) a+A	⑤ EUCENTRIC INVERSION C
δ External	⑥ SYMMETRICAL INTERCHANGE C+C	⑦ ASYMMETRICAL INTERCHANGE D+A
3 Br. 3 Re. ε Internal	⑧ DYSCENTRIC TRANSLOCATION C	⑨ ANNULATION (≈TREBLE DEFIC?) C+A+a etc.
ζ External	⑩ ASYMMETRICAL TRANSLOCATION D+A	⑪ MULTIPLE CHANGE C+c+A etc.

FIG. 159.—Diagram showing certain simple types of primary structural change in relation to the number of breaks and reunions required to make them and the character of the chromosomes produced. (Left) before change; (right) after.

CROSSING-OVER 557

In translocation a segment of one chromosome is never attached to the end of another. He has concluded that true ends cannot rejoin and therefore have a specific non-fusability. Further, the indirect proportion of induced structural changes to X-ray dosage suggests that they sometimes depend on two hits, not, like gene-changes, on one. This at least is true of the changes that survive in breeding experiments. Simple deficiencies from one breakage would not usually survive.

TABLE 78

Analysis of certain Simple Structural Changes in Terms of Breakage and Reunion (*cf.* Table 77).

Events	Internal (C)		Fraternal or External (C + C)	
	Change	Result	Change	Result
1 Br., 0 Re.	Deficiency.	C + A	—	—
2 Br., 1 Re.	1. Deletion.	C + A	—	—
	2. Double deficiency.	c + A + A	—	—
2 Br., 2 Re.	1. Inversion.	C	1. Symmetrical interchange.	C + C
	2. Double deficiency.	c + A	2. Asymmetrical interchange.	D + A
3 Br., 3 Re.	Translocation eucentric or dyscentric.	C	Translocation eucentric or dyscentric.	
			1. Symmetrical	C + C
			2. Asymmetrical	D + A

Catcheside (1935, 1936), on the other hand, explains the reciprocity of structural changes without assuming a specific non-fusability of the ends. He infers that the reciprocity of changes is simply due to the high unlikelihood of an end of one chromosome lying near the point of breakage of another. Stadler considers (from the behaviour of reverse mutation islands) that there is an unlimited time-interval during which breakage-ends could reunite. This view seems untenable in his particular examples, since acentric fragments are nearly always lost at mitosis.

Ring chromosomes in *Zea*, perhaps through interlocking at anaphase, form double dicentric rings which break like the bridges in inversion hybrids. But unlike them (according to Rhoades and McClintock, 1935) these double broken bridges form new rings again. There is no evidence yet of this happening to broken bridges after meiosis. In these it must be assumed undoubtedly where the bridges are single that the new ends become normal ends. Their survival as ends explains the occurrence of lateral chiasmata. There is, therefore, some difficulty in supposing that delayed re-fusion occurs. And hence there is also a difficulty in assuming a specificity in ends without the additional obnoxious assumption that breakage-ends change and become true ends later.

Catcheside, on the other hand, considers that breakage-ends must reunite immediately. His two assumptions of reciprocity and immediate fusion make structural change strictly analogous to crossing-over, as it is described above, and thus simplify the whole problem.

In order to understand the apparent non-fusability and consequent permanence of ends it is necessary to consider the mechanics of breakage and reunion. The process like crossing-over demands movement: the breakage ends must move towards one another before they can reunite. This movement may be due to attraction between the ends or to change within the broken chromosomes as in the case of crossing-over. The first explanation agrees with the view that the breakage-ends have a power of attraction that the true ends lack. The second explanation agrees with, and indeed itself explains, non-fusability of the ends, in this way: the relic spirals show a state of stress of all parts of the chromosome except the ends, for these are free to uncoil. Two breakage-ends are at once released from this stress and will fly apart from one another. They are expected to move when broken, although true ends would not.

Whatever the mechanical conditions, however, their consequences are now clear. We have to consider the type of change produced in relation to its survival at mitosis, and in growth and reproduction.

Survival at mitosis depends on whether they are *efficient* chromosomes, *i.e.* on whether they have one centromere and two ends, or are *inefficient*, *i.e.* having more or fewer centromeres or ends. The permanence of centromeres and ends, however, has two important evolutionary consequences, and ones that can be tested. First, any organism that has more chromosomes than any particular ancestor must have had a centromere and has probably had two ends reduplicated. Secondly, these centromeres and ends will be associated with groups of genes in the same linear order up to the point at which the first structural or gene changes have occurred. In other words, the chance of finding homologous series of genes within a haploid set is greatest next to the ends and next to the

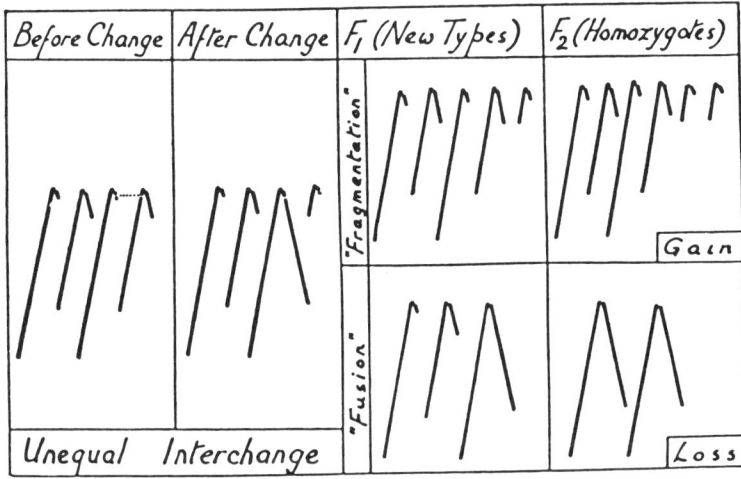

Fig. 160.—Diagram to show how " fusion " and " fragmentation " can arise by unequal interchange followed by loss and gain respectively of the smaller product.

centromeres and diminishes as we proceed away from them. Catcheside (1933) in haploid *Œnothera* and Nishiyama (1934) in triploid *Avena* have found pairing within the haploid set, but only at or near the ends. Incidentally we may notice that if by the chances of breakage a centromere becomes an end it will remain an end.

If the centromeres are permanent structures not arising *de novo* and are also necessary for the perpetuation of chromosomes at mitosis, an important conclusion follows. Increase and decrease of chromosome numbers, such as has been loosely attributed to "fragmentation" and "fusion" in the past, can arise only by gain and loss of centromeres. And since, so far as we know, all such changes apart from simple loss or gain of chromosomes depend on structural changes occurring at random in the chromosomes, we have to consider how they could arise without fatal inconvenience to the organism. It seems that this would be achieved most probably by *unequal interchange*. Thus two rod-shaped chromosomes with sub-terminal centromeres interchanging would give a single V-shaped chromosome and a very short fragment which might then be lost, and the chromosome number reduced. On the other hand if the fragment were retained along with the unchanged types the chromosome number would be increased. The first would depend on the region near the centromere being inert; the frequent occurrence of inert supernumerary fragments shows that this is often the case. In *Fritillaria*, where such fragments are formed, the expected concomitant changes in number and form of the chromosomes are also found in comparing different species (D., 1935 g). It seems, therefore, that the direction of change of chromosome number must be conditioned by the genetic activity or inertness of the proximal parts of the chromosomes. Evolutionary stability of basic chromosome numbers means activity of genes near the centromere and the ends. Instability means their inertness.

(*d*) *The Molecular Basis of Cell Mechanics.* In considering the internal and external mechanics of the chromosomes we find a wide range of movements which are co-ordinated in time and therefore necessarily related in cause. In order to understand their causation we must pursue the study to lower levels of integration, to colloidal and molecular dimensions. Having inferred movements from changes in structure, we must infer structure from differences in movement.

Let us take first the internal changes in the chromosomes. They require, as we have seen, changes in a molecular spiral. Long molecules of a chain structure must be twisted spirally. Such changes of shape are characteristic of certain protein molecules and may be determined in various ways, the most obvious being by a change in the surface electrical conditions, which will necessarily change the internal spatial equilibrium of the molecule (*cf.* Koltzoff,

1928; Astbury, 1933; Caspersson and Hammarsten, 1935; Wrinch, 1936).

Turning to the external movements of the chromosomes we have already seen that they must be determined by surface charges —or double electrical layers—of different intensities on different cell-organs—chromosomes, centromeres, centrosomes, nuclear membranes and cell-walls. Haldane (unpublished) calculates that two spheres, like the centromeres, of diameter 2500 A° at a distance apart of 10,000 A° (1 μ) in a medium of dielectric constant 81 and at a potential difference from this medium of 100 mv. would repel one another with a force equal to 67,000 times their weight. Electrical potential differences of an order that is known in living systems will therefore determine a force sufficient to affect chromosomes perceptibly at the distances at which they are seen to move apart, an action that will be reinforced by the special dielectric conditions that we have inferred from the properties of the spindle. The spindle, as we saw, develops its molecular orientation under the influence of an electrical field and has the character of a liquid crystal.

The existence of different and changing surface charges on different cell-organs may be inferred in an entirely different way, from their different and changing fixation-staining reaction. This reaction is conditioned by the surface charge on the bodies concerned, and, as we have seen, it passes through a cycle of changes co-ordinated with the mitotic cycle (*cf.* Yamaha, 1935).

The next question that arises is how a surface charge comes to be developed on certain cell-structures. The reason is that any amphoteric electrolyte, such as a protein, lying in a medium not at its iso-electric point will develop a surface charge (or strictly, a double electric layer). Different cell-organs have different surface-charges because the charge is a function of the difference between the pH of the medium and the iso-electric point of the ampholyte concerned. Hence also the cycle of changes in surface charge on each organ implies a cycle of changes in the pH of the medium. And hence, again, comparable organs, the centromeres and centrosomes, within and without the nucleus, must have a different cycle of changes. Finally, we must recall that the pH cycle inferred from these external movements is also to be inferred from the internal rearrangements implied by the spiralisation cycle.

While the iso-electric point of the surface molecules will determine the surface charge at any one point, other things being equal, the surface charges at different points will, of course, interact and cause a secondary distribution not directly conditioned by surface molecular structure. To this is probably to be ascribed the " precocious condensation " or staining of the proximal parts of chromosomes irrespective of their internal constitution. Nor must we forget that this behaviour is correlated with precocious division of these parts, division being itself controlled by electrical conditions, and as we have seen, in certain cases by the possession of a centromere. Still more important will be the effect of a localised charge on the viscosity of the chromosome thread and its transmission of torsions. The control of the distribution of crossing-over by the centromere must depend partly on this property.

Thus at every phase of their life one particle, the centromere, governs the activities of the chromosomes. It can do so because it is in them but not of them. The centromere agrees with the chromomeres and the chromosome thread in its property of linear arrangement and regulated reproduction. It disagrees with them in its cycle of reproduction and repulsion, its surface charge and staining capacity, its lack of longitudinal cohesion and spiralisation. In all these respects it is related rather to the centrosomes, from which it differs merely in its position on the gene-thread. On these cardinal similarities and differences of structure and behaviour depends that co-ordination of centric and genic reactions which makes reproduction and heredity possible.

Cell genetics led us to investigate cell mechanics. Cell mechanics now compels us to infer the structures underlying it. In seeking the mechanism of heredity and variation we are thus discovering the molecular basis of growth and reproduction. The theory of the cell revealed the unity of living processes; the study of the cell is beginning to reveal their physical foundations.

APPENDIX I

INTERPRETATION

Take the nucleus—what is it ?—apparently no more than a pellicle or skin, a mere bladder containing fluid !

W. B. HARDY, 1925.

THE cytologist investigating chromosomes chiefly depends on the observation of fixed and stained material. He has two means of judging the value of particular preparations for his studies. The first is by the inductive-deductive method. The effect of particular reagents and treatments on particular structures is determined, laws of behaviour are induced, and from these the value of any particular treatment deduced. This method has been attempted earlier (by Fischer, Yamaha and others), but it has never yet been practicable because the character of the organisms treated, the methods of treatment, the reagents used, and the objects pursued, are all so very diverse in relation to the amount of comparative study that can be carried out. He therefore has to resort to a second more empirical method as being of more immediate practical importance.

This method is the method of comparison of observations, and it may be applied in three ways :—

(i) The observations may be compared with those of living material of the same organism at the same stage of development. This test has been usefully applied in the resting nucleus (Schaede), in the prophase of meiosis (Pl. XVI) and elsewhere, but it is necessarily of restricted use with the ordinary microscopic technique.

The ultra-violet microscope has increased resolving power and increased selectivity both in regard to focus and absorption. Thus photographs of living cells taken by ultra-violet light have a sharper definition and much sharper contrast at a higher magnification. Chromosomes particularly have a strongly differential absorptive capacity as compared with cytoplasm, not only during mitosis but—and this may prove of the greatest significance in relation to the theory of continuity—also during the resting stage. It has already been possible to take photographs of living cells agreeing with those hitherto taken of fixed and stained cells and in some respects equally instructive (Pl. XVI). This method evidently has an important future, especially perhaps in the study of stages in the chromosome history of organisms which have not proved amenable to observation by the customary methods (Lucas and Stark, 1931).

In the salivary glands it has been possible with ultra-violet light to discriminate between chromomeres that were not distinguishable in visible light, but whose separate existence was inferred from the occurrence of X-ray breakages (Ellenhorn, Prokofieva and Muller, 1935). Similarly, the use of X-rays for changing genes and for breaking the chromosome thread has superseded direct observation as a means of testing such delicate questions as the time of division of these structures. Direct X-ray photography has not yet been brought to bear on the chromosomes and the spindle (*cf.* Astbury, 1933).

Evidently visible light can no longer be regarded as the final instrument of observation for chromosomes. Our senses have been surpassed by intermediate mechanisms and in this way the old objection to cytology, *tot capita, tot sensus*, together with its stale controversies, will be put on one side.

(ii) The observations may be compared with those of related organisms at the same stage of development (after similar or different treatment). In this way a study of the chromosomes at mitosis and meiosis in material which has satisfied the first kind of test enables us to conclude that such structures as constrictions and chiasmata are present in other material less suitable for study on account of the concomitant conditions found, *e.g.*, an inturned bend in the chromosome at the constriction or a change of plane in a diplotene loop or opening of the chromatids in anaphase at the chiasma. The same criterion has been used to show that multiple ring formation at meiosis is the result of normal development of structurally dissimilar chromosomes, in hybrids.

(iii) The observations may be compared with those of the same material at the stages immediately preceding and following. This method is essential in the interpretation of meiosis in regard to the pairing of the chromosomes, the movements of chiasmata, the separation at anaphase, the frequency of association, and so forth, and, in examining these, quantitative methods are often necessary. Developmental comparison has shown the invalidity of various theories of pairing and crossing-over. Developmental comparison is equally essential for many minor inferences. For example, nucleolar material surviving to metaphase at meiosis in many forms may be mistaken for free small chromosomes produced by fragmentation unless different stages are compared critically.

It is not yet possible to show the whole course of meiosis in all organisms on account of two major difficulties: (i) The occurrence of stages when the chromosomes are collapsed to one side of the nucleus; this is usually at zygotene and diplotene, probably for the reason that the chromosomes at these stages are slenderer than at the stages

PLATE XVI

FIGS. 1–3.—Diakinesis in *Stenobothrus lineatus*. (From Belar, 1927.) Fig. 1. Living cells. Fig. 2. The same cells after fixation with osmium tetroxide vapour and Flemming. Fig. 3. The same cells after staining with Giemsa-Romanovsky.

FIGS. 4–7.—Photographs of living chromosomes of *Melanoplus femur rubrum* (Orthoptera) at meiosis, taken by ultra-violet light. (After Lucas and Stark, 1931.)
Fig. 4. Late telophase before meiosis. × 1800. Fig. 5. Pachytene. × 1800. Fig. 6. First anaphase showing probable dicentric bridge. × 1200. Fig. 7. First telophase. × 1200.

PLATE XVI

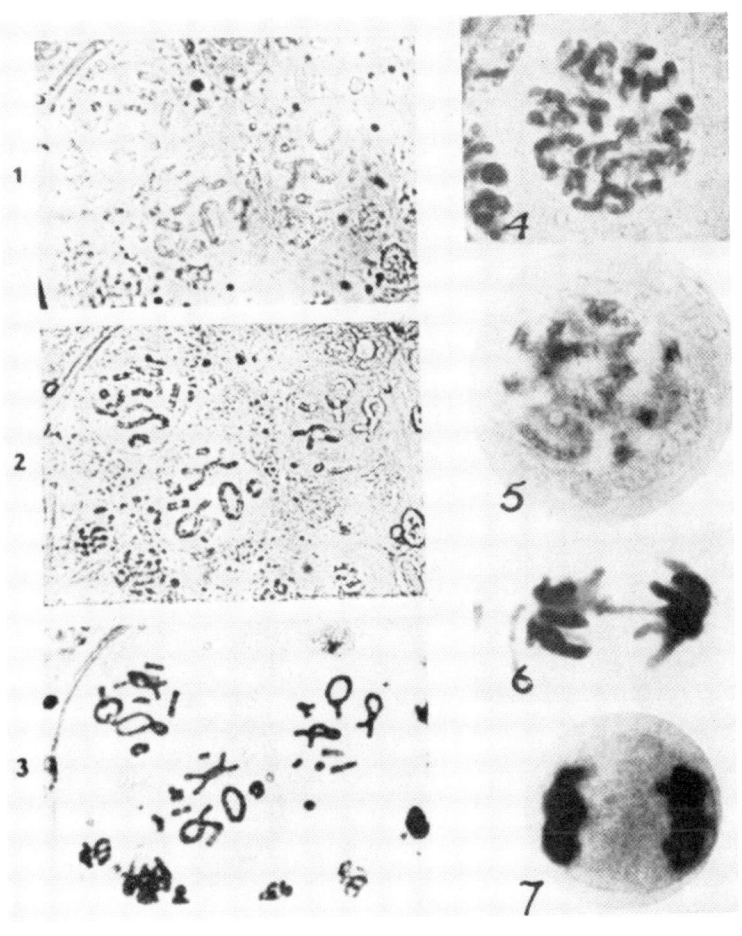

[To face p. 564.

between them and following them, so that sudden dehydration leads to extreme distortion. (ii) The occurrence of a diffuse stage when, as though in a resting stage, the chromosomes are lost to sight. Either of these interruptions is fatal to the study of the succession of stages, and both may occur in the same species (Chick, 1927, on *Benacus*). These difficulties make the comparative method even more important in studying meiosis.

The application of the comparative method enables us to take a more objective view of the nature of artefacts than is commonly done. Artefact has been one of those words that " plainly force and overrule the understanding." An artefact is an appearance that arises from treatment. All living organisms are therefore liable to show artefacts when they are killed, and they may, as Belar (1928, pp. 3-4) pointed out, reveal " vital " artefacts while still showing signs of normal life, *e.g.*, the collapse of the chromosomes referred to above. These follow merely from manipulation. But in ordinary treatment manipulation is followed by fixation (coagulation and hardening) and staining (differential absorption of dyes). Both of these may lead to artefacts. The fact that stains are apparently of three degrees of directness: (i) those which stain cell-structure differentially in the first instance (*e.g.*, Giemsa), (ii) those which stain uniformly and are de-stained differentially (*e.g.*, gentian violet), (iii) those which are de-stained differentially after mordanting (*e.g.*, iron-hæmatoxylin), is probably not of great importance, for the decisive differential reaction is in all cases a single one, and judged by the results, under suitable conditions of fixation, it is of the same degree of specificity. The differences in effect of treatment are therefore almost entirely conditioned by differences in fixation to which staining is secondary.

No appearance of treated material is definitely free from artefact nor is any appearance " pure " artefact. The question is therefore not so much whether an appearance is an artefact but how significant the artefact is. For this purpose comparison enables us to distinguish those that are *characteristic* from those that are not, as follows :

(i) The collapse of the chromosomes to one side of the nucleus is characteristic of two stages of meiosis and is not found in the prophase of the abortive meiosis of parthenogenetic aphides (Morgan, 1912), although it is found in the normal prophase. The reason for this has been shown above. (ii) The chromomere structure of the early prophase chromosome is a characteristic artefact because homologous chromosomes in the same nucleus show similar chromomeres. (iii) The spiral structure seen after squeezing living chromosomes is characteristic, but the fractured structure seen when they are squeezed after fixation is not. (iv) The fibrillar structure of the shrunk spindle is characteristic and is derived from a special arrangement of differently hydrated particles in it. (v) The anastomosed structure seen in the resting nucleus by Sharp and Martens is non-characteristic. It cannot be related to structures seen earlier or later (Ch II).

Having excluded non-characteristic artefacts, we may then consider the significance of different kinds of characteristic artefact. The comparative method at once shows us that some correspond closely with living conditions, *e.g.*, the chromomere structure, while others are definitely abnormal and peculiar to the effects of treatment on certain organisms, *e.g.*, the contraction in the prophase of meiosis.

The artefactual nature of the spiral structure of the chromosomes seen in smear preparations of metaphase, particularly of meiosis, is shown by the optical homogeneity of the living chromosomes. Its characteristic nature is shown by its relationship with uncoiled pre-treated spirals as well as with relic spirals at the following telophase and prophase. It depends, therefore, on a characteristic and artefactual differentiation of the chromosome into staining and non-staining parts. The products of this differentiation have been named " chromonema " and " matrix " and have been given a developmental significance. Such a significance they cannot have in view of their artefactual nature. They depend on the different types of coagulation to which chromosomes, like milk, are subject. The matrix is the whey, which is separated from the curd by fixation. It does not exist morphologically or mechanically until the chromosomes are fixed, however useful its distribution may be in enabling us to determine what the internal structure of the chromosome may be. In its use to fill a morphological and mechanical *rôle* in the living chromosomes the word matrix is nothing more than a myth (D., 1935 *e*).

Genetical inferences are subject to other kinds of error. They are derived from observations of mitosis and meiosis. In mitosis assumptions are made which are correct except for certain exceptional cases.

(i) It is assumed that the chromosome complement seen at one division is characteristic of the individual and even of the species. The exceptions to this rule (based on the theory of permanence) are so few that it can be satisfactorily applied in practice unless there are *a priori* reasons for expecting lack of constancy (as in the examination of mutants or geographical races).

(ii) It is assumed (also on the basis of the theory of permanence) that differences between species in the size of their chromosomes will be maintained in their hybrids and derivatives. The observations described earlier (Ch. III) show that this conclusion is not necessarily true, although experimental evidence is still confined to few and exceptional cases. Since genetic properties influence chromosome size and length where two related species have chromosomes of different size or length but are otherwise comparable, it is rather to be assumed that this difference is due to a genetic difference and that therefore the hybrid will fail to show it. But there is another restriction to this type of inference, viz., the different effects of the same treatment on chromosomes of related species. Such different effects are probably responsible for the differences in the size of chromosomes observed by Farmer and Digby (1914) in diploid and tetraploid *Primula kewensis*, suggesting that the tetraploid was derived by transverse splitting of the chromosomes, and by Tischler (1928) between species of *Ribes*, suggesting that autosyndesis occurred in the hybrid (*cf.* Meurman, 1928; D., 1929 *d*), because large and small pairs of chromosomes were found there (*cf.* Upcott, 1936 *b* on *Æsculus ;* v. Berg, 1931 *b* on *Ægilops*).

In meiosis, erroneous assumptions have been made and still are made with regard to the pairing of chromosomes, such as the following :—

(i) Pairing has been held to be determined by " affinity " between chromosomes (*cf.* Dobzhansky, 1931). It is therefore constant (under constant external conditions), and a few observations are an accurate measure of this " affinity." Chromosomes which do not pair are not " homologous." In this way a simple rule is made for analysing the relationships and even phylogeny of species. The conclusions derived from these imperfect assumptions are of doubtful value (*e.g.*, in *Nicotiana*). Even with quantitative analysis (*e.g.*, in *Triticum-Ægilops* hybrids) it is doubtful whether relationships can be accurately assessed because (*inter alia*) the differences in the chromosomes which reduce their pairing are not necessarily directly related to, and therefore proportional to, the differences which determine genetic differences in the organism (*v.* Ch. V).

(ii) Ring-formation has been held (by adherents of the theory of telosynapsis) to be determined by an undefined " hybridity," a reduced " homology," a genetic factor, and so on. These inferences were loose and unjustifiable and disagreed with genetic evidence. They have been quietly buried (Ch. IX).

Summing up : The mechanical and genetical interpretations put on chromosome behaviour have suffered from two chief sources of error. First, the relative values of direct and comparative inference have not been correctly assessed. Direct inferences have been admitted as " facts " and " observations " or denied as " artefacts " and " misinterpretations " when actually their whole weight depended upon their use for comparative inference. Comparative inference, on the other hand, has too often been put aside as speculative and irrelevant when indeed it has been the only means for any general advance in the understanding of the processes concerned. Hypothesis based on it has often proved more reliable than the " facts " of direct observation.

Secondly, synthesis has preceded complete analysis. " The method of discovery and proof according to which the most general principles are first established, and then intermediate axioms are tried and proved by them, is the parent of error and the curse of all science" (*Novum Organum*, I, 69). This method has been particularly unfortunate in its application to the structural and genetical interpretation of meiosis. The relations between the several stages of any process must be determined before any interpretation of the whole can be attempted.

APPENDIX II

TECHNIQUE

So many coagulations there are in nature; and though we content ourselves with one in the running of milk, yet many will perform the same.

SIR THOMAS BROWNE, *Common Place Book*.

The following is a list of references to authors who describe new methods of preparation.

Microscopy
Belling, 1930.

General Technique
Taylor, 1928.
Belar, 1929 c.
Belling, 1930.
La Cour, 1931.

Fixation
Navashin, 1926 (chrom-formol-acetic).
La Cour, 1931 (various).
Sakamura, 1927 b (hot water).
Kihara, 1924, cf. Huskins, 1928 (Carnoy + Flemming).
Kihara, 1927 (cooling before fixing).
Randolph, 1935.

Smearing
Taylor, 1924 a (Flemming + hæmatoxylin).
Darlington, 1926 (Flemming + gentian violet).
Belling, 1926 a (iron aceto-carmine); cf. de Meijere, 1928, 1930.
Belar, 1929 c (osmic vapour + Flemming).
McClintock, 1929 b; Steere, 1931 (permanent aceto-carmine).
Barrett, 1932 (aceto-hæmatoxylin).
Yasui, 1933.
Sax, K. and H. J., 1933 (endosperm of Conifers).
Backman, 1935.

Staining

Newton and D., 1929 (gentian-violet-iodine).
Belling, 1930 (iron-brazilin).
Belar, 1929, *b* and *c* (Giemsa).
La Cour, 1931 (chromic acid as mordant after gentian violet).
Barrett, 1932.
Hance, 1933.
Ludford, 1933 (vital staining).

Living Material

Chambers, 1925.
Schaede, 1925, 1927.
Sakamura, 1927 *a*.
Belar, 1929, *a*, *b*, and *c*.
Lucas and Stark, 1931
Yamaha, 1926.
Fry, 1930.
Kuwada and Nakamura, 1934 *b*.
Ellenhorn, 1933.

Pre-treatment for Spiral Structure

Nebel, 1932.
Huskins and Smith, 1934.
Kuwada and Nakamura, 1933, 1934 *a*.
Sax and Humphrey, 1934.
La Cour, 1935.

Chemical Tests for Chromatin, etc.

Strugger, 1932.
Bauer, 1932, 1933 (oöcytes).
Shinke and Shigenaga, 1933.
Milovidov, 1933.
Geitler, 1935 (*Spirogyra*).
Yamaha, 1935.
Gardiner, 1935.

Polarised Light

Kuwada and Nakamura, 1934 *c*.

Infra-Red Photography

Prat and Hruby, 1934; Hruby, 1935.

TECHNIQUE

Ultra-Violet Photography
Lucas and Stark, 1931.
Ellenhorn, Prokofieva and Muller, 1935.
Caspersson and Hammarsten, 1935.

Screen Projection
Darlington and Osterstock, 1936.

Mitosis in Tissue Culture
Hearne, 1936.

Centrifuging
Beams and King, 1935.
Kostoff, 1935.

X-ray Treatment

1. *Organisms*

   Stadler, 1931.
   Muller, 1934 ; *et alii*, 1935.
   Timofeeff-Ressovsky, 1934 ; *et alii*, 1935.

2. *Chromosomes*

   Mather and Stone, 1933.
   Mather, 1934.
   Katayama, 1935.
   Catcheside, 1935.
   White, 1935.
   Levan, 1935 *b*.
   Riley, 1936.

APPENDIX III

GLOSSARY

If men shal telle proprely a thing
The word mot cosin be to the werking.
CHAUCER, *Maunciple's Tale*.

Terms not used in the text are enclosed in brackets. The usages given are those adopted in this book and are not always the original ones, since usage necessarily changes. For the sources of many of these the author is indebted to Wilson (1928), Rosenberg (1930), Carothers (1931), and a number of correspondents.

Acentric, of the whole or part of a chromatid or chromosome lacking a centromere.
Allelomorph, one of two dissimilar factors which on account of their corresponding position in corresponding chromosomes are subject to alternative (mendelian) inheritance in a diploid. *Bateson*, 1902.
Allopolyploid, a polyploid whose chromosomes do not usually form multivalents at meiosis but form bivalents as far as their homologies allow them, e.g., a tetraploid with $2x$ bivalents or a triploid with x bivalents and x univalents (*cf.* Autopolyploid, *v*, Text). *Kihara* and *Ono*, 1926.
Allosyndesis, the pairing in a polyploid of chromosomes derived from opposite parents; particularly as opposed to autosyndesis, in a hybrid between allopolyploids. *Ljungdahl*, 1922.
Alternation of Generations, the occurrence of two series of nuclear divisions in the life cycle, a haploid and a diploid.
Ameiosis, the occurrence of one division of the nucleus instead of the two of a normal meiosis, giving non-reduction of the mother-cell.
Amitosis, the division of a nucleus without separation of daughter chromosomes. *Flemming*, 1882.
[**Amphidiploid**], allotetraploid, *v*. allopolyploid.
Amphimixis, the bringing together of elements from two gametes in fertilisation (as opposed to apomixis). *Weismann*, 1891.
Anaphase, the stage at which daughter chromosomes move apart in a nuclear division. *Strasburger*, 1884.
Androgenesis, male parthenogenesis, *q.v.*
Aneuploid, having the different chromosomes of the set present in different numbers, through purely numerical aberration—therefore an unbalanced polyploid. *Täckholm*, 1922.
[**Anisogamy**, heterogamy, *q.v.*]
Apogamy, apomixis involving the replacement of the gametes by unspecialised cells which do not fuse. *De Bary*, 1878; *Winkler*, 1908.
Apomixis, the occurrence of the external form of sexual reproduction with the omission of fertilisation and usually meiosis as well (as opposed to amphimixis). *Winkler*, 1908.

GLOSSARY

Apospory, apomixis involving the replacement of the spores by unspecialised cells which have not undergone meiosis. *Bower*, 1887; *cf. Winkler*, 1908; *Rosenberg*, 1930.

Asynapsis, non-pairing of chromosomes at meiosis. *Beadle*, 1931.

Attachment, [(i) the position of the centromere, *q.v.*;] (ii) the permanent fusion of two chromosomes. *L. V. Morgan*, 1922.

Autosomes, those chromosomes whose segregation does not normally affect the determination of sex (as opposed to sex-chromosomes). *Montgomery*, 1906.

Autosyndesis, the pairing, in a polyploid, of chromosomes derived from the same parent; particularly its exceptional occurrence in an allopolyploid or its hybrid. *Ljungdahl*, 1922.

Azygote, organism arising by haploid parthenogenesis. *Whiting* and *Benkert*, 1934.

Balance, the condition in which the genes are adjusted in proportions which give satisfactory and normal development of the organism (opposite: unbalance). *Bridges*, 1922.

    Secondary ———, a new balance derived by change in the proportion of genes, as in a secondary polyploid, from an old balance, and capable of competing with it. *D.* and *Moffett*, 1930.

Balanced Lethal Factors. *v. Lethal.*

Basic Number. The supposed number of chromosomes found in the gametes of a diploid ancestor of a polyploid, represented by x.

Bivalent (*v.* Univalent).

Bouquet, the zygotene and pachytene stage in those organisms in which the chromosomes lie in loops with their ends near one part of the wall of the nucleus (*cf.* Polarisation). *Eisen*, 1900.

Cell, a unit in the structure of the animal or plant containing one nucleus, or several where these are separated by a uniform substratum. Also a body derived from such a unit. *Leeuwenhoek*, *Schleiden* and *Schwann*, 1840.

Centromere, a particle in the chromosome thread whose special cycles of repulsion and division determine the anaphase and terminalisation movements of the chromosomes (*v.* Table 75).

Centrosome, the self-propagating body which during division in many organisms lies at the two poles of the spindle and appears to determine its orientation. *Boveri*, 1885.

Chiasma, -ta (Hellenistic Greek), an exchange of partners in a system of paired chromatids; observed between diplotene and the beginning of the first anaphase in meiosis. *Janssens*, 1909.

    Comparate and disparate ——— **ta**, pairs of successive chiasmata which compensate, and do not compensate, respectively, for one another in regard to the crossing-over and changes of partner producing them.

    Interstitial ———, where there is a length of chromatid on both sides of the chiasma. *D.*, 1929 *c*.

    Reciprocal and Complementary ——— **ta**, the two kinds of comparate chiasmata according to their crossing-over relationships.

    Terminal ———, where an exchange occurs amongst the end particles of the chromatids, following terminalisation. *D.*, 1929 *c*.

    Multiple ———, a terminal chiasma where three or four pairs of chromatids are engaged. *D.*, 1929 *c*.

    Lateral ———, a chiasma which is terminal as to two chromatids and interstitial as to the others. Of two kinds, symmetrical and asymmetrical. *D.*, 1929 *c*.

    Imperfect ———, where one of the four associations in a chiasma is broken prior to anaphase. *D.*, 1929 *c*.

APPENDIX III

Chiasma Theory of Pairing, the hypothesis that whenever two chromosomes which have been paired at pachytene remain associated until metaphase they do so by virtue of the formation of a chiasma or visible exchange of partners amongst their chromatids. *D.*, 1929, *b* and *c*.

Chiasmatype Theory, (i) the theory that chiasmata are connected with crossing-over either as a cause or as a consequence. *Janssens,* 1909. (ii) The special assumption that chiasmata are determined by crossing-over between two non-sister chromatids of the four involved. *D.*, 1930 *b*.

Chimæra, a plant composed of tissues of two genetically distinct types as a result of mutation, segregation, irregularity of mitosis, or artificial fusion (graft-hybrids), *v.* Mosaic. *Cf. Baur, Winkler.*

Chondriosomes, bodies in the cytoplasm which are believed to be self-propagating, including mitochondria and Golgi-bodies. *Benda,* 1904.

Chromatid, a half chromosome between early prophase and metaphase of mitosis and between diplotene and the second metaphase in meiosis—after which stages, *i.e.*, during an anaphase, it is known as a daughter-chromosome. The separating chromosomes at the first anaphase are known as daughter-bivalents, or, if single chromatids derived from the division of univalents, daughter-univalents. *McClung,* 1900.

—— **Non-Disjunction,** the passing of homologous parts of chromatids to the same pole following crossing-over between homologous differential segments in a multiple interchange hybrid. *E. Sansome,* 1933.

—— **bridge,** dicentric chromatid with centromeres passing to opposite poles at anaphase.

Sister ——s, those derived from division of one and the same chromosome, as opposed to *non-sister chromatids* which are derived from partner chromosomes at pachytene.

Chromatin, the part of the chromosome that stains deeply during mitosis, as opposed to an achromatic part. The distinction is difficult to apply and is often of doubtful validity. *Flemming,* 1879.

Chromocentre, (i) fused prochromosomes (*q.v.*); (ii) body produced by fusion of the centromeres, etc., in nuclei of the salivary glands of certain Diptera. *Painter,* 1935.

Chromomeres, the smallest particles identifiable by their characteristic size and position in the chromosome thread between leptotene and pachytene and in salivary gland nuclei. *Wilson,* 1896.

[Chromonema, -ta], the chromosome thread, *q.v.* (similarly leptonema, pachynema, etc., the chromosome threads at leptotene, pachytene, etc.). *Wilson,* 1896.

Chromosomes, one of the bodies into which the nucleus resolves itself at the beginning of mitosis and from which it is derived at the end of the mitosis. *Waldeyer,* 1888.

Acentric ——, one lacking a centromere.
Dicentric ——, one with two centromeres.
Ring ——, one with no ends.

Chromosome Thread, the thread consisting of centromere, chromomeres, achromatic connective thread and, perhaps, pellicle at prophase ; and constituting, as a spiral, the metaphase chromosome.

Clone, a group of organisms descended by mitosis from a common ancestor.

Coil, *v.* spiral.

Complement, the group of chromosomes derived from a particular nucleus in gamete or zygote, composed of one, two or more sets (*q.v.*).

Condensation or Contraction of the chromosome, the thickening, shortening and spiralisation of the chromatids during prophase (*v.* Text).

GLOSSARY

Configuration, an association of chromosomes at meiosis, segregating independently of other associations at anaphase.
Congression, the movement of chromosomes on to the metaphase plate.
Conjugation, the pairing of the chromosomes, gametes or zygotes, or the fusion of pairs of nuclei.
Constriction, an unspiralised segment of fixed position in the metaphase chromosome. *Agar*, 1911.
    Primary or centric ——, that always associated with the centromere.
    Nucleolar ——, a secondary constriction determined by the organisation of the nucleolus.
    Secondary ——, any acentric constriction. *D.*, 1926.
Co-orientation, the relative orientation of two centromeres on the spindle.
Crossing-over, the exchange of corresponding segments between corresponding chromatids of different chromosomes by breakage and reunion following pairing ; a process inferred genetically from the reassociation of linked factors in mendelian hybrids and cytologically from the reassociation of parts of chromosomes in structural hybrids. *Morgan*, 1911.
    Effective ——, that which is detectable in breeding experiments.
    Illegitimate ——, crossing-over in a haploid or polyploid which is not a structural hybrid, between homologous and reduplicated segments of two chromosomes which being structurally dissimilar as a whole, do not normally pair. Determines secondary structural change. *D.*, 1932.
    Double ——, the production of a chromosome in which crossing-over has occurred twice ; may be reciprocal or non-reciprocal as between chromatids. *Sturtevant*, 1914.
[Cytokinesis], the separation of daughter-cells, usually after nuclear division.
Cytoplasm, the protoplasm apart from the nucleus.
Daughter chromosome (*v.* Chromatid).
Deficiency, loss of a terminal acentric segment of a chromosome from the diploid complement. *Bridges*, 1917.
Deletion, loss of an intercalary acentric segment of a chromosome.
Diakinesis, the last stage in the prophase of meiosis—immediately before the disappearance of the nuclear membrane. *Haecker*, 1897.
Dicentric, of a chromatid or chromosome, having two centromeres.
Differential Affinity, the failure of two chromosomes to pair at metaphase in the presence of a third although they pair in its absence. *D.*, 1928.
Differential Segment, one in respect of which two pairing chromosomes differ in a permanent hybrid (by contrast with a pairing segment). *D.*, 1931 *c*, 1934 *a*.
Diminution, the loss or expulsion of a part of the chromosome complement at mitosis so that a daughter nucleus is formed lacking this part.
Diplochromosome, chromosome which has divided once more often than it normally does in relation to its centromere (*cf.* attached X in *Drosophila melanogaster*). *White*, 1935.
Diploid, (i) the zygotic number of chromosomes ($2n$) as opposed to the gametic or haploid number (n) ; (ii) an organism having two sets of chromosomes ($2x$) as opposed to organisms having one (haploid) three (triploid) or more sets (x, $3x$, etc.) (*v.* Polyploid). *Strasburger*, 1905.
    Functional ——, allopolyploid which behaves as a diploid in segregation. *D.*, 1928.
Diploidisation, process of division of conjugate nuclei in fungi, by which haploid cells or mycelia become diploid. *Buller*, 1930.
[Diplont], organism at the diploid stage of the life-cycle, as opposed to " haplont."
Diplophase, [(i)] the diplotene stage of meiosis, *Belling*, 1928 ; (ii) the diploid

phase of the life-cycle, particularly in the Basidiomycetes where nuclear fusion does not immediately follow cell-fusion. *Goebel.*

Diplotene, the stage of meiosis which follows division of the chromosomes at the pachytene stage. *v. Winiwarter*, 1900.

Disjunction, the separation of chromosomes at anaphase, particularly of the first meiotic division.

Dislocated segments, homologous pairs of segments in a different linear sequence from other segments in a structural hybrid.

[Dislocation], structural change. *Navashin*, 1926, 1932.

Distal, that part of a chromosome which relative to another part is further from the centromere.

Duplication. The occurrence of a segment twice in the same chromosome, or in the same complement or the change by which this condition arises. *Bridges*, 1917.

Dyad, (i) the pair of cells formed after meiosis through irregularity instead of a quadruple " tetrad " (*q.v.*) ; [(ii)] the univalent chromosome, composed of two chromatids, at meiosis.

Embryo-sac, the female gametophyte in the angiosperms.

Environment, those factors external or antecedent to an organism which are related to its development. Its reaction with the genotype determines the phenotype. N.B.—Dissimilar genotypes themselves imply dissimilar environments.

[Ergastic], of materials in the cell : the non-living products of metabolism (*e.g.*, many fats and carbohydrates). *A. Meyer*, 1896.

Factor, the hypothetical determinant of a character in respect of which an organism may show a mendelian difference. It is usually an abstract description of a gene difference. *Bateson* and *Punnett*, 1905.

Fertilisation, the fusion of two germ-cells and of their nuclei, without which their later development is usually impossible.

First Division, the first of the two meiotic divisions, formerly known as the " heterotypic " or " reduction " division.

Gametes, germ cells which are specialised for fertilisation and cannot normally develop without it.

Gametophyte, the haploid plant which produces the gametes where there is an alternation of generations.

Gene, the unit of reproduction and hence of crossing-over in the hereditary material. *Johannsen*, 1911, 1923 ; *Morgan*, 1925. Its change may be inter-genic and structural, or intra-genic and molecular. *D.*, 1932.

Genetic, of a property possessed by an organism by virtue of heredity. *Bateson*, 1906.

[Genome], a chromosome set, *q.v. Winkler*, 1916.

Genotype, the kind or type of the hereditary properties of an individual organism. *Johannsen*, 1911.

Genotypic Control, the control of chromosome behaviour by the hereditary properties of the organism. *D.*, 1932 *a*.

Golgi apparatus, bodies in the cytoplasm, so-called in anticipation of their proving to be an apparatus.

Haploid (*v.* Diploid). *Strasburger*, 1905.

" True ——," an organism having one set of chromosomes (*v.* Text).

Haplo-diploid Sex Differentiation, that in which the sexes are distinguished by the male being haploid, the female diploid (*v.* Text).

Heredity, the process by which like begets like (in sexual reproduction). 1863.

[Heterochromosomes], chromosomes of exceptional form or behaviour, especially at meiosis, such as sex chromosomes, fragments, etc. *Montgomery*, 1904.

GLOSSARY

Hence **Heterochromatin**, parts of the chromosome which stain more deeply or less deeply than the rest during the prophase of mitosis. *Heitz, 1927 et al.*

[**Heterogametic**], producing gametes of two kinds in regard to their properties of sex determination. *Wilson, 1910.*

Heterogamy, [(i)] differentiation of male and female types of gametes ; (ii) the special differentiation by which in the hybrid Œnothera species two kinds of gametes of both sexes are produced, but opposite kinds function in opposite sexes. *Renner, 1919.*

[**Heteropycnosis**], precocious condensation of certain chromosomes in the prophase especially of meiosis (v. Precocity). *Gutherz, 1906.*

Heterozygote, a zygote derived from the union of gametes dissimilar in respect of the constitution of their chromosomes or from mutation in a homozygote. *Bateson, 1902.* Used for narrower categories of hybrid, as : Mendelian, Interchange, Reduplication, Fragmentation, Translocation, and Deficiency Heterozygotes (v. Text).

    Complex ——, one whose gametes have numerous differences which segregate as a unit. *Renner, 1917.*

    Sex —— (a form of complex heterozygote), a zygote of the heterozygous sex, v. Y Chromosome. *D., 1932.*

Heterozygous Sex (v. Y Chromosome).

[**Homogametic Sex**], the homozygous sex, q.v. *Wilson, 1910.*

Homology, the similarity of structures in different organisms which they owe to the common ancestry of the organisms. *Owen.*

Homozygote, a zygote derived from the union of gametes identical in respect of their chromosomes. *Bateson, 1902.*

Homozygous Sex (v. X and Y Chromosomes).

Hybrid (used here *sensu stricto* as a Functional Hybrid), a heterozygote. Used for the broader categories as :

    Numerical ——, one whose parental gametes differed in respect of the number of chromosomes. *D., 1931 c.*

    Structural ——, one whose parental gametes differed in respect of the structure of their chromosomes. *D., 1929 c.* It is Eucentric when its dislocated segments have the same linear sequence with respect to the centromere, Dyscentric when they are relatively inverted. *D., 1936 d.*

Hyperploids, diploids with an extra piece or pieces of chromosome. *Muller, 1932.*

Hypoploids, diploids lacking a piece or pieces of chromosome. *Muller, 1932.*

Hysteresis, a lag in the movement at one level of integration in response to stress at another level, *e.g.*, in the adjustment of the external form of a chromosome to its internal stresses during the spiralisation cycle. *D., 1935.*

Interchange, an exchange of non-homologous terminal segments of chromosomes, v. *Belling, 1927 b.* May be symmetrical or asymmetrical with respect to the centromere. (The term " segmental interchange " is used by Belling to include crossing-over (*q.v.*), but the two are conveniently distinguished, v. Text).

Interference, the property by which the occurrence of one crossing-over reduces the chance of occurrence of another in its neighbourhood. *Muller, 1916.*

    Cytological ——, the same property as between chiasmata. *Haldane, 1931 b.*

    Chromatid ——, the same property with respect to the chromatids taking part in two successive chiasmata. *Haldane, 1931 b ; Mather, 1933.*

Interkinesis (v. Interphase, Resting Stage). *Grégoire, 1905.*

Interphase, the resting stage that may occur between the first and second divisions. Should not be applied to the resting stage in general. *Lundegårdh,* 1912.

Inversion, the reversal of the linear sequence of the genes in one segment of a chromosome relative to the centromere. *Sturtevant,* 1926 (*cf.* Ch. VII).

Isogamy, morphological similarity of fusing gametes.

[Karyokinesis], *Schleicher,* 1878, mitosis, *q.v.*

[Karyology], nuclear cytology.

[Karyomere], a compartment or vesicle in the resting nucleus, usually containing one chromosome.

Leptotene, the single chromosome threads at the early stage of the prophase of meiosis, and by extension the stage itself, (*v.* Zygotene, Pachytene, Diplotene) *v. Winiwarter,* 1900.

Lethal Factors, factors which render inviable an organism or, by extension, a cell possessing them.

    Balanced ———— ————, factors which lie in chromosomes of opposite gametes and do not frequently cross over at meiosis. The chromosomes concerned must be homologues except in a ring-forming interchange heterozygote. *Muller,* 1918, 1930 *b.*

[Linin], a structural component of the nucleus. The term has been applied to the descriptions of various artefacts and has no definite meaning.

Localisation, the genetic property of restriction of pachytene pairing and chiasma formation to one part of the chromosomes—proximal or distal. *D.,* 1931 *c.*

Macronucleus, the large vegetative nucleus of the Infusoria.

[Matrix], the space left round the solid part of a metaphase or anaphase chromosome when it contracts, by syneresis or otherwise, to reveal its characteristic or non-characteristic structure, following fixation. By extension applied to any analogous appearance at prophase or telophase or to the supposed cause of any behaviour that might result from any analogous differentiation. *Sharp,* 1935; *cf. D.,* 1935 *d.*

Matroclinal Inheritance, that in which the offspring resembles the mother more closely than the father (as opposed to Patroclinal).

Maturation, the formation of gametes or spores by meiosis.

Megaspore (Macrospore), a spore having the property of giving gametophytes with only female gametes (the embryo-sac).

Meiosis, a form of mitosis in which the nucleus divides twice and the chromosomes once. The prophase of meiosis is the prophase of the first of the two divisions. *Farmer* and *Moore,* 1905.

Merogony, development of part of an egg with the sperm nucleus, but without the egg nucleus, as in male parthenogenesis.

[Metabolic Stage], an unwarranted physiological description of the resting stage, *q.v.*

Metaphase, the stage of mitosis or meiosis in which the chromosomes lie in a plane at right angles to the axis of the spindle and half-way between the poles. *Strasburger,* 1884.

Microchromosomes, fragment chromosomes in the Hemiptera which do not form chiasmata and do not usually pair until metaphase. *Wilson,* 1905.

Micronucleus, the small, permanent, reproductive nucleus of the Infusoria.

Microspore, the spore which produces a gametophyte bearing only male gametes (the pollen grain).

Mitosis, the process by which daughter chromosomes are separated into two identical groups; the diagnostic property of division of the nucleus. The term may be conveniently used in contradistinction to meiosis. *Flemming,* 1882.

GLOSSARY

[Monoploid], "True Haploid," *q.v.*

Monosomic, a diploid organism lacking one chromosome of its proper complement (*v.* Trisomic, Tetrasomic).

Mosaic, in animals, the equivalent of a chimæra, *q.v.* In plants, a chimæra produced by repeated mutation.

Mother-cell, the cell with a diploid nucleus which by meiosis gives four haploid nuclei, *e.g.*, the spore mother-cell in Bryophyta and Pteridophyta, the microspore or pollen mother-cell and the megaspore or embryo-sac mother-cell in phanerogams. The sperm mother-cell is known as the *spermatocyte*, the egg mother-cell as the *oöcyte*, in animals. Many other terms, used in the lower plants, are unnecessary in general cytology.

Multivalent (*v.* Univalent).

[Non-cellular], unicellular organisms in the Protista. *Dobell.*

Non-disjunction, cytologically, the failure of separation of paired chromosomes at meiosis and their passage to the same pole ; genetically, any results that might be imputed to such an abnormality, although usually arising from the failure of pairing, or from multivalent formation. *Bridges*, 1914.

Non-homologous pairing, association of non-homologous parts of chromosomes at pachytene ; *cf.* torsion-pairing. *McClintock*, 1933 ; *cf.* D., 1935 c.

Nuclear sap, the fluid which is lost by the chromosomes as they contract during prophase and which fills the space of the nucleus.

Nucleolar Organiser, specific chromomere responsible for developing the nucleolus. *McClintock*, 1934.

Nucleolus, a body in the nucleus which disappears and does not resolve itself into chromosomes at mitosis.

Nucleus, a cell body reproducing by mitosis. The most constant constituent of animal and plant cells. *Robert Brown*, 1831.

[Oligopyrene], of spermatozoa, those deficient in chromosomes.

Oöcyte, egg mother-cell.

Orientation, the movement of chromosomes so that their centromeres lie axially with respect to the spindle, either as to their potential halves at mitosis (*auto-orientation*) or as to members of a pair in meiosis (*co-orientation*).

[Orthoploid], polyploid organisms with a number of balanced, complete, chromosome sets. *Cf.* Aneuploid. (Originally used for those with any even number of chromosomes, *v. Winkler*, 1916.)

Pachytene, the double thread (and the stage at which it occurs) produced by pairing of the chromosomes in the prophase of meiosis, *v. Winiwarter*, 1900.

Pairing of Chromosomes, (active) the coming together of chromosomes at zygotene or (passive) the continuance of their association at the first metaphase of meiosis.

    Somatic ———, the lying of homologous chromosomes especially close to one another at metaphase of mitosis.

    Secondary ———, the same phenomenon as seen amongst bivalents at meiosis. *D.*, 1928.

Parasynapsis, the association of chromosomes side by side observed at zygotene, as opposed to their alleged end-to-end association at this stage, or " telosynapsis." *Wilson*, 1912.

Parthenogenesis, a form of apomixis in which the female gamete develops without fertilisation. *Owen*, 1849.

    Diploid ———, that in which meiosis has failed first.

    Haploid ———, that in which fertilisation fails first.

Male ———, that in which the male nucleus develops into the nuclei of the embryo.

Phenotype, the external appearance produced by the reaction of an organism of a given genotype with a given environment. *Johannsen*, 1911.

Polar Bodies, the expelled products of the two divisions of the oöcyte nucleus in animals.

Polarisation, (i) of chromosomes at telophase of mitosis, and later the maintenance of their proximal parts on the polar side of the nucleus; (ii) of chromosomes at zygotene, the movement of their ends towards one part of the nuclear surface; (iii) of centromeres, the initiation of orientated division at metaphase of mitosis.

Polyhaploid, an organism with the gametic chromosome number, arising by parthenogenesis from a polyploid. *Katayama*, 1935.

Polyploid, an organism with more than two sets of homologous chromosomes. The terms used are triploid, tetraploid, pentaploid, hexaploid, heptaploid, octoploid (for octaploid), nonaploid (for enneaploid), decaploid, undecaploid (for hendecaploid), dodecaploid and so on. Higher multiples are best referred to as $14x$, $22x$ and so on (v. Haploid, Diploid and Tetraploid). *Winkler*, 1916.

Secondary ———, a homozygous allopolyploid in which some of the chromosomes in the basic set are present more frequently than others. *D.* and *Moffett*, 1930.

Polysomic (v. Trisomic).

Polytene, of chromosomes in the salivary gland nuclei of Diptera. *Koller*, 1935.

Position Effect, the differences in effect of two or more genes according to their distances apart in the chromosome thread. *Sturtevant*, 1926.

[Post-reduction], the alleged reduction of the number of chromosomes, or segregation of differences between partners, at the second meiotic division, as opposed to pre-reduction, their reduction or segregation at the first division. *Korschelt* and *Heider*, 1903.

Precocity, the property of the nucleus beginning prophase before the chromosomes have divided; characteristic of meiosis. *D.,* 1931, *b* and *c*.

Differential ———, the property of some chromosomes or their parts condensing, dividing or pairing in advance of the rest of the complement during prophase (v. Text).

Pro-chromosome, condensed proximal part of a chromosome, staining during the resting stage. V. Heterochromatin. *Overton*, 1909.

Pro-metaphase, stage between the dissolution of the nuclear membrane and the congression of the chromosomes on the metaphase plate.

Prophase, the stage in mitosis or meiosis from the appearance of the chromosomes to metaphase. *Strasburger*, 1884.

Prothallium, the gametophyte in the Pteridophyta.

[Prothallus], = Prothallium, *q.v.*

Protoplasm, that part of the living organism which is not dead (including the nucleus and cytoplasm). *Purkinje*, 1840; *Strasburger*, 1882.

Protoplast, the protoplasm of one cell. *Hanstein*, 1880.

Proximal, of a chromosome, a part that is nearer the centromere than a particular other part.

Pseudogamy, parthenogenetic development of the female gamete, requiring the stimulation of the male gamete.

Quadrivalent (v. Univalent).

Reduction, the halving of the chromosome number at meiosis and, by extension, its genetical concomitant of segregation. *Weismann*, 1887.

Double ———, the occurrence of a reductional division at both divisions in regard to particular parts of chromosomes; possible in some hybrids and polyploids; hence "equational exceptions." *D.*, 1929 *c*; *Haldane*, 1930 *a*.

    Gametic ———, meiosis immediately before fertilisation.

    Zygotic ———, meiosis immediately after fertilisation.

Reduction Division, formerly applied to the one of the meiotic divisions at which a particular author thought reduction and segregation occurred.

Reductional Division, a separation of homologous parts of chromosomes derived from opposite parents at anaphase of a first or second division (as opposed to *equational*).

Reductional Split, the same at diplotene. *Wenrich*, 1916.

Restitution Nucleus, a single nucleus formed through failure the first division. *Rosenberg*, 1927.

Rings, (i) at mitosis, chromosomes with no ends.

    (ii) at meiosis, chromosomes associated end to end in a ring, usually by terminal chiasmata; especially applied to diploid interchange heterozygotes where more than two chromosomes are so associated.

[Satellite], a trabant, *q.v.*

Segment, a portion of a chromosome which can conveniently be considered as a unit for a given purpose; *e.g.*, Differential segment. *Cf. Belling*, 1927 *b*.

Segregation, the separation of chromosomes of paternal and maternal origin at meiosis and the separation of the differences observed genetically. *Bateson*, 1902.

    Secondary ———, the segregation in an allopolyploid of differences between its ultimate diploid parents. *D.*, 1928.

    Effective ———, that which gives viable gametic or zygotic combinations (especially following multiple association in structural hybrids). *D.*, 1931.

[Semi-heterotypic Division], a first division which gives rise to a restitution nucleus following defective pairing. *Rosenberg*, 1927.

Set of Chromosomes, a minimum complement of chromosomes derived from the gametic complement of a supposed ancestor.

Sex-Chromosome, one which in the heterozygous sex is mated with a dissimilar homologue; the X and Y chromosomes, *q.v.* *Wilson*, 1906.

Sexual Differentiation, the production by an organism or by related organisms of gametes of two sizes such that the larger can fuse only with the smaller.

Sexual Reproduction, that which requires meiosis and fertilisation.

Sperm, Spermatozoon, the male gamete in animals.

Spermatocyte, sperm mother-cell.

Spindle, the axially differentiated part of the cytoplasm within which the centromeres of the chromosomes are held during metaphase and anaphase. Normally bipolar, exceptionally unipolar or multipolar.

[Spindle Attachment], the position of the centromere.

Spiral, a coil of the chromosome thread (chromosome or chromatid), at mitosis or meiosis.

    Internal ———, a coil within a single chromatid between prophase and anaphase.

    Relational ———, coiling of two chromatids or chromosomes round one another.

    Major ———, the larger internal coil at meiosis.

    Minor ———, the smaller internal coil.

    Relic ———, the coiling which survives at telophase and prophase.

    Super- ———, larger coils derived in prophase from the rearrangement of relic spirals.

Molecular ——, the coiling within the chromosome thread which conditions internal and relational spirals. *D.*, 1935.
  (Primary and secondary coils of Kuwada and Nakamura are minor and major spirals).

Spiralisation, the assumption of an internal (but not a relational) spiral by the chromatids in mitosis and meiosis. *D.*, 1932.

[Spireme], the chromosomes during the prophase in mitosis or meiosis, and hence the prophase itself. Particularly used in the expression " continuous spireme." *Flemming*, 1882.

Spore, a cell specialised for reproduction but not for fertilisation. In the higher plants it is always the product of meiosis and gives rise immediately to gametes.

Sporophyte, the spore-producing diploid in the higher plants (*v.* Gametophyte).

Structural Change, change in the genetic structure of the chromosome. May be intra-radial or extra-radial with respect to arms, internal, fraternal or external with regard to chromosomes, symmetrical or asymmetrical with respect to the possession of a centromere, eucentric or dyscentric with respect to the direction of a segment in relation to the centromere. *D.*, 1929 (*cf.* Ch. XII).

  Secondary —— ——, change in structure resulting from crossing-over between two homologous segments in chromosomes which are structurally different on both sides of these segments. *D.*, 1932. (*Cf.* Illegitimate crossing-over.)

Structure, the potentially permanent linear order of the particles, chromomeres or genes in the chromosomes. *Cf.* " Structural hybrid." *D.*, 1929 *c*.

Subsexual reproduction, parthenogenesis following ameiosis with non-reduction, but with segregation owing to crossing-over.

[Synapsis], (i) chromosome pairing at zygotene, also described as syndesis; (ii) contraction of the chromosomes to one side of the nucleus at this stage or at diplotene (an artefact), also described as synizesis.

Syndiploidy, the fusion of nuclei to give a doubled chromosome number, especially in the divisions immediately preceding meiosis. *Strasburger*, 1907.

[Syngamy], fusion of gametes.

Telophase, the last stage of mitosis, after movement of the chromosomes has ceased. *Heidenhain*, 1894.

Terminal Affinity, the property by which chromosomes are held together end to end from diplotene till metaphase or brought together in this way at metaphase. *D.*, 1932.

Terminal Chiasma (*v.* Chiasma.)

Terminalisation, expansion of the association of the two pairs of chromatids on one side of a chiasma at the expense of that on the other side. So called because the resulting " movement " of the chiasma is usually if not always towards the ends of chromosomes. *D.*, 1929 *c*.

  Arrest of ——, stoppage of the movement of a chiasma through the opposite segments distal to it being non-homologous (*v.* Text).

Tetrad, (i) quartet of cells formed by meiosis in a mother-cell; [(ii)] the four chromatids making up a bivalent at meiosis (*v.* Dyad, Hexad). *Nemec*, 1910.

Tetraploid, an organism having four sets of chromosomes (*v.* Polyploid). *Nemec*, 1910.

Tetrasomic, an organism with four chromosomes of one type, usually otherwise a diploid ($2x + 2$).

Torsion-pairing, non-homologous association at pachytene which releases a torsion without satisfying an attraction. *D.*, 1935 *c*.

Trabant (German), a segment of a chromosome, separated from the rest by one long constriction if terminal or two if intercalary ; usually so short that longitudinal contraction reduces it to a sphere before it attains the characteristic chromatid diameter.

Intercalary ——, one with full-sized chromosome limbs on both sides of it.

Lateral ——, where the trabant is a branch of the chromosome.

Translocation, change in position of a segment of a chromosome to another part of the same chromosome or of a different chromosome. *Bridges*, 1917.

Triploid, an organism having three sets of chromosomes. *Nemec*, 1910.

Trisomic, an organism with three chromosomes of one type ; usually otherwise diploid ($2x + 1$). A diploid organism is said to be doubly trisomic, not tetrasomic, when it has the two extra chromosomes of two different types ($2x + 1 + 1$).

Secondary ——, a trisomic organism in which the extra chromosome has two identical ends. *Belling*, 1927 b.

Tertiary ——, a trisomic organism in which the extra chromosome is made up of halves corresponding with the halves of different chromosomes in the normal set. *Belling*, 1927 b.

Trivalent (v. Univalent).

Univalent, a body at the first meiotic division corresponding with a single chromosome in the complement; especially when unpaired. Bivalent, Trivalent, Quadrivalent, Quinquevalent, Sexivalent, Septivalent, Octavalent (for Octovalent), etc., are associations of chromosomes held together between diplotene and metaphase of the first division by chiasmata. Similar associations of more than two including non-homologous chromosomes in structural hybrids should be described as " unequal bivalents," " associations " or " rings " of three, four, etc. The hybrid forms, " monovalent," " tetravalent," " pentavalent," " hexavalent," etc., used in chemistry should be avoided.

[W-chromosome], sometimes used for the X-chromosome where the female is the heterozygous sex, Z being then used for Y.

X-chromosome, with diploid sex differentiation, the sex chromosome in regard to which one sex is homozygous ; this is said to be the homozygous sex. With sex differentiation in the haploid, the sex chromosome of the female.

Y-chromosome, the sex chromosome that is present and pairs with the X in the sex heterozygote.

Zygote, the cell formed by the union of gametes and the individual derived from it (*cf*. Homozygote).

Zygotene, the pairing threads (and the stage at which they occur) in the prophase of meiosis. *v. Winiwarter*, 1900.

APPENDIX IV

BIBLIOGRAPHY

Un homme de bon sens croit toujours ce qu'on luy dit,
et ce qu'il trouve par escrit.

RABELAIS, *Gargantua*. Ch. VI.

NOTE.—Reviews and general works are marked with an asterisk. The earlier works of some authors have been omitted and can be found from their later references. Articles by the present author are referred to in the text by the letter D.

AFZELIUS, K., 1928, Die Embryobildung bei Nigritella nigra. *Sv. Bot. Tidskr.*, **22**, 82–91.
AGAR, W. E., 1911, The spermatogenesis of Leptosiren paradoxa. *Quart. J. Micr. Sci.*, **67**, 1–44.
AGAR, W. E., 1914, Parthenogenetic and sexual reproduction in *Sinocephalus vetulus* and other Cladocera. *J. Genet.*, **3**, 179–194.
*AGAR, W. E., 1920, *Cytology, with special reference to the Metazoan nucleus*. London.
AGAR, W. E., 1923, The Male Meiotic Phase in two Genera of Marsupials (Macropus and Pentauroides). *Quart. J. Micr. Sci.*, **67**, 183–202.
AKEMINE, T., 1935, Chromosome studies on *Hosta* I. The chromosome numbers in various species of *Hosta*. *J. Fac. Sci. Hokkaido* V, **5**, 25–32.
ÅKERLUND, E., 1927, Ein *Melandrium*-Hermaphrodit mit weiblichen Chromosomenbestand. *Hereditas*, **10**, 153–159.
ALLEN, C. E., 1919, The Basis of Sex Inheritance in *Sphærocarpos*. *Proc. Amer. Phil. Soc.*, **58**, 298–316.
ALLEN, C. E., 1926, The Direct Results of Mendelian Segregation. *Proc. Nat. Acad. Sci.*, **12**, 2–7.
*ALLEN, C. E., 1932, Sex-Inheritance and Sex-Determination. *Amer. Nat.*, **66**, 97–107.
ALLEN, C. E., 1934, A diploid female gametophyte of *Sphærocarpos*. *Proc. Nat. Acad. Sci.*, **20**, 147–150.
ALLEN, C. E., 1935 a, The occurrence of polyploidy in *Sphærocarpos*. *Amer. J. Bot.*, **22**, 635–644.
*ALLEN, C. E., 1935 b, The genetics of bryophytes. *Bot. Rev.*, **1**, 269–291.
ALTENBURG, E., 1934, The artificial production of mutations by ultra-violet light. *Amer. Nat.*, **68**, 491–507.
ALVERDES, F., 1912, Die Kerne in den Speicheldrusen der Chironomus-Larve. *Arch. Zellforsch. u. mikr. Anat.*, **9**, 168–204.
ANDERSON, E. G., 1925, Crossing over in a case of attached X chromosomes in *Drosophila melanogaster*. *Genetics*, **10**, 403–417.
ANDERSON, E. G., 1929, Studies on a case of high non-disjunction in *Drosophila melanogaster*. *Z.I.A.V.*, **51**, 397–441.

ANDERSON, E. G., 1935, Chromosome interchanges in maize. *Genetics*, **20**, 70–83.
ANDERSON, E. G., and CLOKEY, I. W., 1934, Chromosomes involved in a series of interchanges in maize. *Amer. Nat.*, **68**, 440–445.
ANDERSSON, I., 1927, Notes on some characters in Ferns subject to Mendelian inheritance. *Hereditas*, **9**, 157–179.
*ANDERSSON-KOTTÖ, I., 1931, The Genetics of Ferns. *Bibliogr. Genet.*, **8**, 269–294.
ANDERSSON-KOTTÖ, I., 1932, Observations on the Inheritance of Apospory and Alternation of Generations. *Sv. Bot. Tidskr.* **26**.
ANDERSSON-KOTTÖ, I., 1936, On the comparative development of alternating generations, with special reference to ferns. *Sv. Bot. Tids.*, **30**, 57–78.
ANDERSSON-KOTTÖ, I., and GAIRDNER, A. E., 1931, Interspecific Crosses in the Genus *Dianthus*. *Genetica*, **13**, 77–112.
ANDERSSON-KOTTÖ, I., and GAIRDNER, A. E., 1936, The inheritance of apospory in *Scolopendrium vulgare*. *J. Genet.*, **32**, 189–228.
ANDREWS, F. M., 1915, Die Wirkung der Zentrifugalkraft auf Pflanzen. *Jahrb. wiss. Bot.*, **56**, 221–253.
*ANKEL, W. E., 1927, 1929, Neuere Arbeiten zur Zytologie der naturlichen Parthenogenese der Tiere. Sammelreferat. *Z.I.A.V.*, **45**, 232–278. **52**, 318–370.
ANKEL, W. E., 1930, Die atypische Spermatogenese von Janthina (Prosobranchia Ptenoglossa). *Zeits. Zellforsch. u. Mikr. Anat.*, **11**, 491–608.
ARAKI, S., 1932, Karyologische Untersuchungen über einen Artbastard zwischen *Potentilla chinensis* (haploid 1 n) und *P. nipponica* (hapl. 2 n). *J. Sci. Hiroshima Univ.* B, **2**, **1**, 103–116.
ARTOM, C., 1928, Diploidismo e Tetraploidismo dell' Artemia salina. *V. Int. Kong. Vererb.*, **1**, 384–386.
ASANA, J. J., and MAKINO, S., 1935, A comparative study of the chromosomes in the Indian dragonflies. *J. Fac. Sci. Hokkaido* VI, **4**, 67–86.
ASTBURY, W. T., 1933, *Fundamentals of Fibre Structure*. Oxford.
AVDOULOV, N. P., 1931, Karyo-systematische Untersuchungen der Familie Gramineen. *Bull. Appl. Bot. Suppt.*, **44**, 1–438.
AVDULOV, N. P., 1933, On the additional chromosomes in Maize. *Bull. Appl. Bot.*, 101–130.
AVDULOV, N. P., and TITOVA, N., 1933, Additional chromosomes in Paspalum stoloniferum Bosco. *Bull. Appl. Bot.*, 165–172.
AVERY, P., 1929, Chromosome Numbers and Morphology in Nicotiana IV. The Nature and Effects of Chromosomal Irregularities in N. alata var. grandiflora. *Univ. Calif. Pub. Bot.*, **11**, 265–284.
AVERY, P., 1930, Cytological Studies of Five Interspecific Hybrids of Crepis Leontodontoides. *Univ. Calif. Pub. Agr. Sci.*, **6**, 135–167.

BABCOCK, E. B., and CAMERON, D. R., 1934, Chromosomes and phylogeny in Crepis II. The relationships of one hundred eight (sic) species. *Univ. Calif. Pub. Agr. Sci.*, **6** (11), 287–324.
BABCOCK, E. B., and CLAUSEN, J., 1929, Meiosis in two species and three hybrids of Crepis and its bearing on taxonomic relationship. *Univ. Calif. Pub. Agr. Sci.*, **2**, 401–432.
*BABCOCK, E. B., and NAVASHIN, M., 1930, The Genus Crepis. *Bibliogr. Genet.*, **6**, 1–90.
BACKMAN, E., 1935, A rapid combined fixing and staining method for plant chromosome counts. *Stain Technology*, **10**, 83–86.
BAEHR, V. B. de, 1919, Recherches surl a maturation des œufs parthénogénétiques dans l'Aphis palmæ. *Cellule*, **30**, 315–353.

BAEHR, W. B. v., 1907, Über die Zahl der Richtungskörper in Parthenogenetisch sich entwickelnden Eiern von *Bacillus rossii*. *Zool. Jahrb. Anat.*, **24**, 175–192.
BALBIANI, E. G., 1881, Sur la structure du noyau des cellules salivaires chez les larves de *Chironomus*. *Zool. Anz.*, **4**, 637–641.
BALTZER, F., 1910, Uber die Beziehung zwischen dem Chromatin und der Entwicklung und Vererbungsrichtung bei Echinodermenbastarden. *Arch. Zellf. u. mikr. Anat.*, **5**, 497–621.
BANTA, A. M., SNIDER, K. G., and WOOD, T. R., 1926, Inheritance in parthenogenesis and in sexual reproduction in a cladoceran. *Proc. Soc. Exp. Biol. N.Y.*, **23**, 621–622.
BANTA, A. M., and WOOD, T. R., 1928, Inheritance in Parthenogenesis and in Sexual Reproduction in Cladocera. *V. Int. Kong. Vererb.*, **1**, 391–396.
BARANETSKY, J., 1880, Die Kerntheilung in den Pollenmutterzellen einiger Tradescantien. *Bot. Zeitg.*, **38**, 241–248, 265–274, 281–296.
BARIGOZZI, C. 1933, Die Chromosomengarnitur der Maulwurfsgrille und ihre systematische Bedeutung. *Zeits. Zellf. u. mikr. Anat.*, **18**, 626–641.
BARIGOZZI, C., 1935, Il legame genetico fra i biotipi partenogenetici di *Artemia salina*. *Archivio zoologico italiano*, **22**, 33–77.
BARRETT, W. C., 1932, Heidenhain's Hæmatoxylin used with the Smear Technic. *Stain Technology*, **7**, 63–64.
BARTLETT, H. H., 1915, Mutation en masse. *Amer. Nat.*, **49**, 129–148.
BAUCH, R., 1926, Untersuchungen über zweisporige Hymenomyceten. I. Haploide Parthenogenesis bei Camarophyllus virgineus. *Zeits. f. Bot.*, **18**, 337–387.
BAUER, H., 1931, Die Chromosomen von Tipula paludosa Meig. in Eibildung und Spermatogenese. *Zeits. Zellforsch. u. mikr. Anat.*, **14**, 138–193.
BAUER, H., 1932, Die Feulgensche Nuklealfärbung in ihrer Anwendung auf cytologische Untersuchungen. *Zeits. Zellf. u. mikr. Anat.*, **15**, 225–247.
BAUER, H., 1933, Die wachsenden Oocytenkerne einiger Insekten in ihrem Verhalten zur Nuklealfärbung. *Zeits. Zellf. u. mikr. Anat.*, **18**, 254–296.
BAUER, H., 1935 *a*, Die Speicheldrüsenchromosomen der Chironomiden. *Naturwiss.*, **23**, 475–476.
BAUER, H., 1935 *b*, Der Aufbau der Chromosomen aus den Speicheldrüsen von Chironomus Thummi Kiefer. *Zeits. Zellf. u. mikr. Anat.*, **23**, 280–313.
BEADLE, G. W., 1930, Genetical and Cytological Studies of Mendelian Asynapsis in Zea Mays. *Cornell Univ. Exp. Sta. (Ithaca) Mem.*, **129**.
BEADLE, G. W., 1931, A gene in maize for supernumerary cell divisions following meiosis. *Cornell Univ. Exp. Sta. (Ithaca) Mem.*, **135**.
BEADLE, G. W., 1932, A gene in *Zea Mays* for failure of cytokinesis during meiosis. *Cytologia*, **3**, 142–155.
BEADLE, G. W., 1932 *a*, Studies of *Euchlæna* and its hybrids with *Zea*. 1. Chromosome behaviour in *Euchlæna mexicana* and its hybrids with *Zea Mays*. *Z.I.A.V.*, **62**, 291–304.
BEADLE, G: W:, 1932 *b*, Genes in maize for pollen sterility. *Genetics*, **17**, 393–412.
BEADLE, G. W., 1932 *c*, The relation of crossing-over to chromosome association in Zea-Euchlænas hybrids. *Genetics*, **17**, 481–501.
BEADLE, G. W., 1932 *d*, A gene in *Zea Mays* for failure of cytokinesis during meiosis. *Cytologia*, **3**, 142–155.
BEADLE, G. W:, 1932 *e*, A gene for sticky chromosomes in *Zea Mays*. *Z.I.A.V.*, **63**, 195–217.
BEADLE, G. W., 1932 *f*, A possible influence of the spindle fibre on crossing-over in Drosophila. *Proc. Nat. Acad. Sci.*, **18**, 160–165.

BIBLIOGRAPHY

BEADLE, G. W., 1933 a, Polymitotic maize and the precocity hypothesis of chromosome conjugation. *Cytologia*, **5**, 118-121.
BEADLE, G. W., 1933 b, Further studies in asynaptic maize. *Cytologia*, **4**, 269-287.
BEADLE, G. W., 1935, Crossing over near the spindle attachment of the X-chromosomes in attached X-triploids of Drosophila melanogaster. *Genetics*, **20**, 179-191.
BEADLE, G. W., and STURTEVANT, A. H., 1935, X-chromosome inversions and meiosis in Drosophila melanogaster. *Proc. Nat. Acad. Sci.*, **21**, 384-390.
BEAL, J. M., 1934, Chromosome behavior in Pinus Banksiana following fertilisation. *Bot. Gaz.*, **95**, 660-666.
BEAL, J. M., 1936, Double interlocking of bivalent chromosomes in Lilium elegans. *Bot. Gaz.*, **97**, 678-680.
BEAMS, H. W., and KING, R. L., 1935, The effect of ultracentrifuging on the cells of the root tip of the bean (Phaseolus vulgaris). *Proc. Roy. Soc.*, B **118**, 264-276.
BEATUS, R., 1934, Der Erbgang der Pentasepalie bei den Zwischenrassen von *Veronica Tournefortii*. *Jahrb. wiss. Bot.*, **79**, 256-295.
BECKER, W. A., 1933, Vitalbeobachtungen über den Einfluss von Methylenblau und Neutralrot auf den Verlauf von Karyo- und Zyto-kinese. Beitrag zur Pathologie der Mitose. *Cytologia*, **4**, 135-157.
BEHRE, K., 1929, Physiologische und zytologische Untersuchungen über Drosera. *Planta* **7**, 208-306.
BELAR, K., 1922, Untersuchungen an *Actinophrys sol* Ehrenberg. I. Die Morphologie des Formwechsels. *Arch. Prot.*, **46**, 1-96.
BELAR, K., 1923, Uber den Chromosomenzyklus von parthenogenetischen Erdnematoden. *Biol. Zbl.*, **43**, 513-518.
BELAR, K., 1925, Der Chromosomenbestand der *Melandrium*-Zwitter. *Z.I.A.V.*, **39**, 184-190.
BELAR, K., 1926 a, Zur Cytologie von *Aggregata eberthi*. *Arch. Prot.*, **53**, 312-325.
BELAR, K., 1926 b, Der Formwechsel der Protistenkerne. *Erg. Fortschr. Zool.*, **6**, 235-652.
BELAR, K., 1927, Beiträge zur Kenntnis des Mechanismus der indirekten Kernteilung. *Naturwiss.*, **36**, 725-733.
BELAR, K., 1928 a, Über die Naturtreue des fixierten Präparats. *V. Int. Kong. Vererb.*, **1**, 402-407.
*BELAR, K., 1928 b, *Die Cytologischen Grundlagen der Vererbung*. Berlin.
BELAR, K., 1929 a, Beiträge zur Kausalanalyse der Mitose II. Untersuchungen an den Spermatocyten von Chorthippus (Stenobothrus) lineatus Panz. *Arch. f. Entwm.*, **118**, 359-484.
BELAR, K., 1929 b, Beiträge zur Kausalanalyse der Mitose III. Untersuchungen an den Staubfaden, Haarzellen und Blattmeristemzellen von *Tradescantia virginica*. *Zeits. Zellforsch. u. mikr. Anat.*, **10**, 73-134.
*BELAR, K., 1929 c, Die Technik der deskriptiven Cytologie. *Meth. wiss. Biol.*, **1**, 638-735.
BELAR, K., 1930, Uber die reversible Entmischung des lebenden Protoplasmas I. *Protoplasma*, **9**, 209-244.
BELAR, K., 1933 (W. Huth), Zur Teilungsantontomie der Chromosomen. *Zeits. Zellforsch. u. mikr. Anat.*, **17**, 51-66.
BELIAJEFF, N. K., 1930, Die Chromosomenkomplexe und ihre Beziehung zur Phylogenie bei den Lepidopteren. *Z.I.A.V.*, **54**, 369-399.
BELLING, J., 1921, The Behavior of homologous Chromosomes in a triploid Canna. *Proc. Nat. Acad. Sci.*, **7**, 197-201.

BELLING, J., 1924 a, The distribution of Chromosomes in the pollen-grains of a triploid Hyacinth. *Amer. Nat.*, **58**, 440–446.
BELLING, J., 1924 b, Detachment (Elimination) of chromosomes in *Cypripedium acaule*. *Bot. Gaz.*, **78**, 458–460.
BELLING, J., 1925 a, Fracture of Chromosomes in Rye. *J. Hered.*, **16**, 360.
BELLING, J., 1925 b, Homologous and similar chromosomes in diploid and triploid hyacinths. *Genetics*, **10**, 59–71.
BELLING, J., 1925 c, The Origin of Chromosomal Mutations in *Uvularia*. *J. Genet.*, **15**, 245–266.
BELLING, J., 1925 d, Chromosomes of Canna and of Hemerocallis. *J. Hered.*, **16**, 465–466.
BELLING, J., 1926 a, The Iron-Acetocarmine Method of Fixing and Staining Chromosomes. *Biol. Bull.*, **50**, 160–162.
BELLING, J., 1926 b, Single and Double Rings at the Reduction Division in *Uvularia*. *Biol. Bull.*, **50**, 355–363.
BELLING, J., 1926 c, The Structure of Chromosomes. *J. Exp. Biol*, **3**, 145–147.
BELLING, J., 1927 a, Configurations of Bivalents of Hyacinthus with regard to Segmental Interchange. *Biol. Bull.*, **52**, 480–487.
BELLING, J., 1927 b, The Attachments of Chromosomes at the Reduction Division in the Flowering Plants. *J. Genet.*, **18**, 177–205.
BELLING, J., 1928 a, A Method for the Study of Chromosomes in Pollen-Mother-Cells. *Univ. Calif. Pub. Bot.*, **14**, 293–299.
BELLING, J., 1928 b, Nodes and Chiasmas in the Bivalents of *Lilium* with regard to segmental interchange. *Biol. Bull.*, **54**, 465–470.
BELLING, J., 1928 c, A Working Hypothesis for Segmental Interchange between Homologous Chromosomes in Flowering Plants. *Univ. Calif. Pub. Bot.*, **14**, 283–291.
BELLING, J., 1928 d, The Ultimate Chromomeres of Lilium and Aloë with Regard to the Number of Genes. *Univ. Calif. Pub. Bot.*, **14**, 307–318.
BELLING, J., 1928 e, Contraction of Chromosomes during Maturation Divisions in Lilium and Other Plants. *Univ. Calif. Pub. Bot.*, **14**, 335–343.
BELLING, J., 1929, Nodes and Internodes of Trivalents of Hyacinthus. *Univ. Calif. Pub. Bot.*, **14**, 379–388.
BELLING, J., 1930, *The Use of the Microscope*. New York. (McGraw Hill.)
BELLING, J., 1931 a, Chromomeres of Liliaceous Plants. *Univ. Calif. Pub. Bot.*, **16**, 153–170.
BELLING, J., 1931 b, Chiasmas in Flowering Plants. *Univ. Calif. Pub. Bot.*, 311–338.
BELLING, J., 1933 a, Crossing-over and gene rearrangement in flowering plants. *Genetics*, **18**, 388–413.
BELLING, J., 1933 b, Critical notes on C. D. Darlington's " Recent Advances in Cytology." *Univ. Calif. Pub. Bot.*, **17**, 75–110.
BELLING, J., and BLAKESLEE, A. F., 1922, The assortment of chromosomes in triploid Daturas. *Amer. Nat.*, **56**, 339–346.
BELLING, J., and BLAKESLEE, A. F., 1923, The Reduction Division in Haploid, Diploid, Triploid and Tetraploid Daturas. *Proc. Nat. Acad. Sci.*, **9**, 106–111.
BELLING, J., and BLAKESLEE, A. F., 1924 a, The Distribution of Chromosomes in Tetraploid Daturas. *Amer. Nat.* **58**, 60–70.
BELLING, J., and BLAKESLEE, A. F., 1924 b, The configurations and sizes of the chromosomes in the trivalents of 25-chromosome Daturas. *Proc. Nat. Acad. Sci.*, **10**, 116–120.

BELLING, J., and BLAKESLEE, A. F., 1926, On the attachment of non-homologous chromosomes at the reduction division in certain 25-chromosome Daturas. *Proc. Nat. Acad. Sci.*, **12**, 7–11.
BELLING, J., and BLAKESLEE, A. F., 1927, The Assortment of Chromosomes in Haploid Daturas. *Cellule*, **37**, 353–365. (*See also :* Blakeslee.)
BERG, K. H. v., 1931 *a*, Ein Bastard mit vier vollständigen haploiden Artgenomen. *Ak. Anz.* (Vienna), **22**.
BERG, K. H. v., 1931 *b*, Autosyndese in *Ægilops triuncialis* L. x *Secale cereale* L. *Zeits. f. Zücht.*, **17**, 55–69.
BERG, K. H. v., 1934 *a*, Cytologische Untersuchungen an den Bastarden des Triticum turgidovillosum und an einer F_1 Triticum turgidum × villosum. *Z.I.A.V.*, **68**, 94–126.
BERG, K. H. v., 1934 *b*, Cytologische Untersuchungen an Triticum turgidovillosum und seinen Eltern. *Z.I.A.V.*, **67**, 342–373.
BERGMAN, B., 1934, Zytologische Studien über sexuelles und asexuelles *Hieracium umbellatum. Hereditas*, **20**, 47–64.
BERGMAN, B., 1935 *a*, Zur Kenntnis der Zytologie der skandinavischen Antennaria-Arten. *Hereditas*, **20**, 214–226.
BERGMAN, B., 1935 *b*, Zytologische Studien über die Fortpflanzung bei den Gattungen Leontodon und Picris. *Sv. Bot. Tids.*, **29**, 155–301.
BERGNER, A. D., and BLAKESLEE, A. F., 1932, Cytology of the Ferox-Quercifolia-Stramonium triangle in *Datura. Proc. Nat. Acad. Sci.*, **18**, 151–159.
BERGNER, A. D., and BLAKESLEE, A. F., 1935, Chromosome ends in Datura discolor. *Proc. Nat. Acad. Sci.*, **21**, 369–374.
BERGNER, A. D., CARTLEDGE, J. L., and BLAKESLEE, A. F., 1934, Chromosome behaviour due to a gene which prevents metaphase pairing in *Datura. Cytologia*, **6**, 19–37.
BERGNER, A. D., SATINA, S., and BLAKESLEE, A. F., 1933, Prime types in Datura. *Proc. Nat. Acad. Sci.*, **19**, 103–115.
BERNAL, J. D., and CROWFOOT, D., 1934, X-ray photographs of crystalline pepsin. *Nature*, **133**, 794.
BLACKBURN, K. B., 1924, The cytological aspects of the determination of sex in diœcious forms of Lychnis. *J. Exp. Biol.*, **1**, 413–430.
BLACKBURN, K. B., 1928, Chromosome Number in Silene and the Neighbouring Genera. *V. Int. Kong. Vererb.*, **1**, 439–446.
BLACKBURN, K. B., 1933 *a*, On the relation between geographic races and polyploidy in *Silene ciliata.* Pourr. *Genetica*, **15**, 49–66.
BLACKBURN, K. B., 1933 *b*, Note on the chromosomes of the Duckweeds (Lemnaceæ) introducing the question of chromosome size. *Proc. Univ. Durham Phil. Soc.*, **9**, 84–90.
BLACKBURN, K. B., and HARRISON, J. W. H., 1921, The Status of the British Rose Forms as determined by their Cytological Behaviour. *Ann. Bot.*, **35**, 159–188.
BLACKBURN, K. B., and HARRISON, J. W. H., 1924, A Preliminary Account of the Chromosomes and Chromosome Behaviour in the *Salicaceæ. Ann. Bot.*, **38**, 361–378.
BLAKESLEE, A. F., 1922, Variations in Datura due to Changes in Chromosome Number. *Amer. Nat.*, **56**, 16–31.
*BLAKESLEE, A. F., 1928, Genetics of *Datura. V. Int. Kong. Vererb.*, **1**, 117–130.
BLAKESLEE, A. F., 1929, Cryptic types in *Datura. J. Hered.*, **20**, 177–190.
BLAKESLEE, A. F., 1931, *Quoted in* An. Rep. of Director, Dept. of Genetics, Carnegie Inst. *Carnegie Inst. Yearbk.*, **29**, 35–68 (v. pp. 35–40, 43, 44).
BLAKESLEE, A. F., 1931, Extra Chromosomes, a Source of Variations in the Jimson weed. *Smithson. Rep.*, 1930, 431–450.

BLAKESLEE, A. F., and BELLING, J., 1924, Chromosomal chimaeras in the Jimson Weed. *Science*, **60**, 19–20.
BLAKESLEE, A. F., BELLING, J., and FARNHAM, A., 1923, Inheritance in Tetraploid Daturas. *Bot. Gaz.*, **76**, 329–373.
BLAKESLEE, A. F., BELLING, J., FARNHAM, A., and BERGNER, D., 1922, A haploid mutant in the Jimson weed, " Datura Stramonium." *Science*, **55**, 646–647.
BLAKESLEE, A. F., and CARTLEDGE, J. L., 1926, Pollen abortion in chromosomal types of Datura. *Proc. Nat. Acad. Sci.*, **12**, 315–323.
BLAKESLEE, A. F., BERGNER, A. D., and AVERY, A. G., 1933, Methods of synthesising pure-breeding types with predicted characters in the Jimson Weed, Datura Stramonium. *Proc. Nat. Acad. Sci.*, **19**, 115–122.
BLAKESLEE, A. F., and CLELAND, R. E., 1930, Circle Formation in Datura and Œnothera. *Proc. Nat. Acad. Sci.*, **16**, 177–183.
BLAKESLEE, A. F., MORRISON, G., and AVERY, A. G., 1927, Mutations in a Haploid Datura, and their bearing on the hybrid-origin theory of Mutants. *J. Hered.*, **18**, 193–199.
BLEIER, H., 1928 a, Zytologische Untersuchungen an seltenen Getreide- und Rübenbastarden. *V. Int. Kong. Vererb.*, **1**, 447–452.
BLEIER, H., 1928 b, Karyologische Untersuchungen an Linsen-Wicken Bastarden. *Genetica*, **11**, 111–118.
BLEIER, H., 1930 a, Untersuchungen über das Verhalten der verchiedenen Kernkomponenten bei der Reduktionsteilung von Bastarden. *Cellule*, **40**, 85–144.
BLEIER, H., 1930 b, Experimentell-cytologische Untersuchungen I. Einfluss abnormale Temperatur auf die Reduktionsteilung. *Zeits. f. Zellf. u. mikr. Anat.*, **11**, 218–236.
BLEIER, H., 1931, Zur Kausalanalyse der Kernteilung. *Genetica*, **13**, 27–76.
BØCHER, T. W., 1932, Beiträge zur Zytologie der Gattung Anemone. *Dansk. Bot. Tids.*, **42**, 183–206.
BØCHER, T. W., 1936, Cytological studies on Campanula rotundifolia. *Hereditas*, **22**, 269–277.
BONNIER, G., 1923, Studies on High and Low Non-Disjunction in Drosophila melanogaster. *Hereditas*, **4**, 81–110.
BORING, A. M., 1907, A study of spermatogenesis of twenty-two species of the Membracidæ, Jassidæ, Ceropidæ and Fulgoridæ, with special reference to the behavior of the odd chromosome. *J. Exp. Zool.*, **4**, 469–512.
BOSTIAN, C. H., 1936, Fecundity of triploid females in Habrobracon juglandis. (Abstract.) *Amer. Nat.*, **70**, 40.
BOUIN, P., 1925, Les cinèses de maturation et la double spermatogenèse chez Scolopendra cingulata L. *Cellule*, **35**, 371–423.
BOVERI, T., 1895, Über das Verhalten der Centrosomen bei der Befruchtung des Seeigel-Eies nebst allgemeinen Bemerkungen über Centrosomen und Verwandtes. *Verh. Phys-Med. Gess. Wurzb.*, N. F., **29**, 1–75.
*BOVERI, T., 1904, Ergebnisse über die Konstitution der chromatischen Substanz des Zellkerns. Jena.
BOVERI, T., 1907, Zellenstudien VI. Die Entwicklung dispermer Seeigel-eier. Ein Beitrag zur Befruchtungslehre und zur Theorie des Kerns. *Jena Zeitschr.*, **37** (43), 1–292.
BOVERI, T., 1909, Die Blastomerenkerne von Ascaris megalocephala und die Theorie der Chromosomenindividualität. *Arch. f. Zellforsch.*, **3**, 181–268.
BOVERI, T., 1911, Über das Verhalten der Geschlechtschromosomen bei Hermaphroditismus. Beobachtungen an Rhabditis nigrovenosa. *Verh. Phys.-Med. Ges. Würzburg*, **41**, 83–97.

BIBLIOGRAPHY

BOWEN, R. H., 1924, Studies on insect spermatogenesis VI. Notes on the formation of the sperm in coleoptera and aptera, with a general discussion of flagellate sperms. *J. Morphol.*, **39**, 351–413.
*BRAGG, W. H., 1933, Liquid crystals. *Proc. Roy. Inst. G. B.*, **28**, 57–92.
BRAUER, A., 1894, Zur Kenntniss der Reifung des parthenogenetisch sich entwickelnden Eies von *Artemia salina*. *Arch. mikr. Anat.*, **43**, 162–222.
BRAUER, A., 1928, Spermatogenesis of Bruchus quadrimaculatus (Coleoptera ; Bruchidæ). *J. Morphol.*, **46**, 217–239.
BREMER, G., 1923, A cytological investigation of some species and species hybrids within the genus *Saccharum*. *Genetica*, **5**, 97–148, 273–326.
BREMER, G., 1924, The cytology of the sugar cane. *Genetica*, **6**, 497–525.
BREMER, G., 1925, The cytology of the sugar cane. Third contribution. *Genetica*, **7**, 293–322.
BREMER, G., 1930, The Cytology of Saccharum. *Proc. 3rd Cong. Int. Soc. Sugar Cane Tech.*, Sourabaya, 408–415.
BRESLAWETZ, L. P., 1925, Polyploide Mitosen bei Cannabis sativa L. *Ber. deuts. bot. Ges.*, **44**, 498–502.
BRESLAWETZ, L., 1929, Zytologische Studien über Melandrium album L. *Planta*, **7**, 444–460.
BRESLAWETZ, L. P., 1935, Abnormal development of pollen in different races and grafts of hemp. *Genetica*, **17**, 154–169.
BRIDGES, C. B., 1914, Direct Proof through Non-disjunction that the Sex-linked Genes of Drosophila are borne by the X-Chromosome. *Science* n.s., **40**, 107–109.
BRIDGES, C. B., 1916, Non-disjunction as a proof of the chromosome theory of heredity. *Genetics*, **1**, 1–52, 107–163.
BRIDGES, C. B., 1917, Deficiency. *Genetics*, **2**, 445–465.
BRIDGES, C. B., 1919, Duplication. *Anat. Rec.*, **15**, 357–358. (Abs. Proc. Amer. Soc. Zool.)
BRIDGES, C. B., 1921 a, Triploid Intersexes in Drosophila melanogaster. *Science*, **54**, 252–254.
BRIDGES, C. B., 1921 b, Genetical and Cytological Proof of Non-disjunction of the Fourth Chromosome of Drosophila melanogaster. *Proc. Nat. Acad. Sci.*, **7**, 186–192.
BRIDGES, C. B., 1922, The Origin of Variations in Sexual and Sex-limited Characters. *Amer. Nat.*, **56**, 51–63.
BRIDGES, C. B., 1923, The translocation of a section of chromosome I upon chromosome III in *Drosophila*. *Anat. Rec.*, **24**, 426–427. (Abs. Proc. Amer. Soc. Zool.)
BRIDGES, C. B., 1925 a, Sex in relation to chromosomes and genes. *Amer. Nat.*, **59**, 127–137.
BRIDGES, C. B., 1925 b, Elimination of Chromosomes due to a mutant (minute-n) in *Drosophila melanogaster*. *Proc. Nat. Acad. Sci.*, **11**, 701–706.
BRIDGES, C. B., 1925 c, Haploidy in *Drosophila melanogaster*. *Proc. Nat. Acad. Sci.*, **11**, 706–711.
BRIDGES, C. B., 1927 a, The Relation of the Age of the Female to Crossing Over in the Third Chromosome of Drosophila melanogaster. *J. Gen. Physiol.*, **8**, 689–700.
BRIDGES, C. B., 1927 b, Constrictions in the Chromosomes of Drosophila melanogaster. *Biol. Zbl.*, **47**, 600–603.
BRIDGES, C. B., 1929, Variation in crossing-over in relation to age of female in *Drosophila melanogaster*. Carn. Inst., Washington, 399.
BRIDGES, C. B., 1935, Salivary chromosome maps. *J. Hered.*, **26**, 60–64.
BRIDGES, C. B., 1936, The Bar " gene " a duplication. *Science*, **83**, 210–211.
BRIDGES, C. B., and ANDERSON, E. G., 1925, Crossing over in the X chromo-

somes of triploid females of *Drosophila melanogaster*. *Genetics*, **10**, 418–441. (*See also*: T. H. Morgan.)
BRIEGER, F., 1928, Über die Vermehrung der Chromosomenzahl bei dem Bastard *Nicotiana tabacum* L. x *N. Rusbyi* Britt. *Z.I.A.V.*, **47**, 1–53.
*BRIEGER, F., 1930 a, Selbststerilität und Kreuzungssterilität. *Monog. Gesamt. d. Phys. Pfl. u. Tiere*, **21** (Berlin).
BRIEGER, F., 1930 b, Über die Bedeutung der Chromosomenverdoppelung für das Problem der Artentstehung. *Ber. Deuts. Bot. Ges.*, **48**, 95–98.
BRIEGER, F., 1934, Ablauf der Meiose bei völliger Asyndese. *Ber. Deuts. Bot. Ges.*, **52**, 149–153.
BRINK, R. A., and COOPER, D. C., 1932 a, A strain of maize homozygous for segmental interchanges involving both ends of the P–Br chromosome. *Proc. Nat. Acad. Sci.*, **18**, 441–447.
BRINK, R. A., and COOPER, D. C., 1932 b, A structural change in the chromosomes of maize leading to chain formation. *Amer. Nat.*, **66**, 310–322.
BRINK, R. A., and COOPER, D. C., 1935, A proof that crossing over involves an exchange of segments between homologous chromosomes. *Genetics*, **20**, 22–35.
BRITTINGHAM, W. H., 1936, Œnothera franciscana mut. tenuifolia, a gene mutation induced by radium treatment. (Abstract.) *Amer. Nat.*, **70**, 41.
BRUNELLI, G., 1910, La spermatogenesi della Tryxalis. IA. Divisioni spermatogoniali. *Mem. Soc. Ital. Sci.*, Ser. III, **16**, 221–236.
BRUNELLI, G., 1911, La spermatogenesi della Tryxalis : Divisioni maturative. *Mem. Acad. Lincei*, Ser. 5, **8**, 634–652.
BRUUN, H., 1930, The cytology of the genus Primula. A preliminary report. *Sv. Bot. Tidskr.*, **24**, 468–475.
BRUUN, H. G., 1932, Cytological studies in Primula. *Symbolæ Botanicæ Upsalienses*, **1**, 1–239.
BUCHNER, P., 1909, Das accessorische Chromosom in Spermatogenese und Ovogenese der Orthopteren u.s.w. *Arch. Zellf. u. Mikr. Anat.*, **3**, 335–430.
BULLER, A. H. R., 1930, The biological significance of conjugate nuclei in *Coprinus lagopus* and other Hymenomycetes. *Nature*, **126**, 686–689.
BULLER, A. H. R., 1931, *Researches on Fungi*, IV. *Observations on the Coprini, together with some investigations on social organisation and sex in the Hymenomycetes*. London.
BURGEFF, H., 1930, Über die Mutationsrichtungen der Marchantia polymorpha, etc. *Z.I.A.V.*, **54**, 239–243.
BURNHAM, C. R., 1930, Genetical and Cytological Studies of Semisterility and Related Phenomena in Maize. *Proc. Nat. Acad. Sci.*, **16**, 269–277.
BURNHAM, C. R., 1932, An interchange in Maize giving low sterility and chain configurations. *Proc. Nat. Acad. Sci.*, **18**, 434–440.
BUXTON, B. H., and DARK, S. O. S., 1934, Hybrids of *Digitalis dubia* and *D. mertonensis*. *J. Genet.*, **29**, 109–122.
BUXTON, B. H., and DARLINGTON, C. D., 1932, Behaviour of a New Species, *Digitalis mertonensis*. *New Phytol.*, **31**, 225–240.
BUXTON, B. H., and NEWTON, W. C. F., 1928, Hybrids of *Digitalis ambigua* and *Digitalis purpurea*, their Fertility and Cytology. *J. Genet.*, **19**, 269–279.
BYTINSKI-SALZ, H., 1934, Verwandtschaftsverhältnisse zwischen den Arten der Gattung Celerio und Pergesa nach Untersuchungen über die Zytologie und Fertilität ihrer Bastarde. *Biol. Zbl.*, **54**, 301–313.
BYTINSKI-SALZ, H., and GÜNTHER, A., 1930, Untersuchungen an Lepidopterenhybriden, I. Morphologie und Cytologie einiger Bastarde der Celerio hybr. galliphorbiae-Gruppe. *Z.I.A.V.*, **53**, 153–234.

BIBLIOGRAPHY

CAMERON, D. R., 1934, The chromosomes and relationship of Crepis syriaca (Bornm.). *Univ. Calif. Pub. Agr. Sci.*, **6** (10), 257–286.

CAPINPIN, J. M., 1930, Chromosome Behaviour of Triploid Œnothera. *Nature*, **126**, 469–470.

CARANO, E., 1921, Nuove richerche sulla embryologia della Asteraceae. *Ann. di Bot.*, **15**, 1–100.

*CARNOY, J. B., 1884, *La biologie cellulaire ; étude comparée de la cellule dans les deux regnes*. I. Lierre.

CAROTHERS, E. E., 1913, The Mendelian ratio in relation to certain orthopteran chromosomes. *J. Morphol.*, **24**, 487–511.

CAROTHERS, E. E., 1917, The Segregation and Recombination of Homologous Chromosomes as found in two genera of Acrididæ (Orthoptera). *J. Morphol.*, **28**, 445–520.

CAROTHERS, E. E., 1921, Genetical behavior of heteromorphic homologous chromosomes of *Circotettix* (Orthoptera). *J. Morphol.*, **35**, 457–483.

CAROTHERS, E. E., 1931, The Maturation Divisions and Segregation of Heteromorphic Homologous Chromosomes in Acrididæ (Orthoptera). *Biol. Bull.*, **61**, 324–349.

CARROLL, M., 1920, An extra dyad and an extra tetrad in the spermatogenesis of Camnula pellucida (Orthoptera) ; numerical variations in the chromosome complex within the individual. *J. Morphol.*, **34**, 375–455.

CARTER, P. W., 1927, The life history of Padina Pavonia. *Ann. Bot.*, **41**, 139–159.

CARTLEDGE, J. L., and BLAKESLEE, A. F., 1934, Mutation rate increased by ageing seeds as shown by pollen abortion. *Proc. Nat. Acad. Sci.*, **20**, 103–110.

CASPERSSON, T., 1936, Über den chemischen Aufbau der Strukturen des Zellkernes. *Skand. Arch. f. Physiol.*, **73**, Sup. 8, 1–151.

CASPERSSON, T., and HAMMARSTEN, E. and H., 1935, Interactions of proteins and nucleic acid. *Trans. Faraday Soc.*, **31**, 367–389.

CASTLE, W. E., 1930, The Quantitative Theory of Sex and the Genetic Character of Haploid Males. *Proc. Nat. Acad. Sci.*, **16**, 783–791.

CATCHESIDE, D. G., 1931 a, Meiosis in a triploid Œnothera. *J. Genet*, **24**, 145–163.

CATCHESIDE, D. G., 1931 b, Critical Evidence of Parasynapsis in *Œnothera*. *Proc. Roy. Soc. B.*, **109**, 165–184.

CATCHESIDE, D. G., 1932, The Chromosomes of a New Haploid Œnothera. *Cytologia*, **4**, 68–113.

CATCHESIDE, D. G., 1933 a, Chromosome catenation in some F$_1$ *Œnothera* hybrids. *J. Genet.*, **27**, 45–70.

CATCHESIDE, D. G., 1933 b, Chromosome configurations in trisomic Œnotheras. *Genetica*, **15**, 177–201.

CATCHESIDE, D. G., 1934, The chromosomal relationships in the swede and turnip groups of Brassica. *Ann. Bot.*, **48**, 601–633.

CATCHESIDE, D. G., 1935, X-ray treatment of Œnothera chromosomes. *Genetica*, **17**, 313–341.

*CATCHESIDE, D. G., 1936 a, Biological Effects of Irradiation. *Sci. J. Roy. Coll. Sci.*, **6**, 71–76.

CATCHESIDE, D. G., 1936 b, The origin, nature and breeding behaviour of trisomics in *Œnothera Lamarckiana*. *J. Genet.*, **33**, 1–23.

CHAMBERS, R., 1925, The physical structure of protoplasm as determined by micro-dissection and injection, in : Cowdry, *General Cytology* (2nd imp. Chicago), 235–309.

CHAMBERS, R., and SANDS, H. C., 1923, A dissection of the chromosomes in

the pollen mother cells of *Tradescantia virginica* L. *J. Gen. Physiol.*, **5**, 815–819.
CHAMPY, C., 1923, La spermatogenèse chez Discoglossus pictus (Otth.) Comparaison avec celles des autres Discoglossides et des Vertébrés en général. *Arch. Zool. Exp. Gen.*, **62**, 1–52.
CHEESMAN, E. E., and LARTER, L. H. N., 1935, Genetical and cytological studies of *Musa* III. Chromosome numbers in the Musaceæ. *J. Genet.*, **30**, 31–52.
CHIARUGI, A., 1926, Aposporia e apogamia in " Artemisia nitida " Bertol. *N. Giorn. Bot. Ital. N.S.*, **33**, 501–626.
CHIARUGI, A., 1929, Diploidismo con amfimissia e tetraploidismo con apomissia in una medesima specie *Nigritella nigra*. Reichb. *Boll. Soc. Ital. Biol. Sper.*, **4(6)**, 1–3.
CHIARUGI, A., and FRANCINI, E., 1930, Apomissia in " Ochna serrulata " Walp. *N. Giorn. Bot. Ital. N.S.*, **37**, 1–250.
CHICKERING, A. M., 1927, Spermatogenesis in the Belostomatidæ. II. *J. Morphol.*, **44**, 541–607.
CHIPMAN, R. H., and GOODSPEED, T. H., 1927, Inheritance in Nicotiana Tabacum, VIII. Cytological features of purpurea haploid. *Univ. Calif. Pub. Bot.*, **11**, 141–158.
CHODAT, R., 1925, La chiasmatypie et la cinèse de maturation dans l'Allium ursinum. *Bull. Soc. Bot. Genève*, 1925, 1–30.
CHIZAKI, Yoshiwo, 1934, Another new haploid plant in *Triticum monococcum* L. *Bot. Mag.*, Tokyo, **48**, 621–628.
CHURCH, G. L., 1929, Meiotic phenomena in certain Gramineæ. I. Festuceæ, Aveneæ, Agrostidæ, Chlorideæ and Phalaridæ. *Bot. Gaz.*, **87**, 608–629.
CLAUSEN, J., 1926, Genetical and Cytological Investigations on *Viola tricolor* L. and *Viola arvensis* Murr. *Hereditas*, **8**, 1–157.
CLAUSEN, J., 1927, Chromosome Number and the Relationship of Species in the Genus *Viola*. *Ann. Bot.*, **41**, 677–714.
CLAUSEN, J., 1929, Chromosome Number and Relationship of some North American Species of *Viola*. *Ann. Bot.*, **43**, 741–764.
CLAUSEN, J., 1930, Male sterility in *Viola Orphanidis*. *Hereditas*, **14**, 53–72.
CLAUSEN, J., 1931 *a*, Genetic Studies in *Polemonium*. III. Preliminary Account of the Cytology of Species and Specific Hybrids. *Hereditas*, **15**, 62–66.
CLAUSEN, J., 1931 *b*, Viola canina L., a cytologically irregular species. *Hereditas*, **15**, 67–88.
CLAUSEN, J., 1931 *c*, Cyto-genetic and Taxonomic Investigations in Melanium Violets. *Hereditas*, **15**, 219–308.
CLAUSEN, J., 1933, Cytological evidence for the hybrid origin of Pentstemon neotericus Keck. *Hereditas*, **18**, 65–76.
CLAUSEN, R. E., 1928, Interspecific Hybridization in Nicotiana VII. The cytology of hybrids of the synthetic species, Digluta, with its parents, Glutinosa and Tabacum. *Univ. Calif. Pub. Bot.*, **11**, 177–211.
CLAUSEN, R. E., 1930, Inheritance in *Nicotiana Tabacum*. X. Carmine-Coral Variegation. *Cytologia*, **1**, 358–368.
CLAUSEN, R. E., 1931 *a*, Inheritance in Nicotiana Tabacum. XI. The Fluted Assemblage. *Amer. Nat.*, **65**, 316–331.
CLAUSEN, R. E., 1931 *b*, Inheritance in Nicotiana tabacum. XII. Transmission features of carmine coral variegation. *Zeits. f. Zücht. A.*, **17**, 108–115.
CLAUSEN, R. E., and GOODSPEED, T. H., 1925, Interspecific Hybridization in *Nicotiana* II. A Tetraploid *Glutinosa-Tabacum* hybrid, an experimental verification of Winge's Hypothesis. *Genetics*, **10**, 279–284.

BIBLIOGRAPHY

CLAUSEN, R. E., and LAMMERTS, W. E., 1929, Interspecific Hybridization in Nicotiana X. Haploid and Diploid Merogony. *Amer. Nat.*, **63**, 279–282.
CLAUSEN, R. E., and MANN, M. C., 1924, Inheritance in Nicotiana Tabacum V. The occurrence of haploid plants in interspecific progenies. *Proc. Nat Acad. Sci.*, **10**, 121–124. (*See also :* Goodspeed.)
CLELAND, R. E., 1922, The reduction divisions in the pollen mother cells of *Œnothera franciscana*. *Amer. J. Bot.*, **9**, 391–413.
CLELAND, R. E., 1926 a, Meiosis in the pollen mother cells of *Œnothera biennis* and *Œnothera biennis sulfurea*. *Genetics*, **11**, 127–162.
CLELAND, R. E., 1926 b, Cytological Studies of Meiosis in anthers of *Œnothera muricata*. *Bot. Gaz.*, **82**, 55–70.
*CLELAND, R. E., 1928, The Genetics of Œnothera in Relation to Chromosome Behavior, with Special Reference to Certain Hybrids. *V. Int. Kong. Vererb.*, **1**, 554–567.
CLELAND, R. E., 1929, Chromosome behavior in the pollen mother-cells of several strains of *Œnothera Lamarckiana*. *Z.I.A.V.*, **51**, 126–145.
CLELAND, R. E., 1935, Cyto-taxonomic studies of certain Œnotheras from California. *Proc. Amer. Phil. Soc.*, **75**, 339–363, 382–384, 405–429.
CLELAND, R. E., and BLAKESLEE, A. F., 1930, Interaction between complexes as evidence for segmental interchange in Œnothera. *Proc. Nat. Acad. Sci.*, **16**, 183–189.
CLELAND, R. E., and BLAKESLEE, A. F., 1931, Segmental Interchange, the Basis of Chromosomal Attachments in *Œnothera*. *Cytologia*, **2**, 175–233.
CLELAND, R. E., and BRITTINGHAM, W. H., 1934, A contribution to an understanding of crossing-over within chromosome rings of Œnothera. *Genetics*, **19**, 62–72.
CLELAND, R. E., and OEHLKERS, F., 1929, New Evidence bearing upon the Problem of the Cytological Basis for Genetical Peculiarities in the Œnotheras. *Amer. Nat.*, **63**, 497–510.
CLELAND, R. E., and OEHLKERS, F., 1930, Erblichkeit und Zytologie verschiedener Œnotheren und ihrer Kreuzungen. *Jahrb. Wiss. Bot.*, **73**, 1–124.
CLEVELAND, L. R., 1934, The wood-feeding roach Cryptocercus, its Protozoa and the symbiosis between Protozoa and roach. *Mem. Amer. Acad. Art. Sci.*, **17**, 187–342.
CLEVELAND, L. R., 1935, The intranuclear achromatic figure of Oxymonas grandis, sp. nov. *Biol. Bull.*, **69**, 54–63.
COLLINS, G. N., and KEMPTON, I. H., 1916, Patrogenesis. *J. Hered.*, **7**, 106–118.
COLLINS, J. L., HOLLINGSHEAD, L., and AVERY, P., 1929, Interspecific Hybrids in *Crepis* III. Constant fertile forms containing chromosomes derived from two species. *Genetics*, **14**, 305–320.
COLLINS, J. L., and MANN, M. C., 1923, Interspecific Hybrids in *Crepis* II. A preliminary report on the results of hybridizing *Crepis setosa* Hall with *C. capillaris* (L.) Wallr. and *C. biennis* L. *Genetics*, **8**, 212–232.
*CONKLIN, E. G., 1925, Cellular Differentiation, in : Cowdry, *General Cytology* (2nd imp. Chicago), 1925, 537–607.
COOPER, D. C., 1931, Microsporogenesis in *Buginvillæa glabra*. *Amer. J. Bot.*, **18**, 337–358.
COOPER, D. C., 1935, Macrosporogenesis and development of the embryo-sac of Lilium Henryi. *Bot. Gaz.*, **97**, 346–355.
COOPER, D. C., and BRINK, R. A., 1931, Cytological evidence for segmental interchange between non-homologous chromosomes in Maize. *Proc. Nat. Acad. Sci.*, **17**, 334–338.
*CORRENS, C., 1928, Bestimmung, Vererbung und Verteilung des Geschlechtes bei den Höhern Pflanzen. *Handbuch der Vererbungswissenschaft*, **2**.

APPENDIX IV

CRANE, M. B., and DARLINGTON, C. D., 1927, The Origin of New Forms in Rubus. *Genetica*, 9, 241-278.
CRANE, M. B., and DARLINGTON, C. D., 1932, Chromatid Segregation in Tetraploid *Rubus*. *Nature*, 129, 869.
CRANE, M. B., and GAIRDNER, A. E., 1923, Species Crosses in *Cochlearia* with a preliminary account of their Cytology. *J. Genet.*, 13, 187-200.
CRANE, M. B., and LAWRENCE, W. J. C., 1929, Genetical and Cytological Aspects of Incompatibility and Sterility in Cultivated Fruits. *J. Pomol.*, 7, 276-301.
CRANE, M. B., and LAWRENCE, W. J. C., 1930, Fertility and Vigour of Apples in Relation to Chromosome Number. *J. Genet.*, 22, 153-163.
CRANE, M. B., and ZILVA, S. S., 1931, The Antiscorbutic Vitamin of Apples, IV. *J. Pomol.*, 9, 228-231.
*CRANE, M. B., and LAWRENCE, W. J. C., 1934, *The genetics of garden plants*. London.
CREIGHTON, H. B., and McCLINTOCK, B., 1931, A Correlation of Cytological and Genetical Crossing-Over in Zea Mays. *Proc. Nat. Acad. Sci.*, 17, 492-497.
CRETSCHMAR, M., 1928, Das Verhalten der Chromosomen bei der Spermatogenese von Orgyia thyellina Btl. und antiqua L. sowie eines ihrer Bastarde. *Zeits. Zellforsch. u. mikr. Anat.*, 7, 290-399.
CREW, F. A. E., 1933, A case of non-disjunction in the fowl. *P. Roy. Soc. Edin.*, 53, 89-100.
CREW, F. A. E., and KOLLER, P. C., 1932, The sex incidence of chiasma frequency and genetical crossing-over in the mouse. *J. Genet.*, 26, 359-382.
CREW, F. A. E., and KOLLER, P. C., 1936, Genetical and cytological studies of the intergeneric hybrid of Cairina moscata and Anas platyrhynca platyrhynca. *Proc. Roy. Soc. Edin.*
CREW, F. A. E., and LAMY, R., 1935, Autosomal colour mosaics in the budgerigar. *J. Genet.*, 30, 233-241.

*DAHLGREN, K. V. O., 1927, Die Befruchtungserscheinungen der Angiospermen, eine monographische Übersicht. *Hereditas*, 10, 169-229.
DAHLGREN, K. V. O., 1930, Zur Embryologie der Saxifragoideen. *Sv. Bot. Tids.*, 24, 429-448.
DARK, S. O. S., 1931, Chromosome Association in Triploid *Primula sinensis*. *J. Genet.*, 25, 91-96.
DARK, S. O. S., 1932 a, Chromosomes of Taxus, Sequoia, Cryptomeria and Thuya. *Ann. Bot.*, 66, 965-977.
DARK, S. O. S., 1932 b, Meiosis in diploid and triploid Hemerocallis. *New Phytol.*, 31, 310-320.
DARK, S. O. S., 1934, Chromosome studies in Scilleæ, II. *J. Genet.*, 29, 85-98.
DARK, S. O. S., 1936, Meiosis in diploid and tetraploid *Pæonia* species. *J. Genet.*, 32, 353-372.
DARLINGTON, C. D., 1926, Chromosome Studies in the Scilleæ. *J. Genet.*, 16, 237-251.
DARLINGTON, C. D., 1927, The Behaviour of Polyploids. *Nature*, 119, 390.
DARLINGTON, C. D., 1928, Studies in *Prunus*, I and II. *J. Genet.*, 19, 213-256.
DARLINGTON, C. D., 1929 a, Ring-formation in Œnothera and Other Genera. *J. Genet.*, 20, 345-363.
DARLINGTON, C. D., 1929 b, Meiosis in Polyploids II. *J. Genet.*, 21, 17-56.
DARLINGTON, C. D., 1929 c, Chromosome Behaviour and Structural Hybridity in the *Tradescantiæ*. *J. Genet.*, 21, 207-286.

BIBLIOGRAPHY 597

*DARLINGTON, C. D., 1929 d, Polyploids and Polyploidy. *Nature*, **124**, 62-64, 98-100.
DARLINGTON, C. D., 1929 e, A comparative study of the chromosome complement in *Ribes*. *Genetica*, **11**, 267-272.
DARLINGTON, C. D., 1929 f, The Significance of Chromosome Behaviour in Polyploids for the Theory of Meiosis. *J.I.H.I. Conference on Polyploidy*, 42-44.
DARLINGTON, C. D., 1930 a, Studies in *Prunus* III. *J. Genet.*, **22**, 65-93.
DARLINGTON, C. D., 1930 b, A Cytological Demonstration of " Genetic " Crossing-Over. *Proc. Roy. Soc.* B, **107**, 50-59.
DARLINGTON, C. D., 1930 c, Chromosome Studies in *Fritillaria*, III. Chiasma Formation and Chromosome Pairing in *Fritillaria imperialis*. *Cytologia*, **2**, 37-55.
DARLINGTON, C. D., 1930 d, Telosynapsis or Structural Hybridity in *Œnothera*? *Nature*, **125**, 743.
DARLINGTON, C. D., 1931 a, Meiosis in Diploid and Tetraploid *Primula sinensis*. *J. Genet.*, **24**, 65-96.
*DARLINGTON, C. D., 1931 b, Cytological Theory in Relation to Heredity. *Nature*, **127**, 709-712.
*DARLINGTON, C. D., 1931 c, Meiosis. *Biol. Revs.*, **6**, 221-264.
DARLINGTON, C. D., 1931, d, The Cytological Theory of Inheritance in *Œnothera*. *J. Genet.*, **24**, 405-474.
DARLINGTON, C. D., 1931 e, The Analysis of Chromosome Pairing in *Triticum* Hybrids. *Cytologia*, **3**, 21-25.
DARLINGTON, C. D., 1931 f, The Mechanism of Crossing-Over. *Science*, **73**, 561-562.
*DARLINGTON, C. D., 1932, *Recent Advances in Cytology*, London, 1st Ed.
*DARLINGTON, C. D., 1932 a, The Control of the Chromosomes by the Genotype and its Bearing on some Evolutionary Problems. *Amer. Nat.*, **66**, 25-51.
*DARLINGTON, C. D., 1932 b, *Chromosomes and Plant Breeding*. London (Macmillan).
DARLINGTON, C. D., 1932 c, The origin and behaviour of chiasmata. V. Chorthippus elegans. *Biol. Bull.*, **63**, 357-367.
DARLINGTON, C. D., 1932 d, The origin and behaviour of chiasmata. VI. Hyacinthus amethystinus. *Biol. Bull.*, **63**, 368-371.
DARLINGTON, C. D., 1933 a, The Behaviour of Interchange Heterozygotes in Œnothera. *Proc. Nat. Acad. Sci.*, **19**, 101-103.
DARLINGTON, C. D., 1933 b, Meiosis in Agapanthus and Kniphofia. *Cytologia*, **4**, 229-240.
DARLINGTON, C. D., 1933 c, The origin and behaviour of chiasmata, VIII. Secale cereale (n, 8). *Cytologia*, **4**, 444-452.
DARLINGTON, C. D., 1933 d, The origin and behaviour of chiasmata, IX. Diploid and tetraploid *Avena*. *Cytologia*, **5**, 128-134.
*DARLINGTON, C. D., 1933 e, Chromosome study and the genetic analysis of species. *Ann. Bot.*, **47**, 811-814.
DARLINGTON, C. D., 1933 f, Studies in Prunus, IV. *J. Genet.*, **28**, 327-328.
DARLINGTON, C. D., 1934 a, Anomalous chromosome pairing in the male Drosophila pseudo-obscura. *Genetics*, **19**, 95-118.
DARLINGTON, C. D., 1934 b, The origin and behaviour of chiasmata, VII. Zea Mays. *Z.I.A.V.*, **67**, 96-114.
DARLINGTON, C. D., 1934 c, Crossing-over of sex chromosomes in Drosophila. *Amer. Nat.*, **68**, 374-377.
DARLINGTON, C. D., 1935 a, The Internal Mechanics of the Chromosomes. I. The nuclear cycle in *Fritillaria*. *Proc. Roy. Soc.* B, **118**, 33-59.
DARLINGTON, C. D., 1935 b, The Internal Mechanics of the Chromosomes.

II. Prophase pairing at meiosis in *Fritillaria*. *Proc. Roy. Soc.* B, **118**, 59–73.
DARLINGTON, C. D., 1935 c, The Internal Mechanics of the Chromosomes. III. Relational coiling and crossing-over in *Fritillaria*. *Proc. Roy. Soc.* B, **118**, 74–96.
DARLINGTON, C. D., 1935 d, The old terminology and the new analysis of chromosome behaviour. *Ann. Bot.*, **49**, 579–586.
*DARLINGTON, C. D., 1935 e, The time, place and action of crossing-over. *J. Genet.*, **31**, 185–212.
DARLINGTON, C. D., 1935 f, Crossing-over and chromosome disjunction. *Nature*, **136**, 835.
DARLINGTON, C. D., 1935 g, The External mechanics of the Chromosomes, I–V. *MS. in the Custody of the Roy. Soc.*
DARLINGTON, C. D., 1936 a, The Limitation of Crossing Over in *Œnothera*. *J. Genet.*, **32**, 343–352.
DARLINGTON, C. D., 1936 b, The analysis of chromosome movements I. *Podophyllum versipelle*. *Cytologia*, **7**, 242–247.
DARLINGTON, C. D., 1936 c, The internal mechanics of the chromosomes. V. Relational coiling of chromatids at mitosis. *Cytologia*, **7**, 248–255.
DARLINGTON, C. D., 1936 d, Crossing-over and its Mechanical Relationships in *Chorthippus* and *Stauroderus*. *J. Genet.*, **33** (in the press).
DARLINGTON, C. D., 1936 e, The sex-determining mechanism of *Drosophila miranda*. *Amer. Nat.*, **70**, 74–75.
DARLINGTON, C. D., and DARK, S.O.S., 1932, The Origin and Behaviour of Chiasmata II. *Stenobothrus parallelus*. *Cytologia*, **3**, 169–185.
DARLINGTON, C. D., and JANAKI-AMMAL, E. K., 1932, The Origin and Behaviour of Chiasmata I. Diploid and Tetraploid *Tulipa*. *Bot. Gaz.*, **93**, 296–312.
DARLINGTON, C. D., and MATHER, K., 1932, The Origin and Behaviour of Chiasmata III. Triploid *Tulipa*. *Cytologia*, **4**, 1–15.
DARLINGTON, C. D., and MOFFETT, A. A., 1930, Primary and Secondary Chromosome Balance in *Pyrus*. *J. Genet.*, **22**, 129–151.
DARLINGTON, C. D., and OSTERSTOCK, H. C., 1936, Projection method for demonstration of chromosomes *in situ*. *Nature*, **138**, 79.
See also : Buxton, Crane, Gairdner, Newton, Koller.
DAVIS, B. M., 1931, Some attempts to obtain haploids from *Œnothera Lamarckiana*. *Amer. Nat.*, **65**, 233–242.
DAVIS, B. M., and KULKARNI, C. G., 1930, The Cytology and Genetics of a Haploid Sport from *Œnothera franciscana*. *Genetics*, **15**, 55–80.
DEARING, W. H., 1934, The material continuity and individuality of the somatic chromosomes of *Amblystoma tigrinum*, with special reference to the nucleolus as a chromosomal component. *J. Morphol.*, **56**, 157–174.
DEDERER, P. H., 1928, Variations in chromosome number in the spermatogenesis of Phylosamia cynthia. *J. Morphol.*, **45**, 599–613.
DELAUNAY, L. N., 1925, The Chromosomes of Ornithogalum L. (English résumé). *Ann. Timiriasev Inst.*, **1**, (II, 3), 1–8.
DELAUNAY, L. N., 1926, Phylogenetische Chromosomenverkürzung. *Zeits. Zellforsch. u. mikr. Anat.*, **4**, 338–364.
DELAUNAY, L. N., 1929, Kern und Art. Typische Chromosomenformen. *Planta*, **7**, 100–112.
DELAUNAY, L. N., 1930, Rönthgenexperimente mit Weizen. *Wiss. Select. Inst. Kiew*, **6**, 3–34.
DEMEREC, M., 1924, A case of pollen dimorphism in maize. *Amer. J. Bot.*, **11**, 461–464.

DEMEREC, M., 1934, Effect of X-rays on the rate of change in the unstable miniature-3 gene of Drosophila virilis. *Proc. Nat. Acad. Sci.*, **20**, 28–31.
DERMEN, H., 1931, Polyploidy in Petunia. *Amer. J. Bot.*, **18**, 250–261.
DETLEFSEN, J. A., and CLEMENTE, L. S., 1923, Genetic Variation in Linkage Values. *Proc. Nat. Acad. Sci.*, **9**, 149–156.
DE WINTON, D., and HALDANE, J. B. S., 1931, Linkage in the tetraploid *Primula sinensis*. *J. Genet.*, **24**, 121–144.
DICKSON, H., 1933, Saltation induced by X-rays in seven species of Chætomium *Ann. Bot.*, **47**, 735–754.
DIGBY, L., 1905, On the Cytology of Apogamy and Apospory. II. Preliminary Note on Apospory. *Proc. Roy. Soc.*, B **76**, 463–467.
DIGBY, L., 1912, The cytology of Primula kewensis and of other related Primula hybrids. *Ann. Bot.*, **26**, 357–388.
DIGBY, L., 1919, On the Archesporial and Meiotic Mitoses in *Osmunda*. *Ann. Bot.*, **33**, 135–172.
DOBELL, C. C., 1925 a, The Life History and Chromosome Cycle of *Aggregata eberthi* (Protozoa, Sporozoa, Coccidia). *Parasitology*, **17**, 1–136.
DOBELL, C. C., 1925 b, The chromosome cycle of the Sporozoa considered in relation to the Chromosome theory of heredity. *Cellule*, **35**, 167–192.
DOBZHANSKY, Th., 1929, Genetical and cytological proof of translocations involving the third and fourth chromosome of Drosophila melanogaster. *Biol. Zbl.*, **49**, 408–419.
DOBZHANSKY, Th., 1930 a, Translocations involving the third and fourth chromosomes of *Drosophila melanogaster*. *Genetics*, **15**, 347–399.
DOBZHANSKY, Th., 1930 b, Cytological map of the second chromosome of Drosophila melanogaster. *Biol. Zbl.*, **50**, 671–685.
DOBZHANSKY, Th., 1931 a, The decrease in crossing-over observed in translocations, and its probable explanation. *Amer. Nat.*, **65**, 214–232.
DOBZHANSKY, Th., 1931 b, Translocations involving the second and fourth chromosomes of Drosophila melanogaster. *Genetics*, **16**, 629–658.
DOBZHANSKY, Th., 1933, On the sterility of the inter-racial hybrids in *Drosophila melanogaster*. *Proc. Nat. Acad. Sci.*, **19**, 397–403.
DOBZHANSKY, Th., 1934, Studies on hybrid sterility. I. Spermatogenesis in pure and hybrid Drosophila pseudo-obscura. *Zeits. Zellf. u. mikr. Anat.*, **21**, 169–223.
DOBZHANSKY, Th., 1935 a, Drosophila miranda, a new species. *Genetics*, **20**, 377–391.
DOBZHANSKY, Th., 1935 b, The Y-chromosome of Drosophila pseudo-obsucra. *Genetics*, **20**, 366–376.
DOBZHANSKY, Th., and BOCHE, R. D., 1933, Intersterile races of *Drosophila pseudo-obscura* Frol. *Biol., Zbl.*, **53**, 314–330.
DOBZHANSKY, Th., and SCHULTZ, J., 1931, Evidence for multiple sex factors in the X-chromosome of Drosophila melanogaster. *Proc. Nat. Acad. Sci.*, **17**, 513–518.
*DODGE, B. O., 1936, Reproduction and inheritance in Ascomycetes. *Science*, **83**, 169–175.
DONCASTER, L., 1910, Gametogenesis of the Gall Fly, *Neuroterus lenticularis* (Spathegaster baccarum) I. *Proc. Roy. Soc.* B, **82**, 88–113.
DONCASTER, L., 1911, Gametogenesis of the Gall Fly, *Neuroterus lenticularis* II. *Proc. Roy. Soc.* B, **83**, 476–489.
DONCASTER, L., 1916, Gametogenesis and sex-determination of the gall-fly, *Neuroterus lenticularis* (Spathegaster baccarum) III. *Proc. Roy. Soc.* B, **89**, 183–200.
*DONCASTER, L., 1920, *An Introduction to the Study of Cytology*. Cambridge.

DOPP, W., 1932 Die Apogamie bei Aspidium remotum Al. Br. *Planta*, **17**, 86–152.
DORSEY, E., 1936, Induced polyploidy in wheat and rye. *J. Hered.*, **27**, 154–160.
DOUTRELIGNE, J., 1933, Chromosomes et nucleoles dans les noyaux du type euchromocentrique. *Cellule*, **42**, 31–100.
DRYGALSKI, U. v., 1935, Über die Entstehung einer tetraploiden, genetisch ungleichmassigen F_2 aus der Kreuzung Saxifraga adscendens L. × S. tridactylites L. *Z.I.A.V.*, **69**, 278–300.
DUBININ, N. P., 1934, Experimental reduction in the number of chromosome pairs in Drosophila melanogaster. *Biol. Jur.*, **3**, 719–736.
DUBININ, N. P., and HEPTNER, M. A., 1935, A new phenotypic effect of the Y-chromosome in *Drosophila melanogaster*. *J. Genet.*, **30**, 423–446.
DUBININ, N. P., and SIDOROFF, B. N., 1934, Relation between the effect of a gene and its position in the system. *Amer. Nat.*, **68**, 377–381.
DUBININ, N. P., *et alii*, 1936, Occurrence and distribution of chromosome aberrations in nature. *Nature*, **137**, 1035–1036.
DU BOIS, A. M., 1933, Chromosome behavior during cleavage in the eggs of Sciara coprophila (Diptera) in relation to the problem of sex determination. *Zeits. Zellf. u. mikr. Anat.*, **19**, 595–614.

EARL, R. O., 1927, The nature of chromosomes. I. Effects of reagents on root-tip sections of *Vicia faba*. *Bot. Gaz.*, **84**, 58–74.
*EAST, E. M., 1929, The concept of the gene. *Proc. Int. Cong. Pl. Sc.*, **1**, 889–895.
EAST, E. M., 1930, The Production of Homozygotes through induced Parthenogenesis. *Science*, **72**, 148–149.
EAST, E. M., 1933, The behavior of a triploid in Nicotiana Tabacum L. *Amer. J. Bot.*, **20**, 269–289.
EAST, E. M., 1934, A novel type of hybridity in Fragaria. *Genetics*, **19**, 167–174.
EDWARDS, CH., 1910, The idiochromosomes in *Ascaris megalocephala* and *Ascaris lumbricoides*. *Arch. Zellf.*, **5**, 422–429.
EISENTRAUT, M., 1926, Die spermatogonialen Teilungen bei Acridiern mit besonderer Berücksichtigung der Überkreuzungsfiguren. *Zeits. wiss. Zool.*, **127**, 141–183.
EISENTRAUT, M., 1928, Über das Auftreten von Chromosomenbläschen in den Reifeteilungen einiger Acridier. *Zeits. wiss. Zool.*, **128**, 253–266.
EKSTRAND, H., 1918, Zur Zytologie und Embryologie der Gattung Plantago (Vorläufige Mitteilung). *Svensk Bot. Tidsskr.*, **12**, 202–206.
ELLENHORN, J., 1933, Experimental-photographische Studien der lebenden Zelle. *Zeits. Zellf. u. mikr. Anat.*, **20**, 288–308.
ELLENHORN, J., 1934, Zytologische Studien über die genetisch bedeutsamen Kernstrukturen, *Zeits. f. Zellf. u. mikr. Anat.*, **21**, 24–41.
ELLENHORN, J., PROKOFJEVA, A., and MULLER, H. J., 1935, The optical dissociation of Drosophila chromomeres by means of ultra-violet light, *C. R. Acad. Sci. U.R.S.S.*, 1935, **1** (4), 234–242.
EMERSON, R. A., 1932, The present status of maize genetics. *Proc. 6th Int. Cong. Genet.*, 141–152.
EMERSON, R. A., BEADLE, G. W., and FRASER, A. C., 1935, A summary of linkage studies in Maize. *Cornell Memoir*, **180**, 1–83.
EMERSON, S. H., 1929, The reduction division in a haploid Œnothera. *Cellule*, **39**, 159–165.
EMERSON, S. [H.], 1931, Parasynapsis and Apparent Chiasma Formation in Œnothera. *Amer. Nat.*, **65**, 551–555.

EMERSON, S. [H.], 1935, The genetic nature of de Vries's mutations in Œnothera Lamarckiana. *Amer. Nat.*, **69**, 545–559.
EMERSON, S. H., 1936, A genetic and cytological analysis of Œnothera pratincola and one of its revolute-leaved mutations. *J. Genet.*, **32**, 315–342.
EMERSON, S. H., 1936, The trisomic derivatives of Œnothera Lamarckiana. *Genetics*, **21**, 200–224.
EMERSON, S. H., and BEADLE, G. W., 1930, A fertile tetraploid hybrid between *Euchlæna perennis* and *Zea Mays*. *Amer. Nat.*, **64**, 190–192.
EMERSON, S. H., and STURTEVANT, A. H., 1931, Genetic and Cytological Studies on Œnothera. III. The Translocation Interpretation. *Z.I.A.V.* **59**, 395–419.
EMME, H., 1928, Karyologie der Gattung *Secale* L. *Z.I.A.V.*, **47**, 99–124.
EMSWELLER, S. L., and JONES, H. A., 1935, A gene for control of interstitial localisation of chiasmata in *Allium fistulosum* L. *Science*, **81**, 543–544.
ERLANSON, E. W., 1929, Cytological conditions and evidences for hybridity in North American wild roses. *Bot. Gaz.*, **87**, 443–506.
ERLANSON, E. W., 1931 *a*, Sterility in wild roses and in some species hybrids. *Genetics*, **16**, 75–96.
ERLANSON, E. W., 1931 *b*, A Group of Tetraploid Roses in Central Oregon. *Bot. Gaz.*, **91**, 55–64.
ERLANSON, E. W., 1931 *c*, Chromosome Organisation in *Rosa*. *Cytologia*, **2**, 256–282.
ERLANSON, E. W., 1933, Chromosome pairing, structural hybridity and fragments in *Rosa*. *Bot. Gaz.*, **94**, 551–566.
ERNST, A., 1913, Embryobildung bei Balanophora. *Flora*, **106**, 129–159.
*ERNST, A., 1918, *Bastardierung als Ursache der Apogamie im Pflanzenreich. Eine Hypothese zur experimentelle Vererbungs- und Abstammungslehre.* Jena.
EVANS, H. M., and SWEZY, O., 1928, A sex difference in chromosome lengths in the Mammalia. *Genetics*, **13**, 532–543.
EVANS, H. M., and SWEZY, O., 1929, The Chromosomes in Man : sex and somatic. *Mem. Univ. Calif.*, **9** (i), 1–41.

FABERGÉ, A. C., 1936, The cytology of the male-sterile *Lathyrus odoratus*. *Genetics* (in the press).
FAGERLIND, F., 1936, Die Chromosomenzahl von Alectorolophus und Saison-Dimorphismus. *Hereditas*, **22**, 189–192.
FANKHAUSER, Gerhard, 1934 *a*, Cytological studies on egg fragments of the salamander Triton. IV. The cleavage of egg fragments without the egg nucleus. *J. Exp. Zool.*, **67**, 351–391.
FANKHAUSER, Gerhard, 1934 *b*, Cytological studies on egg fragments of the salamander Triton, V. Chromosome number and chromosome individuality in the cleavage mitoses of merogonic fragments. *J. Exp. Zool.*, **68**, 1–57.
FARMER, J. B., and DIGBY, L., 1907, Studies in Apospory and Apogamy in Ferns. *Ann. Bot.*, **21**, 161–199.
FARMER, J. B., and DIGBY, L., 1914, On the dimensions of the chromosomes considered in relation to phylogeny. *Phil. Trans. Roy. Soc.*, **205**, 1–26.
FARMER, J. B., and MOORE, J. E. S., 1905, On the Maiotic Phase in Animals and Plants. *Quart. J. Micros. Sci.*, **48**, 489–557.
FASTEN, N., 1914, Spermatogenesis of the American crayfish, *Cambarus virilis* and *Cambarus immunis* (?) with special reference to synapsis and the chromatoid bodies. *J. Morphol.*, **25**, 587–649.
FASTEN, N., 1926, Spermatogenesis of the black-clawed crab, Lophopanopeus bellus (Stimpson) Rathbun. *Biol. Bull.*, **50**, 277–292.

FEDERLEY, H., 1912, Das Verhalten der Chromosomen bei der Spermatogenese der Schmetterlinge *Pygæra anachoreta, curtula* und *pigra* sowie einiger ihrer Bastarde. *Z.I.A.V.*, 9, 1–110.
FEDERLEY, H., 1914, Ein Beitrag zur Kenntnis der Spermatogenese bei Mischlingen zwischen Eltern verschiedener systematischer Verwandtschaft. *Öfv. of Finsk Vetensk Soc. Förh.*, 56 (13), 1–28.
FEDERLEY, H., 1915, Chromosomenstudien an Mischlingen I. Die Chromosomenkonjugation bei der Gametogenese von Smerinthus populi. var. austauti × populi. *Öfv. of Finsk Vetensk Soc. Förh.*, 57 (26), 1–36. II. Die Spermatogenese des Bastards *Dicranura erminea* ♀ × *D. vinula* ♂ ibid. 57 (30), 1–26.
FEDERLEY, H., 1916, Chromosomenstudien an Mischlingen III. Die Spermatogenese des Bastards Chaerocampa porcellus ♀ × elpenor ♂. *Öfv. of Finsk Vetensk Soc. Förh.*, 58 (12), 1–17.
FEDERLEY, H., 1927, Ist die Chromosomenkonjugation eine conditio sina qua non für die Mendelspaltung ? *Hereditas*, 9, 391–404.
FEDERLEY, H., 1931, Chromosomenanalyse der reziproken Bastarde zwischen Pygæra pigra und P. curtula sowie ihrer Ruckkreuzungsbastarde. *Zeits. Zellforsch. u. mikr. Anat.*, 12, 772–816.
FEDERLEY, H., 1932, The conjugation of the chromosomes. *Proc. 6th Int. Cong. Genet.*, 1, 153–164.
FENG, Yen-An, 1934 *a*, Sur la technique employée à la recherche de la présence de centrosomes chez les Lonicera. *C.R. Soc. Biol.*, 115, 1087–1088.
FENG, Yen-An, 1934 *b*, Recherches cytologiques sur la caryocinèse, la spermatogenèse et la fécondation chez les Caprifoliacées (en particulier sur la présence de centrosomes présidant à la caryocinèse dans les Lonicera). *Botaniste*, 26, 1–85.
FERNANDES, A., 1931, Estudos nos Cromosomas das Liliáceas e Amarilidáceas. *Bol. Soc. Brot.* (Coimbra), 7, 1–122.
FEULGEN, R., and IMHAÜSER, K., 1925, Über die für Nuklealreaktion und Nuklealfärbung verantwortlich zu machenden Gruppen nebst Bemerkungen zur Darste.lung des Oxymethylfurfurols. II. *Zeits. f. physiol. Chem.*, 148, 1–16.
FISK, E. L., 1931, The Chromosomes of Lathyrus tuberosus. *Proc. Nat. Acad. Sci.*, 17, 511–513.
FOOT, K., and STROBELL, E. C., 1905, Prophases and metaphases of the first Maturation Spindle of *Allolobophora fœtida*. *Amer. J. Anat.*, 4, 199–243.
FRANDSEN, H. N., and WINGE, Ø., 1932, Brassica napocampestris, a New Constant Amphidiploid Species Hybrid. *Hereditas*, 16, 212–219.
FRANKEL, O. H., 1936, The nucleolar cycle in some species of *Fritillaria*. *Cytologia*, 8. (In the press.)
FRASER, H. C. I., 1908, Contributions to the Cytology of Humaria rutilans, Fries. *Ann. Bot.*, 22, 35–55.
FRIEDMAN, B., and GORDON, M., 1934, Chromosome numbers in Xiphophorin Fishes. *Amer. Nat.*, 68, 446.
FRIES, W., 1910, Die Entwicklung der Chromosomen im Ei von Branchipus Grub. und der parthenogenetischen Generationen von Artemia salina. *Arch. f. Zellforsch.*, 4, 44–80.
FRIESEN, H., 1934, Künstliche Auslösung von Crossing-over bei Drosophila-Männchen, *Biol. Zbl.*, 54, 65–75.
FRIESEN, H., 1936 *a*, Crossing-over in male Drosophila. (Abstract.) *Nature*, 137, 325.
FRIESEN, H., 1936 *b*, Auslösung von crossing-over bei Drosophilamännchen durch röntghenisierung der imago. *Genetica*, 18, 187–192.

BIBLIOGRAPHY

FROLOWA, S., 1912, Idiochromosomen bei Ascaris megalocephala. *Arch f. Zellforsch.*, **9**, 149-167.
FROLOWA, S., 1926, Normale und polyploide Chromosomengarnituren bei einigen Drosophila-Arten. *Zeits. f. Zellforsch. u. mikr. Anat.*, **3**, 682-694.
FROLOWA, S. L., 1929 a, Die Geschlechtschromosomen bei Choaborus plumicornis F. *Zeits. Zellforsch. u. mikr. Anat.*, **9**, 66-82.
FROLOWA, S. L., 1929 b, Die Polyploidie einiger Gewebe bei Dipteren. *Zeits. Zellforsch. u. mikr. Anat.*, **8**, 542-565.
FROLOWA, S. L., 1936, Structure of the nuclei in the salivary gland cells of *Drosophila*. *Nature*, **137**, 319.
FROLOWA, S. L., and ASTAUROW, B. L., 1929, Die Chromosomengarnitur als systematisches Merkmal. *Zeits. Zellforsch. u. mikr. Anat.*, **10**, 201-213.
FROST, H. B., 1925, Tetraploidy in Citrus. *Proc. Nat. Acad. Sci.*, **11**, 535-537.
FROST, H. B., 1927, Chromosome-Mutant Types in Stocks (*Matthiola incana* R. Br.). I. Characters due to Extra Chromosomes. *J. Hered.*, **18**, 474-486.
FROST, H. B., and MANN, M. C., 1924, Mutant forms of *Matthiola* resulting from non-disjunction. *Am. Nat.*, **58**, 569-572.
FRY, H. J., 1930, A critique of the cytological method: determining the structure of living celis from fixed ones. *Anat. Rec.*, **46**, 1-21.
FRY, H. J., and PARKS, M. E., 1934, Studies of the mitotic figure IV. Mitotic changes and viscosity changes in eggs of Arbacia, Cumingia and Nereis. *Protoplasma*, **21**, 473-499.
FYFE, J. L., 1936, The External Forces acting on Chromosomes. *Nature*, **138**, 366.
FUKUSHIMA, E., 1929, Preliminary Report on Brassico-Raphanus Hybrids. *Proc. Imper. Acad.*, **5**, 48-50.
FUKUSHIMA, E., 1931, Formation of Diploid and Tetraploid Gametes in *Brassica*. *Jap. J. Bot.*, **5**, 273-283.

GAINES, E. F., and AASE, H. C., 1926, A haploid wheat plant. *Amer. J. Bot.*, **13**, 373-385.
GAIRDNER, A. E., 1926, *Campanula persicifolia* and its Tetraploid Form, " Telham Beauty." *J. Genet.*, **16**, 341-351.
GAIRDNER, A. E., 1933, Sporangia containing spermatozoids in Ferns. *Nature*, **131**, 621.
GAIRDNER, A. E., and DARLINGTON, C. D., 1931, Ring-Formation in Diploid and Polyploid *Campanula persicifolia*. *Genetica*, **13**, 113-150.
*GAISER, L. O., 1926 a, A list of chromosome numbers in Angiosperms. *Genetica*, **8**, 401-484.
*GAISER, L. O., 1926 b, Chromosome Numbers in Angiosperms, II. *Bibliogr. Genet*, **6**, 171-466.
GAISER, L. O., 1933, Chromosome Numbers in Angiosperms, IV. *Bibliogr. Genet.*, **10**, 105-250.
GALAN, Fernando, 1931, Estudios sobre la espermatogenesis del Coleoptero *Phytodecta variabilis* Ol. *EOS, Revista Española de Entomologia*, **7**, 461-501.
GARDINER, M. S., 1935, The origin and nature of the nucleolus. *Quart. J. Micr. Sci.*, **77**, 523-547.
GATES, R. R., 1907, Pollen development in hybrids of *Œnothera Lata* × *O. Lamarckiana*, and its relation to mutation. *Bot. Gaz.*, **43**, 81-115.
GATES, R. R., 1909, The stature and chromosomes of *Œnothera gigas*, de Vries. *Arch. f. Zellforsch.*, **3**, 525-552.

GATES, R. R., and GOODWIN, K. M., 1930, A new Haploid Œnothera, with some considerations on Haploidy in Plants and Animals. *J. Genet.*, 23, 123–156.
GATES, R. R., and GOODWIN, K. M., 1931, Meiosis in Œnothera purpurata and Œ. blandina. *Proc. Roy. Soc.* B, 109, 149–164.
GATES, R. R., and REES, E. M., 1921, A cytological study of pollen development of two species of Lathraea. *Ann. Bot.*, 35, 365–398.
GATES, R. R., and SHEFFIELD, F. M. L., 1929. Chromosome Linkage in certain Œnothera Hybrids. *Phil. Trans. R.S.* B, 217, 367–394.
GEERTS, J. M., 1909, Beitrage zur Kenntnis der Zytologie und der partiellen Sterilitat von Œnothera Lamarckiana. *Rec. Trav. Bot. Neerl.*, 5, 93–208.
GEINITZ, B., 1915, Über Abweichungen bei der Eireifung von Ascaris. *Arch. f. Zellforsch.*, 13, 588–633.
GEITLER, L., 1927, Somatische Teilung, Reduktionsteilung, Copulation und Parthenogenese bei Cocconeis placentula. *Arch. Prot.*, 59, 506–549.
GEITLER, L., 1929, Zur Zytologie von Ephedra. *Österr. Bot. Zeits.*, 78, 242–250.
GEITLER, L., 1932, Das Verhalten der Nucleolen in einer tetraploiden Wurzel von Crepis capillaris. *Planta*, 17, 801–804.
GEITLER, L., 1933, Das Verhalten der Chromozentren von Agapanthus während der Mitose. *Œsterr. Bot. Zeit.*, 4, 82, 277–282.
*GEITLER, L., 1934, Grundriss der Cytologie, Berlin.
GEITLER, L., 1935 a, Untersuchungen über den Kernbau von Spirogyra mittels Feulgens Nuklealfärbung. *Ber. Deuts. Bot. Ges.*, 53, 270–275.
GEITLER, L., 1935 b, Beobachtungen über die erste Teilung im Pollenkorn der Angiospermen. *Planta*, 24, 361–386.
GEITLER, L., 1935 c, Der Spiralbau somatischer Chromosomen. *Zeits. Zellf. u. mikr. Anat.*, 23, 514–521.
GELEI, J., 1913, Über die Ovogenese von Dendrocoelum lacteum. *Arch. Zellf.*, 11, 51–150.
GELEI, J., 1921, Weitere Studien über die Oogenese des Dendrocœlum lacteum. II. Die Längskonjugation der Chromosomen. *Arch. f. Zellforsch.*, 16, 88–169.
GELEI, J., 1922, Weitere Studien über die Oogenese des Dendrocœlum lacteum. III. Die Konjugationsfrage der Chromosomen in der Literatur und meine Befunde. *Arch. f. Zellforsch.*, 16, 299–370.
GERASSIMOVA, H., 1933, Fertilization in Crepis capillaris. *Cellule*, 42, 103–148.
GERSHENSON, S., 1935, The mechanism of non-disjunction in the ClB stock of Drosophila melanogaster. *J. Genet.*, 30, 115–125.
*GOLDSCHMIDT, R., 1920, Mechanismus und Physiologie der Geschlechtsbestimmung. Berlin.
*GOLDSCHMIDT, R., 1928, The Gene. *Quart. Rev. Biol.*, 3, 307–324.
*GOLDSCHMIDT, R., 1929, Geschlechtsbestimmung im Tier- und Pflanzenreich. *Biol. Zbl.*, 49, 641–648.
*GOLDSCHMIDT, R., 1931, Analysis of Intersexuality in the Gipsy Moth. *Quart. Rev. Biol.*, 6, 125–142.
*GOLDSCHMIDT, R., 1934 a, Lymantria. *Bibliogr. Genet.*, 11, 1–186.
GOLDSCHMIDT, R., 1934 b, The influence of the cytoplasm upon gene-controlled heredity. *Amer. Nat.*, 68, 5–23.
GOLDSCHMIDT, R., 1935, Geographische Variation und Artbildung. *Naturwiss.* 23, 169–176.
GOLDSCHMIDT, R., and KATSUKI, K., 1931, Vierte Mitteilung über erblichen Gynandromorphismus und somatische Mosaikbildung. *Biol. Zbl.*, 51, 58–74.

GOODSPEED, T. H., 1929, Cytological and other features of variant plants produced from X-rayed sex cells of *Nicotiana tabacum*. *Bot. Gaz.*, **87**, 563–582.
GOODSPEED, T. H., 1930, Occurrence of Triploid and Tetraploid Individuals in X-ray Progenies of Nicotiana Tabacum. *Univ. Calif. Pub. Bot.*, **11**, 299–308.
GOODSPEED, T. H., 1934, Nicotiana phylesis in the light of chromosome number, morphology and behavior. *Univ. Calif. Pub. Bot.*, **17**, 369–398.
GOODSPEED, T. H., and AVERY, P., 1929 a, The occurrence of chromosome variants in Nicotiana alata Lk. et Otto. *Proc. Nat. Acad. Sci.*, **15**, 343–345.
GOODSPEED, T. H., and AVERY, P., 1929 b, The occurrence of a *Nicotiana glutinosa* haplont. *Proc. Nat. Acad. Sci.*, **15**, 502–504.
GOODSPEED, T. H., and AVERY, P., 1930, Nature and Significance of Structural Chromosome Alterations Induced by X-rays and Radium. *Cytologia*, **1**, 308–327.
GOODSPEED, T. H., and CLAUSEN, R. E., 1927 a, Interspecific Hybridization in Nicotiana V. Cytological features of the two F_1 hybrids made with Nicotiana bigelovii as a parent. *Univ. Calif. Pub. Bot.*, **11**, 117–125.
GOODSPEED, T. H., and CLAUSEN, R. E., 1927 b, Interspecific Hybridization in Nicotiana VI. Cytological features of sylvestris-tabacum hybrids. *Univ. Cal. Pub. Bot.*, **11**, 127–140.
GOODSPEED, T. H., CLAUSEN, R. E., and CHIPMAN, R. H., 1926, Interspecific Hybridization in Nicotiana IV. Some Cytological Features of the paniculata-rustica hybrid and its Derivatives. *Univ. Calif. Pub. Bot.*, **11**, 103–116.
GOTOH, K., 1924, Über die Chromosomenzahl von *Secale cereale*, L. *Bot. Mag. Tokyo*, **38**, 135–152.
GOTOH, K., 1932, Further investigations on the chromosome number of Secale cereale L. *Jap. J. Genet.*, **7**, 172–182.
GOTOH, K., 1933, Karyologische Studien an Paris und Trillium. *Jap. J. Genet.*, **8**, 197–203.
GOWEN, J. W., 1928, On the Mechanism of Chromosome Behavior in Male and Female Drosophila. *Proc. Nat. Acad. Sci.*, **14**, 475–477.
GOWEN, J. W., 1929, The Cell Division at which Crossing Over takes Place. *Proc. Nat. Acad. Sci.*, **15**, 266–268.
GOWEN, J. W., 1933, Meiosis as a genetic character in Drosophila melanogaster. *J. Exp. Zool.*, **65**, 83–106.
GOWEN, J. W., and GAY, E. H., 1933, Gene number, kind and size in Drosophila. *Genetics*, **18**, 1–31.
GOWEN, M. S., and GOWEN, J. W., 1922, Complete linkage in *Drosophila melanogaster*. *Amer. Nat.*, **56**, 286–288.
GRANATA, L., 1910, Le cinesi spermatogenetiche di Pamphagus marmoratus (Burm.). *Arch. f. Zellforsch*, **5**, 182–214.
GRAUBARD, M. A., 1932, Inversion in *Drosophila melanogaster*. *Genetics* **17**, 81–105.
*GRAY, J., 1931, *Experimental Cytology*, Cambridge.
GRAY, J., 1927, The mechanism of cell-division IV. The effect of gravity on the eggs of Echinus. *J. Exp. Biol.*, **5**, 102–111.
GRAZE, H., 1935, Weitere Chromosomenuntersuchungen bei *Veronica*-Arten der Sektion *Pseudolysimachia* Koch. *Jahrb. wiss. Bot.*, **81**, 609–662.
GRÉGOIRE, V., 1905, Les résultats acquis sur les cinèses de maturation dans les deux règnes. (Premier mémoire.) Revue critique de la littérature. *Cellule* **22**, 219–376.
GRÉGOIRE, V., 1909, La réduction dans le Zoögonus mirus Lss. et le " Primärtypus." *Cellule*, **25**, 243–287.

GRÉGOIRE, V., 1910, Les cinèses de maturation dans les deux règnes. L'unité essentielle du processus méiotique. (Second mémoire.) *Cellule*, **26**, 221–422.
GRÉGOIRE, V., 1931, Euchromocentres et chromosomes dans les végétaux. *Acad. Roy. Belg., Bull. classe de Sci.* (5) **17**, 1435–1448.
GREGOR, J. W., and SANSOME, F. W., 1930, Experiments on the Genetics of Wild Populations. II. *Phleum pratense* L. and the hybrid *P. pratense* L. × *P. alpinum* L. *J. Genet.*, **22**, 373–387.
GROSS, F., 1935, Die Reifungs- und Furchungsteilungen von Artemia salina im Zusammenhang mit dem Problem des Kernteilungsmechanismus. *Zeits. Zellf. u. mikr. Anat.*, **23**, 522–565.
GRÜNEBERG, H., 1935, A new inversion of the X-chromosome in *Drosophila melanogaster*. *J. Genet.*, **31**, 163–184.
GRÜNEBERG, H., 1936, The position effect proved by a spontaneous reversion of the X-chromosome in *Drosophila melanogaster*. *J. Genet.* (In the press.)
*GUILLERMOND, A., MANGENOT, G., and PLANTEFOL, L., 1933, *Traité de cytologie végétale*. Pp. viii + 1195, Paris.
GURWITSCH, A., 1926, Das Problem der Zellteilung physiologisch betrachtet. *Monog. a.d. Gesamtgebiet d. physiol. d. Pfl. u. d. Tiere*, **11**, Berlin.
GUSTAFSSON, A., 1930, Kastrierungen und Pseudogamie bei Rubus. *Bot. Not.* (Lund), 1930, 477–494.
GUSTAFSSON, A., 1931, Sind die Canina-Rosen apomiktisch ? *Bot. Not.* (Lund), 1931, 21–30.
GUSTAFSSON, A., 1932, Zytologische und experimentelle Studien in der Gattung *Taraxacum*. *Hereditas*, **16**, 41–62.
GUSTAFSSON, A., 1932, Spontane Chromosomenzahlerhöhung in Pollenmutterzellen und die damit verbundene Geminibildung. *Hereditas*, **17**, 100–114.
GUSTAFSSON, A., a., 1934, Die Formenbildung der Totalapomikten. *Hereditas*, **19**, 259–283.
GUSTAFSSON, A., b., 1934, Primary and secondary association in Taraxacum. *Hereditas*, **20**, 1–31.
GUSTAFSSON, A., 1935, Studies on the mechanism of parthenogenesis. *Hereditas*, **21**, 1–112.
GWYNNE-VAUGHAN, H. C. I., and WILLIAMSON, H. S., 1931, Contributions to the study of Pyronema confluens. *Ann. Bot.*, **45**, 355–371.
GWYNNE-VAUGHAN, H. C. I., and WILLIAMSON, H. S., 1933, The asci of Lachnea scutellata. *Ann. Bot.*, **47**, 375–383.

HAASE-BESSELL, G., 1916, Digitalisstudien I. *Z.I.A.V.*, **16**, 293–314.
HAASE-BESSELL, G., 1922, Digitalisstudien II. *Z.I.A.V.*, **27**, 1–26.
HAASE-BESSELL, G., 1926, Digitalisstudien III. *Z.I.A.V.*, **42**, 1–46.
HABERLANDT, G., 1921 a, Über experimentelle Erzeugung von Adventivembryonen bei Œnothera Lamarckiana. *Sitzb. preuss. Akad. Wiss.*, **40**, 695–725.
HABERLANDT, G., 1921 b, Die Entwicklungserregung der Eizellen einiger parthenogenetischer Kompositen. *Sitzb. preuss. Akad. Wiss.*, **51**, 861–881.
HABERLANDT, G., 1923, Zur Embryologie von Allium odorum L. *Ber. Deuts. Bot. Ges.*, **41**, 174–179.
HABERLANDT, G., 1927, Zur Cytologie und Physiologie des weiblichen Gametophyten von Œnothera. *Sitzb. preuss. Akad. Wiss.*, **7**, 33–47.
HÄCKER, V., 1892, Die heterotypische Kernteilung im Cyklus der generativen Zellen. *Ber. Naturf. Ges. Freib.*, **6**, 160–193.

HAGA, T., 1934, The comparative morphology of the chromosome complement in the tribe *Parideæ*. *J. Fac. Sci. Hokkaido* V, **3**, 1-32.
HAGA, T., 1935, Sex and chromosomes in *Spinacia oleracea* L. *Jap. J. Genet.*, **10**, 218-222.
HAGERUP, O., 1927, *Empetrum Hermaphroditum* (Lge) Hagerup, A new tetraploid bisexual Species. *Dansk Bot. Ark.*, **5** (2), 1-17.
HAGERUP, O., 1932, Über Polyploidie in Beziehung zu Klima, Ökologie und Phylogenie. Chromosomenzahlen aus Timbuktu. *Hereditas*, **16**, 19-40.
HÅKANSSON, A., 1925, Zur Zytologie der Gattung *Godetia*. *Hereditas*, **6**, 257-274.
HÅKANSSON, A., 1926, Über das Verhalten der Chromosomen bei der heterotypischen Teilung schwedischer *Œnothera Lamarckiana* und einiger ihrer Mutanten und Bastarde. *Hereditas*, **8**, 255-304.
HÅKANSSON, A., 1927, Über das Verhalten der Chromosomen bei der heterotypischen Teilung schwedischer Œnothera Lamarckiana und einiger uhrer Mutanten und Bastarde. *Hereditas*, **8**, 255-304.
HÅKANSSON, A., 1928, Die Reduktionsteilung in den Samenanlagen einiger Œnotheren. *Hereditas*, **11**, 129-181.
HÅKANSSON, A., 1929 a, Chromosomenringe in *Pisum* und ihre mutmassliche genetische Bedeutung. *Hereditas*, **12**, 1-10.
HÅKANSSON, A., 1929 b, Die Chromosomen in der Kreuzung *Salix viminalis* × *caprea* von Heribert-Nilsson. *Hereditas*, **13**, 1-52.
HÅKANSSON, A., 1929 c, Über verschiedene Chromosomenzahlen in *Scirpus palustris* L. *Hereditas*, **13**, 53-60.
HÅKANSSON, A., 1930 a, Die Chromosomenreduktion bei einigen Mutanten und Bastarden von *Œnothera Lamarckiana*. *Jahrb. Wiss. Bot.*, **72**, 385-402.
HÅKANSSON, A., 1930 b, Zur Zytologie trisomischer Mutanten aus *Œnothera Lamarckiana*. *Hereditas*, **14**, 1-32.
HÅKANSSON, A., 1931 a, Über Chromosomenverkettung in *Pisum*. *Hereditas*, **15**, 17-61.
HÅKANSSON, A., 1931 b, Chromosomenverkettung bei Godetia und Clarkia. *Ber. Deuts. Bot. Ges.*, **49**, 228-234.
HÅKANSSON, A., 1933 a, Zytologische Studien an compactoiden Typen von *Triticum vulgare*. *Hereditas*, **17**, 155-196.
HÅKANSSON, A., 1933 b, Die Konjugation der Chromosomen bei einige Salix-Bastarden. *Hereditas*, **18**, 199-214.
HÅKANSSON, A., 1934, Chromosomenbindungen in einigen Kreuzungen zwischen halbsterilen Erbsen. *Hereditas*, **19**, 341-358.
HÅKANSSON, A., 1936, Die Reduktionsteilung in einigen Artbastarden von Pisum. *Hereditas*, **21**, 215-222.
HALDANE, J. B. S., 1919, The Combination of Linkage Values and the Calculation of Distances between the Loci of Linked Factors. *J. Genet.*, **8**, 291-297.
HALDANE, J. B. S., 1922, Sex Ratio and Unisexual Sterility in Hybrid Animals. *J. Genet.*, **12**, 101-109.
*HALDANE, J. B. S., 1927, The comparative genetics of colour in rodents and carnivora. *Biol. Rev.*, **2**, 199-212.
*HALDANE, J. B. S., 1929, The Species Problem in the Light of Genetics. *Nature*, **124**, 514-516.
HALDANE, J. B. S., 1930, Theoretical Genetics of Autopolyploids. *J. Genet.*, **22**, 359-372.
HALDANE, J. B. S., 1931, The Cytological Basis of Genetical Interference. *Cytologia*, **3**, 54-65.
HALDANE, J. B. S., 1932 a, The Time of Action of Genes and its bearing on some Evolutionary Problems. *Amer. Nat.*, **46**, 5-24.

*HALDANE, J. B. S., 1932 c, *The Causes of Evolution*. London (Longmans, Green & Co.).
HALDANE, J. B. S., 1936, A search for incomplete sex linkage in man. *An. Eug.*, **7**, 28–57.
HAMLETT, G. W. D., 1926, The Linkage Disturbance Involved in the Chromosome Translocation I of *Drosophila* and its Probable Significance. *Biol. Bull.*, **51**, 435–442.
HAMMERLING, J., 1932, Entwicklung und Formbildungsvermögen von *Acetabularia mediterranea*. *Biol. Zbl.*, **52**, 42–61.
HANCE, R. T., 1927, Parental chromosome dimensions in Ascaris. A study of the effect of cellular environment on chromosome size. *J. Morphol.*, **44**, 117–125.
HANCE, R. T., 1933, Improving the staining action of Iron Hæmatoxylin. *Science*, **77**, 287.
HARDY, W. B., 1936, *Collected Scientific Papers*. Cambridge.
HARLAND, S. C., 1936, Haploids in polyembryonic seeds of Sea Island Cotton. *J. Hered.*, **27**, 229–231.
HARMAN, M. T., 1920, Chromosome studies in Tettigidæ. II. Chromosomes of Paratettix BB and CC and their Hybrid BC. *Biol. Bull.*, **38**, 213–230.
HARRISON, J. W. H., 1920 a, Genetical studies in the moths of the Geometrid genus *Oporabia* (*Oporinia*) with a special consideration of Melanism in the Lepidoptera. *J. Genet.*, **9**, 195–280.
HARRISON, J. W. H., 1920 b, The inheritance of melanism in the genus *Tephrosia* (*Ectropis*) with some consideration of the Inconstancy of Unit Characters under Crossing. *J. Genet.*, **10**, 61–86.
HARRISON, J. W. H., and DONCASTER, L., 1914, On hybrids between moths of the Geometrid family *Bistoninæ*, with an Account of the Behaviour of the Chromosomes in Gametogenesis in *Lycia* (*Biston*) *Hirtaria*, *Ithysia* (*Nyssa*) *Zonaria* and their Hybrids. *J. Genet.*, **3**, 229–248.
*HARTMANN, M., 1929 a, Fortpflanzung und Befruchtung als Grundlage der Vererbung. *Handb. d. Vererbungswiss.*, **1**.
*HARTMANN, M., 1929 b, Verteilung, Bestimmung und Vererbung des Geschlechtes bei den Protisten und Thallophyten. *Handb. d. Vererbungswiss.*, **2**.
HARTMANN, M., and PROWAZEK, S. v., 1907, Blepharoplast, Caryosom und Centrosome. *Arch. Protist.*, **10**, 306–335.
HASEGAWA, N., 1933, Chromosome studies in diploid and triploid *Disporum sessile*. *Jap. J. Genet.*, **9**, 9–14.
HASEGAWA, N., 1934, A cytological study on 8-chromosome rye. *Cytologia*, **6**, 68–77.
HASEYAMA, S., 1932, On the spermatogenesis of an Orthopteran, *Gampsocleis bürgeri* D. H. *J. Sci. Hiroshima B* (1), **1**, 91–144.
HAUPT, G., 1932, Beitrage zur Zytologie der Gattung *Marchantia* (L.). *Z.I.A.V.*, **62**, 367–428.
HAYDEN, M. A., 1925, Karyosphere Formation and Synapsis in the Beetle Phanæus. *J. Morphol.*, **40**, 261–297.
HEARNE, E. M., 1936, Induced chiasma formation in somatic cells by a carcinogenetic hydrocarbon. *Nature*, **138**, 291.
HEARNE, E. M., and HUSKINS, C. L., 1935, Chromosome pairing in *Melanoplus femur-rubrum*. *Cytologia*, **6**, 123–147.
HEBERER, G., 1924, Die Spermatogenese der Copepoden. I. *Zeits. wiss. Zool.*, **123**, 155–646.
HEBERER, G., 1925, Die Furchungsmitosen von *Cyclops viridis* J. und das Chromosomenindividualitätsproblem. *Zeits. mikr.-anat. Forsch.*, **10**, 169–206.

HEDAYETULLAH, S., 1931, On the Structure and Division of the Somatic Chromosomes in *Narcissus*. *J. Roy. Micr. Soc.*, **51**, 347-386.
*HEGNER, R. W., 1914, *The Germ Cell Cycle in Animals*. New York.
HEIDENHAIN, M., 1894, Neue Untersuchungen über die Centralkörper und ihre Beziehung zum Kern und Zellenprotoplasma. *Arch. mikr. Anat.*, **43**, 423-758.
HEILBORN, O., 1927, Chromosome Numbers in *Draba*. *Hereditas*, **9**, 59-68.
HEILBORN, O., 1935, Reduction division, pollen lethality and polyploidy in apples. *Acta Hort. Berg.*, **11**, 129-184.
*HEILBRUNN, L. V., 1928, *The Colloid Chemistry of Protoplasm*. Berlin.
HEIMANS, J., 1928, Chromosomen und Befruchtung bei Lilium Martagon. *Rec. Trav. Bot. Néerl.*, **25A**, 138-167.
HEITZ, E., 1925, Beitrag zur Cytologie von Melandrium. *Zeits. wiss. Biol.* (E), **1**, 241-259.
HEITZ, E., 1926, Der Nachweis der Chromosomen : Vergleichende Studien über ihre Zahl, Grösse und Form im Pflanzenreich, I. *Zeits. f. Bot.*, **18**, 625-681.
HEITZ, E., 1928, Das Heterochromatin der Moose. I. *Jahrb. wiss. Bot.*, **69**, 762-818.
HEITZ, E., 1929, Heterochromatin, Chromocentren, Chromomeren. *Der Deuts. Bot. Ges.*, **47**, 274-284.
HEITZ, E., 1931 *a*, Die Ursache der gesetzmässigen Zahl, Lage, Form und Grösse pflanzlicher Nucleolen. *Planta*, **12**, 774-844.
HEITZ, E., 1931 *b*, Nukleolen und Chromosomen in der Gattung Vicia. *Planta*, **15**, 495-505.
HEITZ, E., 1932, Die Herkunft der Chromocentren. Dritter Beitrag zur Kenntnis der Beziehung zwischen Kernstruktur und qualitative Verschiedenheit der Chromosomen in ihrer Längsrichtung. *Planta*, **18**, 571-635.
HEITZ, E., 1933 *a*, Über totale und partielle somatische Heteropyknose, sowie strukturelle Geschlechtschromosomen bei Drosophila funebris. *Zeits. Zellf. u. mikr. Anat.*, **19**, 720-742.
HEITZ, E., 1933 *b*, Die somatische Heteropyknose bei Drosophila melanogaster und ihre genetische Bedeutung. *Zeits. Zellf. u. mikr. Anat.*, **20**, 237-287.
HEITZ, E., 1933 *c*, Über α- und β-Heterochromomeren sowie Konstanz und Bau der Chromomeren bei Drosophila. *Biol. Zbl.*, **54**, 588-609.
*HEITZ, E., 1935, Chromosomenstruktur und Gene. *Z.I.A.V.*, **70**, 402-447.
HEITZ, E., 1936, Die Nucleal-Quetschmethode. *Ber. deuts. bot. Ges.*, **53**, 870-878.
HEITZ, E., and BAUER, H., 1933, Beweise fur die Chromosomennatur der Kernschleifen in Knäuelkernen von Bibio hortulanus L. *Zeits. Zellforsch. u. mikr. Anat.*, **17**, 67-82.
HELMS, A., and JORGENSEN, C. A., 1925, Maglemose i Grib Skov. VIII. Birkene paa Maglemose. *Dansk Bot. Tidsskr.*, **39**, 57-135.
HELWIG, E. R., 1929, Chromosomal variations correlated with geographic distribution in Circotettix verruculatus (Orthoptera). *J. Morphol.*, **47**, 1-36.
*HELWIG, E. R., 1933, The effect of X-rays upon the chromosomes of Circotettix verruculatus (Orthoptera). *J. Morphol.*, **55**, 265-312.
HENKING, H., 1890, Untersuchungen über die ersten Entwicklungsvorgänge in den Eiern der Insekten. 2. Über Spermatogenese und deren Beziegung zur Eientwicklung bei Pyrrhocoris apterus, L. *Zeits. wiss. Zool.*, **51**, 685-736.
*HERTWIG, O., 1917, Documente zur Geschichte der Zeugungslehre, etc. *Arch. mikr. Anat.*, **90**, Abt. II, 1-168.

HERTWIG, P., 1920, Abweichende Form der Parthenogenese bei einer Mutation von Rhabditis pellio, eine experimentell cytologische Untersuchung. *Arch. mikr. Anat.*, **94**, 303–337.
*HERTWIG, P., 1927, Vererbungslehre. *Tab. Biol.*, **4**, 114–215.
HIGGINS, E. M., 1930, Reduction Division in a Species of *Cladophora*. *Ann. Bot.*, **44**, 587–592.
HIRAYANAGI, H., 1929, Chromosome Arrangement III. The Pollen Mother Cells of the Vine. *Mem. Coll. Sci. Kyoto*, B **4**, 273–281.
HOAR, C. S., 1927, Chromosome studies in Aesculus. *Bot. Gaz.*, **84**, 156–170.
HOAR, C. S., 1931, Meiosis in Hypericum punctatum Lam. *Bot. Gaz.*, **92**, 396–406.
HOARE, G. V., 1933, Gametogenesis and Fertilisation in Scilla non-scripta. *Cellule*, **42**, 269–292.
HŒPPENER, E. and RENNER, O., 1929. Genetische und Zytologische Œnotherenstudien II. *Bot. Abh.* (Goebel), **15**, 1–86.
HOFELICH, A., 1935, Die Sektion Alsinebe Griseb. der Gattung Veronica in ihren chromosomalen Grundlagen. *Jahrb. wiss. Bot.*, **81**, 541–572.
HOGBEN, L. T., 1920, Studies on Synapsis I and II. I. Oogenesis in the Hymenoptera ; II. Parallel Conjugation and the Prophase Complex in Periplaneta, with Special Reference to the Pre-Meiotic Telophase. *Proc. Roy. Soc.*, B **91**, 268–292, 305–329.
HOLLINGSHEAD, L., 1928 *a*. A preliminary note on the occurrence of haploids in *Crepis*. *Amer. Nat.*, **62**, 282–284.
HOLLINGSHEAD, L., 1928 *b*. Chimaeras in Crepis. *Univ. Calif. Pub. Agr. Sci.*, **2**, 343–354.
HOLLINGSHEAD, L., 1930 *a*, A Cytological Study of Haploid Crepis Capillaris Plants. *Univ. Calif. Pub. Agr. Sci.*, **6**, 107–134.
HOLLINGSHEAD, L., 1930 *b*, Cytological Investigations of Hybrids and Hybrid Derivatives of Crepis capillaris and Crepis tectorum. *Univ. Calif. Pub. Agr. Sci.*, **6**, 55–94.
HOLLINGSHEAD, L., 1932, The occurrence of unpaired chromosomes in hybrids between varieties of *Triticum vulgare*. *Cytologia*, **3**, 119–141.
HOLLINGSHEAD, L. and BABCOCK, E. B., 1930, Chromosomes and Phylogeny in Crepis. *Univ. Calif. Pub. Agr. Sci.*, **6**, 1–53.
HOLMGREN, I., 1919, Zytologische Studien über die Fortpflanzung bei den Gattungen Erigeron und Eupatorium. *K. Sv. Vet. Akad. Handl.*, **59** (7), 1–118.
HOLT, C. M., 1917, Multiple complexes in the alimentary tract of Culex pipiens. *J. Morphol.*, **29**, 607–618.
HONING, J. A., 1928, Canna crosses II. The chromosome number of Canna glauca, C. glauca × C. indica F_1, C. aureo-vittata and C. aureo-vittata gigas. *Med. Landbouwhoogesch. Wageningen*, **32**, 1–14.
HOSONO, S., 1935 *a*, Karyogenetische Studien bei reinen Arten und Bastarden der Emmerreihe. I. Reifungsteilungen. *Jap. J. Bot.*, **7**, 310–322.
HOSONO, S., 1935 *b*, Studien über die 1. Reifungsteilung bei einem tetraploiden *Triticum*-bastard (*T. persicum* Vav. var. *rubiginosum* Zhuk. × *T. ægilopoides* Bal. var. *bœoticum* Perc.). *Jap. J. Genet.*, **11**, 18–29.
HRUBY, K., 1934, Über die Chromosomenstruktur in infraroten Strahlen. *Planta*, **22**, 685–691.
HRUBY, K., 1935 *a*, Durch X-Strahlen hervorgerufene abnorme Plastiden. *Vest. Kral. Ces. Spol. Nauk*, **2**, 1–5.
HRUBY, K., 1935 *b*, Some new Salvia species hybrids. *Studies from the Plant Physiol. Lab., Charles Univ., Prague*, **5**, 1–73.
HUETTNER, A. F., 1930, The spermatogenesis of Drosophila melanogaster. *Zeits. Zellforsch. u. mikr. Anat.*, **11**, 615–637.

BIBLIOGRAPHY

HUGHES-SCHRADER, S., 1927, Origin and Differentiation of the male and female germ-cells in the hermaphrodite of Icerya purchasi (Coccidæ). *Zeits. f. Zellforsch. u. mikr. Anat.*, **6**, 509–540. (*See also*: Schrader.)

HUGHES-SCHRADER, S., 1930, The cytology of several species of iceryine coccids, with special reference to parthenogenesis and haploidy. *J. Morphol.*, **50**, 475–495.

HUGHES-SCHRADER, S., 1931, A study of the chromosome cycle and the meiotic division figure in Llaveia bouvari—a primitive coccid. *Zeits. Zellforsch. u. mikr. Anat.*, **13**, 742–769.

HUGHES-SCHRADER, S., 1935, The chromosome cycle of Phenacoccus (Coccidæ). *Biol. Bull.*, **69**, 462–468.

HUMPHREY, L. M., 1934, Meiotic divisions of haploid, diploid and tetraploid tomatoes. *Cytologia*, **5**, 278–299.

HURST, C. C., 1931, Embryo-sac Formation in Diploid and Polyploid Species of Roseæ. *Proc. Roy. Soc.*, B **109**, 126–148.

HUSKINS, C. L., 1927, On the Genetics and Cytology of Fatuoid or False Wild Oats. *J. Genet.*, **18**, 315–364.

HUSKINS, C. L., 1928, On the Cytology of Speltoid Wheats in relation to their Origin and Genetic Behaviour. *J. Genet.*, **20**, 103–122.

HUSKINS, C. L., 1931 a, The Origin of Spartina Townsendii. *Genetica*, **12**, 531–538.

HUSKINS, C. L., 1931 b, A cytological Study of Vilmorin's Unfixable Dwarf Wheat. *J. Genet.*, **25**, 113–124.

HUSKINS, C. L., 1934, Anomalous segregation of a triploid tomato. *J. Hered.*, **25**, 281–286.

HUSKINS, C. L., and HEARNE, E. Marie, 1933, Meiosis in asynaptic dwarf oats and wheat. *J. Roy. Micr. Soc.*, **53**, 109–117.

HUSKINS, C. L., and HUNTER, A. W. S., 1935, The effect of X-radiation on chromosomes in the microspores of Trillium erectum Linn. *Proc. Roy. Soc. Lond.* B, **117**, 22–33.

HUSKINS, C. L., and SMITH, S. G., 1932, A Cytological Study of the Genus Sorghum Pers., I. The Somatic Chromosomes. *J. Genet.*, **25**, 241–249.

HUSKINS, C. L., and SMITH, S. G., 1934, Chromosome division and pairing in Fritillaria Meleagris; the mechanism of meiosis. *J. Genet.*, **28**, 397–406.

HUSKINS, C. L., and SMITH, S. G., 1935, Meiotic chromosome structure in Trillium erectum L. *Ann. Bot.*, **49**, 119–150.

HUSKINS, C. L., and SPIER, J. D., 1934, The segregation of heteromorphic homologous chromosomes in pollen mother-cells of Triticum vulgare. *Cytologia*, **5**, 269–277.

HUTH, W., 1933, Ophryotrocha-studien. I. Zur Cytologie der Ophryotrochen. *Zeits. Zellforsch. u. mikr. Anat.*, **20**, 309–381.

HÜTTIG, W., 1933, Über physikalische und chemische Beeinflussungen des Zeitpunktes der Chromosomenreduktion bei Brandpilzen. *Zeits. f. Bot.*, **26**, 1–26.

HUXLEY, J. S., 1929, Sexual difference of linkage in Gammarus chevreuxi. *J. Genet.*, **20**, 145–156.

ICHIJIMA, K., 1926, Cytological and Genetic Studies on Fragaria. *Genetics*, **11**, 590–604.

ICHIJIMA, K., 1930, Studies on the Genetics of Fragaria. *Z.I.A.V.*, **55**, 300–347.

ICHIJIMA, K., 1934, On the artificially induced mutations and polyploid plants of rice occurring in subsequent generations. *Proc. Imp. Acad.*, **10**, 388–391.

INARIYAMA, S., 1928, On the Spiral Structure of Chromosomes in *Hosta Sieboldiana* Engl. *Bot. Mag. Tokyo*, **42**, 486–489 (Japanese).
INARIYAMA, S., 1933, Cytological studies in the genus Lycoris I. *Bot. Mag. Tokyo*, **46**, 426–434.
IRIKI, S., 1932, Studies on Amphibian chromosomes 4-8. *Sci. Rep. Tokyo Univ. Sci. Lit.* B, **1**, 61–126.
ISHIKAWA, M., 1911, Cytologische Studien von Dahlien, *Bot. Mag. Tokyo*, **25**, 1–8.
ISHIKAWA, M., 1916, A list of the numbers of chromosomes. *Bot. Mag. Tokyo*, **30**, 404–440.
ISHIKAWA, M., 1918, Studies on the Embryo-Sac and Fertilisation in Œnothera. *Ann. Bot.*, **32**, 279–317.
ITOH, H., 1934, Chromosomal variations in the spermatogenesis of a grasshopper, Locusta danica L. (English summary.) *Jap. J. Genet.*, **10**, 115-134.
IWATA, J., 1935, Chromosome structure in *Lilium*. *Mem. Coll. Sci. Kyoto B*, **10**, 275–288.

JACHIMSKY, H., 1935, Beitrag zur Kenntnis von Geschlechtschromosomen und Heterochromatin bei Moosen. *Jahrb. wiss. Bot.*, **81**, 203–238.
JANAKI-AMMAL, E. K., 1932, Chromosome studies in Nicandra physaloides. *Cellule*, **41**, 89–100.
JANAKI-AMMAL, E. K., 1934, Polyploidy in *Solanum Melongena* Linn. *Cytologia*, **5**, 453–459.
JANAKI-AMMAL, E. K., 1936, Cytogenetic analysis of *Saccharum spontaneum* L. I. Chromosome studies in some Indian forms. *Ind. J. Agr. Sci.*, **6**, 1–8.
JANSSENS, F. A., 1909, Spermatogénèse dans les Batraciens V., La Théorie de la Chiasmatypie, Nouvelle interpretation des cinèses de maturation. *Cellule*, **25**, 387–411.
JANSSENS, F. A., 1924, La Chiasmatypie dans les Insectes. *Cellule*, **34**, 135-359.
JARETSKY, R., 1928, Histologische und karyologische Studien an Polygonaceen. *Jahrb. wiss. Bot.*, **69**, 357–490.
JARETSKY, R., 1930, Zur Zytologie der Fagales. *Planta*, **10**, 120–137.
JENKIN, T. J., and SETHI, B. L., 1932, Phalaris arundinacea, Ph. tuberosa, their F_1 hybrids and hybrid derivatives. *J. Genet.*, **26**, 1–36.
JENKINS, J. A., 1929, Chromosome homologies in wheat and *Ægilops*. *Amer. J. Bot.*, **16**, 238–245.
JENNINGS, H. S., 1923, The numerical relations in the crossing-over of the genes, with a critical examination of the theory that the genes are arranged in a linear series. *Genetics*, **8**, 390–457.
*JOHANNSEN, W., 1911, The Genotype Conception of Heredity, *Amer. Nat.*, **45**, 129–159.
*JOHANNSEN, W., 1923, Some remarks about units in heredity. *Hereditas*, **4**, 133–141.
JOHANSEN, D. A., 1934, Haploids in Hordeum vulgare. *Proc. Nat. Ac. Sci.*, **20**, 98–100.
JONES, D. F., 1934, Unisexual maize plants and their bearing on sex differentiation in other plants and animals. *Genetics*, **19**, 552–567.
JØRGENSEN, C. A., 1923, Studies on Callitrichaceæ. *Dansk Bot. Tidsskr.*, **38**, 81–126.
JØRGENSEN, C. A., 1927, Cytological and Experimental Studies in the Genus *Lamium*. *Hereditas*, **9**, 126–136.
JØRGENSEN, C. A., 1928, The Experimental Formation of Heteroploid Plants in the Genus *Solanum*. *J. Genet.*, **19**, 133–211.

JØRGENSEN, C. A., and CRANE, M. B., 1927, Formation and morphology of *Solanum* chimaeras. *J. Genet.*, **18**, 247-273.
JØRGENSEN, M., 1913, Zellenstudien, I. Morphologische Beiträge zum Problem des Eiwachstums. *Arch. Zellf.*, **10**, 1-126.
JUEL, H. O., 1900, Vergleichende Untersuchungen über typische and parthenogenetische Fortpflanzung bei der Gattung *Antennaria*. *K. Sv. Vet. Akad. Handl.*, **33** (5), 1-59.
JUEL, H. O., 1905, Die Tetradenteilung bei Taraxacum und anderen Cichorieen. *K. Sv. Vet. Akad. Handl.*, **39** (4), 1-21.
JUNKER, H., 1923, Cytologische Untersuchungen an den Geschlechtsorganen der halbzwittrigen Steinfliege *Perla marginata* (Panzer). *Arch. f. Zellforsch.*, **17**, 185-259.

KACHIDZE, N., 1929, Karyologische Studien über die Familie Dipsacaceæ. *Planta*, **7**, 482-502.
KAGAWA, F., 1928, Cytological Studies on *Triticum* and *Ægilops* II. On the genus crosses between *Triticum* and *Ægilops*. *Jap. J. Bot.*, **4**, 1-26.
KAGAWA, F., 1929 a, On the Phylogeny of some Cereals and Related Plants as Considered from the Size and Shape of Chromosomes. *Jap. J. Bot.*, **4**, 363-383.
KAGAWA, F., 1929 b, Cytological Studies on the pollen-formation of the hybrids between *Triticum* and *Ægilops*. *Jap. J. Bot.*, **4**, 345-361.
KAGAWA, F., 1931, Chromosome Studies of a Species Cross in *Ægilops*. *Bull. Utsunomiya Ag. Coll.*, **1**, 57-60.
KAGAWA, F., and NAKAJIMA, G., 1933, Genetical and cytological studies on species hybrids in *Quamoclit*. *Jap. J. Bot.*, **6**, 315-327.
KAHLE, W., 1908, Die Pædogenese des Cecidomyiden. *Zoologica* (Stuttgart), **21** (55), 1-80.
KAPPERT, H., 1933, Erbliche Polyembryonie bei Linum usitatissimum. *Biol. Zbl.*, **53**, 276-307.
KARASAWA, K., 1932, On triploid *Thea*. *Bot. Mag. Tokyo*, **46**, 458-460.
KARASAWA, K., 1933, On the triploidy of *Crocus sativus* L. and its high sterility. *Jap. J. Genet.*, **9**, 6-8.
KARLING, J. S., 1928, Nuclear and cell division in the antheridial filaments of the Characeæ. *Bull. Torrey Bot. Club*, **55**, 11-39.
KARPECHENKO, G. D., 1927 a, The Production of Polyploid Gametes in Hybrids. *Hereditas*, **9**, 349-368.
KARPECHENKO, G. D., 1927 b, Polyploid hybrids of Raphanus sativus L. × Brassica oleracea L. (English summary). *Bull. Appl. Bot.*, **17** (3), 305-410.
KARPECHENKO, G. D., 1928, Polyploid Hybrids of *Raphanus sativus* L. × *Brassica oleracea* L. *Z.I.A.V.*, **48**, 1-85.
KARPECHENKO, G. D., 1929, A contribution to the synthesis of a constant hybrid of three species. (English summary). *Proc. U.S.S.R. Cong. Genet.*, **2**, 277-294.
*KARPECHENKO, G. D., 1935, *Theory of distant hybridisation* (in Russian). Moscow, 1935.
KARPECHENKO, G. D., 1936 a, Experimental Production of Hexagenomic Hybrids (Brassica oleracea × Brassica carinata Czern). *Bull. Appl. Bot. II.*, **7**, (In the press.)
KARPECHENKO, G. D., 1936 b, Increasing the crossability of a species by doubling its chromosome number. *Bull. Appl. Bot. II.*, **7**, 47-50.
KARPECHENKO, G. D., and SHCHAVINSKAIA, S. A., 1929, On sexual incompatibility of tetraploid hybrids. *Proc. U.S.S.R. Cong. Genet.*, **2**, 267-276. (English summary.)

KATAYAMA, Y., 1935 a, Karyological comparisons of haploid plants from octoploid Ægilotricum and diploid wheat. *Jap. J. Bot.*, **7**, 349–380.

KATAYAMA, Y., 1935 b, Haploid plants in the Japanese Morning Glory. *Jap. J. Genet.*, **11**, 279–281.

KATAYAMA, Y., 1935 c, On a chromosomal variant induced by X-ray treatment in *Triticum monococcum*. *Proc. Imper. Acad.*, **11**, 110–111 (Tokyo).

KATAYAMA, Y., 1935 d, Karyogenetic studies on X-rayed sex cells and their derivatives in *Triticum monococcum*. *J. Coll. Agric. Tokyo*, **13**, 333–362.

KATAYAMA, Y., 1935 e, Further investigations on synthesized octoploid Ægilotrichum. *J. Coll. Agric. Tokyo*, **13**, 397–414.

KATAYAMA, Y., 1936, Chromosome studies in some Alliums. *J. Coll. Agric. Tokyo*, **13**, 431–441.

KATER, J. McA., 1926, Chromosomal vesicles and the structure of the resting nucleus in *Phaseolus*. *Biol. Bull.*, **51**, 209–224.

KATTERMANN, G., 1930, Chromosomenuntersuchungen bei Gramineen. *Planta*, **12**, 19–37.

KATTERMANN, G., 1931, Über die Bildung polyvalenter Chromosomenverbände bei einigen Gramineen. *Planta*, **12**, 732–744.

KATTERMANN, G., 1933, Weitere zytologische Untersuchungen an Briza media mit besonderer Berücksichtigung der durch Verbände aus vier Chromosomen ausgezeichneten Pflanzen. *Jahrb. wiss. Bot.*, **78**, 43–91.

KATTERMANN, G., 1934, Die cytologischen Verhältnisse einiger Weizenroggenbastarde und ihrer Nachkommenschaft (F_2). *Züchter*, **6**, 97–107.

KATTERMANN, G., 1935 a, Die cytologischen Verhältnisse bei Primula malacoides II. mitteilung. Die tetraploiden Pflanzen. *Gartenbauwiss*, **9**, 159–174.

KATTERMANN, G., 1935 b, Die Chromosomenverhältnisse bei Weizen-Roggenbastarde der zweiten Generation, mit besonderer Berucksichtigung der Homologiebeziehungen. *Z.I.A.V.*, **70**, 265–308.

KATO, K., 1935, Chromosome behaviour in the interkinesis I. Observation of pollen mother cells in *Tradescantia reflexa*. *Mem. Coll. Sci. Kyoto B*, **10**, 251–272.

KATO, K., and IWATA, J., 1935, Spiral structure of chromosomes in *Lilium Mem. Coll. Sci. Kyoto B*, **10**, 263–273.

KAUFMANN, B. P., 1926 a, Chromosome structure and its relation to the chromosome cycle I. Somatic mitoses in *Tradescantia pilosa* Lehm. *Amer. J. Bot.*, **13**, 59–80.

KAUFMANN, B. P., 1926 b, Chromosome structure and its relation to the chromosome cycle II. *Podophyllum peltatum*. *Amer. J. Bot.*, **13**, 355–363.

KAUFMANN, B. P., 1931, Chromonemata in somatic and meiotic mitoses. *Amer. Nat.*, **65**, 280–283.

KAUFMANN, B. P., 1933, Interchange between X- and Y-chromosomes in attached X females of Drosophila melanogaster. *Proc. Nat. Acad. Sci.*, **19**, 830–838.

KAUFMANN, B. P., 1934, Somatic mitoses of Drosophila melanogaster. *J. Morph.*, **56**, 125–155.

KAWAGUCHI, E., 1928, Zytologische Untersuchungen am Seidenspinner und seinen Verwandten I. Gametogenese von Bombyx mori L. und B. mandarina M. und ihre Bastarde. *Zeits. f. Zellforsch. u. mikr. Anat.*, **7**, 519–552.

KAWAGUCHI, E., 1933, Die Heteropycnose der Geschlechtschromosomen bei Lepidopteren. *Cytologia*, **4**, 339–354.

KAWAKAMI, J., 1930, Chromosome numbers in Leguminosæ. *Bot. Mag. Tokyo,* **44**, 319–328.
KAZAO, N., 1929, Cytological studies on *Iris*. *Sci. Rep. Tohoku Imp. Univ.* IV, **4**, 543–549.
KERKIS, J., 1934, On the mechanism of the development of triploid intersexuality in *Drosophila melanogaster*. *C. R. Acad. Sci. de l'URSS*, 1934, 288–294.
KERKIS, J., 1936, Chromosome conjugation in hybrids between *Drosophila melanogaster* and *Drosophila simulans*. *Amer. Nat.*, **70**, 81–86.
KEUNEKE, W., 1924, Über die Spermatogenese einiger Dipteren. *Zeits. Zell. u. Geweb.*, **1**, 357–412.
KIELLANDER, C. L., 1935, Apomixis bei *Poa serotina*. *Bot. Notiser*, Lund, **1935**, 87–95.
KIHARA, H., 1919, Über cytologische Studien bei einigen Getreidearten I. and II. I. Spezies-Bastarde des Weizens und Weizenroggen-Bastard. II. Chromosomenzahlen und Verwandtschaftsverhaltnisse unter Avena-Arten. *Bot. Mag. Tokyo,* **33**, 17–38, 94–97.
KIHARA, H., 1921, Über cytologische Studien bei einigen Getreidearten. III. *Bot. Mag. Tokyo,* **35**, 19–44.
KIHARA, H., 1924, Cytologische und Genetische Studien bei wichtigen Getreidearten, etc. *Mem. Coll. Sci. Kyoto,* **1**, 1–200.
KIHARA, H., 1927 a, Über das Verhalten der "end-to-end" gebundenen Chromosomen von Rumex acetosella und Œnothera biennis wahrend der heterotypischen Kernteilung. Beitrag zur Frage der Para- und Metasyndese. *Jahrb. wiss. Bot.*, **66**, 429–460.
KIHARA, H., 1927 b, Über die Vorbehandlung einiger pflanzlicher Objecte bei der Fixierung der Pollenmutterzellen. *Bot. Mag. Tokyo,* **41**, 124–128.
KIHARA, H., 1929 a, The Sex chromosomes of *Humulus japonicus*. *Jap. J. Genet.*, **4**, 55–63.
KIHARA, H., 1929 b, A case of linkage of sex chromosomes with autosomes in the pollen mother cell of *Humulus japonicus* (English summary). *Jap. J. Genet.*, **5**, 73–80.
KIHARA, H., 1929 c, Conjugation of homologous chromosomes in the genus hybrids *Triticum* × *Ægilops* and species hybrids of *Ægilops*. *Cytologia*, **1**, 1–15.
KIHARA, H., 1930 a, Karyologische Studien an *Fragaria* mit besonderer Berücksichtigung der Geschlechtschromosomen. *Cytologia*, **1**, 345–375.
KIHARA, H., 1930 b, Genomanalyse bei *Triticum* und *Ægilops* I. KIHARA, H., and NISHIYAMA, I., 1930, Genomaffinitäten in tri-tetra und pentaploiden Weizenbastarden. *Cytologia*, **1**, 263–284.
KIHARA, H., 1931 a, Genomanalyse bei *Triticum* und *Ægilops* II. *Ægilotricum* und *Ægilops cylindrica*. *Cytologia*, **2**, 106–156.
KIHARA, H., 1931 b, Genomanalyse bei *Triticum* und *Ægilops* III. KIHARA, H., and KATAYAMA, Y., 1931, Zur Entstehungsweise eines neuen konstanten oktoploiden *Ægilotricum*. *Cytologia*, **2**, 234–255.
KIHARA, H., 1935 b, Genomanalyse bei Triticum und Ægilops VI. KIHARA, H., and LILIENFELD, F.: Weitere Untersuchungen an *Ægilops* × *Triticum* und *Ægilops* × *Ægilops*-Bastarden. *Cytologia*, **6**, 195–216.
KIHARA, H., and HIRAYOSHI, I., 1932, Die Geschlechtschromosomen von *Humulus japonicus* Sieb. et Zucc. *8th Congr. Jap. Assoc. Adv. Sci.*, 1932, 363–368.
KIHARA, H., and KATAYAMA, Y., 1932, Über das Vorkommen von haploiden Pflanzen bei Triticum monococcum. *Kwagaku*, **2**, 408–410 (Japanese).
*KIHARA, H., and LILIENFELD, F., 1932, Untersuchungen an *Ægilops* × *Triticum* und *Ægilops* × *Ægilops* Bastarden. *Cytologia*, **3**, 384–456.

KIHARA, H., and LILIENFELD, F., 1935, Kerneinwanderung und Bildung syndiploider Pollenmutterzellen bei dem F_1-Bastard *Triticum ægilopoides* × *Ægilops squarrosa*. *Jap. J. Genet.*, 10, 1–28.

KIHARA, H., and NISHIYAMA, I., 1932, Different compatibility in reciprocal crosses of Avena, with special reference to tetraploid hybrids between hexaploid and diploid species. *Jap. J. Bot.*, 6, 245–305.

KIHARA, H., and ONO, T., 1925, The Sex-Chromosomes of *Rumex Acetosa*. *Z.I.A.V.*, 39, 1–7.

KIHARA, H., and ONO, T., 1926, Chromosomenzahlen und systematische Gruppierung der Rumex-Arten. *Zeits. f. Zellforsch. u. mikr. Anat.*, 4, 475–481.

KIHARA, H., and YAMAMOTO, Y., 1931, Karyomorphologische Untersuchungen an *Rumex acetosa* L. und *Rumex montanus* Desf. *Cytologia*, 3, 84–118.

KIHARA, H., and YAMAMOTO, Y., 1935, Chromosomenverhältnisse bei *Aucuba chinensis* Benth. *Agric. and Hort.*, 10, 2485–2496.

KIKKAWA, H., 1935 a, Contributions to the knowledge of non-disjunction of the sex-chromosomes in Drosophila virilis. II. The mode of reduction of the heterochromosomes. *Cytologia*, 6, 177–189.

KIKKAWA, H., 1935 b, Crossing-over in the male of *Drosophila virilis*. *Cytologia*, 6, 190–194.

KING, R. L., and BEAMS, H. W., 1934, Somatic synapsis in Chironomus with special reference to the individuality of the chromosomes. *J. Morphol.*, 56, 577–592.

KIRSSANOW, B. A., 1931, Über die Nachwirkung verschiedener Temperaturen und der X-Strahlen auf das Crossing-over im dritten Chromosom bei Drosophila melanogaster. *Biol. Zbl.*, 51, 529–533.

KLINGSTEDT, H., 1931, Digametie beim Weibchen der Trichoptere Limnophilus decipiens Kol. *Act. Zool. Fenn.*, 10, 1–69.

KLINGSTEDT, H., 1933, Chromosomenstudien an Neuropteren I. Ein Fall von heteromorphen Chromosomenpaaren als Beispiel vom Mendeln der Chromosomen. *Faun. et Flor. Fenn.*, 10, 1–11.

KNAPP, E., 1935 a, Zur Frage der genetischen Aktivität des Heterochromatins, nach Untersuchungen am X-Chromosom von Sphærocarpus Donnellii. *Ber. deuts. Bot. Ges.*, 53, 751–760.

KNAPP, E., 1935 b, Untersuchungen über die Wirkung von Röntgenstrahlen an dem Lebermoos Sphærocarpus, mit Hilfe der Tetraden-Analyse. I. *Z.I.A.V.*, 70, 309–349.

*KNIEP, H., 1928, *Die Sexualität der niederen Pflanzen*. Jena.

*KNIEP, H., 1929, Vererbungserscheinungen bei Pilzen. *Bibliogr. Genet.*, 5, 371–478.

KOBEL, F., 1927, Zytologische Untersuchungen an Prunoideen und Pomoideen. *Arch. Jul.-Klaus-Stift.*, 3, 1–84.

KOLLER, P. C., 1932 a, The relation of fertility factors to crossing-over reduction in the *Drosophila obscura* hybrid. *Z.I.A.V.*, 60, 137–151.

KOLLER, P. C., 1932 b, Constitution of the X-Chromosome in *Drosophila obscura*. *Nature*, 129, 616.

KOLLER, P. C., 1932 c, Further studies in *Tradescantia virginiana* var. *humilis* and *Rhoeo discolor*. *J. Genet.*, 26, 81–96.

KOLLER, P. C., 1934 a, Spermatogenesis in *Drosophila pseudo-obscura* Frolowa. II. The cytological basis of sterility in hybrid males of races A and B. *Proc. Roy. Soc. Edin.*, 54, 67–87.

KOLLER, P. C., 1934 b, The movements of chromosomes within the cell and their dynamic interpretation. *Genetica*, 16, 447–466.

KOLLER, P. C., 1935, The internal mechanics of the chromosomes, IV. Pairing

BIBLIOGRAPHY

and coiling in salivary gland nuclei of Drosophila. *Proc. Roy. Soc. B.*, **810**, 371–397.

KOLLER, P. C., 1936 a, Structural hybridity in Drosophila pseudo-obscura. *J. Genet.*, **32**, 79–102.

KOLLER, P. C., 1936 b, The genetical and mechanical properties of the sex-chromosomes. II. Marsupials. *J. Genet.*, **32**, 451–472.

KOLLER, P. C., and DARLINGTON, C. D., 1934, The genetical and mechanical properties of the sex chromosomes. I. Rattus norvegicus ♂. *J. Genet.*, **29**, 159–173.

KOLTZOFF, N. K., 1928, Physikalisch-chemische Grundlage der Morphologie. *Bot. Zeit.*, **48**, 345–369.

KOLTZOFF, N. K., 1934, The structure of the chromosomes in the salivary glands of Drosophila. *Science*, **80**, 312–313.

KORNHAUSER, S. I., 1914, A comparative study of the chromosomes in the spermatogenesis of Enchenopa binotata (Say) and Enchenopa (Campylenchia Stål) curvata (Fabr). *Arch. f. Zellforsch.*, **12**, 241–298.

KOSSIKOV, K., 1934, The attached X-chromosomes in Drosophila simulans. *C. R. Acad. Sci. U.R.S.S.*, **1934**, 472–475.

KOSSIKOV, K. V., and MULLER, H. J., 1935, Invalidation of the genetic evidence for branched chromonemas (in the case of the pale translocation in Drosophila). *J. Hered.*, **26**, 305–317.

KOSTOFF, D., 1929, An androgenic Nicotiana haploid. *Zeits. f. Zellforsch. u. mikr. Anat.*, **9**, 640–642.

KOSTOFF, D., 1930, Discoid structure of the spireme. *J. Hered.*, **21**, 323–324.

KOSTOFF, D., 1932 (Triticum dicoccum × Triticum monococcum) × Triticum vulgare triple hybrid with 42 chromosomes. *Cytologia*, **3**, 186–187.

KOSTOFF, D., 1933, Cytogenetic studies of the triple fertile hybrid Nicotiana Tabacum × (N. sylvestris × N. Rusbyi) − N. triplex. *Bull. Appl. Bot.* II, **5**, 167–204.

KOSTOFF, D., 1934 a, A contribution to the meiosis of Helianthus tuberosus L. *Zeits. f. Zücht. A.* **19**, 429–438.

KOSTOFF, D., 1934 b, Crossing over in Nicotiana species hybrids. *Cytologia*, **5**, 373–377.

KOSTOFF, D., 1935 a, Conjugation between morphologically different chromosomes in Nicotiana species hybrids. *C. R. Acad. Sci. U.R.S.S.*, March, 1935, **1**, 7–8, 558–560.

KOSTOFF, D., 1935 b, Chromosome alterations by centrifuging. *Z.I.A.V.*, **69**, 301–302.

KOSTOFF, D., 1935 c, On the increase of mutation frequency following interspecific hybridization in Nicotiana. *Current Science*, **3**, 302–304.

KOSTOFF, D., 1935 d, Studies on Polyploid Plants V. Fertile Triticum vulgare-monococcum hybrids. *C. R. Acad. Sci. U.R.S.S.*, **1**, 155–159.

KOZHEVNIKOV, B. T., 1935, Intraspecific isolation artificially obtained through chromosome aberrations. *Amer. Nat.*, **69**, 459–461.

KRAUSE, O., 1931, Zytologische Studien bei den Urticales unter besonderer Berücksichtigung der Gattung Dorstenia. *Planta*, **13**, 29–84.

KUHN, A., 1920, Untersuchungen zur kausalen Analyse der Zellteilung. I. *Arch. Entwickl.*, **46**, 259–327.

KUHN, E., 1929 a, Die Beziehung der Chromocentren zur Chromosomenbildung. *Ber. Deuts. Bot. Ges.*, **47**, 421–430.

KUHN, E., 1929 b, Ein Beweis für die Lebensfähigkeit von Spermatozoen ohne X- und Y-Chromosom bei Drosophila melanogaster. *Z.I.A.V.*, **53**, 26–37.

* KUHN, E., 1930, Pseudogamie und Androgenesis bei Pflanzen. *Züchter*, **2**, 124–136.

KULKARNI, C. G., 1929, Meiosis in Pollen Mother Cells of Strains of Œnothera pratincola Bartlett. *Bot. Gaz.*, **87**, 218–259.

KUNIEDA, H., 1926, On the Mitosis and Fertilisation of *Sargassum Horneri* Ag. *Bot. Mag. Tokyo*, **40**, 545–550.

KUWADA, Y., 1910, A Cytological Study of *Oryza sativa* L. *Bot. Mag. Tokyo*, **24**, 267–281.

KUWADA, Y., 1921, On the So-called Longitudinal Split of Chromosomes in the Telophase. *Bot. Mag. Tokyo*, **35**, 99–105.

KUWADA, Y., 1926, On the Structure of the Anaphasic Chromosomes in the Somatic Mitoses in *Vicia Faba*, with Special Reference to the So-called Longitudinal Split of Chromosomes in the Telophase. *Mem. Coll. Sci. Kyoto B.*, **2**, 1–13.

KUWADA, Y., 1927, On the Spiral Structure of Chromosomes. *Bot. Mag. Tokyo*, **41**, 100–109.

KUWADA, Y., 1928, An Occurrence of Restitution-Nuclei in the Formation of the Embryo sacs in *Balanophora japonica*, Mak. *Bot. Mag. Tokyo*, **42**, 117–129.

KUWADA, Y., 1929, Chromosome Arrangement I. Model Experiments with Floating Magnets and some Theoretical Considerations on the Problem. *Mem. Coll. Sci. Kyoto*, **4**, 199–264.

KUWADA, Y., 1935, Behaviour of chromonemata in mitosis. V. A probable method of formation of the double-coiled chromonema spirals and the origin of the coiling of the chromonemata into spirals. *Cytologia*, **6**, 308–313.

KUWADA, Y., and NAKAMURA, T., 1933, Behaviour of chromonemata in Mitosis I. Observations of pollen mother cells in *Tradescantia reflexa*. *Mem. Coll. Sci. Kyoto Imp. Univ. B*, **9**, 129–139.

KUWADA, Y., and NAKAMURA, T., 1934 *a*, Behaviour of chromonemata in Mitosis. II. Artificial unravelling of coiled chromonemata. *Cytologia*, **5**, 244–247.

KUWADA, Y., and NAKAMURA, T., 1934 *b*, Behaviour of chromonemata in mitosis. III. Observation of living staminate hair cells in *Tradescantia reflexa*. *Mem. Coll. Sci. Kyoto B*, **9** (5), 343–366.

KUWADA, Y., and NAKAMURA, T., 1934 *c*, Behaviour of Chromonemata in Mitosis. IV. Double refraction of chromosomes in *Tradescantia reflexa*. *Cytologia*, **6**, 78–86.

KUWADA, Y., and NAKAMURA, T., 1935, Behaviour of chromonemata in mitosis. VI. Metaphasic and anaphasic longitudinal split of chromosomes in the homotype division in pollen mother cells in *Tradescantia reflexa*. *Cytologia*, **6**, 314–319.

KUWADA, Y., and SAKAMURA, T., 1926, A contribution to the Colloidchemical and Morphological Study of Chromosomes. *Protoplasma*, **1**, 239–254.

KUWADA, Y., and SUGIMOTO, T., 1926, On the Structure of the Chromosomes in *Tradescantia virginica*. *Bot. Mag. Tokyo*, **40**, 19–20.

KUWADA, Y., and SUGIMOTO, T., 1928, On the Staining Reactions of Chromosomes. *Protoplasma*, **3**, 531–535.

LA-COUR, L., 1931, Improvements in Everyday Technique in Plant Cytology. *J. Roy. Micr. Soc.*, **51**, 119–126.

LA-COUR, L., 1935, Technic for studying chromosome structure. *Stain Technology*, **10**, 57–60.

LAMM, R., 1936, Cytological studies on inbred rye. *Hereditas*, **22**, 217–240.

LAMMERTS, W. E., 1929, Interspecific Hybridization in Nicotiana IX. Further Studies of the Cytology of the Backcross progenies of the *Paniculata-Rustica* hybrid. *Genetics*, **14**, 286–304.

BIBLIOGRAPHY

LAMMERTS, W. E., 1931, Interspecific Hybridization in Nicotiana XII. The Amphidiploid *rustica-paniculata* hybrid; its origin and cyto-genetic behavior. *Genetics*, 16, 191–211.
LAMMERTS, W. E., 1932, An experimentally produced secondary polyploid in the genus *Nicotiana*. *Cytologia*, 4, 38–45.
LAMMERTS, W. E., 1934, On the nature of chromosome association in *N. tabacum* haploids. *Cytologia*, 6, 38–50.
LANCEFIELD, D. E., 1929, A genetic study of two races or physiological species of *Drosophila obscura*. *Z.I.A.V.*, 52, 287–317.
LANCEFIELD, R. C., and METZ, C. W., 1922, The sex-linked group of mutant characters in Drosophila Willistoni. *Amer. Nat.*, 56, 211–241.
LANGLET, O. F. I., 1927, Beiträge zur Zytologie der Ranunculazeen. *Sv. Bot. Tidsskr.*, 21, 1–17.
LANGLET, O. F. I., 1932, Über Chromosomenverhältnisse und Systematik der Ranunculaceæ. *Sv. Bot. Tids.*, 26, 381–400.
LANGLET, O. F. I., and SÖDERBERG, E., 1929, Über die Chromosomenzahlen einiger Nymphaeaceen. *Acta Hort. Berg.*, 9, 85–104.
LARTER, L. N. H., 1932, Chromosome variation and behaviour in Ranunculus L. *J. Genet.*, 26.
LARTER, L. N. H., 1935, Hybridism in *Musa*. I. Somatic cytology of certain Jamaican seedlings. *J. Genet.*, 31, 297–316.
LAWRENCE, W. J. C., 1929, The Genetics and Cytology of Dahlia species. *J. Genet.*, 21, 125–159.
LAWRENCE, W. J. C., 1931 a, The Genetics and Cytology of *Dahlia variabilis*. *J. Genet.*, 24, 257–306.
LAWRENCE, W. J. C., 1931 b, Mutation or Segregation in the Octoploid *Dahlia variabilis?* *J. Genet.*, 24, 307–324.
LAWRENCE, W. J. C., 1931 c, The Secondary Association of Chromosomes. *Cytologia*, 2, 352–384.
LAWRENCE, W. J. C., 1931 d, The Chromosome Constitution of *Cardamine pratensis* and *Verbascum phœniceum*. *Genetica*, 13, 183–208. (*See also*: Crane.)
LAWRENCE, W. J. C., 1936, On the origin of new forms in Delphinium. *Genetica*, 18, 109–115.
LAWTON, E., 1932, Regeneration and induced polyploidy in ferns. *Amer. J. Bot.*, 19, 303–333.
LEAGUE, B. B., 1928, The chromosomes of the guinea pig. *J. Morphol.*, 46, 131–142.
LEBEDEFF, G. A., 1934, Genetics of hermaphroditism in Drosophila virilis. *Proc. Nat. Acad. Sci.*, 20, 613–616.
LEBEDEFF, V. N., 1932, The New Phenomenon in Wheat-Rye Hybrids. (Plant Breeding Inst., Belaya Zerkov, Ukraine.) Kiev.
LEBEDEFF, V. N., 1934, Neue Fälle der Formierung von Amphidiploiden in Weizen-Roggen-Bastarden. *Zeits. Zücht. A*, 19, 509–25.
LEBRUN, H., 1902, La cytodiérèse de l'œuf. La vésicule germinative et les globules polaires chez les Anoures. *Cellule*, 19, 315–402.
LESLEY, J. W., 1928, The cytological and genetical study of progenies in triploid tomatoes. *Genetics*, 13, 1–43.
LESLEY, J. W., and LESLEY, M. M., 1929, Chromosome fragmentation and mutation in tomato. *Genetics*, 14, 321–336.
LESLEY, M. M., 1926, Maturation in diploid and triploid Tomatoes. *Genetics*, 11, 267–279.
LESLEY, M. M., and FROST, H. B., 1927, Mendelian Inheritance of Chromosome Shape in *Matthiola*. *Genetics*, 12, 449–460.
LESLEY, M. M., and FROST, H. B., 1928, Two extreme " small " Matthiola

plants; a haploid with one and a diploid with two additional chromosome fragments. *Amer. Nat.*, **62**, 22–33.

LESLEY, M. M., and LESLEY, J. W., 1930, The Mode of Origin and Chromosome Behaviour in Pollen Mother Cells of a Tetraploid Seedling Tomato. *J. Genet.*, **22**, 419–425.

LEVAN, A., 1931, Cytological studies in Allium. A preliminary note. *Hereditas*, **15**, 347–356.

LEVAN, A., 1932, Cytological Studies in *Allium* II., Chromosome morphological contributions. *Hereditas*, **16**, 257–294.

LEVAN, A., 1933 a, Über das Geschlechtschromosom in Sedum Rhodiola DC. *Bot. Notiser*, Lund, **1933**, 195–197.

LEVAN, A., 1933 b, Cytological studies in *Allium* III. *Allium carinatum* and *Allium oleraceum*. *Hereditas*, **18**, 101–114.

LEVAN, A., 1933 c, Cytological studies in *Allium* IV. *Allium fistulosum*. *Sv. Bot. Tidskr.*, **27**, 211–232.

LEVAN, A., 1934, Cytological studies in *Allium* V. *Allium macranthum*. *Hereditas*, **18**, 349–359.

LEVAN, A., 1935 a, Cytological studies in *Allium*, VI. *Hereditas*, **20**, 289–330.

LEVAN, A., 1935 b, Zytologische Studien in Allium Schœnoprasum. *Hereditas*, **22**, 1–128.

LEVAN, A., 1936, Die Zytologie von Allium cepa × fistulosum. *Hereditas*, **21**, 195–214.

LEWITSKY, G. A., 1931, Experimentally induced alterations of the morphology of chromosomes. *Amer. Nat.*, **65**, 564–567.

*LEWITSKY, G. A., 1931, The morphology of the chromosomes. *Bull. Appl. Bot.*, **27**, (1), 19–173.

LEWITSKY, G. A., and ARARATIAN, G. A., 1932, Transformation of Chromosomes under the influence of X-rays. *Bull. App. Bot.*, **27** (1), 265–286.

LEWITSKY, G. A., and BENETZKAIA, G. K., 1929, Cytological Investigations of constant intermediate rye-wheat hybrids. *Proc. U.S.S.R. Cong. in Genet.*, **2**, 345–352 (English summary).

LEWITSKY, G. A., and KUZMINA, N. E., 1927, Karyological investigations on the genus *Festuca*. *Bull. Appl. Bot.*, **17** (3), 3–36.

LEWITSKY, G., SHEPELEVA, H., and TITOVA, N., 1934, Cytology of F_1, F_2 and F_3 of X-rayed Crepis capillaris Wallr. *Plant Industry in U.S.S.R.*, No. 11.

LEWITSKY, G., and SIZOVA, M., 1934, On regularities in chromosome transformations induced by X-rays. *C. R. Acad. Sci. U.R.S.S.*, **1934**, 86–87.

LEWITSKY, G. A., and TRON, E. J., 1930, Zur Frage der karyotypische Evolution der Gattung Muscari Mill. *Planta*, **9**, 760–775.

LI, J.-C., 1927, The Effect of chromosome aberrations on development in *Drosophila melanogaster*. *Genetics*, **12**, 1–58.

LIDFORSS, B., 1914, Résumé seiner Arbeiten über Rubus. *Z.I.A.V.*, **12**, 1–13.

LILIENFELD, F. A., 1933, Karyologische und genetische Studien an Fragaria. Ein tetraploider Bastard zwischen *F. nipponica* (n = 7) und *F. elatior* (n = 21). *Jap. J. Bot.*, **6**, 425–458.

LILLIE, R. S., 1905 a, On the conditions determining the disposition of the chromatic filaments and chromosomes in mitosis. *Biol. Bull.*, **8**, 193–204.

LILLIE, R. S., 1905 b, The physiology of cell-division, I. Experiments in the conditions determining the distribution of chromatic matter in mitosis. *Amer. J. Physiol.*, **15**, 46–84.

LINDEGREN, C. C., 1933, The genetics of Neurospora III. The development of pure-bred stocks and crossing-over in *N. crassa*. *Bull. Torr. Bot. Club.*, **60**, 133–154.

BIBLIOGRAPHY

LINDEGREN, C. C., 1936, A six-point map of the sex-chromosome of *Neurospora crassa*. *J. Genet.*, **32**, 243-256.
LINDENBEIN, W., 1927, Beiträge zur Cytologie der Charales. *Planta*, **4**, 437-466.
LINDSTROM, E. W., 1929, A haploid mutant in the tomato. *J. Hered.*, **20**, 23-30.
LINDSTROM, E. W., and HUMPHREY, L. M., 1933, Comparative cyto-genetic studies of tetraploid tomatoes from different origins. *Genetics*, **18**, 193-209.
LINDSTROM, E. W., and KOOS, K., 1931, Cytogenetic investigations of a haploid tomato and its diploid and tetraploid progeny. *Amer. J. Bot.*, **18**, 398-410.
LITARDIÈRE, R. de, 1925, Sur l'existence de figures didiploides dans le meristème radiculaire de Cannabis sativa L. *Cellule*, **35**, 19-25.
LJUNGDAHL, H., 1922, Zur Zytologie der Gattung *Papaver*. Vorläufige Mitteilung. *Sv. Bot. Tidsskr.*, **16**, 103-114.
LJUNGDAHL, H., 1924, Über die Herkunft der in der Meiosis konjugierenden Chromosomen bei Papaver-Hybriden. *Sv. Bot. Tidsskr.*, **18**, 279-291.
LONGLEY, A. E., 1924, Chromosomes in Maize and Maize Relatives. *J. Agric. Res.*, **28**, 673-681.
LONGLEY, A. E., 1927, Supernumerary Chromosomes in Zea Mays. *J. Agr. Res.*, **35**, 769-784.
LORBEER, G., 1927, Untersuchungen über die Reduktionsteilung und Geschlechtsbestimmung bei Lebermoosen. *Z.I.A.V.*, **44**, 1-109.
LORBEER, G., 1930, Geschlechtsunterschiede im Chromosomensatz und in der Zellgrösse bei Sphareocarpus Donnellii Aust. *Zeits. f. Bot.*, **23**, 932-956.
LORBEER, G., 1934, Die Zytologie der Lebermoose mit besonderer Berücksichtigung allgemeiner Chromosomenfragen. *Jahrb. wiss. Bot.*, **80**, 567-817.
LOTSY, J. P., 1905, Die X-Generation und die 2X-Generation. Eine Arbeitshypothese. *Biol. Zbl.*, **25**, 97-117.
*LOTSY, J. P., 1925, *Evolution considered in the light of hybridization*. Christchurch, N.Z.
LUCAS, F. F., and STARK, M. B., 1931, A Study of Living Sperm Cells of Certain Grasshoppers by means of the Ultraviolet Microscope. *J. Morphol.*, **52**, 91-107.
LUDFORD, R. J., 1933, Vital staining in relation to cell physiology and pathology. *Biol. Revs.*, **8**, 357-369.
LUTZ, A. M., 1907 a, A preliminary note on the chromosomes of Œnothera Lamarckiana and one of its mutants, O. gigas. *Science*, **26**, 151-152.
LUTZ, A. M., 1907 b, A study of the chromosomes of Œnothera Lamarckiana, its mutants and hybrids. *Proc. Int. Zool. Cong.*, Boston, 352-354.

MCCLINTOCK, B., 1929 a, A cytological and genetical study of triploid maize. *Genetics*, **14**, 180-222.
MCCLINTOCK, B., 1929 b, A method for making Aceto-carmin smears permanent. *Stain Tech.*, **4**, 53-56.
MCCLINTOCK, B., 1929 c, A 2n-1 Chromosomal Chimaera in Maize. *J. Hered.*, **20**, 218.
MCCLINTOCK, B., 1931 a, A Cytological Demonstration of the Location of an Interchange between the non-homologous chromosomes of Zea Mays. *Proc. Nat. Acad. Sci.*, **16**, 791-796.
MCCLINTOCK, B., 1931 b, The order of the genes C Sh and Wx in Zea Mays with reference to a Cytologically Known Point in the Chromosome. *Proc. Nat. Acad. Sci.*, **17**, 485-491.

McCLINTOCK, B., 1932, A correlation of ring-shaped chromosomes with variegation in Zea Mays. *Proc. Nat. Acad. Sci.*, 18, 677–681.

McCLINTOCK, B., 1933, The association of non-homologous parts of chromosomes in the mid-prophase of meiosis in Zea Mays. *Zeits. Zellf. u. mikr. Anat.*, 19, 191–237.

McCLINTOCK, B., 1934, The relation of a particular chromosomal element to the development of the nucleoli in Zea Mays. *Zeits. Zellf. u. mikr. Anat.*, 21, 294–328.

McCLUNG, C. E., 1905, The chromosome complex of orthopteran spermatocytes. *Biol. Bull.*, 9, 304–340.

McCLUNG, C. E., 1914, A comparative study of the chromosomes in orthopteran spermatogenesis. *J. Morphol.*, 25, 651–749.

McCLUNG, C. E., 1917, The multiple chromosomes of Hesperotettix and Mermiria. *J. Morphol.*, 29, 519–605.

McCLUNG, C. E., 1927 a, Synapsis and Related Phenomena in Mecostethus and Leptysma. (Orthoptera.) *J. Morphol. Physiol.*, 43, 181–264.

*McCLUNG, C. E., 1927 b, The Chiasmatype Theory of Janssens. *Quart. Rev. Biol.*, 2, 344–366.

McCLUNG, C. E., 1928 a, The Generic Constancy of a Particular Chromosome in Mecostethus Gracilis, Lineatus and Grossus. *Arch. Biol.*, 38, 503–528.

McCLUNG, C. E., 1928 b, Differential Chromosomes of *Mecostethus gracilis*. *Zeits. f. Zellforsch. u. mikr. Anat.*, 7, 756–778.

McCRAY, F. A., 1932, Another haploid Nicotiana Tabacum plant. *Bot. Gaz.*, 93, 227–230.

MÄCKEL, H. G., 1928, Zur Cytologie einiger Saprolegniaceen. *Jahrb. wiss. Bot.*, 69, 517–548.

McNABB, J. W., 1928, A study of the chromosomes in meiosis, fertilization and cleavage in the grasshopper egg (Orthoptera). *J. Morphol.*, 45, 47–93.

MAEDA, T., 1928, The Spiral Structure of Chromosomes in the sweet-pea (*Lathyrus odoratus* L.). *Bot. Mag. Tokyo*, 42, 191–195.

MAEDA, T., 1930 a, The Meiotic Divisions in the Pollen Mother Cells of the Sweet-pea (Lathyrus odoratus L.) with special reference to the Cytological Basis of Crossing-over. *Mem. Coll. Sci. Kyoto B.*, 5, 89–123.

MAEDA, T., 1930 b, On the Configurations of Gemini in the Pollen Mother Cells of *Vicia Faba*, L. *Mem. Coll. Sci. Kyoto B*, 5, 125–137.

MAEDA, T., and KATO, K., 1929, The Pollen Mother Cells of *Spinacia oleracea*, Mill. and *Vicia Faba*, L. *Mem. Coll. Sci. Kyoto B*, 4, 327–345.

MAHONY, K. L., 1935, Morphological and cytological studies on *Fagopyrum esculentum*. *Amer. J. Bot.*, 22, 460–475.

MAINX, F., 1924, Versuche über die Beeinflussung der Mitose durch Giftstoffe. *Zool. Jahrb.*, 41, 553–580.

MAINX, F., 1931, Physiologische und genetische Untersuchungen an Œdogonien I. *Zeits. f. Bot.*, 24, 481–526.

MAKINO, S., 1932, An unequal pair of idiochromosomes in the Tree-cricket, Œcanthus longicauda Mats. *J. Fac. Sci. Hokkaido*, VI, 1, 1–35.

MAKINO, S., 1934, The chromosomes of the Sticklebacks, *Pungitius tymensis* (Nikolsky) and *P. pungitius* (Linnæus). *Cytologia*, 5, 155–168.

MAKINO, S., 1935, The chromosomes of Cryptobranchus allegheniensis. *J. Morphol.*, 58, 573–583.

MAKINO, S., 1936, The spiral structure of chromosomes in the Meiotic Divisions of *Podisma* (Orthoptera). *J. Fac. Sci. Hokkaido* VI., 5, 29–40.

MALAN, D. E., 1918, Ergebnisse anatomischer Untersuchungen an STANDFUSS'schen Lepidopteren-Bastarden. *Mitt. Entom. Zurich*, 4, 201–260.

MANGELSDORF, P. C., and REEVES, R. G., 1931, Hybridization of Maize, Tripsacum and Euchlæna. *J. Hered.*, **22**, 329–343.
MANGELSDORF, P. C., and REEVES, R. G., 1935, A trigeneric hybrid of Zea, Tripsacum and Euchlæna. *J. Hered.*, **26**, 128–140.
MANTON, I., 1932 a, Contributions to the Cytology of Apospory in *Osmunda regalis* L. *J. Genet.*, **25**, 423–430.
MANTON, I., 1932 b, Introduction to the Cytology of the Cruciferæ. *Ann Bot.*, **46**, 509–556.
MANTON, I., 1935, Some new evidence on the physical nature of plant nuclei from intra-specific polyploids. *Proc. Roy. Soc.* B, **118**, 522–547.
MARCHAL, EL., and MARCHAL, EM., 1911, Aposporie et Sexualité chez les Mousses III. *Bull. Acad. Roy. Belg.*, 9–10, 1911, 750–778.
MARCHAL, EM., 1912, Recherches cytologiques sur le genre " *Amblystegium.*" *Bull. Soc. Roy. Bot. Belg.*, **51**, 189–203.
MARSDEN-JONES, E. M., and TURRILL, W. B., 1930, The History of a Tetraploid Saxifrage. *J. Genet.*, **23**, 83–92.
MARTENS, P., 1928, Le cycle du chromosome somatique dans les phanérogames. III. Recherches expérimentales sur la cinèse dans la cellule vivante. *Cellule*, **38**, 69–174.
MATHER, K., 1932, Chromosome Variation in *Crocus*, I. *J. Genet.*, **26**, 129–142.
MATHER, K., 1933 a, Interlocking as a demonstration of the occurrence of genetical crossing-over during chiasma-formation. *Amer. Nat.*, **67**, 476–479.
MATHER, K., 1933 b, The relation between chiasmata and crossing-over in diploid and triploid Drosophila melanogaster. *J. Genet.*, **27**, 243–260.
MATHER, K., 1934 a, The behaviour of meiotic chromosomes after X-irradiation. *Hereditas*, **20**, 303–322.
MATHER, K., 1935 a, Meiosis in Lilium. *Cytologia*, **6**, 354–380.
MATHER, K., 1935 b, Chromosome behaviour in a triploid wheat hybrid. *Zeits. Zellf. u. mikr. Anat.*, **23**, 117–138.
MATHER, K., 1935 c, Reductional and equational separation of the chromosomes in bivalents and multivalents. *J. Genet.*, **30**, 53–78.
MATHER, K., 1935 d, Crossing-over and chromosome conjugation in triploid Drosophila. *J. Genet.*, **30**, 481–485.
MATHER, K., 1936 a, Competition between bivalents during chiasma-formation. *Proc. Roy. Soc.*, B, **120**, 208–227.
MATHER, K., 1936 b, Segregation and linkage in autotetraploids. *J. Genet.*, **32**, 287–314.
MATHER, K., 1936 c, The determination of position in crossing-over, I. *Drosophila melanogaster*. *J. Genet.* (in the press).
MATHER, K., 1936 d, The determination of position in crossing-over, II. The length—chiasma-frequency relationship. *Cytologia* (in the press).
MATHER, K., and LAMM, R., 1935, The negative correlation of chiasma frequencies. *Hereditas*, **20**, 65–70.
MATHER, K., and STONE, L. H. A., 1933, The effect of X-radiation upon somatic chromosomes. *J. Genet.*, **28**, 1–24.
MATSUDA, H., 1928, On the origin of big pollen grains with an abnormal number of chromosomes. *Cellule*, **38**, 215–243.
MATSUDA, H., 1935, Cytological studies of the genus Petunia. *Cytologia*, **6**, 502–522.
MATSUMOTO, K., 1933, Zur Kritik der Kryptogonomerie-Theorie von Bleier *Mem. Coll. Agr. Kyoto*, **25**, 1–10.
MATSUURA, H., 1935 a, On the relation of chromosomes to nucleoli. *Botany and Zoology* (Japan), **3**, 1589–1594.

MATSUURA, H., 1935 b, A cytological study on *Phacellanthus tubiflorus*. Sieb. et Zucc. I. *J. Fac. Sci. Hokkaido* V, **3**, 169–187.
MATSUURA, H., 1935 c, Chromosome studies on *Trillium kamtschaticum* Pall. I. The number of coils in the chromonema of the normal and abnormal meiotic chromosomes and its relation to the volume of chromosomes. *Cytologia*, **6**, 270–280.
MATSUURA, H., 1935 d, Chromosome studies on Trillium kamtschaticum Pall. II. The direction of coiling of the chromonema within the first meiotic chromosomes of the PMC. *J. Fac. Sci. Hokkaido*, V, **3**, 233–250.
MATSUURA, H., 1935 e, On karyo-ecotypes of Fritillaria camschatcensis (L) Ker-Gawler. *J. Fac. Sci. Hokkaido* V, **3**, 219–232.
MATSUURA, H., 1935 f, On the secondary association of meiotic chromosomes in *Tricyrtis latifolia* Max. and *Dicentra spectabilis*. *J. Fac. Sci. Hokkaido* V, **3**, 251–260.
MATSUURA, H., 1935 g, A karyological investigation of Mitrastemon Yamamotoi Mak., with special reference to the so-called " diffuse stage " in Meiosis. *J. Fac. Sci. Hokkaido* V, **3**, 189–204.
MATSUURA, H., and GONDO, A., 1935, A karyological study on *Peziza subumbrina* Boud., with special reference to a heteromorphic pair of chromosomes. *J. Fac. Sci. Hokkaido* V, **3**, 205–217.
MATSUURA, H., and SUTÔ, T., 1935, Contributions to the idiogram study in phanerogamous plants. I. *J. Fac. Sci. Hokkaido* V, **5**, 33–75.
MATTHEY, R., 1929, Les Chromosomes de la Vipère mâle (Vipera aspis, Lin.). *Biol. Zbl.*, **49**, 35–43.
MATTHEY, R., 1931, Chromosomes de Reptiles Sauriens, Ophidiens et Cheloniens. L'Évolution de la formule chromosomiale chez les Sauriens. *Rev. suisse Zool.*, **38**, 117–186.
MATTHEY, R., 1933, Nouvelle contribution a l'étude des chromosomes chez les Sauriens. *Rev. suisse de Zool.*, **40**, 281–316.
MATTHEY, R., 1936, Le problème des hétérochromosomes chez les Mammifères. *Arch. Biol.*, **47**, 319–383.
MEIJERE, see MEYERE.
MELBURN, M. C., 1929, Heterotypic prophases in the absence of chromosome pairing. *Canad. J. Res.*, **1**, 512–527.
MELBURN, M. C., and THOMPSON, W. P., 1927, The cytology of a tetraploid wheat hybrid (*Triticum spelta* × *T. monococcum*). *Amer. J. Bot.*, **14**, 327–333.
MESSERI, A., 1930, Il numero dei cromosomi dell' *Allium roseum* v. *bulbilliferum* e dell' *A.* confr. *odorum* e nuovo esempi di rapporti fra apomissia e poliploidismo. *N. Gior. Bot. Ital.*, **37**, 276–277.
MESSERI, A., 1931, Ricerche embriologiche e cariologiche sopra i generi " Allium " e " Nothoscordon." *Nuov. Giorn. Bot. Ital.*, **38**, 409–441.
METCALF, M. M., 1928, Trends in Evolution : A discussion of data bearing upon "orthogenesis." *J. Morphol.*, **45**, 1–45.
METZ, C. W., 1916, Chromosome Studies on the Diptera, II. The Paired Association of Chromosomes in the Diptera and its Significance. *J. Exp. Zool.*, **21**, 213–262.
METZ, C. W., 1922, Association of homologous chromosomes in tetraploid cells of Diptera. *Biol. Bull.*, **43**, 369–373.
METZ, C. W., 1926, Observations on Spermatogenesis in Drosophila. *Zeits. f. Zellforsch. u. mikr. Anat.*, **4**, 1–28.
METZ, C. W., 1933, Monocentric mitosis with segregation of chromosomes in Sciara and its bearing on the mechanism of mitosis. I. The normal monocentric mitosis. II. Experimental modification of the monocentric mitosis. *Biol. Bull.*, **54**, 333–347.

METZ, C. W., 1934, Evidence indicating that in Sciara the sperm regularly transmits two sister sex chromosomes. *Proc. Nat. Acad. Sci.*, **20**, 31–36.

METZ, C. W., 1935, Structure of the salivary gland chromosomes in Sciara. *J. Hered.*, **26**, 177–188.

METZ, C. W., MOSES, M. S., and HOPPE, E. N., 1926, Chromosome Behavior and Genetic Behavior in Sciara (Diptera). I. *Z.I.A.V.*, **42**, 237–270.

METZ, C. W., and NONIDEZ, J. F., 1921, Spermatogenesis in the fly, Asilus sericeus. Say. *J. Exp. Zool.*, **32**, 165–185.

METZ, C. W., and NONIDEZ, J. F., 1923, Spermatogenesis in *Asilus notatus*, Wied. (Diptera). *Arch. f. Zellforsch.*, **17**, 438–449.

METZ, C. W., and NONIDEZ, J. F., 1924, The behavior of the nucleus and chromosomes during spermatogenesis in the robber fly, *Lasiopogon bivittatus*. *Biol. Bull.*, **46**, 153–164.

METZNER, R., 1894, Beiträge für Granularlehre I. *Arch. Anat. Physiol.* (Phys. Abt.), **1894**, 309–348.

MEURMAN, O., 1925, The Chromosome Behaviour of some Diœcious Plants and their Relatives with Special Reference to the Sex Chromosomes. *Comm. Biol. Soc. Sci. Fenn.*, **2** (3), 1–104.

MEURMAN, O., 1928, Cytological Studies in the Genus Ribes L. *Hereditas*, **11**, 289–356.

MEURMAN, O., 1929 a, Association and Types of Chromosomes in *Aucuba japonica*. *Hereditas*, **12**, 179–209.

MEURMAN, O., 1929 b, Prunus laurocerasus L., a Species showing High Polyploidy. *J. Genet*, **21**, 85–94.

MEURMAN, O., 1933. Chromosome morphology, somatic doubling and secondary association in Acer platanoides L. *Hereditas*, **18**, 145–173.

MEVES, F., 1907, Die Spermatocytenteilungen bei der Honigbiene (Apis mellifica L.), etc. *Arch. mikr. Anat.*, **70**, 414–491.

MEVES, F., 1915, Über die Mitwirkung der Plasmosomen bei der Befruchtung des Eies von *Filaria papillosa*. *Arch. mikr. Anat.*, **87**, 12–46.

MEYER, K. J., 1925, Parthenogenesis bei Thismia javanica im Lichte der Haberlandtschen Anschauung. *Ber. Deuts. Bot. Ges.*, **43**, 193–197.

MEYERE, J. H. C. de, 1928, Über haltbare rasche Färbung vermittels Azetokarmin. *Zeits. wiss. Mikr.*, **46**, 189–195.

MEYERE, J. H. C. de, as MEIJERE, J. C. H. de, 1930, Über einige europäische Insekten, besonders günstig zum Studium der Reifungsteilungen, nebst einigen Zusätzen zur Azetokarminmethode. *Zool. Anzeig.*, **88**, 209–219.

MICHAELIS, P., 1926, Über den Einfluss der Kälte über die Reduktionsteilung von Epilobium. *Planta* **1**, 569–582.

MICHAELIS, P., 1930, Über experimentell erzeugte, heteroploide Pflanzen von Œnothera Hookeri. *Zeits. f. Bot.*, **23**, 288–308.

MICZYNSKI, K., 1931. Genetic studies in the genus Ægilops II. The morphology and cytology of the interspecific hybrids. *Bull. Acad. Pol.* (B), **1**, 51–82.

MILOVIDOV, P. F., 1933, Independence of chondriosomes from nuclear matter. *Cytologia*, **4**, 158–173.

MINOUCHI, O., 1929 a, Chromosome Arrangement, VI. The Behaviour of Chromosomes from the Moment of Disappearance of the Nuclear Membrane up to the Formation of the Equatorial Plate in the First Spermatocyte Division of the Albino Rat. *Mem. Coll. Sci. Kyoto* B, **4**, 323–326.

MINOUCHI, O., 1929 b, On the Spermatogenesis of the Racoon Dog (Nyotereutes viverrinus), with special reference to the Sex Chromosomes. *Cytologia*, **1**, 88–108.

MIYAJI, Y., 1929, Studien über die Zahlenverhältnisse der Chromosomen bei der Gattung *Viola*. *Cytologia*, 1, 28–58.
MODILEWSKI, J., 1930, Neue Beiträge zur Polyembryonie von *Allium odorum*. *Ber. Deuts. Bot. Ges.*, 48, 285–294.
MODILEWSKI, I., 1932, New observations on the nucleolus in the nucleus of higher plants. *Bull. Kiev Bot. Gard.*, 14 (in Russian).
MODILEWSKI, I., 1934, Über den Nucleolus (weitere Beiträge). *J. Inst. bot. Acad. Sci. Ukraine*, 3 (11), 1–15.
MOFFETT, A. A., 1931 a, A preliminary account of chromosome behaviour in the Pomoideæ. *J. Pomol.*, 9, 100–110.
MOFFETT, A. A., 1931 b, The Chromosome Constitution of the Pomoideæ. *Proc. Roy. Soc. B*, 108, 423–446.
MOFFETT, A. A., 1932 a, Studies on the Formation of Multinuclear Giant Pollen Grains in *Kniphofia*. *J. Genet.*, 25, 315–337.
MOFFETT, A. A., 1932 b, Chromosome Studies in *Anemone*, I. A new type of chiasma behaviour. *Cytologia*, 4, 26–37.
MOFFETT, A. A., 1934, Cytological studies in pears. *Genetica*, 15, 511–518.
MOFFETT, A. A., 1936, The origin and behaviour of chiasmata. XIII. Diploid and tetraploid Culex pipiens. *Cytologia*, 7, 184–197.
MOHR, O. L., 1915, Sind die Heterochromosomen wahre Chromosomen? Untersuchungen über ihr Verhalten in der Ovogenese von *Leptophyes punctatissima*. *Arch. f. Zellforsch.*, 14, 151–176.
MOHR, O. L., 1916, Studien über die Chromatin-reifung der männlichen Geschlechtszellen bei *Locusta viridissima*. *Arch. Biol.*, 29, 579–752.
MOHR, O., 1932, Genetical and cytological proof of somatic elimination of the fourth chromosome in *Drosophila melanogaster*. *Genetics*, 17, 60–80. (Cf. *Arch. Biol.*, 42, 365–373.)
MOL, W. E. de, 1921, De l'éxistence de variétés hétéroploides de l'Hyacinthus orientalis L. dans les cultures hollandaises. *Arch. Néerl. Sci.* III B, 4, 18–117.
MOL, W. E. de, 1923, Duplication of generative nuclei by means of physiological stimuli and its significance. *Genetica*, 5, 225–272.
MOL, W. E. de, 1928, Nucleolar number and size in Diploid, Triploid and Aneuploid Hyacinths. *Cellule*, 38, 1–65.
MOL, W. E. de, 1929, The Originating of Diploid and Tetraploid Pollen Grains in Duc van Thol-Tulips (Tulipa suaveolens) dependent on the method of culture applied. *Genetica*, 11, 119–212.
MOL, W. E. de, 1933, Die Entstehungsweise anormaler Pollenkörner bei Hyazinthen, Tulpen und Narzisse. *Cytologia*, 5, 31–65.
MONTGOMERY, T. H., 1901, A Study of the chromosomes of germ cells of Metazoa. *Trans. Amer. Phil. Soc.*, 20, 154–236.
MONTGOMERY, T. H., 1911, The Spermatogenesis of an hemipteron, *Euchistus*. *J. Morphol.*, 22, 731–799.
MORGAN, L. V., 1922, Non-criss-cross inheritance in Drosophila melanogaster. *Biol. Bull.*, 42, 267–274.
MORGAN, L. V., 1933, A closed X-chromosome in Drosophila melanogaster. *Genetics*, 18, 250–283.
MORGAN, T. H., 1911, Random Segregation versus Coupling in Mendelian Inheritance. *Science* n.s., 34, 384.
MORGAN, T. H., 1912, The elimination of the sex chromosomes from the male-producing eggs of Phylloxerans. *J. Exper. Zool.*, 12, 479–498.
MORGAN, T. H., 1915, The predetermination of sex in phylloxerans and aphids. *J. Exper. Zool.*, 19, 285–321.
*MORGAN, T. H., 1922, On the Mechanism of Heredity. *Proc. Roy. Soc. B*, 94, 162–197.

*MORGAN, T. H., 1926, *The Theory of the Gene.* New Haven.
MORGAN, T. H., 1927, The Constitution of the Germ Material in Relation to Heredity. *Carneg. Inst. Yearbk.*, **26**, 284–288.
MORGAN, T. H., and BRIDGES, C. B., 1919, The origin of gynandromorphs. *Carneg. Inst. Pub.*, **278**, 1–122.
*MORGAN, T. H., BRIDGES, C. B., and STURTEVANT, A. H., 1925, The Genetics of Drosophila. *Bibliogr. Genet.*, **2**, 1–262.
MORGAN, W. P., 1928, A comparative study of spermatogenesis of five species of earwigs. *J. Morph.*, **46**, 241–273.
MORINAGA, T., 1929, Interspecific Hybridization in *Brassica* II. The Cytology of F_1 Hybrids of *B. cernua* and Various other Species with 10 Chromosomes. *Jap. J. Bot.*, **4**, 277–289.
MORINAGA, T., 1934 a, Cyto-genetical studies on *Oryza sativa* L. I. Studies on the haploid plant of *Oryza sativa* L. *Jap. J. Bot.*, **7**, 73–106.
MORINAGA, T., 1934 b, Interspecific hybridization in *Brassica*. VI. The cytology of F_1 hybrids of B. juncea × B. nigra. *Cytologia*, **6**, 62–67.
MORINAGA, T., and FUKUSHIMA, E., 1931, Preliminary Report on the Haploid Plant of Rice, *Oryza sativa* L. *Proc. Imper. Acad.*, Tokyo, **7**, 383–384.
MORINAGA, T., and FUKUSHIMA, E., 1933, Karyological studies on a spontaneous haploid mutant of *Brassica Napella. Cytologia*, **4**, 457–460.
MORINAGA, T., and FUKUSHIMA, E., 1935, Cyto-genetical studies on *Oryza sativa* L. II. Spontaneous autotriploid mutants in Oryza sativa L. *Jap. J. Bot.*, **7**, 207–225.
MORINAGA, T., FUKUSHIMA, E., KANTO, T., MARUYAMA, Y., and YAMASAKI, Y., 1929, Chromosome Numbers of Cultivated Plants II. *Bot. Mag. Tokyo*, **43**, 589–592.
MÜLLER, H. A. C., 1912, Kernstudien an Pflanzen, I und II. *Arch. f. Zellforsch.*, **8**, 1–51.
MULLER, H. J., 1916, The Mechanism of Crossing-Over, II. *Amer. Nat.*, **50**, 193–221, 284–305, 350–366, 421–434.
MULLER, H. J., 1918, Genetic variability, twin hybrids, and constant hybrids in a case of balanced lethal factors. *Genetics*, **3**, 422–499.
MULLER, H. J., 1925 a, The regionally differential effect of X-rays on crossing-over in autosomes of *Drosophila*. *Genetics*, **10**, 470–507.
MULLER, H. J., 1925 b, Why Polyploidy is rarer in Animals than in Plants. *Amer. Nat.*, **59**, 346–353.
MULLER, H. J., 1927, Artificial transmutation of the gene. *Science*, **66**, 84–87.
MULLER, H. J., 1928, The Production of Mutations by X-Rays. *Proc. Nat. Acad. Sci.*, **14**, 714–726.
*MULLER, H. J., 1930 a, Radiation and Genetics. *Amer. Nat.*, **64**, 220–251.
MULLER, H. J., 1930 b, Types of Visible Variations Induced by X-Rays in *Drosophila*. *J. Genet.*, **22**, 299–334.
MULLER, H. J., 1930 c, Œnothera-like Linkage of Chromosomes in *Drosophila*. *J. Genet.*, **22**, 335–357.
*MULLER, H. J., 1932 a, Some Genetic Aspects of Sex. *Amer. Nat.*, **66**, 118–138.
MULLER, H. J., 1932 b, Further studies on the nature and causes of gene mutations. *Proc. 6th Int. Cong. Genet.*, **1**, 213–255.
*MULLER, H. J., 1934, The effects of Roentgen Rays upon the hereditary material. *The Science of Radiology*, 1934, 305–318.
MULLER, H. J., 1935 a, A viable two-gene deficiency. *J. Hered.*, **26**, 469–478.
MULLER, H. J., 1935 b, The origination of chromatin deficiencies as minute deletions subject to insertion elsewhere. *Genetica*, **17**, 237–252.
MULLER, H. J., 1935 c, On the dimensions of chromosomes and genes in dipteran salivary glands. *Amer. Nat.*, **69**, 405–411.

MULLER, H. J., and GERSHENSON, S. M., 1935, Inert regions of chromosomes as the temporary products of individual genes. *Proc. Nat. Acad. Sci.*, 21, 69-75.
MULLER, H. J., and PAINTER, T. S., 1932, The differentiation of the sex chromosomes of Drosophila into genetically active and inert regions. *Z.I.A.V.*, 62, 316-365.
MULLER, H. J., and PROKOFJEVA, A. [A.], 1934, Continuity and discontinuity of the hereditary material. *C. R. Acad. Sci. U.R.S.S.*, 1934, 4 (1-2), 74-83. (Oct. 11, 1934.)
MULLER, H. J., and PROKOFYEVA, A. A., 1935, The individual gene in relation to the chromomere and the chromosome. *Proc. Nat. Acad. Sci.*, 21, 16-26.
MULLER, H. J., PROKOFYEVA, A., and RAFFEL, D., 1935, Minute intergenic rearrangement as a cause of apparent " Gene Mutation." *Nature*, 135, 253.
MULSOW, K., 1912, Der Chromosomencyclus bei Ancyracanthus cystidicola Rud. *Arch. f. Zellforsch.*, 9, 63-72.
MÜNTZING, A., 1927, Chromosome Number, Nuclear Volume and Pollen Grain Size in *Galeopsis*. *Hereditas*, 10, 241-260.
MÜNTZING, A., 1928, Pseudogamie in der Gattung *Potentilla*. *Hereditas*, 11, 267-283.
MÜNTZING, A., 1930 a, Outlines to a Genetic Monograph of the Genus *Galeopsis*. *Hereditas*, 13, 185-341.
MÜNTZING, A., 1930 b, Über Chromosomenvermehrung in *Galeopsis*-Kreuzungen und ihre phylogenetische Bedeutung. *Hereditas*, 14, 153-172.
MÜNTZING, A., 1931, Note on the Cytology of some Apomictic *Potentilla* Species. *Hereditas*, 15, 166-178.
MÜNTZING, A., 1932, Cyto-Genetic Investigations on Synthetic *Galeopsis Tetrahit*. *Hereditas*, 16, 105-154.
MÜNTZING, A., 1933 a, Quadrivalent formation and aneuploidy in Dactylis glomerata. *Bot. Not.* (Lund), 1933, 189-205.
MÜNTZING, A., 1933 b, Apomictic and sexual seed formation in Poa. *Hereditas*, 17, 131-154.
MÜNTZING, A., 1933 c, Studies on meiosis in diploid and triploid Solanum tuberosum L. *Hereditas*, 17, 223-245.
MÜNTZING, A., 1934, Chromosome fragmentation in a Crepis hybrid. *Hereditas*, 19, 284-302.
*MÜNTZING, A., 1936 a, The evolutionary significance of autopolyploidy. *Hereditas*, 21, 263-378.
MÜNTZING, A., 1936 b, The chromosomes of a giant *Populus tremula*. *Hereditas*, 21, 383-393.
MURBECK, S., 1902, Über Anormalien des Nucellus und des Embryosackes bei parthenogenetischen Arten der Gattung *Alchemilla*. *Lunds Univ. Arsskr.*, 38 (2).

NABOURS, R. K., 1919, Parthenogenesis and Crossing-over in the Grouse locust Apotettix. *Amer. Nat.*, 53, 131-142.
NABOURS, R. K., and ROBERTSON, W. R. B., 1933, An X-ray induced chromosomal translocation in Apotettix eurycephalus Hancock (Grouse Locusts). *Proc. Nat. Acad. Sci.*, 19, 234-239.
NACHTSHEIM, H., 1913, Cytologische Studien über die Geschlechtsbestimmung bei der Honigbiene (Apis mellifica L.). *Arch. f. Zellforsch.*, 11, 169-241.
NAGAO, S., 1929, Karyological studies of the Narcissus plant. I. Somatic Chromosome Number of Some Garden Varieties and Some Meiotic Phases of a Triploid Variety. *Mem. Coll. Sci. Kyoto* B, 4, 175-198.

NAGAO, S., 1933, Number and behaviour of chromosomes in the genus Narcissus. *Mem. Coll. Sci. Kyoto B*, **8**, 81-200.
NAGAO, S., 1935, Distribution of pollen grains in certain triploid and hypertriploid *Narcissus* plants. *Jap. J. Genet.*, **11**, 1-5.
NAKAMURA, T., 1929, Chromosome Arrangement IX. The Pollen Mother Cells in *Cycas revoluta* Thunb. *Mem. Coll. Sci. Kyoto B*, **4**, 353-369.
NAKAMURA, K., 1931, Studies on Reptilian Chromosomes, II. On the chromosomes of *Eumeces latiscutatus* (Hallowell), a Lizard. *Cytologia*, **2**, 385-400.
NAVASHIN, M., 1926, Variabilität des Zellkerns bei Crepis-Arten in Bezug auf die Artbildung. *Zeits. f. Zellforsch. u. mikr. Anat.*, **4**, 171-215.
NAVASHIN, M., as NAWASCHIN, 1927 a, Ein Fall von Merogonie infolge Artkreuzung bei Compositen. *Ber. Deuts. Bot. Ges.*, **45**, 115-126.
NAVASHIN, M., 1927 b, Über die Veränderungen von Zahl und Form der Chromosomen infolge der Hybridisation. *Zeits. f. Zellforsch. u. mikr. Anat.*, **6**, 195-233.
NAVASHIN, M., 1930 a, Unbalanced Somatic Chromosomal Variation in Crepis. *Univ. Calif. Pub. Agr. Sci.*, **6**, 95-106.
NAVASHIN, M., 1930 b, *Zacintha verrucosa* Gärtner : another plant with six Somatic Chromosomes. *Nature*, **126**, 604.
NAVASHIN, M., 1931 a, Spontaneous Chromosome Alterations in *Crepis tectorum* L. *Univ. Calif. Pub. Agr. Sci.*, **6**, 201-206.
NAVASHIN, M., 1931 b, Chromatin Mass and Cell Volume in related species. *Univ. Calif. Pub. Agr. Sci.*, **6**, 207-230.
NAVASHIN, M., 1931 c, A Preliminary Report on some Chromosome Alterations by X-Rays in *Crepis*. *Amer. Nat.*, **65**, 243-252. (*See also*: Babcock.)
NAVASHIN, M., 1932, The dislocation hypothesis of evolution of chromosome numbers. *Z.I.A.V.*, **63**, 224-231.
NAVASHIN, M. (as Nawaschin), 1933, Altern der Samen als Ursache der Chromosomenmutationen. *Planta*, **20**, 233-243.
NAVASHIN, M., 1934, Chromosome alterations caused by hybridisation and their bearing upon certain general genetic problems. *Cytologia*, **5**, 169-203.
NAVILLE, A., 1923, Recherches sur la constance numérique des Chromosomes dans la lignée germinale mâle de *Helix pomatia* L. *Rev. suisse Zool.*, **30**, 353-385.
NAVILLE, A., 1931, Les Sporozoaires (cycles chromosomiques et sexualité). *Mem. Soc. Phys. Hist. Nat. Genève*, **41**, 1-223.
NAVILLE, A., 1932, Les bases cytologiques de la théorie du " Crossing-over." Etude sur la spermatogénèse et l'ovogénèse des Calliphorinæ. *Zeits. Zellforsch. u. mikr. Anat.*, **16**, 440-470.
NAVILLE, A., and BEAUMONT, J. de, 1933, Recherches sur les chromosomes des Neuroptères. *Arch. Anat. mic.*, **29**, 199-243.
NAVILLE, A., and BEAUMONT, J. de, 1934, Les chromosomes des panorpes. *Bull. Biol.*, **68**, 98-107.
NAWASCHIN, S., 1912, Über den Dimorphismus der Zellkerne in den somatischen Zellen von *Galtonia candicans*. *Bull. Acad. Imp. Sci. Petersb. VI.*, **6**, 373-385.
NEBEL, B. R., 1929, Zur Cytologie von Malus und Vitis. *Gartenbauwiss.*, **1**, 549-592.
NEBEL, D. R., 1932, Chromosome studies in the Tradescantiæ II. The direction of coiling of the chromonema in *Tradescantia reflexa* Raf., *T virginiana* L. *Zebrina pendula* Schnizl, and *Rhœo discolor* Hance. *Zeits. Zellforsch. u. mikr. Anat.*, **16**, 285-304.
NEBEL, B. R., 1933, Chromosome numbers in aneuploid apple seedlings. *Tech. Bull. N.Y. Agric. Expt. Sta.*, **209**, 12.

NEBEL, B. R., 1933 a, Chromosome structure in Tradescantiæ IV. The history of the chromonemata in Tradescantia reflexa Raf. *Cytologia*, 5, 1–14.

NEBEL, B. R., 1933 b, Chromosome structure in Tradescantiæ V. Optical analysis of a somatic telophase chromosome. *Geneva Tech. Bull.*, 220.

NEBEL, B. R., 1935, Chromosomenstruktur VI. Ein Ausschnitt. *Züchter*, 7, 132–136, 155.

*NEMEC, B., 1910, *Das Problem des Befruchtungsvorgänge*. Berlin.

NEMEC, B., 1926, Multipolare Teilungsfiguren und vegetative Chromosomenreduktion. *Biol. Gen.*, 2, 96–103.

NEMEC, B., 1929 a, Über Struktur und Aggregatzustand des Zellkerns. *Protoplasma*, 7, 423–443.

NEMEC, B., 1929 b, Multipolare Teilungen in chloralisierten Wurzeln. *Mem. Soc. Roy. Sci. Bohème* (Prague), 2 (4), 1–7.

NEUHAUS, M. J., 1936, Crossing-over between the X- and Y-chromosomes in the female of Drosophila melanogaster. *Z.I.A.V.*, 71, 265–275.

NEWELL, W., 1914, Inheritance in the honey bee. *Science*, 41, 218–219.

NEWTON, W. C. F., 1924, Studies on somatic chromosomes, I. Pairing and segmentation in *Galtonia*. *Ann. Bot.*, 38, 197–206.

NEWTON, W. C. F., 1927, Chromosome Studies in *Tulipa* and Some Related Genera. *J. Linn. Soc.* (Bot.), 47, 339–354.

NEWTON, W. C. F., and DARLINGTON, C. D., 1929, Meiosis in Polyploids I. *J. Genet*, 21, 1–16.

NEWTON, W. C. F., and DARLINGTON, C. D., 1930, *Fritillaria Meleagris*: Chiasma-Formation and Distribution. *J. Genet*, 22, 1–14.

NEWTON, W. C. F., and PELLEW, C., 1929, *Primula kewensis* and its derivatives. *J. Genet.*, 20, 405–467.

NIIYAMA, H., 1935, The chromosomes of the edible crab, *Paralithodes camtschatica* (Tilesius). *J. Fac. Sci. Hokkaido* VI., 4, 59–65.

NISHIYAMA, I., 1928, Reduction Division in *Lycoris*. *Bot. Mag. Tokyo*, 42, 509–513.

NISHIYAMA, I., 1929, The Genetics and Cytology of Certain Cereals, I. Morphological and Cytological Studies on Triploid, Pentaploid and Hexaploid *Avena* Hybrids. *Jap. J. Genet.*, 5, 1–48.

NISHIYAMA, I., 1931, The Genetics and Cytology of Certain Cereals, II. Karyo-Genetic Studies of Fatuoid Oats with special reference to their origin. *Jap. J. Genet.*, 7, 49–102.

NISHIYAMA, I., 1933 a, The genetics and cytology of certain cereals, I. On the occurrence of an unexpected diploid in the progeny of pentaploid Avena hybrids. *Cytologia*, 5, 146–148.

NISHIYAMA, I., 1933 b, The genetics and cytology of certain cereals, IV. Further studies on fatuoid oats. *Jap. J. Genet.*, 8, 107–124.

NISHIYAMA, I., 1934, The genetics and cytology of certain cereals, VI. Chromosome behavior and its bearing on inheritance in triploid *Avena* hybrids. *Mem. Coll. Agr. Kyoto*, 32, 157.

NISHIYAMA, I., 1935, The genetics and cytology of certain cereals, VII. Genetical significance of the c-chromosome in hexaploid *Avena* species. *Jap. J. Bot.*, 7, 453–469.

NONIDEZ, J. F., 1921, The meiotic phenomena in the spermatogenesis of *Blaps* with special reference to the X-complex. *J. Morphol.*, 34, 69–117.

OEHLKERS, F., 1933, Crossing over bei Œnothera. *Zeits. f. Bot.*, 26, 385–430.

OEHLKERS, F., 1935, Untersuchungen zur Physiologie der Meiosis I. *Zeits. f. Bot.*, 29, 1–53.

OFFERIJNS, F. J. M., 1935, *Meiosis in the pollen mother cells of some Cannas*. Thesis. The Hague, 1935, 1–60.

OGUMA, K., 1921, The idiochromosomes of the mantis. *J. Coll. Agr. Imp. Univ. Hokkaido*, **10**, 1–27.
OGUMA, K., 1934, A new type of the mammalian sex chromosome, found in a field mouse, *Apodemus speciosus*. *Cytologia*, **5**, 460–471.
OGUMA, K., 1935, The chromosomes of four wild species of Muridæ. *J. Fac. Sci. Hokkaido*, VI., **4**, 35–57.
OHMACHI, F., 1929 a, A Short Note on the Chromosomes of *Gryllus campestris*, L., in Comparison with those of *Gryllus mitratus*, Burm. *Proc. Imp. Acad. Tokyo*, **5**, 357–359.
OHMACHI, F., 1929 b, A Short Note on the Chromosomes of *Gryllotalpa africana* Pal. *Proc. Imp. Acad. Tokyo*, **5**, 360–363.
OHMACHI, F., 1929 c, Preliminary Note on a Case of Facultative Parthenogenesis in Loxoblemmus frontalis, Shir. (Gryllidæ). *Proc. Imp. Acad. Tokyo*, **5**, 367–369.
OHMACHI, F., 1934, A comparative study of chromosome complements in the Gryllodea in relation to taxonomy. *Bull. Imp. Coll. Agric. and Forest*, V., 1–45.
OKABE, S., 1928, Zur Zytologie der Gattung *Prunus*. *Sc. Rep. Tohoku Imp. Univ.*, Ser. 4, **3**, 733–743.
OKABE, S., 1929 a, Über eine tetraploide Gartenrasse von *Psilotum nudum*, Palisot de Beauvois (= *P. triquetum* Sw.) und die tripolige Kernteilung in ihren Sporenmutterzellen. *Sc. Rep. Tohoku Imp. Univ.*, Ser. 4, **4**, 373–380.
OKABE, S., 1929 b, Meiosis im Oogonium von *Sargassum Horneri* (Turn.) Ag. *Sc. Rep. Tohoku Imp. Univ.*, Ser. 4, **4**, 661–669.
OKABE, S., 1932, Parthenogenesis bei *Ixeris dentata* Nakai. *Bot. Mag. Tokyo*, **46**, 518–523.
OKURA, E., 1933, A haploid plant in Portulaca grandiflora. *Jap. J. Genet.*, **8**, 251–260.
OLMO, H. P., 1934, Prophase association in triploid *Nicotiana Tabacum*. *Cytologia*, **5**, 417–431.
O'MARA, J., 1932, Chromosome Pairing in *Yucca flaccida*. *Cytologia*, **3**, 66–76.
ONO, H., and SATO, D., 1935, Intergenra hibridigo en Cichorieæ, II. (In Esperanto.) *Jap. J. Genet.*, **11**, 169–179.
ONO, T., 1927, Reducing Division in Triploid *Primula*. A preliminary note. *Bot. Mag. Tokyo*, **41**, 601–604. (English summary.)
ONO, T., 1928, Further investigations on the cytology of *Rumex*. *Bot. Mag. Tokyo*, **42**, 524–533. (English summary.)
ONO, T., 1930 a, Further investigations on the cytology of *Rumex*. V–VII. *Bot. Mag. Tokyo*, **44**, 168–176. (English summary.)
ONO, T., 1930 b, Chromosomenmorphologie von *Rumex Acetosa*. *Sc. Rep. Tohoku Imp. Univ.*, Ser. 4, **5**, 415–422.
ONO, T., 1935, Chromosomen und Sexualität von Rumex Acetosa. *Sci. Rep. Tohoku Imp. Univ.* IV., **10**, 41–210.
ONO, T., and SHIMOTOMAI, N., 1928, Triploid and Tetraploid intersex of *Rumex Acetosa*, L. *Bot. Mag. Tokyo*, **42**, 266–270.
OSAWA, J., 1913, Studies on the cytology of some species of Taraxacum. *Arch. f. Zellforsch.*, **10**, 450–469.
OSAWA, J., 1920, Cytological and Experimental Studies in Morus with special reference to triploid mutants. *Bull. Imp. Ser. Stu. Tohyo*, **1**, 317–369.
OSTENFELD, C. H., 1910, Further studies on the Apogamy and Hybridisation of the Hieracia. *Z.I.A.V.*, **3**, 241–285.
OSTENFELD, C. H., 1921, Some Experiments on the Origin of New Forms in the genus *Hieracium* sub-genus *Archieracium*. *J. Genet.*, **11**, 117–122.

OSTENFELD, C. H., and ROSENBERG, O., 1907, Experimental and Cytological Studies in the Hieracia II. O. Rosenberg : Cytological studies on the apogamy in *Hieracium*. *D. Bot. Tidsskr.*, **28**, 143-170.

OVEREEM, C. VAN, 1921, Über Formen mit abweichender Chromosomenzahl bei Œnothera. *Beih. Bot. Zbl.*, **38**, 73-113.

OVEREEM, C. VAN, 1922, Über Formen mit abweichender Chromosomenzahl bei *Œnothera*. Fortschritt. *Beih. Bot. Zbl.*, **39**, 1-80.

OVERTON, J. B., 1904, Über Parthenogenesis bei Thalictrum purpurascens. *Ber. deuts. bot. Ges.*, **22**, 274-283.

OVERTON, J. B., 1909, On the Organisation of the Nuclei in the Pollen-Mother-Cells of certain plants, with special reference to the Permanence of the Chromosomes. *Ann. Bot.*, **23**, 19-61.

PACE, L., 1913, Apogamy in Atamasco. *Bot. Gaz.*, **56**, 376-394.

PAINTER, T. S., 1921, Studies in reptilian spermatogenesis I. The spermatogenesis of lizards. *J. Exper. Zool.*, **34**, 281-327.

PAINTER, T. S., 1923, Studies in mammalian spermatogenesis. II. Spermatogenesis of man. *J. Exper. Zool.*, **37**, 291-321.

PAINTER, T. S., 1924, Studies in mammalian spermatogenesis III. The fate of the chromatin-nucleolus in the opossum. *J. Exper. Zool.*, **39**, 197-247.

PAINTER, T. S., 1925, A Comparative Study of the Chromosomes of Mammals. *Amer. Nat.*, **59**, 385-409.

PAINTER, T. S., 1927, The Chromosome Constitution of Gates' "Nondisjunction" (v-o) Mice. *Genetics*, **12**, 379-392.

PAINTER, T. S., 1934 a, A new method for the study of chromosome aberrations and the plotting of chromosome maps in Drosophila melanogaster. *Genetics*, **19**, 175-188.

PAINTER, T. S., 1934 b, The morphology of the X-chromosomes in salivary glands of Drosophila melanogaster and a new type of chromosome map for this element. *Genetics*, **19**, 448-469.

PAINTER, T. S., 1935, The morphology of the third chromosome in the salivary gland of Drosophila melanogaster and a new cytological map of this element. *Genetics*, **20**, 301-326.

PAINTER, T. S., and MULLER, H. J., 1929, Parallel Cytology and Genetics of induced translocations and deletions in Drosophila. *J. Hered.*, **20**, 287-298.

PAINTER, T. S., and STONE, W., 1935, Chromosome fusion and speciation in Drosophila. *Genetics*, **20**, 327-341.

PARISER, K., 1927, Die Zytologie und Morphologie der triploiden Intersexe des rückgekreuzten Bastards von *Saturnia pavonia* L. und *Saturnia pyri* Schiff. *Zeits. f. Zellforsch. u. mikr. Anat.*, **5**, 415-447.

PARNELL, F. R., 1921, Note on the Detection of Segregation by examination of the pollen of rice. *J. Genet.*, **11**, 209-212.

PÄTAU, K., 1935, Chromosomenmorphologie bei Drosophila melanogaster und Drosophila simulans und ihre genetische Bedeutung. *Naturwiss*, **23**, 537-543.

PATTERSON, J. T., 1933, The mechanism of mosaic formation in *Drosophila*. *Genetics*, **18**, 32-52.

PATTERSON, J. T., 1935, The question of delayed breakage in the chromosomes of Drosophila. *J. Exp. Zool.*, **70**, 233-242.

PATTERSON, J. T., and HAMLETT, G. W. D., 1925, Haploid males in Paracopidosomopsis. *Science*, **61**, 89.

PATTERSON, J. T., and MULLER, H. J., 1930 d, Are "Progressive" Mutations produced by X-rays ? *Genetics*, **15**, 495-578.

PATTERSON, J. T., and SUCHE, M. L., 1934, Crossing-over induced by X-rays in Drosophila males. *Genetics*, **19**, 223–236.
PAYNE, F., 1910, The chromosomes of *Acholla multispinosa*. *Biol. Bull.*, **18**, 174–179.
PAYNE, F., 1914, Chromosomal variations and the formation of the spermatocyte chromosomes in the European earwig, *Forficula* sp. *J. Morphol.*, **25**, 559–585.
*PEACOCK, A. D., 1925, Animal Parthenogenesis in relation to Chromosomes and Species. *Amer. Nat.*, **59**, 218–224.
PEACOCK, A. D., and SANDERSON, A. R., 1931, Cytological evidence of male haploidy and female diploidy in a sawfly. (Hymen. Tenthred.) *Proc. 2nd Int. Cong. Sex Research.*
PEARSON, N. E., 1927, A Study of Gynandromorphic Katydids. *Amer. Nat.*, **61**, 283–285.
PELLEW, C., and SANSOME, E. RICHARDSON, 1931, Genetical and Cytological Studies on the Relations between European and Asiatic Varieties of *Pisum sativum*. *J. Genet.*, **25**, 25–54.
PERCIVAL, J., 1930, Cytological Studies of some Hybrids of *Ægilops* sp. x Wheats, and of some Hybrids between different Species of *Ægilops*. *J. Genet.*, **22**, 201–278.
PETO, F. H., 1930, Cytological Studies in the Genus *Agropyron*. *Canad. J. Res.*, **3**, 428–448.
PETO, F. H., 1934, The Cytology of certain Intergeneric Hybrids between *Festuca* and *Lolium*. *J. Genet.*, **28**, 113–156.
PETO, F. H., 1935, Associations of somatic chromosomes induced by heat and chloral hydrate treatments. *Canad. J. Res. C.*, **13**, 301–314.
PHILIP, U., 1935, Crossing-over between X- and Y-chromosomes in *Drosophila melanogaster*. *J. Genet.*, **31**, 341–352.
PHILP, J., 1933, The genetics and cytology of some interspecific hybrids of *Avena*. *J. Genet*, **27**, 133–179.
PHILP, J., 1934 *a*, Note on the cytology of *Saxifraga granulata* L., *S. rosacea* Mœnch, and their hybrids. *J. Genet.*, **29**, 197–201.
PHILP, J., 1934 *b*, Aberrant albinism in polyploid oats. *J. Genet.*, **30**, 267–302.
PHILP, J., and HUSKINS, C. L., 1931. The Cytology of *Matthiola incana* R. Br., especially in relation to the Inheritance of Double Flowers. *J. Genet.*, **24**, 359–404.
PIECH, K., 1928 *a*, Zytologische Studien an der Gattung Scirpus. *Bull. Acad. pol. Sci.* B, 1928, 1–43.
PIECH, K., 1928 *b*, Über die Entstehung der generativen Zelle bei Scirpus uniglumis Link durch freie Zellbildung. *Planta*, **6**, 96–117.
PIETSCHMANN, Käthe, 1929, Untersuchungen an Vahlkampfia tachypoda Gläser. *Arch. Prot.*, **65**, 379–425.
PINNEY, E., 1908, Organisation of the chromosomes in *Phrynotettix*. *Kansas Univ. Sci. Bull.*, **4**.
PLOTNIKOWA, T. W., 1932, Zytologische Untersuchung der Weizen-Roggen-Bastarde. I. Anormale Kernteilung in somatischen Zellen. *Planta*, **16**, 174–177.
PLOUGH, H. H., 1917, The Effect of Temperature on Crossing-over in *Drosophila*. *J. Exper. Zool.*, **24**, 147–209.
PODDUBNAJA-ARNOLDI, V., 1933 *a*, Spermazellen in der Familie der Dipsacaceæ. *Planta*, **21**, 381–386.
PODDUBNAJA-ARNOLDI, W. A., 1933 *b*, Geschlechtliche und ungeschlechtliche Fortpflanzung bei einigen Chondrilla-Arten. *Planta*, **19**, 46–86.
PODDUBNAJA-ARNOLDI, V., and DIANOWA, V., 1934, Eine zytoembryologische Untersuchung einiger Arten der Gattung Taraxacum. *Planta*, **23**, 19–46.

PODDUBNAJA-ARNOLDI, W., STESHINA, N., and SOSNOVETZ, A., 1935, Der Character und die Ursachen der Sterilität bei Scorzonera tan-saghys Lipsch. et Bosse. *Beiheft. Bot. Zbl.*, 53, A., 309–339.

POOLE, C. F., 1931, The Interspecific Hybrid Crepis rubra × C. fœtida and Some of its Derivatives, I. *Univ. Calif. Pub. Agr. Sci.*, 6, 169–200.

POOLE, C. F., 1932, The Interspecific Hybrid, Crepis rubra × C. fœtida, and Some of its Derivatives. II. Two selfed generations from an amphidiploid hybrid. *Univ. Calif. Pub. Agr. Sci.*, 6, 231–255.

POPOFF, W. W., 1933, Über die Zahl der Chromosomen und über die Heterochromosomen des Haushuhnes. *Zeits. Zellforsch. u. mikr. Anat.*, 17, 341–346.

PRAT, S., and HRUBY, K., 1934, Die Mikrophotographie mit infraroten Strahlen im Dienste der Zellforschung.   Sammelreferate mit neuen Beobachtungen. *Protoplasma*, 22, 145–152.

PRELL, H., 1923. Der Vererbungstheoretische Character der Parthenogenese. *Genetica*, 5, 191–208.

PROKOFIEWA, A., 1933, Vergleichend-karyologische Studien von elf Arten der Familie Corixidæ (Hemiptera, Heteroptera). *Zeits. Zellf. u. mikr. Anat.*, 19, 1–27.

PROKOFIEWA, A. A., 1935 a, Morphologische Struktur der Chromosomen von Drosophila melanogaster. *Zeits. Zellf. u. mikr. Anat.*, 22, 255–262.

PROKOFIEVA, A., 1935 b, On the chromosome morphology of certain Amphibia. *Cytologia*, 6, 148–164.

PROKOFYEVA (BELGOVSKAYA), A. A., 1935 c, The structure of the Chromocenter. *Cytologia*, 6, 438–443.

PROKOFJEVA (BELGOVSKAJA), A. A., 1935 d, Y chromosome in salivary glands of Drosophila. *C. R. Acad. Sci. U.R.S.S.*, 1935, 3, 365–366.

PROPACH, H., 1934, Cytologische Untersuchungen an Limnanthes Douglasii R. Br. *Zeits. Zellf. u. mikr. Anat.*, 21, 357–375.

PROPACH, H., 1935, Studien über heteroploide Formen von Antirrhinum majus L. II. *Planta*, 23, 349–357.

*PUNNETT, R. C., 1925, *Lathyrus odoratus*. *Bibliograph. Genet.*, 1, 69–82.

RAMAER, H., 1935, Cytology of Hevea. Thesis. The Hague, 1935, pp. 193–236.

RAMIAH, K., et alii, 1934, A haploid plant in rice. *J. Indian Bot. Soc.*, 13, 153–164.

RANCKEN, G., 1934, Zytologische Untersuchungen an einigen wirtschaftlich wertvollen Wiesengräsern, etc. *Acta Agral. Fenn.*, 29, 1–92.

RANDOLPH, L. F., 1928, Chromosome Numbers in Zea Mays L. *Cornell Univ. Exp. Sta.* (Ithaca) *Mem.*, 117.

RANDOLPH, L. F., 1932, Some effects of high temperature on polyploidy and other variations in Maize. *Proc. Nat. Acad. Sci.*, 18, 222–229.

RANDOLPH, L. F., 1935, A new fixing fluid and a revised schedule for the paraffin method in plant cytology. *Stain Technol.*, 10, 95–96.

RANDOLPH, L. F., and McCLINTOCK, B., 1926, Polyploidy in Zea Mays L. *Amer. Nat.*, 60, 99–102.

RAO, T. R., 1933, Chromosomal aberrations occurring in un-irradiated grasshoppers. *J. Mysore Univ.*, 7, 1–12.

REDFIELD, H., 1930, Crossing-over in the third chromosomes of triploids of *Drosophila melanogaster*. *Genetics*, 15, 205–252.

RENNER, O., 1919 a, Zur Biologie und Morphologie der männlichen Haplonten einiger Önotheren. *Zeits. f. Bot.*, 11, 305–380.

RENNER, O., 1919 b, Über Sichtbarwerden der Mendelschen Spaltung im Pollen von Önotherabastarden. *Ber. Deuts. Bot. Ges.*, 37, 129–135.

RENNER, O., 1921, Heterogamie im weiblichen Geschlecht und Embryosackentwicklung bei den Önotheren. *Zeits. f. Bot.*, **13**, 609–621.
RENNER, O., 1925, Untersuchungen über die faktorielle Konstitution einiger komplexheterozygotischer Önotheren. *Biblioth. Genet.*, **9**, 1–168.
RENNER, O., 1927, Über eine aus *Œnothera suaveolens* durch Bastardierung gewonnene homozygotische *lutescens*-Form. *Hereditas*, **9**, 69–80.
*RENNER, O., 1929, Artbastarde bei Pflanzen. *Handb. d. Vererbungswissenschaft*, **2**. (*See also :* Hœppener.)
RENNER, O., 1933, Zur Kenntnis der Letalfaktoren und des Koppelungswechsels der Œnotheren. *Flora*, **27**, 215–250.
RENNER, O., 1934, Die pflanzlichen Plastiden als selbstandige Elemente der genetischen Konstitution. *Ber. Math.-Phys. Kl. Acad. Wiss. Leipzig*, **86**, 241–266.
*REUTER, E., 1930, Beiträge zu einer einheitlichen Auffassung gewisser Chromosomenfragen. *Act. Zool. Fenn.*, **9**, 1–487.
RHOADES, M. M., 1931, The frequencies of homozygosis of factors in attached-X females of *Drosophila melanogaster*. *Genetics*, **16**, 375–385.
RHOADES, M. M., 1933 *a*, A cytological study of a reciprocal translocation in *Zea*. *Proc. Nat. Acad. Sci.*, **19**, 1022–1031.
RHOADES, M. M., 1933 *b*, An experimental and theoretical study of chromatid crossing over. *Genetics*, **18**, 535–555.
RHOADES, M. M., 1933 *c*, A secondary trisome in maize. *Proc. Nat. Acad. Sci.*, **19**, 1031–1038.
*RHOADES, M. M., and McCLINTOCK, B., 1935, The cytogenetics of maize. *Bot. Rev.*, **1**, 292–325.
RICHARDSON, M. [M.], 1934, The origin and behaviour of chiasmata. X. *Spironema fragrans*. *Cytologia*, **5**, 337–354.
RICHARDSON, M. M., 1935 *a*, Setcreasia brevifolia, a further example of polyploidy and structural hybridity in the Tradescantiæ. *Bot. Gaz.*, **97**, 400–407.
RICHARDSON, M. M., 1935 *b*, Meiosis in *Crepis*. I. Pachytene association and chiasma behaviour in *Crepis capillaris* (L.) Wallr. and *C. tectorum* L. II. Failure of pairing in *Crepis capillaris* (L.) Wallr. *J. Genet.*, **31**, 101–117, 119–143.
RICHARDSON, M. M., 1936, Structural hybridity in *Lilium Martagon album* × *L. Hansonii*. *J. Genet.*, **32**, 411–450.
RILEY, H. P., 1936, The effect of X-rays on the chromosomes of Tradescantia gigantea. *Cytologia*, **7**, 131–142.
ROBERTSON, M., 1929, Life cycles in the Protozoa. *Biol. Revs.*, **4**, 152–179.
ROBERTSON, W. R. B., 1915, Chromosome Studies III. Inequalities and deficiencies in homologous chromosomes : their bearing upon synapsis and the loss of unit characters. *J. Morphol.*, **26**, 109–141.
ROBERTSON, W. R. B., 1916, Chromosome Studies I. Taxonomic relationships shown in chromosomes of Tettigidæ and other subfamilies of the Acrididæ: V-shaped chromosomes and their significance in Acrididæ, Locustidæ, and Gryllidæ : chromosomes and variation. *J. Morphol.*, **27**, 179–332.
ROBERTSON, W. R. B., 1930, Chromosome Studies V. Diploidy and persistent chromosome relations in partheno-produced Tettigidæ (Apotettix eurycephalus and Paratettix texanus). *J. Morphol.*, **50**, 209–257.
ROBERTSON, W. R. B., 1931 *a*, Hybrid Vigor—a Factor in Tettigid Parthenogenesis ? *Amer. Nat.*, **65**, 165–172.
ROBERTSON, W. R. B., 1931 *b*, A split in chromosomes about to enter the spermatid (Paratettix texanus). *Genetics*, **16**, 349–352.

ROBERTSON, W. R. B., 1931 c, On the origin of partheno-produced males in Tettigidæ Apotettix and Paratettix. *Genetics*, **16**, 353–356.
ROBERTSON, W. R. B., 1931 d, Chromosome Studies II. Synapsis in the Tettigidæ, with special reference to the presynapsis split. *J. Morphol.*, **51**, 119–139.
ROHWEDER, H., 1929, Über Kernuntersuchungen an *Dianthus*-Arten (Vorl. Mitt.). *Ber. Deuts. Bot. Ges.*, **47**, 81–86.
ROHWEDER, H., 1934, Beiträge zur Systematik und Phylogenie der Genus Dianthus. *Bot. Jahrb.*, **66**, 249–368.
ROSENBERG, O., 1906, Über die Embryobildung in der Gattung Hieracium. *Ber. Deuts. Bot. Ges.*, **24**, 157–161.
ROSENBERG, O., 1909 a, Cytologische und morphologische Studien an Drosera longifolia × rotundifolia. *K. Sv. Vet. Handl.*, **43** (11), 1–64.
ROSENBERG, O., 1909 b, Über den Bau des Ruhekerns. *Sv. Bot. Tidsskr.*, **3**, 163–173.
ROSENBERG, O., 1912, Über die Apogamie bei *Chondrilla juncea*. *Svensk Bot. Tidsskr.*, **6**, 915–919.
ROSENBERG, O., 1917, Die Reduktionsteilung und ihre Degeneration in Hieracium. *Svensk Bot. Tidsskr.*, **11**, 145–206.
ROSENBERG, O., 1927, Die Semiheterotypische Teilung und ihre Bedeutung für die Entstehung verdoppelter Chromosomenzahlen. *Hereditas*, **8**, 305–338.
*ROSENBERG, O., 1930, Apogamie und Parthenogenesis bei Pflanzen. *Handb. d. Vererbungswissenschaft*, **2**. (See also : Ostenfeld.)
RÜCKERT, J., 1892, Zur Entwicklungsgeschichte des Ovarialeies bei Selachiern. *Anat. Anz.*, **7**, 107–158.
RUDLOFF, F., 1930, Oe. pachycarpa Renner. Genetische und cytologische Untersuchungen. *Gartenbauwissenschaft*, **3**, 499–526.
RUDLOFF, F., 1931, Zur Polarisation in der Reduktionsteilung heterogamer Œnotheren I. Die Embryosack-Entwicklung und ihrer Tendenzen. *Z.I.A.V.*, **58**, 422–433.
RUTTLE, M. L., 1927, Chromosome numbers and morphology in *Nicotiana* I. The somatic chromosomes and non-disjunction in *N. alata* var. *grandiflora*. *Univ. Calif. Pub. Bot.*, **11**, 159–176.
RUTTLE, M. L., 1928, Chromosome numbers and morphology in *Nicotiana* II. Diploidy and Partial Diploidy in root-tips of *Tabacum* haploids. *Univ. Calif. Pub. Bot.*, **11**, 213–232.
RYBIN, V. A., 1927 a, On the number of chromosomes observed in the somatic and reduction divisions of the cultivated apple in connection with pollen sterility of some of its varieties. *Bull. Appl. Bot.* **17** (3), 101–120.
RYBIN, V. A., 1927 b, Polyploid Hybrids of Nicotiana Tabacum L. × Nicotiana rustica L. *Bull. Appl. Bot.*, **17** (3), 191–240.
RYBIN, V. A., 1929, Über einen allotetraploiden Bastard von *Nicotiana tabacum* × *Nicotiana sylvestris*. *Ber. Deuts. Bot. Ges.*, **37**, 385–394.
RYBIN, W. A., 1936, Spontane und experimentell erzeugte Bastarde zwischen Schwarzdorn und Kirschpflaume und das Abstammungsproblem der Kulturpflaume. *Planta*, **25**, 22–58.

SAEZ, F. A., 1930, Investigaciones sobre los cromosomas de algunos Ortópteros de la America del Sur. *Rev. Mus. La Plata*, **32**, 317–361.
SAEZ, F. A., ROJAS, P., and ROBERTS, E. de, 1936, Investigaciones sobre las células sexuales de los anfibios anuros : el proceso meiótico en *Bufo arenarum* Hensel. *Inst. Mus. Univ. Nat. La Plata. Obra Cincuentenario* **11**, 95–143.

SAKAI, K.-I., 1935, Chromosome association in Oryza sativa L. I. The secondary association of the meiotic chromosomes. *Jap. J. Genet.*, 11, 145–156.
SAKAMURA, T., 1915, Über die Einschnürung der Chromosomen bei *Vicia Faba* L. *Bot. Mag. Tokyo*, 29, 287–300.
SAKAMURA, T., 1920, Experimentelle Studien über die Zell- und Kernteilung, etc. *J. Coll. Sci. Imp. Univ. Tokyo*, 39 (ii), 1–221.
SAKAMURA, T., 1927 a, Chromosomenforschungen an frischem Material. *Protoplasma*, 1, 537–565.
SAKAMURA, T., 1927 b, Fixierung von Chromosomen mit siedendem Wasser. *Bot. Mag. Tokyo*, 41, 59–64.
SAKAMURA, T., and STOW, I., 1926, Über die experimentell veranlasste Entstehung von keimfähigan Pollenkörner mit abweichenden Chromosomenzahlen. *Jap. J. Bot.*, 3, 111–137.
SANDS, H. C., 1923, The structure of the chromosomes in *Tradescantia virginica* L. *Amer. J. Bot.*, 10, 343–360.
SANSOME, E. R., 1929, A chromosome ring in *Pisum*. *Nature*, 124, 578.
SANSOME, E. R., 1932, Segmental Interchange in *Pisum*. *Cytologia*, 3, 200–219.
SANSOME, E. R., 1933, Segmental interchange in *Pisum* II. *Cytologia*, 5, 15–30.
SANSOME, F. W., 1931 a, Saxifrage Crosses. *Nature*, 127, 59–60.
SANSOME, F. W., 1931 b, Graft Hybrids and Induction of Polyploids in Solanum. *Proc. 9th Int. Hort. Cong.*, 92–99.
SANSOME, F. W., 1933, Chromatid segregation in *Solanum Lycopersicum*. *J. Genet.*, 27, 105–132.
*SANSOME, F. W., and PHILP, J., 1932, *Recent Advances in Plant Genetics*. London, Churchill.
SANTOS, J. K., 1923, Differentiation among chromosomes in *Elodea*. *Bot. Gaz.*, 75, 42–59.
SANTOS, J. K., 1924, Determination of sex in Elodea. *Bot. Gaz.*, 77, 353–376.
SAPĚHIN, A. A., 1928, Hylogenetic Investigations of the Vulgare group in Triticum. *Bull. Appl. Bot.*, 19 (1), 127–166.
SAPĚHIN, L. A., 1931, Über die faktorielle Natur der Unterschiede im Verlaufe der Reduktionsteilung. *Ber. Deuts. Bot. Ges.*, 48, 443–457.
SASS, J. E., 1929, The cytological basis for homothallism and heterothallism in the Agaricaceæ. *Amer. J. Bot.*, 16, 663–701.
SASS, J. E., 1934, Chromosome fragmentation in *Lilium tigrinum* Ker. *Amer. Nat.*, 68, 471–475.
SATINA, S., and BLAKESLEE, A. F., 1935, Cytological effects of a gene in Datura which causes dyad formation in sporogenesis. *Bot. Gaz.*, 96, 521–532.
SATO, D., 1934 a, Chiasma studies in plants, I. Chromosome pairing and chiasma behaviour in *Allium Moly*. *Jap. J. Genet.*, 10, 155–159.
SATO, D., 1934 b, Chiasma studies in plants, II. Chromosome pairing and chiasma behaviour in Yucca, Scilla and Urginea, with special reference to Interference. *Bot. Mag. Tokyo*, 48, 823–846.
SATO, D., and SINOTO, Y., 1935, Chiasma studies in plants, III. Chromosome pairing and chiasma behaviour in the male Rumex acetosa with special reference to the tripartite sex-chromosome. *Jap. J. Genet.*, 11, 219–226.
SATO, I., 1932, Chromosome behaviour in Urodele Amphibia. *Dismyctulus pyrrhogaster* (Boie). *J. Sci. Hitoshima Univ.*, B, I., 2, 33 47.
SAX, H. J., 1933, Chiasma formation in Larix and Tsuga. *Genetics*, 18, 121–128.
SAX, K., 1922, Sterility in Wheat Hybrids II. Chromosome behavior in partially sterile hybrids. *Genetics*, 7, 513–552.

SAX, K., 1929, Chromosome Behavior in Sorbopyrus and Sorbaronia. *Proc. Nat. Acad. Sci.*, **15**, 844–845.
SAX, K., 1931, The Mechanism of Crossing Over. *Science*, **74**, 41–42.
SAX, K., 1935, Chromosome structure in the meiotic chromosomes of Rhœo discolor Hance. *J. Arn. Arb.*, **16**, 216–224.
SAX, K., 1935, Variation in chiasma frequencies in Secale, Vicia and Tradescantia. *Cytologia*, **6**, 289–293.
SAX, K., and ANDERSON, E., 1932, Segmental interchange in chromosomes of Tradescantia. *Genetics*, **18**, 53–67.
SAX, K., and HUMPHREY, L. M., 1934, Structure of meiotic chromosomes in microsporogenesis of Tradescantia. *Bot. Gaz.*, **96**, 353–362.
SAX, K., and SAX, H. J., 1924, Chromosome behaviour in a genus cross. *Genetics*, **9**, 454–464.
SAX, K., and SAX, H. J., 1933, Chromosome number and morphology in the conifers. *J. Arn. Arb.*, **14**, 356–375.
SCHAEDE, R., 1925, Untersuchungen über Zelle, Kern und ihre Teilung am lebenden Objekt. *Beitr. Biol. Pfl.*, **14**, 231–260.
SCHAEDE, R., 1927, Vergleichende Untersuchungen über Cytoplasma, Kern und Kernteilung im lebenden und im fixierten Zustand. *Protoplasma*, **3**, 145–190.
SCHAEDE, R., 1929 a, Kritische Untersuchungen über die Mechanik der Karyokinesis. *Planta*, **8**, 383–397.
SCHAEDE, R., 1929 b, Die Kolloidchemie des pflanzlichen Zellkernes in der Ruhe und in der Teilung. *Erg. d. Biol.*, **5**, 1–28.
SCHAEDE, R., 1930, Über die Struktur des ruhenden Kernes. *Ber. Deuts. Bot. Ges.*, **48**, 342–348.
SCHAFER, B., and LA COUR, L., 1934, A chromosome survey of Aconitum. I. *Ann. Bot.*, **48**, 693–713.
SCHAFFSTEIN, Gerhard, 1935, Untersuchungen über den Feinbau der Prophasechromosomen in der Reduktionsteilung von Lilium martagon. *Zeits. Zellf. u. mikr. Anat.*, **22**, 275–281.
SCHIEMANN, E., 1929, Zytologische Beiträge zur Gattung Aegilops. Chromosomenzahlen und Morphologie. (III Mittlg.) *Ber. Deuts. Bot. Ges.*, **47**, 164–181.
SCHKWARNIKOW, P. K., and NAWASCHIN, M. S., 1934, Über die Beschleunigung des Mutationsvorganges in ruhenden Samen unter dem Einfluss von temperaturerhöhung. *Planta*, **22**, 720–736.
*SCHNARF, K., 1931, *Vergleichende Embryologie des Angiospermen.* Berlin.
SCHRADER, F., 1923 a, Haploidie bei einer Spinnmilbe. *Arch. mikr. Anat.*, **97**, 610–622.
SCHRADER, F., 1923 b, A study of the chromosomes in three species of Pseudococcus. *Arch. Zellforsch.*, **17**, 45–62.
SCHRADER, F., 1925, The cytology of pseudo-sexual eggs in a Species of Daphnia. *Z.I.A.V.*, **40**, 1–27.
*SCHRADER, F., 1928, *The Sex Chromosomes.* Berlin.
SCHRADER, F., 1929, Experimental and cytological investigations of the life-cycle of Gossiparia spuria (Coccidæ) and their Bearing on the Problem of Haploidy in Males. *Zeits. wiss. Zool.*, **134**, 149–179.
SCHRADER, F., 1932, Recent hypotheses on the structure of spindles in the light of certain observations in Hemiptera. *Zeits. wiss. Zool.*, **142**, 520–539.
SCHRADER, F., 1934, On the reality of spindle fibres. *Biol. Bull.*, **57**, 519–533.
SCHRADER, F., 1935, Notes on the mitotic behavior of long chromosomes. *Cytologia*, **6**, 422–430.
*SCHRADER, F., and HUGHES-SCHRADER, S., 1931, Haploidy in Metazoa. *Quart. Rev. Biol.*, **6**, 411–438.

SCHRADER, F., and STURTEVANT, A. H., 1923, A Note on the theory of sex determination. *Amer. Nat.*, 57, 379–381.
SCHREINER, A., and SCHREINER, K. E., 1906 a, Neue Studien über die Chromatinreifung der Geschlechtszellen. I. Die Reifung der männlichen Geschlechtszellen von Tomopteris onisciformis, Eschscholz. *Arch. Biol.*, 22, 1–69.
SCHREINER, A., and SCHREINER, K. E., 1906 b, Neue Studien über die Chromatinreifung der Geschlechtszellen. II. Die Reifung der männlichen Geschlechtszellen von Salamandra maculosa (Laur.) Spinax niger (Bonap.) und Myxine glutinosa (L.). *Arch. Biol.*, 22, 419–492.
SCHULTZ, E. S., 1927, Nuclear division and spore-formation in the ascus of *Peziza domiciliana*. *Amer. J. Bot.*, 14, 307–322.
SCHUSSNIG, B., 1930, Der Chromosomencyclus von *Cladophora Suhriana*. *Œsterr. Bot. Zeits.*, 79, 273–278.
SCHUSSNIG, B., 1931, Die somatische und heterotype Kernteilung bei Cladophora Suhriana Kützing. *Planta*, 13, 474–528.
SCHWARTZ, H., 1932, Der Chromosomenzyklus von Tetraneura ulmi de Geer. *Zeits. Zellforsch. u. mikr. Anat.*, 15, 645–687.
SCHWARZENBACH, M., 1926, Regeneration und Aposporie bei Anthoceros. *Arch. J. Klaus Stift.*, 2, 91–140.
SCHWEIZER, J., 1923, Polyploidie und Geschlechter-verteilung bei *Splachnum sphaericum* (Linn. fil.) Swartz. *Flora*, 116, 1–72.
SCHWEMMLE, J., 1926, Vergleichend zytologische Untersuchungen an Onagraceen. II. Die Reduktionsteilung von Eucharidium concinnum. *Jahrb. wiss. Bot.*, 65, 778–818.
SCHWEMMLE, J., 1928, Genetische und zytologische Untersuchungen an Eu-Œnotheren. *Jahrb. wiss. Bot.*, 67, 849–876.
SEILER, J., 1914, Das Verhalten der Geschlechtschromosomen bei Lepidopteren. Nebst einem Beitrag zur Kenntnis der Eireifung, Samenreifung und Befruchtung. *Arch. Zellforsch.*, 13, 159–269.
SEILER, J., 1920, Geschlechtschromosomenuntersuchungen an Psychiden I. *Arch. Zellf.*, 15, 249–268.
SEILER, J., 1923, Geschlechtschromosomen-Untersuchungen an Psychiden. IV. Die Parthenogenese der Psychiden. *Z.I.A.V.*, 31, 1–99.
SEILER, J., 1925, Ergebnisse aus Kreuzungen von Schmetterlingsrassen mit verschiedener Chromosomenzahl. *Arch. Jul. Klaus-Stift.*, 1, 63–117.
SEILER, J., 1927, Ergebnisse aus der Kreuzung parthenogenetischer und zweigeschlechtliche Schmetterlinge. *Biol. Zbl.*, 47, 426–446.
SEILER, J., and HANIEL, C. B., 1921, Das verschiedene Verhalten der Chromosomen im Eireifung und Samenreifung von Lymantria monacha L. *Z.I.A.V.*, 27, 81–103.
SEITZ, F. W., 1935, Zytologische Untersuchungen an tetraploiden Œnotheren. *Zeits. f. Bot.*, 28, 481–542.
SELIM L. A. G., 1930, A Cytological Study of *Oryza sativa* L. *Cytologia*, 2, 1–26.
SENJANINOVA, M., 1927, Beitrag zur vergleichend-karyologische Untersuchung des Linneons Valeriana officinalis L. (sensu-lato). *Zeits. Zellforsch. u. mikr. Anat.*, 5, 675–679.
SEREBROVSKY, A. S., 1929, A General Scheme for the Origin of Mutations. *Amer. Nat.*, 63, 374–378.
SHARP, L. W., 1920, Spermatogenesis in *Blasia*. *Bot. Gaz.*, 69, 258–268.
*SHARP, L. W., 1925, The factorial interpretation of sex determination. *Cellule*, 35, 195–235.
SHARP, L. W., 1929, Structure of Large Somatic Chromosomes. *Bot. Gaz.*, 88, 349–382.

*SHARP, L. W., 1934, *Introduction to Cytology*. N.Y. (3rd edition).
SHEFFIELD, F. M. L., 1929, Chromosome Linkage in Œnothera, with special reference to some F_1 hybrids. *Proc. Roy. Soc.* B., **105**, 207–228.
SHEN, T. H., 1936, Beiträge zum Studium der Geschlechts bestimmung bei Dinophilus apatris. *Zool. Jahrb.*, **56**, 219–238.
SHIMAMURA, T., 1935, Über die Bestäubung und Befruchtung bei *Ginkgo biloba* L. *Jap. J. Genet.*, **11**, 180–184.
SHIMOTOMAI, N., 1927, Über Störungen der meiotischen Teilungen durch niedrige Temperatur. *Bot. Mag. Tokyo*, **41**, 149–160.
SHIMOTOMAI, N., 1928, Karyokinese im Oogonium von *Cystophyllum sisymbrioides*, J. Ag. *Sci. Rec. Tohoku Imp. Univ.* (4), **3**, 577–579.
SHIMOTOMAI, N., 1930, Chromosomenzahlen und Phylogenie bei der Gattung Potentilla. *J. Sci. Hiroshima* B2, **1**, 1–11.
SHIMOTOMAI, N., 1931 *a*, Bastardierungsversuche bei *Chrysanthemum*, I. *J. Sci. Hiroshima* B2, **1**, 37–54.
SHIMOTOMAI, N., 1931 *b*, Über die abnorme Reduktionsteilung in Pollenmutterzellen die einen riesigen oder überzähligen Zwergkerne enthalten. *Bot. Mag. Tokyo*, **45**, 356–363.
SHIMOTOMAI, N., 1933, Zur Karyogenetik der Gattung *Chrysanthemum*. *J. Sci. Hiroshima Univ.*, B. 2, **2**, 1–100.
SHIMOTOMAI, N., 1935, Zur Kenntnis der Pseudogamie bei Potentilla. *Proc. Imper. Acad. Tokyo*, **11**, 338–339.
SHIMOTOMAI, N., and KOYAMA, Y., 1932, Geschlechtschromosomen bei *Pogonatum inflexum* Lindb. und Chromosomenzahlen bei einigen anderen Laubmoosen. *J. Sci. Hiroshima Univ.*, B, 2, **1**, 95–101.
SHINJI, O., 1931, The evolutional significance of the chromosomes of the Aphididæ. *J. Morphol. and Physiol.*, **51**, 373–433.
SHINKE, N., 1929, Chromosome arrangement IV. The Meiotic Divisions in Pollen Mother Cells of *Sagittaria Agniashi*, Makino and *Lythrum salicaria*, L. var. *vulgare*, DC. subvar. *genuina*, Koehne. *Mem. Coll. Sci. Kyoto* B, **4**, 283–308.
SHINKE, N., 1930, On the Spiral Structure of Chromosomes in some Higher Plants. *Mem. Coll. Sci. Kyoto* B, **5**, 239–245.
SHINKE, N., 1934, Special structures of chromosomes in meiosis in *Sagittaria Aginashi*. *Mem. Coll. Sci. Kyoto* B, **9**, 367–392.
SHINKE, N., and SHIGENAKA, M., 1933, A histochemical study of plant nuclei in rest and mitosis. *Cytologia*, **4**, 189–221.
SHULL, A. F., and WHITTINGHILL, M., 1934, Crossovers in male Drosophila melanogaster induced by heat. *Science*, **80**, 103–104.
SIMONET, M., 1931, Etude génétique et cytologique de l'hybride Iris pallida Lan. × I. tectorum Max. *C. R. Acad. Sci.*, **193**, 1214–1216.
SIMONET, M., 1934 *a*, Nouvelles recherches cytologiques et génétiques chez les Iris. *Ann. des Sci. Nat., Bot.*, 10th ser., 1934, 231–383.
SIMONET, M., 1934 *b*, Sur la valeur taxonomique de l'Agropyron acutum Roehm. et S.—Controle cytologique. *Bull. Soc. Bot. France*, **81**, 801–814.
SINÉTY, R. de, 1901, Recherches sur la Biologie et l'Anatomie des Phasmes. *Cellule*, **19**, 119–278.
SINOTÔ, Y., 1929, Chromosome Studies in Some Dioecious Plants, with special reference to the Allosomes. *Cytologia*, **1**, 109–191.
SINOTO, Y., and ONO, H., 1934, Intergenra hibridigo en Chicorieæ, I. Hibridoj de *Crepis capillaris* kaj *Taraxacum platycarpum* (in Esperanto). *Jap. J. Genet.*, **10**, 160–164.
SKALINSKA, M., 1935, Cytogenetic investigations of an allotetraploid Aquilegia. *Bull. Acad. Pol. Sci.* B (1), 33–63.

SKOVSTED, A., 1929, Cytological investigations of the genus *Æsculus* L. *Hereditas*, **12**, 64–70.
SKOVSTED, A., 1933, Cytological studies in cotton. I. The mitosis and the meiosis in diploid and triploid Asiatic cotton. *Ann. Bot.*, **47**, 227–251.
SKOVSTED, A., 1934, Cytological studies in cotton. II. Two interspecific hybrids between Asiatic and New World cottons. *J. Genet.*, **28**, 407–424.
SMITH, B. G., 1929, The history of the chromosomal vesicles in the segmenting egg of Cryptobranchus allegheniensis. *J. Morphol.*, **47**, 89–133.
SMITH, F. H., 1933, The relation of the satellites to the nucleolus in *Galtonia candicans*. *Amer. J. Bot.*, **20**, 188–195.
SMITH, F. H., 1935, Anomalous spindles in *Impatiens pallida*. *Cytologia*, **6**, 165–176.
SMITH, L., 1936, Cytogenetic studies of Triticum monococcum and T. ægilopoides. (Abstract.) *Amer. Nat.*, **70**, 66.
SMITH, S. G., 1931, Cytology of *Anchusa* and its Relation to the Taxonomy of the Genus. *Nature*, **128**, 493–494.
SMITH, S. G., 1935, Chromosome fragmentation produced by crossing-over in *Trillium erectum* L. *J. Genet.*, **30**, 227–232.
SOROKIN, H., 1927 a, Cytological and morphological investigations on gynodimorphic and normal forms of *Ranunculus acris* L. *Genetics*, **12**, 59–83.
SOROKIN, H., 1927 b, A study of meiosis in *Ranunculus acris*. *Amer. J. Bot.*, **14**, 76–84.
SOROKIN, H., 1929, Idiograms, Nucleoli, and Satellites of certain Ranunculaceæ. *Amer. J. Bot.*, **16**, 407–420.
SPEICHER, K. G., 1934, Impaternate females in Habrobracon. *Biol. Bull.*, **67**, 277–293.
SPIER, J. D., 1934, Chiasma frequency in species and species hybrids of *Avena*. *Canad. J. Res.*, **11**, 347–361.
SPRUMONT, G., 1928, Chromosomes et satellites dans quelques espèces d'Ornithogalum. *Cellule*, **38**, 269–292.
STADLER, L. J., 1929, Chromosome Number and the Mutation Rate in Avena and Triticum. *Proc. Nat. Acad. Sci.*, **15**, 876–881.
STADLER, L. J., 1930 a, Recovery following Genetic Deficiency in Maize. *Proc. Nat. Acad. Sci.*, **16**, 714–720.
STADLER, L. J., 1930 b, Some genetic effects of X-rays in plants. *J. Hered.*, **21**, 3–19.
*STADLER, L. J.. 1931, The Experimental Modification of Heredity in Crop Plants. I. Induced Chromosomal Irregularities. *Sci. Agric.*, **11**, 557–572.
*STADLER, L. J., 1932, On the genetic nature of induced mutations in plants. *Proc. 6th Int. Cong. Genet.*, 274–294.
STADLER, L. J., 1933, On the genetic nature of induced mutations in plants. II. A haplo-viable deficiency in maize. *Univ. Missouri Agr. Exp. Sta. Research Bull.*, **204**, 1–29.
STADLER, L. J., and SPRAGUE, G. F., 1936, Genetic effects of ultra-violet radiation in maize. (Abstract.) *Amer. Nat.*, **70**, 69.
STÄHLIN, A., 1929, Morphologische und zytologische Untersuchungen an Gramineen. *Wiss. Arch. Landw.* A (Pflanzenbau), **1**, 330–398.
STEBBINS, G., 1932, Cytology of Antennaria II. Parthenogenetic species. *Bot. Gaz.*, **94**, 322–345.
STEERE, W. C., 1931, A New and Rapid Method for making permanent acetocarmin Smears. *Stain Technol.*, **6**, 107–111.
STEERE, W. C., 1932, Behavior in triploid *Petunia* hybrids. *Amer. J. Bot.*, **19**, 340–356.
STEIL, W. N., 1919, A study of apogamy in *Nephrodium hirtipes*. *Ann. Bot.*, **33**, 109–132.

STENAR, H., 1928, Zur Embryologie der *Veratrum-* und *Anthericum-*Gruppen. *Bot. Notiser,* 1928, 357-378.
STERN, C., 1927, Über Chromosomenelimination bei der Taufliege. *Naturwiss.,* **36,** 740-746.
*STERN, C., 1928, Fortschritte der Chromosomentheorie der Vererbung. *Ergeb. d. Biol.,* **4,** 205-359.
STERN, C., 1929 a, Über Reduktionstypen der Heterochromosomen von *Drosophila melanogaster. Biol. Zbl.,* **49,** 718-735.
STERN, C., 1929 b, Untersuchungen über Aberrationen des Y-Chromosoms von *Drosophila melanogaster. Z.I.A.V.,* **51,** 253-353.
STERN, C., 1931, Zytologisch-genetische Untersuchungen als Beweise für die Morgansche Theorie des Faktorenaustauschs. *Biol. Zbl.,* **51,** 547-587.
STERN, C., 1934, On the occurrence of translocations and autosomal non-disjunction in Drosophila melanogaster. *Proc. Nat. Acad. Sci.,* **20,** 36-39.
STERN, C., 1935, Further studies on somatic crossing-over and segregation. (Abstract.) *Amer. Nat.,* **69,** 81-82.
STERN, C., and OGURA, S., 1931, Neue Untersuchungen über Aberrationen des Y-chromosoms von *Drosophila melanogaster. Z.I.A.V.,* **58,** 81-121.
STERN, C., and SEKIGUTI, K., 1931, Analyse eines Mosaikindividuums bei *Drosophila melanogaster. Biol. Zbl.,* **51,** 194-199.
STEVENS, N. M., 1905, Studies in spermatogenesis with special reference to the accessory chromosome. *Carneg. Inst. Pub.,* **36** (1).
STEVENS, N. M., 1909 a, Further Studies on the Chromosomes of the Coleoptera. *J. Exp. Zool.,* **6,** 101-113.
STOMPS, T. J., 1911, Kernteilung und Synapsis bei *Spinacia oleracea* L. *Biol. Zbl.,* **31,** 257-309.
STOMPS, T. J., 1930 a, Über Parthenogenesis infolge Fremdbefruchtung bei *Œnothera. Z.I.A.V.,* **54,** 243-245.
STOMPS, T. J., 1930 b, Über parthenogenetische Œnothera. *Ber. Deuts. Bot. Ges.,* **48,** 119-126.
STOMPS, T. J., 1931, Weiteres über Parthenogenesis bei Œnothera. *Ber. Deuts. Bot. Ges.,* **49,** 258-266.
STONE, L. H. A., 1933, The effect of X-radiation on the meiotic and mitotic divisions of certain plants. *Ann. Bot.,* **41,** 815-826.
STONE, L. H. A., and MATHER, K., 1932, The origin and behaviour of chiasmata IV. Diploid and triploid *Hyacinthus. Cytologia,* **4,** 16-25.
STONE, W., 1934, Linkage between the X and IV chromosomes in Drosophila melanogaster. *Genetica,* **16,** 506-519.
STONE, W., and THOMAS, I., 1935, Cross-over and disjunctional properties of X-chromosome inversions in Drosophila melanogaster. *Genetica,* **17,** 170-184.
STOW, I., 1927, A Cytological Study on Pollen Sterility in *Solanum tuberosum* L. *Jap. J. Bot.,* **3,** 217-238.
STOW, I., 1930, Experimental Studies on the Formation of the Embryosac-like Giant Pollen-Grains in the Anther of *Hyacinthus orientalis. Cytologia,* **1,** 417-439.
STOW, I., 1933, On the female tendencies of the embryo sac-like giant pollen grains of *Hyacinthus orientalis. Cytologia,* **5,** 88-108.
STRANGEWAYS, T. P., 1922, Observations on the Changes seen in Living Cells during Growth and Division. *Proc. Roy. Soc.* B, **94,** 137-141.
STRANGEWAYS, T. S. P., and HOPWOOD, F. L., 1927, The effects of X-rays upon mitotic cell division in tissue cultures *in vitro. Proc. Roy. Soc.* B., **100,** 283-293.
STRASBURGER, E., 1907, Apogamie bei Marsilia. *Flora,* **97,** 123-191.

STRASBURGER, 1910, Sexuelle und Apogame Fortpflanzung bei Urticaceen. *Jahrb. Wiss. Bot.*, **47**, 245–288.
STRAUB, J., 1936, Untersuchungen zur Physiologie der Meiosis II. *Zeits. f. Bot.*, **30**, 1–57.
STRUGGER, S., 1932, Über das Verhalten des pflanzlichen Zellkernes gegenüber Anilinfarbstoffen. Ein Beitrag zur Methodik der Bestimmung des isoelektrischen Punktes der Kernphasen. *Planta*, **18**, 561–570.
STURTEVANT, A. H., 1925, The effects of unequal crossing-over at the bar locus in *Drosophila*. *Genetics*, **10**, 117–147.
*STURTEVANT, A. H., 1926 a, Renner's Studies on the Genetics of *Œnothera*. *Quart. Rev. Biol.*, **1**, 283–288.
STURTEVANT, A. H., 1926 b, A cross-over reducer in Drosophila melanogaster due to inversion of a section of the third chromosome. *Biol. Zbl.*, **46**, 697–702.
STURTEVANT, A. H., 1929, The Genetics of *Drosophila simulans*. *Carnegie Inst. Wash. Pubn.*, **399**, 1–62.
STURTEVANT, A. H., 1929, The claret type of mutant of *Drosophila simulans* : a study of chromosome elimination and of cell-lineage. *J. wiss. Zool.*, **135**, 323–356.
STURTEVANT, A. H., 1931, Known and Probable Inverted Sections of the Autosomes of *Drosophila melanogaster*. *Carneg. Inst. Wash. Pub.*, **421**, 1–27.
STURTEVANT, A. H., and DOBZHANSKY, T., 1930, Reciprocal translocations in Drosophila and their bearing on Œnothera cytology and genetics. *Proc. Nat. Acad. Sci.*, **16**, 533–536.
STURTEVANT, A. H., and PLUNKETT, C. R., 1926, Sequence of corresponding third chromosome genes in *Drosophila melanogaster* and *D. simulans*. *Biol. Bull.*, **50**, 56–60. (*See also* : Morgan.)
SUESSENGUTH, K., 1923, Über die Pseudogamie bei *Zygopetalum Mackayi* Hook. *Ber. Deuts. Bot. Ges.*, **41**, 16–23.
SUGIURA, T., 1925, Meiosis in *Tropæolum majus*, L. *Bot. Mag. Tokyo*, **39**, 47–54.
SUGIURA, T., 1927, Some observations on the meiosis of the pollen mother cells of *Carica papaya*, *Myrica rubra*, *Aucuba japonica*, and *Beta vulgaris*. *Bot. Mag. Tokyo*, **41**, 219–224.
SUGIURA, T., 1928, Cytological Studies on *Tropæolum* II. *Tropæolum peregrinum*. *Bot. Mag. Tokyo*, **42**, 553–556.
SUOMALAINEN, E., 1933, Der Chromosomencyclus von Macrosiphum pisi Kalt. (Aphididæ). *Zeits. Zellf. u. mikr. Anat.*, **19**, 583–594.
SUTTON, E., 1935, Half-disjunction in an association of four chromosomes in Pisum sativum. *Ann. Bot.*, **49**, 689–698.
SUTTON, W. S., 1902, On the morphology of the chromosome group in Brachystola magna. *Biol. Bull.*, **4**, 24–39.
SWESCHNIKOWA, I., 1928, Die Genese des Kerns im Genus *Vicia*. *Verh. V. Int. Kong. Vererb.*, **2**, 1415–1421.
SWESCHNIKOWA, I., 1929 a, *Vicia sativa* L. and *Vicia Cracca* L. *Ann. Timir. Ag. Acad.* (Moscow), **4**, 1–22.
SWESCHNIKOWA, I., 1929 b, Reduction Division in the Hybrids of *Vicia* (in Russian). *Proc. U.S.S.R. Cong. Genet.*, **2**, 447–452.
SWEZY, O., 1929, Maturation of the Male Germ Cells in the Rat. *J. Morphol.*, **48**, 433–443.

TÄCKHOLM, G., 1922, Zytologische Studien über die Gattung *Rosa*. *Acta Hort. Berg.*, **7**, 97–381.

TAHARA, M., and SHIMOTOMAI, N., 1927, Bastardierung als eine Ursache für die Entstehung der Chromosomenpolyploidie. I. Bastard zwischen *Chrysanthemum marginatun* und *C. lavandulaefolium*. *Sci. Rep. Tohoku Imp. Univ.*, Ser. 4, **2**, 293–299.

TAKAGI, F., 1928, The Influence of Higher Temperature on the Reduction Division of the Pollen Mother-Cells of *Lychnis Sieboldii*, van Houtte. *Sci. Rep. Tohoku Imp. Univ.*, Ser. 4, **3**, 461–466.

TAKAGI, F., 1935, Karyogenetical studies on Rye. I. A trisomic plant. *Cytologia*, **6**, 496–501.

TAKAMINE, N., 1927, Some observations on the chromosomes of *Najas major* All. *Bot. Mag. Tokyo*, **41**, 118–122.

TAKENAKA, Y., 1929, Karyological Studies in *Hemerocallis*. *Cytologia*, **1**, 76–83.

TAKENAKA, Y., 1930 a, On the Sex-Chromosomes of *Rumex montanus* Desf. *Bot. Mag. Tokyo*, **44**, 176–185.

TAKENAKA, Y., 1930 b, On the Chromosomes of *Lycoris squamigera* Maxim. *J. Nat. Hist. Soc. Korea*, **10**, 54–56. Abstr. *Jap. J. Bot.*, **5** (80).

TAKENAKA, Y., and NAGAMATSU, S., 1930, On the Chromosomes of *Lilium tigrinum* Ker. *Bot. Mag. Tokyo*, **44**, 386–391.

TAN, C. C., 1935, Salivary gland chromosomes in the two races of Drosophila pseudo-obscura. *Genetics*, **20**, 392–402.

TATUNO, S., 1933, Geschlechtschromosomen bei einigen Lebermoosen. I. *Bot. Mag. Tokyo*, **47**, 30–44. II. *Bot. Mag. Tokyo*, **47**, 438–445. III. *Bot. Mag. Tokyo*, **47**, 715–720.

TATUNO, S., 1936, Geschlechtschromosomen bei einigen Lebermoosen. II. *J. Sci. Hiroshima* B, II, **3**, 1–9.

TAUSON, A., 1927, Die Spermatogenese bei Asplanchna intermedia Huds. *Zeits. Zellforsch. u. mikr. Anat.*, **4**, 652–681.

TAYLOR, W. R., 1924 a, The smear method for plant cytology. *Bot. Gaz.*, **78**, 236–238.

TAYLOR, W. R., 1924 b, Cytological Studies on Gasteria I. Chromosome Shape and Individuality. *Amer. J. Bot.*, **11**, 51–59.

TAYLOR, W. R., 1925 a, The Chromosome Morphology of Veltheimia, Allium and Cyrtanthus. *Amer. J. Bot.*, **12**, 104–115.

TAYLOR, W. R., 1925 b, Cytological Studies on Gasteria II. A Comparison of the Chromosomes of Gasteria, Aloë and Haworthia. *Amer. J. Bot.*, **12**, 219–223.

TAYLOR, W. R., 1925 c, Chromosome constrictions as distinguishing characteristics in plants. *Amer. J. Bot.*, **12**, 238–244.

*TAYLOR, W. R., 1928, General Botanical Microtechnique (*Microscopical Technique*, ed. C. E. McClung). (Paul Hoeber.)

TAYLOR, W. R., 1931, Chromosome Studies on Gasteria III. Chromosome structure during microsporogenesis and the postmeiotic mitosis. *Amer. J. Bot.*, **18**, 367–386.

THOMAS, P. T., 1936, Genotypic control of chromosome size. *Nature*, **138**, 402.

THOMPSON, W. C., 1926 a, Chromosome behavior in a cross between wheat and rye. *Genetics*, **11**, 317–332.

THOMPSON, W. C., 1926 b, Chromosome Behaviour in Triploid Wheat Hybrids. *J. Genet.*, **17**, 43–48.

THOMPSON, W. C., 1930, Shrivelled endosperm in species crosses in wheat, its cytological causes and genetical effects. *Genetics*, **15**, 99–113.

THOMPSON, W. C., 1931, Cytology and Genetics of Crosses between Fourteen- and Seven-Chromosome Species of Wheat. *Genetics*, **16**, 309–324.

THOMSEN, M., 1927, Studien über die Parthenogenese bei einigen Cocciden und Aleurodiden. *Zeits. Zellforsch. u. mikr. Anat.*, **5**, 1–116.

*Timoféeff-Ressovsky, N. W., 1931, Die bisherigen Ergebnisse der Strahlengenetik. *Ergeb. midiz. Strahlenforsch.*, 5, 131–228.
*Timoféeff-Ressovsky, N. W., 1934, The Experimental Production of Mutation. *Biol. Rev.*, 9, 411–457.
Timoféeff-Ressovsky, N. W., 1935, Auslösung von Vitalitätsmutationen durch Röntgenbestrahlung bei Drosophila melanogaster. *Nachr. v. d. Gesellschaft der Wiss. z. Göttingen* n.f., 1, 163–180.
*Timoféeff-Ressovsky, N. W., Zimmer, K. G., and Delbrück, M., 1935, Über die Natur der Genmutation und der Genstruktur. Berlin, *Math.-Phys. Kl., Fachgruppe VI*, 1, 189–245.
Tinney, F. W., 1935, Chromosome structure and behavior in *Sphærocarpos*. *Amer. J. Bot.*, 22, 543–558.
Tischler, G., 1918, Untersuchungen über den Riesenwuchs von *Phragmites communis* var. *pseudodonax*. *Ber. Deuts. Bot. Ges.*, 36, 549–558.
*Tischler, G., 1922, Allgemeine Pflanzenkaryologie. *Hb. d. Pflanzenanatomie* II., Berlin.
*Tischler, G., 1927, Pflanzliche Chromosomenzahlen. *Tabulae Biol. Period.*, 4, 1–83.
Tischler, G., 1928, Über die Verwendung der Chromosomenzahl für phylogenetische Probleme bei den Angiospermen. *Biol. Zbl.*, 48, 321–345.
Tischler, G., 1929, Verknüpfungsversuche von Zytologie und Systematik bei den Blütenpflanzen. *Ber. Deuts. Bot. Ges.*, 47, (30)–(49).
Tischler, G., 1930, Über die Bastardnatur des persischen Flieders. *Zeits. f. Bot.*, 23, 150–162.
*Tischler, G., 1931, Pflanzliche Chromosomenzahlen. *Tab. Biol. Period.*, 7, 109–226.
*Tischler, G., 1935, Die Bedeutung der Polyploidie für die Verbreitung der Angiospermen. *Bot. Jahrbucher*, 67, 1–36.
*Tischler, G., 1936, Pflanzliche Chromosomen-Zahlen. *Tab. Biol. Per.*, 11, 281–304, and 12, 1–115.
Torvik, M. M., 1931, Genetic Evidence of biparental males in Habrobracon. *Biol. Bull.*, 61, 139–156.
Torvik-Greb, M., 1935, The chromosomes of Habrobracon. *Biol. Bull.*, 68, 25–34.
Trankowski, D. A., 1930, Leitkörperchen der Chromosomen bei einigen Angiospermen. *Zeits. Zellforsch. u. mikr. Anat.*, 10, 736–743.
Trankowski, D. A., 1931, Zytologische Beobachtungen über die Entwicklung der Pollenschläuche einiger Angiospermen. *Planta*, 12, 1–18.
Tschermak, E., 1930, Neue Beobachtungen am fertilen Artbastard Triticum turgidovillosum. *Ber. Deuts. Bot. Ges.*, 48, 400–407.
Tschermak, E. v. and Bleier, H., 1926, Über fruchtbare Ægilops-Weizenbastarde. *Ber. Deuts. Bot. Ges.*, 44, 110–132.
Tuan, H–C., 1931, Unusual Aspects of Meiotic and Post-Meiotic Chromosomes of *Gasteria*. *Bot. Gaz.*, 92, 45–65.
Turesson, G., 1930, Studien über *Festuca ovina* L. II. Chromosomenzahl und Viviparie. *Hereditas*, 13, 177–184.
Turesson, G., 1931, Über verschiedene Chromosomenzahlen in *Allium schoenoprasum* L. *Bot. Not. (Lund)*, 1931, 15–20.
Tuschnjakowa, M., 1929 a, Embryologische und Zytologische Beobachtung über Listera ovata (Orchidaceae). *Planta*, 7, 29–44.
Tuschnjakowa, M., 1929 b, Untersuchungen über die Kernbeschaffenheit einiger diözischer Pflanzen. *Planta*, 7, 427–443.
Tuschnjakowa, M., 1930, Über einen eigenartigen dreifachen Chromosomenkomplex in der Reduktionsteilung der Pollenmutterzellen von Humulus Japonicus. S. et Z. *Planta*, 10, 597–610.

UCHIKAWA, I., 1934, Genetisch-cytologische Studien an Weizenspeltoiden, I. Speltoide der C-Serie. *Cont. Lab. Genet. Kyoto*, **43**, 851-864.
UPCOTT, M. B., 1935, The cytology of triploid and tetraploid *Lycopersicum esculentum*. *J. Genet.*, **31**, 1-19.
UPCOTT, M. B., 1936 a, The origin and behaviour of chiasmata, XII. *Eremurus*. *Cytologia*, **7**, 118-130.
UPCOTT, M. B., 1936 b, The parents and progeny of *Æsculus carnea*. *J. Genet.*, **33**, 135-150.
UPCOTT, M. B., 1936 c, The Mechanics of Mitosis in the Pollen Tube of *Tulipa*. *Proc. Roy. Soc. B.*, **121**, 207-220.
UPCOTT, M. B., 1936 d, Timing unbalance at meiosis in the pollen-sterile *Lathyrus odoratus*. *Cytologia* (in the press).
UPCOTT, M. B., and LA COUR, L., 1936, The Genetic Structure of Tulipa, I. A Chromosome Survey. *J. Genet.*, **33**, 237-254.

VAKAR, B. A., 1930, Cytological investigation of hybrids between Triticum persicum Vav. and other wheat species. *Proc. U.S.S.R. Cong. Genet.*, 1929, **2**, 187-196 (in Russian).
VAKAR, B. A., 1935, Cytological analysis of wheat—couch grass hybrids (Russian, with Eng. summary, p. 27). *Omsk, Inst. of Agriculture*.
*VANDEL, A., 1927 a, La Cytologie de la Parthenogénèse Naturelle. *Bull. Biol.*, **61**, 93-125.
VANDEL, A., 1927 b, Gigantisme et Triploidie chez l'Isopode *Trichoniscus (Spiloniscus) provisorius* Racovitza. *C. R. Soc. Biol.*, **97**, 106-108.
VANDEL, A., 1934, La Parthenogénèse géographique, II. Les mâles triploides d'origine parthenogénétique : *Trichoniscus (Spiloniscus) elisabethæ* de Herold. *Bull. Biol.*, **68**, 419-463.
VILMORIN, R. de, 1929, Etude cytologique du *Solanum Commersonii*. *Arch. d'Anat. micr.*, **25**, 382-387.
VILMORIN, R. de, and SIMONET, M., 1928, Recherches sur le Nombre des Chromosomes chez les Solanées. *Proc. V. Int. Kong. Vererb.*, **2**, 1520-1536.
VRIES, H. de, 1918, Mass Mutations and Twin Hybrids in Oenothera grandiflora Ait. *Bot. Gaz.*, **65**, 377-422.
VRIES, H. de, 1919, Oenothera rubrinervis, a half mutant. *Bot. Gaz.*, **67**, 1-26.

WADA, B., 1933, Mikrodissektion der Chromosomen von *Tradescantia reflexa* (VM). *Cytologia*, **4**, 222-227.
WADA, B., 1935, Mikrurgische Untersuchungen lebender Zellen in der Teilung, II. *Cytologia*, **6**, 381-406.
WAKAKUWA, SH., 1931, A karyological study on the triploid hybrid of the genus *Celosia*. *Jap. J. Genet.*, **7**, 17-23.
WAKAYAMA, K., 1930, Contributions to the Cytology of Fungi. I. Chromosome Number in Agaricaceæ. II. Cytological studies in Morchella deliciosa. *Cytologia*, **1**, 369-388, **2**, 27-36.
WALTON, A. C., 1924, Studies on nematode gametogenesis. *Zeits. Zell. u. Geweb.*, **1**, 167-239.
*WANSCHER, J. H., 1934 a, The basic chromosome number of the higher plants. *New Phytol.*, **33**, 101-126.
WANSCHER, J. H., 1934 b, Secondary (Chromosome) associations in Umbelliferæ and Bicornes. *New Phytol.*, **33**, 58-65.
WARTH, G., 1925, Zytologische, histologische und stammesgeschichtliche Fragen aus der Gattung Fuchsia. *Z.I.A.V.*, **38**, 200-257.

WATKINS, A. E., 1924, Genetic and Cytological studies in Wheat I. *J. Genet.*, **14**, 129–171.
*WATKINS, A. E., 1930, The Wheat Species : A Critique. *J. Genet.*, **23**, 173–263.
*WATKINS, A. E., 1932, Hybrid Sterility and Incompatibility. *J. Genet.*, **25** 125–162.
WEBBER, J. M., 1930, Interspecific Hybridization in Nicotiana XI. The cytology of a sesquidiploid hybrid between Tabacum and sylvestris. *Univ. Calif. Pub. Bot.*, **11**, 319–354.
WEBBER, J. M., 1933, Cytological features of Nicotiana glutinosa haplonts. *J. Agric. Res.*, **47**, 845–867.
WEDEKIND, G., 1927, Zytologische Untersuchungen an Barrouxia Schneideri (Gametenbildung, Befruchtung und Sporogonie) zugleich ein Beitrag zum Reduktionsproblem (Coccidienuntersuchungen I). *Zeits. Zellforsch. u. mikr. Anat.*, **5**, 505–595.
WEINEDEL-LIEBAU, F., 1928, Zytologische Untersuchungen an *Artemisia*-Arten. *Jahrb. wiss. Bot.*, **69**, 636–686.
*WEISMANN, A., 1892, *Das Keimplasma :* eine Theorie der Vererbung. Jena.
WENRICH, D. H., 1916, The Spermatogenesis of Phrynotettix magnus, with special reference to synapsis and the individuality of the chromosomes. *Bull. Mus. Comp. Zool. Harvard*, **60**, 57–135.
WENRICH, D. H., 1917, Synapsis and chromosome organisation in Chorthippus (Stenobothrus) curtipennis and Trimeropteris suffusa (Orthoptera). *J. Morphol.*, **29**, 471–516.
WERNER, O. S., 1931, The Chromosomes of the Domestic Turkey. *Biol. Bull.*, **61**, 157–164.
WETTSTEIN, F. v., 1920, Künstliche haploide Parthenogenese bei *Vaucheria* und die geschlechtliche Tendenz ihrer Keimzellen. *Ber. Deuts. Bot. Ges.*, **38**, 260–266.
WETTSTEIN, F. v., 1924, Morphologie und Physiologie des Formwechsels der Moose auf genetische Grundlage, I. *Z.I.A.V.*, **33**, 1–236.
*WETTSTEIN, F. v., 1927, Die Erscheinung der Heteroploidie, besonders im Pflanzenreich. *Ergeb. Biol.*, **2**, 311–356.
WETTSTEIN, F. v., 1928 a, Über plasmatische Vererbung und über das Zusammenwirken von Genen und Plasma. *Ber. Deuts. Bot. Ges.*, **46** (32)–(49).
WETTSTEIN, F. v., 1928 b, Morphologie und Physiologie des Formwechsels der Moose, II. *Biblioth. Genet.*, **10**, 1–216.
WHITE, M. J. D., 1932, The chromosomes of the domestic chicken. *J. Genet.*, **26**, 345–350.
WHITE, M. J. D., 1933, Tetraploid spermatocytes in a locust, *Schistocerca gregaria*. *Cytologia*, **5**, 135–139.
WHITE, M. J. D., 1934, The influence of temperature on chiasma frequency. *J. Genet.*, **29**, 203–215.
WHITE, M. J. D., 1935, The effects of X-rays on mitosis in the spermatogonial divisions of Locusta migratoria L. *Proc. Roy. Soc. B*, **119**, 61–84.
WHITE, M. J. D., 1936 a, Chiasma-localisation in Mecostethus grossus L. and Merioptera brachyptera L. (Orthoptera). *Zeits. Zellf. u. mikr. Anat.*, **24**, 128–135.
WHITE, M. J. D., 1936 b, The chromosome cycle of Ascaris megalocephala. *Nature*, **137**, 783.
*WHITING, P. W., 1933, Selective fertilisation and sex determination in Hymenoptera. *Science*, **78**, 537–538.
WHITING, P. W., 1935, Sex determination in bees and wasps. *J. Hered.*, **26**, 263–278.

WHITING, P. W., and BENKERT, L. H., 1934, Azygotic ratios in Habrobracon. *Genetics* **19**, 237–267.
WHITNEY, D. D., 1929, The Chromosome cycle in the rotifer Asplanchna amphora. *J. Morphol.*, **47**, 415–433.
WILSON, E. B., 1905 a, Studies on Chromosomes I. The Behavior of the Idiochromosomes in Hemiptera. *J. Exp. Zool.*, **2**, 371–405.
WILSON, E. B., 1905 b, The Paired Microchromosomes, Idiochromosomes and Heterotropic Chromosomes in Hemiptera. *J. Exp. Zool.*, **2**, 507–545.
WILSON, E. B., 1909, Studies on Chromosomes V. The Chromosomes of Metapodius. A Contribution to the Hypothesis of the Genetic Continuity of the Chromosomes. *J. Exp. Zool.*, **6**, 147–205.
WILSON, E. B., 1911, Studies on the Chromosomes VII. A Review of the Chromosomes of Nezara; with some more general considerations. *J. Morphol.*, **22**, 71–110.
WILSON, E. B., 1912, Studies on Chromosomes VIII. Observations on the maturation phenomena in certain hemiptera and other forms, with considerations on synapsis and reduction. *J. Exp. Zool.*, **13**, 345–448.
*WILSON, E. B., 1914, The Bearing of Cytological Research on Heredity. *Proc. Roy. Soc.* B, **88**, 333–352.
*WILSON, E. B., 1928, *The Cell in Development and Heredity*. 3rd Ed. New York. (Macmillan.)
*WINGE, Ø., 1917, The chromosomes. Their numbers and general importance. *C.R. Trav. Carlsb.*, **13**, 131–275.
WINGE, Ø., 1922 a, A Peculiar Mode of Inheritance and its Cytological Explanation. *J. Genet.*, **12**, 137–144.
WINGE, Ø., 1922 b, Crossing over between the X- and the Y-chromosome in Lebistes. *J. Genet.*, **13**, 201–217.
WINGE, Ø., 1923 a, One-sided Masculine and Sex-linked Inheritance in *Lebistes reticulatus*. *J. Genet.*, **12**, 145–162.
WINGE, Ø., 1923 b, On sex chromosomes, sex-determination and preponderance of females in some dioecious plants. *C. R. Trav. Carlsb.*, **15** (5), 1–25.
WINGE, Ø., 1924, Zytologische Untersuchungen über Speltoide und andere mutantenähnliche Aberranten beim Weizen. *Hereditas*, **5**, 241–286.
WINGE, Ø., 1925, Contributions to the knowledge of chromosome numbers in plants. *Cellule*, **35**, 303–324.
WINGE, Ø., 1926, Das Problem der Jordan—Rosen'schen *Erophila*-Kleinarten. *Beitr. z. Biol. d. Pflanz.*, **14**, 313–334.
WINGE, Ø., 1929, On the nature of the sex chromosomes in Humulus. *Hereditas*, **12**, 53–63.
WINGE, Ø., 1930, On the occurrence of XX males in Lebistes, with some remarks on Aida's so-called " non-disjunctional " males in *Aplocheilus*. *J. Genet.*, **23**, 69–76.
WINGE, Ø., 1931, X- and Y-Linked Inheritance in *Melandrium*. *Hereditas*, **15**, 127–165.
WINGE, Ø., 1932, The nature of sex chromosomes. *Proc. 6th Int. Cong. Genet.*, **1**, 343–355.
WINGE, Ø., 1933, A case of amplidiploidy within the collective species *Erophila verna*. *Hereditas*, **18**, 181–191.
WINGE, Ø., 1934, The experimental alteration of sex chromosomes into autosomes and *vice versâ*, as illustrated by Lebistes. *C. R. Lab. Carlsberg, Ser. Physiol.*, **21** (1), 1–49.
WINGE, Ø., 1935, On haplophase and diplophase in some Saccharomycetes. *C. R. Carlsb.*, **21**, 77–112.
WINGE, Ø., 1936, Linkage in Pisum. *C.R. Carslb.*, **21** (15), 271–393.

BIBLIOGRAPHY

WINIWARTER, H. de, 1927. Étude du cycle chromosomique chez diverses races de *Gryllotalpa gryll*. (L). *Arch. Biol.*, **37**, 515-572.
WINIWARTER, H. de, 1931, Evolution de l'Hétérochromosome chez Tettigonia (Decticus) albifrons (Fab.). *Arch. d. Biol.*, **42**, 201-228.
WINIWARTER, H. de, 1934, La formula chromosomiale chez diverses races de chats (Felis domestica). *Acad. R. Belg., Bull. Cl. Sci.*, 5th ser., **20**, 512-518.
WINKLER, H., 1908, Über Parthenogenesis und Apogamie im Pflanzenreiche. *Prog. rei Bot.*, **2**, 293-454.
WINKLER, H., 1916, Über die experimentelle Erzeugung von Pflanzen mit abweichenden Chromosomenzahlen. *Zeits. f. Bot.*, **8**, 417-531.
WINKLER, H., 1921, Über die Entstehung von genotypischer Verschiedenheit innerhalb einer reinen Linie (Report of paper read to Deuts. Ges. f. Vererb.). *Z.I.A.V.*, **27**, 244-245.
WITSCHI, E., 1924, Die Entwicklung der Keimzellen der Rana temporaria L. I. Urkeimzellen und Spermatogenese. *Zeits. Zell. u. Geweb.*, **1**, 523-561.
*WITSCHI, E., 1929, Bestimmung und Vererbung des Geschlechtes bei Tieren. *Hdb. d. Vererbungswiss.*, **2**.
WITSCHI, E., 1935, The chromosomes of hermaphrodites. I. Lepas anatifera L. *Biol. Bull.*, **68**, 263-267.
WOLFSON, A. M., 1925, Studies on aberrant forms of *Sphærocarpos Donnellii*. *Amer. J. Bot.*, **12**, 319-326.
WOLTERECK, R., 1898, Zum Bildung und Entwicklung des Ostrakoden-Eies. Kerngeschichtliche und biologische Studien an parthenogenetischen Cypriden. *Zeits. Wiss. Zool.*, **64**, 596-623.
WOODRUFF, L. L., and ERDMAN, R., 1914, A normal periodic reorganisation process without cell fusion in Paramœcium. *J. Exp. Zool.*, **17**, 425-502.
WOODRUFF, L. L., and SPENCER, H., 1924, Studies on *Spathidium spathula*, II. The significance of conjugation. *J. Exper. Zool.*, **39**, 133.
WOOLSEY, C. I., 1915, Linkage of Chromosomes correlated with reduction in number among the species of a genus, also within a species of the Locustidæ. *Biol. Bull.*, **28**, 163-186.
WRINCH, D. M., 1936, On the molecular structure of chromosomes. *Protoplasma*, **25**, 550-569.

YAMAHA, G., 1926, Über die Lebendbeobachtung der Zellstrukturen, nebst dem Artefactproblem im Pflanzenzytologie. *Bot. Mag. Tokyo*, **40**, 172-197.
YAMAHA, G., 1927, Experimentelle zytologische Beiträge, III. Mitteilung. Über die Wirkung einiger Chemikalien auf die Pollenmutterzellen von *Daphne odora*, Thunb. *Bot. Mag. Tokyo*, **41**, 181-211.
YAMAHA, G., 1935, Über die pH-Schwankung in der sich teilenden Pollenmutterzelle einiger Pflanzen. *Cytologia*, **6**, 523-526.
YAMAHA, G., and SINOTO, Y., 1925, On the Behaviour of the Nucleolus in the Somatic Mitosis of Higher Plants, with Micro-chemical Notes. *Bot. Mag. Tokyo*, **39**, 205-226.
YAMAMOTO, Y., 1933, Karyotypes in *Rumex Acetosa* L. and their geographical distribution. *Jap. J. Genet.*, **8**, 264-274.
YAMAMOTO, Y., 1934 a, Reifungsteilungen bei einer asynaptischen Pflanze von Rumex acetosa L. *Botany and Zoology*, **2**, 1160-1168.
YAMAMOTO, Y., 1934 b, Karyogenetische Untersuchungen bei der Gattung Rumex I. Hetero- und Euploidie bei Rumex acetosa L. *Cytologia*, **5**, 317-336.

YAMAMOTO, Y., 1935 a, Karyogenetische Untersuchungen bei der Gattung Rumex. *Cytologia*, 6, 407–412.
YAMAMOTO, Y., 1935 b, Karyogenetische Untersuchungen bei der Gattung Rumex. II–III. *Jap. J. Genet.*, 11, 6–17.
YAMANOUCHI, S., 1908, Apogamy in Nephrodium. *Bot. Gaz.*, 45, 289–318.
YAMANOUCHI, S., 1912, The Life History of Cutleria. *Bot. Gaz.*, 54, 441–502.
YAMAURA, A., 1933, Karyologische und embryologische Studien über einige Bambusa Arten. *Bot. Mag. Tokyo*, 47, 551–555.
YARNELL, S. H., 1929, Meiosis in a triploid Fragaria. *Proc. Nat. Acad. Sci.*, 15, 843–844.
YARNELL, S. H., 1931 a, Genetic and Cytological Studies on *Fragaria*. *Genetics*, 16, 422–454.
YARNELL, S. H., 1931 b, A study of certain Polyploid and Aneuploid Forms in *Fragaria*. *Genetics*, 16, 455–489.
YASUI, K., 1921, On the Behavior of Chromosomes in the Meiotic Phase of some Artificially Raised *Papaver* Hybrids. *Bot. Mag. Tokyo*, 35, 154–168.
YASUI, K., 1927, Further Studies on Genetics and Cytology of Artificially Raised Interspecific Hybrids of *Papaver*. *Bot. Mag. Tokyo*, 41, 235–261.
YASUI, K., 1931, Cytological Studies in Artificially Raised Interspecific Hybrids of *Papaver* III. Unusual Cases of Cytokinesis in Pollen Mother-Cells in an F_1 Plant. *Cytologia*, 2, 402–419.
YASUI, K., 1933, Ethyl alcohol as a fixative for smear materials. *Cytologia*, 5, 140–145.
YASUI, K., 1935, Cytological studies in diploid and triploid *Hosta*. *Cytologia*, 6, 484–491.

ZIRKLE, C., 1928, Nucleolus in root tip mitosis in Zea mays. *Bot. Gaz.*, 86, 402–418.
ZIRKLE, C., 1931, Nucleoli of the Root Tip and Cambium of *Pinus Strobus*. *Cytologia*, 2, 85–105.

INDEX

Words in capitals appear in the glossary.
Page references to the text are in roman type, to tables in italics, to figures in clarendon, and to plates, in roman numerals.

Acanthocystis, 535
Acarina, 376
ACENTRIC FRAGMENT, see fragment.
Acer, 64, *242*
Acetabularia, 21
Acholla, 58, *368*
Aconitum, 37, 232 et sqq.
Acrididæ, 77
Acridium, **261**
Acrocladium, 222
Actinophrys, 15
adaptation, 474 et sqq. See also evolution.
Ægilotriticum, see Triticum-Ægilops
Æsculus, 56, *234*, *274*, *567*, VIII
univalents, 530
AFFINITY, 129
differential, 171, 185 et sqq., 198 et sqq., 328
terminal, 370, 517
See also attraction, pairing.
Agapanthus, *306*, *308*, *498*, 514, XV
centromere, 538
pachytene, IV
Agar, 575
Agaricaceæ, 82
Agave, 81, *218*
Aggregata, 28, 43, **48**, 82
centrosome, 533
chiasmata, 110
segregation, 249
Agropyrum, 218, *234*
Akemine, 523
Alchemilla, 437, 453
Alectorolophus, *145*
Aleurodes, 377
ALLELOMORPH, 334
Allen, 16, 357
Allium, hybrid, *162*
interchange, 151
interlocking, 256
localisation, 110

Allium—(contd.).
mitotic chromosomes, **37**, 39
pollen, 317
spindle failure, 400, *405*
spirals, *491*
terminalisation, 113
tetraploid, 119, *125*, 129
X-rayed, XII
Allium Moly, XII
odorum, 401, 439, 441
Schœnoprasum, 226
ursinum, 46
Allolobophora, XIV
ALLOPOLYPLOID, 183 et sqq., 248, *338*
evolution of, 329
secondary structural change, 277
ALLOSYNDESIS, 199 et sqq.
Aloë, *31*
Alstrœmeria, XV
ALTERNATION OF GENERATIONS, 6, 435 et sqq., 449
Alydus, 89, 142, 519
Amblystegium, 217, *448*
Amblystoma, *306*, 541
AMEIOSIS, 296, 379, **406**, 473
AMITOSIS, 47, 49
Amœba, 21, 50
Amorpha, *174*
ANAPHASE, forces, 531 et sqq.
meiosis, 113, **114**
mitosis, *23*, 28
multivalents, **130**
univalents, 410 et sqq.
Anas, 171, 535
Anasa, 373, 519
Ancyracanthus, 9, 45, 363
ANDROGENESIS, 442
Anemone, *110*, 227
ANEUPLOID, 61, 68, 126, *127*. See also monosomic, trisomic, tetrasomic.

Angiostomum, *404*
ANISOGAMY, 17
Annelida, 67, 469
Antennaria, *438*
antheridia, 436
Anthoceros, 223
Anthoxanthum, *159*, *161*
Antirrhinum, 319
Aphides, 373, 383
Aphis, 91, *373*, *376*
Apis (bee), 20, 376
Apodemus, *368*, *370*
APOGAMY, 435 et sqq.
APOMIXIS, 18, 434 et sqq.
APOSPORY, 435 et sqq., 448
Apotettix, 383
Aquilegia, *190*
archegonia, 437
Arphia, 313
artefact, 25, 34, 236, 565
Artemia, 221, 438 et sqq., *469*, 544
Artemisia, *438*, *452*
Ascalaphus, 58
Ascaris, centromeres, 537
 chromosome aggregation, 57, 59, 364
  number, 82, *83*
  persistence, 43
 diminution, 57, 329, 431, **432**
 non-pairing, *404*
 non-reduction, 41
 triploidy, 67
Ascomycetes, 6, 10, 248
asexual reproduction, 4, 271
Asilus, 358
Asplanchna, 377
ASYNAPSIS, 200, 291
Athyrium, 436
ATTACHMENT, of chromosomes, see fusion
 spindle, see centromere
 of X and Y, see *Drosophila melanogaster*
attachment chromomere, see centromere
attraction (specific), 481 et sqq.
 and crossing-over, 548
 in twos, 121
 primary, 121, 303, 492
 secondary, 178, 314, 493, 497
Aucuba, 38, **158**, *159*, *161*
Aulacantha, 82, *83*
auto-orientation, see orientation
autopolyploid, *125*, 183 et sqq., 226
 apomixis, 451

AUTOSOME, 356
 Drosophila, **371**, 372
AUTOSYNDESIS, 198 et sqq., *208*
Avena, conditional precocity, 314
 hybrid, *193*, 276
 triploid, 559
 trisomic, *319*
Avena sativa, 278
 asynaptic, 200
 non-pairing, 400 et sqq.
 unbalance, 323
 X-rayed, 429
Avery, **163**

Bacillus, *450*
BALANCE, 219, 240, 315 et sqq.
 secondary, 326 et sqq.
 and sex, 385
 See also theory.
BALANCED LETHAL, see lethal
 balanced variation, 61 et sqq.
Balanophora, *437*, *450*
Balbiani, 175
Baranetsky, *491*
BASIC NUMBER, 62, 75, 239
Basidiomycetes, 6, 8, 248
basidium, **8**
Bateson, 183, 572 et sqq.
Beadle, **283**, *284*, 397, 420
de Beaumont, **374**, **357**
bee, see *Apis*
Belar, 21, **31**, 43, 47, 480, **520**, 540, 565, XVI
Bellevalia, **31**, **92**
Belling, **31**, 156, 253, 569
Belostoma, 89
Benacus, 565
van Beneden, 532
Berberideæ, 84
Beta, 205
Betula, *208*
Bibio, 178
Biscutella, *46*
BIVALENT, 87, 106 et sqq.
 non-congression, **526**
 ring, **108**
 unequal, *161*, 247, 260, **261**
Blakeslee, *321*
Blaps, *145*, 363
blepharoplast, 9
Bombyx, 292, 370
 hybrid, 140
 mosaic, 448
 triploid, *216*
Bonellia, 15

INDEX

BOUQUET, see polarisation.
Boveri, 41, 315, **432**, 573
brachymeiosis, 263
Brachystola, *161*
Bragg, 522
Brassica, 66, 103
  basic number, 241
  hybrids, 192, *418*
  secondary pairing, 242
  See also *Raphanus*.
breakage ends, 555 et sqq.
  See also chromosome breakage.
breeding, as opposed to cytological method, 298, 302
bridge, 268
Bridges, 278, 285, 385, **386**
Brink, **262**
Brochymene, 366
Bromus, 215, *403*
Bryophyta, 6, 246, 248, 356
Buller, 8, 248, 575

Callitriche, 82, *227*
callus, 63
Camarophyllus, 442, **457**
Cambarus, 82, *83*
cambium, 84
Camnula, 314
Campanula persicifolia, VI
  haploid, *444*
  interchange, 150, *159*, *161*
  interlocking, 256
  non-reduction, 192
  terminalisation, **104**, **506**, *513*
Canna, 217, 308
Cannabis, 66, *405*
Cardamine, 78, *214*
Cassia, 228
Catcheside, 241, 354, 557
CELL, 2
  elimination, 555
  glandular, 175
  multinucleate, 65
  selection, 430
cell-size, 84, 221, 515
cell-wall, action on spindle, 532
  and centromere, 540, **545**
  embryo-sac, 437
centric reaction, 524
centric segment, 69
centrifuging, 19, 20, 46, 571
  spindle, 522
CENTROMERE, **23**, 26, 28 et sqq., 315, 535 et sqq., **562**, XV
  abnormal division, 411

CENTROMERE—(*contd.*).
  in *Ascaris*, 58
  attractions, 498
  branching of chromosome, 74
  cohesion, 552
  and crossing-over, 260, 551
  cycle, **486**, 561
  and differential segments, **346**, **369**
  doubleness, 431
  effect on condensation, 308 et sqq.
  equilibrium, **528**, **529**
  function, 540
  and interchange, 350
  interphase, **117**
  inversion relative to, 265
  leptotene, 90
  in multivalents, 129
  and non-disjunction, 153
  pachytene, **98**, XV
  pairing, 372
  permanence, 273, 424
  priority, 552
  repulsion, 152, 235, 519
  salivary glands, **176**, 177
  second metaphase, **117**
  spindle reaction, 193, 530
  in structural change, 69
  synchronisation, **72**, 396
  and terminalisation, 105 et sqq., 507 et sqq.
  trivalents, **529**
  See also chromatid and chromosome, dicentric and acentric.
CENTROSOME, 9, 26, **49**, 523
  continuity, 533 et sqq.
  division-cycle, 399, **538**
  and parthenogenesis, 464
  X-rayed, 427
centrosphere, **49**, 534
Ceresa, 373
Chambers, 43, 521
Chara, 435, 442, 458, *469*
Charophyta, 6
Chiarugi, **219**, 438 et sqq., 452
CHIASMA, between sex chromosomes, 370
  classification, **504**
  coiling at, 95 et sqq., **549**
  comparate, 116, **255**, 268 et sqq.
  complementary, 268
  and crossing-over, 252 et sqq., 550
  disparate, 116, **255**, 268 et sqq.
  failure of, 405 et sqq.

CHIASMA—(contd.).
 frequency, 99, 129, 398
  of fragments, 144, *146*
  in hybrids, 138, 400
  of polyploids, 218 et sqq.
  statistics, **169, 290, 294**
  after X-raying, 429
 imperfect, 495
 interchange heterozygote, 151
 interstitial, 105, 131, 495
  in sex chromosomes, 366
 lateral, **142,** 148, **261,** 262, **504,** 558
 localisation, 99 et sqq., *110,* 289, 391
 at mitosis, 397
 movement, see terminalisation.
 multiple, **142,** 350
 origin, 550
 in polyploids, 120 et sqq.
 position, **147,** *110*
 pseudo-, 421, 426, 433, 527, XII
 random, *110*
 reciprocal, 268, **499**
 terminal, 105, *110,* 170, 495, **504,** 517
 theory of pairing, see theory.
  See also terminalisation, crossing-over.
CHIASMATYPE THEORY, see theory.
Chilodon, 21
CHIMÆRA, 71, 443
Chironomus, 175 et sqq., *178,* VII
Chlamydophrys, 21
Chlorophyceæ, 6
chloroplasts, 221
Chondrilla, 438
Chorthippus, **91, 95,** III, IX, XI
 inversion, 265
 mitosis, **53**
 non-pairing, *402*
 relational coiling, 260
 See also *Stenobothrus.*
CHROMATID, 23
 acentric, 420 et sqq., 487, 524
 breakage, 425 et sqq.
 bridge, see dicentric.
 coiling, see coiling.
 dicentric, 268, 420 et sqq., IX, XII
 at diplotene, 549 et sqq.
 interference, see interference.
 interlocking, **116**
 at meiosis, 100 et sqq., **114**
  anaphase, 130

CHROMATID—(contd.).
 non-disjunction, see non-disjunction.
 ring, IX, XII
 segregation, 285 et sqq.
 sister, **117**
  See also crossing-over.
CHROMATIN, 570
 diminution, 329
 nucleoli, 22, 311
CHROMOCENTRE, 45
CHROMOMERE, 24, 87, 423, 566
 salivary gland, 177, VII
CHROMOSOME, 22, **23**
 acentric, 71, 424 et sqq., 541, *554*
 aggregation, 57, **59**
 arm, 35
 branched, 74, 424, *554*
 breakage, 425 et sqq., 547 et sqq.
 clonal, 478
 coiling, see coiling.
 complement, 40
 compound, 364
 constants, 52
 dicentric, 71, 424 et sqq., *554,* IX, XII
 differential condensation, 307 et sqq.
 diminution, 57, 431
 direction of split, 75
 distribution, 524
 division, of, 47, 425, **486,** 501 et sqq., 549
 elimination of, 429
 individuality, 40 et sqq., 60
 inert, 63, 329 et sqq., 560
  sex chromosomes, 177, 370 et sqq., 391
 length, 56
 linear differentiation, 25, 42, 245, 332
 natural types, **37**
 nucleolar, 40
 number, 60 et sqq., 220
 polycentric, 537, *554*
 qualitative differentiation, 167, 316
 reunion, 557 et sqq.
 rigidity, 521, 530
 ring, see ring chromosome.
 semi-mitotic, 408
 set, 60, 207
 size, 55, 78 et sqq., 112, 566
  genotypic control of, **399**
  mutation, 54

INDEX

CHROMOSOME—(contd.).
  structural differentiation, 303 et sqq.
  structure, 32
  supernumerary, 142
  survival, 71
  theory, see theory.
  thread, 22, 34
  unpaired, 98, 376
  vesicle, 45, 89
Chrysanthemum, clonal forms, 319
  hybrids, 171, *193*, *208*
  non-pairing, *404*
  non-reduction, 192, 416
  polyploidy, *191*
  syndiploidy, 66
Chrysopa, 367, **374**
Circotettix, 77, 247, **261**
Citrus, 217, 440
  somatic doubling, 66
Cladocera, 471
Cladophora, 6, *111*
cleavage, 41, 45
  division, 533, 541
  surface, 484
CLONE, 218, *319*, 323
  old, 421
  See also species.
Coccidæ, 377, 379, **381**
  spindle, 543
Cocconeis, 442, 450
Cochlearia, 220
coefficient, of hybridity, 271
  terminalisation, 505, *513*
COILING, direction, 488
  relational, chromatid, 25, 27, 271, 483
  chromosome, 93 et sqq., 100 et sqq., 166, **177**, 428
  and crossing-over, 259, 548 et sqq.
  in salivary glands, **177**
  and structural change, 558
  See also spiral, spiralisation.
Coleoptera, 367
Coleotrype, 80
Commelina, 80
comparative method, 563 et sqq.
competition, in chiasmata, 402
  between embryo sacs, see Renner effect.
  in nucleolus-formation, 305
  in pairing, 199, **206**
  in spindle formation, 523
  between zygotes, 318 et sqq.

COMPLEMENT, see chromosome, balance.
complex heterozygote, see heterozygote.
complex hybrid, see hybrid.
CONDENSATION, see differential condensation.
CONGRESSION, 523
conjecture, see theory.
CONJUGATION, of gametes, see fertilisation.
  distance, *367*, **374**, 375, *528*
conjugate nuclei, 10
CONSTRICTION, 34 et sqq., *536*
  nucleolar, **23, 27, 59**, 305, 485
CO-ORIENTATION, see orientation.
Coregonus, XIV
correlation, negative, chiasma frequency, 293
  fertility, 196
Correns, 384
Cotoneaster, 242
Cratægus, 212, 217
Crepis, autosyndesis, 202, 208
  chromosome length, 56
  chromosome numbers, 220, **231** et sqq.
  haploid, *444*
  hybrids, 184 et sqq., 276
    chromosome size, 57
    pairing in, **163**, *172*, *174*, *208*, *400*
    trabants, 305
  inversions, *274*
  non-pairing, *400*
  nucleoli, *306*
  phylogenetics, 172
  progeny of triploids, *320*
  somatic doubling, 64
  trisomic, *319*
Crepis artificialis, 242
Crepis capillaris, chromosome shapes, **44**, I
  chromosome length, **57**
  chromosome number, 82
  haploid, *416*, *418*
  polyploidy, **62**, *191*
  population analysis, 70
  syndiploidy, 66
  X-rayed, 430
  X-strain, *400*, *404*
Crepis capillaris × *tectorum*, 199
Crepis tectorum, chromosome shapes, **44**
  old seed, 420
  population analysis, 70

Crepis tectorum—(contd.).
  ring chromosomes, **73**
  tetraploid, 222
  trisomic, 69
Crocus, 76, 82, *160*, XII
  chromosome numbers, **231** et sqq.
  polyploidy, **228**
CROSSING-OVER, 87, 245 et sqq., 250
  with apomixis, 477
  complementary, **255**
  dyscentric, 268 et sqq.
  four-strand, 285
  frequency, 281
  illegitimate, 275
  inference, 249, 298
  and internal spirals, 488
  map, **180**, **283**, 293 et sqq.
  mechanism, 547 et sqq.
  non-homologous, 494
  reciprocal, **255**, 370
  in sex chromosomes, **359**, 389 et sqq.
  somatic, 553
  suppression, 333, 337
  theory of, see theory.
  unit, 333
  and X-rays, 553
Cruciferæ, *306*
Cryptobranchus, 45
Cryptomonas, **48**
crystal, liquid, 522, 546
Culex, 110, 376
Cutleria, 442
Cyanophyceæ, 2
Cyanotis, **80**
Cycads, 533
Cycas, *115*
Cyclops, 45, 522
Cydonia, VIII
Cynips, 450
Cyperaceæ, 7
Cypris, 450
CYTOPLASM, 2, 193, 520

Dactylis, *125*
Dahlia, 235, VIII
  secondary pairing, **238**
  polyploidy, **239** et sqq., 326
Daphnia, 436, *469*
Dark, **92**, **96**, **403**
Dasychira, 78, 82
Dasyurus, 94
Datura, ameiosis, 399 et sqq.
  fragment, 144

Datura—(contd.).
  haploid, *416*, *444*, 447
  interchange, 150 et sqq.
  monosomic, 316
  old seed, 420
  pollen abortion, *321*
  progeny of triploids, *320*
  somatic doubling, 64
  tetraploid, *185*, *194*
  trisomic, *61*, 148, *319*
DEFICIENCY, **59**, 69, 431, 554 et sqq.
  heterozygote, **152**, *161*, 260
  inert, 330
Delaunay, **57**
DELETION, **181**, 554 et sqq.
Delphinium, *191*
Dendrocœlum, 31, 256
Dermaptera, *367*
detachment (sex chromosomes), 370
DIAKINESIS, 87, 101, 508
Dianthus, 228
  chromosome size, 55, 84
  hybrids, *208*
diatom, **49**
Dicentra, 242
DICENTRIC CHROMATID, see chromatid.
DICENTRIC CHROMOSOME, see chromosome.
Dicranura, 140, 163
DIFFERENTIAL AFFINITY, see affinity.
differential condensation, 98, **282**, 307, 500
DIFFERENTIAL SEGMENT, 264, 390
  Œnothera, 338, **346**, **347**, 349 et sqq.
  Pisum, 263
  sex chromosomes, 365
  Zea, 262
differentiation, 3
  between chromosome sets, 227
  qualitative, 197, 245, 316 et sqq.
  of sex chromosomes, 389 et sqq.
  See also chromosome.
diffuse stage, *94*
Digitalis, haploid, 201, *444*
  hybrids, *172*, *173*, *185* et sqq., *208*
Digitalis mertonensis, *185*
Dinophilus, 15
diœcism, 18, 335
DIPLOCHROMOSOME, 427, 531, 554

INDEX

DIPLOID, 5, 6
  constitution, 61
  functional, 184 et sqq., 214
DIPLOIDISATION, **8, 10**
DIPLOPHASE, 10
DIPLOTENE, 87 et sqq., **259, 287,** 550 et sqq.
Diptera, 7, 367, 497
  salivary glands, 175 et sqq.
DISJUNCTION, 113 et sqq., 132
distribution, see chromosomes.
Dobzhansky, 371, **386**, 389, 398, **410** et sqq.
donation, 430
Dorstenia, 402
Draba, 227
  See also *Erophila*.
Drosera, 83, 417
  hybrid, 205
Droseraceæ, 79, 84
Drosophila, VII
  abnormal meiosis, 372
  coiling, 492
  crossing-over, 250, 252, 289 et sqq
  heterochromatin, **309**
  intersexes and supersexes, 385
  inversions, 265, 272 et sqq.
  irradiation, 422 et sqq.
  nucleoli, 305, 306
  salivary glands, **176, 177,** *178,* **180, 181**
  sex chromosomes, 358 et sqq.
  somatic pairing, 236, 497 et sqq.
Drosophila funebris, 330
Drosophila melanogaster, attached X, 285
  attached XY, 277
  bar eye, 333
  cell size, 221
  chromosome map, 295
  diminution, 58
  gynandromorph, 68
  interchange, *160*
  interference, 551
  monosomic, 316
  mosaic, 68, 425, 448
    haploid, 385
  nucleoli, 40
  polyploid cells, 63
  reduplications, 317
  ring chromosome, 74
  sex chromosomes, 330
  tetraploidy, 66
  triploidy, 67, 216, 284
  trisomic, *319*

Drosophila melanogaster—(*contd.*).
  types of complement, **386**
  viability of deficiencies and reduplications, *321*
Drosophila miranda, 367, 389
Drosophila pseudo-obscura, 79
  geographic races, 181
  hybrid, **181**
  meiosis, **371**
  non-pairing, 400, *405*
  nucleoli, 40
  polymitosis, 398, 535
  sex chromosomes, 181
  tetraploid meiosis, *125*
Drosophila simulans, 179, **180,** 388
  abnormal mitosis, 396
Drosophila virilis, 388
Drosophyllum, 83
DUPLICATION, 141, 181, 272, 278
  and balance, 327
  buffer effect, 317
  haploid, 398
DYSCENTRIC HYBRID, see structural hybrid.
Dzierzon, 249, 376

Echinodermata, 434, 466
  double fertilisation, 64
  polyploidy, 63
Ectocarpus, 17
egg-cell, 7, 26, 94, 103
  unreduced, 189 et sqq.
Elatostema, 451
electric field, 522
electric potential differences, 561
electric surface charge, 315, **499** et sqq., **503,** 519 et sqq.
  of sex chromosomes, 375
  and spiralisation, 560
elimination, of cells, 555
  of gametes, 219, 316 et sqq.
  at mitosis, 555
  of zygotes, 240, 318 et sqq.
  See also chromosome.
embryonic membranes, 50, 63
EMBRYO-SAC, 10 et sqq., **11,** 437, **454**
  competition, 340
  differential viability, 316 et sqq.
Empetrum, 67, *242*, 364
Enchenopa, 313, 368
endogamous group, 478
endomixis, 50
endosperm, *11*, 50, 62, 67, 438, 441
  X-rayed, 431

ENVIRONMENT, 12
epigenesis, 324
Epilobium, 535
equational exceptions, 285
equational separation, 252, **261**
equilibrium, between chiasmata, 514
  chromatid, 497, **499**
  hybridity, 271
  metaphase plate, 497, 524, **529**
  molecular, 561
  pachytene, 549
  repulsion, 235, 531, 546
Eremurus, *111*, *258*, 306, III
Erlanson, 224, 460, **508**
Ernst, 434
Erophila, *190*, 228
EUCENTRIC HYBRID, see structural hybrid.
Euchlæna, *173*, *191*, *446*
  See also *Zea*.
euchromatin, see heterochromatin.
Eucomis, *81*
Euglypha, 21, *31*, 49
Euphorbia, **238**
Euschitus, 358
evolution, and apomixis, 470
and chromosome numbers, 560
and chromosomes, 75 et sqq.
of genes, 328, 332
and haplo-diploidy, 379
and meiosis, 134
and mutation, 246
and permanent hybrids, 337
and polyploid species, 220 et sqq.
and secondary polyploids, 243
and sex-determination, 388 et sqq.
Evotomys, 368

FACTOR, 245
See also gene.
Felis, 78
female, 18, 356 et sqq.
female gamete, see egg-cell.
ferns, see Pteridophyta.
fertility, and crossing-over, 291
in *Œnothera*, 338
of tetraploids, 186
of triploids, 319
See also sterility.
FERTILISATION, 5
double, 11, 64
failure of, 434 et sqq.
Festuca, 228
interchange, 157

Festuca—(*contd.*).
polyploid, *191*
See also *Lolium*.
Feulgen reaction, 20, 46, 307, 312
Filaria, 9, 45
Fischer, 563
fixation, staining reaction, 307, 315, 563 et sqq., 569
Flagellata, 4
flagellum, 9
Flemming, 22, 574
forces, 482 et sqq.
  inter-molecular, 493
Forficula, 161
Fourcræa, 306
Fragaria, haploid, *445*, *446*
  hybrid, *190*, *193*, 209
  interchange, 157
  sex, 358, 364
fragment, *27*, *71*, **72**, *73*, *143*
  acentric, 268, 420 et sqq.
  elimination, 322 et sqq.
  inert, 330
  pairing of —s, *143*, 287, 519
  supernumerary —s, 276
fragmentation, 59, *70*, *77*, **559**
  heterozygote, *140*, *162*
Fritillaria, V
  coiling, 487, *492*
  complement, 77
  crossing-over, 260
  diplotene, **259**, 287, *552*
  fragments, *72* et sqq., *143*
  pairing of, *146*
  interrupted pairing, **100, 102**
  meiosis, 92 et sqq.
  nucleoli, **98**, *306*
  pollen grains, *396*, *525*
  structural change, 560
  terminalisation, **506**
  triploid, *216*
Fritillaria imperialis, **98, 143, 287**
Fritillaria Meleagris, *110*
frog, see *Rana*.
Frolowa, **176**
Frullania, 314, *366*
Fujii, *101*, 485
Funaria, 466
  cell size, *221* et sqq.
  polyploid, 69, *195*
fungi, mitosis, 47
  spindle, 543
fusion, of chromosomes, 77, 239, 363, **559**
  heterozygote, *140*, *162*

INDEX

fusion—(*contd.*).
  of nuclei, see fertilisation and diploidisation.
  nucleus, 12, 438

Gagea, *418*
Galeopsis, **128,** *190*, *234*
Gallus, *368*
Galtonia, *306*, XV
GAMETE, 5
  deficient, 316 et sqq.
  nucleoli, 21
  unreduced, 157, 188, 296, 414 et sqq.
gametic reduction, 6
GAMETOPHYTE, 6
  differential viability, 316 et sqq.
gamma rays, 422
Gasteria, interphase, 118, 487, *492*
GENE, 167, 179, 245
  differentiation of, 303
  division, 501
  evolution, 332
  inert, 331 et sqq.
  mutable, 422
  mutation, 422 et sqq.
  theory of, see theory.
generative nucleus, 8, 10
genetic isolation, 79
GENOTYPE, 12
GENOTYPIC CONTROL, 53, 164
  of apomixis, 470 et sqq.
  of crossing-over, 288 et sqq.
    between sex chromosomes, 390
  of meiosis, 399 et sqq.
  of mitosis, 395 et sqq.
  of mutation rate, 332
  of nucleolar organiser, 305
  of structural change, 419
gentian violet, 565
germ-plasm, 13
Giemsa, 565
gigantism, 222 et sqq.
Godetia, *159*
Goldschmidt, 384
Gomphocerus, 65
Gossyparia, 310, **381**
Gossypium, 56, *191*, 243
  twin haploid, *444*
Gregarina, 534
growth rate, differential, 325
Gryllotalpa, 367
Gryllus, 78
Guignard, 11, 534
Gustafsson, 438, 465, 475

Gymnosperms, 79, 84
gynandromorph, 68

Haberlandt, 441, 465
Habrobracon, 377, **378,** 459
  cell-size, 222
  mosaic, 448
  triploid, 216
hæmatoxylin, 565
Haldane, 372, 391, 555
half-mutant, see *Œnothera*.
HAPLO-DIPLOID SEX DETERMINATION, see sex determination.
haplo-diploid twin seedlings, 447
HAPLOID, 5, *61*
  chiasmata, 254
  crossing-over, 262, 278
  differentiation (of sex), **336**
  mosaics, 385
  origin, *444* et sqq.
  pairing in, 328
  pollen grains of, **195,** *416*
  polyhaploid, 199 et sqq.
  progeny, *417*
  size, 222 et sqq.
  somatic doubling, 64
haplophase, 10
Hardy, 493, 522, 563
Hartmann, 17
Heitz, **40,** 176, 307, **309,** 330
Helianthus, *214*
Helix, 469
Hemerobius, 57, XI
Hemerocallis, *125*, *217*
Hemiptera, 366, *368*, 376
Hepaticæ, *310*, 366
HEREDITY, 1, 14, 224
hermaphroditism, 17, 18
Hertwig, O., 5, 85
Hesperotettix, 78, **141,** 364
HETEROCHROMATIN, 307, 330, 358
  See also precocity.
HETEROGAMY, 18
Heteroptera, *368*
HETEROPYCNOSIS, 309
heterosis, 136
heterothallism, 17
HETEROZYGOTE, 136 et sqq.
  complex, 335 et sqq.
  fragmentation, 140
  See also hybrid.
HETEROZYGOUS SEX, 136
Hevea, *404*, 405

HEXAPLOID, 62, 184, 187
Hieracium, 438 et sqq.
  apomixis, 473
  fragment, *145*
Hollingshead, **44, 62**
HOMOLOGY, 328, 567
  change of, 115, 199, 499
    and terminalisation, 510, *511*
  and pairing, 164
Homoptera, 368
HOMOZYGOTE, *Œnothera*, 342
Hordeum, 421
Hosta, 523, 524
Humulus, 358 et sqq.
Hyacinthus, IV
  abnormal pollen, 396
  chiasmata, **127**, 128
  chromosome size, 82, *83*
  clones, *319*
  mitotic chromosomes, **37**
  polyploid meiosis, 119
  triploid, *216*
    pollen, 318
  unbalance, 325
HYBRID, classification, 137
  complex, 138
  fertility of, *187* et sqq., **196**
  intersexual, 384
  at mitosis, 44, 45
  numerical, 137
  permanent, 335
  pollen grains of, **195**
  polyploid, 138, 183 et sqq., 208 et sqq.
  structural, see structural hybrid.
  in systematics, 135
  twin, 339
  undefined, 137, 168 et sqq.
  See also heterozygote.
hybridisation, 17
hybridity, and apomixis, 440, 467, 472
  coefficient, 271
  equilibrium, 271
  and non-reduction, 297
  of old clones, 421
hydration, of nucleus, 544
  of spindle, **520**
hydrogen ion concentration, 32, 547, 561
Hymenoptera, polyploid cells, 63
  salivary glands, 175
  segregation, 248
Hypericum, 160, 338
hypothesis, see theory.
hysteresis (lag), 483, 550

Iberis, 235
Icerya, 6, 24, 311, 397, 449
  sex, 377
Iguanidæ, 65
Impatiens, 405
inbreeding, 279, 378
incompatibility, 12, 17
inertness, see chromosome, gene.
inference, of crossing-over, 249, 298
  of forces, 482 et sqq.
  of movement of chromosomes, 503 et sqq.
infra-red photography, 570
Infusoria, 50
insertion-region, see centromere.
INTERCHANGE, 69, **124**, 555 et sqq.
  and chromosome form, 59
  heterozygote, **149, 150**, 263, 286 et sqq., VI
  internal, 156
  Œnothera, 342
  sex chromosomes, 360
  time of, **71**
  Zea Mays, **282, 304**
INTERFERENCE, between chiasmata, 292
  between chromosomes, 293, 551
  between crossings-over, 252, 287, 292
  chromatid, 292, 551
  at zygotene, 165
interlocking, at meiosis, 156, 255
  at mitosis, 74, 502
INTERPHASE, **117**, 118, 487
intersex, 384
interstitial chiasma, see chiasma.
interstitial segment, 263
INVERSION, 69, 554 et sqq.
  crossing-over in heterozygote, 265 et sqq.
  and the gene, 333
  heterozygote, **152**, IX
  reversal, 280
  salivary glands, 180, 181
irradiation, see X-ray.
iso-electric point, 315, 561
ISOGAMY, 18
Ixeris, 438, **454**

Jamaicana, 140
Janssens, 250, 253, 573
Johannsen, 245, 576
Jørgensen, 195, 442

INDEX

Karpechenko, *187, 192, 223, 418*
karyosphere, 94
Kaufmann, 277
kern-plasma ratio, 20
Kihara, **204**, *361, 411, 572*
Klingstedt, XI
Kniphofia, *175*, *402, 409*, XIV
  somatic doubling, 66
Koller, **177, 179, 181**, VII, XI, XIV
Kuwada, **483**, *533, 570*, XIII

Lacertilia, *65*
Lachenalia, 115
La Cour, **232**, *569, 570*, II, III, IV, V, VI, VIII, XII, XV
lagging, *115*
Lamarck, 52
Lamm, **407**
Lathyrus, 110, 111, 195, 405
  terminalisation, **512**
Lawrence, 237, VIII
Lebistes, *333, 388*
Lecanium, *310, 380, 450*
Leguminosæ, 281
Lens esculenta, *446*
Leontodon, *437, 450*
Lepas, 55
Lepidoptera, 78, 91, 171 et sqq.
  sex chromosomes, *358*
Leptophyes, *358*
LEPTOTENE, 87 et sqq., 501
Lesley, *319, 404*
LETHAL, balanced, 339
  cell, 555
  factor, Œnothera, 354 et sqq.
  through unbalance, *322*
Lewitsky, 37, 83, *430*
life-cycle, 3 et sqq., 435 et sqq.
  and elimination, 318 et sqq.
Liliaceæ, 67
Lilium, 9
  chromosome number, 84
  embryo-sac, *437*
  interlocking, *258*
  inversions, *274*
  length of chromosomes, *31*
  minor spirals, *487*
  pachytene, 93
  triploid, *125, 126*
Lilium Henryi, 145
  tigrinum, 125, 126
Lillie, 503
Limnophilus, *368*
Lindegren, 248
linear arrangement, see chromosome.

linkage, 245 et sqq.
Liriope, *418*
living cells, 22, 27, XV
Llaveia, **381**, *519*
LOCALISATION, see pairing, chiasma, zygotene.
Locusta, III
  fragment, *145*
  X-rayed, *427*
Lolium, chromosome size, 55, **399**
  hybrid, *173, 191*
Lucretius, 1, 244
Lychnis, *418*
Lycia, 162
Lycopersicum esculentum, see *Solanum lycopersicum*.
Lycoris, 216
Lygæus, *358, 519*
Lymantria, *384, 388*
Lythrum, 228

McClintock, 166, 265, 276, **304** et sqq., *431*
McClung, **141**, *363*
Macronemurus, *111*, **357**, *512*
MACRONUCLEUS, 50
Macropus, *358, 364*
Mæda, **169**
maize, see *Zea Mays*.
male, 18, 356 et sqq.
  gamete, see spermatozoon and generative nucleus.
  haploid, *376*
Mammalia, *368*
Man, 55, *83, 368*, XI
Manton, 46, *449*
Marchal, *448*
Marsilia, *436*
Marsupials, *93, 310*
Mather, **76, 231, 257**, *292*, **426**
MATRIX, *566, 578*
Matthiola, **59**, *111*, *146*
  deficiency, *316*
  haploid, *416*
  non-pairing, 400 et sqq.
  trisomic, *148, 319, 321*
mechanics, 479 et sqq.
  of cell-wall, **541**
  external, of chromosomes, **492** et sqq.
  internal, of chromosomes, **483** et sqq.
  of meiosis, 132
  ultra-, **547** et sqq.
Mecoptera, *367*

Mecostethus, 261, XIV
  chiasma frequency, 294
  localisation, 99, *110*
  precocity, 309
  spiralisation, 28
MEGASPORE, see embryo-sac.
MEIOSIS, 5, 88 et sqq.
  abnormal, 372, 380, 460 et sqq.
  displaced, 377, 397, 448
  first metaphase, 106 et sqq.,
    525 et sqq., IV, VI, VIII, XIV
  and parthenogenesis, 451 et sqq.
  theory of, see theory.
  X-rayed, 428
Melandrium, 367, X
  chromosome length, 56
  chromosome size, 55
Melanoplus, *110*, 261, 292, XVI
Meleagris, 81
Melopsittacus, 68
Mendel, 135, 245, 250
Mendelian principles, 138, 247, 335
Menexenus, 364
Mermiria, 140, 364
Mespilus, *242*
METAPHASE, of meiosis, 106 et
    sqq., 525 et sqq.
  of mitosis, 23
  plate, 28, 235 et sqq., 497, **523**, **525**
    accessory, **526**
Metapodius, 142, 310, 313, *368*, 392
Metzner, **536**
Meurman, **158**
Miastor, 58
MICROCHROMOSOME, see
    fragment.
microdissection, 35, 43
MICRONUCLEUS, 19, 50
microscopy, 569
MICROSPORE, see pollen.
Miltochrista, 78
mitochondria, 9
MITOSIS, 2, 22, **23**
  abnormal, 395 et sqq., 420
  duration of, 30
  failure of, 63, 395 et sqq.
  gonial, 28
  and meiosis, 134
  multipolar, 64
  pollen-grain, II, III
  spirals at, XIII
  X-rayed, 427, XII
mitotic constants, 52
Mitrastemon, 94
Moffett, **238**

molecular change, see gene.
molecular orientation, 546
molecular spiral, see spiral.
molecular structure of spindle, 532,
    545
Mollusca, 7, 15, 67, 469
monaster, *535*, 543
Monocystis, **48**
MONOSOMIC, 61, 316
  Zea, X
Morgan, 250, 575
MOSAIC, 12, 68, 246
  haploid, 385
  sex, 385
  X-ray, 425
Mosses, see Bryophyta, *Funaria*.
MOTHER-CELL, 5
Muller, 252, 286, 330 et sqq., 339,
    387, 422
multipolar mitosis, see mitosis.
MULTIVALENT, 120 et sqq., *196*
    219
  anaphase, 411
  co-orientation, 131
  See also trivalent, quadrivalent,
    etc.
Mus musculus, 115
Muscari, 37
mutation, 54, 167
  in apomictics, 475
  from crossing-over, 279, 345
  mass, 348, **352**
  in *Œnothera*, 338
  in parthenogenesis, 298
  rate, 332
  and unbalance, 325
  unit, 333
  X-ray, 422 et sqq.

Narcissus, 217, *318*, *319*, *402*
Nasturtium, 227
Natrix, 313
natural population, see population.
Navashin, M., 62, 71, **73**, 420, 424, I
Naville, 357, 374
Nawaschin, S., 11, 35
negative correlation, see correlation.
Nematoda, *367*
Neottia, 307
Nephrodium, 436
Neuroptera, *367*
  salivary glands, 175
Neuroterus, 376
Newton, 120, **189**, V
Nezara, 358

INDEX

Nicandra, 509
Nicotiana, balance, 327
  basic number, 241
  haploid, 417, *444* et sqq.
  non-reduction, 298
  origin of polyploids, *190*, *191*
  triploids, 210
Nicotiana, alata, 77, 402
  digluta, *190*, 225
  glutinosa-Tabacum, *187*
  Lansdorffii, 77
  longiflora, 239
  rustica-paniculata, *187*
  suaveolens, *125*
Nicotiana Tabacum, autosyndesis, 214
  non-pairing, *403*
  pentasomic, *321*
  secondary polyploid, 241
  structural change, 278
Nigritella, 469
Nishiyama, **324**
NON-DISJUNCTION, **152** et sqq., 353
  chromatid, 263, 350
NON-HOMOLOGOUS PAIRING, see pairing.
non-reduction, see ameiosis.
Nothoscordum, 79, *217*, 440.
nucellar embryony, 440 et sqq.
nuclear cycle, 481, 561
nuclear division, multiple, 49
  See also mitosis, amitosis, meiosis, ameiosis.
nuclear membrane, see nucleo-cytoplasmic surface.
NUCLEAR SAP, 24, 46, 521
nucleo-cytoplasmic surface, 19, 25, 503, 544
nucleolar constriction, see constriction.
NUCLEOLUS, 20, **23**, **98**, 102
  movement of, 46, 92
  organisation of, 228, 304
  origin of, 39
  salivary glands, **176**
  and sex chromosomes, *306*, *312*, 501
NUCLEUS, 2, 9, 18 et sqq.
  division of, see mitosis, meiosis.
  multiple division, 49
  size of, 20, 103
  solid and vesicular, *46*
Nyctereutes, 313

Ochna, *218* et sqq., *438* et sqq.

OCTOPLOID, *215*
Odonata, 367
Œcanthus, 367
Œdogonium, 5
Œnothera, 335 et sqq., VI
  crossing-over, 288
  half-mutants, 277, 345
  haploid, 262, 328, 417, *444*, 559
  monosomic, 316
  multiple chiasmata, 122
  mutations, 279
  non-pairing, 151
  nucleoli, 103
  progeny of triploids, *320*
  segregation, 248
  terminalisation, 112, *511*
  tetraploid, *194*
  triploid, 157
  trisomic, 353 et sqq.
  X-ray, *160*
Œnothera Lamarckiana, *194*
  trisomic, *319*
Oligochæta, 534
OLIGOPYRENE, 10
Oncopeltus, *358*
OÖCYTE, crossing-over, 272
  meiosis, XIV
oögenesis, nucleoli, *306*, 312
Ophryotrocha, 390, 534
organiser, see nucleolus.
Orgyia, 78, 162
ORIENTATION, auto-, 539
  centromere, 35, 523
  co-, **131**, 539
  failure of, **526**
  and hydration, 544
  multivalents, **120**, **529**
  univalents, 411 et sqq.
Ornithogalum, 37, *81*
Orthoptera, 367, 373
Oryza, *125*, 242, 248, 429
  haploid, *444*
  triploid, *216*
Osmunda, 448

PACHYTENE, 87 et sqq., 166
  in allopolyploids, **206**
  coiling, 489
  and crossing-over, 548 et sqq.
  forces, 492
  interlocking, **257**
  Œnothera, 346
  sex chromosomes, 368, **369**
  structural hybrid, 262, 266, 270
  in tetraploids, *120*

Pæonia, 162, 262
  inversion crossing-over, 268, IX
Painter, 176, 368
PAIRING, block, 121, 127, 199
  chiasma theory of, 126 et sqq.
  failure of, 126 et sqq., 399 et sqq.
  illegitimate, 214, 225
  internal, 199
  intra-haploid, 278
  localisation, 99
  loose, 166
  at meiosis, 87 et sqq.
  non-homologous, 166, **304**, 406, 489
    in differential segments, 265, 368
    in structural hybrids, 493
  principles of, 163 et sqq.
  secondary, 236 et sqq., *242*, VIII
  segment, 338
  somatic, 234 et sqq., 553
  torsion, see non-homologous pairing.
  touch and go, *366*, 375
  in undefined hybrids, 168
  See also zygotene, auto-syndesis, allo-syndesis, chiasma, attraction, time-limit.
Pallavicinia, 366
Panorpa, 367
Pantala, 367
Papaver, 202, 228, 239, 413
  hybrids, *208*
Paracopidosomopsis (sic), *377*
Paramæcium, 6, 50
Paratettix, 383, 464
Paris, 110, II, XIII
PARTHENOGENESIS, 85, 434 et sqq.
  cyclical, 383, 402
  in haplo-diploids, 376
  male, 193
  of monosomics, 316
  and segregation, 247, 298
  and spindle failure, 409
Pasteur, 549
Patau, **180**
Pellia, 228, 308, 366
PENTAPLOID, **215**, *218*, **219**
Pentstemon, **234**, VIII
Perla, 311 et sqq., *363*, *418*
permanence, theory of, see theory.
Petunia, 318
  hybrid, *193*
Peziza, 261
Phæophyceæ, 6
Phacellanthus, 228

Phalaris, 228
Phanæus, 94
Pharbitis, 445
Phenacoccus, 380
Philocamia, 58, 63
Phleum, 187
Photinus, 373
Phragmatobia, **59**, 78, X
  aggregation, 58
  chromosome length, 56
  sex chromosomes, 358, **359**
Phrynotettix, 89, 93, *162*, **261**
  resting stage, 45
  terminalisation, 105
Phycomycetes, 6
Phylloxera, 402
Physaloptera, 115
Phytodecta, 367
Piscicola, 533
Pisum, 151 et sqq., VI
  crossing-over, **263**
  hybrid, *160*
  inversions, 274
  terminalisation, *511*
plastids, 2, 9, 427
Platypæcilus, 172
Plecoptera, *367*
Poa, 319, 440
Podisma, 492
Podophyllum, acentric fragment, 271, 426, IX
  chiasmata, *110*
  non-congression, 526
  spindle, **532**
  spirals, **117**, 487
Pogonatum, 366
POLAR BODY, 7, 41, 85, 459, 476
polar granule, see centromere.
POLARISATION, of centromere, 413
  leptotene, 90, 500
  telophase, 44
  of univalents, 461
polarised light, 570
pole, see spindle, centrosome.
Polemonium, *159*, *160*
pollen, abortion, 320, 321
  deficient, 316
  grain, 9 et sqq., II, III, XIII
  of haploids, **195**
  of hybrids, **195**
  inactive, 340
  mitosis, 30, 396 et sqq., **542**
  of polyploids, **195**
  segregation, 248

INDEX

pollen—(contd.).
 size, 221, *224*
 spindle, 532
 sub-haploid, 317
 tube, **542**, XIII
 unreduced, 189 et sqq.
Polychæta, 390
polyembryony, 447
Polygonatum, *306*
POLYHAPLOID, see haploid.
polymitosis, 279, 397, **535**, X
POLYPLOID animals, 66
 Bryophyta, 448
 gland cells, 175 et sqq.
 hybrids, 202 et sqq., *208* et sqq.
 interchange heterozygotes, 156
 meiosis, 119 et sqq.
 resting nuclei, 46
 segregation, 247
 species, 197
polyploids, 60, *61*, 75
 and apomixis, *468*, *469*
 classification of, 183 et sqq.
 failure of pairing, 165
 induction of, 64
 inference of, 229
 origin of, 63, *190*, *191*, 396, 416 et sqq.
 pollen grains of, **195**
 secondary, 202, **238**
  See also triploid, tetraploid, etc.
polysomy, 60, 68
POLYTENE, 178
Pomoideæ, 79, 239
population, natural, 299
 in *Crepis*, **70**
 hybridity of, 271
Portulaca, 228, *445*
POSITION EFFECT, 280, 333
POST-REDUCTION, see reduction.
Potentilla, 454, *465*
PRECOCITY, of anaphase separation, 115
 of condensation, 308, 364, 562
 of sex chromosomes, 380, 390
 of metaphase, 413
 of prophase, 309, 398
 theory, see theory.
pre-reduction, see reduction.
Primula, tetraploids, 217
 triploids, 217
Primula kewensis, *190*, *193*, 248
 differential affinity, 198
 diploid, *172*
 pairing, 184 et sqq.

Primula kewensis—(contd.).
 secondary segregation, 278
 unbalance, 323, 400
Primula malacoides, *125*
Primula obconica, 224
Primula sinensis, 105, 185
 polyploid, 121 et sqq.
Pristiurus, 94, 103
PRO-CHROMOSOMES, 45, 307
progeny tests, *298*
PRO-METAPHASE, 236, 527, **545**
PROPHASE, 503 et sqq.
 of meiosis, 87, 89 et sqq.
 of mitosis, 22 et sqq.
Protenor, 358
PROTHALLIUM, 6
Protista, 2 et sqq., 13, 85
 centrosome, 533, 534
 mitosis, 47
 nucleoli, 21
PROTOPLASM, 1
Protortonia, 379, **381**
Prunus, 79, VIII
 autosyndesis, **203**
 chiasmata, 508
 hybrids, *172*, *193*
 multivalents, **128**, 132
 new species, *234*
Prunus avium syndiploidy, 66
Prunus cerasus, 242
Prunus laurocerasus, 197
Pseudococcus, 310, 379
PSEUDOGAMY, 454, 464
Psilotum, 535
Pteridonea, 377
Pteridophyta, 7, 246, 436
Pungitius, 536
Puschkinia, **27**, III
Putorius, XI
Pygæra, *173*, 210 et sqq.
 hybrids, *418*
 non-pairing, 400, *404*
 segregation, 247
 tetraploids, 183, *191*
 triploids, 67
Pyrus, **128, 202, 238**
 balance, 326

QUADRIVALENT, **120, 128,** 183 et sqq.
 crossing-over, **518**
 Hyacinthus, **127**
 segregation, **196**
 types, **124**

qualitative differentiation, see differentiation.
quantitative data, see also statistics.
　nuclear size, 46
　spiralisation, *31*
　timing of division, *31*
quart, in a pint pot, 84

race, **374**
　geographic, 228, 361, 384
　in *Drosophila pseudo-obscura*, 181
Rana, 15, *393*, 466
Ranunculaceæ, 84
Ranunculus, 115, **402**
　fragments, 72 et sqq.
Raphanus-Brassica, 181 et sqq., 190, *193*
　pairing, *174*, *210* et sqq.
　polyploid gametes, 416
　segregation, 248
　syndiploidy, 66
Rattus, *110*, 310, *368*, 370, XI
REDUCTION, 85, 249
　division, see meiosis.
　double, **518**
　failure of, 188
　numerical, 252
reductional separation, 252, **261**
REDUPLICATION, see duplication.
regeneration, 448
Renner, 248, 338, 577
　effect, 340
reproduction, gene, 182, 331
　semi-clonal, 463
　sexual, 4, 219, 300, 434
　sub-sexual, 477
Reptilia, *82*
repulsion, 103 et sqq., 481 et sqq., 503 et sqq.
　between centromeres, 152, 235
　body, 524, **528**
　centrosome, 525, 531
resting nucleus, *23*, *34*, *42* et sqq., **426**
RESTITUTION NUCLEUS, 297, 414, 436, **454**
Rhabditis, 311 et sqq., *404*, *450*, *464*
Rhodites, 450
Rhodophyceæ, 6, 7
Rhœo, **80**, 151, *160*, 338, VI
　coiling, *492*
　ring formation, **154**
　terminalisation, 112
Rhogostoma, 31
Richardson, 268
Ribes, 56, 79, *173*, 567

RING chromosomes, at meiosis,
　bivalents, **108**
　interlocking, 256
　multivalents, **124**, **128**, 149 et sqq.
　and segregation, 248
　at mitosis, 24, 272, 424, 502, 554
　in *Crepis*, **73**
　in *Zea*, 74, 557
Ring formation, with interchange, 149, 340
Rosa, conditional precocity, 314
　interchange, 156, *159*, *161*
　pollen grains, 224
　polyploidy, 227
　terminalisation, **508**, *511*
　univalents, *418*
Rosa canina, 460 et sqq.
Rosaceæ, 326
Rosenberg, 45, 413, 453, **456**
Rotifera, 376, 459
Roux, 22
Rubus, hybrids, **194**, *208*
　polyploidy, 189, *190*, *191*
　pseudogamy, 465
　secondary pairing, **238**
Rumex, basic number, 239
　intersexes, 387
　trisomic, 364
Rumex acetosa, 79, 199, 364
　fragments, *145*
Rumex acetosella, *161*, 214
Rumex Hydrolapathum, 197, 225

Saccharomyces, 10
Saccharum, *191*, 314, 402
Sagittaria, *110*, 487
Sakamura, IV
Salamandra, 28, 536
salivary glands, 175 et sqq., 308, 396, VII
Salvia, *173*
Sansome, **263**, VI
Saprolegnia, 82, *83*, 442
SATELLITE, see trabant.
Saturnia, 191
Sax, 487, 570
Saxifraga, 190
Schistocerca, *82*, *125*, 129
Schizophyllum, 442
Schrader, 381, 431, 543
Sciara, *178*, 543
　chromosome diminution, 58
scientific method, 298, 302, 333, 479 et sqq., 563

INDEX 667

Scilla, **403**, 437
Scolopendra, 103
Scolopendrium, 436, 448
Secale, fragment, 144, **147**, 528
  interchange, 71, **147**, *160*
  interference, 293
  inversions, *274*, 279
  semi-precocity, 400, **407**
  trisomic, *319*
second division, of meiosis, 89, 119
secondary balance, see balance.
SECONDARY PAIRING, see pairing.
SECONDARY POLYPLOID, see polyploid.
Sedum, *367*
seed mutations, 421
SEGMENT, see interstitial pairing, differential, interchange.
segmental interchange, see interchange.
segmentation, see cleavage.
SEGREGATION, 135, 245 et sqq.
  in allotetraploids, *186*
  in autotetraploids, *186*
  chromatid, 285
  in complex heterozygotes, **336** et sqq.
  and mutation, 339
  secondary, 205, 225
  and sex determination, **336**
  without reduction, 296
Seiler, 359
semi-permeable membrane, 19
SET, see chromosome.
seta, 38
Setcreasia, *111*
sex, determination of, 249, 335, 356 et sqq., 383 et sqq.
  haplo-diploid, 376
SEX CHROMOSOME, **141**, 356 et sqq., X, XI
  differential affinity, 199
  inert parts, 330 et sqq.
  loss at mitosis, 68
  mitosis, **53**
  precocity, 308 et sqq.
  in salivary glands, 177
  unpaired, 373
  Y chromosome, 330, 356 et sqq., 478
sex linkage, 356
  partial, 372
SEXUAL DIFFERENTIATION, 14
  and polyploidy, 67

sexual reproduction, see reproduction.
Sharp, 9, 578
Silene, 227
Simuleum, *178*
Smerinthus, 400
Solanum, basic number, 241
  pollen grains, 195
Solanum Lycopersicum, chiasmata, *111*
  fragments, *145*, 148, 330
  haploid, **417**
  polyploid, 121 et sqq.
  progeny of triploids, *320*
  somatic doubling, 64
  tetraploid, *185*
  trabants, **37**
  trisomic, *319*, *321*
Solanum nigrum, haploid, 195, *201*, 442 et sqq.
  tetraploid, 185
Solenobia, 450, 467, *469*, 470
somatic doubling, 63 et sqq., *187*, 235
SOMATIC PAIRING, see pairing.
Sorghum, 404
Spartina, *234*
species, 79
  chromosome numbers, 229 et sqq.
  clonal, 218, 227
  of known origin, *234*
  polyploid, 197 et sqq., 220
  polyploidy within, 226
specificity, see attraction, differentiation.
spermatid, 396, **410**
spermatozoid, 7, 449, 533
SPERMATOZOÖN, 9 et sqq.
  non-functional, 382
Sphærocarpus, 15
  cell size, 55
  intersexes, 387
  sex chromosomes, **308**, 331, 357
  triploid, *216*
  trisomic, 69
Sphæromyxa, 82, *83*
Spilosoma, 78
Spinacia, *115*, 235, 390
SPINDLE, bent, 413, 530
  central, 28, 542
  centromere, 531
  in Coccidæ, 380, 543
  failure of, *405*, 408 et sqq.
  hollow, 397, **525**
  intra-nuclear, 47
  at mitosis, **23**, 25

SPINDLE—(contd.).
movement of, 29
multipolar, 523, **535**
in polyploids, 129
structure, 520 et sqq.
tri-polar, *405*
SPINDLE ATTACHMENT, see centromere.
spindle fibre, 26, 521
SPIRAL, major, 101, 485, IV, XIII
minor, 101, 485
molecular, 483
relic, 33, 93, 311, 483, II
See also coiling.
spiral structure, 31
See also coiling.
SPIRALISATION, 25, 52, 483 et sqq., 562
See also coiling.
SPIREME, 24
continuous, 311
Spirogyra, 20, 21, 307
Spironema, 80, *145*, 507
split, see chromosome division.
SPORE, 5, 6, 449
segregation, 248
SPOROPHYTE, 6
Sporozoa, 6
Stadler, 422, 555
staining reaction, 307, 315
centromere, 536, XV
centrosome, 533
statistics, of chiasma frequency, **169**, 281 et sqq., *400*
of chromosome numbers, 229
of crossing-over, 281 et sqq.
of fragment pairing, *146*
of gametes of triploids, *318*
of natural populations, 70
of numbers of species, **230, 231**
of pairing frequency, **219**, *400*
of pairing in hybrids, *159*, *174*, *175*
of polyploid pairing, 126
of terminalisation, **508, 510**, *513*
of types of ring, 349
See also correlation.
Stauroderus, 261
Stenobothrus, **96**, XV
chiasma frequency, 99, 294
chromosome size, 83
coiling, 487
crossing-over, 553
deficiency, *162*
differential condensation, *310*
duration of meiosis, *31*

Stenobothrus—(contd.).
terminalisation, *513*
See also *Chorthippus*.
sterility, cross-, 135
self-, see incompatibility.
of tetraploids, 187
of triploids, 422
Stern, 277
Strasburger, 572, 581
STRUCTURAL CHANGE, 59, 77, 167
classification, 69, 554, **556**
dyscentric, 557
eucentric, 557
and genetic change, 197
induced, 422 et sqq.
inter-genic, 333
mechanism of, 554 et sqq.
natural, 419 et sqq.
secondary, 273, 275, *280*, 351, 421
and spiralisation, 489
symmetrical, 555, **556**, 557
time of, 71
See also structural hybrid.
STRUCTURAL HYBRID, 135 et sqq., 260 et sqq.
crossing-over, 286
dyscentric, 265
and sex, 363
Sturtevant, 180, 252
substrate, 315, 331
effect, 497, 498, 546
surface charge, see electric field.
Surirella, 48
SYNDIPLOIDY, 65
See also somatic doubling.
Syrbula, 294
Syringa, 417
systematics, 75

Täckholm, 460, 572
Talaporia, 368
tapetum, 63
Taraxacum, 438 et sqq., 457
secondary pairing, *242*
syndiploidy, 66
Taxus, *146*, 390
technique, 569
for spirals, 487 et sqq.
teleology, 197, 253
TELOPHASE, meiosis, 305; see also interphase.
mitosis, **23**, 30, 33
spirals, *491*
Tenebrio, 307

Tenodera, 364
Tephrosia, 467
TERMINAL AFFINITY see affinity.
TERMINALISATION, 103 et sqq., 503 et sqq.
  and coiling, 488
  in fragments, 143
  and interlocking, 257
  in *Œnothera*, 342
  in polyploids, **120**
  in sex chromosomes, **357**
  in unequal bivalents, **261**
Tetranychus, 377
TETRAPLOID, 63, 183
  animals, 66
  configurations at meiosis, **124**
  constitution, 61
  hybrids, 203 et sqq.
  interchange heterozygote, **158**
  meiosis, **120** et sqq.
  segregation, 286
  species, 217, **230**
    See also polyploid, auto- and allo-polyploid.
TETRASOMIC, 61, **322**, 325
  segregation, 249
Tettigidæ, 281
Tettigidea, *145*
Tettigonia, 82, 311, *491*
Thalictrum, 453
Thallophyta, 6, 441
Thea, 217
theory, balance, of mitosis, 544, **545**
  of body repulsions, 514, 524, 527
  chiasma, of pairing, 126 et sqq., 164, 236, 300, 337
  chiasmatype, 250, **516**, **518**
  chromosome, 244
  co-orientation, **538**, **539**
  crossing-over, 250 et sqq., 547 et sqq.
  of differential segment, **346** et sqq., 349, 365
  of evolution, see evolution.
  of the gene, 245, 332, 502
  of genetic balance, 315, 385
  of genotypic control, 53, 399
  of heterochromatin, 307
  of heterosis, 136
  of hysteresis, 483, 550
  of illegitimate crossing-over, 275, 575
  of inert genes, 329
  of localisation, 99

theory—(*contd.*).
  Mendelian, 245, 250
  of molecular spirals, 483 et sqq.
  of mutation, 328, 345
  of negative correlation, 293
  of old clones, 421
  of origin of apomixis, 434 et sqq.
  of permanence of chromosomes, 40
    of centromeres, 536
    of centrosomes, 534
  of pre- and post-reduction, 262
  of precocity of meiosis, *113*, 132 et sqq., 433, 473 et sqq., 481 et sqq., **486**
  of reciprocal chiasmata, 371
  of relational coiling, 489, 548
  of secondary segregation, 225
  of specific attraction, 492
  of structural change, 554
    secondary, 275
  of sub-sexual reproduction, 477
  of terminal affinity, 517
  of terminalisation, 495
  of the time-limit, 99 et sqq., 123, 134, 166, 279
  of torsion pairing, 493
Thismia, 438
Thomas, **399**
Thyanta, **358**
thymo-nucleic acid, 20
Thysanoptera, 376
time limit, in pairing, see theory.
Timoféeff-Ressovsky, 422
Tinney, 308
tissue culture, 571
tomato, see *Solanum Lycopersicum*.
TORSION, *179* et sqq.
  pairing, 166
    See also non-homologous pairing.
TRABANT, 37, 38, 305, **499**
  lateral, 424
Tradescantia
  centromere, **528**, XV
  chromosome size, **54**
  coiling, 487 et sqq., IV, XIII
  fragments, 71, **142**
  mitosis, 30, *31*
  nucleoli, *306*
  pollen grains, *221*, **532**, III
    of triploid, *318*
  somatic interchange, 421
  terminalisation, *111*, 113
  X-ray breakage, 425
TRANSLOCATION, **59**, 69, 555 et sqq.

TRANSLOCATION—*(contd.)*.
heterozygote, 265
lateral, 74
reciprocal, see interchange.
Trialeurodes, 377, *450*
Trichoniscus, 216, *469*
Trichoptera, *94*, 358, *368*
Tricyrtis, III
Trillium, 110, *274*, 487
Trimerotropis, 153, **261**, **264**
TRIPLOID, and apomixis, 470
constitution, 61
crossing-over, 284, 551
Habrobracon, 379
hybrid, 187 et sqq.
interchange, 157
inversions, 271, 279
origin of, 67
pachytene, 493
progeny, 317 et sqq., *326*, *327*
Tripsacum, 191, *208*, *446*
See also *Zea*.
Trirhabda, 358
TRISOMIC, *319*
constitution, *61*
mutation, 69
Œ*nothera*, 338, 353, 448
origin, 68
Rumex acetosa, 364
secondary, 148
secondary change, 279
sex chromosomes, 385, **386**
tertiary, 156
true breeding, 353 et sqq.
Zea Mays, **282**
Triticum, haploid, *201*, *417*, *444*
interchange, 157, *160*, *161*
spindle, 530
failure, 408 et sqq.
triploid progeny, 317 et sqq.
Triticum hybrids, *172*, *173*, **204**
meiosis in, 414
non-reduction, *417*
pairing, 209 et sqq.
Triticum durum, **170**
monococcum, 311, *417*
turgidum, **170**
vulgare, 200, 278
unbalance, 323
Triticum-Ægilops hybrids, 161, 169, *174*
haploid, *201*, *418*, *445*
tetraploid, 185 et sqq., *190*
Triton, 523, *535*
chromosome size, 55

TRIVALENT, **120**, **124**, **128**
Hyacinthus, **130**
inversions, 272
orientation, **131**, *529*
Tulipa, **122**
Tryxalis, 311
Tulipa, V, XII
chromosome number, 84
doubling in embryo-sac, 67
first metaphase, **122**
inversions, 265, IX
pollen grains, *396*
spirals, *491*, XIII
terminalisation, *513*
triploid meiosis, 119
triploids, *216*, *217*
univalents, 411
Tulipa Clusiana, **215**, **226**
Tulipa galatica, 63
fragments, **72**, *145*, 531
Turbellaria, 7
twin seedlings, 447

Ulothrix, 458
ultra-violet irradiation, 422
ultra-violet photography, 180, 334, 423, *563*, XV
unbalance, see balance.
UNIVALENT, 123, 132, 164 et sqq., 410 et sqq.
and cell-wall, **545**
division cycle, **538**
Hyacinthus, **127**
metaphase plate, 530
Rosa canina, **462**
secondary pairing, *242*
Urechis, **535**
Uredineæ, 458
Urginea, 129
Uvularia, 69, **114**, 221
sub-haploid, 317

Vahlkampfia, 50, 531
Valeriana, 228
Vallisneria, **224**, 234
Vanduzea, 373
Vaucheria, 458, 466
vegetative reproduction, 4
Veronica, 217
viability, differential, 316 et sqq.
Vicia hybrids, 55, 140, 148, **155**
trisomic, *319*
Vicia Cracca, **59**, *228*

INDEX

Vicia Faba, XII
  chiasma frequency, *169*, 288
  mitotic chromosomes, **37**
  non-reduction, *418*
  nucleoli, **40**
  X-rayed, 428
Vicia narbonensis, **40**
Vicia sativa, *446*
Vilmorin, L. de, 13
Viola, chromosome length, 56
  hybrids, 171, *209*
  non-pairing, 400, *404*
  polyploid species, *228*
Viola Kitaibeliana, 227, 228
Viviparus, 256
vivipary, 440
Vries, H. de, 338

W-CHROMOSOME, 583
Weismann, 5, 85, 244, 302, 477
Wenrich, 105
Wettstein, F. v., 222, 448, 466
White, 427, 431
Whiting, 378
Wilson, 142, **358**, 574
Winge, 229, 388
Winiwarter, de, 572 et sqq.
Winkler, 434, 572

X-CHROMOSOME, see sex
  chromosome.
X-ray, 331, 422 et sqq.
  and crossing-over, 553
  effects, 71, 555 et sqq., XII
  and gene analysis, 334, 564
  photography, 564
  on pollen, 316
  technique, 571

Yamamoto, **365**
Y-CHROMOSOME, see sex
  chromosome.
Yucca, *81*, 294

Zacintha, 82
Z-CHROMOSOME, 583
Zea Mays, asynapsis, 291, 399 et
  sqq., *418*
  B chromosomes, **314**, 493, 498
  centromere, *536*
  chiasmata, **282**
  crossing-over map, **283**
  diœcious, 388
  endosperm, 12
  fragment, *145*
  haploid, *445*
  hybrid, *191*, *208*
  interchange, **262**, **282**, 430
  inversion, 265 et sqq.
  irradiation, 422 et sqq.
  monosomic, 61, 316, X
  non-homologous pairing, 166
  pachytene, 93, 166, **262**, **304**
  polymitosis, 279, 397, X
  progeny of triploid, *320*
  ring chromosome, 74, 557
  segregation, 248
  " sticky " chromosomes, 420
  terminalisation, 111, **510**, *511*
  triploid, 119, *216*
  trisomic, **282**, *319*
Zea-Euchlæna hybrids, 284, 494
Zebrina, 80
Zephyanthes, 438, 453
Zygopetalum, 440
ZYGOTE, 5
ZYGOTENE, 87 et sqq.
  polyploid, 119 et sqq., **120**, V
  zygotic reduction, 6